COBE Differential Microwave Radiometers
FULL SKY MICROWAVE MAP

53 GHz   5.7 mm

−6.6   +6.6

mK

Launch (November 1989) thru May 1990

PRIMORDIAL
NUCLEOSYNTHESIS
AND EVOLUTION OF
EARLY UNIVERSE

ASTROPHYSICS AND SPACE SCIENCE LIBRARY

A SERIES OF BOOKS ON THE RECENT DEVELOPMENTS
OF SPACE SCIENCE AND OF GENERAL GEOPHYSICS AND ASTROPHYSICS
PUBLISHED IN CONNECTION WITH THE JOURNAL
SPACE SCIENCE REVIEWS

PROCEEDINGS
VOLUME 169

PRIMORDIAL NUCLEOSYNTHESIS AND EVOLUTION OF EARLY UNIVERSE

PROCEEDINGS OF THE INTERNATIONAL CONFERENCE "PRIMORDIAL
NUCLEOSYNTHESIS AND EVOLUTION OF EARLY UNIVERSE"
HELD IN TOKYO, JAPAN, SEPTEMBER 4–8 1990

Edited by

K. SATO
University of Tokyo, Tokyo, Japan

J. AUDOUZE
Institut d'Astrophysique, Paris, France

SPRINGER SCIENCE+BUSINESS MEDIA, B.V.

Library of Congress Cataloging-in-Publication Data

Primordial nucleosynthesis and evolution of early universe:
  proceedings of an international conference held in Tokyo, Japan.
  September 4-8, 1990 / edited by K. Sato. J. Audouze.
      p.   cm. -- (Astrophysics and space science library : v. 169)
    Sponsored by the International Union of Pure and Applied Physics
  and the University of Tokyo.
    ISBN 978-0-7923-1193-5   ISBN 978-94-011-3410-1 (eBook)
    DOI 10.1007/978-94-011-3410-1
    1. Cosmology--Congresses. 2. Nucleosynthesis--Congresses.
  I. Sato, K. II. Audouze, Jean. III. International Union of Pure
  and Applied Physics. IV. Tokyo Daigaku. V. Series.
  QB980 . P75 1991
  523.1 -- dc20   91-11099

ISBN 978-0-7923-1193-5

Contents

Preface   xiii
Opening Address   xv

1. STANDARD MODEL OF PRIMORDIAL NUCLEOSYNTHESIS AND OBSERVATIONS OF LIGHT ELEMENTS   1

Standard Model of Primordial Nucleosynthesis: A Few General Remarks
Audouze, J.   1

The Abundances of D, He and Li Test and Constrain
the Standard Model of Cosmology
Steigman, G.   3

Lithium, Beryllium and Boron: Observational Constraints
on Primordial Nucleosynthesis
Beckman, J.   23

The Evolution of the Galactic Lithium Abundance
Hobbs, L.   39

Chemical Evolution of Galaxies
Pagel, B.   45

The Effect of Some Nonequilibrium Processes on the Primordial Nucleosynthesis
Dolgov, A. and Kirilova, D.   55

Analysis of the Reaction $^7Li(d,n)^8Be$ at Subcoulomb Energies
Rauscher, T., Grün, K., Krauss, H. and Oberhummer, H.   61

Experimental Study of the Key Reaction to the Nucleosynthesis in the Inhomogeneous
Big Bang Models
Kubono, S., Boyd, R., Peterson, R., Ikeda, N., Tanaka, M., Yasue, M.,
Yun, S., Toyokawa, H., Yosoi, M., Ohnuma, H., Tanihata, I. and Kajino, T.   63

Primordial Black Holes and Big Bang Nucleosynthesis
Arai, K., Hashimoto, M. and Futamase, T.   65

Constraints from Primordial Nucleosynthesis on Neutrino Degeneracy
Kang, H.   67

2. QCD PHASE TRANSITION AND NUCLEOSYNTHESIS IN INHOMOGENEOUS UNIVERSES   69

Strange Quark Matter in Physics and Astrophysics
Madsen, J.   69

Primordial Nucleosynthesis in Inhomogeneous Universe
Sato, K.   79

Sterile Neutrinos in the Early Universe
Malaney, R. and Fuller, G.   91

Could Cosmic QCD Phase Transition Produce Strange Quark Matter
Which Survives until the Present Time?
Kajino, T. and Sumiyoshi, K.   95

Multi-Zone Calculation of Nucleosynthesis in Inhomogeneous
Universe and Be-9 Abundance
Terasawa, N.   101

Signatures of Inhomogeneity in the Early Universe
Fowler, W., Kawano, L., Malaney, R. and Kavanagh, R.   107

Diffusion Coefficients of Nucleons in the Inhomogeneous Big Bang Model
Banerjee, B. and Chitre, S.   117

Reactions on Carbon-14
Kawano, L.   121

Survival of Strange Matter Lumps Formed in the Early Universe
Lee, C. and Lee, H.   123

Measurement of the Cross Section of the $^{12}C(n,g)^{13}C$ Reaction
at Stellar Energy
Nagai, Y.   125

Inhomogeneous Universes in the Framework of Lattice QCD
Hackel, M., Faber, M., Markum, H. and Oberhummer, H.   127

3. INFLATION AND VERY EARLY UNIVERSE   129

The Beginning of the Universe
Hawking, S.   129

Extended Inflationary Cosmology: A Primer
Steinhardt, P.   141

The Inflation Sector of Extended Inflation
Kolb, E.   153

Inflation in Generalized Einstein Theories
Maeda, K., Berkin, A. and Yokoyama, J.   167

Baryogenesis in the Universe
Dolgov, A.   173

Formation of Topological Defects in the Inflationary Universe
Yokoyama, J. and Nagasawa, M.   187

Non-Zel'dovich Fluctuations from Inflation
Primack, J.   193

Magnetic Theory of Gravitation
Cho, Y.   203

Chaotic Inflation and the Omega Problem
Mathews, G., Graziani, F., and Kurki-Suonio, H.   213

Late-Time Cosmological Phase Transitions
Schramm, D.   225

False-Vacuum Decay in Generalized Extended Inflation
Holman, R.   243

Reconciling a Small Density Parameter to Inflation
Fujii, Y. and Nishioka, T.   249

Soft Inflation: A Model for Easing Constraints
Berkin, A., Maeda, K. and Yokoyama, J.   251

Stochastic Inflation Lattice Simulations: Ultra-Large Scale Structure
of the Universe
Salopek, D.   253

Purely Quantum Derivation of Density Fluctuations in the Inflationary Universe
Nambu, Y. and Sasaki, M.   265

Constraints on the Coupling of Weakly-Interacting Particles to Matter
from Stellar Evolution
Iwamoto, N.   267

Formation and Evolution of Domain-Wall-Networks
Kubotani, H., Ishihara, H. and Nambu, Y.   269

Catastrophe of Spacetime in the Early Universe
Cao, S.   271

(2+1)-Dimensional Quantum Gravity
Hosoya, A.   273

A Stringy Universe Scenario
Mashino, M. and Minakata, H.   275

The Constant-Mean-Curvature Slicing of
the Schwarzschild-de Sitter Space-Time
Nakao, K.   277

Schwarzschild-de Sitter Type Wormhole and Cosmological Constant
Kim, S., Kim, S. and Yang, J.   279

4. BACKGROUND RADIATION   281

COBE: New Sky Maps of the Early Universe
Smoot, G.   281

Large Scale Cosmic Instability
Hogan, C.   295

Gas-Induced Primary and Secondary CMB Anisotropies
Bond, J. and Myres, S.   305

Cosmic X-Ray Background
Inoue, H.   325

Large Scale Anisotropy of the CMB in an Open Universe and
Constraints on the Models of Galaxy Formation
Gouda, N., Sugiyama, N. and Sasaki, M.   335

5. DARK MATTER   337

The Best-Fit Universe
Turner, M.   337

LEP Physics and the Early Universe
Krauss, L.   351

A Search for Dark Matters in the Kamiokande II
Suzuki, Y.   363

Baryonic Dark Matter
Silk, J.   371

Phenomenological Dark Matter Detection Rate-form WIMP to SIMP-
Sato, H.   381

6. GALAXIES AND AGN   387

X-Ray Iron Line of Cluster of Galaxies
Koyama, K.   387

Dynamical Evolution of Compact Groups of Galaxies
Kodaira, K., Okumura, S., Makino, J., Ebisuzaki, T. and Sugimoto, D.   393

Correlations of Spin Angular Momenta of Galaxies
Iye, M.   403

Formation of Bipolar Radio Jets and Lobes from Accreton Disk
around Forming Blackhole at the Center of Protogalaxies
Uchida, Y., Matsumoto, R., Hirose, S. and Shibata, K.   409

An Evolutionary Unified Scheme for Radio-Loud Quasars and Blazars
Vagnetti, F., Giallongo, E. and Cavaliere, A.   417

Magnetohydrodynamical Energy Extraction from a Kerr Black Hole
Yokosawa, M. 419

Spherical Symmetric Model for Calculating Large Peculiar Velocities
of Galaxies
Xiang, S., Cheng, F. and Liu, J. 421

On the Origin of Cosmological Magnetic Fields
Tajima, T., Cable, S. and Shibata, K. 423

7. *LARGE SCALE STRUCTURE* 425

The Hawaii Deep Survey-Implications for Cosmology and
Galaxy Formation
Cowie, L. 425

Analysis of the Large Scale Structure with
Deep Pencil Beam Surveys
Szalay, A., Ellis, R., Koo, D. and Broadhurst, T. 435

Distance to the Coma Cluster and the Value of H_0
Okamura, S. and Fukugita, M. 451

Cosmological Implications of HI Absorption systems
Ikeuchi, S. 461

Gravitational Lens Effect on Cosmological Distance
Sasaki, M. 469

Apparent and Biased Effects in Oscillating Universe
Morikawa, M. 475

Oscillating Physics and Periodic Universes
Steinhardt, P. 483

Probes of the High-Redshift Universe
Rees, M. 487

Large Scale Coherent Structures in the Universe
Kofman, L. 495

Non-Gaussian Density Field in Nonlinear Gravitational Clustering
Suto, Y. 509

United Ways of Inflation in Generalized Gravities
Amendola, L., Capozziello, S., Litterio, M. and Occhionero, F. 517

Reviving Massive Neutrinos for Large-Scale Structure
Scherrer, R. 525

Cellular Distribution of the Universe and Dark Matter
Calzetti, D., Giavalisco, M. and Ruffini, R. 533

The Large-Scale Peculiar Velocity at High Redshift
Zhu, X. and Chu, Y.   543

An Evidence of Multiply-Connected Topology Model of the Universe
Zhu, X. and Chu, Y.   545

Non-Gaussian Initial Conditions in CDM Model Simulations
Lucchin, F.   547

Evolution of Galaxy Cluster and Primordial Nucleosynthesis
Chuvenkov, V., Glukhov, A. and Vainer, B.   549

Constraints on Universe Models with Cosmological Constant
from Cosmic Microwave Background Anisotropy
Sugiyama, N., Gouda, N. and Sasaki, M.   553

Gravitational Lens Effect and Distance Measure
Watanabe, K., Sasaki, M. and Tomita, K.   555

Constraints on the Cosmic String Scenario from the
Large-Scale Distribution of Galaxies
Miyoshi, S. and Hara, T.   557

Clustering and Velocity Field of Galaxies in Cold Dark Matter Scenario
Suginohara, T. and Suto, Y.   559

Formation of Galactic Halos from Seeded Hot Dark Matter
van Dalen, A.   561

Gravitational Clustering of Galaxies: Comparison between
Thermodynamic Theory and N-body Simulations
Itoh, M.   563

The Hydrodynamical Interaction of Galaxies
Umemura, M. and Yoshioka, S.   565

Velocity Fields around Superclusters: The Effect of Asphericity
Watanabe, T.   567

Evolution and Statistics from the Number Counts of X-Ray Cluster and Intergalactic
Medium
Hanami, H.   569

A Sample of Ultraluminous IRAS Galaxies in Southern Sky
Zou, Z., Xia, X. and Deng, Z.   571

Backreactions of Inhomogeneities on Expanding Universes
Bildhauer, S. and Futamase, T.   573

Multi-Level Feature in the Large Scale Distribution of Galaxies
Deng, Z., Xia, X., Liu, Y. and Zou, Z.   575

Fragmentation and Clustering in the Wake Formed by Cosmic String
Hara, T. and Miyoshi, S.   577

Perturbations of the One-Dimensional Structure in the Universe
Bildhauer, S., Aso, O., Kasai, M. and Futamase, T.   579

Three Points Spatial Correlation Function on the Self Similar
Observer Homogeneous Structure
Lee, H. and Ruffini, R.   581

Limits from Cosmological Nucleosynthesis on the Leptonic Numbers
of the Universe
Bianconi, A., Lee, H. and Ruffini, R.   583

Inhomogeneities and Non-Friedmann Light Propagation
Linder, E.   587

On the Generation of Non-Gaussian Fluctuations During
Chaotic Inflation
Vishniac, I. and Mineshige, S.   589

A Unified Picture of HI Absorption-Line Systems by Minihalos
Murakami, I. and Ikeuchi, S.   591

Gravitational Lens Effect on the Images of High Redshift Objects
Tomita, K. and Watanabe, K.   593

8. COSMOLOGICAL PARAMETERS   595

Gravitational Lens Determinations of Cosmological Parameters
Turner, E.   595

Global Cosmological Parameters
Fukugita, M.   599

Gravitational Lens Effects on the Redshift-Volume Test of W
Omote, M. and Yoshida, H.   615

9. Summary   617

Summary of the Conference
Hayakawa, S.   617

10. List of Participants   623

Preface

The International Conference "Primordial Nucleosynthesis and Evolution of Early Universe" was held in the presence of Prof. William Fowler on 4 - 8 September 1990 at the Sanjo Conference Hall, the University of Tokyo. This conference was co-sponsored by IUPAP, the International Union of Pure and Applied Physics, and by the University of Tokyo.

The number of participants was 156, 58 from 15 foreign countries and 98 from Japan. About 120 contributions were submitted orally or as posters.

Originally this conference was planned as a small gathering on Primordial Nucleosynthesis as indicated in the title, since primordial nucleosynthesis is the most important probe of the early stage of the universe. As is well known, light element abundances strongly depend on the time evolution of temperature and density. In this sense we can say that primordial nucleosynthesis is both the thermometer and speedometer of the early universe. Moreover, recently it has been claimed that primordial nucleosynthesis is an indicator of inhomogeneity of the early universe too. Now research of the primordial nucleosynthesis is in a boom.

We, however, decided to include observational cosmology, taking into account the recent remarkable results of observations. Nowadays, to reveal the large scale structure of the universe and discover its origin is a main subject in cosmology. We invited distinguished scientists from all over the world, and very fortunately almost all these people accepted to attend this conference.

This proceedings includes about thirty reviews by active and distinguished researchers, which cover the most recent developments in each field nicely. We believe this proceedings is very useful for people who want to know recent issues in cosmology.

Financially, this conference was also supported by some science foundations. Here we sincerely thank the Nishina Memorial Foundation, Inoue Foundation for Science, Kajima Foundation and the Commemorative Association for the Japan World Exposition for their support.

Katsuhiko Sato
Jean Audouze

January 1991

OPENING ADDRESS

Professor Fowler, distinguished guests, ladies and gentlemen,

I am very happy to have the opportunity to address you this morning.

I am as a nuclear physicist very much interested in this topical conference on Primordial nucleosynthesis and evolution of early universe. In recent years astrophysics has been rapidly developing because of the development of high energy and nuclear physics as well as of that of experimental methods.

This is the physics which requires cooperations among various fields of physics and also international cooperation.

I hope that this conference will provide to be successful and fruitful.

To all of our foreign guests, I wish you pleasant stay in Tokyo.

Akito Arima

President, the University of Tokyo

OPENING ADDRESS

Distinguished guests, ladies and gentlemen,

It is my great pleasure to make the opening address for the International Conference on Primordial Nucleosynthesis and Evolution of Early Universe. This conference is sponsored by many organizations, one of which is IUPAP. I am one of the Vice-Presidents of IUPAP.

On behalf of IUPAP, I am very happy to welcome all participants and their accompanying persons.

May I speak from the view of high energy community, to which I belong?

The relation between physics, particularly particle physics, and astrophysics has recently become quite intimate each other. Particle theorists, after grand unified theories, trials to unify all forces in nature, are now discussing many issues on cosmology, some of which are the subjects of this conference, and quantum theory of gravity. Experimentally, high energy physics contributed to detect supernova neutrinos and solar neutrinos opening the new field, the neutrino astronomy. We, high energy physicists, now hold very much at home feeling to astrophysics.

I am sure that reciprocal by astrophysicists will have a similar altitude to particle or high energy physics.

Reflecting such a status, six years ago, IUPAP has established the Commission on Astrophysics, C19, to stimulate better relation between astrophysics and physics. This commission, chaired by S. Hayakawa, president of Nagoya University, was very much active in the past three years. One of the vivid outcomes of activities of this commission is indeed this Conference.

I wish that this Conference will be very successful by cooperation of all of you.

Finally but not the least, I wish very much that people coming here from abroad would have enjoyable stay in Japan, beside hot scientific debates.

Thank you for your attention.

Yoshio Yamaguchi

Vice President, IUPAP

(Tokai University)

I UPAP Conference
Primordial Nucleosynthesis
and
Evolution of Early Universe

Scientific Organizing Committee; Chairman K. Sato(Japan)

J. Audouze (France), J. R. Bond(Canada), E. Kolb(USA), P. J. E. Peebles(USA),
M. Rees(UK), H. Sato(Japan), D. N. Schramm(USA), J. Silk(USA),
A.A. Starobinski(USSR), G.Steigman(USA), R. A. Sunyaev(USSR),
A. Szalay(Hungary)

Local Organizing Committee;

J. Arafune (Institute For Cosmic Ray Research, The University of Tokyo)
M. Fukugita (Research Institute For Fundamental Physics, Kyoto University)
S. Hayakawa (President, Nagoya University)
S. Ikeuchi (National Astronomical Observatory)
K. Kodaira (National Astronomical Observatory)
K. Maeda (Department of Physics, Waseda University)
M. Morita (Department of Physics, Osaka University)
S. Okamura (Institute of Astronomy, The University of Tokyo)
H. Sato (Department of Physics, Kyoto University)
K. Sato (Department of Physics, The University of Tokyo) ;   Chairman
Y. Tanaka (Institute of Space and Astronautical Science)
Y. Uchida (Department of Astronomy, The University of Tokyo)
T. Yamazaki (Institute For Nuclear Study, The University of Tokyo)
M. Yosimura (Department of Physics, Tohoku University)

Secretariat and Treasurer

J. Yokoyama (Department of Physics, The University of Tokyo)

Host Organizations

International Union of Pure and Applied Physics
The University of Tokyo

Acknowledgement

This conference has been executed with the grant of the Commemorative
Association for the Japan World Exposition. This conference was also
supported by Nishina Memorial Foundation, Inoue Foundation for Science
and Kajima Foundation.

STANDARD MODEL OF PRIMORDIAL NUCLEOSYNTHESIS: A FEW GENERAL REMARKS

Jean AUDOUZE
Institut d'Astrophysique du CNRS
Paris, France

Primordial nucleosynthesis plays indeed a major role both in cosmology and in particle and nuclear physics. The comparison between the observations of the abundances of the light elements (D, ^{3}He, ^{4}He and ^{7}Li) and the predictions coming from the Standard Model provides information on the following facts :

- it indicates that the baryonic density of the Universe is at most 10% of the critical density. This means that most of the dark matter should be of non-baryonic nature.

- There is only room for three different flavours of neutrinos and leptons which is consistent both with the simplest Grand-Unification schemes and with the results of the LEP and SLC particle physics experiments.

Many attempts have been made, and some are reported in this book, to analyze more complicated models: inclusions of massive unstable non-baryonic particles, inhomogeneous cosmologies, etc... . It is fair to say that these models are satisfactory to reproduce the currently observed abundances of the light elements. So, researchers in this field can adopt two different attitudes:

- pursue the study of non standard cosmologies. A lot can be learnt on the quark-baryon transition, the antimatter suppression at very early epochs or the possible effect of still unknown non baryonic particles. This is discussed in detail in these proceedings and I shall not elaborate more on it here.

- push the Standard Model to its extreme: First one still needs better abundance determinations. The agreement between the presently observed values and the model predictions is good, may be because the abundance uncertainties are much too large and this is true for the four relevant elements (D, ^{3}He, ^{4}He, ^{7}Li).

Second more work should be devoted to a careful study of the chemical evolution of these four elements.

To sum up, both observers and specialists in chemical evolution of galaxies should be encouraged to pursue their analyses along these lines. In the present situation, the Standard Models of primordial nucleosynthesis still provide the best answer to basic problems such as the number of neutrino flavours, the open nature of the Universe or the non baryonic nature of dark matter.

THE ABUNDANCES OF DEUTERIUM, HELIUM AND LITHIUM TEST AND CONSTRAIN THE STANDARD MODEL OF COSMOLOGY

GARY STEIGMAN
Physics Department
The Ohio State University
174 West 18th Avenue
Columbus, OH 43210
USA

ABSTRACT. The standard hot big bang model with input from the experimentally tested standard model of particle physics (e.g., three flavors of neutrinos) is remarkably successful in accounting for the abundances of the light elements D, ^3He, ^4He and ^7Li. Accurate observations of the abundances of these light elements, along with the means to extrapolate them to their primordial values, provide an invaluable tool for testing the predictions of the standard cosmological model and for constraining the baryon density of the Universe. The current data on the abundances of the light elements is reviewed and is supplemented by model independent and some model dependent (i.e. galactic chemical evolution models) estimates to derive their primordial values. The inferred primordial abundances are compared with the predictions of nucleosynthesis in the standard cosmological model to test the consistency of the model and to constrain the allowed range of the baryon-to-photon ratio (η). The preferred value of $\eta = 4 \times 10^{-10}$ points to a very low baryon density ($\Omega_B \lesssim 0.1$) and strengthens the case for nonbaryonic dark matter.

1. Introduction

The existence of the microwave background radiation and of the light elements D, ^3He, ^4He and ^7Li are the only predictions of the standard hot big bang cosmological model which have been subjected to direct, observational tests. The beautiful data of the COBE satellite (Mather et al. 1990) dramatically confirms the hot, dense infancy of the Universe. It is, then, inevitable that some nucleosynthesis would have occurred during the early evolution of the Universe. The predictions of primordial nucleosynthesis in the context of the standard model of cosmology have been compared with the observed abundances of the light elements, establishing the consistency of the standard model (Yang et al. 1984; Boesgaard and Steigman 1985; Walker et al. 1991). The relic cosmic background radiation and the relic abundances of the light elements provide the only direct tests of the standard model of cosmology and, offer observational constraints on non-standard cosmologies.

In this contribution I will review the current observational status of the abundances of $D, {}^3He, {}^4He$ and ${}^7Li$ and, their inferred primordial values. Whenever possible, it is preferrable to employ model (galactic evolution) independent *bounds* to the primordial abundances rather than model dependent *determinations*. Here, I will emphasize the former but, I will comment on the implications of the latter. Since the "theory" of the synthesis of the light elements during the first few minutes in the evolution of the Universe is well-known and reviewed in many places (c.f., Yang et al. 1984; Boesgaard and Steigman 1985; Walker et al. 1991), this overview will concentrate on the comparison of the predicted and observed abundances; the interested reader is invited to consult the above-cited references for the theoretical background. The predicted abundances of $D, {}^3He$ and ${}^7Li$ are sensitive to the nucleon density at the epoch of nucleosynthesis. Therefore, the abundances of these light elements provide a test of the standard model and offer the possibility of constraining the present ratio, η, of nucleons (baryons) to photons. Since virtually all the neutrons present when primordial nucleosynthesis commences are incorporated in ${}^4He$, the helium-4 abundance is relatively insensitive to η but, its consistency with observations provides a crucial test of the standard model and, offers constraints on exotic new elementary particles (Steigman, Schramm and Gunn 1977). The discussion here, therefore, will commence with deuterium and helium-3, proceed to lithium-7 and, conclude with helium-4. After the data is reviewed and the primordial abundances (or, bounds thereto) inferred, theory and observations will be compared. In a concluding section the cosmological consequences of the confrontation between theory and observation will be discussed.

2. The Observed Abundances

Ideally, we would like to know the "Universal" abundances of the light elements. For ${}^4He$, the second most abundant element in the Universe, this is possible. To date, with one notable exception to be discussed shortly, this has proven impossible for $D, {}^3He$ and ${}^7Li$. For these latter elements/isotopes, we are restricted to solar system and/or Galactic data. Nonetheless, it is often possible to use such "local" data, in a (nearly) model independent fashion, to derive bounds to the primordial abundances of these species. Further, if one is willing to sacrifice model independence, models of Galactic chemical evolution (which are subjected to other observational constraints) can be used to infer "the" primordial abundances from the local data. In this section I review the status of the observed abundances; subsequently, their primordial values will be inferred and compared to the predictions.

2.1 DEUTERIUM

The solar system and the local interstellar medium provide virtually our only information on the deuterium abundance (Boesgaard and Steigman 1985). The Copernicus data (Rogerson and York 1973; York and Rogerson 1976; Vidal-Madjar et al. 1977; Laurent et al. 1979; Ferlet et al. 1980; Vidal-Madjar et al. 1982; York 1983) comparing the UV absorption due to interstellar atomic deuterium and hydrogen along the lines-of-sight to nearby hot bright stars reveals a linear correlation between the D and H column densities (Steigman, unpublished). The resulting ratio,

$$(D/H)_{ISM} = 1.6^{+0.9}_{-0.6} \times 10^{-5} \tag{1}$$

provides a measure of the present deuterium abundance in the interstellar gas of the Galaxy.

In principle, similar observations could provide information on the D to H ratio in external galaxies. In practice, since this would require a large UV telescope above the absorbing effects of the atmosphere, such data doesn't exist. The Hubble Space Telescope and the proposed FUSE project of NASA could fill this observational gap and provide valuable information on the abundance of deuterium in the Universe. Indeed, data on the deuterium abundance in "metal-poor" systems could bring us much closer to the primordial deuterium abundance. In this context, it may be possible to search the "Lyman-alpha forest" (the hydrogen absorption features in front of high redshift QSOs) for deuterium absorption. The problem here is that for those features with relatively high hydrogen column density, the deuterium absorption will be buried in the wings of the saturated hydrogen features whereas, for those low column density hydrogen absorbers a deuterium abundance of $10^{-5} - 10^{-4}$ would be unobservably small. There is, therefore, only a small subset of $Ly\alpha$ absorbers with column densities appropriate to such a search. This subset is culled further by the requirement that the absorbing cloud be "cool and quiet" lest the dominant hydrogen absorption overwhelm any deuterium absorption which may be present. Difficult as such a search may be, the rewards could be great.

Although there exists data on molecular deuterium in the Galaxy, its analysis is complicated by uncertainties in interstellar chemistry (c.f., Boesgaard and Steigman 1985; Dalgarno and Lepp 1984). It is, however, fair to say that the deuterium abundances inferred from studies of deuterated molecules are consistent with those of atomic deuterium.

The solar system provides a sample of the interstellar gas in the Galaxy some 4.6 Gyr ago. Although the metallicity of the galactic disk has apparently not changed dramatically in that time interval, nevertheless, the solar system data complements that from the present ISM and takes us back to an earlier epoch in the chemical evolution of the Galaxy. The deuterium abundance in the presolar nebula may be inferred from studies of deuterated molecules in the atmospheres of the giant planets (Kunde et al. 1982; Encrenaz and Combes 1982). Here, too, however, the inferred presolar abundances are complicated by considerations of the chemistry of the nebula and by physical processes occurring in the atmospheres of the giant planets. The conclusion, that $(D/H)_\odot \approx 1 - 3 \times 10^{-5}$, is consistent with the meteoritic data to be reviewed next.

The meteorites provide a valuable sample of solar system material but, care must be taken when inferring abundances in the presolar nebula from the meteoritic data. For example, the deuterium abundance in the Earth and the meteorites are substantially higher – by factors of 4-40 – than those measured in the giant planets and inferred (to be described next) from the ^3He observed in the meteorites and the solar wind. Therefore, I will postpone the discussion of the presolar deuterium abundance to include it along with the analysis of the ^3He abundance from the meteoritic data.

2.2 HELIUM-3

During the approach of the Sun to the main sequence any presolar deuterium would have been burned to ^3He. Thus, the solar wind ^3He abundance should provide a measure of the presolar abundance of D plus ^3He. From studies of the gas-rich meteorites, lunar soil and breccias and the solar wind (Jeffrey and Anders 1970; Black 1972; Geiss et al. 1970), the ratio of D plus ^3He to ^4He is determined. The 95 % confidence level range is,

$$3.65 \leq 10^4 \left[(D + {}^3He)/{}^4He \right]_\odot \leq 4.41. \tag{2}$$

In contrast, in the carbonaceous chondrites – believed to provide a sample of the most primitive solar system material – there is a component of ^3He which Black (1971, 1972) and Geiss and Reeves (1972) have interpreted as representative of presolar ^3He uncontaminated by ^3He from the burning of presolar D. The data of Frick and Moniot (1977) and of Eberhardt (1978) yield a 95 % CL range of

$$1.44 \leq 10^4 ({}^3He/{}^4He)_\odot \leq 1.60. \tag{3}$$

If, indeed, this interpretation of the two components of ^3He in the meteorites is correct, then (2) and (3) may be used to infer the presolar abundance of deuterium.

$$2.05 \leq 10^4 (D/{}^4He)_\odot \leq 2.97. \tag{4}$$

In order to find the presolar D/H and $^3He/H$ ratios, it is necessary to know the solar helium-4 abundance. From observations of solar prominences, Heasley and Milkey (1978) found $(^4He/H)_\odot = 0.100 \pm 0.025$. Recent, "best" solar models (Bahcall and Ulrich 1988; Turck-Chièze et al. 1988; Guenther, Jaffe and Demarque 1989; Sackmann, Boothroyd and Fowler 1990) are consistent with $0.27 \leq Y_\odot \leq 0.28$ (where Y is the helium-4 mass fraction) which, for a solar "metal" abundance of $Z_\odot = 0.02$, corresponds to $0.095 \leq (^4He/H)_\odot \leq 0.100$. To be cautious, I will arbitrarily adopt a "95 % CL" range of $0.26 \leq Y_\odot \leq 0.30$ which corresponds to $0.090 \leq (^4He/H)_\odot \leq 0.110$. In concert with (2)-(4), this leads to our inferred, presolar abundances of D, ^3He and $D + {}^3He$,

$$1.85 \leq 10^5 (D/H)_\odot \leq 3.27, \tag{5a}$$

$$1.30 \leq 10^5 (^3He/H)_\odot \leq 1.76, \tag{5b}$$

$$3.29 \leq 10^5 \left[(D + {}^3He)/H \right]_\odot \leq 4.85. \tag{5c}$$

Note that the deuterium abundance inferred from the meteoritic data is entirely consistent with that inferred from deuterated molecules in the atmospheres of the giant planets. The presolar deuterium abundance is also consistent with that in the ISM; however, there is a hint that the deuterium abundance may have decreased in the 4.6 Gyr since the formation of the solar system.

Often, it is useful to express the abundances as mass fractions. For deuterium and helium-3, the meteoritic ratios can be used along with the solar helium-4 mass fraction to infer,

$$X_{2\odot} = 1/2 \, Y_\odot \, (D/^4He)_\odot, \tag{6a}$$

$$X_{3\odot} = 3/4 \, Y_\odot \, (^3He/^4He)_\odot, \tag{6b}$$

$$X_{23\odot} = 1/2 \, Y_\odot \left[(D + {}^3He)/^4He + 1/2(^3He/H) \right]_\odot, \tag{6c}$$

where $X_{23} \equiv X_2 + X_3$. From the results summarized in (2)-(4) and, for $0.26 \leq Y_\odot \leq 0.30$, 95 % CL ranges for the presolar mass fraction may be derived.

$$2.7 \leq 10^5 \, X_{2\odot} \leq 4.5, \tag{7a}$$

$$2.8 \leq 10^5 \, X_{3\odot} \leq 3.6, \tag{7b}$$

$$5.7 \leq 10^5 \, X_{23\odot} \leq 7.8. \tag{7c}$$

The present abundance of ^3He in the ISM has been probed by studying the hyperfine line of singly ionized ^3He in Galactic HII regions (Rood, Wilson and Steigman 1979; Rood, Bania and Wilson 1984; Bania, Rood and Wilson 1987). ^3He has been detected in 9 of 17 HII regions observed and the derived abundances range from $^3He/H = 1.2 \times 10^{-5}$ to $^3He/H = 14.7 \times 10^{-5}$. This range in the inferred ^3He abundance would appear to confirm the prediction of Rood, Steigman and Tinsley (1976) that stellar production of ^3He (in $\sim 1-2\ M_\odot$ stars) should have increased the interstellar abundance of ^3He since the formation of the solar system 4.6 Gyr ago. Although the highest ^3He abundance is found for $W3$ which lies in the outer Galaxy where it might have been supposed there was less chemical processing than in the solar vicinity, Fich and Silkey (1991) have found "normal" metal abundances for a sample of such outer Galaxy HII regions. The data of Bania, Rood and Wilson (1987) are consistent with the conclusion that $(^3He/H)_{ISM} \gtrsim (^3He/H)_\odot$ but, it is difficult – and dangerous! – to try to infer the primordial abundance of ^3He from this data. Therefore, it is those results summarized in (5) and (7) which will be used to infer bounds to the primordial abundances of D and ^3He. But first, we turn to a discussion of ^7Li.

2.3 LITHIUM-7

In complementary observational approaches, lithium is observed on the surfaces of stars and in the interstellar gas in the Galaxy. Because of its extremely low abundance, lithium has yet to be observed outside the Galaxy. Each approach to the lithium abundance (stars/gas) has its assets and liabilities.

Stars offer the possibility of tracking the evolution of lithium through a comparison of its abundance in metal-rich, solar type stars (Pop I) with that inferred from observations of metal-poor, halo stars (Pop II). However, ^7Li is a very fragile nucleus which burns at the relatively low temperature of $\sim 2 \times 10^6 K$. Thus, lithium in stellar interiors is burned away and only that lithium which is unmixed from the surface to the interior may survive. Indeed, for the Sun, the surface lithium abundance is observed to be depleted by more than a factor of 100 from the meteoritic value of $(^7Li/H)_{met} = 2 \times 10^{-10}$ (Grevesse and Anders 1989) suggesting that during 4.6 Gyr of solar evolution, surface lithium has been convected to the interior and destroyed. It is, therefore, necessary to restrict attention to those hotter stars where convection, and its consequent destruction of surface lithium, may be minimized. In the hotter stars, the observed lithium abundance may provide a fair sample of the lithium abundance in the gas out of which the star formed. However, it is neutral lithium that provides the optically observed lines and, in hotter stars less of the lithium is neutral and the lines become unobservably weak. Furthermore, the warmer stars are more massive and evolve more quickly so that the only remaining Pop II stars are cool and convection may have depleted their surface lithium. It is in this context that the extraordinary importance of the discovery by the Spites (1982 a,b) of lithium in extreme Pop II halo stars can, perhaps, be best appreciated. We return to the comparison and implications of the Pop I and Pop II lithium observations shortly.

Although the interstellar observations offer the lithium abundance at present, there are many problems here too. As with the stellar observations, it is neutral lithium which is observed so that a large correction for unseen LiII must be made. When this is done, it is found that along all lines-of-sight in the Galaxy for which lithium has been observed, the derived lithium abundance falls short of the solar system lithium abundance. This deficiency is interpreted as evidence for depletion of lithium onto grains or in molecules. By "correcting" to the solar system value, of course we learn nothing about the "true" interstellar abundance. Even if the interstellar abundance of lithium were known, it would only provide a snapshot – here and now – of the overall lithium evolution. To learn if lithium has been increasing or decreasing since the big bang, the lithium abundance in a galaxy at a different evolutionary stage than our own would be of immense value. SN 1987A, which illuminated the interstellar gas in the Large Margellanic Cloud, provided just such an opportunity. But first, the stellar lithium data.

The stellar data reviewed by Boesgaard and Steigman (1985) has been supplemented by much recent work on lithium in field stars and in Pop I clusters ranging in age from 50-70 Myr for α Per and the Pleiades to 630-760 Myr for the Hyades and Praesepe to 1.7 Gyr for NGC 752 (Hobbs and Pilachowski 1986a; Pilachowski, Booth and Hobbs 1987; Boesgaard 1987a,b; Boesgaard and Budge 1988; Boesgaard, Budge and Burck 1988; Boesgaard, Budge and Ramsay 1988). The previously noted trend to higher lithium abundances (less convection and nuclear burning) in the hottest stars is confirmed and, the data are consistent with

$$(Li/H)_{\text{PopI}} \approx 1 - 2 \times 10^{-9}. \tag{8}$$

The older Pop I clusters M67 (Hobbs and Pilachowski 1986b) and NGC188 (Hobbs and Pilachowski 1988) have lower maximum lithium abundances, presumably – but, not necessarily – because they lack the hottest ($\gtrsim 6700$ K) stars. The derived interstellar lithium abundances are all *below* this Pop I value showing that lithium is depleted in the ISM of the Galaxy (Steigman 1991).

As already mentioned, the observed low abundance of lithium in relatively cool ($\lesssim 6000$ K) Pop I stars, led to the expectation that in such cool, but older, Pop II stars, lithium would have been unobservable. In stark contrast, the initial discovery of Pop II lithium (Spite and Spite 1982a,b) has been supplemented by many subsequent observations (Spite, Maillard and Spite 1984; Spite and Spite 1986; Beckman, Rebolo and Molaro 1986; Hobbs and Duncan 1987; Hobbs and Pilachowski 1988; Rebolo, Molaro and Beckman 1988; Spite et al. 1989). The 35 Pop II stars with

$[Fe/H] \leq -1.3$ and $T_{eff} \geq 5500$ K , all have lithium abundances in the range: $-10.16 \leq log(Li/H)_{\text{PopII}} \leq -9.68$. The cooler Pop II stars *do* show the depletion pattern familiar from the Pop I stars. That the "Spite plateau" in Li vs. Fe abundances is so narrow, is in support of the suggestion (Spite and Spite 1982a,b) that the plateau value is a measure of the primordial abundance of lithium. For the 35 stars with $T_{eff} \geq 5500$ K and $[Fe/H] \leq -1.3$, the inferred Pop II lithium abundance is (Walker et al. 1991),

$$12 + log(Li/H)_{\text{PopII}} = 2.08 \pm 0.07, \tag{9}$$

where the dispersion is the 2σ error in the mean.

The interstellar observations of lithium can complement the stellar data. The problems of correcting for unobserved LiII and for lithium depletion have already been mentioned. Indeed, since it is not known (quantitatively) how to correct for depletion along individual lines-of-sight, it is in principle impossible to derive the *absolute* lithium abundance from interstellar absorption data. However, it is possible to compare lithium and potassium data to learn about *relative* lithium abundances (Steigman 1991). Both species are observed in their neutral state but, are predominantly singly ionized in the ISM. In correcting the observed LiI/KI ratio to the Li/K ratio, the uncertainties in the photoionization and recombination rates tend to cancel. Furthermore, the *relative* lithium and potassium depletions may be similar due to their similar condensation temperatures (Field 1974) and/or their similar first ionization potentials (Snow 1975). Thus, although, neither the lithium nor the potassium intrstellar abundance is well-known, the observed ratio may be representative of the true interstellar ratio. For the seven lines-of-sight in the Galaxy for which observations of LiI and KI are available, the Li/K ratio, corrected for ionization (Pequignot and Aldrovandi 1986), is (Steigman 1991)

$$log(Li/K)_{\text{ISM}} = -1.80 \pm 0.11. \tag{10}$$

This compares remarkably well with the solar system value (Grevesse and Anders 1989) of

$$log(Li/K)_{\odot} = -1.82 \pm 0.07, \tag{11}$$

suggesting – but not proving – that the true interstellar Li/K ratio has the solar value. This "coincidence" is precisely what would be expected if the majority of the interstellar lithium (and, all of the observed potassium) has been produced in stars.

As alluded to earlier, SN87A "lit up" the interstellar medium of the LMC, permitting a search for lithium in the LMC (Vidal-Madjar et al. 1987; Baade and Magain 1988; Sahu, Sahu and Pottasch 1988; Malaney and Alcock 1989). No lithium absorption feature associated with gas in the LMC was seen. From the data of Baade and Magain (1988), the limit is tantalizingly close to the solar system and Galactic ISM values,

$$log(Li/K)_{LMC} \leq -1.87. \tag{12}$$

The implications of this limit will be discussed when we try to assess the primordial abundances in the next section.

2.4 HELIUM-4

Finally, we review the data on ^4He determinations in extragalactic HII regions. This has been the major concern of several recent papers (Pagel, Terlevich and Melnick 1986; Pagel 1988; Pagel and Simonson 1989; Pagel 1990; Walker et al. 1991) so that here I shall simply summarize the conclusions.

Helium, the second most abundant element in the Universe can be observed in stars and in ionized gas (HII regions, planetary nebulae, etc.) throughout the Universe. In proceeding from observations to the inferred primordial abundance, several complications intrude which guide us towards extragalactic HII regions as the objects of choice. In the first place, ^4He is produced as stars burn their hydrogen. To minimize the contamination from the debris of stellar evolution, it is advisable to concentrate on low "metal-abundance" objects. Accurate helium determinations in the cool, Pop II stars are difficult. Extragalactic HII regions of low heavy element abundance provide the possibility of tracking the ^4He abundance to its primordial value (Peimbert and Torres-Peimbert 1974; Lequeux et al. 1979). In HII regions the helium is ionized and recombination lines from $^4He^+$ and $^4He^{++}$ are observed but, neutral helium is unobserved. To minimize the correction for $^4He^o$, it is advisable to concentrate on those high-excitation HII regions ionized by very hot stars; again, this points to low Z, extragalactic HII regions. Since helium is abundant and its emission lines strong, it is possible to determine the helium abundance to a much higher accuracy than is common for abundance determinations in ionized nebulae. It is, therefore, necessary to consider systematic uncertainties which ordinarily could be ignored (Davidson and Kinman 1985; Davidson et al. 1989; Pagel and Simonson 1989). In general, to achieve the accuracy offered by the signal to noise of the data requires that, in addition to the data on ^4He and H, the HII region temperature (where the helium emission is produced), electron density, dust to gas ratio as well as the spectra of the ionizing stars be determined. At the few percent level, the systematic effects of neutral helium, (Dinnerstein and Shields 1986), collisional excitation (Berrington and Kingston 1987; Clegg 1987) and dust (Cota and Ferland 1988) can not be ignored. Conversely, at the few percent level, the uncertainties may well be mainly systematic. This remark is intended as a caution in interpreting the sigmas quoted in the determination of the primordial abundance of helium-4; the uncertainties are unlikely to be gaussian distributed.

It is clear that to minimize the extrapolation from the observed ^4He abundance (more precisely, the ^4He abundance derived from observational data), it is best to restrict attention to the lowest heavy element (typically oxygen or nitrogen) abundance HII regions. On the other hand, if only one or a few of the most metal poor HII regions are considered, the statistical uncertainties will dominate the inferred primordial abundance. Pagel (1988; 1990) and Walker et al. (1991) have considered the ^4He abundance in roughly three dozen of the most oxygen/nitrogen-poor, extragalactic HII regions. Here, I quote the results of the analysis of Walker et al. (1991), which are in agreement with the conclusions of Pagel (1990). The data set in Walker et al. (1991) consists of 36 HII regions with $O/H \leq 2 \times 10^{-4}$, $N/H \leq 1 \times 10^{-5}$; recall that the solar abundances are: $(O/H)_\odot = 8.5 \times 10^{-4}$, $(N/H)_\odot = 1.1 \times 10^{-4}$ (Grevesse and Anders 1989).

The helium/oxygen data set is strongly correlated; the probability is less than 10^{-3} that they are drawn from an uncorrelated parent population. With equal weights ($\sigma_Y = \pm 0.009$) the fit which minimizes χ^2 (reduced $\chi^2 = 0.95$) is

$$Y = 0.229 \pm 0.004 + (1.3 \pm 0.3) \times 10^2 (O/H). \tag{13}$$

In (13), Y is the ^4He mass fraction; for very low Z HII regions, $Y = 4y(1 + 4y)^{-1}$ where $y = {}^4He/H$. Alternately, for the 10 most oxygen-poor HII regions $(O/H \leq 8 \times 10^{-5}) : \langle Y \rangle = 0.236 \pm 0.009$.

Alternately, the helium/nitrogen data set is also very well correlated $(P < 10^{-4})$; indeed, this is to be expected since the nitrogen/oxygen data is also highly correlated. A linear fit with equal weights ($\sigma_Y = \pm 0.009$) which minimizes χ^2 (reduced $\chi^2 = 0.81$) is

$$Y = 0.231 \pm 0.003 + (2.8 \pm 0.7) \times 10^3 (N/H). \tag{14}$$

For the 10 most nitrogen-poor HII regions $(N/H \leq 2.4 \times 10^{-6}) : \langle Y \rangle = 0.234 \pm 0.009$. The inference of the primordial helium abundance (or, a bound thereto) is postponed to the next section.

3. The Inferred Primordial Abundances

The data assembled in the previous section may be used to infer the primordial abundances of the light elements. However, to derive the primordial abundances from the observational data requires some knowledge/model of galactic chemical evolution. Since the model dependent uncertainties of such extrapolations may overwhelm the observational uncertainties, caution (conservatism?) suggests we content ourselves with (nearly) model independent bounds to the primordial abundances. Although this is the approach to be adopted here, I will comment on some work in progress with M. Tosi (Bologna) in which the Galactic evolution of D and ^3He have been followed.

3.1 DEUTERIUM AND HELIUM-3

It is useful to consider D and ^3He together. The reason is that when gas is incorporated into stars, deuterium is quickly burned to helium-3 but, some ^3He survives subsequent stellar nucleosynthesis (Yang et al. 1984; Dearborn, Schramm and Steigman 1986). Hence, if there were extensive processing of the interstellar gas in the Galaxy prior to the formation of the solar system, primordial deuterium may have been destroyed only to reappear as enhanced ^3He.

Since deuterium is only destroyed in the course of galactic chemical evolution (Epstein, Lattimer and Schramm 1976), the solar system deuterium abundance provides a lower bound to the primordial value. From (5a),

$$(D/H)_P \geq (D/H)_\odot \geq 1.8 \times 10^{-5}. \tag{15}$$

To obtain an upper bound, we use D plus ^3He. Any gas cycled through a generation of stars will have all of its deuterium but only a fraction, g_3, of its helium-3 destroyed (Yang et al. 1984). Ignoring production of ^3He in low mass stars (Iben 1967), the primordial abundances are related to the later abundances (e.g., in the presolar nebula) by,

$$y_{23P} + (g_3^{-1} - 1)(y_{3P}/y_{2P})y_2 \leq y_{23} + (g_3^{-1} - 1)y_3. \tag{16}$$

In (16), $y_2 \equiv D/H$, $y_3 \equiv {}^3He/H$ and $y_{23} \equiv y_2 + y_3$; the inequality reflects the neglect of stellar produced ^3He. Since the second term on the left-hand side of (16) is positive definite, it may be ignored; this reinforces the inequality and relates the primordial abundance (y_{23P}) to the observed abundances (Yang et al. 1984).

$$y_{23P} < y_{23} + (g_3^{-1} - 1)y_3. \tag{17}$$

Although the "true" value of y_{23P} may be *smaller* than that derived from (17) because of the neglect of the term from (16), it could also be *larger* because these equations represent only "one-cycle" of stellar processing; any gas which has been through several cycles will have had more of its ^3He destroyed. The only ways to assess the importance of the neglect of many cycles are to compare these results with those derived from evolution models and/or from the "instantaneous recycling approximation" (IRA). Since the bound from (17) is in agreement with that derived from the IRA (Walker et al. 1991), I will restrict myself here to (17). From Dearborn, Schramm and Steigman (1986), $g_3 \gtrsim 1/4$; using the upper bounds to the solar system values of y_3 ($y_{3\odot} \leq 1.76 \times 10^{-5}$) and $y_{23}(y_{23\odot} \leq 4.85 \times 10^{-5})$, the sum of the primordial abundances of D plus ^3He is bounded by

$$y_{23P} \leq 1.0 \times 10^{-4}. \tag{18}$$

3.2 LITHIUM-7

The oldest and most metal-deficient Pop II stars which are sufficiently warm to have – presumably – minimized convection (Deliyannis, Demarque and Kawaler 1990), have their surface lithium abundance constrained to a very narrow range,

$$(Li/H)_{Pop\ II} = 1.2 \pm 0.2 \times 10^{-10}. \tag{19}$$

If, indeed, this were representative of the abundance of lithium in the gas out of which these stars formed, then we would bound the primordial abundance of lithium: $(Li/H)_P \geq (Li/H)_{Pop\ II}$. However, some destruction of lithium may have occurred even in these extreme Pop II stars (Deliyannis et al. 1990). With allowance for this effect (which, in any case, is small) we choose,

$$(Li/H)_P \leq 2.3 \times 10^{-10}. \tag{20}$$

3.3 HELIUM-4

The HII region data on ^4He – and its correlation with oxygen and nitrogen – summarized in (13) and (14) is consistent with a primordial abundance whose formal, 2σ ($\sim 95\ \%$ CL) range is encompassed by,

$$Y_P = 0.230 \pm 0.007. \tag{21}$$

The upper bound to Y_P from the mean value of Y for the 10 most oxygen/nitrogen-poor HII regions, $Y_P \leq 0.235 \pm 0.009(1\sigma)$, is consistent with (21) but, less constraining. As was emphasized, however, systematic uncertainties may well dominate the small statistical uncertainties. *Guessing* that the systematic uncertainties are comparable to the statistical ones (if the systematic corrections were known, they would have been made) and adding them in quadrature, we obtain for our best estimate of the 95 % CL range for the primordial abundance of ^4He,

$$0.220 \leq Y_P \leq 0.240. \tag{22}$$

The reader is strongly cautioned to regard with skepticism, the third significant figure.

3.4 MODEL DEPENDENT BOUNDS TO PRIMORDIAL DEUTERIUM AND HELIUM-3

Before proceeding to compare the bounds to the primordial abundances of $D, {}^3He, {}^4He$ and ${}^7Li$ inferred in this section from the observational data summarized in the previous section, I will present the results of some work in progress with M. Tosi (Bologna) which uses published models of galactic chemical evolution (Tosi 1988) to follow the evolution of D and ${}^3He$.

Of the large number of models considered by Tosi (1988), models # 1 and # 25 provide the best fit to the many observational constraints. Although we are considering a wider class of models (Steigman and Tosi 1991), the results presented here from # 1 and # 25 are typical in their predictions of the evolution of D and ${}^3He$. In these models, for which $Y_P = 0.23$ and the present age of the Galaxy is 13 Gyr, at the time of the formation of the solar system 4.6 Gyr ago, the primordial deuterium mass fraction had been reduced by a factor f_2^{-1} (where $X_{2\odot} = f_2 X_{2P}$) which is

$$f_2^{-1}(\# \ 1) = 1.5, \tag{23a}$$

$$f_2^{-1}(\# \ 25) = 1.8. \tag{23b}$$

For $X_{2\odot} \gtrsim 2.7 \times 10^{-5}$ (see eq. 7a), this corresponds to a *lower* bound to the primordial abundance of

$$X_{2P} \gtrsim 4.0 \times 10^{-5} \tag{24}$$

which, for $0.22 \leq Y_P \leq 0.24$, is equivalent to

$$y_{2P} \gtrsim 2.6 \times 10^{-5}. \tag{25}$$

In following the D plus ${}^3He$ evolution, we find that consistency with the solar system mass fraction $X_{23\odot} \lesssim 7.8 \times 10^{-5}$ (see eq. 7c) requires that

$$X_{23P} \lesssim 9.7 \times 10^{-5}. \tag{26}$$

These, model dependent, constraints on the primordial abundances of D and ${}^3He$ are entirely consistent with – albeit tighter than – the less model dependent bounds derived earlier in eq. (15) and eq. (18).

4. Comparison With The Predictions Of The Standard Model

Having reviewed the observational data and the bounds to the primordial abundances inferred therefrom, we are now in a position to compare with the predictions of nucleosynthesis in the standard, hot big bang cosmological model (Boeasgaard and Steigman 1985 and references therein). The comparison here will be to the predictions of Walker et al. (1991) who have updated the nuclear reaction network.

The *lower* bound to the primordial abundance of D, (eq. 15), leads to an *upper* bound on the nucleon-to-photon ratio ($\eta_{10} \equiv 10^{10}\eta$; $\eta \equiv n_N/n_\gamma$)

$$\eta_{10}(D) \leq 6.8. \tag{27}$$

The *upper* bound to primordial D plus ^3He, eq. (18), yields a *lower* bound to η,

$$\eta_{10}(D + {}^3He) \geq 2.8. \tag{28}$$

Thus, the standard model is consistent with the inferred bounds to the primordial abundances of deuterium and helium-3 provided that the true nucleon-to-photon ratio lies in the narrow range: $2.8 \leq \eta_{10} \leq 6.8$. Is this range consistent with the abundances of the remaining light elements, ^7Li and ^4He?

Using the "Pop II Plateau" value of the lithium abundance and, allowing, for some modest possible destruction in these stars, we were led to the bound (eq. 20) on the primordial abundance of lithium. The standard model prediction is consistent with this bound provided that,

$$1.5 \leq \eta_{10}(^7Li) \leq 4.2. \tag{29}$$

Comparing (27)-(29), reveals that consistency among D, ^3He and ^7Li is achieved for

$$2.8 \leq \eta_{10}(D, {}^3He, {}^7Li) \leq 4.2. \tag{30}$$

What of consistency with ^4He?

For the standard case of 3 light neutrino flavors ($N_\nu = 3$) and, for the neutron lifetime in the (95 % CL) range: $882 \leq \tau \leq 896$ min., with η_{10} bounded as in (30), the predicted primordial mass fraction of ^4He (Walker et al. 1991) lies in the range,

$$0.236 \leq Y_P \leq 0.243. \tag{31}$$

There is, indeed, consistency between the inferred primordial abundance of ^4He (see eq. 22) and that predicted by the standard, hot big bang model. Although the reader may have noted an apparent discrepancy between the *central* values of the observed ($Y_P = 0.23$) and predicted ($Y_P =$

0.24) primordial helium-4 mass fractions, he ignores the observational and theoretical uncertainties at his own peril.

The narrow overlap in consistency between the prediction (eq. 31) and the data (eq. 22) for the standard $N_\nu = 3$ case, leads to a very strong bound to the number of "new", light, weakly-interacting particles (Steigman, Schramm and Gunn 1977). For $\Delta N_\nu \equiv N_\nu - 3 \neq 0$, the predicted primordial ^4He mass fraction changes by (Walker et al. 1991),

$$\Delta Y_P = 0.012 \Delta N_\nu. \tag{32}$$

The predicted *lower* bound will exceed the inferred *upper* bound to Y_P unless $N_\nu \leq 3.3$ (Walker et al. 1991).

5. Closing In On The Universal Density Of Nucleons

The bounds to the primordial abundances of the light elements inferred from observational data were derived and compared to the predictions of nucleosynthesis in the standard hot big bang model ($N_\nu = 3$). Theory and observation are consistent provided that the nucleon-to-photon ratio is restricted to the narrow range given in eq. (30). It was noted, however, that if a chemical evolution model (Tosi 1988) was used to follow the evolution of the abundances of D and ^3He (Steigman and Tosi 1991), then the bounds on primordial D and ^3He could be restricted further (see eqs. 24 and 26). These tighter bounds will result in a correspondingly smaller range in η allowed for consistency with standard big bang nucleosynthesis. Therefore, it is worth seeing if consistency can be preserved with the abundances of all four light elements. Comparing the predictions of Walker et al. (1991) with the lower bound to the primordial deuterium mass fraction (eq. 24),

$$\eta_{10}(D) \leq 5.5, \tag{33}$$

and, with the upper bound to that of D plus ^3He (eq. 26),

$$\eta_{10}(D + {}^3He) \geq 4.0. \tag{34}$$

Although we have now relied on a model for galactic chemical evolution, the range of consistency has been restricted considerably. Indeed, comparing with the – unmodified – bound from ^7Li (eq. 29), we find we have "zeroed-in" on the cosmic abundance of nucleons.

$$4.0 \leq \eta_{10}(D, {}^3He, {}^7Li) \leq 4.2. \tag{35}$$

For this narrow range in η and, for $N_\nu = 3$ and $882 \leq \tau \leq 896$ min., the primordial abundance of helium is predicted to be restricted to the range,

$$0.240 \leq Y_P \leq 0.243. \tag{36}$$

Since the *lower* bound in eq. 36 and the observational *upper* bound (eq. 22) now coincide, to the extent that the third significant figure can be claimed to be known, there is no room for any further equivalent neutrino species: $N_\nu \leq 3.0$!

6. Cosmological Consequences

The baryon density parameter Ω_B, the ratio of the present values of the baryon and critical mass densities, is proportional to the nucleon-to-photon ratio η.

$$\Omega_B = 3.73 \times 10^{-3} \, \eta_{10} h^{-2} T_{2.75}^3. \tag{37}$$

In eq. 37, h is the Hubble constant in units of 100 kms^{-1} Mpc^{-1} and $T_{2.75}$ is the cosmic background radiation temperature in units of 2.75 K. For $2.75 \leq T \leq 2.79$ K (Walker et al. 1991) and the "conservative" range of η from eq. 30,

$$0.010 \leq \Omega_B h^2 \leq 0.016. \tag{38}$$

The largest residual uncertainty in bounding Ω_B comes from the uncertainty in the Hubble constant H_0. Sandage and Tammann (1990) claim a "99 % CL" range of $40 \leq H_0 \leq 76$ kms^{-1}Mpc^{-1}. For $h \geq 0.4$, $\Omega_B \lesssim 0.10$ and the Universe fails to be closed by nucleons by an order of magnitude. If the lower bound in eq. 38 is compared to the density of luminous matter in galaxies (Pagel 1989), $\Omega_{LUM} \approx 0.07$, it is revealed that most of the baryons in the Universe are "dark".

$$\Omega_B/\Omega_{LUM} \gtrsim 1.5 \, h^{-2}. \tag{39}$$

The more restrictive range of allowed η values (eq. 35) reinforces the above conclusion about dark nucleons: $\Omega_B/\Omega_{LUM} \gtrsim 2.1 h^{-2}$, and maintains the conclusion that nucleons can contribute – at most – only 10% of the critical density. That this range is so narrow ($4.0 \leq \eta_{10} \leq 4.2$) suggests that further observational work could provide a crucial test of the overall consistency of the standard hot big bang model. Already, from this analysis of primordial nucleosynthesis, the universal density of baryons has been bounded with an accuracy which make it one of the best determined of all cosmological parameters.

ACKNOWLEDGEMENTS

I have great pleasure in thanking Professor K. Sato and his hard-working students and colleagues for having been such excellent hosts of such a successful conference. I also wish to acknowledge my colleagues K. Olive, D.

Schramm, M. Tosi and T. Walker and my student H. Kang for permitting me to quote liberally from our joint work. I also owe an incalculable debt to B. Pagel from whom I've learned much about the art of abundance determinations. This research has been supported at the Ohio State University by the DOE. This manuscript was prepared at the Max-Planck-Institut für Physik in Munich where I am an Alexander von Humboldt Awardee and I am pleased to thank the Humboldt Stiftung for support and the MPI for hospitality.

REFERENCES

Baade, D. and Magain, P. (1988) Astron. Astrophys. **351**, 31.

Bahcall, J.N. and Ulrich, R.K. (1988) Rev. Mod. Phys. **60**, 297.

Bania, T.M., Rood, R.T. and Wilson, T.L. (1987) Ap. J. **323**, 30.

Beckman, J.E., Rebolo, R. and Molaro, P. (1986) in Advances in Nuclear Astrophysics (Editions Frontières) p. 29.

Berrington, K.A. and Kingston, A.E. (1987) J. Phys. **B20**, 6631.

Black, D.C. (1971) Nature Phys. Sci. **234**, 148.

Black, D.C. (1972) Geochim. Cosmochim. Acta **36**, 347.

Boesgaard, A.M. (1987a) Ap. J. **321**, 967.

Boesgaard, A.M. (1987b) P.A.S.P. **99**, 1067.

Boesgaard, A.M. and Budge, K.G. (1988) Ap. J. **322**, 410.

Boesgaard, A.M., Budge, K.G. and Burck, E.E. (1988) Ap. J. **325**, 749.

Boesgaard, A.M., Budge, K.G. and Ramsay, M.E. (1988) Ap. J. **327**, 389.

Boesgaard, A.M. and Steigman, G. (1985) Ann. Rev. Astron. Astrophys. **23**, 319.

Clegg, R.E.S. (1987) MNRAS **229**, 31.

Cota, S.A. and Ferland, G. (1988) Ap. J. **326**, 889.

Dalgarno, A. and Lepp, S. (1984) Ap. J. Lett. **287**, L47.

Davidson, L. and Kinman, T.D. (1985) Ap. J. Suppl. **58**, 321.

Davidson, K. Kinman, T.D. and Friedman, S.D. (1989) Astron. J. **97**, 1591.

Dearborn, D.S.P., Schramm, D.N. and Steigman, G. (1986) Ap. J. **302**, 35.

Deliyannis, C.P., Demarque, P. and Kawaler, S.D. (1990) Ap. J. Suppl. **73**, 21.

Dinnerstein, H.L. and Shields, G.A. (1986) Ap. J. **311**, 45.

Eberhardt, P. (1978) Proc. 9th Lunar Planet. Sci. Conf., p. 1027.

Encrenaz, T. and Combes, M. (1982) Icarus **52**, 54.

Epstein, R.I., Lattimer, J.M. and Schramm, D.N. (1976) Nature **263**, 198.

Ferlet, R., Vidal-Madjar, A., Laurent, C. and York, D.G. (1980) Ap. J. **242**, 576.

Fich, M. and Silkey, M. (1991) Ap. J. **366**, 107.

Field, G.B. (1974) Ap. J. **187**, 453.

Frick, U. and Moniot, R.K. (1977) Proc. 8th Lunar Planet. Sci. Conf., p. 229.

Geiss, J., Eberhardt, P., Bühler, F., Meister, J. and Signer, P. (1970), J. Geophys. Res. 75, 5972.

Geiss, J. and Reeves, H. (1972) Astron. Astrophys. 18, 126.

Grevesse, N. and Anders, E. (1989) AIP Conf. Proc. 183, 1.

Guenther, D.B., Jaffe, A. and Demarque, P. (1989) Ap. J. 345, 1022.

Hobbs, L.M. and Duncan, D. (1987) Ap. J. 317, 796.

Hobbs, L.M. and Pilachowski, C. (1986a) Ap. J. Lett. 309, L17.

Hobbs, L.M. and Pilachowski, C. (1986b) Ap. J. Lett. 311, L37.

Hobbs, L.M. and Pilachowski, C. (1988) Ap. J. 334, 734.

Iben, I. (1967) Ap. J. 147, 624, 650.

Jeffrey, P.M. and Anders, E. (1970) Geochim. Cosmochim. Acta 34, 1175.

Kunde, V., Hanel, R. et al. (1982) Ap. J. 263, 443.

Laurent, C., Vidal-Madjar, A. and York, D.G. (1979) Ap. J. 229, 923.

Lequeux, J., Peimbert, M., Rayo, J.F., Serrano, A. and Torres-Peimbert, S. (1979) Astron. Astrophys. 80, 155.

Malaney, R.A. and Alcock, C.R. (1990) Ap. J. 351, 31.

Mather, J.C., Cheng, E.S. et al. (1990) Ap. J. Lett. 354, L37.

Pagel, B.E.J. (1988) in A Unified View of the Macro- and the Micro-Cosmos (World Scientific) p. 399.

Pagel, B.E.J. (1989) in Baryonic Dark Matter (Klumer) p. 237.

Pagel, B.E.J. (1990) Nordita Preprint 90/47A.

Pagel, B.E.J., Terlevich, R.J. and Melnick, J. (1986) P.A.S.P. 98, 1005.

Pagel, B.E.J. and Simonson, E.A. (1989) Rev. Mex. Astron. Astrof. 18, 153.

Peimbert, M. and Torres-Peimbert, S. (1974) Ap. J. 193, 327.

Pequignot, D. and Aldrovandi, S.M.V. (1986) Astron. Astrophys. 161, 169.

Pilachowski, C.A., Booth, J. and Hobbs, L.M. (1987) P.A.S.P. 99, 1288.

Rebolo, R., Molaro, P. and Beckman, J. (1988) Astron. Astrophys. 192, 192.

Rogerson, J.B. and York, D.G. (1973) Ap. J. Lett. 186, L97.

Rood, R.T., Steigman, G. and Tinsley, B.M. (1976) Ap. J. Lett. 207, L57.

Rood, R.T., Wilson, T.L. and Steigman, G. (1979) Ap. J. Lett. 227, L97.

Rood, R.T., Bania, T.M. and Wilson, T.L. (1984) Ap. J. 280, 629.

Sackmann, I.-J., Boothroyd, A.J. and Fowler, W.A. (1990) Preprint (submitted to the Ap. J.).

Sahu, K.C., Sahu, M. and Pottasch, S.R. (1988) Astron. Astrophys. 107, L1.

Sandage, A. and Tammann, G.A. (1990) Ap. J. 365, 1.

Snow, T.P. (1975) Ap. J. Lett. 202, L87.

Spite, F. and Spite, M. (1982a) Astron. Astrophys. 115, 357.

Spite, F. and Spite, M. (1982b) Nature 297, 483.

Spite, F. and Spite, M. (1986) Astron. Astrophys. 163, 140.

Spite, F., Maillard, J.P. and Spite, M. (1984) Astron. Astrophys. **141**, 56.

Spite, F. Spite, M., Peterson, R.C. and Chaffee, F.H. (1987) Astron. Astrophys. **172**, L9.

Steigman, G. (1991) "Cosmic Lithium: Going Up Or Coming Down? (In Preparation).

Steigman, G. and Tosi, M. (1991) "Galactic Evolution of Deuterium And Helium-3" (in Preparation).

Tosi, M. (1988) Astron. Astrophys. **197**, 33.

Turck-Chièze, S., Cahen, S., Cassé, M. and Doom, C. (1988), Ap. J. **335**, 415.

Vidal-Madjar, A., Laurent C., Bonnet, R.M. and York, D.G. (1977) Ap. J. **211**, 91.

Vidal-Madjar, A., Ferlet, R., Laurent, C. and York, D.G. (1982) Ap. J. **260**, 128.

Vidal-Madjar, A., Andreani, P., Cristiani, S., Ferlet, R., Lanz, T. and Vladilo, G. (1978) Astron. Astrophys. **177**, L17.

Walker, T.P., Steigman, G., Schramm, D.N., Olive, K.A. and Kang, H.-S. (1991) Ap. J. (In Press, July 20, 1991)

Yang, J., Turner, M.S., Steigman, G., Schramm, D.N. and Olive, K.A. (1984) Ap. J. **281**, 493.

York, D.G. (1983) Ap. J. **264**, 172.

York, D.G. and Rogerson, J.B. (1976), Ap. J. **203**, 378.

LITHIUM, BERYLLIUM AND BORON: OBSERVATIONAL CONSTRAINTS ON PRIMORDIAL NUCLEOSYNTHESIS

J E BECKMAN
Instituto de Astrofísica de Canarias
38200 - La Laguna
Tenerife
Spain

ABSTRACT. Here I summarize the evidence bearing on the derivation of the primordial ^{7}Li abundance from observations of extremely metal deficient stars in the galactic halo (Population II). Arguments, primarily observational, but supported theoretically, are presented to show that the ^{7}Li observed in warm halo subdwarfs has neither been supplemented by ^{7}Li production in the early galaxy, nor significantly depleted during the lifetime of the stars themselves, and gives a good approximation to the primordial value. The absence of ^{9}Be or ^{10}B+^{11}B in these objects at present limits of observational accuracy tends to support the primordiality of their ^{7}Li. Observing ^{9}Be, and heavier nuclides in the most metal deficient stars, does offer a possible route to distinguish between standard (SBB) and inhomogeneous (IBB) primordial cosmological models.

1. Introduction

The remarkable success of the Standard Hot Big Bang model in predicting, at least to first order, the fractions of light nuclides so widely different in their cosmic abundances as ^{4}He($\sim$25%) and ^{7}Li($\sim10^{-10}$) has been a major factor in its acceptance as the point of departure for all serious modern cosmology. It is accepted that, at least in the Standard (homogeneous) model, there are four primordially produced nuclides: D, ^{3}He, ^{4}He, and ^{7}Li, whose abundances can in principle be used to constrain the parameters of the model. These parameters can, in their turn, be translated into two physical observables: the mean life of the neutron τ_n, and the number of different two-component massless neutrino flavours N_ν, plus two astrophysical observables: the Hubble parameter H_o, and the mean baryon density Ω_b. In this article I will review the observational evidence pertinent to the establishment of an agreed value for the primordial abundance of one of the four nuclides cited above, viz. ^{7}Li. Starting with a description of the most relevant observational material, I will discuss the barriers, chiefly astrophysical, to reaching an unchallengeable value of log N(^{7}Li)$_p$, the primordial abuundance of ^{7}Li. I will explain how

23

24

present-day observations of the nuclides ^{7}Li, ^{9}Be, ^{10}B and ^{11}B, which in a homogeneous hot Big Bang were not produced in significant quantities, may be used to help isolate the fraction of N(^{7}Li) produced primordially, and outline the arguments between those who believe that the ^{7}Li that we see today in the oldest stars is representative of the gas mixture from which these were formed (and hence primordial), and those who feel that the bulk of the ^{7}Li with which these stars began their existence has been destroyed during their main sequence lifetimes. Combining what is assessed as the most probable value of log N(^{7}Li)$_p$ with corresponding estimates (especially) ^{4}He, also ^{3}He and D, limits may be set on Ω_b in a standard Big Bang context. In the final section the consequences of non-standard models, based on inhomogeneities arising at the end of the supposed inflationary epoch, during the quark-hadron phase transition, for the light nuclides, are briefly reviewed. I will show that although the light element abundances cannot yet offer a critical test to distinguish standard from non-standard Big Bang scenarios, in some scenarios ^{9}Be could be the key tracer.

2. The ^{7}Li in Halo stars

The major step in demonstrating that measuring log N(^{7}Li)$_p$ is an attainable goal was taken by Spite and Spite (1982) who made observations of the λ 6708Å resonance doublet of Li in solar neighbourhood stars of low metallicity. They found that the lowest

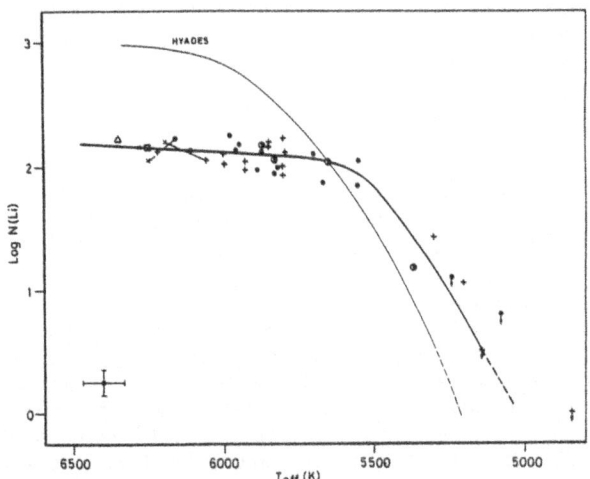

Fig 1: Observed Li abundance v. effective temperature for solar neighbourhood stars with extreme metal deficiency ([Fe/H] $\lesssim$ -1.4), from Rebolo et al. (1988) showing the "Li plateau", at logN (^{7}Li) = 2.15. The curve for a young cluster, the Hyades, is shown for comparison.

metallicity stars with surface temperatures higher than some 5800 K showed a remarkably uniform ^{7}Li abundance, with a value log N(^{7}Li) of 2.1 (on the conventional scale where log N(H)=12). This result was placed on a wider basis of stellar observations subsequently by the Spites themselves (1984, 1986), by Hobbs and Duncan (1987), and by Beckman et al. (1987) and Rebolo et al. (1987, 1988). Figure 1, taken from the last paper cited, gives a summary of measurements for stars with metallicities, [Fe/H] less than -1.4 (where [Fe/H]$_\odot$ = 0), which are presumed to have been formed very early in the galaxy lifetime and which have, in general, kinematics characteristic of the halo. In Fig. 1 we see, as well as the abundance "plateau" discovered by the Spites, a fall-off in log N(^{7}Li) to lower surface temperatures, produced (as is generally agreed) by depletion of ^{7}Li in cooler stars whose sub-surface convective zones are deep enough to transport the ^{7}Li towards the "combustion" level, where at 2.4x10^6 K, it is destroyed. It is tempting to infer from the existence of the plateau that the stars with T_{eff} > 5800 K (and the [Fe/H] < -1.4) have not depleted the ^{7}Li with which they were formed, and to take the abundance value of 2.1, i.e. ^{7}Li/H = 1.25x10^{-10} in these stars as primordial. The fact that this value fits well, in the context of the predictions of standard Big Bang nucleosynthesis (SBBN), the observed value of Y_P, the primordial ^{4}He abundance, with a mutually consistent range of Ω_b between 0.01 and 0.1, seen in Fig. 2, gives support to this hypothesis. However, with such basic parameters at stake one should not take any result for granted, and there have been several reasonable grounds for doubting whether the present-day halo value of N(^{7}Li) is a true measure of the primordial abundance. In this article I will order the discussion by

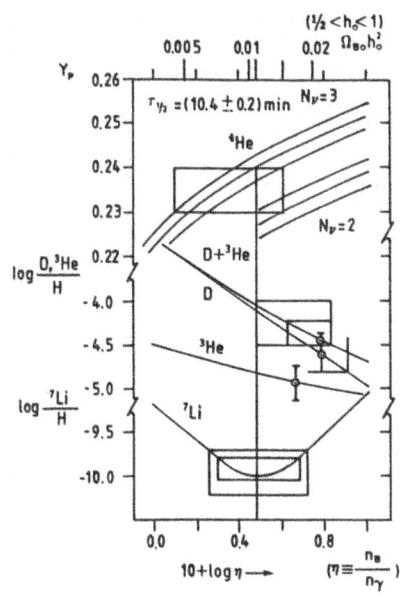

Fig. 2: Standard Big Bang nucleosynthesis production curves for ^{4}He, B, ^{3}He and ^{7}Li, plotted against η, the baryon to photon ratio, also converted to baryon density Ω_b (using T_{BB} = 2.74 K). Note: (1) We can now count on N_ν = 3 from laboratory data. (2) Values of log (^{7}Li/H) higher than -9.5 are inconsistent with the measured ^{4}He abundance in standard (though not necessarily in inhomogeneous) models.

asking whether N(^{7}Li) in Pop. II stars represents "the primordial abundance, the whole primordial abundance and nothing but the primordial abundance" to paraphrase the Anglo-American legal system in circumstances where the substantive questions differ, but where comparable rigour in presenting the evidence is, and should be, required.

3. The primordial ^{7}Li abundance

The question whether at least some important component of the ^{7}Li in population II stars is primordial is readily answered affirmatively. We know that low metallicity stars, traditionally those with low [Fe/H] (but more clearly in analytic terms those with low [O/H]) were formed at an early epoch in the galaxy, and that going back below [Fe/H] = -1, i.e. below 10% of the solar iron metallicity, stars in the solar neighbourhood have velocity vectors which differ markedly from those with higher metallcities. These stars move on orbits which are not, generally, in the plane of the Galactic disc, and are elliptically elongated rather than nearly circular. They were formed during an early contracting phase of the galaxy, before centrifugal support had become important. While it is not at all trivial to discover an accurate chronology for them, these halo stars were formed before most of the heavy element material now observed in the galaxy, within the first 10^9 years of Galactic evolution, as a conservative limit. There is in any case no possibility that any of their observed ^{7}Li could have been produced in normal "steady" stellar nucleosynthesis, since it is well known that fragile ^{7}Li does not survive the temperatures present in the inner zones of stars where nuclear reactions occur. We will examine in Section 4 other possible specific processes which might have contributed to the ^{7}Li in the material from which these metal deficient stars were formed, but in fact it is a problem to produce any ^{7}Li in processes which could have occurred between the Big Bang and the formation of Population II, the halo stars, and for this reason a primordial origin for this ^{7}Li is by far the most probable cause for its presence. This argument is strengthened by the presence of ^{7}Li with an abundance log N(^{7}Li) of ~2.1 in the most metal deficient dwarfs known (Rebolo, Molaro, Beckman, 1987) with [Fe/H] some 3 orders of magnitude below the solar. This is clearly demonstrated in Fig. 3 where the Li abundance in all dwarf and subdwarf field stars (which means stars within a few hundred pc. of the sun which are not in known clusters) with $T_{eff} >$ 5500 K, where it has been measured, is plotted against iron metallicity [Fe/H]. The trend towards a constant value at decreasing [Fe/H] is evidence for an origin for the Li which is not the same as that for the Fe or for other heavy elements which decrease monotonically with Fe, and primordial nucleosynthesis offers a natural explanation.

4. The whole primordial ^{7}Li abundance

Although primordial ^{7}Li is the main candidate for the observed Pop. II abundance, the very reasonable question has been raised about whether, during the long lifetimes of order 10^{10} years, or even more, of these stars on the main sequence, they might not have consumed some, even most of the original ^{7}Li, i.e. whether significant ^{7}Li depletion has occurred. The question springs naturally from two observations. In main sequence stars in open clusters and in the field, the level of ^{7}Li in the warmer objects: F stars with temperatures between 6000 K and 7000 K reach values up to log N(^{7}Li)>3. On the other hand, in even young open clusters there is clear evidence for depletion at T_{eff} below 5500 K, and in older open clusters this depletion reaches to higher temperatures, as illustrated in the comparison between the Hyades (age ~10^8 years) and M67 (age ~5×10^9 yrs); should not the oldest stars of all show the greatest depletion? Perhaps instead of interpreting the Pop. II abundance of log N(^{7}Li)~2 as primordial, we should consider the "young cluster" abundance of log N(^{7}Li)~3 as the true primordial abundance, and the pop. II abundance as the result of the 10^{10} years of depletion. This scenario supposes that no significant ^{7}Li has been produced during the galaxy lifetime, an idea which is compatible with the absence of ^{7}Li production in normal stars.

Protagonists of this "high" primordial ^{7}Li scenario have included Boesgaard (1985), and Hobbs and Pilachowski (1988), who are amongst the most practised observers of stellar Li abundances. A further set of observations which could be considered as somewhat supportive of the "high" primordial value is the presence of a ^{7}Li abundance "gap" in the surface temperature range from 6300 K to 7000 K, i.e. mid-F stars in fairly young clusters, such as the Hyades (see Fig. 4), first found by Boesgaard and Trippico (1986) and systematically investigated by Boesgaard and her coworkers (1986b, 1987, 1988) and by Hobbs and Pilachowski (1986a, b, 1988). This gap does not appear in the youngest clusters such as the Pleiades and α Per (Boesgaard et al., 1989), which implies that the depletion occurs while the stars are on the main sequence rather than in the pre-main sequence phase. The depletion in the Hyades Li gap reaches factors of two orders of magnitude. Clearly anyone supporting the "low" primordial ^{7}Li value needs to explain how Pop. II stars have avoided the depletion effects which lead to this gap, while a supporter of the "high" primordial value must show how the Li depletion which leads to a "gap" in young clusters gives rise to a "plateau" in Pop. II.

To distinguish whether the "low" or "high" ^{7}Li abundance (the "Pop. II" or the "Pop. I" value) is closer to the primordial value is made difficult by the limitations to our knowledge of the depletion mechanisms and how they operate in stars with different masses. If we had a clear and accepted theory for Li depletion we would use it to see, for example, whether the Pop. II "plateau" in Fig. 1 was produced by depletion from a population with an initial Pop. I Li abundance or, on the contrary, could only be the result of an initial abundance from which very little depletion has occurred. In order for the second case to hold, it would be necessary for the depletion rates in metal deficient stars to be much less than in stars with "normal" (near-solar) metallicities. On general physical grounds this is not unlikely, given that the lower opacity of stellar material of low metallicity leads naturally to solutions for the energy transport in stellar interiors which have shallower convection layers. While the various serious attempts to model Li depletion over a range of stellar masses have given results not always consistent among themselves (see e.g. Schatzman, 1977; Michaud, 1986; Vauclair, 1988) the most comprehensive attempt to model the Li abundance in low metallicity main sequence stars, by Deliyannis et al. (1989) is able to match the observed "plateau" rather well. If we go along with the results of the "best fit" model of Deliyannis et al. represented in Fig. 4, we can see that in the T_{eff} range of the Pop. II plateau of ^{7}Li, there has been very little main sequence depletion of this nuclide, i.e. the present Pop. II abundance represents virtually the whole of the ^{7}Li which went into these stars initially, i.e. "the whole primordial abundance".

Advocates of the thesis that the Pop. I abundance is the true primordial value have not, at this point, exhausted the evidence for the defence of their proposal. They have pointed out that the abundance of ^{7}Li measured in open clusters takes the value $\log N(^7Li) \sim 3$, a value persisting back to the oldest clusters. For example, Hobbs and Pilachowski (1988) measured log $N(^7Li)$ in NGC 188, finding a value close to 3 in the T_{eff} range showing least depletion. They estimated the age of NGC 188 as 10^{10} years, using isochrones of Vandenberg (1985). Although a more recent re-examination of the age by Twarog and Twarog (1989) puts it at 6×10^9 years, a yet more recent study by Hobbs et al. (1990) gives 8×10^9 years. Whatever the precise value, the underlying idea is the same: if log $N(^7Li)$ were not the primordial value, how could the galactic ^{7}Li abundance possibly have risen so rapidly between the epoch of formation of the halo stars, and that of the disc stars in the open clusters? In the most unfavourable case, this time interval might not have been greater than a couple of Gyr. Given that stellar mechanisms for ^{7}Li production are uncertain, this could be taken as important evidence for believing that log $N(^7Li) \sim 3$ is the primordial value.

Even before tackling the problem of sources of ^{7}Li in the Galactic disc, however, we can examine the above argument more critically in the light of the observations. Plotting metallicity (e.g. [Fe/H]) against age for open clusters, following Garcia Lopez et al. (1989) we see that there has been little measureable change over an epoch of 5×10^9 years, comparable to the lifetime of the sun and this epoch can be extended to the age of NGC 188 itself. If Fe has maintained a constant abundance over this long period, it need not be so surprising that ^{7}Li has done the same. No-one would try to argue that this constant Fe abundance in the disc implies that Fe is primordial. Explanations for this unchanging Fe abundance have been the province of chemical evolution modelling of the local Galaxy. It is generally agreed that the reason is the constant infall of unprocessed material to the galactic plane (Lynden-Bell, 1975; Clayton, 1988) which has maintained a rough balance between the enrichment of the ISM due to stellar nucleosynthesis and its dilution by this metal-poor inflow.

The most complete models in fact claim that the metallicities of new stars have tended to fall very slowly with epoch over the past 5×10^9 years, but even if this were not the case infall models, which account well for the paucity of low metallicity stars in the solar neighbourhood, account nicely for the observed long-term stability of [Fe/H].

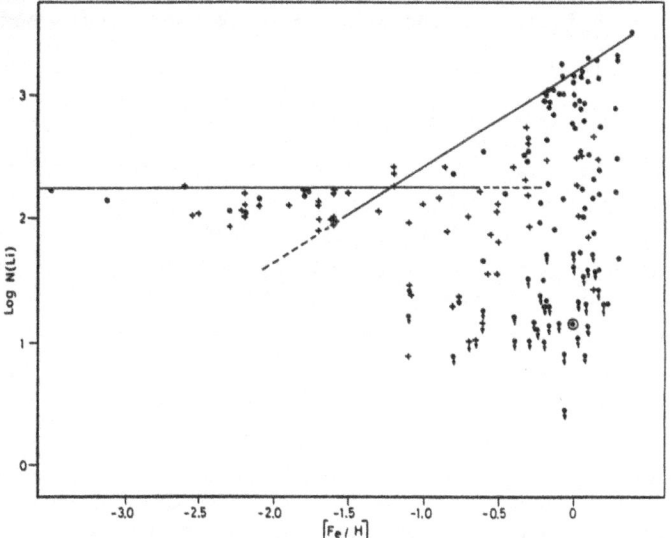

Fig. 3: Observed ^{7}Li abundance against metallicity for local field stars with T_{eff} > 5500 K. Note the lack of scatter for [Fe/H] < -1.4. The horizonal line is the hypothetical primordial abundance, and the sloping envelope shows a presumed galactic production curve (points below this curve imply depleted ^{7}Li).

A further piece of evidence is that offered in Fig. 3, presenting the ^{7}Li abundances against Fe. We can see three important features in Fig. 3: the low-scatter "plateau" in ^{7}Li for [Fe/H] < -1.4, already noted in section 3, the rising upper envelope between [Fe/H] = -1.4 and [Fe/H] > 0, and the widening scatter in log N(^{7}Li) below this envelope. These features have a natural explanation if the rising envelope is caused by ^{7}Li production in the Galactic disc, while the scatter below it is due to differing degrees of ^{7}Li depletion (note that the stars in Fig. 3 cover a range of T_{eff} and of initial rotational velocity, so one would expect a range of ^{7}Li depletion). The fact that the upper envelope of the plot in Fig. 3 between [Fe/H] = -1.4 and [Fe/H] = 0.2 is close to linear implies that the production mechanism for galactic ^{7}Li closely tracks the mechanism for Fe. Plausible sites suggested for this ^{7}Li production include the atmospheres of red giants, (Cameron and Fowler, 1971), novae (Starrfield et al., 1987) and more recently type II supernovae (Dearborn et al., 1989); it is known that ^{7}Li is also produced by spallation of C, N and O in the interstellar medium by galactic cosmic rays (Reeves et al., 1970; Meneguzzi et al., 1971), but that this cannot account for more than 10% of the observed Pop. I abundance. The linear ^{7}Li envelope dependence of Fe in Fig. 3 tells against supernovae as the prime sources of galactic ^{7}Li, because in that case the ^{7}Li abundance would track O rather than Fe, and it is known that O/Fe rises as Fe/H falls, in the solar neighbourhood (Abia and Rebolo, 1990).

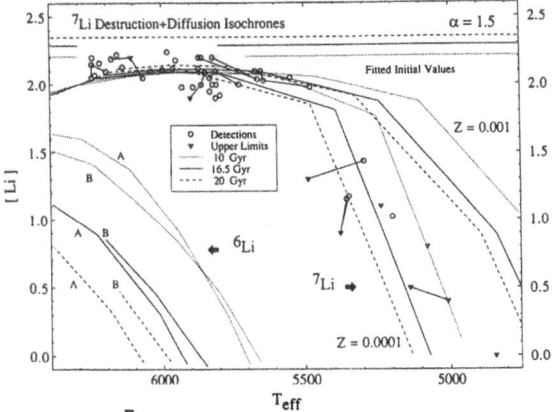

Fig. 4: Predicted ^{7}Li in Pop. II stars from one of the models of Delyannis et al. (1989). These models, the most complete in the literature, suggest that the Pop. II ^{7}Li abundance for $T_{eff} > 6000$ K could show little or no depletion.

It is also very difficult, if not impossible, to reproduce this envelope from a model in which all stars start from a uniform, log N=3, abundance of ^{7}Li. Obtaining a curve in which the ^{7}Li depletion mimics the rise in Fe abundance requires "forcing" of the depletion to a degree which is hard to believe.

The fact that below the rising envelope of Fig. 3 there is such a wide depletion scatter, while below the plateau, i.e. for [Fe/H] < -1.4, there is no such scatter, speaks in favour of a qualitatively different depletion regime for the extremely metal - weak stars. Combining this evidence with the theoretical modelling of Delyannis et al. (1989) gives grounds for believing that there has really been very little ^{7}Li depletion in these stars, in spite of their ages. We can indeed assume that the stars in this Pop.II plateau show "the whole primordial abundance". In fact when the plateau is represented as a function of T_{eff}, as in Fig. 1, the graph shows a tendency to rise slightly between $T_{eff} > 6000$ K and $T_{eff} > 6350$ K, the highest temperature at which a near-MS Pop. II star has been measured for ^{7}Li (see Rebolo et al., 1987). This, too, is consistent with Delyannis et al.'s modelling and suggests that for the best primordial ^{7}Li abundance we should use Log N(^{7}Li)$_p$ = 2.2, i.e. (^{7}Li/H)$_{primordial}$ = 1.6x10^{-10}, the value found at the warm end of the plateau.

5. Nothing but the primordial ^{7}Li abundance

There is still a possible qualification to the identification of Pop. II ^{7}Li as primordial. If some of this ^{7}Li had been produced in sources active between the epoch of the Big Bang and the formation of the halo stars, the value for Log $N(^7$Li)$_p$ quoted in section 4 would have to be treated as aun upper limit rather than the true primordial abundance. Sources for this ^{7}Li which have been suggested are: (i) a supposed Population III, of pre-galactic objects which had been invoked either to explain the supposed non-Planckian excesses in the Background radiation, or even the whole Background spectrum, as an alternative to the Big Bang, and (ii) an early burst of relatively massive stars formed at the beginning of the lifetime of the galaxy, which could give rise to ^{7}Li by spallation in the interstellar medium, since they would be sources both of C, N, O nuclei and of high energy cosmic rays after exploding as supernovae. Since the beautiful measurement of the spectrum of the Background radiation by COBE, the Population III hypothesis, always somewhat speculative and hard to quantify, has been shown to be unnecessary, and need no longer be taken seriously as a source of ^{7}Li. The early massive galactic starburst hypothesis needs further examination.

One virtue of this hypothesis is that is makes testable predictions about early spallogenic light element production. Walker et al. (1985) derive the abundance ratios of the isotopes ^{6}Li, ^{7}Li, ^{9}Be, ^{10}B and ^{11}B which would be formed in these circumstances. If some, or in the limiting case all of the Pop. II ^{7}Li were spallogenic, these predictions allow us to compute the abundances of the other nuclides which would have been present at the formation of the Pop. II stars.

Observations of the abundances of ^{6}Li, ^{9}Be and ^{10}B+^{11}B in Pop. II can then be used to measure, or place upper limits on the fraction of non-primordial ^{7}Li in the halo stars. ^{6}Li is the least helpful of these nuclides because it is subject to depletion in a similar fashion to ^{7}Li but at a greatly enhanced rate, due to its lower thermonuclear combustion temperature. If it were detected in halo stars it would not only delimit the non-primordial ^{7}Li, but would imply categorically that no ^{7}Li depletion whatever has occurred in these objects. However, its non-detection does not imply the contrary, i.e. does not mean that ^{7}Li depletion has in fact occurred. Careful theoretical estimates for ^{6}Li as well as ^{7}Li depletion are available for metal poor stars in Delyannis et al. (1989) which bear out these last statements. A recent measurement by Hobbs et al. of an upper limit to the ^{6}Li/^{7}Li ratio in HD 84937 ([Fe/H] = -2.1; log N(Li) = 2.1; T_{eff} = 6200K) gives ^{6}Li/^{7}Li <0.1. This does not in itself exclude non-primordial sources for some of the ^{7}Li in Pop. II but taken together with the ^{9}Be and ^{10}B+^{11}B observations it does offer extra evidence that they are unimportant. As we shall see in section 6, this limit does give information about some non-standard Big Bang nucleosynthesis schemes.

Only two literature observations exist for B in metal poor objects, both by Molaro (1987, 1988) using the BI 2496 Å resonance doublet, for a blend between ^{10}B and ^{11}B from which their sum may be inferred. In HD 76932 ([Fe/H] = -1.1) Molaro placed an upper limit to log N(B) of

2.3, and in HD 140283 ([Fe/H] = -3.0) an upper limit to log N(B) of 1.3. These limits show, qualitatively that there was no excess B in the early galactic material (the solar and Pop. I stellar values are in the range 2.3 to 2.6), telling clearly against any model of galactic evolution with an early "burst" of massive objects (cf. Quirk and Tinsley, 1973). The stronger limit for HD 140283, used in conjunction with the theoretical ratios of Walker et al. (1985) limits strictly the possible fraction of non-primordial ^{7}Li in this star. The ^{7}Li/B ratio observed is in fact 25 times higher than if all the ^{7}Li were spallogenic (computations by Walker et al.) and in this case depletion could only strengthen the case against spallogenic ^{7}Li, since the B depletion rate is lower than that for ^{7}Li. The maximum spallogenic fraction of ^{7}Li in this star is 4%.

^{9}Be was the first nuclide to be used in this way to limit spallogenic early galactic ^{7}Li in Pop. II. Molaro and Beckman (1984), Rebolo et al. (1988b) and Beckman et al. (1989) observed the region of the ^{9}Be II resonance doublet at 3130Å using IUE and also the 2.5m Isaac Newton Telescope on La Palma, in a sequence of objects of low and intermediate metallicities. A summary of these results is that for Pop. II objects with [Fe/H] < -1.4 there is no detectable ^{9}Be down to levels of log N(^{9}Be) = -0.3 (i.e. ^{9}Be/H < 5x10^{-13}). This conclusion was strengthened recently by Ryan et al. (1990) who obtained limits reaching down as far as log N(^{9}Be) = -2.2 (i.e. ^{9}Be/H < 6x10^{-15}) in the giant HD 47147 where [Fe/H] = -3.2, and log N(^{9}Be) = -1.2 (i.e. ^{9}Be/H <6x10^{-14}) in the subdwarf HD 140283, the same object that Molaro used for B, and which offers a more valid test than the giant, where ^{9}Be depletion might have been much greater. This limit in HD 140283, taken together with cosmic ray spallation theory, shows that no more than 1% of the ^{7}Li in Pop. II could be non-primordial in origin. We are apparently very close to be able to say that the ^{7}Li in halo stars with [Fe/H]<-1.4 is "nothing but the primordial ^{7}Li", and hence to use its abundance in model calculations.

A word of caution is still required. This reasoning which leads to the conclusion that virtually none of the ^{7}Li in the halo stars could have been of galactic origin is based on the specific spallogenic model for the production of ^{7}Li, ^{9}Be, ^{10}B and ^{11}B, in which the production ratios of these nuclides are specified. If the galactic source of these were different if, for example, some or all of the ^{9}Be in the galaxy were produced directly in supernovae, it would not be possible to obtain such a clean quantitative upper limit to the non-primordial ^{7}Li in Pop. II. Even in this case, however, the fact that, as seen clearly in Fig. 1, the ^{7}Li abundance in extremely metal deficient stars holds constant between [Fe/H] = -3.5 and [Fe/H] = -1.4, i.e. over a range of two orders of magnitude in Fe abundances, implies that whatever process was producing the rise in Fe during that epoch was not affecting the ^{7}Li abundance. If the ^{9}Be were produced in supernovae, it would track the Fe, and we can use this part of Fig. 1, together with the limit on ^{9}Be for stars with [Fe/H] ~-1.4 to again infer, although more circumstantially, that less than 1% of the ^{7}Li in the halo stars is of non-primordial origin.

6. Non-standard Big Bang models

During the latter part of the 1980's a series of variants on the standard homogeneous Hot Big Bang cosmology have been produced (Alcock et al., 1987; Applegate et al., 1988; Malaney and Fowler, 1988) which have as a basic premise $\Omega_b = 1$, i.e. they are inflationary models, closed by baryons alone without recourse to undiscovered particles. In order to conform to the constraints of the observed primordial abundances of, especially ^{4}He and ^{7}Li, but also of D and ^{3}He, these models use the fluctuations in density (Witten, 1984) and chemical separation (Applegate and Hogan, 1985) which could occur when the universe made the phase transition from a quark-dominated to a hadron-dominated state, at 10^{-35} sec after the Planck epoch. Most of these first inhomogeneous (IBB) models could reproduce the observed D, ^{3}He, and ^{4}He abundances without forcing, but over-produced Li by an order of magnitude. This is one of the reasons why the issue of whether the Pop. II or the Pop. I ^{7}Li abundance is the primordial one has become such a critical matter. As the astrophysical evidence has seemed to tilt somewhat towards the Pop. II value, the theorists have looked with increasing refinement at the ^{7}Li production in IBB models, to see whether an abundance of $\sim 10^{-10}$ rather than 10^{-9} for Li/H could be readily achieved. Malaney and Fowler (1988) used the time lag between ^{4}He and ^{7}Li production, and the tendency for the fluctuations to smooth out by back-diffusion, to reduce the ^{7}Li to a Pop. II value. This has been criticised as needing a highly specific part of a large parameter space. Some workers (Kurki-Suonio et al., 1988; 1989) have claimed that $\Omega_b = 1$ is in fact excluded by the ^{7}Li observations, while others (Mathews et al., 1988, 1990; Terasawa and Sato, 1989) feel that there is still room for inflationary baryonic IBB models yielding the observed primordial abundances, including that of ^{7}Li. Clearly, although it is very difficult to pindown the situation now, if IBB were eventually established as valid, the specificity of the ^{7}Li prediction would in fact be an advantage, enabling us to fix the parameters of our cosmology within a narrow range.

Since IBB models are "tailored" to reproduce the primordial D, He and ^{7}Li abundances it is ipso facto difficult to use these abundances alone to distinguish between IBB and standard Big Bang (SBB), and we must look for observables capable of offering this distinction. One of these is a spatial frequency sepctrum of the cosmic background fluctuations, a theme I have dealt with in some detail elsewhere (Beckman, 1990) and which I will not deal with further here, except to say that present observations are not yet able to distinguish between IBB and SBB. Others are the abundances of the light nuclides ^{6}Li, ^{9}Be, and others of increasing atomic weight, which are predicted as having negligible abundances in SBB, but may be produced in measurable quantities in the context of one of other IBB model. In this context, ^{9}Be has been singled out, since the reaction ^{7}Li$(^3$H,n$)^9$Be was shown by Boyd and Kajino (1989) to be capable of yielding ^{9}Be abundances much higher in IBB models with a specified parameter range than in SBB, and sufficiently high to be close to the currently observable limits (Rebolo et al., 1988; Beckman et al., 1989; Ryan et al., 1990). The signature of an IBB model would, in this case, be the presence of ^{9}Be at a level somwhere between 5×10^{-13} and 10^{-14} for ^{9}Be/H, but with abundance independent of Fe/H for the range [Fe/H] < -1.5 (if the ^{9}Be were galactic in this range, it would rise with increasing Fe/H, and the functional relation would help to indicate its production mechanism). The recent paper by Kajino and Boyd (1990) places the predicted SBB and IBB production of ^{9}Be in the context of the galactic evolution of this nuclide, and does the same for ^{10}B and ^{11}B, showing these to be predicted at levels $10^{-16} < {}^{10}\text{B/H}_{IBB} < 3 \times 10^{-13}$ and $3 \times 10^{-16} < {}^{11}\text{B/H}_{IBB} < 3 \times 10^{-14}$ respectively. These values, as well as being very low for measurement purposes, contain uncertainties due to uncertain cross-sections in their production reactions, and similar uncertainties hold for ^{9}Be. Thus our conclusion must be that if no ^{9}Be, ^{10}B or ^{11}B are found in Pop. II stars at levels at or just below 10^{-14}, the levels attainable in the next couple of years, this would not rule out IBB models, on the other hand measureable abundances of one or more of ^{9}Be and ^{10}B+^{11}B in Pop. II would offer, if invariant with Fe/H, evidence decisively in favour of IBB. A parallel situation has already been reached for ^{6}Li, which was not found (at a level of ~10^{-11}) in the halo star HD 84937 by Pilachowski et al. (1990). The conclusion that this excludes the non-standard Big Bang model by Dimopoulos et al. (1988) in which Li is produced by the slow decay of a postulated massive particle, such as the gravitino, is subject to our understanding of the differential depletion of ^{6}B and ^{7}Li in metal deficient stars during the galaxy lifetime. Although modelling of this process by Delyannis et al. (1989) is now on a thorough basis it is still true to claim that while detection of ^{6}Li in Pop. II stars would be capable of giving definite information either about non-standard Big Bang models, or about depletion in Pop. II stars, or both, its absence is not yet a definitive test of either.

To round off this discussion of IBB predictions, it is worth mentioning that other light nuclides heavier than ^{11}B, and their ratios could in principle be used to test primordial nucleosynthesis predictions. Ratios such as C^{13}/C^{12}, $^{17}O/^{16}O$, $^{21}Ne/^{20}Ne$, and $^{25}Mg/^{24}Mg$ are predicted to be different in the Big Bang from their values in stellar nucleosynthesis, and different in SBB from IBB. Similarly, the C/N and N/O ratios would also be distinctive. However, a search for this type of signatures is very challenging observationally. In the range of mass numbers between 10 and 14, for example, a recent set of predictions by Kajino et al. (1990) gives IBB, $\Omega_b = 1$ abundances which are between 3 and 5 orders of magnitude below the values observed by Bessel and morris (1984) in metal deficient stars. Although in principle measurements of C, N and O (also Ne and Mg) as functions of Fe down to extremely low Fe abundance could be of interest, the practical difficulties entailed in reaching cosmologically interesting levels seem rather prohibitive. Similar considerations apply to r-process elements, whose synthesis emerges from the products of the light element chains. Thus although one should certainly encourage observations of the nuclides touched on in this paragraph, it is probable that their differential behaviour at low metallicity will give us information about their sites of stellar genesis, and about the early galaxy, before offering any clear cosmological information.

6. Conclusions

In this article I have reviewed the evidence relevant to establishing a reliable value for the primordial ^{7}Li abundance. In spite of the fact that, since the birth of the galaxy, processes have acted to destroy ^{7}Li within most stars, and to produce ^{7}Li in novae, giants and in the interstellar medium, there is good reason to believe that the ^{7}Li observed in the halo dwarfs in the solar neighbourhood with surface temperatures in the range 6000K to 6350K represents an abundance close to the primordial figure. Summarizing, the invariance of this measured abundance over the temperature range cited, and over two orders of magnitude in metallicity (below [Fe/H] = -1.5), with very low scatter in both cases, is easy to explain if this ^{7}Li is primordial but difficult to explain quantitatively using alternative hypotheses, notably taking the primordial abundance as that seen in young stars, and assuming that the halo dwarf abundance is heavily depleted. The value, log N(^{7}Li) = 2.2 (i.e. ^{7}Li/H = 1.25x10^{-10}) fits with the observed primordial ^{4}He abundance; Y_p =0.23, yielding a universal baryon density Ω_b between 0.03 and 0.1 in the framework of standard Big Bang (SBB) nucleosynthesis. In this context, the low abundances of ^{9}B and ^{10}B+^{11}B in the halo dwarfs point to an absence of spallogenic ^{7}Li production in the early galaxy, and even if this circumstantial evidence is not binding, the measured invariance of ^{7}Li with Fe/H between [Fe/H] = -1.5 and -3.5 strongly implies the absence of significant galactic sources of ^{7}Li prior to the formation of the halo.

In the inhomogeneous Big Bang (IBB) models which have been invoked to account for the observed ^{4}He and ^{7}Li with the condition that Ω = Ω_b = 1, there exists the possibility of using the dependence of ^{9}Be (and at lower levels of ^{10}B+^{11}B, and even C, N, O, Ne and Mg and their isotopes) on [Fe/H] at low metallicities to distinguish between IBB and SBB. Detection and measurement of one or more of these dependences could even be used to isolate the parameter space in the IBB model, specifying the fraction of neutron or proton rich matter, and its relative concentration. Although beyond the reach of current experimental limits this use of the light elements could, in principle, give us an observational insight as far back as the quark-hadron phase transition, some 10^{-35} seconds into the lifetime of the Universe.

7. References

Abia, C., Rebolo, R. (1990) Astron. Astrophys. 347, 186
Alcock, C.R., Fuller, G.M., Mathews, G.J. (1987) Astrophys. J. 320, 439
Applegate, J.H., Hogan, C.J., Scherver, R.J. (1988) Astrophys. J. 329, 72
Beckman, J.E., Abia, C., Rebolo, R. (1989) Astrophys. Spc. Sci. 157, 41.

Beckman, J.E., Rebolo, R., Molaro, P. (1987) in Advances in Nuclear
    Astrophysics (eds E. Vangioni-Flam et al.) Editions Frontières 29
Boesgaard, A.M. (1987) Astrophys. J. 321, 967
Boesgaard, A.M., Budge, K.G., Ramsay, M.E. (1988) Astrophys. J. 327, 389
Boesgaard, A.M., Budge, K.G. (1989) Astrophys. J. 338, 375
Boesgaard, A.M., Trippico, M.J. (1986a) Astrophys. J. Lett. 302, L49
Boesgaard, A.M., Trippico, M.J. (1986b) Astrophys. J. 303, 374
Boyd, R.N., Kajino, T. (1989) Astrophys. J. Lett. 336, L55
Cameron, A., and Fowler, W. (1971). Astrophys. J. 167, 221.
Clayton, D.D. (1988) Mon. Not. Roy. Astr. Soc. 234, 1
Dearborn, P.S.P., Schramn, D.N., Steigman, G., Truran, J. (1989)
    Astrophys. J. 347, 455.
Deliyannis, C.P., Demarque, P., Kawaler, S. (1990) Astrophys. J. Suppl.
    Ser. 73, 21
Dimapoulos, S., Esmailzadeh, R., Hall, L.J., Starkman, G.D. (1988) Phys.
    Rev. Lett. 60, 7
Garcia Lopez, R.J., Rebolo, R., Beckman, J.E. (1988) P.A.S.P. 100, 1489
Hobbs, L.M., Pilachowski, C. (1986a) Astrophys. J. Lett. 309, L17
Hobbs, L.M., Pilachowski, C. (1986b) Astrophys. J. Lett. 311, L37
Hobbs, L.M., Duncan, D.K. (1987) Astrophys. J. 290, 284
Hobbs, L.M., Pilachowski, C. (1988) Astrophys. J. 334, 734
Hobbs, L.M., Thorburn, L.A. Rodriguez-Bell, T. (1990) Astrophys. J. 100,
    710
Kajino, T., Boyd, R.N. (1990) Astrophys. J. 359, 267
Kurki-Suonio, H., Matzner, R.A., Centrella, J.M., Rothman, T., Wilson,
    J.R. (1988) Phys. Rev. D. 38, 1091
Kurki-Suonio, H., Matzner, R.A. (1989) Phys. Rev. D. 39, 1046
Lynden-Bell, D. (1975) Vistas in Astronomy 19, 299
Malaney, R.A., Fowler, W.A. (1988) Astrophys. J. 333, 14
Mathews, G.J., Fuller, G.M., Alcock, C.R., Kajino, T. (1988) in
    Darkmatter, eds. J. Audouze & J. Tran Thanh Van, Editions
    Frontières 319
Mathews, G.J., Meyer, B.S., Alcock, C., Fuller, G.M. (1990) Astrophys.
    J. 358, 36
Meneguzzi, M., Audouze, J., Reeves, H. (1971) Astron. Astrophys. 15, 337
Michaud, G. (1986) Asrtrophys. J. 302, 650
Molaro, P. (1987) Astron. Astrophys. 183, 241
Molaro, P. (1988) in The Impact of very high S/N Spectroscopy on Stellar
    Physics, Eds. G. Cayrel de Strobel & M. Spite, 511
Molaro, P., Beckman, J.E. (1984) Astron. Astrophys. 139, 394
Pilachowski, C., Hobbs, L.M., De Young, D.S. (1990) Astrophys. J. Lett.
    (In Press)
Quirk, W.J., Tinsley, B.M. (1973) Astrophys. J. 179, 69
Rebolo, R., Beckman, J.E., Molaro, P. (1987) Astron. Astrophys. Lett.
    172, L17
Rebolo, R., Molaro, P., Beckman, J.E. (1988a) Astron. Astrophys. 192,
    192
Rebolo, R., Molaro, P., Abia, C., Beckman, J.E. (1988b) Astron.
    Astrophys. 193, 193
Reeves, H, Fowler, W.A., Hoyle, F. (1976) Nature 226, 727
Schatzmann, E. (1977) Astron. Astrophys. 56, 211

Spite, F., Spite, M. (1982) Astron. Astrophys. 115, 357

Spite, M., Maillard, J.P., Spite, F. (1984) Astron. Astrophys. 141, 56

Spite, F., Spite, M. (1986) Astron. Astrophys. 163, 340

Starrfield, S., Truran, J.W., Sparks, J.W., Arnould, M. (1987)
    Astrophys. J. 222, 660

Teresawa, N., Sato, N. (1989) Proy. Theor. Phys. 81, 254

Truran, J.W. Cameron, A.G.W. (1971) Astrophys. Spa. Sci. 14, 179

Twarog, B.A., Anthony-Twarog, D.J. (1989) Astron. J. 97, 759

Vandenberg, D.A. (1985) Astrophys. J. Suppl. 58, 711

Vauclair, S. (1988) Astrophys. J. 335, 971

Walker, T.R., Mathews, G.J., Viola, V.E. (1985) Astrophys. J. 299, 745

Witten, E. (1984) Phys. Rev. D. 30, 272

THE EVOLUTION OF THE GALACTIC LITHIUM ABUNDANCE

L. M. HOBBS
University of Chicago
Yerkes Observatory
Williams Bay, WI 53191-0258
USA

ABSTRACT. Studies of the lithium abundances in Population I and Population II stars have been used to determine the Li/H fraction produced in Big Bang nucleosynthesis. A summary of the principal results derived from the 6707 Å line measured in the spectra of various stars is presented. The value Li/H $\approx$ 1 x 10^{-10} uniformly found for about 40 relatively hot, extreme halo stars appears to have survied, nearly unaltered, from that early epoch.

1. Introduction

During much of the 1980s, measurements of cosmic lithium abundances have been of central interest in many studies of Big Bang nucleosynthesis (BBN). One of several primary reasons for this interest is that the evolution of the Li/H ratio in the Galactic gas over much of the full age of the Galaxy can be directly investigated observationally. In particular, a dependable determination of the initial lithium abundance in the Galaxy, produced by BBN, may therefore be possible from stellar spectra alone.

The key point was discovered by Spite and Spite [1]. Despite the very low cosmic abundance of lithium, the weak 6707 Å spectral line of Li I can be detected in the spectra of the oldest, essentiallly unevolved Galactic stars, which necessarily are solar-like stars of relatively low mass. In contrast, no abundance measurements of ^{2}H, ^{3}He, or ^{4}He have been obtained in such old, unevolved Galactic objects, although these are the three other isotopes which so far have been of primary importance to BBN studies.

Table 1 summarizes the types and ages of Galactic objects in which lithium has been measured.

TABLE 1. The ages and lithium abundances of various objects.

type	object	age (Gyr)	Li/H (10^{-9}) present	initial
1	interstellar medium	0	$\approx$ 1.	--
2	7 young star clusters	< 2	$\lesssim$ 1.	1.
3a	Sun	5	0.01	1.
3b	some meteorites	5	1.	1.
4	2 old star clusters	5 - 8	$\lesssim$ 0.5	$\approx$ 1.
5	$\approx$ 45 Population II stars with [Fe/H] $\leq$ -1.4	$\approx$ 12	$\lesssim$ 0.1	$\gtrsim$ 0.1

The last two columns of the table give the approximate Li/H fractions, by number, which are derived from the observations of the various objects today, as well as the inferred Li/H fractions with which they probably were born. The latter are the data of interest, because they can reveal the evolutionary history and the primordial abundance of lithium in the Galactic gas. In the diverse collections of stars which constitute groups 2, 4, and 5, we need to identify the respective subsets of stars which have atmospheres in which the currently observable Li/H ratio has remained virtually unmodified since the births of the stars. The quantity $[X/H] = \log [(X/H)/(X/H)_\odot]$ appearing in the table is the logarithm of the abundance of element X relative to hydrogen, by number, after normalization to the corresponding solar ratio.

2. Population II Stars

The oldest Galactic stars presumably are those most directly pertinent to observational studies of BBN. The Galactic abundances of heavy elements such as iron are believed to have increased steadily with time, as a result of ongoing stellar nucleosynthesis. The oldest stars therefore are identified primarily by the requirement that $[Fe/H] \le -1.4$, or $(Fe/H)/(Fe/H)_\odot \le 0.04$. These wil be referred to as extreme halo (EH) stars. At the present time, the EH stars in which Li/H has been measured include nine with $[Fe/H] \le -2.6$, in particular; these extreme examples even among the EH stars are more metal deficient than, and therefore apparently older than, any well studied globular cluster.

For main sequence EH stars in whose spectra the 6707 Å line has been measured, Figure 1 shows the derived lithium abundances as a function of effective temperature, or, approximately, of mass [2, 3, 4].

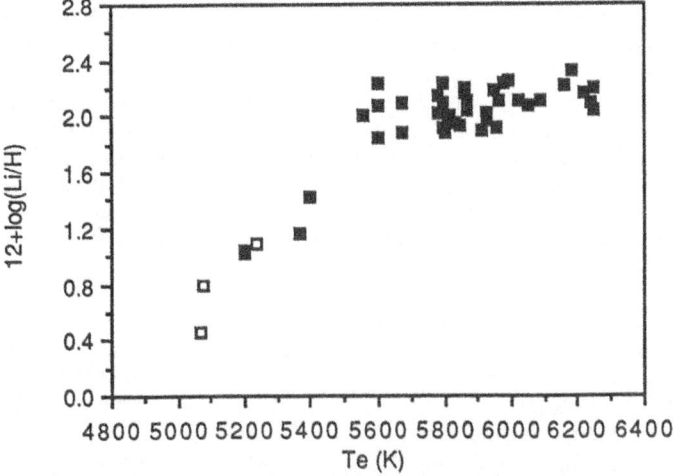

Figure 1. Lithium abundance as a function of effective temperature, for 45 extreme halo stars. The three open symbols show upper limits, only, on Li/H.

The abundances are expressed in the usual logarithmic form, $N(Li) = 12 + \log (Li/H)$, and two principal results are seen. (1) At $T_e \ge 5500$ K, all individual abundances agree fairly closely with an asymptotic value, $N(Li) \approx 2.1$ or $Li/H \approx 1 \times 10^{-10}$. (2) At $T_e \le 5500$

K, the surface lithium abundances decrease rapidly with decreasing temperature and mass, i. e. with increasing depth of the convective envelope which exists in these stars. Because lithium is rapidly destroyed by the $^7Li(p,\alpha)^4He$ reaction in stellar interiors at relatively low temperatures of $T \gtrsim 2.5 \times 10^6$ K, it is likely that the reduced lithium abundances are caused by the deeper convective circulation of material in the cooler stars [5].

The crucial question then is whether the present asymptiotic value $Li/H \approx 1 \times 10^{-10}$ has been similarly reduced from a higher value present at the birth of these EH stars. At least three arguments suggest that very little lithium destruction has occurred over the lifetimes of these slightly hotter, more massive stars and therefore that the asymptotic abundance observed today indeed is the nearly unaltered Li/H fraction produced in BBN, i.e. that $(Li/H)_{BBN} \approx 1 \times 10^{-10}$.

1) As is discussed briefly in section 3 below, lithium abundances have also been measured for a considerably larger number of main sequence stars with $-1.4 \lesssim [Fe/H] \lesssim 0$ and $T_e \gtrsim$ 5500 K. These more metal rich stars constitute a younger population of stars, born more recently in Galactic history. The lithium abundances of these stars and of the EH stars can be plotted together as a function of [Fe/H], rather than effective temperature. At low metallicities, the asymptotic plateau occupied by the EH stars is necessarily recovered. At higher metallicities, however, the abundances spread progressively and fairly smoothly both upward and downward from $N(Li) \approx 2.1$, occupying a wedge shaped region [4]. The upper envelope of these abundances can be interpreted as evidence for later "contamination" of the initial, primordial lithium by Galactic lithium, which apparently was cumulatively produced in eventually detectable quantities by some process of stellar nucleosynthesis. The abundances which range downward from this upper envelope may reflect a complex competition between this stellar production of lithium and the stellar destruction of lithium which is expected as higher metallicities cause deeper surface convective zones to compensate for reduced radiative energy transport.
2) The flatness of the asymptotic plateau requires the effectiveness of any proposed destruction process to vary almost negligibly with stellar mass, over the range represented by the plateau [1]. In contrast, most mechanisms advocated in this connection would be expected to show a strong dependence on mass.
3) Standard stellar evolutionary models of EH stars can fit the data of Figure 1 in a natural way. Such models which represent stars on the asymptotic plateau do not generally destroy much lithium [5]. As a further verification, it will be desirable eventually to have a similar set of models calculated under the same assumptions, in order to test whether such models also can satisfactorily explain the more complex behavior of surface lithium in the Population I clusters.

A final point concerns the primordial $^6Li/^7Li$ isotope ratio. In the spectrum of HD 84937, which is a relatively bright EH star with $[Fe/H] = -2.1$ and $T_e = 6200$ K, an upper limit $^6Li/^7Li < 0.1$ is derived from high-resolution observations of the 6707 Å line profile [6]. Because ^6Li is burned in (p, α) reactions at even lower temperatures than ^7Li, the original value of this ratio in HD 84937 was, almost certainly, appreciably higher than the one now observed about 12 Gyr later. If the corrected, original ratio nevertheless was $^6Li/^7Li \lesssim 1$, as recent stellar evolutionary models indicate [5, 7], then some non-standard theoretical models of BBN may be disproved.

3. Population I Clusters

A different observational approach to determining the primordial Li/H ratio is to measure the lithium abundances in a sequence of open clusters with relatively well known ages which span nearly the full age of the Galactic disk. The relation determined in this way might, for example, decline fairly obviously to the value $(Li/H)_{BBN} \approx 1 \times 10^{-10}$ estimated from the EH stars, when the varying Galactic lithium abundance is extrapolated back over the relatively brief interval from the age of the oldest open clusters, about 8 Gyr, to the age of the Galaxy, about 12 Gyr. The 6707 Å line of Li I was measured for this purpose in the spectra of a number of main sequence stars in each of four progressively older open clusters which are listed in Table 2 [8]. Their ages range from < 0.1 Gyr to about 8 Gyr [9]. Data for a fifth relevant cluster, the Hyades, also are available [10, 11].

TABLE 2. Five open clusters in which lithium abundances have been measured in main sequence stars.

cluster	age (Gyr)	max T_e (K)	d (pc)	"V_o"
Pleiades	< 0.1	> 7300	125	10
Hyades	0.8	> 7300	45	8
NGC 752	1.7	6900	400	13
M67	5.	6300	800	14
NGC 188	8.	5950	1600	15.5

At present, there are no observational data accessible in the age gap between NGC 188 and the EH stars, at about 12 Gyr. The last column of the table gives the apparent visual magnitude of a solar-like star in each cluster; in NGC 188, the very weak 6707 Å line is near the limits of the observational capabilities of current telescopes and spectrographs. Figure 2 shows the results of the observations.

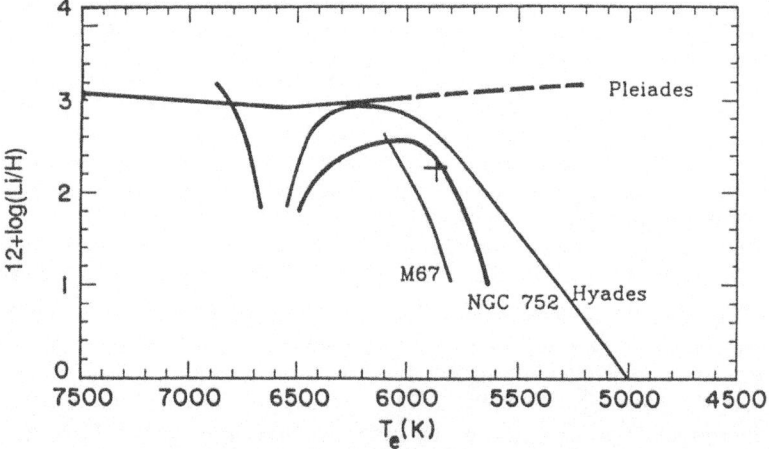

Figure 2. Lithium abundances in the five open clusters of Table 2. Smooth curves have been drawn through the individual data points,

and the cross indicates an average value for five similar stars in NGC 188. The steep curve near 6800 K represents the Hyades and, roughly, NGC 752.

No individual abundances significantly exceed N(Li) ≈ 3.1. However, a complex variation of Li/H with mass and time, which is not yet satisfactorily understood theoretically [12, 13] and which is similar in part to that seen previously in the EH stars, is found in the stars of lower mass, at T_e ≲ 6800 K. On the other hand, an asymptotic abundance N(Li) ≈ 3.1, which is consistent with that measured in the present interstellar medium, is apparently found at hotter temperatures, in seven clusters younger than 2 Gyr old. From a comparison in Figure 2 with three of the younger clusters, the main sequence stars originally found at T_e ≲ 6800 K in the old clusters M67 and NGC 188 also probably showed N(Li) ≈ 3, before they long ago evolved away from the ZAMS.

The final interpretation is that each of the five clusters in Table 2, as well as four additional open clusters not older than the Hyades [14], formed from gas which had essentially the same lithium fraction, N(Li) ≈ 3.1 or Li/H ≈ 1×10^{-9}, as that still found in the interstellar gas today. Hence, the Galactic lithium abundance has been nearly invariant at this "high" value for at least the last 8 Gyr, since the birth of NGC 188.

4. Conclusions

With respect to the Li/H fraction produced in BBN, perhaps the two simplest alternative conclusions which can be drawn from the various observations are the following.

1) $(Li/H)_{BBN} \approx \underline{1 \times 10^{-10}}$.

The lithium now present in the atmospheres of sufficiently hot, metal-poor dwarfs must be an essentially unmodified product of the Big Bang which has avoided significant stellar processing over the subsequent 12 Gyr. A Galactic mechanism [15, 16] not yet clearly identified must produce, in the first few Gyr, about 90% of the lithium now found in the interstellar gas, while producing amounts of ^6Li, Be, and B which are smaller by an order of magnitude. This rapid, early net production of lithium must also have ceased permanently at least 8 Gyr ago.

2) $(Li/H)_{BBN} \approx \underline{1 \times 10^{-9}}$.

In this view, the abundance of lithium in the Galactic gas must have been approximately invariant at $(Li/H)_{BBN} \approx 1 \times 10^{-9}$. The surface Li/H ratio in hot, metal-poor halo dwarfs must have been reduced over the lifetime of the Galaxy to about 10% of its initial value, at an average rate which is therefore about 20 times slower than that in the Sun and in 1 $M_\odot$ stars in M67, but is not negligible.

The first of these two conclusions is favored by at least the three arguments listed above in section 2. If the standard theoretical model of BBN is further assumed to be correct, a priori, then the lower abundance, $(Li/H)_{BBN} \approx 1 \times 10^{-10}$, also is not inconsistent in this model with the primordial fractions of $^2H/H$, $^3He/H$, $^4He/H$ presently inferred from the pertinent observations. The ratio of baryons to photons in the universe then is required to be η ≈ 4×10^{-10} [17].

44

5. References

1) Spite, F., and Spite, M. 1982, <u>Astr. Ap.</u> **115**, 357.
2) Spite, M., Maillard, J. P., and Spite, F. 1984, <u>Astr. Ap.</u> **141**, 56.
3) Hobbs, L. M., and Duncan, D. K. 1987, <u>Ap. J.</u> **317**, 396.
4) Rebolo, R., Molaro, P., and Beckman, J. E. 1988, <u>Astr. Ap.</u> **192**, 192.
5) Deliyannis, C. P., Demarque, P., and Kawaler, S. D. 1990, <u>Ap. J. Suppl</u>. **73**, 21.
6) Pilachowski, C., Hobbs, L. M., and DeYoung, D. S. 1989, <u>Ap. J. Letters</u> **345**, L39.
7) Brown, L., and Schramm, D. N. 1988, <u>Ap. J. Letters</u> **329**, L103.
8) Hobbs, L. M., and Pilachowski, C. 1988 <u>Ap. J.</u> **324**, 734.
9) Hobbs, L. M., Thorburn, J. A., and Bell, T. 1990, <u>A. J.</u> **100**, 710.
10) Cayrel, R., Cayrel de Stroebel, G., Campbell, B., and Dappen, W. 1984, <u>Ap. J.</u> **283**, 205.
11) Boesgaard, A. M, and Tripicco, M. 1986, <u>Ap. J. Letters</u> **302**, L49.
12) Michaud, G. 1986, <u>Ap. J.</u> **302**, 650.
13) Hobbs, L. M., Iben, I., and Pilachowski, C. 1989, <u>Ap. J.</u> **347**, 817.
14) Boesgaard, A. M., Budge, K. G., and Ramsey, M. 1988, <u>Ap. J.</u> **327**, 389.
15) Smith, V. V., and Lambert, D. L. 1989, <u>Ap. J. Letters</u> **345**, L75.
16) Dearborn, D. S. P., Schramm, D. N., Steigman, G., and Truran, J. 1990, preprint.
17) Boesgaard, A. M., and Steigman, G. 1985, <u>Ann. Rev. Astr. Ap.</u> **23**, 319.

CHEMICAL EVOLUTION OF GALAXIES *

B.E.J. PAGEL
NORDITA
Blegdamsvej 17
DK - 2100 Copenhagen Ø
Denmark

ABSTRACT

Initial conditions are probably set by results of Big Bang nucleosynthesis (BBNS) without intervening complications affecting the composition of visible matter so that extrapolation of observed abundances to BBNS products seems fairly secure. Primordial helium and deuterium abundances deduced in this way place upper and lower limits on baryonic density implying that both baryonic and non-baryonic dark matter exist and predicting no more than 3 neutrino flavours as recently confirmed in accelerator experiments. The validity of simple galactic chemical evolution models assumed in extrapolating back to the Big Bang is examined in the light of the frequency distribution of iron or oxygen abundances in the Galactic halo, bulge and disk.

1. INTRODUCTION

Chemical evolution of galaxies (GCE) involves many very uncertain factors, arising from stellar nucleosynthesis and especially from a state of ignorance about the origin and evolution of galaxies (cf. Larson 1990). Big Bang nucleosynthesis (BBNS) in itself is actually rather well understood from a theoretical point of view, but to reach primordial abundances resulting from BBNS, by way of observations that we can make now, it is necessary to cut through a dense forest of GCE effects. In what follows I shall try to convince you that quite simple models of GCE fit relevant data well enough to justify equally simple extrapolations, but I begin with some comments on the primordial abundances themselves.

2. PRIMORDIAL ABUNDANCES

Fig. 1 shows the present state of helium abundance determinations in extragalactic H II regions (Pagel and Simonson 1989; Simonson 1990; Pagel 1991) following the method originally suggested by Peimbert and Torres-Peimbert (1974, 1976) and extended to consider nitrogen as well as oxygen (Pagel, Terlevich and Melnick 1986). A relationship evidently exists, though with marginally significant scatter or nonlinearity in the case of oxygen. The relation with nitrogen shows no significant departure from a linear regression within the limited range of abundances considered here and leads to the maximum-likelihood solution

$$Y = 0.229 + 3310 \quad (N/H) \tag{1}$$
$$\pm 4 \quad \pm 940 \quad (1\sigma)$$

giving (after making reasonable allowance for systematic errors) 95 per cent confidence that the primordial helium mass fraction Y_p does not exceed 0.24. Figure 2 shows implications of this together with a neutron half life of at least 10.1 minutes (10.25 according to Mampe et al. 1989 and 10.32 according to Byrne et al. 1990) and a primordial $(D+ {}^3He)/H$ ratio not exceeding 10^{-4} (Yang et al. 1984; Boesgaard and Steigman 1985) which seems to me to be quite a generous upper limit (Pagel 1990). Tall vertical lines show corresponding limits to the density parameter in a homogeneous BBNS model (Yang et al. 1984) and shorter vertical stripes indicate roughly how these limits might be extended in a reasonable inhomogeneous one

---

* Invited talk to IUAP Conference on Primordial Nucleosynthesis and Evolution of Early Universe, Tokyo, 4-8 September 1990. K. Sato (ed.). Kluwer publication.

(Kurki-Suonio et al. 1990). In either case more than 3 neutrino flavours are ruled out as has been clear for some time (cf. Pagel 1991) and confirmed by accelerator measurements of the height and width of the Z^0 resonance (e.g. Ellis, Salati and Shaver 1990), although the implications are not identical in the two cases. Furthermore, the homogeneous model would be in trouble if it could be established that Y_p is significantly less than 0.24. Lithium is discussed by J.E. Beckman in this volume.

An additional implication of the upper limit on primordial deuterium is that Ω_{b0}, the fraction of the closure density supplied today by baryons, is at least $0.008\ h_0^{-2}$ (where h_0 is the Hubble constant in units of 100 km $s^{-1}\ Mpc^{-1}$ and is generally thought to lie in the range 0.7 ± 0.2), implying the existence of about as much dark baryonic matter as the amount of dark matter inferred from flat 21 cm rotation curves of spiral galaxies. The (inhomogeneous) upper limit, $\Omega_{b0}\ h_0^2 < .032$, is barely consistent with $\Omega_0 \simeq 0.1$ to 0.15 deduced from several dynamical investigations if $h_0 = 0.5$, but such a low value of h_0 now seems unlikely (Tully 1990). The homogeneous BBNS upper limit to Ω_{b0} is 0.05 even if $h_0 = 0.5$. Thus unless Ω_0 is as little as 0.05 or so, there is still a need for some non-baryonic matter as well.

3. INGREDIENTS OF GCE MODELS

The word "ingredients", reminiscent of cookery, is used here advisedly and I spell them out because only too often some of them are taken for granted. They are

1   Initial conditions, which should involve primordial abundances if we are to deduce these from observation. Models with an initial production spike (Truran and Cameron 1971) - whatever their validity may be for certain purposes - clearly cannot be used to extrapolate back to the Big Bang.

2   End-results of stellar evolution including the role of close binaries and many other details which are still uncertain.

3   The initial mass function (Scalo 1986), which together with 2. fixes the mass-fraction of a generation of newly formed stars that is returned to the diffuse interstellar medium (the "return fraction") and the yields of various elements (defined as the mass of a nuclear species freshly synthesised and ejected by a generation of stars, divided by the mass locked up in long-lived stars and compact remnants). It is not yet clear how, or whether, the IMF varies under different ambient conditions (Scalo 1989).

4   Star formation laws, which depend (from a theoretical point of view) on still largely unknown physics and cannot be deduced from observation independently of biases resulting from variations in ambient conditions. Non-monotonic time-dependence of the star formation rate is evident in starburst galaxies (Thuan et al. 1987) and apparently even in the solar neighbourhood (Barry 1988). Fortunately some results of GCE theory are quite insensitive to the actual past rates of star formation.

5   The galactic context, which includes the different circumstances of formation of various stellar populations, dynamical relaxation and collapse with or without dissipation, presence or absence of exchange of material between the region of interest and its environment and possible effects of mergers and other complications (see Gilmore, Wyse & Kuijken 1989). Three stellar populations can be distinguished in our own Galaxy: that of the disk, typified by the solar neighbourhood, which will be schematically treated as a cylindrical shell perpendicular to the Galactic plane and including the Sun, and where the stars are predominantly rotationally supported and have a rather narrow distribution of abundances; that of the Population II spheroid or halo, very sparsely represented in the solar neighbourhood, where the stars are predominantly "pressure" supported, very old and deficient in carbon and heavier elements; and that of the bulge, within a kpc or so of the Galactic centre, in which the stars are mostly old and have a broad range of chemical compositions extending up to a few times solar abundances (Frogel 1986; Rich 1988). An index of stellar chemical composition is the "metallicity" which I shall define as [Fe/H], i.e. the logarithm of the iron abundance in units of the solar value. Metallicity (together with effective temperature) governs the colours and spectral appearance of the cooler stars because of the prominence of absorption lines due to iron and can be taken as a first indication of their chemical composition; but careful measurements indicate that other elements (some of which, like oxygen, are still more abundant) can display significant variations relative to iron which are important clues to GCE (cf. Wheeler, Sneden and Truran 1989).

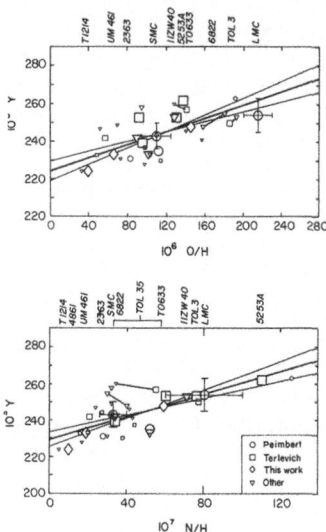

Fig. 1
Relation between helium, oxygen and nitrogen abundances in extragalactic H II regions (Simonson 1990; Pagel 1991). Sizes of symbols increase with the weights of the data and a few typical error bars are shown.

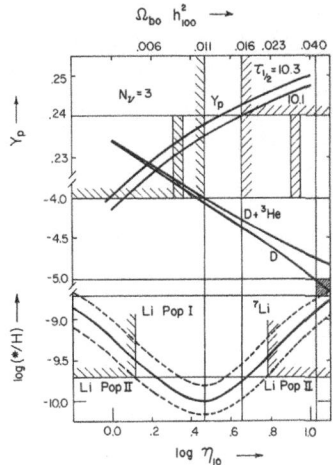

Fig. 2
Theoretical dependence of primordial abundances on the baryon: photon ratio η and the baryonic density parameter. Some upper limits inferred from observational data are shown by horizontal lines.

4. THE (INSTANTANEOUS) SIMPLE MODEL

The instantaneous recycling approximation (Schmidt 1963) assumes that the stars can be divided into two classes: big stars which instantaneously (on the time scale of galactic evolution) return gas and freshly synthesised elements to the interstellar medium; and small stars which merely serve to lock up diffuse material. In this approximation the return fraction and the yields are fixed by the IMF and by the evolution of its component stars at each mass, and the former are simple constants if the latter do not change. This approximation is often quite good for oxygen and other elements synthesised by massive stars which become core-collapse supernovae, but less good for carbon, nitrogen and s-process elements resulting from thermal pulses in intermediate-mass stars and for iron which (in disk stars at any rate) is thought to have a substantial contribution from Type 1a supernovae with an evolution time of the order of 1 Gyr (Tinsley 1979; Matteucci and Greggio 1986). The great advantages of the instantaneous recycling approximation are analytical simplicity and insensitivity to past star formation rates when stable elements are considered.

The "Simple" model assumes that GCE takes place in a single well-mixed, isolated zone with constant yields starting from pure gas with primordial abundances. In this model the abundance of a primary element (i.e. one for which the yield is independent of chemical composition) is given in the instantaneous recycling approximation by the well-known equation

$$z \equiv \frac{Z}{p} = \ln (m/g) \ or \ \frac{g}{m} = e^{-z} \tag{2}$$

(Searle and Sargent 1972), where Z is the mass fraction of that element in the diffuse medium (and in newly formed stars), z the same thing in units of the yield p and g/m is the gas fraction which declines from 1 to some smaller value in the course of time. The final limb of equation (2) leads directly to the differential frequency distribution function of abundances among the constituent long-lived stars of such a system:

$$\frac{ds}{d \log z} \propto z e^{-z} \ ; \ z \leq \ln (m/g) \tag{3}$$

$$= 0 \ ; \ z > \ln (m/g)$$

This function has a peak at $z = 1$ (provided $\ln (m/g) > 1$). In a modified version of the Simple Model in which gas is lost to the system at a rate of Λ times the net star formation rate (by mass), the distribution function takes the same form but with a lower "effective" yield $p/(1 + \Lambda)$ (Hartwick 1976), as opposed to the true yield p that is fixed by stellar evolution and the IMF.

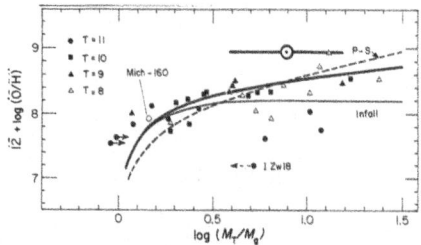

Fig. 3
Relation between oxygen abundance and ambient gas fraction in extragalactic H II regions and the Sun, adapted from Axon et al. (1988). Symbols represent the morphological types of the host galaxies. Full-drawn curves represent a "Simple" model with an oxygen yield of 0.2 times solar abundance and its modification by infall of primordial gas at a constant rate. The broken line represents a model in which the yield increases with metallicity (Peimbert and Serrano 1982).

5. SOME TESTS OF THE SIMPLE MODEL

5.1 IRREGULAR GALAXIES

The gas-fraction dependence in equation (2) can in principle be tested in gas-rich disk galaxies, especially irregular and blue compact galaxies where no significant abundance gradients are observed, although there are difficulties in practice in determining the relevant masses, especially the total mass of stars-plus-gas in the presence of probable dynamical effects of dark matter. Figure 3 shows an attempt at such a comparison, based on data by Axon et al. (1988), which gives a fair fit to the Simple model but with a low effective yield, a factor of 3 or so below what it must be in the solar neighbourhood. This modest success of the Simple Model is of importance in assessing the reliability of using the same class of objects to extrapolate to primordial helium. At the same time, the significance of the low effective yield is unclear; one factor is probably the escape of hot, enriched gas from small systems with shallow gravitational wells (Larson 1974; Yoshii and Arimoto 1987) and indeed there appears to be a high correlation between gas-phase oxygen abundance and luminosity among irregular galaxies similar to that between mean stellar metallicity and luminosity among ellipticals (Skillman, Kennicutt and Hodge 1979). In spirals there is a relation with local surface density deduced from rotation curves (McCall 1982; Edmunds and Pagel 1984).

5.2 THE GALACTIC HALO

Fig. 4 shows the metallicity distribution function of globular clusters and field stars in the halo, with a "Simple Model" distribution shown by the smooth curve (with broken-line extension) in the left part of the diagram. While there is marginal evidence for a lower cutoff in the case of the globular clusters, the Simple Model (as modified by Hartwick) gives a perfect fit to the field stars within the statistics, down to [Fe/H] = -3, with an effective yield 1/40 the iron abundance or 1/13 the solar oxygen abundance if [O/Fe] = 0.5 in the halo (Wheeler, Sneden and Truran 1989; Ryan, Norris and Bessell 1990; but cf. also Abia and Rebolo 1989). Whether or not this implies that the halo formed in a single collapse process (Eggen, Lynden-Bell and Sandage 1962) or could have formed by the merger of smaller fragments (Searle and Zinn 1978), the important point is that there is no evidence whatever for a "floor" in stellar metallicities or indeed to stellar abundances of any elements other than the BBNS products helium and lithium (cf. Pagel 1991 and references therein).

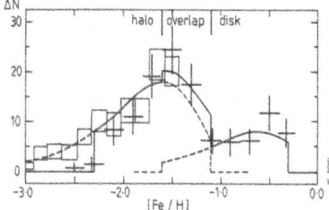

Fig. 4
Metallicity distribution function in globular clusters (crosses) and field stars of the halo (boxes). For details see Pagel (1990) or Pagel (1991).

5.3 THE GALACTIC BULGE

Fig. 5 shows that here again a "Simple Model" gives a good fit to the data, as previously noted by Rich (1988) and by Matteucci and Brocato (1990), but now with a remarkably large effective yield, especially if [O/Fe] > 0 for these stars which is not yet known but could be expected if they are indeed very old. There is a possibility that this high effective yield is a universal true yield (cf. Yoshii and Arimoto 1987) with the lower effective yields found elsewhere resulting from mass loss or other effects, but clearly this is not the only one. The Galactic bulge population is especially interesting from the point of view of cosmology because it may well be representative of stellar populations in nuclear regions of other large galaxies.

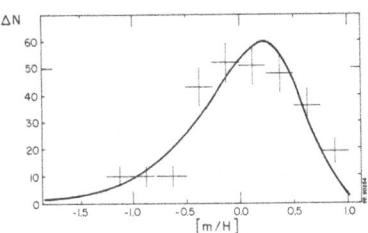

Fig. 5
Metallicity distribution function in the Galactic bulge after Geisler and Friel (1990) as reported by Grenon (1990). The curve gives the prediction of a "Simple" model with a metal yield 1.7 times solar abundance.

5.4 THE SOLAR NEIGHBOURHOOD

In the solar neighbourhood the instantaneous recycling approximation cannot be applied to iron, but it seems reasonable to study the distribution function of oxygen abundances which can be deduced from the more widely available metallicities using an empirical relationship between [Fe/H] and [O/H] (Pagel 1989a). Fig. 6 shows the latest version of this distribution function for G dwarf stars in the solar neighbourhood, corrected to the solar cylinder (Sommer-Larsen 1990ab), which illustrates the much-discussed "G-dwarf problem" (van den Bergh 1962; Schmidt 1963; Pagel and Patchett 1975; Lynden-Bell 1975; Pagel 1989a), i.e. the lack of low-metallicity stars compared to predictions of the Simple Model (dotted curve). The other curves, which give more adequate fits to the data, assume that the disk formed by gradual accretion of unprocessed gas (Larson 1976; Lynden-Bell 1975; Clayton 1985; Pagel 1989b; Sommer-Larsen 1990ab). Another kind of solution, first proposed by Ostriker and Thuan (1975), assumes that star formation in the disk was preceded by prior enrichment due to ejecta from stars in the halo (cf. also Gilmore and Wyse 1986). The difficulty with this is that such prior enrichment, in an otherwise "Simple" model, would need to make [O/H] ≥ -0.6 requiring (with constant IMFs and yields of the order of solar abundance) a mass of stars in the halo of at least 1/4 that of those in the disk, which is too large by an order of magnitude relative to star counts. However, with the large effective yield now seen to be present in the bulge, a substantial "initial enrichment" due to prior activity in the bulge and halo becomes a serious possibility after all (Köppen and Arimoto 1990). A third view of the G dwarf problem is that it is meaningless because the solar cylinder is not sufficiently representative of all the stars in the disk (Grenon 1989, 1990), but this seems to me rather extreme; the "silent majority" of stars not contained in parallax and proper motion catalogues seems to follow an exponential distribution in surface density with galactocentric radius and a fairly mild gradient in mean abundances (Lewis and Freeman 1989) implying a correspondingly shallow variation of scale length with metallicity. If this is so, then the solar neighbourhood is quite typical of the disk as a whole.

Of course, the models discussed here all assume that metallicity has increased, on average, with time, although by no means at the same rate in the different populations. Thus we have super metal-rich stars in the bulge that are probably older than quite metal-poor stars in the disk and the populations can be spatially mixed. For the halo and bulge, enrichment in the course of time has to be taken on faith, since the systematic metallicity-dependent errors in age dating exceed 1 Gyr. In the case of the disk there is evidence for an age-metallicity relation (Nissen, Edvardsson and Gustafsson 1985), but with a large scatter and with remaining doubts as to the effect of selection biases. There is no evidence for enrichment of the local interstellar medium in oxygen or iron since the formation of the Solar System. On the other hand, the oldest disk stars are significantly metal-deficient, on average, and absorption-line systems with large H I column density (damped Lyman-α systems) at large red-shifts, which may be proto-disks (Wolfe 1986; Briggs et al. 1989), display very low metallicities at red-shifts greater than 2 (Pettini, Boksenberg and Hunstead 1990; Rauch et al. 1990) and apparently low rates of star formation (Smith et al. 1989) consistent with inflow models of the sort illustrated in Fig. 6.

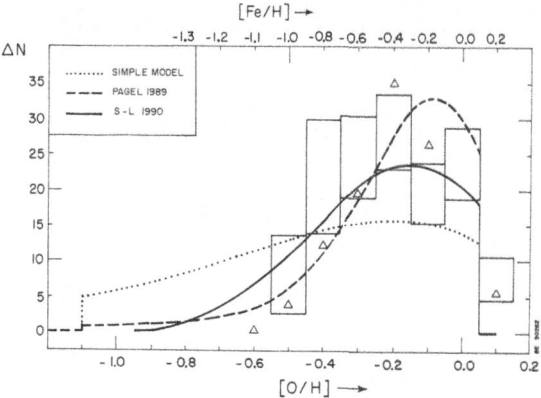

Fig. 6

Distribution function of oxygen abundance in G dwarfs of the solar neighbourhood, corrected to the Solar cylinder after Pagel (1989a, triangles) and after Sommer-Larsen (1990ab, boxes representing bin widths and error bars). Model curves are as follows: "Simple" model with plausible enrichment from the halo (dotted); inflow model of Pagel (1989b; broken-line curve); and model of Sommer-Larsen (1990b; solid curve). The data have been corrected for an assumed Gaussian dispersion σ ([O/H]) = 0.075 owing to errors and cosmic dispersion in the age-metallicity relation.

6. CONCLUSION

Some understanding of galactic chemical evolution is important to cosmology in at least two ways, one being the need for assurance in extrapolating observed abundances back to the Big Bang and the other being the desirability of knowing the numbers, intrinsic colours, luminosities and ages of distant galaxies used as cosmological probes (e.g. Guiderdoni and Rocca-Volmerange 1990). The latter is a complicated subject not touched on in this brief article, but the former appears to be reasonably well established; there is no reason to doubt that the composition of H II regions and stars that we can see today smoothly approaches primordial abundances as the amount of elements synthesised in stars according to conventional ideas tends to zero.

REFERENCES

Abia C and Rebolo R (1989) "Oxygen abundances in unevolved metal-poor stars: interpretation and consequences" **Astrophys J,** 347, 186 - 194

Axon D J, Stavely-Smith L, Fosbury R A E, Danziger J, Boksenberg A and Davies R D (1988) "Spectrophotometric and neutral hydrogen observations of Michigan 160" **Mon Not R Astr Soc,** 281, 1077 - 1090

Barry D C (1988) "The chromosperic age dependence of the birthrate, composition, motions and rotation of late F and G dwarfs within 25 parsecs of the Sun" **Astrophys J,** 334, 436 - 448

Boesgaard A and Steigman G (1985) "Big Bang nucleosynthesis: theories and observations" **Ann Rev Astr Astrophys,** 23, 319 - 378

Briggs F H, Wolfe A M, Liszt H M, Davis M M and Turner K L (1989) "The spatial extent of the z = 2.04 absorber in the spectrum of PKS 0458 - 020" **Astrophys J,** 341, 650 - 657

Byrne J et al (1990) "Measurement of the neutron lifetime by counting trapped protons" **Phys Rev Let** 65, 289 - 292

Clayton D D (1985) "Galactic chemical evolution and nucleocosmochronology: a standard model" in W D Arnett and J W Truran (eds) **Nucleosynthesis: Challenges and New Developments** Univ of Chicago Press, 65 - 88

Edmunds M G and Pagel B E J (1984) "On the composition of H II regions in southern galaxies - III. NCG 2997 and 7793" **Mon Not R Astr Soc,** 211, 507 - 519

Eggen O J, Lynden-Bell D and Sandage A R (1962) "Evidence from the motions of old stars that the Galaxy has collapsed" **Astrophys J,** 136, 748 - 766

Ellis J, Salati P and Shaver P (eds) (1990) Proceedings of the ESO - CERN Topical Workshop on **LEP and the Universe,** Geneva: CERN - TH 5709/90

Frogel J A (1988), "The Galactic nuclear bulge and the stellar content of spheroidal systems" **Ann Rev Astr Astrophys** 26, 51 - 92

Geisler D and Friel E D (1990) "Abundance distribution of Baade's window giants'" ESO/CTIO Workshop: **Bulges of Galaxies,** B J Jarvis & D M Terndrup (eds), Garching: ESO Conference and Workshop Proceedings no 35, 77 - 82

Gilmore G and Wyse R F G (1986), "The Chemical Evolution of the Galaxy" **Nature,** London, 322, 806-807

Gilmore G, Wyse R F G and Kuijken K (1989), "Kinematics, Chemistry and Structure of the Galaxy", **Ann Rev Astr Astrophys,** 27, 555-627

Grenon M (1989), "The chemical evolution of the Galactic disk from the kinematics and metallicities of proper motion stars" **Astrophys Sp Sci,** 156, 29-37

Grenon M (1990) "From halo to SMR stars" in ESO/CTIO Workshop: **Bulges of Galaxies,** B J Jarvis & D M Terndrup (eds), Garching: ESO Conference and Workshop Proceedings no 35, 143-152

Guiderdoni B and Rocca-Volmerange B (1990), "Constraints on the evolution of high red-shift galaxies and on q_0 from faint galaxy counts" **Astr Astrophys,** 227, 362-378

Hartwick F D A (1976), "The chemical evolution of the Galactic halo" **Astrophys J** 209, 418-423

Köppen J and Arimoto N (1990) "A "standard" sequence of chemical evolution models for disk galaxies" in F Ferrini, J Franco and F Matteucci (eds) **Chemical and Dynamical Evolution of Galaxies** Pisa, Giardini Editore, in press

Kurko-Suonio H, Matzner R A, Olive K A and Schramm D N (1990) "Big Bang nucleosynthesis and the quark-hadron transition" **Astrophys J,** 353, 406-410

Larson R B (1974) "Effects of supernovae on the early evolution of galaxies" **Mon Not R Astr Soc** 169, 229-245

Larson R B (1976) "Models for the formation of disc galaxies" **Mon Not R Astr Soc** 176, 31-52

Larson R B (1990) "Galaxy building" **Pub Astr Soc Pacific** 102, 709-722

Lewis J R and Freeman K C (1989) "Kinematics and chemical properties of the old disk of the Galaxy" **Astr J**, 97, 139-162

Lynden-Bell D (1975) "The chemical evolution of galaxies" **Vistas in Astr** 19, 299-316

Mampe W et al (1989) "Neutron lifetime measured with stored ultracold neutrons" **Phys Rev Lett,** 63, 593-596

Matteucci F and Brocato E (1990) "Metallicity distribution and abundance ratios in the stars of the Galactic bulge" submitted to **Astrophy J Lett,**

Matteucci F and Greggio L (1986) "Relative roles of type I and II supernovae in the chemical enrichment of the interstellar gas" **Astr Astrophys,** 154, 279-287

McCall M L (1982) **Thesis** University of Texas, Austin

Nissen P E, Edvardsson B and Gustafsson B (1985) "Oxygen and α-element abundances in Galactic disk stars as a function of age" in I J Danziger, F Matteucci and K Kjär (eds), **Production and distribution of the C, N, O elements** ESO, Garching, 131-149

Ostriker J B and Thuan T X (1975) "Galactic evolution II. Disk galaxies with massive halos" **Astrophys J**, 202, 353-364

Pagel B E J (1989a) "The G-dwarf problem and radio-active cosmochronology" in J E Beckman and B E J Pagel (eds), **Evolutionary Phenomena in Galaxies,** Cambridge University Press, 201-223

Pagel B E J (1989b) "An analytical model for the evolution of primary elements in the Galaxy" **Rev Mex Astr Astrofis,** 18, 161-172

Pagel B E J (1990) "Baryonic dark matter and the chemical evolution of galaxies" in D Lynden-Bell and G Gilmore (eds), **Baryonic Dark Matter** Kluwer, Dordrecht, 237-256

Pagel B E J (1991) "Big Bang nucleosynthesis - observational aspects" **Phys Scripta** in press

Pagel B E J and Patchett B E (1975) "Metal abundances in nearby stars and the chemical history of the solar neighbourhood" **Mon Not R Astr Soc** 172, 13-40

Pagel B E J and Simonson E A (1989) "Helium in three H II galaxies and the primordial helium abundance" **Rev Mex Astr Astrofis,** 18, 153-159

Pagel B E J, Terlevich R J and Melnick J (1986) "New measurements of helium in H II galaxies" **Pub Astr Soc Pacific,** 98, 1005-1008

Peimbert M and Serrano A (1982) "Chemical evolution of galaxies - II. Variation of the heavy element yield with Z" **Mon Not R Astr Soc** 198, 563-572

Peimbert M and Torres-Peimbert S (1974) "Chemical composition of H II regions in the Large Magellanic Cloud and its cosmological implications" **Astrophys J,** 193, 327-333

Peimbert M and Torres-Peimbert S (1976) "Chemical composition of H II regions in the Small Magellanic Cloud and the pregalactic helium abundance" **Astrophys J,** 203, 581-586

Pettini M, Boksenberg A and Hunstead R W (1990) "Metal enrichment, dust and star formation in galaxies at high redshifts I. The z = 2.3091 absorber towards PHL 957" **Astrophys J,** 348, 48-56

Rauch M, Carswell R F, Robertson J G, Shaver P A and Webb J K (1990) "The heavy element abundance in the z = 2.076 absorption system towards the QSO 2206-199N" **Mon Not R Astr Soc,** 242, 698-703

54

Rich R M (1988), "Spectroscopy and abundances of 88 K giants in Baade's window" **Astr J,** 95, 828-865

Ryan S, Norris J and Bessell M S (1990) "Element abundance ratios in extremely metal-deficient stars" **Astrophys J,** in press

Scalo J M (1986) "The stellar initial mass function" **Fund Cosm Phys,** 11, 1-278

Scalo J (1989) "Top-heavy IMFs in starburst galaxies" in A Renzini, G. Fabbiano and J S Gallagher (eds) **Windows on Galaxies,** Dordrecht: Kluwer

Schmidt M (1963) "The rate of star formation II. The rate of formation of stars of different mass" **Astrophys J,** 137, 758-769

Searle L and Sargent W L W (1972) "Inferences from the composition of two dwarf blue galaxies" **Astrophys J,** 173, 25-33

Searle L and Zinn R (1978) "Composition of the halo clusters and the formation the Galactic halo" **Astrophys J,** 225, 357-379

Simonson E A (1990) **Thesis** Sussex University

Skillman E D, Kennicutt R C and Hodge P W (1979) "Oxygen abundances in nearby dwarf irregular galaxies" **Astrophys J,** 347, 875-882

Smith H E, Cohen R D, Burns J E, Moore D J and Uchida B A (1989) "Ly α emission from disk absorption systems at high redshift: star formation in young galaxy disks" **Astrophys J,** 347, 87-95

Sommer-Larsen J (1990a) "On the G-dwarf abundance distribution in the solar cylinder" **preprint**

Sommer-Larsen J (1990b) "The formation and chemical evolution of the Galactic disk" submitted to **Mon Not R Ast Soc**

Thuan T X, Montmerle T and Van J T T (eds) (1987), **Starbursts and Galaxy Evolution,** Paris: Ed Frontières

Tinsley B M (1979) "Stellar lifetimes and abundance ratios in chemical evolution" **Astrophys J,** 229, 1046-1056

Truran J W and Cameron A G W (1971) "Evolutionary models of nucleosynthesis in the Galaxy" **Astrophys Space Sci,** 14, 179-222

Tully R B (1990) "The Hubble constant" in E Vangioni-Flam, M Cassé, J Audouze and J T T Van (eds) **Astrophysical Ages and Dating Methods"** Paris: Ed Frontières, 3-13

van den Bergh S (1962) "The frequency of stars with different metal abundances" **Astr J,** 67, 486-490

Wheeler J C, Sneden C and Truran J W (1989) "Abundance ratios as a function of metallicity" **Ann Rev Astr Astrophys,** 27, 279-349

Wolfe A M (1986) "New evidence from the Lyman-alpha forest concerning the formation of galaxies" **Phil Trans R Soc,** London, A, 320, 503-515

Yang J, Turner M S, Steigman G, Schramm D S and Olive K A (1984) "Primordial nucleosynthesis: a critical comparison of theory and observation" **Astrophys J** 281, 493-511

Yoshii Y and Arimoto N (1987) "Spheroidal systems as a one-parameter family of mass at their birth" **Astr Astrophys** 188, 13-23

THE EFFECT OF SOME NONEQUILIBRIUM PROCESSES ON THE PRIMORDIAL NUCLEOSYNTHESIS

A. D. DOLGOV
ITEP, Moscow

D. P. KIRILOVA[†]
JINR, Dubna

ABSTRACT.   The   effect   of   nonequilibrium   processes,   namely nonequilibrium   decays   of   light   particles   and   nonequilibrium   neutrino oscillations,   on   primordial   nucleosynthesis   is   investigated.   First,   a modification   of   the   standard   big-bang   nucleosynthesis   scenario   with additional   light   (m_x=O(MeV))   quasistable   particles   (t ~ 1 s)   decaying into $\nu\bar{\nu}$ is considered. It is proved that in case the decay products do not   thermalize   they   change   the   kinetics   of   the   neutron-proton transitions   and   respestively   the   abundance   of   the   light   elements produced   primordially.   In   case   of   primordial   nucleosynthesis   with neutrino   oscillations   a   concrete   model   of   oscillations   between   active and   nonthermalized   sterile   neutrinos   is   investigated.   The   effect   on the   production   of   helium-4   has   been   proved   considerable   for   a   certain range   of   oscillation   parameters.   Thus,   stringent   constraints   on   the oscillation parameters were obtained.

We investigated two modifications of the standard model of the primordial nucleosynthesis [1], namely primordial nucleosynthesis with nonequiliblrium decaying light particles X and primordial nucleosynthesis with nonequilibrium neutrino oscillations.

1.We studied the effect of massive, quasistable particles X decaying during or before the freezing of the neutron-to-proton ratio [2]. Nonequilibrium interactions of the nonthermalized decay products with the nucleons and their effect on He-4 production have been investigated. The total effect of the eventual existence of these particles includes both the change of the thermal history of the Universe due to the presence of additional massive particles and the direct influence on the kinetics of the neutron-proton transitions by the interactions of the decay products with nucleons. We considered in detail the case of the dominant decay mode X→νν̄. The equation defining the evolution of decaying X-particles in the expanding Universe reads

$$\partial n_x/\partial t = H\, p\, (\partial n_x/\partial p) - (m_x/E_x)\, \Gamma_x\, n_x \qquad (1)$$

where E_x is the total energy of the X-particles, Γ_x is the decay width. The first term describes the expansion, and the second term is due to the decays. We assume that the X-particles get out of the thermal

---

[†] Permanent address: Sofia 1309, P.B. 15, Sofia's Astronomical Observatory

contact with the plasma at a temperature $T_F^x > 1$ MeV. We solved the equation (1) for two different possibilities for the mass-to-freezing temperature ratio: $m_x/T_F^x > 1$ and $m_x/T_F^x \leq 1$. The initial condition for eq. (1) i.e. the number densities of X- particles at T_F is $n_x^F = \exp(-E_x/T_F^x)$.

The evolution of the decay products of X-particles is described by the equation:

$$\partial n_\nu/\partial t = Hp(\partial n_\nu/\partial p) + (m_x/E_\nu^2)\Gamma_x \int_{E_x^{min}}^{\infty} dE_x n_x(p) \qquad (2)$$

where $E_x^{min} = (m^2 + 4E_\nu^2)/(4E_\nu)$. The two terms on the right-hand side represent the diluting effect of the expansion and the effect of the creation of ν from X-decays.

The kinetic equation governing the neutron number density can be written as

$$\dot{N}_n = \dot{N}_n^{standard} + \delta\dot{N}_n \qquad (3)$$

where $\delta\dot{N}_n = -(G_F^2/2\pi^3)(1+3\alpha^2)T^5(N_n-N_p)\int d. \quad \lambda(\lambda+\Delta/T)^2\lambda^2 n_\nu$, $\Delta = m_n - m_p$

The last term of the equation (3) describes the effect of the $\nu n \rightarrow pe$ and $\bar{\nu}p \rightarrow ne^+$ nonequilibrium reactions with ν and $\bar{\nu}$ from X- decays. The equilibrium number densities of neutrons and protons at high energy T>3 MeV are the initial conditions for the integration of the kinetic equation. For the case $m_x/T_F^x > 1$ numerical calculations have been performed for $T_F^x > 1.5$ MeV, $m_x > 2(\Delta+m_e)$, $\Gamma \in (1 \div 500)$ sec^{-1}. An overproduction of He-4 is achieved in this case. There also exists a range of values of m_x, Γ_x and T_F^x, for which the effect is negligible, e.g. the model reduces to the standard one for He- 4 abundance. In the second case $m_x/T_F^x \leq 1$ the equations were integrated for $\Gamma_x t \leq 1$, at $t \sim 1$ sec and $m_x \in (3 \div 14)$MeV. An underproduction of He-4 is possible in this case for a certain mass interval $m_x < 7$ MeV. It is evident that the result reflects the expected physical picture: below definite energies of the neutrinos $E_\nu \sim (\Delta+m_e)$, e.g. $m_x \sim 2(\Delta+m_e)$, the p-producing reaction becomes predominant and a decrease of n_n and the final He-4 can be achieved. The comparison of the primordial production of helium-4 with the observations puts stringent limits to the allowed range of parameters of the model.

2. We studied a possible modification of the standard nucleosynthesis scenario allowing neutrino oscillations [3]. A concrete model of oscillations between active and nonthermalized sterile neutrinos ν_s is investigated; ν_s denotes right-handed neutrinos, i.e. SU(2)xU(1)

singlets. Oscillations between sterile neutrinos and the active ones are possible, i.e. we study M+D mixing scheme:

$$\nu_e = c \, \nu_1 - s \, \nu_2$$

$$\overline{\nu}_{eR} = s \, \nu_1 + c \, \nu_2 \tag{4}$$

where $c=\cos(\theta)$, $s=\sin(\theta)$, and θ is the mixing angle in the electron sector. ν_1 and ν_2 are Majorana particles with masses m_1 and m_2 respectively. Sterile neutrinos decouple from the plasma much earlier than the active ones. So, they are not heated by the numerous annihilations of particles in the process of plasma's cooling from the temperature of decoupling of the sterile particles T_s^F till the temperature of decoupling of the electron neutrino $T(\nu_L)_F \sim 3$ MeV: $T(\nu_R)_F \equiv T_s^F \gg T(\nu_L)_F \sim 3$ MeV. So, particle number densities of the sterile neutrinos $n_s \sim T_s^3$ at the nucleosynthesis period will be considerably lower than electron neutrino particle densities $n_e \sim T_e^3$. We discuss the case when neutrino oscillations become effective after the decoupling of ν_e. The reactions of active neutrinos with the plasma are the source of thermalization for the ν_s in case of oscillations $\nu_e \leftrightarrow \nu_s$. If neutrino oscillations become effective after the decoupling of ν_e the sterile neutrinos will not be thermalized. The essential point in our research is the element of nonequilibrium introduced by the nonthermalized sterile neutrinos, after the temperature falls below 3 MeV and oscillations become effective. We estimated the effect of M+D vacuum oscillations of nonthermalized neutrinos on the primordial production of He-4. The total effect of neutrino oscillations consists of the following: As a result of oscillations the number density of electron neutrinos will decrease in comparison to their standard equilibrium value. Oscillations also change the energy distribution of active neutrinos. These changes influence the kinetics of reactions of neutrinos with nucleons. As a result of the mixing the number of the relativistic degrees of freedom during nucleosynthesis changes and the expansion rate changes correspondingly. All this leads to a change in the production of the light elements during nucleosynthesis. In case of nonequilibrium oscillations it is necessary to work in terms of density matrix ρ for the neutrinos, because in this case ρ essentially differs from its equilibrium form $\rho_{ij}^{eq} = \delta_{ij} \exp(\mu - E/T)$ [4]. Using the approach of refs. [4,5] we write the equation for the density matrix of the oscillating neutrinos in the primeval plasma in the expanding Universe

$$\partial \rho / \partial t = H \, p_\nu \, \partial \rho / \partial p_\nu + i \, [H_o, \rho] +$$

$$+ \int d\Omega(\overline{\nu}, e^+, e) \left[n_e \, n_{\overline{e}} \, A \, A^+ - 0.5 \, \{\rho, \, A^+ \, \overline{\rho} \, A\}_+ \right] +$$

$$+ \int d\Omega(e, \nu', e') \left[n'_e \, B \, \rho' \, B^* - 0.5 \, \{B^+ \, B, \, \rho\}_+ \, n_e \right] + \tag{5}$$

$$+ \int d\Omega(e^+, \nu', e^{+'}) \left[n_{e^-}' C \rho' C^+ - 0.5 \{C^+ C, \rho\}_+ n_{e^-} \right]$$

where $\quad d\Omega(a,b,c) = \dfrac{(2\pi)^4}{2E_\nu} \dfrac{d^3 p_a}{(2\pi)^3 2E_a} \dfrac{d^3 p_b}{(2\pi)^3 2E_b} \dfrac{d^3 p_c}{(2\pi)^3 2E_c} \delta^4(p_a + p_b - p_c - p_\nu)$, $\quad p$ and

E_ν are respesctively the momentum and the energy of neutrino, ρ_{ij} is the (ij)-element in the density matrix of the massive Majorana neutrinos. n stands for the density number of the interacting particles. H_0 is the free neutrino hamiltonian, A is the amplitude of the process $e^+ e^- \rightarrow \nu_i \nu_j$, B is the amplitude of the process $e^- \nu_i \rightarrow e^- \nu_j$, C is that of the process $\nu_i e^+ \rightarrow \nu_j e^+$. They can be expressed through the known amplitudes $A_e(e^+ e^- \rightarrow \nu_e \bar\nu_e)$, $B_e(e\nu_e \rightarrow e\nu_e)$ and $C(e^+ \nu_e \rightarrow e^+ \nu_e)$: $A_{ij} = \alpha_{ij} A_e$, $\quad B_{ij} = \alpha_{ij} B_e$, $\quad C_{ij} = \alpha_{ij} C_e$, $\quad \alpha_{ij} = U_{ie}^* U_{je}$, $\quad \nu_i = U_{il} \nu_l, l = e, eR$. The first term in the equation describes the effect of expansion, the second one is responsible for oscillations $\nu_e \leftrightarrow \nu_{eR}$ and the last describes the weak reactions $e^+ e^- \rightarrow \nu_i \bar\nu_j$, $e' \nu_j' \rightarrow e \nu_i$, $e^{+'} \nu_j' \rightarrow e^+ \nu_i$. Kinetic equation for the case T<3 MeV after the decoupling of ν_e reduces to:

$$\partial \rho / \partial t = H p_\nu \, \partial \rho / \partial p_\nu + i \, [H_0, \rho] \qquad (6)$$

The initial condition reads $\rho^0 = \begin{pmatrix} c^2 & -cs \\ -cs & s^2 \end{pmatrix}$ at $T = T_0 = 3$ MeV. For the density number of the electron neutrinos $n_e \equiv \rho_{LL}$ we obtained

$$\rho_{LL} = \exp(-E_\nu / T) \left[1 - 2c^2 s^2 + 2c^2 s^2 \cos[DT(T^{-3} - T_0^{-3}) / E_\nu] \right] \qquad (7)$$

where $D = 0.1 M_{Pl} k^{-1/2} |m_1^2 - m_2^2|$ and $T = 1.56 k^{-1/4} t^{-1/2}$. The changes in the number density of the active neutrinos and in their energy distribution due to the oscillations lead to a change in the reactions of neutrinos with nucleons. The kinetic equation describing the evolutin of the neutron number density n_n reads

$$\partial n_n / \partial t = H p_n \, \partial n_n / \partial p_n + \int d\Omega(e^-, p, \nu) |A(e^- p \rightarrow \nu_e n)|^2 (n_e n_p - n_n \rho_{LL}) -$$

$$- \int d\Omega(e^+, p, \bar\nu) \, |A(e^+ n \rightarrow p\bar\nu)|^2 (n_e n_n - n_p \bar\rho_{LL}) \qquad (8)$$

Numerical integration showed that for a certain range of the model parameters a considerable increase in the He-4 production in comparison with the standard value can be observed. The relative increase in He-4 may reach 10÷20%. Consequently the observational data on primordial He-4 can be used to limit the possible oscillation parameters. The cosmological bound on the oscillation parameters for the case of mixing only in the electron sector, i.e. $N_\nu = 4$, reads

$$\delta m^2 < 10^{-9} \text{ eV}^2 \;, \qquad \theta > \pi/15 \;.$$

The kinetic effect of oscillations plays a considerable role in overproduction of He-4 in this case.

References

1. A. Boesgaard and G. Steigman, Ann. Rev. Astr. Astrophys. 23(1985)319
2. A. Dolgov and D. Kirilova, Int. J. Mod. Phys. A3(1988)267
3. D. Kirilova, JINR E2-88-301, Dubna 1988; Proc. School-Seminar "Foundations of Physics", Sochi, 1989
4. A. Dolgov, Yad. Phys. 33(1981)1309
5. A. Sapar, Derivation of the Generalized Equations of Kinetics for Photons and Particles and Their Use in Astrophysics, Valgus, Tallin, 1985

ANALYSIS OF THE REACTION ^{7}Li(d,n)^{8}Be AT SUBCOULOMB ENERGIES

T. RAUSCHER, K. GRÜN, H. KRAUSS AND H. OBERHUMMER
Institut für Kernphysik, TU Wien
Wiedner Hauptstr. 8-10
A-1040 Wien
Austria

ABSTRACT: The reaction ^{7}Li(d,n)^{8}Be is analyzed in the energy range 50 keV to 2 MeV using the DWBA formalism. The astrophysical S-factor for this reaction is calculated and compared to experimental data.

The transfer reaction ^{7}Li(d,n)^{8}Be is necessary for the calculation of the primordial abundance of ^{7}Li. There has been renewed interest in this reaction because the reaction ^{8}Li(d,n)^{9}Be has to be included in the reaction network for the inhomogenious big-bang nucleosynthesis. Due to the instability of ^{8}Li it was not possible up till now to obtain experimental data for the reaction ^{8}Li(d,n)^{9}Be. Therefore the same reaction rate as for ^{7}Li(d,n)^{8}Be was assumed [1-4].

It has been shown recently that the direct reaction mechanism can be dominant for transfer reactions involving light nuclei at subCoulomb energies [5-8]. In the present work we calculated the astrophysical S-factor for the above reaction in zero range for energies between 50 keV and 2 MeV using the DWBA-code TETRA [9]. This code is especially designed for subCoulomb energies. We used double folding and folding type optical potentials in the entrance and exit channel, respectively [10,11]. The strength of the double folding potential in the entrance channel was adjusted to reproduce the 5/2$^-$ resonance (E_{Res}= 680 keV, Γ=195 keV [12,13]). The potential in the exit channel was 55.7 MeV deep. The depth of the bound-state potential was determined by the separation energy of the proton in ^{8}Be. The geometrical parameters of the bound-state potential are similar to those given in [14]. For the zero-range normalization constant the usual value of D_o = -125.2 MeV·fm$^{3/2}$ was employed [15]. The spectroscopic factor used for the ground state transition is S = 2.9 [16].

In fig. 1 the results of our calculations for the astrophysical S-factor are compared to experimental data points at 400 keV and 1.98 MeV deduced from [17,18]. These data points and the 5/2$^-$ resonance are reproduced simultaneously by our DWBA calculation. Furthermore the angular distributions given in [17,18] at 400 keV and 1.98 MeV could also be reproduced fairly well. For the astrophysical S-factor we obtain the values S(0) = 26 MeV·barn and $\dot{S}$(0) = -48 barn.

The resonance energy and width of the 7/2$^+$ resonance (E_{Res}= 980 keV, Γ=195 keV [12,13]) could also be reproduced. However, to obtain the 5/2$^-$ and 7/2$^+$ resonance simultaneously a parity dependent optical potential in which the depths of the odd-even potentials differ by about 15% has to be applied in the entrance channel. This will be investigated in future work. In any case, the above resonances can be explained by our direct DWBA model showing that these resonances can be regarded as optical or potential resonances.

References:

[1] Malaney, R.A. and Fowler, W.A. (1987), in Mathews, G.J. (ed.), 'The origin and distribution of the elements', World Scientific, Singapore, p.76.
[2] Malaney, R.A. and Fowler, W.A. (1988), Ap.J. 333, 14.
[3] Malaney, R.A. and Fowler, W.A. (1989), Ap.J. 345, L5.
[4] Malaney, R.A. (1990), in Thomson, W.J. and Carney, B.W. (eds.), 'Primordial Nucleosynthesis', World Scientific, Singapore, in press.

[5] Oberhummer, H., Herndl, H., Leeb H. and Staudt, G. (1989), Kerntechnik 53, 211.
[6] Raimann, G., Bach, B., Grün, K., Herndl, H., Oberhummer, H., Engstler, S., Rolfs, C., Abele, H., Neu, R. and Staudt, G. (1990), Phys. Lett. B 249, 191.
[7] Grün, K., Krauss, H., Rauscher, T., Oberhummer, H. and Raimann, G. in Muon Catalyzed Fusion, in press.
[8] Herndl, H., Abele, H., Staudt G., Bach, B., Grün, K., Oberhummer, H., Raimann, G., (1991), submitted to Phys. Rev. C.
[9] Bach, B., Grün, K. and Raimann, G., code TETRA, not published.
[10] Satchler, G.R. and Love, W.G. (1978), Phys.Rep. 55, 183.
[11] Kobos, A.M., Brown, B.A., Lindsay, R. and Satchler, G.R. (1984), Nucl.Phys. A425, 205.
[12] Baggett, L.M. and Bame, S.J. (1952), Phys. Rev. 85, 434.
[13] Slattery, J.C., Chapman, R.A. and Bonner, T.W. (1957), Phys. Rev. 108, 809.
[14] Cecil, F.E., Peterson, R.J. and Kunz, P.D. (1985) Nucl. Phys. A441, 477.
[15] Knutson, L.D. (1977), Ann. Phys.(NY) 106, 1.
[16] Cohen, S. and Kurath, D. (1967), Nucl. Phys. A101, 1.
[17] Johnson, C.H. and Trail, C.C. (1964), Phys. Rev. B133, 1183.
[18] Galloway, R.B. and Ghazarian, A.M. (1984), Phys. Rev. C29, 2349.

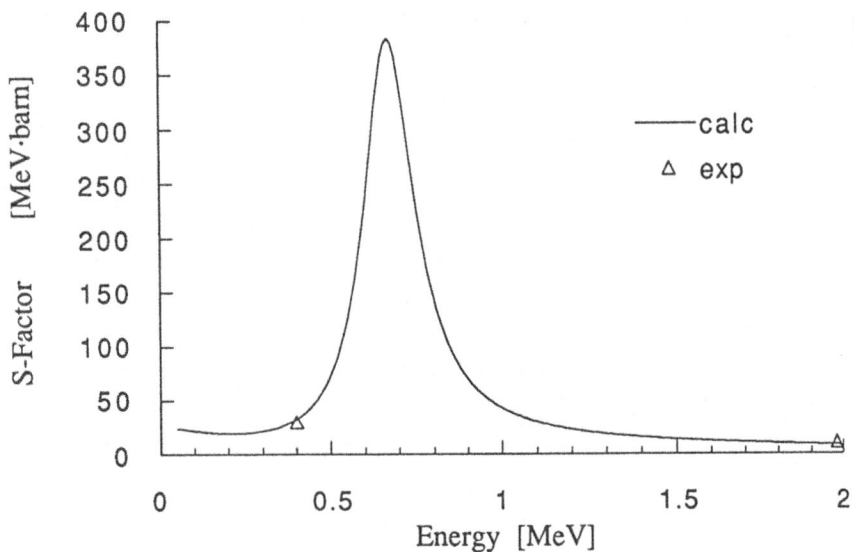

Fig. 1: Astrophysical S-factor in the energy range of 50 keV to 2 MeV. The experimental data is taken from [17,18], the full line gives the result of the DWBA-calculation.

Experimental Study of the Key Reaction to the Nucleosynthesis in the Inhomogeneous Big Bang Models

S. Kubono, R. Boyd[a], R. J. Peterson[b], N. Ikeda, M. H. Tanaka, M. Yasue, S. Yun[e], H. Toyokawa[e], M. Yosoi[e], H. Ohnuma[e], I. Tanihata[f], and T. Kajino[g]
Institute for Nuclear Study, University of Tokyo, Tanashi, Tokyo, 188 Japan
[a] Ohio State University, [b] University of Colorado, [c] Tohoku University, [d] Yamagata University, [e] Tokyo Institute of Technology, [f] RIKEN, [g] Tokyo Metropolitan University

Inhomogeneous Big Bang models[1-4], which introduced inhomogeneities in baryon density in the quark-hadron phase transition, have succeeded in predicting synthesis of finite metallicity. Among the postulated dominant flow path[3], which bypasses the A = 8 gap in the inhomogeneous models, the most unknown process is $^{8}Li(\alpha,n)^{11}B$. An experiment[5] for this process was recently reported which used the reverse reaction $^{11}B(n,\alpha)^{8}Li$. Some resonances were clearly observed, although crucial physical parameters such as total level widths and the spin-parities were not determined for the resonances. We have studied the property of these resonant states in ^{12}B by using the direct reaction $^{9}Be(\alpha,p)^{12}B$.

A 64.21-MeV α beam was obtained from the sector-focusing cyclotron of the Institute for Nuclear Study, University of Tokyo. A ^{9}Be target of 510 $\mu g/cm^2$ was bombarded, and the reaction products protons were momentum analyzed by a QDD-type high-resolution magnetic spectrograph system. The overall energy resolution obtained was about 40 keV.

The proton spectrum from the $^{9}Be(\alpha,p)^{11}B$ reaction obtained at 15° for the high-lying states at 7 - 14 MeV was calibrated by the $(\alpha,^{3}He)$ reactions on $^{24,25,26}Mg$ and $^{12,13}C$. See table 1. The differential cross sections were measured at ten angles for angular distributions. The states identified here are well corresponding in energy to the resonances observed in the $^{11}B(n,\alpha)^{8}Li$ reaction[5] except for the lowest state which was on the edge of their measurement. The most important state for the $^{8}Li(\alpha,n)^{11}B$ reaction at around $T_9 = 1$ ($T_9 = T/(10^9$ K)) is the one at 10.572 MeV, which is very clearly excited here. The observed level width of this state is not larger than 20 ($\pm$ 10) keV, which is much smaller than 200 keV assumed in ref. 5. The shape of the (α,p) angular distribution for this state is fitted reasonably well by exact

Table 1   Experimental results on the resonances in ^{12}B.

Present				$^{11}B(n,\alpha)^{8}Li$[5]		
Ex. (MeV)	ΔEx (keV)	J^π	Γ_{tot} (keV)	Ex. (MeV)	J^π	Γ_{tot} (keV)
10.115	11		180			
10.418	11	(1^-)	130	0.38		
10.572	11	(1^-)	$\leq$ 20	0.58	$\cong 1^-$	$\cong 200$
11.346	11		140	1.34		
12.226	11		160	2.24		

64

finite-range Distorted Wave Born Approximation calculation assuming $J^\pi = 1^-$. This spin assignment is consistent with that assumed in ref. 5.

For the reaction rate calculation of the $^8\text{Li}(\alpha,n)^{11}\text{B}$ process, we adopted $\Gamma_{tot} = 20$ keV for the state to calculate the upper bound for the problem, and the same branching ratios for particle decays as was assumed in ref. 5. The result is lower than the previous inhomogeneous model predictions[1-4] by more than an order of magnitude at around T9 = 1 and even more at other temperature region. The calculation with a constant $S(E)$[5] gives roughly the same rate as the previous predictions.

Figure 1 displays the mass fraction of ^{12}C calculated by the inhomogeneous model based on the present experimental data. The ordinary inhomogeneous model predictions are very close to that of "Paradellis" calculated based on the recent data[5]. The dotted line indicates the calculated result without the $^8\text{Li}(\alpha,n)^{11}\text{B}$ process. Since the present result gives the upper bound of this process, it is suspicious if this $^8\text{Li}(\alpha,n)^{11}\text{B}$ process is really the dominant flow path in the inhomogeneous Big Bang models.

REFERENCES:
1) J. H. Applegate, C. J. Hogan and R. J. Scherrer, Phys. Rev. D35(1987) 1151 and Astrophys. J. 329 (1988) 592.
2) C. Alcock, G. M. Fuller and G. J. Mathews, Astrophys. J. 320 (1987) 439.
3) R. A. Malaney and W. A. Fowler, Origin and Distribution of the Elements, World Scientific, 1988, p. 76, and Astrophys. J. 333 (1988) 14.
4) T. Kajino, G. J. Mathews, and G. M. Fuller, Heavy Ion Physics and Nuclear Astrophysical Problems, ed. S. Kubono et al., World Scientific, 1988, p. 51, and Astrophys. J. 364 (1990) 226.
5) T. Paradellis, et al., Z. Phys. A, to be published.

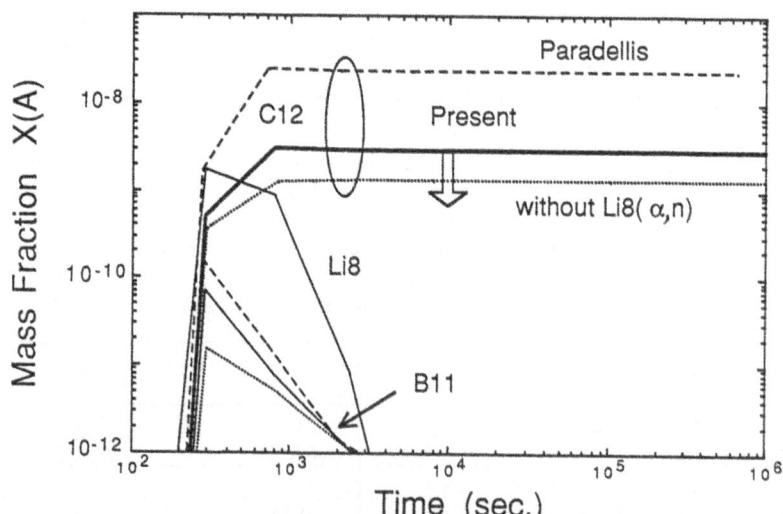

Fig. 1   The Inhomogeneous Big Bang Model calculation for heavy elements based on the present experimental result (solid line), and a constant $S(E)$-factor in ref. 5 (dashed line). The dotted line is the calculation by switching off the $^8\text{Li}(\alpha,n)^{11}\text{B}$ process.

PRIMORDIAL BLACK HOLES AND BIG BANG NUCLEOSYNTHESIS

Kenzo Arai, Masa-aki Hashimoto[*] and Toshifumi Futamase[**]
Department of Physics, Kumamoto University,
Kumamoto 860, Japan
[*] Department of Physics, College of General Education,
Kyushu University, Fukuoka 810, Japan
[**] Department of Physics, Hirosaki University,
Hirosaki 036, Japan

ABSTRACT. Primordial nucleosynthesis is examined with evaporating black holes. A low density model of the universe is preferable because of the heating by the evaporation.

1. Introduction

According to the scenario proposed by Hawking (1975), primordial black holes might have evaporated in the early stage of the universe. The black holes with mass 10^{10} g would have evaporated at the epoch of about 10^2 s which corresponds just to the stage of big bang nucleosynthesis. The emitted energy had been thermalized and used in rising up the matter temperature. It is expected that physical conditions during the epoch of nucleosynthesis deviate from those in the standard model. Miyama and Sato (1978) have investigated the nucleosynthesis in case of monochromatic mass spectrum of holes.

We follow the evolution of the abundances of elements synthesized in the early universe with the evaporating black holes. It is assumed that the emitted energy is thermalized immediately and the universe is Friedmann apart from small density fluctuations which give rise to the primordial black holes. For simplicity the mass spectrum is assumed to have a form of power law.

2. Basic Equations

The evaporation of a black hole is given by Hawking as (1975)

$$\frac{dm}{dt} = -\alpha \frac{\hbar c^4}{G^2} \frac{1}{m^2} \quad ,$$

where $\alpha = 10^{-4}$. The number density of primordial black holes with mass in the range m and m+dm is

$$n(m) \, dm = f \, m^{-\gamma} \, dm \quad .$$

The energy density emitted by the black holes during the time Δt is

$$\Delta \rho_{bh} = f \alpha \frac{\hbar c^4}{G^2} m^{-(\gamma+1)} \Delta t \quad .$$

66

3. Evolution of the Universe
Figure 1 shows the density-temperature diagram in the early universe with the present matter density $\rho_0 = 1 \times 10^{-29}$ g cm^{-3}. The dotted line refers to the Friedmann (FR) model. The solid line indicates model G3, which has $\gamma = 3$ and $n_1 = 3.17 \times 10^{-5}$ cm^{-3}, the number density of the holes at the epoch 1 s. In model G3, effective contribution of evaporating black holes takes place at the epoch t 10^6 s. Therefore, the early stage of G3 has lower T at fixed ρ, and hence higher ρ at fixed T, than that of FR. On the other hand, model G4, which has $\gamma = 4$ and $n_1 = 1.0 \times 10^{-4}$ cm^{-3}, does not deviate appreciably from FR model, because the effective evaporation takes place as early as 10^{-3} s.

4. Primordial Nucleosynthesis
Once temperature and density are specified, we can calculate the abundance of elements. The nuclear reaction rates are taken from Caughlan and Fowler (1988). The final abundances by mass fraction are presented in Figure 2. Since model G3 takes a higher density path than FR model does, the abundances produced in model G3 (solid lines) with lower density are nearly same as in FR model (dotted lines).

The observed abundances of light elements are summarized by Boesgaard and Steigman (1985). A quantitative comparison of the calculated results and the observational data yields the permitted range of the present matter density:
$(1.2 - 4.0) \times 10^{-31}$ g cm^{-3} for FR,
$(0.31 - 1.2) \times 10^{-31}$ g cm^{-3} for G3.

References
Boesgaard, A.M. and Steigman, G. 1985,
    Ann. Rev. Astr. Ap. **23**, 319.
Caughlan, G.R. and Fowler, W.A. 1988,
    Atomic Nucl. Data Tables **40**, 283.
Hawking, S.W. 1975, Comm. Math. Phys.
    43, 199.
Miyama, S. and Sato, K. 1978, Prog.
    Theor. Phys. **59**, 1012.

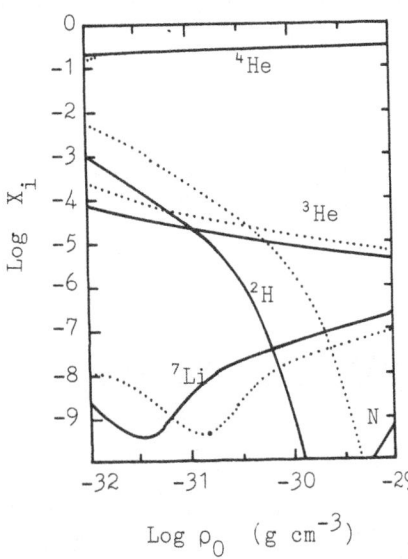

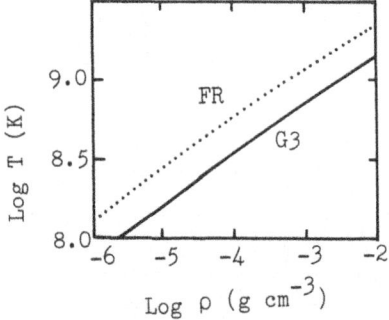

Fig. 1. Density-temperature diagram in the early universe.

Fig. 2. Final abundunces produced in model G3 with evaporating black holes. The dotted lines refer to the Friedmann model.

CONSTRAINTS FROM PRIMORDIAL NUCLEOSYNTHESIS
ON NEUTRINO DEGENERACY

HO-SHIK KANG

Department of Physics
The Ohio State University
Columbus, OH 43210-1106
U. S. A.

Abstract: We present the permissible ranges of, and most stringent constraints on, the lepton asymmetry(degeneracy), particulary the electron-neutrino asymmetry, for different baryon asymmetries by comparing calculated primordial abundances of the light elements(D, ^3He, ^4He, ^7Li) in Big Bang Nucleosynthesis(BBN) with those abundances inferred from current observational data. Furthermore, we show that the Universe can be **closed** by baryons only ($\Omega_b = 1$) if neutrinos are sufficiently degenerate. Finally, we study the behaviour of the abundances of the light elements for different baryon and lepton asymmetries.

1. Degenerate BBN: Since synthesis of the light elements based on BBN could be modified in the presence of a lepton asymmetry(neutrino degeneracy) it is possible to obtain a large present ratio of nucleons-to-photons, η, therefore, a large baryon mass density[1),2)], even $\Omega_b = 1$, without destroying the consistency of the comparison of the predicted primordial abundances of the light elements with observations. One can show that the degeneracy parameters ξ (defined by the ratio of the chemical potential to k_BT) of $\mu^\pm$ and $\tau^\pm$ are zero because μ and τ decay at the time of nucleosynthesis, and that due to the charge neutrality of the Universe, $\xi_e \approx \mathcal{O}(B)$ which is negligible. Further nonelectron neutral leptons(neutrinos) have cosmologically same effects on the Universe; i.e., increase the total energy density and hence expand the Universe faster, which can be expressed through an equivalent speed-up factor, $S(\xi_{\nu_\mu}, \xi_{\nu_\tau})$ defined by[3),4)]:

$$S(\xi_{\nu_\mu}, \xi_{\nu_\tau}) \equiv \frac{H'}{H} = \left(\frac{\varrho'_{tot}}{\varrho_{tot}}\right)^{1/2} = \left(\frac{g'_{eff}}{g_{eff}}\right)^{1/2}, \tag{1a}$$

$$g'_{eff}(T) = \sum_B g_B \left(\frac{T_B}{T_\gamma}\right)^4 + \frac{7}{8} \sum_F g_F \left(\frac{T_F}{T_\gamma}\right)^4 \left(1 + \frac{30}{7}\left(\frac{\xi_{\nu_\ell}}{\pi}\right)^2 + \frac{15}{7}\left(\frac{\xi_{\nu_\ell}}{\pi}\right)^4\right), \tag{1b}$$

where the primes are for the case of degenerate neutrinos and the unprimed quantities are for the standard case of 3 species of light, nondegenerate neutrinos, $H(t)$ is the Hubble parameter, g_{eff} is the total effective number of degrees of freedom defined by $\varrho_{tot}/\varrho_\gamma \equiv g_{eff}/2$, B(F) indicates bosons(fermions), $g_{B(F)}$ is the number of boson(fermion) degrees of freedom, $T_{B(F)}$ is the temperature, and the subscript ℓ denotes all possible flavors of leptons ($\ell = e, \mu, \tau, \cdots$).

For the electron-neutrino degeneracy, there is a more important effect on the neutron-to-proton ratio. In thermal equilibrium,

$$\frac{n(n)}{n(p)} = exp(-\frac{\Delta m}{T} - \xi_{\nu_e} + \xi_e) \simeq exp(-\frac{\Delta m}{T} - \xi_{\nu_e}), \tag{2}$$

where $\Delta m = 1.29 MeV$ is the neutron-proton mass difference. Since the cosmological role of electron-neutrinos is quite different from that of nonelectron-neutrinos, we shall separate ξ_{ν_e} from $(\xi_{\nu_\mu}, \xi_{\nu_\tau}, \cdots)$ in our analysis. We will therefore explore an extended range of η, ξ_{ν_e}, $S(\xi_{\nu_\mu}, \xi_{\nu_\tau})$ parameter space, and probe in detail the possibility of an $\Omega_b = 1$ Universe. We have taken for the neutron mean-life $\tau_n = 882$ sec. and assumed 3 flavors of light neutrinos.

2. Results of Degenerate BBN: We adopt the following set of the abundances of the light elements as the acceptable values inferred from observational data in order to derive constraints on the three parameters $S(\xi_{\nu_\mu}, \xi_{\nu_\tau})$, ξ_{ν_e}, and η by comparing with the primordial abundances calculated in BBN: $D/H > 1.0 \times 10^{-5}$; $(D +^3 He)/H < 1.0 \times 10^{-4}$; $0.21 < Y_p < 0.25$; $1.0 \times 10^{-10} <^7 Li/H < 3.0 \times 10^{-10}$, where Y_p is the ^4He mass fraction and other abundances are the number ratios to hydrogen.

We plot the abundance curves of the light elements as a function of η, ξ_{ν_e}, and $S(\xi_{\nu_\mu}, \xi_{\nu_\tau})$ respectively in order to study the η, ξ_{ν_e}, and $S(\xi_{\nu_\mu}, \xi_{\nu_\tau})$ contributions to the production of the light elements, and find that large negative values of ξ_{ν_e} ($\lesssim -1.0$) can not satisfy all observational constraints simultaneously because the abundances of D and ^3He do not decrease rapidly enough to satisfy the observations as ξ_{ν_e} becomes more negative than $\xi_{\nu_e} \simeq -1$ in contrast with those of Y_p and ^7Li and also the window of large positive values of ξ_{ν_e} ($\gtrsim 2.0$) can be closed. Furthermore, we plot isoyield curves of the abundances of the light elements in the $(\xi_{\nu_e}, S(\xi_{\nu_\mu}, \xi_{\nu_\tau}))$ plane for each different $\eta_{10}(= 10^{10}\eta)$ values from $\eta_{10} = 1.0$ to $\eta_{10} = 200$, focussing on whether or not there is any allowed region in the plane to satisfy all observational data and on whether or not, if there is some allowed region, the corresponding η_{10} value is large enough to close the Universe. Indeed, it is possible to find an allowed region for each different η_{10} values if $\eta_{10} \gtrsim 3$. We present all of the allowed regions in the plane with corresponding values of η_{10} in Fig. (1), and then summarize them numerically in Table 1.

3. Conclusion: From the agreement of the primordial abundances of the light elements calculated in BBN and those abundances inferred from observational data, firstly, we are able to obtain the $\Omega_b = 1$ Universe if neutrinos are sufficiently degenerate. At the same time we are able to derive constraints on particulary the electron-neutrino degeneracy and nonelectron-neutrino degeneracies in terms of the speed-up factor, at a given η_{10} value; see Table 1.

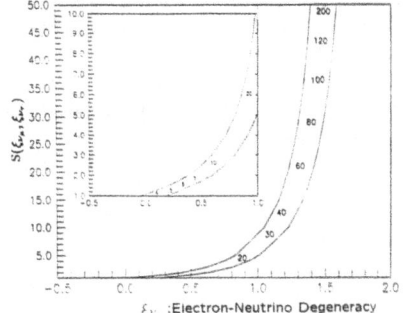

η_{10}	$D_p h^1$	$S(\xi_{\nu_\mu}, \xi_{\nu_\tau})$	ξ_{ν_e}
200	0.73	$\gtrsim 48.0$	$\gtrsim 1.41$
120	0.45	$\gtrsim 26.8$	$1.20 \sim 1.66$
100	0.37	$72.0 \sim 43.0$	$1.28 \sim 1.55$
80	0.30	$17.3 \sim 33.3$	$1.20 \sim 1.51$
60	0.22	$13.3 \sim 24.6$	$1.13 \sim 1.44$
40	0.15	$9.1 \sim 16.3$	$1.02 \sim 1.38$
30	0.11	$6.9 \sim 11.6$	$0.94 \sim 1.20$
20	0.076	$3.8 \sim 7.4$	$0.74 \sim 1.10$
10	0.037	$1.86 \sim 3.75$	$0.39 \sim 0.80$
7	0.026	$1.40 \sim 2.36$	$0.31 \sim 0.60$
6	0.022	$1.08 \sim 2.18$	$0.08 \sim 0.61$
5	0.019	≤ 1.77	$-0.04 \sim 0.49$
4	0.015	≤ 1.36	$-0.08 \sim 0.29$
3	0.011	≤ 1.08	$-0.08 \sim 0.14$

Figure 1: A summary of all allowed regions with each different η_{10} values, which satisfy all of the light elements abundances inferred from observational data. Note $40 \lesssim \eta_{10} \lesssim 120$ is relevant for the $\Omega_b = 1$ Universe.

Table 1: Constraints on $S(\xi_{\nu_\mu}, \xi_{\nu_\tau})$ and ξ_{ν_e} for different η_{10} values.

1. Kang, H.-S. and Steigman, G. (1990), to be submitted to *Nucl. Phys.* B, Dec. 1990.
2. David, Y. and Reeves, H., (1980), *Phil. Trans. Roy. Soc.*, **A296**, 415.
3. Steigman, G., (1979), *Ann. Rev. Nucl. Part. Sci.*, **29**, 313.
4. Steigman, G., (1985), *Nucleosynthesis and Its Implications on Nuclear and Particle Physics*, in J. Audouze and N. Mathieu (eds.), NATO ASI Series, vol. 163, p. 45.

STRANGE QUARK MATTER IN PHYSICS AND ASTROPHYSICS

JES MADSEN
Institute of Astronomy
University of Aarhus
DK-8000 Aarhus C
Denmark

ABSTRACT. The physical properties of stable strange quark matter/quark nuggets are described. Laboratory searches are briefly discussed, but the main emphasis of the review is on the astrophysical consequences of strange matter. In particular on the possible formation and survival of quark nuggets in the cosmological quark-hadron phase transition, where new results regarding boiling of strange matter are mentioned. Also discussed are the rather stringent astrophysical limits, that can be derived on the galactic flux of quark nuggets. Finally it is pointed out, that much of the strange matter-discussion, including the flux-limits, can be generalized to other non-topological solitons carrying baryon number.

1. Introduction

Iron may not be the most tightly bound state of strongly interacting matter at zero temperature and pressure. For surprisingly large ranges of values for the strange quark mass, m_s, the bag constant, B, and the strong fine-structure constant, α_s, bulk systems of roughly equal numbers of u, d and s quarks have an energy per baryon below that of iron (930.4 MeV) (Witten 1984; Farhi and Jaffe 1984), making so-called strange matter the ground state of the strong interactions. This possibility has far-reaching consequences for astrophysics and cosmology, as well as for cosmic ray research and heavy ion collisions. In the following a brief review will be given of the physics of strange matter and some of its consequences. It will also be pointed out how astrophysical arguments can limit the parameter space available for strange matter (and other similar non-topological solitons such as baryonic Q-balls) quite significantly.

2. Physics of Strange Matter

At low temperature and pressure quarks appear only in color-neutral systems, and only in combinations of either quark-antiquark, or three quarks (antiquarks). It is believed (and to some extent supported by calculations and experiments) that only systems with at most one unit of baryon number are stable. In particular, a bulk phase of macroscopic numbers of u and d quarks is expected to be unstable. However, bulk systems of equal numbers of massless u, d and s quarks have a lower energy per baryon due to the availability of an extra Fermi sea. Taking into account that m_s may in fact be as large as 50-200 MeV makes the energetics less favorable for strange matter, but for large ranges of parameters, strange matter in bulk *is* stable.

Farhi and Jaffe (1984) investigated this possibility in detail. Their results showed that stability might be possible for baryon numbers, A, in the range $10-10^3 \lesssim A \lesssim 2 \times 10^{57}$, where the upper bound corresponds to the Chandrasekhar mass (a few times the solar mass, $M_\odot$, where gravity quenches the system into a black hole), and the lower limit is due to the destabilizing effect of surface tension and "shell structure" (the $A=2$ dilambda may be a stable exception).

The large stability window for the baryon number makes strange matter interesting in many contexts. The lower limit of $A \approx 100$ is part of the reason why iron seems to be the most stable system in everyday life. A more important reason is, that stability of strange matter requires roughly one out of three quarks to be strange. Thus the transformation of an iron nucleus would require simultaneous conversion of roughly 50 u and d quarks, the probability of which is zero. As further explored below, this creates problems for strange matter formation in the laboratory.

Due to the s-quark mass, strange matter is slightly deficient in strange quarks relative to ups and downs, and charge neutrality requires a small fraction of electrons (typically 10^{-5} relative to the baryon number for strange matter in bulk). (For limited ranges of parameters the net quark charge is negative, so that positrons are required; the existence of such systems would be disastrous due to their ability to absorb nuclei; in the following we shall limit the discussion to systems with positive quark charge). These electrons are bound electrostatically, and thus extend beyond the sharp surface of the strongly bound quark phase. For $A \lesssim 10^7$ the system resembles a super-massive atom with a low Z/A. For $A \gtrsim 10^7$ most electrons are inside the bulk quark phase, but a small electron atmosphere leads to an electrostatic potential at the quark surface of typically +10 MeV.

The positive surface potential means that quark nuggets (as lumps of strange matter are often called) repel protons and nuclei in much the same way as ordinary nuclei (though the height of the barrier may vary over a larger range). Neutrons however will most likely be absorbed and dissolved into quarks, one of which transforms to a strange quark, thus simply adding one unit of baryon number to the nugget. This property of quark nuggets will be explored in different contexts below.

3. Strange Matter in the Laboratory

An important aspect of ultrahigh energy heavy ion collision experiments is to find evidence for the quark-hadron phase transition. One might even hope to be able to form lumps of strange matter, but several problems exist. First, the baryon number involved in these collisions may be insufficient to overcome the lower limit on A for strange matter to be stable, and second, the weak interaction is much too slow ($\approx 10^{-7}$ seconds) to transform one third of the up and down quarks in the colliding nuclei to strange quarks during the short time of collision ($\approx 10^{-22}$ seconds). On the other hand, even meta-stable, low-A nuggets would be of great interest, and furthermore s-quarks are created due to formation of thermal quark-antiquark pairs. The $\bar{s}$'s are removed in the form of kaons, leaving a net excess of s-quarks (Greiner, Koch, and Stöcker 1987). So in spite of difficulties attempts of direct formation of strange matter in the laboratory are being pursued by several groups.

Quark nuggets of cosmic origin may have been deposited in the upper parts of the Earth's surface layers, which are able to sustain nuggets with $A \lesssim 10^{14}$, if they are brought to rest by electrostatic scatterings (heavier lumps will be dragged to the center of the Earth by gravity; lumps with $A \gtrsim 10^{23}$ penetrate the Earth without suffering significant energy loss; c.f. De Rujula and Glashow 1984). Such deposits have been looked for in the laboratory by Rutherford backscattering experiments, the idea being, that the high mass of a nugget would be clearly revealed by its ability to completely backscatter the projectiles. This technique enabled Brügger *et al.* (1989) to place upper limits on the concentration of nuggets with baryon number $4 \times 10^2 < A < 10^7$.

Other techniques suggested include heavy-ion activation utilizing a high multiplicity, isotropic burst of photons as signature of interactions with strange matter (Farhi and Jaffe 1985), energy shifts of atomic x-rays due to the unusual mass-to-charge ratio of quark nuggets (Pacheco and Sanudo 1989), etcetera. So far those techniques actually tried have yielded limits only.

De Rujula and Glashow (1984) suggested to look for effects of low-energy cosmic nuggets penetrating the Earth or its atmosphere. The electrostatic scattering would stop low-mass nuggets as discussed above, and the slow-down itself might be revealed as unusual meteorite events, earthquakes with special signatures, burnt-out tracks in ancient mica, cosmic ray detectors etcetera. No characteristic nugget events have been detected, and the limits so obtained have ruled out certain ranges of flux for cosmic nuggets. We shall return to these limits below, and also comment on a recent possible detection of high-energy nuggets.

4. Strange Matter in Cosmology

The cosmic quark-hadron phase transition taking place at $T \approx 100$ MeV, 10^{-4} seconds after the Big Bang, may be first order, although this is by no means proven yet. If it is first order, a rich zoo of events may make cosmology even more interesting. One of the more fascinating possibilities is of course the inhomogeneous Big Bang nucleosynthesis treated by many authors in these proceedings, and the possible formation of quark nuggets is in a way the ultimate version of that scenario.

As the Universe cools below 100 MeV it becomes energetically favorable for hadrons to form in bubbles nucleating throughout the quark phase. In a first order transition however, surface tension requires a certain minimum radius of the hadron bubbles, so some supercooling takes place before large enough bubbles can form and start expanding by converting quark phase into hadron phase. The latent heat released reheats the Universe and the formation of further bubbles is suppressed. Instead the expanding bubbles percolate, leaving shrinking bubbles of the quark phase inside the hadron phase. If entropy (mainly in the form of neutrinos) is transmitted slowly from the quark phase compared to the transmission of baryon number, the quark lumps will be completely transformed into hadrons, leaving density inhomogeneities which due to later neutron diffusion gives rise to non-standard Big Bang nucleosynthesis (Applegate and Hogan 1985; Bonometto, Marchetti and Matarrese 1985). If entropy is transmitted sufficiently fast from the quark phase compared to the relatively slow transmission of baryon number, the central parts of the quark bubbles may reach nuclear density. If strange matter is stable, these dense bubbles are nothing but quark nuggets (Witten 1984).

The efficiency of entropy transport relative to baryon number transport is an unsettled issue, but it is fair to say, that the possibility of nugget formation remains open. Limiting the discussion to causal processes, an upper limit on A for cosmically formed nuggets is around 10^{49}, which is the baryon number within the horizon at $T=100\text{MeV}$. Estimates of "typical" bubble sizes most likely lead to somewhat smaller baryon numbers, but a fair part of the possible stability window seems to be within reach.

But do nuggets survive even if they are formed in the cosmic quark-hadron phase transition? In the hot environment emission of nucleons and boiling are two potential dangers. Nucleon emission in equilibrium was shown by Alcock and Farhi (1985) to destroy all causally formed cosmic nuggets. Madsen, Heiselberg and Riisager (1986) argued that flavor equilibration and reabsorption reduced the evaporation efficiency so that nuggets with high baryon numbers might survive after all, and Sumiyoshi et al. (1990) (see also Kajino, these proceedings) have recently stressed, that non-equilibrium effects may further reduce the importance of evaporation (and at the same time increase the likelihood of formation).

Boiling in the sense of spontaneous formation of hadron bubbles inside a hot quark nugget is more efficient than evaporation if the total surface area of the bubbles exceeds the surface area of the nugget itself. As discussed by Alcock and Olinto (1989) this ultimately depends on the surface tension, σ, of the strange matter and the pressure difference between the two phases, that together determines the rate of formation of hadron bubbles. The authors assumed an ideal gas equation of state for the hadron phase and concluded that nuggets would boil at cosmological temperatures unless the surface tension was unrealistically high.

However, a recent reanalysis (Madsen and Olesen 1990) essentially reaches the opposite conclusion, namely that boiling seems rather unimportant for nuggets that are otherwise stable at low temperatures. The major new ingredients in this analysis are the use of Walecka's mean field equation of state for an interacting npe-gas to describe the hadron phase (Serot and Walecka 1985; Reinhardt and Dang 1988), and a self-consistent calculation of pressures, surface tension and bag constant assuming equal baryon chemical potentials, μ, in the two phases:

$$\mu_q(T,P_q) = \mu_h(T,P_h) \tag{1}$$

where indices q and h refer to the quark and hadron phase respectively. At a given temperature boiling turns out to be extremely sensitive to the values of B and σ, so either nuggets boil regardless of A, or no nuggets boil at all. Typically boiling is unimportant for $B^{1/4} \lesssim (135\text{-}160)\text{MeV}$, corresponding to $\sigma^{1/3} \lesssim (40\text{-}60)\text{MeV}$ (a *lower* rather than *upper* bound on σ for boiling to take place). Care should be taken when comparing the actual numbers given here with experimental or theoretical inferences at low T. It is not so obvious how to interpret the bag constant at non-zero temperature. At any rate the investigation indicates, that nuggets may survive boiling, and proper inclusion of non-equilibrium effects (i.e. deviations from Eq.(1)) and strange hadrons would seem to strengthen this conclusion. (Related results are presented by Lee in these proceedings).

Thus quark nuggets may form during the cosmic quark-hadron phase transition, and at least the larger ones seem capable of surviving in spite of the hot environment. This still leaves open the possibility, that quark nuggets could be the dynamically inferred dark matter in galaxies and the universe at large, but it is rather difficult to make any predictions regarding the actual amount of dark nuggets. Such a calculation is however possible in principle, which distinguishes quark nuggets from almost all other dark matter candidates. Witten (1984) argued that 90-99% of the mass in the Universe might be in quark nuggets. These nuggets would behave as cold dark matter in the context of galaxy formation.

Nuggets present during Big Bang nucleosynthesis will absorb neutrons and repel protons, thereby reducing the production of ^{4}He relative to the standard model. The neutron absorption rate per neutron is proportional to $T^{7/2} A^{-1/3} \Omega_Q$, where Ω_Q is the corresponding present-day mean-density of nuggets in units of the closure density. This means that small nuggets (due to a larger area-to-volume) are the most efficient neutron absorbers. Under the (rather crude) assumption of homogeneous nucleosynthesis one finds (Madsen and Riisager 1985; Olesen 1990), that the formation of a reasonable amount of primordial helium requires, that $A \gtrsim 10^{27} \Omega_Q^3$. However, this limit must be taken with a grain of salt because formation of nuggets inevitably leads to inhomogeneities in the distribution of nucleons surrounding nuggets.

Some further limits can be put on the galactic flux/density of quark nuggets and their contribution to the dark matter from direct searches on the Earth as well as from astrophysical arguments. We shall return to these limits after a brief discussion of strange stars.

5. Strange Stars

The early Universe is a possible origin of quark nuggets due to the high temperature, that is above that of the quark-hadron phase transition. Similarly, neutron star interiors may reach densities that opens the possibility for formation of a u-d quark phase, which is likely to lead to strange matter formation via weak interactions. Once formed, a strange matter core will grow, eating up the exterior neutrons. Similarly any seed of strange matter reaching regions with free neutrons will lead to transformation of the whole star.

The resulting structures are called *strange stars* (Haensel, Zdunik and Schaeffer 1986; Alcock, Farhi and Olinto 1986).

There is a continuos transition from quark nuggets, bound by strong interactions alone, to strange stars, where gravity plays an important role. For masses below a few percent of solar ($A < 10^{56}$) the density in the quark phase is constant, and the systems obey a mass-radius relation of the form $M \sim R^3$, in contrast to the well-known $M \sim R^{-3}$ for neutron stars. This, and the related differences in moment of inertia, would make it possible to distinguish low-mass strange stars from neutron stars. However, Nature favors the formation of compact remnants with $M \approx 1.4 M_\odot$ as results of supernova explosions, and in this mass range gravity is so important, that the global properties of strange stars resemble ordinary neutron stars. Some differences in optical properties may occur if strange stars reveal a bare quark surface, but the electrostatic surface potential is capable of sustaining a thin layer of ordinary material without transforming it into quark matter, so with a little pollution it seems most likely, that the surface will look much like an ordinary neutron star.

Neutrino cooling of strange stars is more rapid than that predicted for most other neutron star equations of state, but stars with kaon condensates or a quark matter core formed due to the high density may mimic the signature of a star made of stable quark matter. (See Alcock and Olinto 1988, and Haensel 1987 for reviews of strange star properties).

Only the properties of pulsars seem to make a possible distinction for strange stars with masses near $1.4 M_\odot$. First of all, strange stars may be able to rotate more rapidly than allowed by most models of ordinary neutron stars. The limit seems to be near the speed of the once-believed 0.5 millisecond pulsar in SN1987A (e.g. Glendenning 1989; Frieman and Olinto 1989; Zdunik and Haensel 1990). And secondly, it seems impossible for strange stars to explain the phenomenon of *pulsar glitches* observed in well-known objects like the Crab and Vela pulsars.

Glitches are events where the otherwise monotonically decreasing rotation velocity of pulsars suddenly speeds up before relaxing back towards a slow-down parallel to the previous one. Although no detailed model of glitches can be said to be proven for ordinary neutron stars, the event clearly involves a restructuring of a coherent component containing a fair fraction of the stars moment of inertia (the star wants to become more spherical when it slows down). In ordinary neutron stars this part is played by the crust and mantle, but nothing similar exists for a strange star. The electrostatic surface potential can only sustain a thin shell of ordinary material (no more than around $10^{-5} M_\odot$), because if it becomes too massive it will touch the quark phase and be converted, and if the density exceeds neutron drip density ($4 \times 10^{11} \mathrm{gcm}^{-3}$), neutrons will be absorbed in the quark phase. No coherent structure with a large enough moment of inertia seems possible. This indicates, that pulsars with glitches, such as Crab and Vela, are in fact ordinary neutron stars rather than strange stars (see also Alpar 1987). Since a single seed of strange matter absorbed in a neutron star or being absorbed in the supernova progenitor prior to neutron star formation would result in the presence of a strange star rather than a neutron star, this opens the perspective of using pulsars and their progenitors as huge cosmic detectors for a background flux of quark nuggets, coming either from the Big Bang production described above, or from the inevitable pollution created when neutron stars/strange stars in binary systems collide. As described below the use of stars as large surface area detectors with long integration times allows significant improvements in the limits one may put on the local galactic flux of quark nuggets.

6. Limits on the Galactic Nugget Abundance

Two completely different sources may have polluted our local galactic neighborhood with quark nuggets: The cosmic quark-hadron phase transition and violent phenomena including strange stars, such as collisions in binary systems, that may release a few percent of a solar mass in the form of nuggets with an unknown distribution of baryon number.

Cosmologically produced quark nuggets would cluster in galaxies like other cold dark matter candidates, and have typical speeds of around 250kmsec^{-1} given by the depth of the galactic gravitational potential. The flux of nuggets hitting the Earth would be

$$F = 6.0 \times 10^5 A^{-1} \rho_{24} v_{250} \text{ cm}^{-2}\text{sec}^{-1}\text{sr}^{-1}, \tag{2}$$

with density $\rho \equiv 10^{-24}\text{gcm}^{-3}\rho_{24}$ and speed $v \equiv 250\text{kmsec}^{-1}v_{250}$. For nuggets to be the dark matter would require $\rho_{24} \approx 1$, but much smaller densities may be probed as discussed below.

Strange star collisions, supernova explosions, and other violent events involving strange matter in our own galaxy almost inevitably leads to a background flux of quark nuggets. A single event releasing $0.1M_\odot$ would give a mean density in the galactic disk of 10^{-35}gcm^{-3} if the nuggets were spread evenly. Unfortunately the production rate, the distribution of baryon number, and the velocity distribution of the fragments are almost impossible to predict, but fluxes within observational reach are not *a priori* excluded. In fact a recent reanalysis of a 1981-balloon experiment studying cosmic-ray nuclei (Saito *et al.* 1990) finds indications of two possible quark nugget detections, with $A \approx 370$, $Z \approx 14$, and 450MeV/nucleon energy, corresponding to a flux of $6 \times 10^{-9}\text{cm}^{-2}\text{sec}^{-1}\text{sr}^{-1}$. Interestingly, the candidate events follow a relation $Z \approx 2A^{1/3}$, close to the prediction of Farhi & Jaffe (1984) for low-baryon number nuggets. A more sensitive follow-up experiment is under construction.

It has been suggested that even higher-energy cosmic rays, such as Centauro events may be related to quark nuggets (e.g. Witten 1984; see also Halzen and Liu 1985 for a discussion of strange quark matter formation when heavy primary cosmic rays collide with the atmosphere), and searches are certainly worth pursuing.

Low-energy nuggets, with speeds given by the depth of the galactic potential, may be of either cosmological or local origin. Search strategies from ground and space were described by De Rujula & Glashow (1984), including special-looking meteors, earthquakes, tracks etched in cosmic ray detectors and ancient mineral deposits, etcetera. The basic ingredient in these suggestions is the fact that low-energy nuggets will experience a braking force due to electrostatic scattering off ambient atoms. The collisions will be elastic, and the cross-section will be the geometric cross-section of the electron cloud, which is approximately given by the nugget radius itself for $A > 10^{15}$, and by the Bohr radius for $A < 10^{15}$. Nuggets are stopped after sweeping up a mass comparable to their own. This happens for $A < 6 \times 10^{-6} D^3$ (for $A > 10^{15}$), and $A < 2 \times 10^8 D$ (for $A < 10^{15}$), where D is the column density in gcm^{-2}. Nuggets with $A \gtrsim 10^{10}$ can penetrate the Earth's atmosphere, and for $A \gtrsim 10^{23}$ they may pass through the Earth completely. For $A < 10^{14}$ nuggets can be trapped and sustained by the upper surface layers, making a direct search possible.

Several experiments have been analyzed to limit the flux of nuggets near the Earth (see Price 1988 for a compilation that shows, that nuggets with $3 \times 10^7 \lesssim A \lesssim 5 \times 10^{25}$ have fluxes significantly below that required for dark matter). There have also been attempts to find nuggets trapped in mineral deposits by means of Rutherford scattering. This has yielded upper limits on the concentration of nuggets with $A \lesssim 10^7$ (Brügger et al. 1989), but it is quite difficult to relate this to flux limits. A number of experiments are presently being planned or tested.

A powerful method for significant improvement of the above mentioned limits rests on the use of stars as giant, long-lived nugget detectors (Madsen 1988; 1990a; 1990b). As alluded to previously, a single nugget present during or after formation of a neutron star will transform the star to strange matter. Pulsar glitches seem to be inconsistent with the strange star hypothesis, so the flux of nuggets reaching glitching pulsars like Crab and Vela and their progenitors must have been such, that no quark nuggets were trapped. For a $10 M_\odot$ main sequence star this limits the nugget flux to at most 3×10^{-42} cm^{-2}sec^{-1}sr^{-1} for $A \lesssim 10^{30}$, 20-50 orders of magnitude (!) below the dark matter flux, and several orders of magnitude below even the mean flux corresponding to a single binary star collision, as discussed earlier. The excluded range of nugget baryon numbers providing dark matter may be extended as high as 10^{33}-10^{37} depending on the details of nugget capture after the main sequence phase (see references above).

In view of the extreme sensitivity of the "glitch argument" as a detector of quark nuggets, a few comments are in order: First, the potentially observed nugget flux with $A \approx 370$ (Saito et al. 1990) is not directly excluded by the pulsar limit, since this limit only works for low-energy nuggets. High-energy cosmic-ray nuggets will interact strongly and probably explode due to the heat released by the collisions. But of course it does require significant fine-tuning to have a high-energy flux like that observed without producing any low-energy nuggets. Also, no fragment of a high-energy nugget may survive inelastic collisions with nuclei in stars. Tough requirements, but unfortunately detailed calculations are quite difficult.

Another important comment is, that the imagination of future theorists may after all come up with a model for pulsar glitches in strange stars. If so (however unlikely it seems at present) the flux limits just described can be inverted to a prediction, that (given the stability of strange stars) all neutron stars are actually strange, simply because it seems impossible to avoid polluting the galactic disk with nuggets at a flux level above 10^{-42} cm^{-2}sec^{-1}sr^{-1}.

Of course one must also keep an open mind to the possibility that strange matter at low pressure takes a different form than assumed above, such as the stable quark complexes (*quark alphas*) proposed by Michel (1988).

7. Conclusion

The existence of stable quark matter in the form of strange matter/quark nuggets is possible for wide ranges of strong interaction parameters. If it exists it will have significant implications for heavy ion collisions, cosmic ray research, the astrophysics of neutron stars (which may actually be strange stars), the cosmic quark-hadron phase transition, Big Bang nucleosynthesis and the dark matter problem. Several experimental and astrophysical limits on quark nugget abundances have been discussed above, but at present many possibilities remain open, including the cosmic production and survival of nuggets, which now seems more likely than just one or two years ago. In view of the fact that the issue at stake is nothing less than the ground state of strongly interacting matter, experimental as well as theoretical investigations in this exciting field should be strongly pursued. Either the existence of strange matter will be demonstrated, or at least we will obtain information of strong interaction parameters that are otherwise difficult to probe.

Finally it is worth mentioning that much of the discussion above can be (and to some extent has been) generalized to other non-topological solitons, i.e. stable solutions of classical field theories with a conserved quantum number carried by fields confined to a finite spatial region (see e.g. Frieman *et al.* 1989). In particular generalizations are possible when the conserved quantum number is the baryon number, as in the case of strange quark matter. Candidates include baryonic Q-balls (e.g. Bahcall, Lynn and Selipsky 1990) and strange baryon matter (e.g. Lynn, Nelson and Tetradis 1990; Nelson 1990). A discussion of the astrophysical constraints on baryonic Q-balls and similar neutron absorbers is given in Madsen (1990b) with a parametrization that also includes strange quark matter as a special case.

References

Alcock, C., and Farhi, E. 1985, *Phys. Rev.*, D32, 1273.

Alcock, C., Farhi, E., and Olinto, A. 1986, *Ap. J.*, **310**, 261.

Alcock, C., and Olinto, A. 1988, *Ann. Rev. Nucl. Part. Sci.*, 38, 161.

Alcock, C., and Olinto, A. 1989, *Phys. Rev.*, D39, 1233.

Alpar, M. A. 1987, *Phys. Rev. Letters*, 58, 2152.

Applegate, J. H., and Hogan, C. J. 1985, *Phys. Rev.*, D31, 3037.

Bahcall, S., Lynn, B.W., and Selipsky, S.B. 1990, *Nucl. Phys.*, **331**, 67.

Bonometto, S. A., Marchetti, P. A., and Matarrese, S. 1985, *Phys. Letters*, 157B, 216.

Brügger, M., *et al.* 1989, *Nature*, 337, 434.

De Rujula, A., and Glashow, S. L. 1984, *Nature*, **312**, 734.

Farhi, E., and Jaffe, R. L. 1984, *Phys. Rev.*, **D30**, 2379.

Farhi, E., and Jaffe, R. L. 1985, *Phys. Rev.*, **D32**, 2452.

Frieman, J. A., and Olinto, A. 1989, *Nature*, **341**, 633.

Frieman, J. A., Olinto, A. V., Gleiser, M., and Alcock, C. 1989, *Phys. Rev.*, **D40**, 3241.

Glendenning, N. K. 1989, *Phys. Rev. Letters*, **63**, 2629.

Greiner, C., Koch, P., and Stöcker, H. 1987, *Phys. Rev. Letters*, **58**, 1825.

Haensel, P. 1987, *Prog. Theor. Phys. Suppl.*, **91**, 268.

Haensel, P., Zdunik, J. L., and Schaeffer, R. 1986, *Astr. Ap.*, **160**, 121.

Halzen, F., and Liu, H. C. 1985, *Phys. Rev.*, **D32**, 1716.

Lynn, B.W., Nelson, A.E., and Tetradis, N. 1990, *Nucl. Phys.*, **B345**, 186.

Madsen, J. 1988, *Phys. Rev. Letters*, **61**, 2909.

Madsen, J. 1990a, in *XXIVth Rencontre de Moriond - The Quest for the Fundamental Constants in Cosmology*, ed. J. Audouze and J. Tran Thanh Van (Editions Frontieres), p. 119.

Madsen, J. 1990b, *Phys. Letters*, **246B**, 135.

Madsen, J., Heiselberg, H., and Riisager, K. 1986, *Phys. Rev.*, **D34**, 2947.

Madsen, J., and Olesen, M. L. 1990, *Phys. Rev.*, **D** (to be published).

Madsen, J., and Riisager, K. 1985, *Phys. Letters*, **158B**, 208.

Michel, F. C. 1988, *Phys. Rev. Letters*, **60**, 677.

Nelson, A. E. 1990, *Phys. Letters*, **240B**, 179.

Olesen, M. L. 1990, *M.Sc. thesis*, University of Aarhus.

Pacheco, A. F., and Sanudo, J. 1989, *Phys. Rev.*, **A39**, 380.

Price, P. B. 1988, *Phys. Rev.*, **D38**, 3813.

Reinhardt, H., and Dang, B. V. 1988, *Phys. Letters*, **202B**, 133.

Saito, T., Hatano, Y., Fukada, Y., and Oda, H. 1990, *Phys. Rev. Letters*, **65**, 2094.

Serot, B. D., and Walecka, J. D. 1985, *Advances in Nuclear Physics*, **16**, 1.

Sumiyoshi, K., Kajino, T., Alcock, C. R., and Mathews, G. J. 1990, *Phys. Rev.*, **D** (to be published).

Witten, E. 1984, *Phys. Rev.*, **D30**, 272.

Zdunik, J. L., and Haensel, P. 1990, *Phys. Rev.*, **D42**, 710.

Primordial Nucleosynthesis in Inhomogeneous Universe

Katsuhiko Sato

Department of Physics, Faculty of Science, The University of Tokyo
Bunkyo-ku, Tokyo 113, Japan

Abstract

From recent studies of nucleosynthesis in inhomogeneous universe, It has been pointed out that 1) $\Omega_B = 1$ model becomes consistent with light element observations, 2) heavy elements including r-process elements are abundantly synthesized, and 3) ^{9}Be plays a role of the indicator of inhomogeneity since ^{9}Be is also abundantly synthesized . These suggestions were, however, based on a simple two-zone model in which back diffusion of neutrons are neglected during the nucleosynthesis. In this short review, It is shown that 1)the upper limit of Ω_B which is consistent with light element observations is almost the same as that of homogeneous case, 2) the ^{9}Be abundance in the simple two-zone model is overestimated at least two orders of magnitude in plausible parameter space and 3) the abundances of CNO elements are very small in the proper range of baryon/photon ratio consistent with the abundances of light elements.

1. Introduction

It has been suggested, recently, that large baryon density fluctuations are formed by QCD phase transition which occurs in early universe (Witten, 1984), then primordial nucleosynthesis is greatly modified and $\Omega_B = 1$ model becomes consistent with light elements abundance since overproduction of ^{4}He is suppressed and D is abundantly synthesized (Applegate and Hogan, 1985). Stimulated by this pioneering work, nucleosynthesis in the inhomogeneous universe was investigated in detail and following two important possibilities were pointed out. The first one is that ^{9}Be is synthesized abundantly and ^{9}Be may provide a test of inhomogeneous model of nucleosynthesis (Boyd and Kajino 1989, Malaney and Fowler,1990, Kawano et al,1990a, 1990b). Efforts to find metal-poor population-II stars with lower ^{9}Be abundance were made (Ryan et al. 1990).

The second is that heavy elements are abundantly synthesized in low density neutron rich regions (Sale and Mathews, 1986, Applegate et al., 1987, Malaney and Fowler, 1989, Kawano et al., 1990a, 1990b). More over, it was suggested by Applegate (1988), Applegate et al.(1988) and Fuller et al (1988) that the r-process also works in the low density neutron rich regions. If this is the case, the none-zero abundances of r-process elements in metal-poor stars might be explained by the cosmic r-process.

These important suggestions, however, are based on a simple two-zone model in which neutron diffusion during nucleosynthesis is completely neglected. Importance of neutron diffusion during nucleosynthesis was emphasized by several authors including us (Kurki-Suonio et al. 1988, 1990; Malaney and Fowler 1988; Terasawa and Sato 1989a,b,c, 1990; Mathews et al., 1990). Neutrons once diffused out of the high density zones diffuse back as soon as nucleosynthesis in the high density zone begins since it takes place earlier than that in the low density zone. Hence the neutron number density in the low density region is overestimated in the simplified two-zone models due to cutting off the neutron back diffusion during the nucleosynthesis, and the abundances of neutron-rich nuclei such as ^{9}Be may not be so large as expected. Two-zone model is, furthermore, too simplified to evaluate the effects of neutron diffusion on the abundances of elements quantitatively and we need to perform more realistic multi-zone calculations. Until now some multi-zone calculations (Kurki-Suonio et al. 1988, 1990, Mathews et al., 1990) have been performed, but heavy elements and neutron rich nuclei were not included in their calculation. In order to make clear these suggestions, it is necessary to perform realistic multi-zone calculations with large nuclear reaction network including very neutron rich nuclei and heavy elements.

2. Multi-zone Calculations with Large Nuclear Reaction Network

Recently Terasawa and Sato (1990) carried out multi-zone calculations with large nuclear reaction network. In this work, spherical symmetry was assumed as a model of the density inhomogeneity for simplicity. The diffusion equation for neutrons, $dn/dt = \nabla(D_n \nabla n)$, coupled with nucleosynthesis was solved in a completely implicit manner. For the initial ($T = 10^{11}$K) density profile, a top-hat profile that the density of inner shells is R times higher than that of surrounding low-density shells was assumed. The parameter f_v refers to the volume fraction of the initial high-density shells in the whole region. Another parameter $d(T = 1 MeV)$ refers to the radius of the site in problem normalized by the size at T=1MeV. We calculated in the ranges 10^5cm $\leq d(T = 1 MeV) \leq 10^8$cm, $1 \leq R \leq 1000$, , $0 \leq f_v \leq 0.5$, and $4 \times 10^{-10}(\Omega_B = 0.014h^{-2} = 0.056h_{50}^{-2}) \leq \eta_0 \leq 7 \times 10^{-9}(\Omega_B = 0.25h^{-2} = 1.0h_{50}^{-2})$.

Our reaction network contains almost 70 species of nuclei including very neutron-rich nuclei such as ^{8}Li ,^{9}Li, ^{11}Be ,^{12}Be, ^{13}B ,^{14}B, ^{15}C ,^{16}C, ^{17}C ,^{17}N, ^{18}N ,^{19}N, ^{19}O

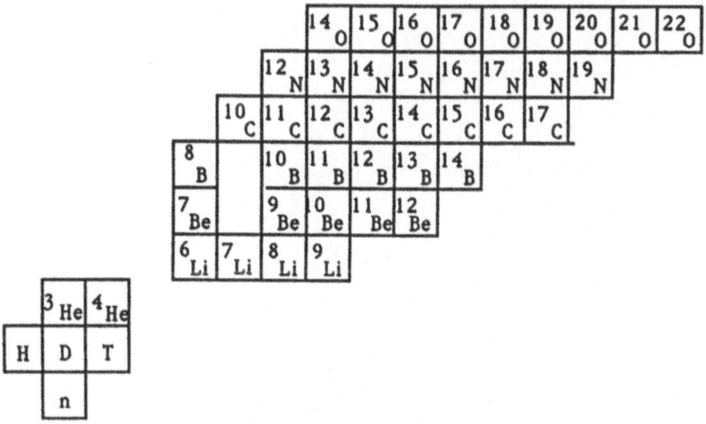

Figure 1: A part of nuclear reaction network ($Z \leq 8$). A lot of neutron-rich nuclei such as ^{8}Li ,^{9}Li, ^{11}Be ,^{12}Be, ^{13}B ,^{14}B, ^{15}C ,^{16}C, ^{17}C ,^{17}N, ^{18}N ,^{19}N, ^{19}O ,^{20}O, ^{21}O ,^{22}O are included.

,^{20}O, ^{21}O ,^{22}O. The lifetime of beta decay of the most neutron rich isotopes is about 1 sec. If the r-process nucleosynthesis occurs in the neutron rich side, the process is included in our network.

Reaction rates were updated according to Caughlan and Fowler (1988), Thielemann and Wiescher (1990), and those relating to ^{9}Be are added and updated by the table given in Malaney and Fowler (1990). The lifetime of neutrons is taken to be 887.6 sec (Mampe et al. 1989) and the number of neutrino species is 3.

In order to demonstrate the importance of back diffusion of neutrons during nucleosynthesis, we show, in Fig. 2, the time evolution of the local mass fractions of neutron in each shell for $d(T = 1\text{MeV}) = 10^5\text{cm}, R = 1000, f_v = 0.11$, and $\eta_0 = 4 \times 10^{-10}$. Neutrons begin to diffuse out of the inner high-density shells into the surrounding space and spread over the low-density zone around $t = 0.1\text{sec}$. Since the density contrast among the high density shells and the low density shells is very large in this case, the neutron fraction becomes completely dominant in the low density shells and the baryon density in the outer low-density shells grows simultaneously. Thus the ratio of the baryon density in the innermost shell to that in the outermost shell, which is 1,000 at first, finally reduces to 107. However, neutrons begin to diffuse back into the inner high-density shells to keep the neutron density distribution uniform when nucleosynthesis starts and the neutron density decreases in the high-density shells. It results in the decrease of baryon density of the outer low-density shells. Due to this effect, the baryon density contrast, which once reduces to 38 at $t \sim 100\text{sec}$, increases again to ~ 100. This example clearly shows that we cannot neglect the neutron diffusion, especially the back diffusion

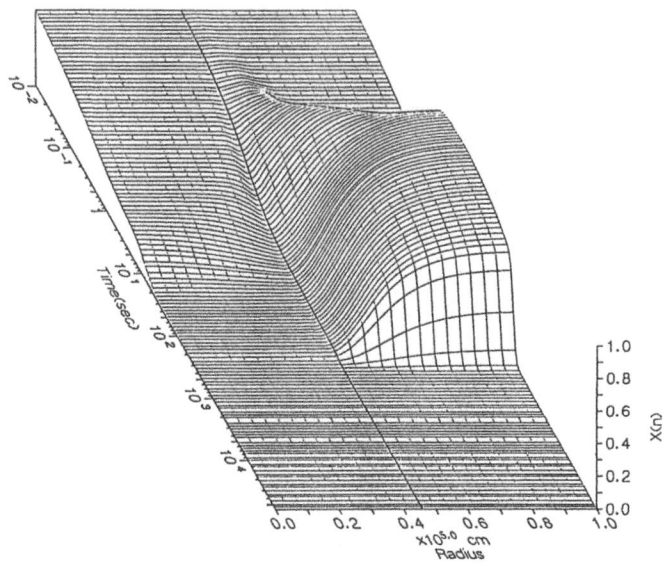

Figure 2: Time evolution of the local mass fraction of neutron in the case that $d(\mathrm{T} = 1\mathrm{MeV}) = 10^5$cm, $R = 1000$, $f_v = 0.11$, and $\eta_0 = 4 \times 10^{-10}$.

out of the low-density region into the high-density region, during nucleosynthesis.

3. Constraints on Ω_B

In our previous two-zone calculation with back diffusion of neutrons (Terasawa and Sato, 1989), we concluded that the region $\Omega_B h_{50}^2 \leq 0.8$ becomes consistent with observation of light elements if we make fine-tuning of inhomogeneity parameters, R, d and f_v. However, we took old and conservative values as for the observational limit of light elements abundance;

$$^4\mathrm{He} \ < 0.26, \tag{1}$$

$$\mathrm{D/H} > 1 \times 10^{-5}, \tag{2}$$

$$(\mathrm{D} + {}^3\mathrm{He})/\mathrm{H} < 1 \times 10^{-4} \tag{3}$$

and

$$^7\mathrm{Li/H} < 1 \times 10^{-9}. \tag{4}$$

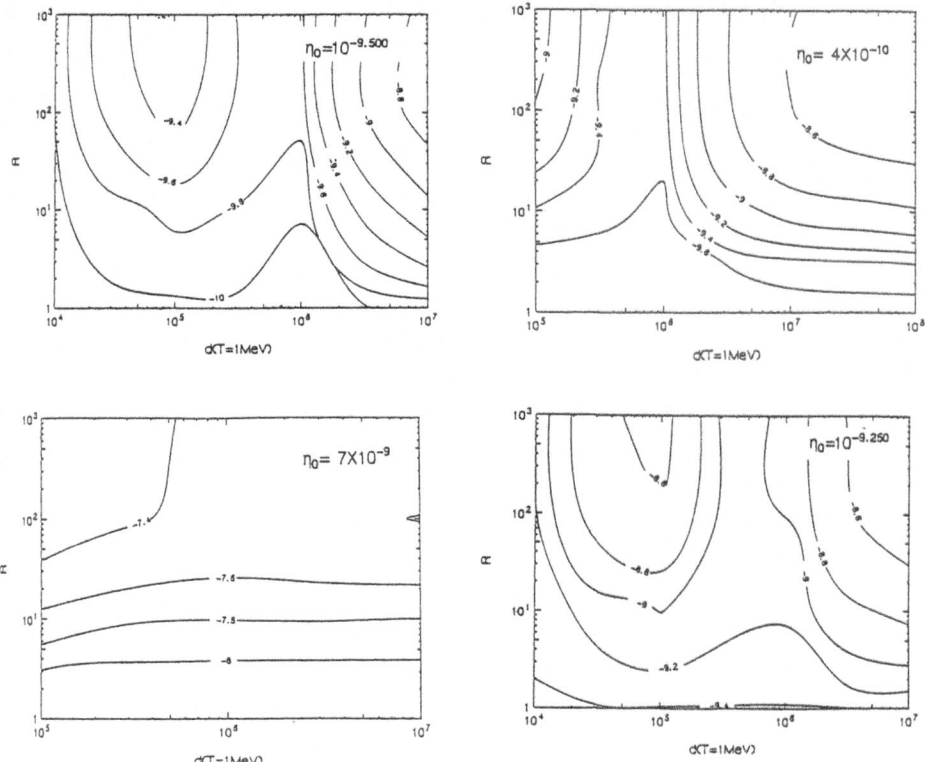

Figure 3: The average abundance of ^{7}Li is shown on $R-d$ plane for various values of η_0, but for $f_v = 0.11$

As discussed in detail by Pagel (1991, in this volume), observational limits of these abundance have become more severe by remarkable development of observations. If we take the stringent limit;

$$^4\text{He} < 0.24, \tag{5}$$

and

$$^7\text{Li/H} < 10^{-9.6} (= 2.5 \times 10^{-10}), \tag{6}$$

the constraint on Ω_B turns to $\Omega_B h_{50}^2 \leq 0.08$. We are now searching allowed regions in parameter space R, d and f_v in multi-zone calculation. As an example, the average abundance of ^{7}Li/H is shown on $R-d$ plane for various values of η_0 in Fig. 2. As seen from this figure, models with $\eta_0 \geq 5.6 \times 10^{-10}(\Omega_B = 0.078 h_{50}^{-2})$ is ruled out, since there exist no regions which satisfy the observational limit ^{7}Li/H $\leq 10^{-9.6}$ (see the review of Pagel in this volume). We searched regions where the constraint

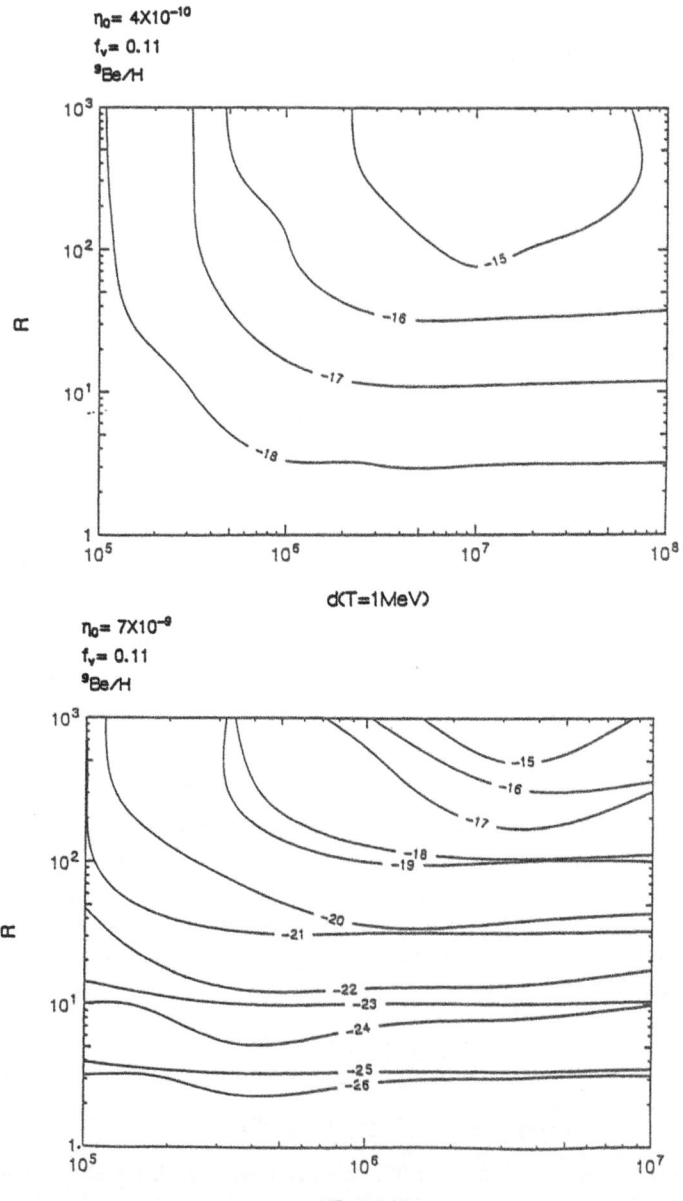

Figure 4: The contour of the average abundances of ^{9}Be on the $d - R$ plane for $\eta_0 = 4 \times 10^{-10}$(top) and $\eta_0 = 7 \times 10^{-9}$(bottom). Contour levels are log(^{9}Be/H).

^{7}Li/H $\leq 10^{-9.6}$ is satisfied in very wide range parameter space d, R and f_v, but no regions could not be found for $\Omega_B \geq 0.06h_{50}^{-2}$. This conclusion is essentially consistent with Kurki-Suonio et al. (1990) and Mathews et al. (1990).

4. ^{9}Be abundance

^{9}Be is produced in the low density and neutron rich region as discussed by Boyd and Kajino (1989). In the high-density regions, ^{9}Be once produced is burnt into heavier elements and its abundance turns out to be very small. The dependence of the ^{9}Be abundance on the scale of fluctuations is shown in Figure 4 as contour on the $d - R$ plane for $f_v = 0.11$. The ^{9}Be abundance increases with increasing R and take the maximum value around $d(\mathrm{T} = 1\mathrm{MeV}) = 10^{7}$cm. Though the f_v dependence of the ^{9}Be abundance is different with R, the ^{9}Be abundances takes the maximum value at $f_v \sim 0.1$ for large R. In the calculated ranges of parameters, the highest value of the ^{9}Be abundance is ^{9}Be/H $= 4 \times 10^{-15}$ ($d(\mathrm{T} = 1\mathrm{MeV}) = 10^{7}$cm, $R = 1000, f_v = 0.11$) for $\eta_0 = 4 \times 10^{-10}$ and ^{9}Be/H $= 3 \times 10^{-15}$ ($d(\mathrm{T} = 1\mathrm{MeV}) = 10^{6.5}$cm, $R = 1000, f_v = 0.11$) for $\eta_0 = 7 \times 10^{-9}$. These values are two orders smaller than those calculated in the simplified two-zone model. The reason for the discrepancy between these results is the overestimate of the neutron diffusion in the latter model. To confirm this point, we cut off neutron diffusion artificially when the neutron density distribution became almost uniform and calculated the abundances of elements in the case of $d(\mathrm{T} = 1\mathrm{MeV}) = 10^{5.5}$cm, $R = 1000$, $f_v = 0.11$, and $\eta_0 = 4 \times 10^{-10}$. The ^{9}Be abundance in this case is ^{9}Be/H $= 1.0 \times 10^{-13}$, which is as large as that obtained in the simplified two-zone model and higher by orders of magnitude than that in the case that neutron diffusion is not cut off with the same parameters; ^{9}Be/H $= 8.3 \times 10^{-18}$.

The recent observations of several population-II stars have placed the upper limit on the primordial ^{9}Be abundance to $\log {}^{9}$Be /H $= -13.7$ (Ryan et al. 1990), which is two orders of magnitude greater than the predicted abundance in the present work.

The constraints on the parameters from other light elements, D, ^{3}He, ^{4}He, and ^{7}Li are more stringent than those placed by the ^{9}Be abundance. The abundances of these elements, of course, depend on the scale of density fluctuations and they reduce to the uniform abundances in the limit of small scale. However, in the range of the fluctuation scale so long that the ^{9}Be abundance is larger than the uniform abundance; $d(\mathrm{T} = 1\mathrm{MeV}) \gtrsim 10^{6}$cm, the ^{7}Li abundance increases monotonically with R and R is constrained to $R \lesssim 10$ even when the very loose bound on the primordial abundance of the ^{7}Li abundance; ^{7}Li/H $< 1 \times 10^{-9}$, is taken. The constraints placed by the upper bound on the sum of the D abundance and the ^{3}He abundance are also stringent: R should be smaller than ~ 50 in the range, $d(\mathrm{T} = 1\mathrm{MeV}) \gtrsim 10^{6}$cm. Hence, the constraints from the light elements except for ^{9}Be, that is, D, ^{3}He, ^{4}He, and ^{7}Li are still more stringent and powerful to examine the inhomogeneous models

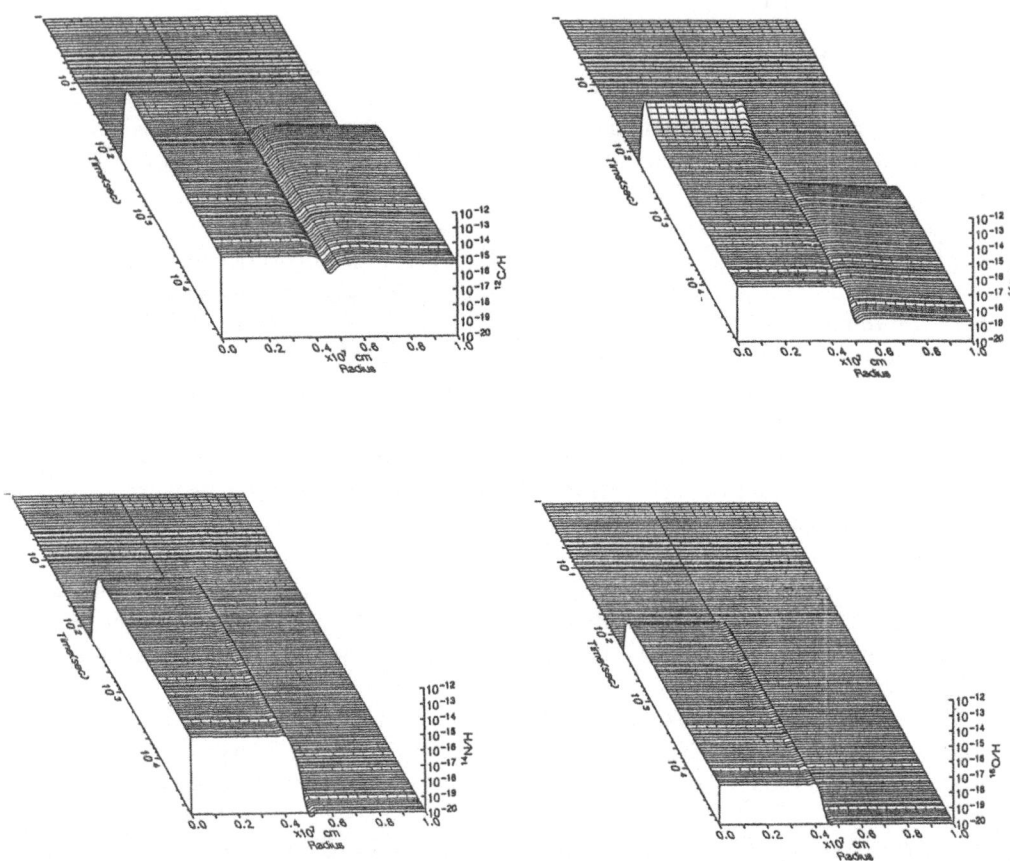

Figure 5: The time evolution of the local abundance (mass fraction) of ^{12}C, ^{12}C, ^{14}N and ^{16}O for $d(\mathrm{T} = 1\mathrm{MeV}) = 10^7\mathrm{cm}, R = 1000, f_v = 0.11$, and $\eta_0 = 4 \times 10^{-10}$

of nucleosynthesis in the present stage.

5. Abundance of Heavy Elements and r-process

In Fig. 5, time evolution of spatial distribution of ^{12}C, ^{14}C, ^{14}N and ^{16}O are shown for the model with $\eta_0 = 4. \times 10^{-10}$, $R = 1,000$, $f_v = 0.11$ and $d = 10^7$ cm. As shown in this figure, ^{12}C is synthesized not only in the high density region, but also in the low density regions. This was synthesized by usual 3α reaction in high density region, but was synthesized via the path of neutron rich nuclei, ^{7}Li$(n, \gamma)^8$Li$(\alpha, n)^{11}$B$(n, \gamma)^{12}$B$(\beta, \nu)^{12}$C in low density regions as suggested by Kawano et. al.(1990). However, the abundance is at most ^{12}C/H $= 10^{-15}$. As shown in Fig. 5, the abundances of ^{14}N and ^{16}O in the low density (neutron rich) regions are less than 10^{-20}. This shows that heavy elements synthesis via r-process like process does not work because the neutron density becomes small by the effect of back diffusion.

In Fig. 6, contour of the total abundance of heavy elements ($A > 7$) is shown for $\eta_0 = 4. \times 10^{-10}(\Omega = 0.014h^{-2} = 0.056h_{50}^{-2})$ and $f_v = 0.11$. The abundance is at most $\sim 10^{-13}$ by mass fraction. The abundance increases to $\sim 10^{-11}$ if we take $\eta_0 = 7 \times 10^{-9}(\Omega = 0.25h^{-2} = 1.0h_{50}^{-2})$ which is dominated by ^{11}B and CNO elements produced in the inner high density shells. The abundances of CNO elements in the low-density neutron-rich shells are at most $X(^{12}$C$) \sim 10^{-14}$, $X(^{14}$N$) \sim 10^{-20}$, and $X(^{16}$O$) \ll 10^{-20}$ for $\eta_0 = 4 \times 10^{-10}$. For $\eta_0 = 7 \times 10^{-9}$, $X(^{12}$C$) \sim 10^{-11}$, $X(^{14}$N$) \sim 10^{-16}$, and $X(^{16}$O$) \sim 10^{-17}$. Thus the abundances of CNO elements in the proper range of baryon/photon ratio, which is consistent with the abundances of light elements, are too small to be the seeds of cosmological r-process. For the large value of baryon/photon ratio as to close the universe only by baryons, the abundances of these elements are significantly large. However, the fact that such high baryon density models are inconsistent with the abundances of light elements should not be ignored.

Recently, Thielemann, Applegate and Wiescher(1990) investigated heavy element synthesis in a two zone model with and without back diffusion. In a $\Omega_B = 1$ model without back diffusion, heavy elements ($A \leq 100$) are considerably synthesized. In a model with back diffusion, however, the abundance of each element whose mass number is greater than 30 is less than 10^{-20}. They also investigated r-process nucleosynthesis up to $A \leq 300$. In a model without back diffusion, the mean abundance of r-process elements was $\sim 10^{-5}$ of the solar system abundance or $\sim 10^{-2}$ of the Pop-II star abundance. In a model with back diffusion, however, nothing was synthesized. Their conclusion is very consistent with our results.

6. Conclusion and Remarks

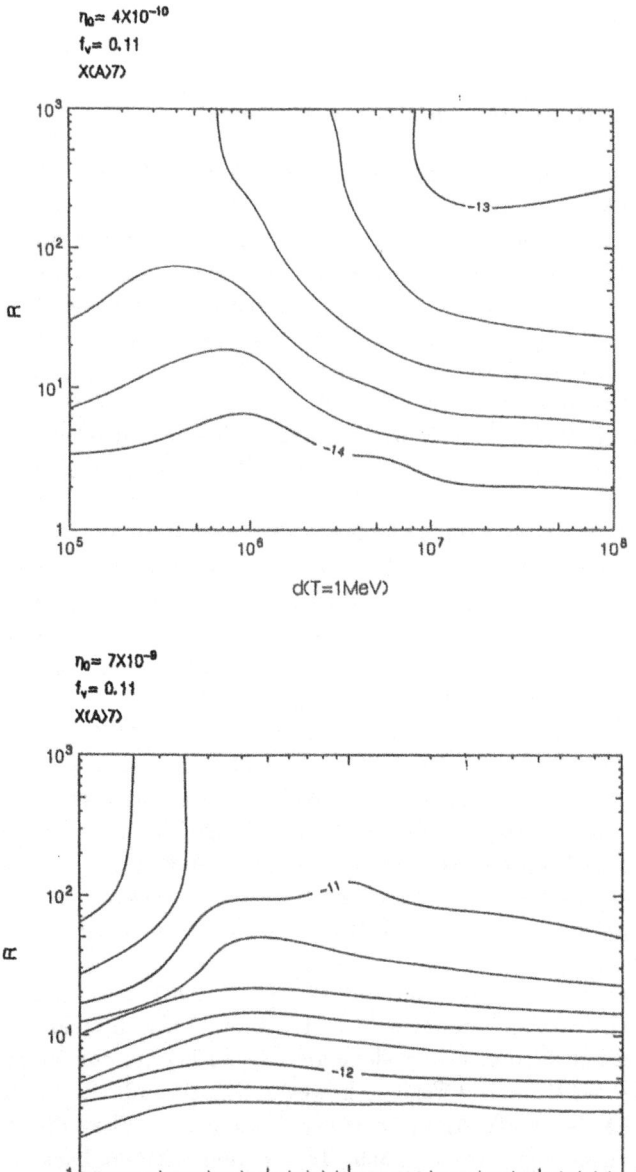

Figure 6: The contour of the average abundances of heavy elements $(A \geq 7)$ on the $d - R$ plane for $\eta_0 = 4 \times 10^{-10}$(top) and $\eta_0 = 7 \times 10^{-9}$(bottom). Contour levels are $\log(X(A(\geq 7)/H)$.

As shown in preceding sections, we carried out systematic multi-zone calculations with large nuclear reaction network. The result is summarized as follows.

If we take Pop-II abundance as the observational limit of light elements, ^{4}He $<$ 0.24and ^{7}Li/H $< 2.5 \times 10^{-10}$, i)the upper limit of $\Omega_b h_{50}^2$ is 0.06 even if we make fine tuning of parameters of inhomogeneity, R, f_v and d. This result is consistent with Kurki-Suonio et al. (1990) and Mathews et al.(1990) though the value depends on authors a little. This limit is almost the same value in the homogeneous universe (see, for example, Steigman (1991). ii) If we take $\Omega_b h_{50}^2 \leq 0.08$, ^{9}Be/H is less than 10^{-17}. This value is 10^{-3} of the upper limit of observation (Ryan et al., 1990). iii) Heavy elements are not synthesized in the low-density and neutron-rich regions because the neutron number density is reduced by back diffusion. No r-process elements are also synthesized. iv) From the upper limit of ^{7}Li abundance, ^{7}Li/H $< 2.5 \times 10^{-10}$, the density contrast must be smaller than 10, $R < 10$.

If we take Pop-I abundance as the observational limit of ^{7}Liabundance, ^{7}Li/H $<$ 1. $\times 10^{-9}$, the upper limit of $\Omega_b h_{50}^2$ can become as large as ~ 0.3. (Kurki-Suonio et al., 1990). Though heavy element abundance $X(C, N, O)$ can become as much as 10^{-12}, r-process elements are not synthesized.

Recently Alcock et al.(1990) proposed an mechanism which might reduce ^{7}Li overproduction. When the mean free path of photons equals to the characteristic size of high density regions, the high density regions expand and the density contrast disappear. This occurs when the cosmic temperature becomes 20 kev for typical case. They suggested significant reduction of ^{7}Li is expected by the chain reaction ^{7}Be$(n, p)^7$Li$(p, \alpha)^4$He if mixing between high and low density regions occurs. Though this is interesting suggestion, we don't know any mechanism for mixing at present. It seems that there is no possibility of turbulence, since Reynols number is much smaller than the critical value (Alcock et al., 1990).

ACKNOWLEDGMENT

Present review is based on the collaboration with N. Terasawa. Author thanks him for stimulating discussion. This work was supported in part by the Grant-in-Aid for Scientific Research Fund from the Ministry of Education, Science and Culture No. 01540308.

Reference

Alcock, C. R., Dearborn, D. S., Fuller, G. M., Mathews, G. J. and Meyer, B. S., 1990, *Phys. Rev. Lett.*, **64**, 2607.

Applegate, J. H., Hogan, C. J., 1985, *Phys. Rev.*, **D31**, 3037.

Applegate, J. H., Hogan, C. J., and Scherrer, R. J., 1988, *Ap. J.*, **329**, 572.

Applegate, J. H., 1988, *Phys. Rep.*, **163**, 141.

Boyd, R. N. and Kajino, T. 1989, *Ap. J.(Letters)*, **336**, L55.

Fuller, G. M., Mathews, G. J. and Alcock, C. R.,1988, *Phys. Rev.*, **D37**, 1380.

Kawano, L. H. , Fowler, W. A. and Malaney, R. A. and, 1990, *Ap. J.*, to be published.

Kawano. L. H., Fowler, W. A. , Kavanagh, R. W. and Malaney, R. A. 1990, *Ap. J.*, to be published.

Kurki-Suonio, H., Matzner, R. A., Centrella, J. M., Rothman, T. and Wilson, J. R., 1988, *Phys. Rev. D*, **38**, 1091.

Kurki-Suonio, H. and Matzner, R. A., 1989, *Phys. Rev. D*, **39**, 1046.

Kurki-Suonio, H., Matzner, Olive, K. A. and Schramm, D. N., 1990, *Ap. J.*, **353**, 406.

Kurki-Suonio, H., Matzner, R. A., 1990, *Phys. Rev. D*, **42**, 1047.

Malaney, R. A. and Fowler, W. A. , 1988, *Ap. J.*, **333**, 14.

Malaney, R. A. and Fowler, W. A. , 1990, *Ap. J.*, to be published.

Mampe, W., Ageron, P., Bates, C., Pendlebury, J. M. and Steyerl, A., 1989, *Phys. Rev. Lett.*, **63**, 593.

Mathews, G. J., Meyer, B. S.,Alcock, C. R. and Fuller, G. M.,1990, *Ap. J.*, **358**, 36.

Pagel, B. J. E., 1991, in this volume.

Ryan, S. G., Bessell, M. S., Sutherland, R. S., Norris, J. E., 1990, *Ap. J. (Letters)*, **348**, L57.

Sale, K. E. and Mathews,1986, *Ap. J.*, **309**, L1.

Steigman, G., 1991, in this volume.

Terasawa, N. and Sato, K., 1989a, *Prog. Theor. Phys.* **81**,254.

Terasawa, N. and Sato, K., 1989b, *Phys. Rev. D*, **39**, 2893.

Terasawa, N. and Sato, K., 1989c, *Prog. Theor. Phys. (Letters)* **81**,1085.

Terasawa, N. and Sato, K., 1990, *Ap. J. Letters*, **362**, L47.

Thielemann, F. K. and Wiescher, M. Proceedings of Primordial Nucleosynthesis Symposium, Chapel Hill, Noth Carolina, 1989 to be published.

Thielemann, F. K., Appligate, J. H. and Wiescher, M., 1990, private communication.

Witten, E., 1984, *Phys. Rev. D*, **30**, 272.

STERILE NEUTRINOS IN THE EARLY UNIVERSE

ROBERT A. MALANEY
Institute of Geophysics and Planetary Physics,
University of California,
Lawrence Livermore National Laboratory.

and

GEORGE M. FULLER
Physics Department
University of California, San Diego.

ABSTRACT. We discuss the role played by right-handed sterile neutrinos in the early universe. We show how the well known ^{4}He constraint on the number of relativistic degrees of freedom at early times limits the equilibration of the right-handed neutrino sea with the background plasma. We discuss how this allows interesting constraints to be placed on neutrino properties. In particular, a new limit on the Dirac mass of the neutrino is presented.

1. Introduction

If the presence of right-handed (RH) "sterile" neutrinos, ν_i^R, $(SU(2)_L \otimes U(1)_Y$ singlets) are allowed, then interesting scenarios can arise when mechanisms for their conversion into left-handed (LH) "active" neutrinos, ν_i^L, are introduced. Hereafter when we refer to RH neutrinos we mean both RH neutrinos and LH anti-neutrinos. By "sterile" here we mean that they do not transform under the standard model gauge group.

In the Standard Big Bang (SBB) model RH neutrinos play no role. Even if they exist they are assumed to have decoupled at $T > 100$ GeV. The neglect of RH neutrinos in nucleosynthesis calculations is to a large extent justified from consideration of the statistical weight, g, of the relativistic particles present before and after the QCD phase transition. Since the evolution of the universe through phase-transition and particle-annihilation epochs is characterized by constant *co-moving* entropy density, the amount of heating of the LH sea relative to the RH sea is given by $T_{LH}/T_{RH} = (g_1/g_2)^{1/3}$, where the subscripts 1 and 2 refer to the times before and after the annihilation epoch, respectively, and g_1 and g_2 are the effective statistical weights of relativistic particles at these times. Prior to confinement the statistical weight in relativistic particles is $g_1 \approx 56.5$; whereas, after the quarks are confined the only nearly relativistic strongly interacting particles will be pions, so that $g_2 \approx 17.25$ and $T_{LH}/T_{RH} \approx 1.5$. This implies that at the time of nucleosynthesis the relative energy densities in a LH and RH neutrino species will be $\rho_{LH}/\rho_{RH} \approx 5$, so that a sterile neutrino species counts as less than 1/5 of an additional neutrino flavor. Muons and pions should annihilate or drop out of equilibrium at roughly $T \approx 100$ MeV, and if we add their statistical weight to the differential heating of the LH and RH seas then a RH neutrino species would count for only about 0.1 of an additional flavor.

SBB nucleosynthesis predicts that the number of relativistic neutrino species be constrained by [1]

$$N_\nu \leq 3.4 + 20\left(\frac{Y_p - 0.240}{0.240}\right), \tag{1}$$

where Y_p is the primordial ^{4}He mass fraction. If we take $Y_p = 0.23 \pm 0.01$ from observation [1-2] then $N_\nu \leq 3.4$. We stress here that our result will be sensitive to the adopted upper limit of $Y_p < 24\%$. A value of Y_p larger by even 0.2% would weaken our constraint. If we assume three light Dirac neutrinos and that the RH components of one of them did not decouple until after the QCD epoch, then those RH components would count for an extra 0.7 neutrino flavor. If the decoupling of the RH sea is after muon and pion annihilation then the RH components count for almost a full extra neutrino flavor. Even an extra 0.7 neutrino flavor is incompatible with the limit $N_\nu \leq 3.4$, thus the RH components of the neutrino (ν_μ or ν_τ) would have to decouple prior to the QCD epoch, which we define to occur at a temperature $T = T_{QCD}$.

It is important to note here that because this argument is based on added degrees of freedom contributed by the RH neutrinos we must assume that neutrinos are purely Dirac particles. Any Majorana contribution to the neutrino mass term will invalidate the argument.

A particle species, x, is decoupled from the background plasma when

$$\sum_i n_i \sigma << H \ , \tag{2}$$

where σ is the cross section of the interaction processes involving x, and H is the Hubble parameter which can be written as

$$H = \left(\frac{4\pi^3}{45}\right)^{1/2} m_{pl}^{-1} g^{1/2} T^2, \tag{3}$$

where m_{pl} is the Planck mass.

Several studies have used the necessity that any RH neutrino species be decoupled prior to the QCD phase transition to impose constraints on the mass, electromagnetic, oscillation and weak interaction properties of neutrinos.

2. Neutrino Mass

Consider a neutrino of mass m_ν and energy E_ν. One can show that the cross section for helicity-flip in any scattering event is given by [3-4]

$$\sigma_R \sim \sigma_L (\frac{m_\nu}{E_\nu})^2 \quad , \tag{4}$$

where $\sigma_L \sim G_F^2 T^2$ is the cross section for the normal non-flip scattering event (G_F is the Fermi coupling constant). Demanding that eq. (3) be valid at $T = T_{QCD}$ and adopting statistical weights relevant to the epoch prior to quark confinement, leads to the following constraint on the neutrino mass [5]

$$m_\nu << 150 \text{ keV} (\frac{100 \text{ MeV}}{T_{QCD}})^{1/2} \quad . \tag{5}$$

Assuming $T_{QCD} \sim 100$ MeV, we find a cosmological limit of $m_\nu << 150$ keV. The present experimental upper limits for the muon and tau neutrino are 250 keV and 35 MeV, respectively [6].

Finally, we note that if the tau neutrino is too massive ($m_\nu \gtrsim 1$ MeV) then it can become non-relativistic during nucleosynthesis [7]. In such circumstances the universe is matter dominated as it passes through nucleosynthesis and the nucleosynthesis yields can be perturbed from their SBB values. This effect should modify somewhat the limit of eq. (5).

3. Other Constraints

The RH neutrino sea can be populated through other effects such as electromagnetic interactions, oscillations, or new forms of weak interactions. For example, similar arguments to those given above have been previously adopted in order to constrain the neutrino magnetic moment, μ_ν, viz., $\mu_\nu < 2 \times 10^{-11} \mu_B$ [8], where μ_B is the Bohr magnetron. Similarly, the neutrino charge radius, r, is constrained by $< r^2 > \lesssim 10^{-32}$cm^2 [9]. The ^{4}He observation coupled with simple extensions of the standard electroweak model, leads to the following constraint on G_R, the RH-coupling constant,

$$G_R \lesssim G_F T_{QCD}^{-3/2} \quad . \tag{6}$$

For $T_{QCD} \sim 100$ MeV results in $G_R \lesssim 10^{-3} G_F$ [10-11]. In addition, if $G_R \propto M_{W_R}^{-2}$, where M_{W_R} is the mass of W_R (the RH counterpart of the LH intermediate vector boson W_L), then this leads to the constraint

$$M_{W_R} \gtrsim M_{W_L} \left(\frac{T_{QCD}}{1 \text{ MeV}}\right)^{3/4} , \tag{7}$$

which for $T_{QCD} \sim 100$ MeV results in $M_{W_R} \gtrsim 2.5$ TeV [11]. The importance of RH neutrinos in the context of neutrino oscillations has also been previously discussed [12].

4. Conclusions

We have discussed the constraints imposed on Dirac neutrinos derived from considerations of their spin flipping into sterile RH neutrino states. Such flipping is severely constrained by the inferred primordial ^{4}He abundance. We showed in particular, how this astronomical observation imposes a mass limit of order 150 keV on the Dirac neutrino.

This work performed under the auspices of the U. S. Department of Energy by the Lawrence Livermore National Laboratory under contract W-7405-ENG-48. This work was partially supported by NSF grant PHY-8914379 and IGPP grant LLNL 90-08 at UCSD.

5. References

[1] K. A. Olive, D. N. Schramm, G. Steigman and T. P. Walker, Phys. Lett. B236 (1990) 454.

[2] B. E. J. Pagel in: "A Unified View of the Macro-and Micro-Cosmos" eds. A. DeRujula, D. V. Nanopoulos and P. A. Shaver (World Scientific, Singapore, 1988) p. 399.

[3] S. L. Shapiro, S. A. Teukolsky, and I. Wasserman, Phys. Rev. Lett. 45 (1980) 669.

[4] A. Perez and R. Gandhi, preprint Fermilab-Pub-89/156-A, (1989).

[5] G. M. Fuller and R. A. Malaney, Phys. Rev. D. submitted (1990).

[6] M. Aguilar-Benitez, et al. (Particle Data Group), Phys. Lett. B204 (1988) 1.

[7] E. W. Kolb and R. J. Scherrer, Phys. Rev. D25 (1982) 1481.

[8] J. A. Morgan, Phys. Lett. B102 (1981) 247.

[9] J. A. Grifols and E. Massó, Mod. Phys. Lett. A2, (1987) 205.

[10] J. R. Bond, G. Efstathiou and J. Silk, Phys. Rev. Lett. 45 (1980) 1980.

[11] K. A. Olive, D. N. Schramm and G. Steigman, Nuc. Phys. B180 (1981) 497.

[12] M. Yu. Khlopov and S. T. Petcov, Phys. Lett. B99 (1981) 117.

COULD COSMIC QCD PHASE TRANSITION PRODUCE STRANGE QUARK MATTER WHICH SURVIVES UNTIL THE PRESENT TIME?

T. KAJINO and K. SUMIYOSHI
Department of Physics
Tokyo Metropolitan University
Setagaya, Tokyo 158
Japan

ABSTRACT.   Microscopic calculations using a chromoelectric flux tube model provide a low baryon penetrability through the phase boundary between the quark-gluon plasma and the hadron gas. The result is applied to the study of the evolution of baryon number density during the cosmic phase transition in quantum chromodynamics (QCD). Remarkable inhomogeneities of baryon number density result from a plausible range of QCD parameters constrained from lattice simulations. If the QCD phase transition is weakly first order, the strange quark matter could have been formed and survived evaporation and resolution in the hot early universe.

1.   INTRODUCTION

A first order QCD phase transition provides interesting physics as inhomogeneous cosmologies [1-3] or the creation of primordial strange quark matter [4]. All these studies are strongly motivated by an inflationary cosmology [5,6] which requires marginally closed universe in order to resolve many fundamental problems of hot big-bang expansion. However, since neither homogeneous nor inhomogeneous nucleosynthesis models seem to close the universe by baryonic mass alone, dark metter search is important. Primordial strange quark matter is one of the strongest candidates for baryonic dark matter. If the strange quark matter is observed today, this will show a clear evidence of cosmological phenomena of first order QCD phase transition or neutron star collision. In this paper we confine ourselves to the discussion of primordial strange quark matter.

It is known that the specific dynamics of a first order cosmic phase transition may allow a strong quark concentration [7,8]. If the concentration is strong, strange quark matter is to be formed. If not very strong, we can still find moderately inhomogeneous universe of fluctuating baryon distribution which may lead to inhomogeneous nucleosynthesis. The QCD physics thus controls the late time evolution of baryons. However, all previous studies of the primordial nucleosynthesis and strange quark matter are subject to the troubles of unknown fluctuation shapes of baryon inhomogeneities. The first purpose of this paper is to discuss quantitatively how large baryon inhomogeneities are attained at the end of the cosmic phase transition from the physics of QCD. Having calculated a possibility of finding strange quark matter at this epoch, we secondly explore the survival condition of the strange quark matter until the present universe.

2. BARYON INHOMOGENEITIES IN THE EARLY UNIVERSE

The purpose of this section is to show the possible fluctuation shapes of baryon number density distribution as a function of microscopic parameters characterizing QCD [8]. These parameters are the coexistence temperature of the qurak-gluon plasma and the hadron gas, T_C, the surface tension of the phase boundary, σ, and the flux of baryons penetrating through the boundary, J_B.

2.1. Baryon penetrability

Excess quark concentration during the cosmic QCD phase transition depends strongly on the baryon penetrability through the phase boundary. However, this significant quantity was taken to be a free parameter in the previous studies. We here calculate J_B in the chromoelectric flux tube model [9].

This model is an effective model for the confinement of quarks in QCD, which has enjoyed a success for jet phenomenology of high energy e^+e^- and proton-antiproton collision experiments. The hadronization of the quark-gluon plasma is described in this model by taking account of the Schwinger mechanism [10] similar to that in QED and the dynamics around the surface. As a quark passes through the phase boundary, a tube of color electric field is built up behind it in the shortest distance. A meson is produced by fission of the tube induced by quark-antiquark pair creation [11]. Extending the same mechanism to the baryon production via diquark-antidiquark pair creation, the flux of baryons evaporating from the quark-gluon plasma is calculated [12] by

$$J_B = \sum_q \frac{n_q}{Z_q} \int d^3k_0 \exp\left(-\frac{E_0}{T}\right)\frac{k_{z0}}{E_0}$$

$$\int dk_z^B \int dE^B \frac{1}{k_C^2}\exp\left[-\frac{E_0}{k_C^2}\left\{(k_{z0}-k_z^B)+\frac{1}{2E_{z0}}\left((k_z^B E_z - k_{z0}E_{z0})+m_q^2\ln\frac{k_z^B+E_z}{k_{z0}+E_{z0}}\right)\right\}\right], \quad (1)$$

where Z_q is the partition function and E_0 is the averaged thermal energy of an initial quark having a mass m_q and number density n_q at a temperature T in the plasma. The integrand of the right hand side of eq. (1) is the evaporation probability of baryon with an energy E^B and momentum k_z^B for an given initial quark momentum k_0 and k_{z0} when it passes through the boundary. Integral is over the threshold energy and momentum satisfying the kinematical conditions. k_C is a typical linear momentum

$$k_C = \sqrt{\frac{3\sigma_0^3}{2\pi\alpha_C p}}, \quad (2)$$

which is lost before the baryon evaporation and include unknown quantity p of diquark-antidiquark pair creation and fission in the tube. We adopted the p-values determined from the jet phenomenology [13]. Here $\sigma_0 = 0.177$ GeV^2 is an energy density per unit length of the flux tube which was deduced from Regge trajectories of hadrons, and $\alpha_C = 2$ is QCD coupling constant. The temperature in question is of an order of 100 MeV. This determines only the quark and antiquark distribution function, but the typical energy which contributes to the baryon evaporation is ≈ 1 GeV or higher due to a kinematical thresold of baryon masses. High energy jet phenomenology is applied successfully to the cosmic QCD phase transition.

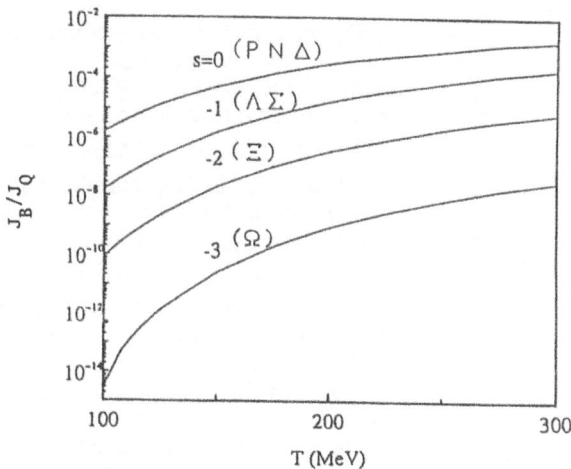

Figure 1 displays the calculated flux of baryons evaporating from the quark-gluon plasma normalized to the quark flux [12]. Non-strange baryons have the largest flux among the baryon octet and decuplet. Their values are two or three orders of magnitude smaller than those assumed in the equlilibrium conditions of the previous calculations.

Difference of this quantity among different baryons do not affect very much the dynamics of the evolution of baryon number density in the weak equilibrium of the early universe.

Fig. 1   Calculated baryon number flux penetrating through the phase boundary divided by the quark flux.

2.2. Evolution of baryon number density during the QCD phase transition

Having known the baryon penetrability, we can calculate the evolution of baryon number densities of the two phases at the coexistence epoch $T = T_C$;

$$\frac{dn_B^q}{dt} = -\lambda n_B^q + \lambda' n_B^h - \left\{ \frac{\dot{V}}{V} + \frac{\dot{f_V}}{f_V} \right\} n_B^q, \tag{3a}$$

$$\frac{dn_B^h}{dt} = \frac{f_V}{1 - f_V} \left\{ -\lambda' n_B^h + \lambda n_B^q + \frac{\dot{f_V}}{f_V} n_B^h \right\} - \frac{\dot{V}}{V} n_B^h, \tag{3b}$$

where n_B^q and n_B^h are the net baryon number densities in the quark-gluon plasma phase and hadron phase. V is the horizon volume and f_V is the volume fraction of quark-gluon plasma, which are known functions of time as solution of Einstein equation [14]. In these equations, λ is the baryon number transfer rate from the quark-gluon plasma phase to hadron phase defined by

$$\lambda = \frac{4\pi r^2 N_n V_i}{f_V V} J_B/n_B{}^q, \qquad\qquad (4)$$

and λ' is determined from the detailed balance condition [8]. In this equation $r(t)$ is the average bubble radius and N_n is the density of nucleation sites. Given T_C and σ, these quantities are calculated in isothermal fluctuation theory, assuming monochromatic spectrum of spherical bubbles [7].

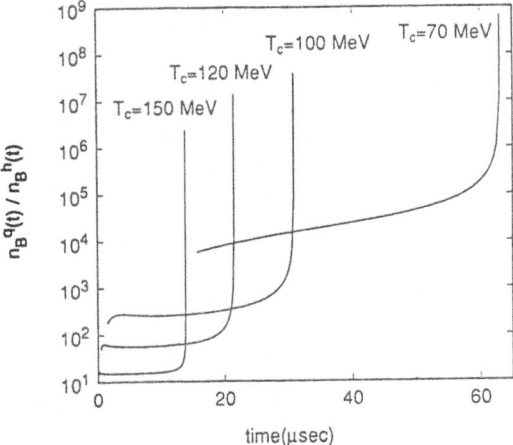

Fig. 2 Baryon number density ratio $n_B{}^q/n_B{}^h$ for various QCD parameters. Time = 0 refers to the epoch when T_C is first reached.

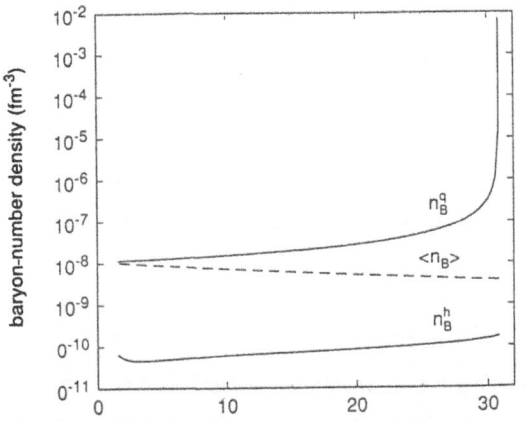

Fig. 3 Evolution of baryon number densities in the two phases.

Figure 2 displays the evolution of the baryon number density ratio during the cosmic QCD phase transition at various temperatures T_C = 70, 100, 120 and 150 MeV with σ = 10^6 MeV^3 [8]. The ratio increases strongly from the initial equilibrium value because of the increasing density $n_B{}^q$. This behavior is observed for all T_C and σ, but the final density ratio is strongly dependent upon T_C.

Figure 3 displays an example of the time evolution of the baryon number density in the quark-gluon plasma phase $n_B{}^q$ and the hadron phase $n_B{}^h$ [8]. In this calculation we set T_C =100 MeV and σ =10^6 MeV^3. The value of $n_B{}^q$ increases up to the order of 10^{-2} fm^{-3} near the end of the phase transition. This quantity even reaches nuclear matter density if $\sigma \leq 10^6$ MeV^3.

In the present analysis we assume that when the velocity of the phase boundary necessary to maintain the universal temperature at T_C exceeds the velocity of sound then the release of latent heat will no longer be sufficient to prevent the universe from cooling. New bubbles of the hadron phase will then nucleate and the baryon number density in the quark-gluon plasma will appear in the hadron phase as a density enhancement. We therefore conclude that the strange quark matter is able to have been formed in the cosmic QCD phase transition if 80 MeV $\leq T_C \leq$ 150 MeV and $\sigma \leq 10^6$ MeV^3 [15].

This is a hopeful result because the degree of the first order QCD phase transition, if it is first order, is suggested to be very weak in recent lattice simulations including the effect of dynamical quarks [16,17]. The weaker first order corresponds to the smaller surface tension $\sigma \leq 10^7$ MeV^3, where the upper bound is set from pure gauge simulations. If σ has a value of order 10^6 - 10^7 and $T_C \leq$ 150 MeV, then the strange matter is not probably formed but the nucleosynthesis in the baryon inhomogeneous universe may be very different from that of the standard model. It is highly desirable to determine these microscopic parameters from the theoretical studies of QCD and the ultra-relativistic heavy-ion collision experiments [18].

3.   PRIMORDIAL STRANGE QUARK MATTER

It has been a general consensus that the primordial strange quark matter (strangelet) cannot survive evaporation into nucleons in the hot early universe 10 MeV $\leq T_U \leq$ 100 MeV [19-21]. This tentative conclusion arises from the calculations using a large baryon penatrability through the surface. However, we found in the previous sections that the baryon evaporation is not very efficient. Let us reexamine the survival condition of the strangelet.

Rate equation of the evaporation of strangelet A is written as

$$\frac{dA}{dt} = \left\{ \frac{m_B T_S^2}{2\pi^2} \exp(-\frac{I_0}{T_S}) - \frac{11\pi^2}{360 T_S}(T_U^4 - T_S^4)\sqrt{\frac{T_S}{2\pi m_B}} \right\} x_0 A^{2/3}, \qquad (5)$$

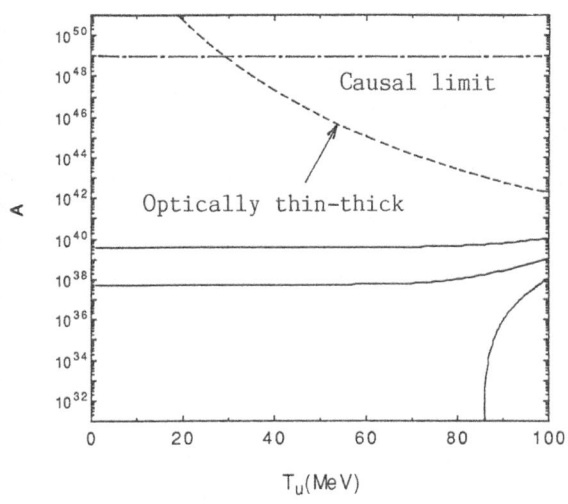

Fig. 4   Evolution of A with T_U for the strangelets.

where T_U is the universal (neutrino) temperature and T_S is that at the surface of strangelet which is lower than T_U when the medium is optically thin to neutrinos. A is the baryon number contained in the strangelet, and m_B is the mass of nucleon. The parameters I_0 and x_0 are determined from the calculations of baryon penetrability using Fermi distribution of quarks inside the strangelet at temperature T_S [22]. In order to estimate the lowest number A for which the strangelet can survive until the present time, we neglected the second term of eq. (5), i.e. $T_U = T_S$. This gives the strongest constraint on the survival condition.

We show the evolution of evaporating strangelets in the T_U-A plane in Fig. 5. Dashed curve shows a transition from diffusive neutrino heating to optically thin neutrino heating, $1 = G_F^2 T_U^5 r_S$. Since most of the evaporation occurs in the optically thin region, the computations were started from $T_U = 100$ MeV. The strangelets having A greater than 10^{39} can survive evaporation in the hot early universe [22]. This critical value $A = 10^{39}$ is much smaller than the causal limit 10^{49} and in the predicted range from phenomenological QCD calculations [15].

We also calculated the mean separation distance between the primordial strangelets. The present scale is of the order 10^{-4} pc. The number density of these strange quark matter may be so high that the observation is not impossible today, though the mass is lighter than that of Jupiter, by using some revolutionary method of using gravitational microlensing effect.

4. CONCLUSION

Dynamical process of generating baryon inhomogeneities during the cosmic QCD phase transition is studied in an effective phenomenological model of QCD. It is found that the primordial strange quark matter could be formed at this epoch if the phase transition is weakly first order and satisfy 80 MeV $\leq T_C \leq$ 150 MeV and $\sigma \leq 10^6$ MeV^3. The strange quark matter formed can survive evaporation into nucleons in the hot early universe if it includes baryon number greater than 10^{39}.

REFERENCES

[1]   J. H. Applegate and C. J. Hogan, Phys. Rev. **D31** (1985) 3037.
[2]   C. R. Alcock, G. M. Fuller and G. J. Mathews, Ap. J. **320** (1987) 439.
[3]   R. A. Malaney, and W. A. Fowler, Ap. J. **333** (1988) 14.
[4]   E. Witten, Phys. Rev. **D30** (1984) 272.
[5]   A.H. Guth, Phys. Rev. **D23** (1981) 347.
[6]   K. Sato, Mon. Not. Roy. Astron. Soc. **195** (1981) 467.
[7]   G. M. Fuller, G. J. Mathews and C. R. Alcock, Phys. Rev. **D37**, 1380 (1988).
[8]   K. Sumiyoshi, T. Kajino, C. Alcock and G. Mathews, Phys. Rev. **D42** (1990) 3963.
[9]   A. Casher, H. Neuberger and S. Nussinov, Phys. Rev. **D20** (1979) 179.
[10] J. Schwinger, Phys. Rev. **82** (1951) 664.
[11] B. Banerjee, N. K. Glendenning and T. Matsui, Phys. Lett. **127B** (1983) 453.
[12] K. Sumiyoshi, K. Kusaka, T. Kamio and T. Kajino, Phys. Lett. **225B** (1989) 10.
[13] B. Andersson, G. Gustafson and T. Sjostrand, Nucl. Phys. **B197** (1982) 45.
[14] K. Kajantie and H. Kurki-Suonio, Phys. Rev. **D34** (1986) 1719.
[15] T. Kajino, K. Kusaka and K. Sumiyoshi, preprint of Tokyo Metro. Univ. (1990).
[16] M. Fukugita, H. Mino, M. Okawa and A. Ukawa, Phys. Rev. Lett. **65** (1990) 816.
[17] F.R. Brown, F.P. Butler, H.C. Chen, N.H. Christ, Z. Dong, W. Schaffer, L.I. Unger and A. Vaccarino, Phys. Rev. Lett. **65** (1990) 2491.
[18] T. Kajino, Phys. Rev. Lett. **66** (1991) 125.
[19] C. R. Alcock and E. Farhi, Phys. Rev. **D32** (1985) 1273.
[20] J. Madsen, H. Heiselberg and K. Riisager, Phys. Rev. **D34** (1986) 2947.
[21] C. Alcock and A. Olinto, Phys. Rev. **D39** (1989) 1233.
[22] T. Kajino, K. Sumiyoshi, C. Alcock and G. Mathews, in preparation.

MULTI-ZONE CALCULATION OF NUCLEOSYNTHESIS IN INHOMOGENEOUS UNIVERSE AND ^{9}Be ABUNDANCE

N. Terasawa
The Institute of Physical and Chemical Research
Hirosawa 2-1
Wako-shi 351-01
Japan

ABSTRACT. While ^{9}Be is expected to provide a test of inhomogeneous models of primordial nucleosynthesis, we show that the previous estimate by the simplified two-zone model of neutron diffusion overestimated the ^{9}Be abundance at least two orders of magnitude. The abundances of CNO elements expected to be the seeds of cosmic r-process are very small in the proper range of parameters consistent with the abundances of light elements.

1. Introduction

It has been suggested, recently, that ^{9}Be may provide a test of inhomogeneous model of nucleosynthesis (Boyd and Kajino 1989), and efforts to find metal-poor population-II stars with lower ^{9}Be abundance were made (Ryan et al. 1990). Their estimate of the ^{9}Be abundance, however, is based on the simple two-zone model in which neutron diffusion during nucleosynthesis is completely neglected. Importance of neutron diffusion during nucleosynthesis was emphasized by several authors including us (Kurki-Suonio et al. 1988; Malaney and Fowler 1988; Terasawa and Sato 1989a,b). For density fluctuations with small scale, neutrons once diffused out of the high density zones diffuse back as soon as nucleosynthesis in the high density zone begins since it takes place earlier than that in the low density zone. For larger scale of fluctuations the time scale of neutron diffusion is much longer than the cosmic expansion scale in the time of nucleosynthesis and the amount of neutrons diffuse out into low density zones is small. Hence the effect of the neutron diffusion is not incorporated correctly in the simplified two-zone models cutting off the neutron diffusion during the nucleosynthesis and the abundances of neutron-rich nuclei such as ^{9}Be may not be so large as expected. Malaney and Fowler(1990) suggested that ^{9}Be can also provide the test for inhomogeneous cosmological models with high baryon density as $\Omega_B = 1$ according to their results incorporating the neutron diffusion during nucleosynthesis. However the dependence of the ^{9}Be abundance on the scale of density fluctuations are not clear in their works. The effects of neutron diffusion strongly depend on the scale of density fluctuations and this dependence is the very thing which we are interested in because the scale of density fluctuations is decided by the physical processes, which generate density fluctuation, such as the QCD phase transition and it is expected to give some informations about the process. Two-zone model is, furthermore, too simplified to evaluate the effects of neutron diffusion on the abundances of elements quantitatively and we need to perform more realistic multi-zone calculations.

Another interest in the inhomogeneous models of nucleosynthesis is the cosmological r-process suggested by Applegate, Hogan, and Scherrer(1988). Although the high baryon density universe with $\Omega_B = 1$ was rejected even in inhomogeneous models(Kurki-Suonio and Matzner 1989; Kurki-Suonio et al. 1989, Terasawa and Sato a,b), speculations that the none-zero abundances of r-process elements in metal-poor stars are explained by the cosmic r-process are often discussed neglecting the crucial problem of the abundance of ^{7}Li. Many reaction rates, however, concerning to the flows to heavy elements are poorly determined and we cannot completely exclude the possibility that significant amount of r-process elements is produced for lower baryon density consistent with the abundances of light elements.

In this talk, we show the abundances of ^{9}Be and CNO elements in the inhomogeneous universe calculated with extended network including elements up to ^{22}Ne in the multi-zone scheme. Some neutron-rich nuclei which are important to estimate the abundances of these elements are added to the network. As it will be shown bellow, the ^{9}Be abundance is two-orders of magnitude smaller than that calculated in the simplified two-zone model. Furthermore, heavy elements are produced mainly in the high-density zones and the abundances of CNO elements, which are seeds of a cosmic r-process, in the low-density and neutron-rich zones are fairly small.

2. The Computational Model

Spherical symmetry is assumed for the space which is divide into 25 shells with same thickness. The diffusion equation for neutrons,

$$\frac{dn}{dt} = \nabla(D_n \nabla n) \tag{1}$$

coupled with nucleosynthesis is solved in a completely implicit manner. The simple version of a flux-limiter is applied for the diffusion coefficient to avoid the overestimate of the neutron diffusion (Terasawa and Sato1989b). Though the diffusion coefficient has uncertainties arising from uncertainties in nuclear data, they reduce to the changes in the dependence of the abundances of elements on the length scale in problem. At the boundary of the region, the neutron flux is taken to be zero. It is consistent with the assumption that the universe consists of repeated sites with the same scale and the same density profile.

For the initial ($T = 10^{11}$K) density profile, we assume a top-hat profile; the density of inner shells is R times higher than that of surrounding low-density shells. The parameter f_v refers to the volume fraction of the initial high-density shells in the whole region. Another parameter $d(T = 1\text{MeV})$ refers to the radius of the site in problem normalized by the size at T=1MeV. We calculated in the ranges 10^5cm $\leq d(T = 1\text{MeV}) \leq 10^8$cm, $1 \leq R \leq 1000$, and $0 \leq f_v \leq 0.5$, for $\eta_0 = 4 \times 10^{-10}(\Omega = 0.014h^{-2} = 0.056h_{50}^{-2})$ and $\eta_0 = 7 \times 10^{-9}(\Omega = 0.25h^{-2} = 1.0h_{50}^{-2})$.

Some neutron-rich elements such as ^{9}Li are added to the network which includes the elements up to ^{22}Ne. Reaction rates are updated according to Caughlan and Fowler 1988 and those relating to ^{9}Be are added and updated by the table given in Malaney and Fowler 1990. The lifetime of neutrons is taken to be 887.6 sec (Mampe et al. 1989) and the number of neutrino species is 3.

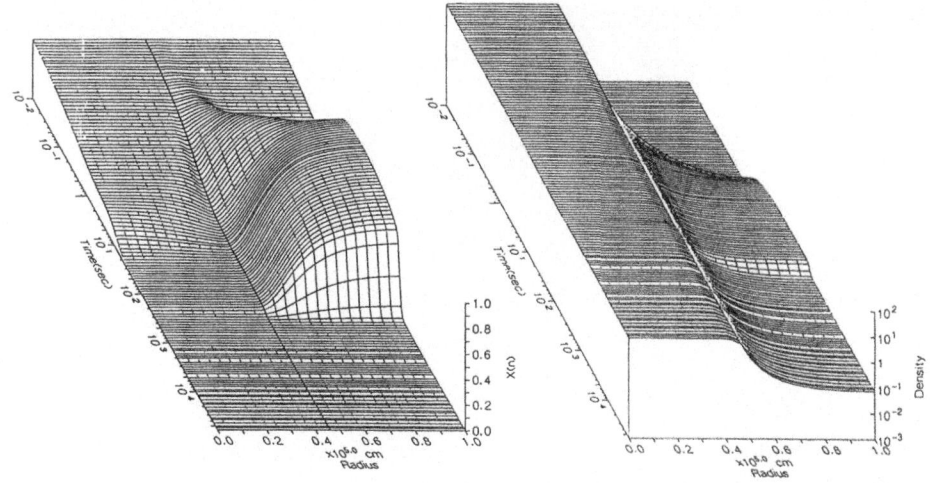

Figure 1. Time evolution of the local mass fraction of neutron (left) and the time evolution of the local baryon density(right) in the case that $d(T = 1\text{MeV}) = 10^5\text{cm}, R = 1000, f_v = 0.11$, and $\eta_0 = 4 \times 10^{-10}$. The local baryon density is denoted by the ratio to the average density.

3. The Numerical Results

Figure 1 displays the time evolution of the local mass fractions of neutron in each shell and the evolution of the ratio of local baryon density to the mean density for $d(T = 1\text{MeV}) = 10^5\text{cm}, R = 1000, f_v = 0.11$, and $\eta_0 = 4 \times 10^{-10}$. Neutrons begin to diffuse out of the inner high-density shells into the surrounding space and spread over the low-density zone around $t = 0.1\text{sec}$. Since the density contrast among the high density shells and the low density shells is very large in this case, the neutron fraction becomes completely dominant in the low density shells and the baryon density in the outer low-density shells grows simultaneously. Thus the ratio of the baryon density in the innermost shell to that in the outermost shell, which is 1,000 at first, finally reduces to 107. However, neutrons begin to diffuse back into the inner high-density shells to keep the neutron density distribution uniform when nucleosynthesis starts and the neutron density decreases in the high-density shells. It results in the decrease of baryon density of the outer low-density shells. Due to this effect, the baryon density contrast, which once reduces to 38 at $t \sim 100\text{sec}$, increases again to ~ 100. This example clearly shows that we cannot neglect the neutron diffusion, especially the back diffusion out of the low-density region into the high-density region, during nucleosynthesis.

^{9}Be is produced in the outer shells as shown in Figure 2 ($d(T = 1\text{MeV}) = 10^7\text{cm}, R = 1000, f_v = 0.11$, and $\eta_0 = 4 \times 10^{-10}$). In the high-density shells, ^{9}Be once produced is burnt into heavier elements and its abundance turns out to be very small.

The dependence of the ^{9}Be abundance on the scale of fluctuations is shown in Figure 3 as contour on the $d-R$ plane for $f_v = 0.11$. The ^{9}Be abundance increases with increasing R and take the maximum value around $d(T = 1\text{MeV}) = 10^7\text{cm}$. Though the f_v dependence of the

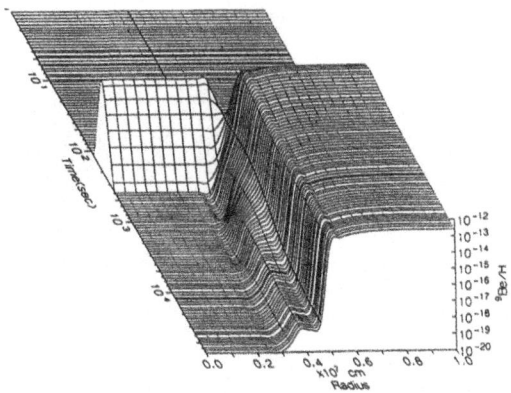

Figure 2. The time evolution of the local ^{9}Be abundance (mass fraction) for $d(T = 1\text{MeV}) = 10^7$cm, $R = 1000$, $f_v = 0.11$, and $\eta_0 = 4 \times 10^{-10}$.

^{9}Be abundance is different with R, the ^{9}Be abundance takes the maximum value at $f_v \sim 0.1$ and decreases with increasing f_v in the range $f_v > 0.1$ for large R. In the calculated ranges of parameters, the highest value of the ^{9}Be abundance is ^{9}Be/H $= 4 \times 10^{-15}$ $(d(T = 1\text{MeV}) = 10^7$cm, $R = 1000$, $f_v = 0.11)$ for $\eta_0 = 4 \times 10^{-10}$ and ^{9}Be/H $= 3 \times 10^{-15}$ $(d(T = 1\text{MeV}) = 10^{6.5}$cm, $R = 1000$, $f_v = 0.11)$ for $\eta_0 = 7 \times 10^{-9}$. These values are two orders smaller than those calculated in the simplified two-zone model. The reason for the discrepancy between these results is incomplete treatments of the neutron diffusion in the latter model. To confirm this point, we cut off neutron diffusion artificially when the neutron density distribution became almost uniform and calculated the abundances of elements in the case of $d(T = 1\text{MeV}) = 10^{5.5}$cm, $R = 1000$, $f_v = 0.11$, and $\eta_0 = 4 \times 10^{-10}$. The ^{9}Be abundance in this case is ^{9}Be/H $= 1.0 \times 10^{-13}$, which is as large as that obtained in the simplified two-zone model and higher by orders of magnitude than that in the case that neutron diffusion is not cut off with the same parameters; ^{9}Be/H $= 8.3 \times 10^{-18}$.

4. Comparison with Observations and Heavy Elements

The recent observations of several population-II stars have placed the upper limit on the primordial ^{9}Be abundance to $\log B/H = -13.7$ (Ryan et al. 1990), which is two orders of magnitude greater than the predicted abundance in the present work.

The constraints on the parameters from other light elements, D, ^{3}He, ^{4}He, and ^{7}Li are more stringent than those placed by the ^{9}Be abundance. The abundances of these elements, of course, depend on the scale of density fluctuations and they reduce to the uniform abundances in the limit of small scale. However, in the range of the fluctuation scale so long that the ^{9}Be abundance is larger than the uniform abundance; $d(T = 1\text{MeV}) \gtrsim 10^6$cm, the ^{7}Li abundance increases monotonically with R and R is constrained to $R \lesssim 10$ even when the very loose bound on the primordial abundance of the ^{7}Li abundance; ^{7}Li/H $< 1 \times 10^{-9}$,

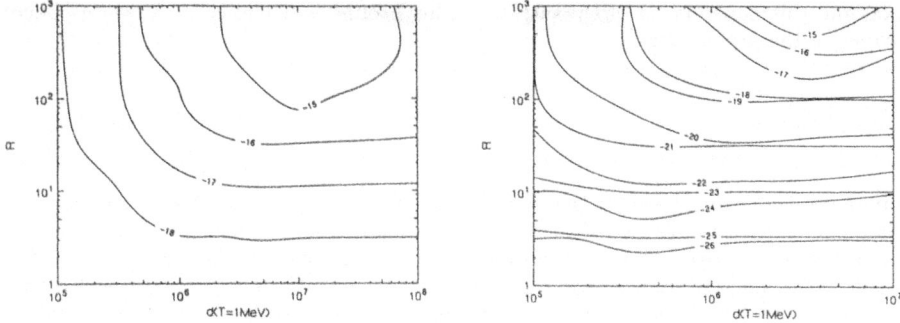

Figure 3. The contour of the average abundances of ^{9}Be on the $d - R$ plane for $\eta_0 = 4 \times 10^{-10}$(left) and $\eta_0 = 7 \times 10^{-9}$(right). Contour levels are log(^{9}Be/H).

is taken. The constraints placed by the upper bound on the sum of the D abundance and the ^{3}He abundance are also stringent: R should be smaller than ~ 50 in the range, $d(T = 1\text{MeV}) \gtrsim 10^6$cm. Hence, the constraints from the light elements except for ^{9}Be, that is, D, ^{3}He, ^{4}He, and ^{7}Li are still more stringent and powerful to examine the inhomogeneous models of nucleosynthesis in the present stage.

The total abundance of heavy elements ($A > 7$) is at most $\sim 10^{-13}$ by mass fraction for $\eta_0 = 4 \times 10^{-10}(\Omega = 0.014h^{-2} = 0.056h_{50}^{-2})$ and $\sim 10^{-11}$ for $\eta_0 = 7 \times 10^{-9}(\Omega = 0.25h^{-2} = 1.0h_{50}^{-2})$ in the ranges of parameter calculated in the present study which is dominated by ^{11}B and CNO elements produced in the inner high density shells. The abundances of CNO elements in the low-density neutron-rich shells are at most $X(^{12}\text{C}) \sim 10^{-14}$, $X(^{14}\text{N}) \sim 10^{-20}$, and $X(^{16}\text{O}) \ll 10^{-20}$ for $\eta_0 = 4 \times 10^{-10}$. For $\eta_0 = 7 \times 10^{-9}$, $X(^{12}\text{C}) \sim 10^{-11}$, $X(^{14}\text{N}) \sim 10^{-16}$, and $X(^{16}\text{O}) \sim 10^{-17}$. Thus the abundances of CNO elements in the proper range of baryon/photon ratio, which is consistent with the abundances of light elements, are too small to be the seeds of cosmological r-process. For the large value of baryon/photon ratio as to close the universe only by baryons, the abundances of these elements are significantly large. However, the fact that such high baryon density models are inconsistent with the abundances of light elements should not be ignored.

5. Discussion

Since we did not survey all over the parameter spaces, we cannot necessarily exclude the possibility that heavy elements and ^{9}Be are synthesized abundantly in the inhomogeneous universe. The important point is, however, the consistencies with the abundances of light elements. In the ranges of parameters consistent with the light element abundances, the abundances of heavy element and ^{9}Be are imposed stringent upper limits. This point is confirmed also by additional calculations in the ranges of small f_v and large R ($10^{-4} \leq f_v \leq 10^{-1}$, $10^3 \leq R \leq 10^6$). Anyway, the range of parameter space excluded only by the ^{9}Be abundance is so limited even if it exists as pointed out by Malaney and Fowler(1989).

The effects of neutron diffusion may depend on the initial density profile and we need to examine this dependance quantitatively. This profile dependance is, however, beyond the range of the present work.

6. Conclusions

In summary, we have calculated the abundances of elements up to ^{22}Ne in the inhomogeneous universe with the extended network in the multi-zone scheme. The ^{9}Be abundance is two orders of magnitude smaller than that calculated in the simplified two-zone model in the calculated ranges of parameters. Further observational efforts seem to be needed to say something about the inhomogeneous cosmological models basing on the ^{9}Be abundance. Cosmic r-process proposed previously seems not to work in the proper range of parameters consistent with the light element abundances.

This talk is based on the collaboration with K. Sato. The author acknowledges the support by Special Researchers' Basic Science Program of the Institute of Physical and Chemical Research in Japan.

References

Applegate, J. H., Hogan, C. J., and Scherrer, R. J., *Ap. J.*, **329**, 572.

Boyd, R. N. and Kajino, T. 1989, *Ap. J.(Letters)*, **336**, L55.

Kurki-Suonio, H., Matzner, R. A., Centrella, J. M., Rothman, T. and Wilson, J. R., 1988, *Phys. Rev. D*, **38**, 1091.

Kurki-Suonio, H. and Matzner, R. A., 1989, *Phys. Rev. D*, **39**, 1046.

Malaney, R. A. and Fowler, W. A. , 1988, *Ap. J.*, **333**, 14.

Malaney, R. A. and Fowler, W. A. , 1990, *Ap. J.*, to be published.

Mampe, W., Ageron, P., Bates, C., Pendlebury, J. M. and Steyerl, A., 1989, *Phys. Rev. Lett.*, **63**, 593.

Ryan, S. G., Bessell, M. S., Sutherland, R. S., Norris, J. E., 1990, *Ap. J. (Letters)*, **348**, L57.

Terasawa, N. and Sato, K., 1989a, *Prog. Theor. Phys. (Letters)* **81**,1085.

Terasawa, N. and Sato, K., 1989b, *Phys. Rev. D*, **39**, 2893.

SIGNATURES OF INHOMOGENEITY IN THE EARLY UNIVERSE

WILLIAM A. FOWLER (with L. H. KAWANO, R. A. MALANEY, and R. W. KAVANAGH)
W. K. Kellogg Radiation Laboratory
California Institute of Technology
Pasadena, CA 91125

ABSTRACT. We have made a systematic study of the production of elemental CNO in inhomogeneous nucleosynthesis, investigating a much larger region of parameter space than previously studied. We have determined abundances of CNO elements and ascertained the main channels to their production. We have focussed in particular on the role played by the $^7\text{Li}(n,\gamma)^8\text{Li}(\alpha,n)^{11}\text{B}(n,\gamma)^{12}\text{B}(\beta^-\nu)^{12}\text{C}$ reaction sequence: in models with $\Omega_b = 1$, we show that this sequence provides the main channel to CNO element production of which there is a significant amount; for lower values of Ω_b there is competition from $^7\text{Li}(\alpha,\gamma)^{11}\text{B}$ but here there is a concurrent decline in CNO production. From these determinations, CNO element production emerges as a distinct signature of an $\Omega_b = 1$ inhomogeneous universe. Beyond ^{12}C, a series of neutron captures is thought to lead to the production of ^{14}C. We have studied the branching at ^{14}C and have deduced the importance of the $^{14}\text{C}(d,n)^{15}\text{N}$ reaction which may be the dominant channel for the production of heavier elements.

I. Introduction

Ever since the work of Applegate and Hogan (1985), there has been keen interest and fruitful research in the study of inhomogeneous nucleosynthesis. In this model of nucleosynthesis, density fluctuations produced in the quark-hadron phase transition lead to the formation of two distinct regions: a high-density region (which made a late transition from the quark state to the hadron state) and a low-density region (which made the transition earlier). The environment is thus characterized by the fractional volume of each region (given as f_v and $1 - f_v$ respectively) and the density contrast between the two regions (given as $R = \Omega_h/\Omega_l$ in which Ω_h is the ratio of the baryon density to the critical density in the high-density region and Ω_l in the low-density region; Ω_h and Ω_l correspond to Ω_q and Ω_h respectively in Malaney and Fowler [1988]). The mean baryon density in the universe, Ω_b, is given as the weighted sum of the densities of the high-density region and low-density region,

$$\Omega_b = f_v\Omega_h + (1 - f_v)\Omega_l \ . \tag{1.1}$$

Once these regions are formed, however, the density inhomogeneities are soon characterized by a chemical inhomogeneity as neutrons stream out of the high-density regions because of their relatively long mean-free-path (compared with the length scale of the high-density regions). Thus, the high-density regions become proton-rich and the low-density regions become neutron-rich. The baryon densities for the two regions are then given by (Alcock, Fuller and Mathews 1987)

$$\Omega_b^{(p)} = X_n\Omega_b + X_p\Omega_h \tag{1.2}$$

$$\Omega_b^{(n)} = X_n\Omega_b + X_p\Omega_l \tag{1.3}$$

with X_n and X_p the neutron and proton mass fractions before the neutron diffusion takes place. The neutron contributions to the baryon densities in each region are given in terms of an overall neutron density because of the assumption of complete neutron homogenization. Thus, after the diffusion of neutrons, the new neutron mass fractions are

$$X_n^{(p)} = \frac{\Omega_b}{\Omega_b^{(p)}}X_n \tag{1.4}$$

$$X_n^{(n)} = \frac{\Omega_b}{\Omega_b^{(n)}}X_n \tag{1.5}$$

with $X_n^{(p)}$ the mass fraction in the proton-rich regions and $X_n^{(n)}$ in the neutron-rich regions. Equation 1.4 (Equation 1.5) says that the neutron density in the proton-rich region (neutron-rich region) is the same as the overall average. Substituting for Ω_h and Ω_l using equations 1.2 and 1.3, the overall baryon density is now given by

$$\Omega_b = f_v \Omega_b^{(p)} + (1 - f_v)\Omega_b^{(n)} \ . \tag{1.6}$$

The two regions, each characterized by its own baryon density and neutron mass fraction, undergo a standard homogeneous nucleosynthesis (e.g., see Fig. 1a and 1b), so that to obtain the total elemental abundances produced, one takes the mass fractions from each region ($X_i^{(p)}$ and $X_i^{(n)}$ for each element i) and does the following average:

$$X_i^{avg} = f_v X_i^{(p)}(\frac{\Omega_b^{(p)}}{\Omega_b}) + (1 - f_v)X_i^{(n)}(\frac{\Omega_b^{(n)}}{\Omega_b}) \ . \tag{1.7}$$

The authors of the original papers made the claim that the averaged abundances so obtained for $\Omega_b = 1$ were within observational bounds for the inferred primordial abundances for D, ^{3}He, and ^{4}He (the computed abundances for ^{7}Li came out somewhat high) so that baryonic nucleosynthesis could be reconciled with a universe of critical density. Moreover, Malaney and Fowler (1988) pointed out that late-time diffusion of neutrons back into the high-density regions could reduce the amount of lithium produced to an acceptable level if the diffusion was late enough (so as not to perturb the other light elements); however, more recent and more sophisticated studies (Kurki-Suonio et al. 1988, Kurki-Suonio and Matzner 1989) questioned the efficiency of such late-time diffusion and raised the issue of ^{4}He overproduction through diffusion. Other groups have presented a more optimistic picture (Mathews et al. 1988; Terasawa and Sato 1989a, b; Alcock et al. 1990); thus, while it remains plausible that the quark-hadron transition has left its mark on nucleosynthesis, it is quite uncertain whether $\Omega_b = 1$ can be sustained. Besides the possibility of having $\Omega_b = 1$, there was the prospect that inhomogeneous nucleosynthesis might provide significant primordial abundances of CNO isotopes (Applegate, Hogan, and Scherrer 1987). It was subsequently determined that the bulk of the CNO production came through the ^{7}Li$(n,\gamma)^8$Li$(\alpha,n)^{11}$B reaction sequence (Malaney and Fowler 1987; Applegate, Hogan, and Scherrer 1988). In this work, we have followed up this suggestion and have looked at this reaction sequence for a much wider range of parameters than hitherto studied, examining the relative importance of the ^{7}Li$(n,\gamma)^8$Li reaction and the ^{7}Li$(\alpha,\gamma)^{11}$B reaction in the production of CNO elements. We discuss this in detail in part II. We have also examined the path of heavy-element production at ^{14}C, looking at the relative importance of the ^{14}C$(n,\gamma)^{15}$C and ^{14}C$(p,\gamma)^{15}$N reactions as well as the ^{14}C$(d,n)^{15}$N reaction which we have included for the first time. This work is described in Part III. Finally, we summarize and give our conclusions in Part IV.

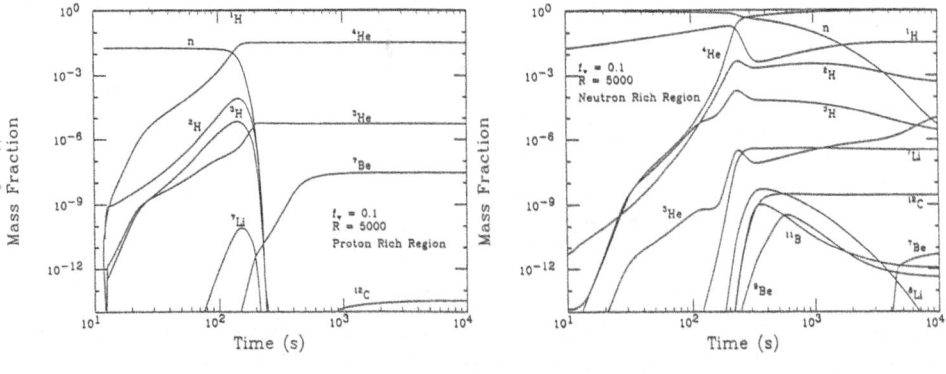

Figure 1a.   Figure 1b

Fig. 1. Evolution of isotopic abundances (in mass fraction), for the parameters indicated, in the a) proton-rich regions; b) neutron-rich regions.

II. The Production Channels

Malaney and Fowler (1987) suggested the following reaction sequence for the formation of CNO elements:

$$^4\text{He}(t,\gamma)^7\text{Li} \begin{cases} (n,\gamma)^8\text{Li} \begin{cases} (\alpha,n)^{11}\text{B}(n,\gamma)^{12}\text{B}(\beta^-\nu)^{12}\text{C} \\ (n,\gamma)^9\text{Li}(\beta^-\nu)^9\text{Be}(p,\alpha)^6\text{Li} \\ (d,n)^9\text{Be}(p,\alpha)^6\text{Li} \end{cases} \\ (\alpha,\gamma)^{11}\text{B}(n,\gamma)^{12}\text{B}(\beta^-\nu)^{12}\text{C} \end{cases}$$

with the first and the last chains the main contributors to CNO element production, the latter a weak (15%) branch. The middle two chains siphon off ^{8}Li to reduce the production from the first chain by up to 50%. However, Malaney and Fowler (1987) made these observations for a limited region of parameter space; furthermore, new measurements have been made on some key reactions. In light of this, we have made an examination of the importance of the $^7\text{Li}(n,\gamma)^8$Li reaction relative to the $^7\text{Li}(\alpha,n)^{11}$B reaction for a wide range of parameters, incorporating the new experimental results.

In our investigations, we have utilized the nucleosynthesis code of Robert Malaney: a description of the code, the method of solution, and the reaction network used are given in detail elsewhere (Malaney 1986, Malaney and Fowler 1987). The updated reactions rates we use in the network can be found in Caughlan and Fowler (1988), Malaney and Fowler (1989), and Table 1 of this paper. The code assumes 3 neutrino families and a Hubble expansion value of 50 kms^{-1}Mpc^{-1} in computing abundances in the proton-rich and neutron-rich regions and utilizes eqns (1.1) through (1.7) to average abundances in the two regions. The results are characterized by parameters f_v, R, and Ω_b. These parameters were allowed to vary over a large range: we used values of the fractional volume of the proton-rich region as small as 10^{-3} to as large as 0.9, focussing on the intermediate values of 0.1, 0.223, and 0.5; the values of the density contrast R were varied from 50 to 5000; and three values of Ω_b, 0.1, 0.4, and 1.0, were used. The parameter space under investigation was enlarged due to recent controversy concerning appropriate values for the parameters. The value for R has been the subject of much debate: Kurki-Suonio et $al.$ (1989) give $R \leq 100$ whereas Fuller, Mathews, and Alcock (1988) consider much larger values ($R \lesssim 10^6$). We tried to cover most of this ground, going from a low value of 50 to as high as 5000; we do not go any higher than this as the simple neutron diffusion model begins to break down at these values – the density in the high-density regions prevents the diffusion of neutrons outward. With regard to Ω_b, Kurki-Suonio et $al.$ (1989) conclude that $\Omega_b \leq 0.3$, contrary to original claims that $\Omega_b = 1$ can be accommodated by inhomogeneous nucleosynthesis. We have therefore allowed Ω_b to vary as a parameter, going from the standard low value of 0.1 to as high as the closure density, $\Omega_b = 1$. Thus, rather than limiting ourselves to regions of parameter space which have been proclaimed compatible with lithium abundances, we extend the parameter ranges seeking a signal of inhomogeneity distinct from the lithium issue.

Table 1

Nuclear Reaction Rates (cm^{3}s^{-1}mole^{-1})

Reaction	Value
$^{14}\text{C}(d,n)^{15}\text{N}$	$4.27 \times 10^{13}T_9^{-2/3}\exp(-16.939/T_9^{1/3})$
$^{14}\text{C}(p,\gamma)^{15}\text{N}$	$1.09 \times 10^8 T_9^{-2/3}\exp(-13.71/T_9^{1/3} - (T_9/4.694)^2)$
	$(1 + 3.04 \times 10^{-2}T_9^{1/3} + 0.105T_9^{2/3}$
	$+ 2.24 \times 10^{-2}T_9 + 0.109T_9^{4/3} + 5.94 \times 10^{-2}T_9^{5/3})$
	$+ 5.36 \times 10^3 T_9^{-3/2}\exp(-3.811/T_9)$
	$+ 9.82 \times 10^4 T_9^{-1/3}\exp(-4.739/T_9)$
$^{14}\text{C}(n,\gamma)^{15}\text{C}$	$3240T_9$
$^7\text{Li}(n,\gamma)^8\text{Li}$	$3144 + 4.26 \times 10^3 T_9^{-3/2}\exp(-2.576/T_9)$

There have been a number of recent experiments pertinent to some of the reactions in the production sequence given above. A new investigation of the reaction rate for $^7\text{Li}(n,\gamma)^8\text{Li}$ has been reported by Wiescher, Steininger and Käppeler (1989) in which they report a smaller cross section than previously used. They give a nonresonant rate of

$$N_A < \sigma v >_{nr} = 3144 \text{ cm}^3\text{s}^{-1}\text{mole}^{-1}$$

(in place of 6015 cm^{3}s^{-1}mole^{-1}) and a resonant rate of

$$N_A < \sigma v >_r = 4.26 \times 10^3 T_9^{-3/2}\exp(-2.576/T_9) \text{ cm}^3\text{s}^{-1}\text{mole}^{-1}$$

(in place of $1.014 \times 10^4 T_9^{-3/2}\exp(-2.576/T_9)$ cm^{3}s^{-1}mole^{-1}). We have incorporated this new rate into our calculations; its effect is to reduce the production of CNO elements through the $^7\text{Li}(n,\gamma)^8\text{Li}(\alpha,n)^{11}\text{B}$ channel by about 40% for most of parameter space for $\Omega_b = 1$.

A very recent result involves the important reactions $^8\text{Li}(\alpha,n)^{11}\text{B}$ and $^8\text{Li}(d,n)^9\text{Be}$ (T. Paradellis, *et al.* 1990). The first reaction was measured to have an $S(E)$ factor dominated by a resonance at 0.60 MeV (which lies near the center of the Gamow energy range for $T_9 = 1$) with a value of $S = 8400$ MeV-b at the resonance peak (coincidentally, the same value as the estimate for the continuum as given by Malaney and Fowler 1989). This S factor was measured for the n_0 channel only but this was assumed to be dominant. The second reaction was found to have an $S(E)$ factor at a constant level of 1.0 MeV-b at energies near the Gamow peak for $T_9 = 1$. We have not analyzed the full implications of the first reaction rate due to the fact that the new rate is not significantly different from the rate previously used. Investigations of the change in the second reaction rate (which is a factor of 34 lower than the previous estimate for $S(E)$) indicate little effect on the Li, Be, and CNO abundances implying that the $^8\text{Li}(d,n)^9\text{Be}$ reaction is relatively unimportant for most of parameter space. Our study of the branching at ^8Li showed that the $^8\text{Li}(p,n)^8\text{Be}$ reaction (the rate given by Wagoner [1969]) is dominant, destroying up to 50% of the ^8Li for large regions of parameter space whereas the $^8\text{Li}(n,\gamma)^9\text{Li}$ and the $^8\text{Li}(d,n)^9\text{Be}$ have proved to be generally negligible.

To measure the relative importance of the $^7\text{Li}(n,\gamma)^8\text{Li}$ channel to that of $^7\text{Li}(\alpha,\gamma)^{11}\text{B}$, three runs were made for a particular point in parameter space: one standard run served as the control; one run was made with the $^7\text{Li}(\alpha,\gamma)^{11}\text{B}$ reaction turned off; one run was made with the $^7\text{Li}(n,\gamma)^8\text{Li}$ reaction turned off. The importance of each channel was thus reflected in the reduction of CNO elemental production. The ratios of this reduction for the $^7\text{Li}(n,\gamma)^8\text{Li}$ over the $^7\text{Li}(\alpha,\gamma)^{11}\text{B}$ reaction are shown in Figs. 2a, 2b, and 2c for three different values of Ω_b.

It is evident from our results that for $\Omega_b = 1$ the $^7\text{Li}(n,\gamma)^8\text{Li}$ channel dominates for most of parameter space, having nine times greater effect than the $^7\text{Li}(\alpha,\gamma)^{11}\text{B}$ reaction at large values of R; before the new rate from Wiescher, Steininger and Käppeler (1989) was included, the effect was sixteen times greater. For $\Omega_b = 0.4$, the $^7\text{Li}(n,\gamma)^8\text{Li}$ reaction dominates less of parameter space and by a smaller amount. Finally, for $\Omega_b = 0.1$, the situation reverses and the $^7\text{Li}(\alpha,n)^{11}\text{B}$, reaction becomes more important. However, at such low values of Ω_b, the production of CNO elements is correspondingly less (Figs. 3a, 3b, and 3c) and therefore, the $^7\text{Li}(n,\gamma)^8\text{Li}$ channel emerges as the most important path to CNO production.

In Fig. 4, we have plotted the CNO total mass fraction as a function of Ω_b. Aside from regions of particularly low values of f_v and R (represented by the curves a, b, and c), the CNO elemental production scales with Ω_b to a maximum mass fraction of about 10^{-10} without much sensitivity to the values of the parameters (curve d). Any detection of primordial CNO abundances material will give a strong signal of a high Ω_b universe (as well as inhomogeneity), independent of the parameters. Although earlier work had clearly shown that significant CNO production could take place, this dependence of the production on the value of Ω_b was not transparent. The use of specific CNO isotopic ratios as a signature of inhomogeneity has recently been discussed by Kajino, Mathews, and Fuller (1990).

III. Reactions on ^{14}C

The production sequences discussed in section II result in the creation of significant amounts of ^{12}C. At that point, transmutation of the ^{12}C by neutrons via $^{12}\text{C}(n,\gamma)^{13}\text{C}(n,\gamma)^{14}\text{C}$ produces ^{14}C. Applegate, Hogan, and Scherrer (1988) proposed that heavy elements were created by the further reactions

$$^{14}\text{C}(\alpha,\gamma)^{18}\text{O}(n,\gamma)^{19}\text{O}(\beta^-)^{19}\text{F}(n,\gamma)^{20}\text{F}(\beta^-)^{20}\text{Ne}.$$

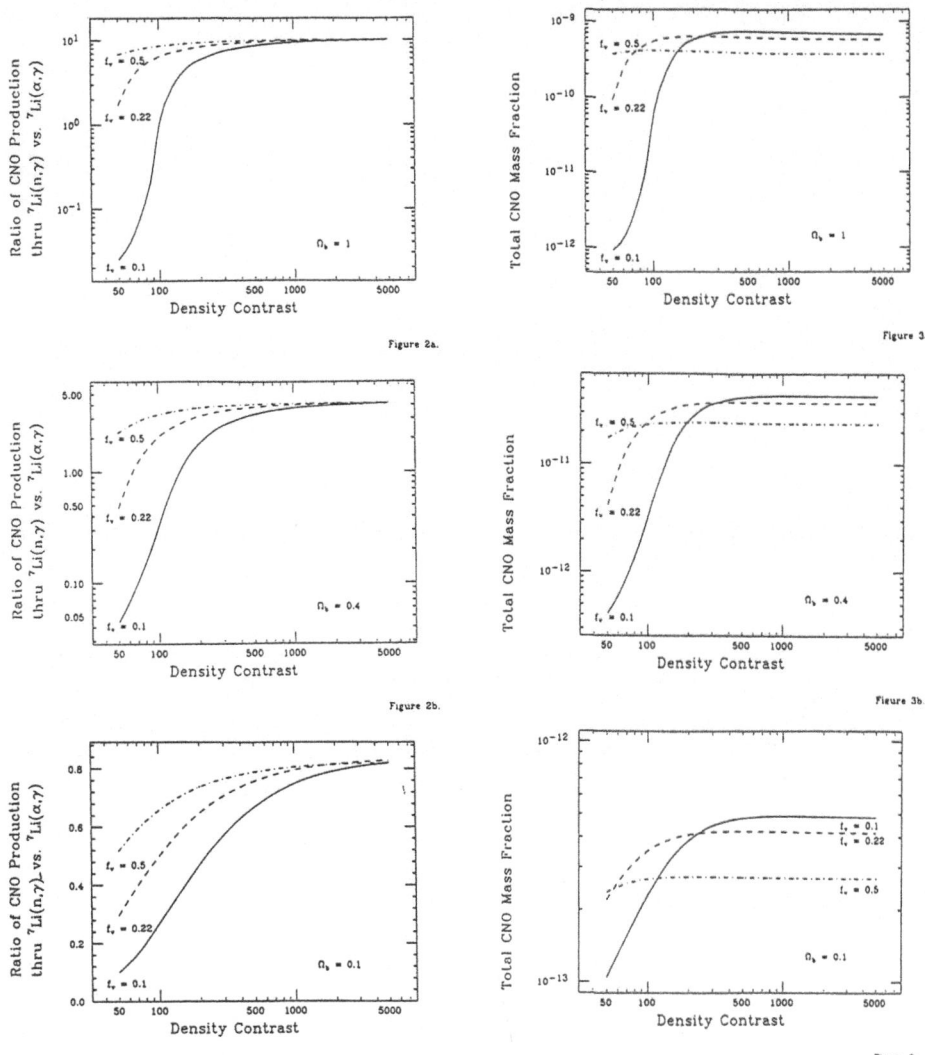

Fig. 2. Ratio of CNO element production effectiveness of $^7Li(n, \gamma)^8Li$ to $^7Li(\alpha, \gamma)^{11}B$ as a function of the density contrast for 3 values of f_v at a value of $\Omega_b =$ a) 1.0; b) 0.4; c) 0.1.

Fig. 3. CNO element (A = 12-16) total mass fraction as a function of the density contrast for 3 values of f_v at a value of $\Omega_b =$ a) 1.0; b) 0.4; c) 0.1.

Kajino, Mathews, and Fuller (1989) pointed out the significance of the $^{14}C(n, \gamma)^{15}C$ reaction and proposed the alternative route:

$$^{14}C(n, \gamma)^{15}C(\beta^-)^{15}N(n, \gamma)^{16}N(\beta^-)^{16}O(n, \gamma)^{17}O(n, \gamma)^{18}O(n, \gamma)^{19}O$$

$$^{19}O \begin{cases} (n, \gamma)^{20}O(n, \gamma)^{21}O(\beta^-)^{21}F(\beta^-)^{21}Ne \\ (\beta^-)^{19}F(n, \gamma)^{20}F(\beta^-)^{20}Ne \end{cases}$$

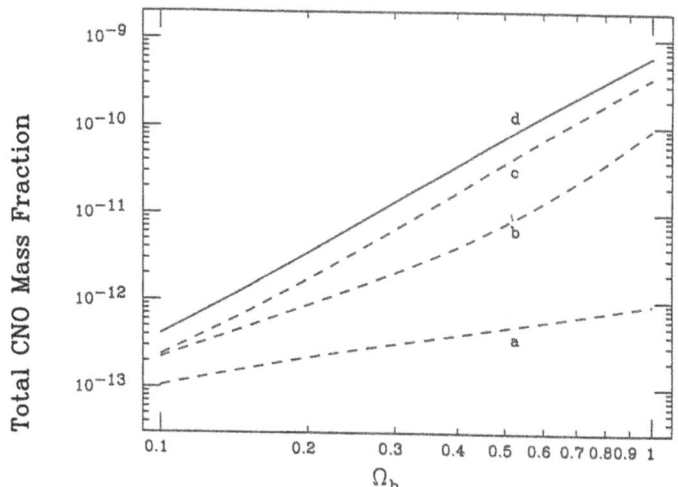

Fig. 4. CNO element total mass fraction as a function of Ω_b for various values of f_v and R: a) R = 50, f_v = 0.1; b) R = 50, f_v = 0.22; c) R = 50, f_v = 0.5; d) R = 200, f_v = 0.22.

with a branching which depends on parameters fixing the neutron abundance. More recently, Wiescher, Görres, and Thielemann (1990) have proposed that ^{14}C is comparably consumed by the ^{14}C$(p, \gamma)^{15}$N reaction. We suggest here that the ^{14}C$(d, n)^{15}$N reaction is the fastest of all: the deuterium abundance is within several orders of magnitude of the neutron and proton abundances and because it is a strong interaction, it will have a larger cross section.

We first incorporated rates for ^{14}C$(p, \gamma)^{15}$N and ^{14}C$(n, \gamma)^{15}$C as given by Wiescher, Görres, and Thielemann (1990). For the ^{14}C$(p, \gamma)^{15}$N reaction, they make a fit to an experimentally determined S-factor curve and obtain

$$N_A < \sigma v >_{nr} = 1.09 \times 10^8 T_9^{-2/3} \exp[-13.71/T_9^{1/3} - (T_9/4.694)^2]$$

$$\times (1 + 3.04 \times 10^{-2} T_9^{1/3} + 0.105 \times T_9^{2/3} + 2.24 \times 10^{-2} T_9 + 0.109 T_9^{4/3} + 5.94 \times 10^{-2} T_9^{5/3}) \text{ cm}^3 \text{s}^{-1} \text{mole}^{-1}$$

for the nonresonant component of the reaction. Their resonant rate agrees well with that of Caughlan and Fowler (1988) which is given as

$$N_A < \sigma v >_r = 5.36 \times 10^3 T_9^{-3/2} \exp(-3.811/T_9)$$

$$+ 9.82 \times 10^4 T^{-1/3} \exp(-4.739/T_9) \text{ cm}^3 \text{s}^{-1} \text{mole}^{-1}$$

and which we utilized. For the ^{14}C$(n, \gamma)^{15}$C reaction, they find resonant contributions small compared to the nonresonant part which they formulate as

$$N_A < \sigma v >_{nr} = 1.08 \times 10^8 (S) T_9 \text{ cm}^3 \text{s}^{-1} \text{mole}^{-1}$$

with the S-factor determined to be slowly varying with energy with a value of about 3×10^{-5}, which we adopt.

Noting the substantial abundance of deuterons produced in the neutron-rich region during the inhomogeneous nucleosynthesis, we decided to look into the possibility that there might be enough deuterons to make the ^{14}C$(d, n)^{15}$N reaction important. Using equation C60 from Fowler and Hoyle (1964), we derived the following expression for the rate of the ^{14}C$(d, n)^{15}$N reaction:

$$N_A < \sigma v >_{nr} = 4.27 \times 10^{13} T_9^{-2/3} \exp(-16.939/T_9^{1/3}) \text{ cm}^3 \text{s}^{-1} \text{mole}^{-1}.$$

There is about a factor 3 uncertainty in these results.

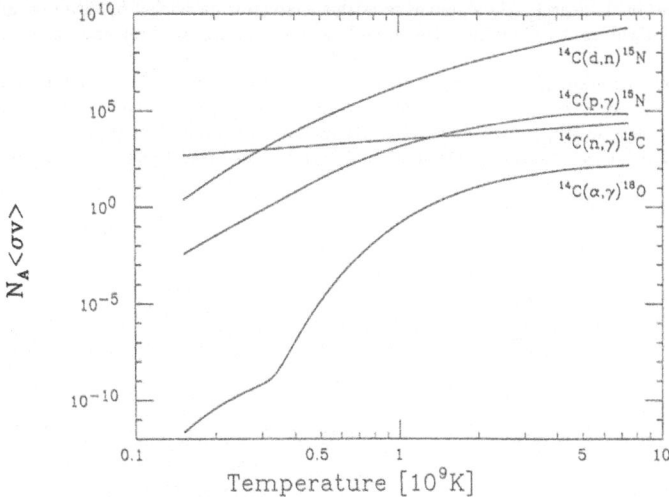

Fig. 5. The reaction rates (in units of cm^3 s^{-1} mole^{-1}) for four of the primary reactions on ^{14}C as functions of the temperature (in units of 10^9 K).

For comparison, the reaction rates for these three reactions and for the ^{14}C$(\alpha, \gamma)^{18}$O reaction are plotted versus the temperature in Fig. 5. As expected, the ^{14}C$(d, n)^{15}$N interaction dominates over the other reactions. Among these other reactions, the proton and neutron reactions dominate over that for the alpha. Fig. 5 essentially duplicates what is given in Fig. 3 of Wiescher, Görres, and Thielemann (1990) with the exception of our curve for ^{14}C$(d, n)^{15}$N reaction.

As Wiescher, Görres, and Thielemann (1990) did in their paper, we derived destruction rates of ^{14}C with respect to these reactions. With the destruction rate given as $\lambda = (X/A)N_A< \sigma v >\rho$, X the mass fraction and A the atomic number of the light nuclide involved, and ρ the baryon density, it was necessary to compute the elemental mass fractions and the baryon density at various temperatures and fold them with the reaction rates. The ^{14}C is produced mainly in the neutron-rich region and thus we can restrict ourselves to the elemental abundances there. The elemental abundances in the neutron-rich region are given as functions of temperature in Fig. 6 for 3 different points in parameter space. The parameters for Fig. 6a are the same as for Fig. 1b in Malaney and Fowler (1988) and Fig. 6c here is very similar to our Fig. 1b as variation in parameter R does not make much of a difference in this range. With these abundances, the destruction rate of ^{14}C was computed and the results are given in Figs. 7. The figures show the destruction rates for the 4 reactions discussed as well as for the ^{14}C beta decay. One sees that the ^{14}C$(d, n)^{15}$N reaction dominates for the temperature range most important for nucleosynthesis in all three points in parameter space. Although there is some uncertainty in our computed ^{14}C$(d, n)^{15}$N rate, this reaction – overlooked until now – certainly has an important role to play in the destruction of ^{14}C.

IV. Conclusions

These investigations of CNO elemental production have given results for a large region of parameter space in the model of inhomogeneous nucleosynthesis. The inhomogeneous nucleosynthesis produces significant amounts of CNO material at high values of Ω_b which should be observable as a "cosmic floor" in metal-poor halo stars. Since this result occurs unambiguously for much of parameter space, an observation of primordial CNO isotopes is perhaps the clearest indication of inhomogeneous nucleosynthesis. Astrophysical observations of CNO abundances will thus provide an independent probe into this volume of parameter space and allow constraints to be put on the nature of the quark-hadron phase transition.

The ^{7}Li$(n, \gamma)^8$Li$(\alpha, n)^{11}$B channel has proven to be the dominant path to CNO production in regions of parameter space in which there is significant production. This reaction sequence dominates for most values of f_v and R at values of Ω_b near 1. The dominance of this channel is paralleled by CNO production, thus making observation of significant primordial CNO abundances a clear signal of inhomogeneous $\Omega_b \sim 1$ universe.

There have been new measurements affecting both the $^7\text{Li}(n,\gamma)^8\text{Li}$ and the $^8\text{Li}(\alpha,n)^{11}\text{B}$ reactions. The result reported by Wiescher, Steininger, and Käppeler (1989) regarding the former reaction has been incorporated into the reaction network, reducing the CNO production somewhat. The recent report on the $^8\text{Li}(\alpha,n)^{11}\text{B}$ and the $^8\text{Li}(d,n)^9\text{Be}$ reactions (Paradellis *et al.* 1990) does not lead to any dramatic change.

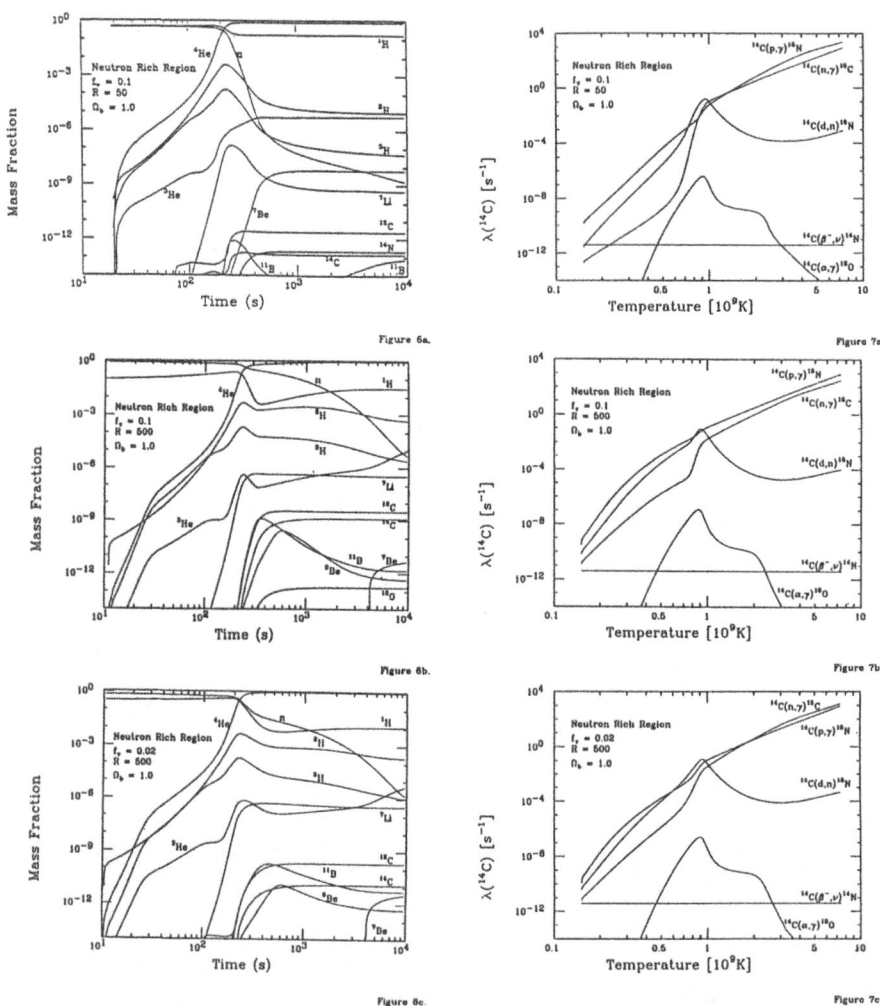

Fig. 6. Mass fractions of elements as functions of the temperature for parameter values of $\Omega_b = 1.0$, a) $f_v = 0.1$ and $R = 50$; b) $f_v = 0.1$ and $R = 500$; c) $f_v = 0.02$ and $R = 500$.

Fig. 7. The destruction rate of ^{14}C (in units of inverse seconds) with respect to four reactions and one decay channel as a function of the temperature (in units of 10^9 K): for $\Omega_b = 1.0$, a) $f_v = 0.1$ and $R = 50$; b) $f_v = 0.1$ and $R = 500$; c) $f_v = 0.02$ and $R = 500$.

There has been a general neglect of the study of deuterium reactions in the field. A case in point is ^{11}B, the termination point of the both the ^{7}Li$(n, \gamma)^8$Li$(\alpha, n)^{11}$B and ^{7}Li$(\alpha, \gamma)^{11}$B sequences that we have studied. It has been thought that a neutron-capture and a beta-decay lead to ^{12}C; however, some preliminary work we have done indicates that the ^{11}B$(d, n)^{12}$C reaction may very well be competitive. Another instance is further up in the chain with ^{14}C which we have discussed in detail in this paper. We have introduced a new reaction, ^{14}C$(d, n)^{15}$N, and have deduced that it competes favorably with the reactions ^{14}C$(n, \gamma)^{15}$C and ^{14}C$(p, \gamma)^{15}$N in the representative portion of parameter space sampled. Eventually, more work will have to be done to put this result into proper perspective. Some preliminary work has shown that: 1) the ^{12}C$(d, n)^{13}$N and ^{12}C$(p, \gamma)^{13}$N reactions are comparable to the ^{12}C$(n, \gamma)^{13}$C reactions, suggesting alternative paths from ^{12}C which may bypass ^{14}C; 2) the ^{15}N$(p, \alpha)^{12}$C reaction is several orders of magnitude more important than the ^{15}N$(n, \gamma)^{16}$N reaction given as part of the heavy-element production sequence mentioned above, leading to the possibility that processing through ^{14}C merely cycles the network back to ^{12}C unless reactions such as ^{15}N$(d, n)^{16}$O and ^{15}N$(d, \alpha)^{13}$C prove comparable.

Before any of these aspects can be thoroughly investigated, however, the current two-zone model of inhomogeneous nucleosynthesis will have to be modified. Recent work by Terasawa and Sato (1990) has indicated possible important consequences for heavy-element production as a result of neutron diffusion back into high-density regions. In their work the low-density regions are depleted of their neutrons, effectively short-circuiting the neutron-capture chains vital to the heavy-element production and thus severely reducing their production. While heavy-element production may still be a reasonable possibility, future investigations of heavy-element production will have to take neutron-diffusion effects into account.

We would like to thank Karlheinz Langanke, Martin Savage, Charles Barnes, Carl Brune, Richard Boyd, Toshitaka Kajino, and Michael Wiescher for their invaluable suggestions and assistance. This work was supported in part by the DoE (at Lawrence Livermore National Laboratory under contract W-7405-ENG-48), and by the NSF (at Caltech under grant PHY88-17296).

References

Alcock, C. R., Dearborn, D. S. P., Fuller, G. M., Mathews, G. J., and Meyer, B. 1990, *Phys. Rev. Letters*, **64**, 2607.

Alcock, C. R., Fuller, G. M., and Mathews, G. J. 1987, *Ap. J.*, **320**, 439.

Applegate, J. H., and Hogan, C. J. 1985, *Phys. Rev. D*, **31**, 3037.

Applegate, J. H., Hogan, C. J., and Scherrer, R. J. 1987, *Phys. Rev. D*, **35**, 1151.

———. 1988, *Ap. J.*, **329**, 572.

Caughlan, G. R., and Fowler, W. A. 1988, *Atomic Data Nucl. Data Tables*, **40**, 283.

Fowler, W. A. and Hoyle, F. 1964, *Ap. J. Suppl.*, No. 91, **9**, 201.

Fuller, G. M., Mathews, G. J., and Alcock, C. R. 1988, in *Dark Matter, Proc. XXIIIrd Rencontre de Moriond*, ed. J. Audouze and J. Tran Thanh Van (Gif-sur-Yvette: Éditions Frontières), p. 303.

Kajino, T., Mathews, G. J., and Fuller, G. M. 1989, in *Heavy Ion Physics and Astrophysical Problems*, ed. S. Kubono (Singapore: World Scientific), p. 51.

Kajino, T., Mathews, G. J., and Fuller, G. M. 1990, *Ap. J.*, (submitted).

Kurki-Sounio, H., and Matzner, R. A. 1989, *Phys. Rev. D*, **39**, 1046.

Kurki-Sounio, H., Matzner, R. A., Centrella, J. M., Rothman, T., and Wilson, J. R. 1988, *Phys. Rev. D*, **38**, 1091.

Kurki-Sounio, H., Matzner, R. A., Olive, K. A., and Schramm, D. N. 1989, *Ap. J.*, **353**, 406.

Malaney, R. A. 1986, *Mon. Not. R. Astro. Soc.*, **223**, 683.

Malaney, R. A. and Fowler, W. A. 1987, in *The Origin and Distribution of the Elements*, ed. G. J. Mathews (Singapore: World Scientific), p. 76.

———. 1988, *Ap. J.*, **333**, 14.

———. 1989, *Ap. J. (Letters)*, **345**, L5.

Mathews, G. J., Fuller, G. M., Alcock, C. R., and Kajino, T. 1988, in *Dark Matter, Proc. XXIIIrd Rencontre de Moriond*, ed. J. Audouze and J. Tran Thanh Van (Gif-sur-Yvette: Éditions Frontières), p. 319.

Paradellis, T., Kossionides, S., Doukellis, G., Aslanoglou, X., Assimakopoulos, P., Pakou, A., Rolfs, C., and Langanke, K. 1990, *Z. Phys. A*, in press.

Terasawa, N., and Sato, K. 1989*a*, *Phys. Rev. D*, **39**, 2893.

————. 1989*b*, *Prog. Theor. Phys.*, **81**, 1085.

————. 1990, *Ap. J.*, in press.

Wagoner, R. V. 1969, *Ap. J. Suppl.*, No. 162, **18**, 247.

Wiescher, M., Görres, J., and Thielemann, F. K. 1990, *Ap. J.*, in press.

Wiescher, M., Steininger, R., and Käppeler, F. 1989, *Ap. J*, **334**, 464.

Diffusion Coefficients of Nucleons in the Inhomogeneous Big Bang Model

B.Banerjee and S.M.Chitre

Tata Institute of Fundamental Research,Homi Bhabha Road,Bombay 400005, India

There is a strong possibility that a first order QCD phase transition from the quark-gluon plasma to the confined hadronic matter had occurred in the early universe when the temperature was about 100 MeV. This phase transition might have produced isothermal baryon number fluctuations [1,2,3]. Applegate and Hogan [2,3] have suggested that the characteristic size of these fluctuations could be such that protons would not be able to diffuse across them before the onset of nucleosynhthesis , but the neutrons would as they have no electrical charge and therefore suffer less scattering. Their calculations have shown that the difference in the mean free paths of neutrons and protons and the resultant diffusive segregation influences the formation of the light elements very significantly.In ref. [3] the diffusion coeffficients were calculated using a mobility formula and the Einstein relation between mobility and the diffusion coefficient [4]. We calculate the diffusion coefficients in the framework of relativistic kinetic theory [5] in the temperature range $10^8 \leq T \leq 5.10^9 \, {}^0K$, assuming all particles to be classical. For these temperatures neutrons and protons are no longer in equilibrium with respect to weak interactions and as a result they retain their identity for diffusive segregation to take place.Neutrons are scattered by electrons through the interaction of their magnetic moments and by protons due to nuclear interaction.Protons ,on the other hand , undergo Coulomb scattering by electrons and are also scattered by neutrons.With these elementary cross–sections as input we calculate the neutron–electron and neutron–proton diffusion coefficients using the first order Chapman–Enskog expressions [5].

(a) Neutron–Electron

The diffusion coefficient of non–relativistic neutrons moving through a gas of relativistic electrons is given by,

$$D_{ne} = \frac{3}{8} \sqrt{\frac{\pi}{2}} \frac{c}{n\sigma_{ne}} \frac{1 - x_n}{x_e} \frac{1}{z_e^{1/2}} \frac{K_2(z_e)}{K_{\frac{5}{2}}(z_e)}, \tag{1}$$

where n is the total density of the system consisting of neutrons,protons, electrons and photons,

$$n = n_n + n_p + n_e + n_\gamma,$$

and

$$x_i = \frac{n_i}{n},$$

$$z_i = m_i c^2 / k_B T$$

$K_\nu(z_i)$ is the modified Bessel function of order ν. The cross–section σ_{ne} is ,

$$\sigma_{ne} = 3\pi\kappa^2 \left(\frac{\alpha \hbar c}{m_n c^2} \right)^2 \tag{2}$$

where κ is the anomalous magnetic moment of the neutron in nuclear magnetons.

117

It is interesting to observe that for non–relativistic electrons ($k_B T \ll m_e c^2$) one obtains the same expression for D_{ne} as given by the mobility formula [4]. The mobility b of the neutrons moving through the electron gas of density n_e is given by ,

$$b^{-1} = \frac{8}{3}\sqrt{\frac{2}{\pi}}(m_e c^2 k_B T)^{1/2}\frac{n_e \sigma_{ne}}{c} \tag{3}$$

We now use the Einstein relation between b and the diffusion coefficient ,

$$D_{ne} = b k_B T, \tag{4}$$

and obtain

$$D_{ne} = \frac{3}{8}\sqrt{\frac{\pi}{2}}\frac{c}{n_e \sigma_{ne}}\left(\frac{k_B T}{m_e c^2}\right)^{1/2} \tag{5}$$

which is the same as Eq. (1) in the limit of large z_e for a dilute gas of neutrons ($x_n \ll 1$) diffusing through electrons.

(b) Neutron–Proton

In this case both the particles are non–relativistic and the scattering cross–section is given by the triplet and singlet scattering lengths a_t and a_s,

$$\sigma_{np} = 3\pi a_t^2 + \pi a_s^2 \tag{6}$$

We get for the diffusion coefficient in this case,

$$D_{np} = \frac{3\sqrt{\pi}}{4}\frac{c}{n\sigma_{np}}\frac{1 - x_n}{x_p}\frac{1}{z_p^{1/2}}. \tag{7}$$

We have calculated the numerical values of the diffusion coefficients D_{ne} and D_{np} . The densities n_γ, n_p, n_e and n_n are calculated using the following expressions valid in the radiation era [6,7]

$$n_\gamma = 20.3 \, T^3$$
$$n_p = n_{e-} = 1.5 \times 10^{-7}\Omega T^3\{1 - 0.17exp(-2.10^{20}T^{-2}/t_n)\} \tag{8}$$
$$n_n = 1.5 \times 10^{-7}\Omega T^3(0.17exp(-2.10^{20}T^{-2}/t_n))$$

where Ω and t_n are respectively the density parameter and neutron life-time. We take $\Omega = 0.2$ and $t_n = 887.6$ sec. For D_{ne} our results differ widely from those of Applegate et al [3]. This is largely because their diffusion coefficient does not depend on the electron density ,whereas it depends very strongly on the temperature–D_{ne} is essentially proportional to $exp(m_e c^2/k_B T)$ .For D_{np} our expression Eq.(6) is the same,except for numerical factors,as the elementary kinetic theory expression $\frac{1}{3}v\lambda$,which has been used in ref. [3].Here v is the velocity of the neutron and $\lambda = 1/n_p \sigma_{np}$,is its mean free path. The agreement of numerical values of the two calculations is to within an order of magnitude.

We should like to emphasise that our expressions for the diffusion coefficients have explicit dependence on the densities of the diffusing particles and we recover the correct expressions in the non–relativistic limit. An important feature that emerges from the calculations is the equivalence between the expression for the diffusion coefficient D_{ne} given by the mobility formula and that derived from the kinetic theory for a dilute neutron gas diffusing through electrons at low temperatures. We have thus shown that the relativistic kinetic theory can be used to calculate the various diffusion coefficients needed for the inhmogeneous nucleosynthesis model as long as the classical approximation is valid. This approximation can be checked by calculating the degeneracy parameter $\left(h^2/2\pi mk_BT\right)^{3/2}n$ which is $\ll 1$ throughout the density and temperature range that we have considered. We expect our values for the diffusion coefficients, which differ from the values used so far, to have significant influence on the abundance of light elements in an inhomogoneous cosmological model. However ,this can only be shown by detailed calculations which are in progress [8].

References

[1] E. Witten,Phys.Rev. D 30 (1984) 272.

[2] J. H. Applegate and C. J. Hogan,Phys.Rev. D 31 (1985) 3037.

[3] J. H. Applegate, C. J. Hogan and R. J. Scherrer,Phys.Rev. D 35 (1987) 1151. (1987) 439.

[4] E. M. Lifshitz and L. P. Pitaevskii, *Physical Kinetics* (Pergamon Press, Oxford, 1981).

[5] S. R. de Groot, W. A. van Leeuwen and Ch. G. van Weert,*Relativistic Kinetic Theory* (North Holland,Amsterdam,1980)

[6] S. Weinberg,*Gravitation and Cosmology* (Wiley,New York, 1972).

[7] E. R. Harrison, Ann.Rev.Astron.Astrophys. 11 (1973) 155.

[8] D. N. Schramm— private communication.

REACTIONS ON CARBON-14

LAWRENCE. H. KAWANO
W. K. Kellogg Radiation Laboratory
California Institute of Technology
Pasadena, CA 91125

ABSTRACT. In inhomogeneous nucleosynthesis, a series of neutron captures is thought to lead to the production of ^{14}C. We have studied the branching at ^{14}C and have deduced the importance of the ^{14}C$(d,n)^{15}$N reaction which may be the dominant channel for the production of heavier elements.

There is the possibility of heavy element production in inhomogeneous nucleosynthesis and as discussed in Kawano *et al.* (1991), the production chain

$$^{4}\text{He}(t,\gamma)^{7}\text{Li}(n,\gamma)^{8}\text{Li}(\alpha,n)^{11}\text{B}(n,\gamma)^{12}\text{B}(\beta^{-},\nu)^{12}\text{C}$$

results in the creation of significant amounts of ^{12}C. At that point, transmutation of the ^{12}C by neutrons via ^{12}C$(n,\gamma)^{13}$C$(n,\gamma)^{14}$C produces ^{14}C. Applegate, Hogan, and Scherrer (1988) proposed that heavy elements were created by the further reactions

$$^{14}\text{C}(\alpha,\gamma)^{18}\text{O}(n,\gamma)^{19}\text{O}(\beta^{-})^{19}\text{F}(n,\gamma)^{20}\text{F}(\beta^{-})^{20}\text{Ne} \ .$$

Kajino, Mathews, and Fuller (1989) pointed out the significance of the ^{14}C$(n,\gamma)^{15}$C reaction and proposed the alternative route to ^{19}O:

$$^{14}\text{C}(n,\gamma)^{15}\text{C}(\beta^{-})^{15}\text{N}(n,\gamma)^{16}\text{N}(\beta^{-})^{16}\text{O}(n,\gamma)^{17}\text{O}(n,\gamma)^{18}\text{O}(n,\gamma)^{19}\text{O} \ .$$

More recently, Wiescher, Görres, and Thielemann (1990) have proposed that ^{14}C is comparably consumed by the ^{14}C$(p,\gamma)^{15}$N reaction. We suggest here that the ^{14}C$(d,n)^{15}$N reaction is the fastest of all: the deuterium abundance is within several orders of magnitude of the neutron and proton abundances and because it is a strong interaction, it will have a larger cross section.

We first looked to see if the (d,n) reaction rate was fast enough to make the reaction reasonably interesting. Using equation C60 from Fowler and Hoyle (1964), we derived the following expression for the

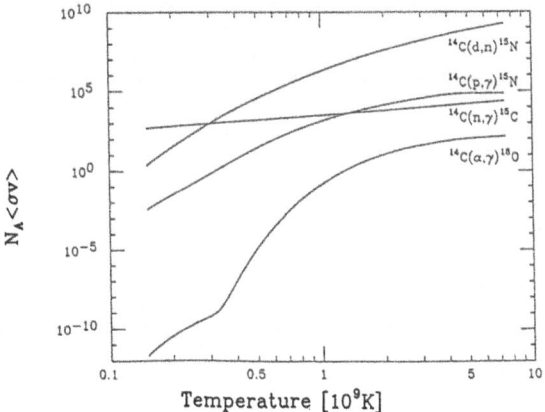

Fig. 1. The reaction rates (in units of cm^{3} s^{-1} mole^{-1}) for four of the primary reactions on ^{14}C as functions of the temperature (in units of 10^{9} K).

rate of the $^{14}C(d,n)^{15}N$ reaction:

$$N_A <\sigma v>_{nr} = 4.27 \times 10^{13} T_9^{-2/3} \exp(-16.939/T_9^{1/3}) \text{ cm}^3 \text{s}^{-1} \text{mole}^{-1}.$$

There is about a factor 3 uncertainty in these results.

For comparison, the reaction rates for this reaction, $^{14}C(n,\gamma)^{15}C$ (rate from Wiescher et al. 1990), $^{14}C(p,\gamma)^{15}N$ (rate from Wiescher et al. 1990), and $^{14}C(\alpha,\gamma)^{18}O$ (rate from Caughlan and Fowler 1988) are plotted versus the temperature in Fig. 1. As expected, the $^{14}C(d,n)^{15}N$ interaction dominates over the other reactions. Among these other reactions, the proton and neutron reactions dominate over that for the alpha.

We next derived destruction rates of ^{14}C with respect to these reactions. With the destruction rate given as $\lambda = (X/A)N_A <\sigma v>\rho$, X the mass fraction and A the atomic number of the light nuclide involved, and ρ the baryon density, it was necessary to compute the elemental mass fractions and the baryon density at various temperatures and fold them with the reaction rates. This was done for a number of points in the space of parameters f_v, the volume fraction of the high-density region and R, the density contrast between the high- and low-density regions. With these abundances, the destruction rate of ^{14}C was computed and the results for a representative point in parameter space are given in Fig. 2. The figure shows the destruction rates for the 4 reactions discussed as well as for the ^{14}C beta decay. One sees that the $^{14}C(d,n)^{15}N$ reaction dominates for the temperature range most important for nucleosynthesis in all three points in parameter space. Although there is some uncertainty in our computed $^{14}C(d,n)^{15}N$ rate, this reaction – overlooked until now – certainly has an important role to play in the destruction of ^{14}C.

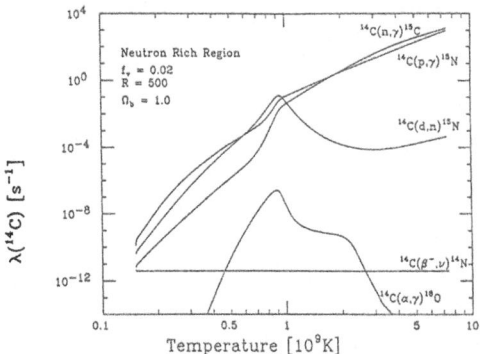

Fig. 2. The destruction rate of ^{14}C (in units of inverse seconds) with respect to four reactions and one decay channel as a function of the temperature (in units of 10^9 K): for $\Omega_b = 1.0$, $f_v = 0.02$ and $R = 500$.

I would like to thank my co-authors Ralph Kavanagh, William Fowler, and Robert Malaney for their contributions to this research. This work was supported by the NSF (at Caltech under grant PHY88-17296).

References

Applegate, J. H., Hogan, C. J., and Scherrer, R. J. 1987, Phys. Rev. D, 35, 1151.

Caughlan, G. R., and Fowler, W. A. 1988, Atomic Data Nucl. Data Tables, 40, 283.

Fowler, W. A. and Hoyle, F. 1964, Ap. J. Suppl., No. 91, 9, 201.

Kajino, T., Mathews, G. J., and Fuller, G. M. 1989, in Heavy Ion Physics and Astrophysical Problems, ed. S. Kubono (Singapore: World Scientific), p. 51.

Kawano, L. H., Fowler, W. A., Kavanagh, R. W., and Malaney, R. A., 1991, Ap. J., to be published February 1991.

Malaney, R. A. and Fowler, W. A. 1988, Ap. J., 333, 14.

Wiescher, M., Görres, J., and Thielemann, F. K. 1990, Ap. J., in press.

SURVIVAL OF STRANGE MATTER LUMPS FORMED
IN THE EARLY UNIVERSE

CHUL H. LEE and HYUN KYU LEE
Department of Physics
Hanyang University
Seoul 133-791, Korea

ABSTRACT. Strange matter lumps are supposed to form during the quark-hadron coexistence period at T_c. We discuss the conditions in which the lumps may avoid dissolution into hadron matter through boiling and survive until the preent time.

Strange matter lumps can be dissolved by evaporation[1,2] and boiling[3]. Evaporation means baryons escaping from the surface and boiling refers to hadron bubbles forming and growing throughout the interior. However, if a strange matter lump is surrounded by a hadron shell with high baryon number density evaporation is much suppressed, and boiling remains an important mechanism for the dissolution of strange matter lumps. Boiling can occur only if strange matter becomes unstable compared to hadron matter as the temperature goes down. However, the equilibrium between strange matter and hadron matter depends not only on the temperature but also on chemical potentials. Let us consider how to determine the equilibrium condition at a given T. If the electron chemical potential in the strange matter is ignored for simplicity, the equilibrium about the weak reactions requires that the chemical potentials of u, d, and s quarks are the same; $\mu_u = \mu_d = \mu_s \equiv \mu$. Then the quark-gluon contribution to the pressure of strange matter is

$$P_Q = T^4 \{ \frac{37\pi^2}{90} + (\frac{\mu}{T})^2 + \frac{1}{2\pi^2}(\frac{\mu}{T})^4 \}$$
$$+ \frac{1}{\pi^2} \int_{m_s}^{\infty} dE (E^2 - m_s^2)^{\frac{3}{2}} \{ \frac{1}{exp(\frac{E-\mu}{T})+1} + \frac{1}{exp(\frac{E+\mu}{T})+1} \} - B \qquad (1)$$

Here m_s is the strange quark mass which we take to be 150 MeV and B is the bag constant for which we consider the three different values, $60, 100$ and $250 MeV/fm^3$, in this work. With the electron chemical potential in the hadron matter also ignored, the baryon chemical potential at equilibrium is three times the quark chemical potential ($\mu_b = 3\mu$) and the meson chemical potential is zero. For the pressure of the hadron gas, we first treat the baryons as free point particles summing contributions from all the low-lying hadrons listed in the particle data book. The final result is then obtained by applying the proper-volume corrections[4],

$$P_H = P_H^{pt} / (1 + \frac{\rho^{pt}}{4B}) \qquad (2)$$

Here P^{pt} and ρ^{pt} are the pressure and energy density of hadrons treated as free point particles.

124

To obtain the equilibrium chemical potential, μ_{eq}, as a function of T, we solve the equation,

$$P_Q(T,\mu) = P_H(T,\mu) \tag{3}$$

with P_Q and P_H obtained in Eqs.(1) and (2) respectively. The result is given in Fig. 1 for the three different values of B. Strange matter is self-bound for $B = 60$ MeV/fm^3 while it is not in the other two cases of $B = 100$ and $250 MeV/fm^3$. If the quark chemical potential, and thus the pressure, inside the strange matter lump is maintained higher than the equilibrium values, the lump can avoid boiling.

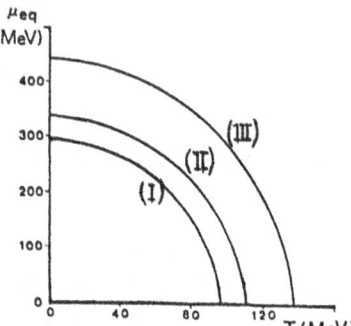

Fig.1 Equilibrium chemical potentials. (I)$B = 60$, (II)$B = 100$, (III)$B = 250$ MeV/fm^3.

Let us now consider how the high pressure inside a strange matter lump can be maintained. As conversion of strange matter into hadron matter proceeds at the critical temperature T_c, the baryon number density in a strange matter lump is likely to become greater and greater[4]. This means the succesive hadron shells forming outside the lump become densere and denser in baryons[5]. The baryon chemical potential becomes larger and larger as one proceeds from the background space through the hadron shells and into the strange matter lump. The energy densities of the strange matter lump and the hadron shells surrounding it are greater than the background energy density. Therefore the gravitational force acts towards the center of the lump and, for gravitational stability, this force has to be balanced out by the pressure gradient. A larger pressure difference between the strange matter core and the background requires a thicker hadron shell. The curves in Fig.1 imply that in the case of $B = 60 MeV/fm^3$ the equilibrium pressure is much smaller than in the other two cases, and therefore those strange matter lumps at the core of much thinner hadron shells can survive boiling. Another interesting aspect in the case of $B = 60 MeV/fm^3$ is that the equilibrium pressure approaches zero as the temperature approaches zero. This means that the hadron shells are eventually reabsorbed into the strange matter lumps and the lumps will be exposed to the background.

This work is supported in part by the Korea Science and Engineering Foundation (No. 891-0201-00902) and in part by the Korea Ministry of Education through the Research Institute for Basic Science of Hanyang University.

REFERENCES
1. C. Alcock and E. Farhi, Phys. Rev. **D32** (1985) 1273
2. J. Madsen, H. Heiselberg and K. Riisager, Phys. Rev. **D34** (1986) 2947
3. C. Alckock and A. Olinto, Phys. Rev. **D39** (1989) 1233
4. E. Witten, Phys. Rev. **D30** (1984) 272
5. G. Fuller, G.Mathews and C. Alcock, Phys. Rev. **D37** (1988) 1380

MEASUREMENT OF THE CROSS SECTION OF THE ^{12}C(n,γ)^{13}C REACTION AT STELLAR ENERGY

Y. Nagai
Department of Applied Physics,Faculty of Science,
Tokyo Institute of Technology,Oh-okayama,Meguro-ku,
Tokyo 152,Japan

ABSTRACT:
    The cross section of the ^{12}C(n,γ)^{13}C reaction at stellar energy is important for discussing the nucleosynthesis of intermediate mass nuclei in inhomogeneous big-bang models. By detecting prompt γ-rays from a captured state the reaction cross section was measured; 16.8 2.1 μb was obtained. This value favours the nucleosynthesis of intermediate-mass nuclei in inhomogeneous big-bang models.

    In the low density neutron-rich regions formed by the isothermal baryonic density fluctuations, caused by the phase transition from the quark-gluon plasma to the hadronic phase, a sufficient production of heavy elements (A 12) has been believed possible through a primordial rapid-process nucleosynthesis[1]. Since the possibility can be tested quantitatively by comparing the calculated abundances of elements with the astronomical observations, it is very important to know the precise nuclear reaction rates to produce the heavy elements. The main reaction sequence to synthesize the heavy elements is: ^{1}H(n,γ)^{2}H(n,γ)^{3}H(d,n)^{4}He(t,γ)^{7}Li(n,γ)^{8}Li(α,n)^{11}B ^{11}B(n,γ)^{12}B(β^{-})^{12}C(n,γ)^{13}C(n,γ)^{14}C etc.
    Since the capture rate of the ^{12}C(n,γ)^{13}C reaction at stellar energy is uncertain, the rate was measured accurately by using pulsed neutrons with the most probable energy of 30 keV and detecting prompt γ-rays from a captured state. The pulsed neutrons were produced by the ^{7}Li(p,n)^{7}Be reaction using the 1.5 ns bunched proton beam, provided by the 3.2-MV Pelletron accelerator at Tokyo Institute of Technology. Capture γ-rays were detected by a 7.6 cm   x 15.2 cm NaI(Tl) detector,

surrounded by an annular NaI(Tl) detector with the size
of 25.4 cm   x 28 cm (2). γ-ray events were stored in
a minicomputer as two-dimensional data of TOF vs pulse
height, where digital gates were set to obtain the prompt
γ-ray spectrum as well as the background spectrum. As a
sample a natural carbon disk of 5.4 cm in diameter and
1.5 cm in thickness was positioned on the proton beam
axis. A ^{197}Au sample was used for normalization
of the absolute capture cross section of ^{12}C.

Several discrete γ-rays from the captured state of
^{13}C to lower excited states were observed. The intensity
of each γ-ray peak was extracted by a stripping method,
using the response function of the γ-ray detector.
Partial capture cross sections were derived by using
these γ-ray intensities from the captured state.
Finally, the total capture cross section is obtained as
16.8(21) μ-barn by adding these partial cross sections.
Corrections for the neutron self-shielding, the neutron
multiple scattering in the sample and the γ-ray
absorption in the sample and the dependence of the γ-ray
detection efficiency on the γ-ray source position were
made.

The present value is about five-times larger than
that estimated from a thermal neutron capture cross
section by using the 1/v law and favours the
nucleosynthesis of intermediate-mass nuclei in
inhomogeneous big-bang models. It should also affect the
discussion concerning the baryon number fluctuation
amplitude and the average baryon density during the
transition from the quark-gluon plasma to the hadronic
phase, since the isotopic abundance ratio ^{13}C/^{12}C
relative to solar has been suggested to determine these
important quantities, independent of the fraction of the
closure density contributed by baryons(3).

References:
(1) Applegate, J., Hogan, C., Scherrer, R. J. (1987)
    'Cosmological baryon diffusion and nucleosynthesis',
    Phys. Rev. D35, 1151-1160, Terasawa, N. Sato, K.
    (1989), Prog. Theor. Phys. 81, 254
(2) Igashira, M., Kitazawa, H., Yamamuro, N. (1986)
    Nucl. Instr. Meth. A245, 432-437
(3) Kajino, T., Mathews, G., J., Fuller., G. M.(1990)
    ' Primordial nucleosynthesis of intermediate-mass
    elements in baryon-number inhomogeneous big-bang
    models: obsevational tests' Ap. J. in press

* Present work has been carried out in collaboration with
  M. Igashira, K. Takeda, N. Mukai, S. Motoyama,
  F. Uesawa, H. Kitazawa and T. Fukuda

INHOMOGENEOUS UNIVERSES IN THE FRAMEWORK OF LATTICE QCD

M. HACKEL, M. FABER, H. MARKUM, H. OBERHUMMER
Institut für Kernphysik, Technische Universität Wien
A-1040 Vienna, Austria

ABSTRACT. There exist two calculations for the interface tension α of coexisting quark and hadron matter with two different methods. We perform an independent computation of α for time extension $N_t = 2$ and 4 and find that (i) the interface tension vanishes for our volume within error bars and (ii) there are systematic deviations between the differential and integral method.

QCD thermodynamics on space-time lattices has demonstrated that matter can exist in a confining hadron phase and a free quark-gluon plasma-phase separated by a first order phase transition implying that the two different phases can coexist at the critical temperature. This opens up the possibility of the creation of an inhomogeneous universe which can now be studied from the first principles of lattice QCD. The most important observable under consideration is the surface energy α between the quark-gluon plasma-state and the hadronic bubbles. The numerical value of α is an important quantity for the inhomogeneity of the universe and represents an input parameter for the probability of nucleation and the average distance between nucleation centers, and further effects the nucleosynthesis of light elements.

To extract α we consider the thermodynamics of a two-phase system with an interface on the lattice. We start with the the relation for the free energy $F(T,V,A) = -T \ln Z(T,V,A)$ as a function of the temperaure T, volume V and interface size A . The partition function $Z(T,V,A) = \int D[U] \exp(-S(U))$ is expressed by a path integral over the lattice action $S(U)$ being a sum over all plaquettes $P_{\mu\nu}$ defined as the gluonic gauge field $U_{\mu\nu}(n)$ at site n around an elementary unit square.

The evaluation of thermodynamical expressions demands to differentiate the partition function with respect to the temperature, volume and interface size. This can be realized on the lattice by directly summing over plaquettes at fixed couplings (differential method) or by integrating the sum of plaquettes over the coupling (integral method). Although the asymptotic value of α at the critical coupling seems to be the same for both methods the slopes turn out to be different [1-3]. In order to shed more light on this problem we perform an independent analysis for $N_t = 2$ and 4 relying on the differential method. There one can proceed straightforward to derive the thermodynamical observables of interest and give the formula for the surface energy α explicitly [1]

$$\alpha(n_z)\frac{A}{T} = \langle A\frac{\partial S(U)}{\partial A}\big|_{T,V}\rangle = \sum_{n_x,n_y,n_t} (\frac{1}{g^2} + \frac{1}{2}(c_t - c_s))(2P_{03} - P_{01} - P_{02} - 2P_{12} + P_{13} + P_{23}) \quad (1)$$

where g is the gluon coupling and c_s, c_t are the Karsch coefficients. To obtain the physical expectation values one has to take the limit of the two-phase system towards the critical point.

We approximated the path integral by 25000 Monte Carlo iterations around $\beta_c = 6/g_c^2$ for four different values of $\Delta\beta = (0.25,) 0.20, 0.15, 0.10, 0.05$ on an 8x8x16 hypercubic lattice with $N_t = 2$ (4) using periodic boundary conditions.

128

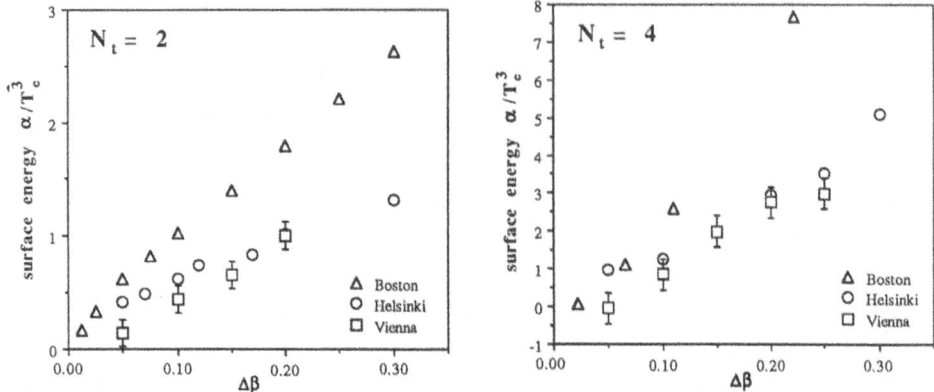

Figure 1. Compilation of all available results for the surface energy α/T_c^3 for an 8x8xN$_z$ lattice with $N_t = 2$ and $N_t = 4$ [1,2]. The error bars denote mean standard deviation with 100 configurations taken to be correlated.

To get the surface energy α we have to integrate equation (1) along the z axis. In figure 1 the surface energy normalized to physical units is compared with some other recent data [1,2]. Our computations yield a value of α compatible with zero for both time extension $N_t = 2, 4$ lattice units. Our data points agree within error bars with the Helsinki data [1] but are systematically lower (especially for $\Delta\beta = 0.05$). Since both analyses rely on the differential method the deviation is a consequence of the shorter z elongation of our system. Comparing our data with the Boston data [2] one clearly observes the different slopes. An explanation might be that the vacuum contributions are handled differently in the the integral and differential method [3]. Previous computations [1,2] for $N_t = 2$ have yielded $\alpha/T_c^3 = 0.24 \pm 0.06$ definitely non-zero whereas α/T_c^3 turned out to be compatible with zero for $N_t = 4$ both by Boston and Helsinki group although with different slopes. Because the difference between the space like and time like plaquettes decreases with increasing elongation in time the interface tension becomes more difficult to be extracted. A way out might be to incorporate the Polyakov loop into the action to stabilize the interface.

Let us draw the conclusions. There is a systematic difference between the differential and the integral method. It is rather certain that $\alpha/T_c^3 \approx 0.1$ is an upper bound for the interface tension which is important for astrophysics. At the moment we make great effort to study the effect of dynamical quarks to the chiral order parameter profile and interface tension.

[1] Kajantie, K., Kärkkäinen, L. and Rummukainen, K. (1990) 'Interface tension in QCD matter', Helsinki Preprint HU-TFT-89-29.
    Kajantie, K., Kärkkäinen, L. and Rummukainen, K., private communication.
[2] Huang, S., Potvin, J., Rebbi, C. and Sanielevici, S. (1989) 'Surface tension in finite temperature quantum chromodynamics', Boston Preprint BUHEP-89-34.
[3] Potvin, J. and Rebbi, C. (1990) 'The surface tension in QCD at finite temperature and study of the surface tension in spin models', Nucl.Phys.B(Proc.Suppl.)17, 223-229.
Supported by 'Fonds zur Förderung der wissenschaftlichen Forschung' under Contract No. P7510.

The Beginning Of The Universe

S. W. HAWKING
Department of Applied Mathematics and Theoretical Physics
University of Cambridge, Silver Str., UK

In this lecture, I would like to discuss whether the universe has a beginning in time, and whether it will have an end. All the evidence seems to indicate, that the universe has not existed for ever, but that it had a beginning about 15 billion years ago. This is probably the most remarkable discovery of modern cosmology. We are not yet certain whether the universe will have an end, but I can re-assure anyone who is nervous about their investments, that it is a bit early to sell: the end will not come for at least ten billion years. By that time, maybe trade relations between Japan and America, will be sorted out.

The time scale for movement of the position of the stars, is very long compared to a human life time. It was therefore not surprising that until recently, the universe was thought to be essentially static, and unchanging in time. On the other hand, it must have been obvious, that society is evolving in culture and technology. This indicates that the present phase of human history, can not have been going for more than a few thousand years. It was therefore natural to believe that the human race, and maybe the whole universe, had a beginning in the fairly recent past. However, many people were unhappy with the idea that the universe had a beginning, because it seemed to imply the existence of a supernatural being, who created the universe. They prefered to believe that the universe, and the human race, had existed for ever, but that there had been periodic floods, or other natural disasters, which repeatedly set back the human race to a primitive state.

This argument about whether or not the universe had a beginning, persisted into the 19th and 20th centuries. It was conducted mainly on the basis of theology and philosophy, with little consideration of observational evidence. This may have been reasonable, given the notoriously unreliable character of cosmological observations, until fairly recently. As Eddington once said: Don't worry if your theory doesn't agree with the observations, because they are probably wrong. But if your theory disagrees with the Second Law of Thermodynamics, it is in bad trouble. In fact, the theory that the universe has existed forever, is in serious difficulty with the Second Law of Thermodynamics. This states that entropy, or disorder, always increases with time. Like the argument about human progress, it suggests that there must have been a beginning. Otherwise, the universe would have reached thermal equilibrium by now, and everything would be at the same temperature. The night sky would have been as bright as the surface of the Sun, unless, for some reason, the stars did not shine before a certain time.

In a universe that was essentially static, there would not have been any dynamical reason, why the stars should have suddenly turned on, at some time. Any such "lighting up time" would have to be imposed by an intervention from outside the universe. The situation was different, however, when it was realized that the universe is not static, but expanding. Galaxies are moving steadily apart from each other. This means that they were closer together in the past. One can plot the separation of two galaxies, as a function of time. If there were no acceleration due to gravity, the graph would be a straight line. It would go down to zero separation, about twenty billion years ago. One would expect gravity, to cause the galaxies to accelerate towards each other. This will mean that the graph of the separation of two galaxies, will bend downwards, below the straight line. So the time of zero separation, would have been less than twenty billion years ago.

At this time, the Big Bang, all the matter in the universe, would have been on top of itself. The density would have been infinite. It would have been what is called, a singularity. At a singularity, all the laws of physics would have broken down. This means that the state of the universe, after the Big Bang, will not depend on anything that may have happened before the Big Bang. The universe will evolve from the Big Bang, completely independently of what it was like before. Even the amount of matter in the universe, can be different to what it was before the Big Bang, as the Law of Conservation of Matter, will break down at the Big Bang.

Since events before the Big Bang have no observational consequences, one may as well cut them out of the theory, and say that time began at the Big Bang. Events before the Big Bang, are simply not defined, because there's no way one could measure what happened at them. This kind of beginning to the universe, is very different to the beginnings that had been considered earlier. These had to be imposed on the universe by some external agency: there is no dynamical reason why the motion of bodies in the solar system, can not be extrapolated back in time, far beyond four thousand and four BC. This is the traditional date for the creation of the universe, according to the book of Genesis. By contrast, the Big Bang is a beginning, that is required by the dynamical laws that govern the universe. It is therefore intrinsic to the universe, and is not imposed on it from outside.

Although the laws of science seemed to predict the universe had a beginning, they also seemed to predict, that they could not determine how the universe would have begun. This was obviously very unsatisfactory. so there were a number of attempts to get round the conclusion, that there was a singularity of infinite density in the past. One suggestion was to modify the law of gravity, so that it became repulsive. This could lead to the graph of the separation between two galaxies, being an exponential curve, which didn't pass through zero, at any finite time in the past. Instead, the idea was that, as the galaxies moved apart, new galaxies were formed in between, from matter that was supposed to be continually created. This was the Steady State theory, proposed by Bondi, Gold, and Hoyle.

The Steady State theory, was what Karl Popper would call, a good scientific theory: it made definite predictions, which could be tested by observation, and possibly falsified. Unfortunately for the theory, they were falsified. The first trouble came with the Cambridge observations, of the number of radio sources of different strengths. On average, one would expect that the fainter sources, would also be the more distant. One would therefore expect them to be more numerous than bright sources, which would tend to be near to us. However, the graph of the number of radio sources, against there strength, went up much more sharply at low source strengths, than the Steady State theory predicted.

There were attempts to explain away this number count graph, by claiming that some of the faint radio sources, were within our own galaxy, and so did not tell us anything about cosmology. This argument didn't really stand up to further observations. But the final nail in the coffin of the Steady State theory, came with the discovery of the microwave background radiation, in 1965. This radiation is the same in all directions. It has the spectrum of radiation in thermal equilibrium at a temperature of 2 point 7 degrees above the Absolute Zero of temperature. There doesn't seem any way to explain this radiation in the Steady State theory. I believe that Hoyle still claims that it could be generated by iron needles, distributed throughout inter-galactic space, and heated by ultra violet light. However, the recent microwave background observations, by the Cosmic Background Explorer sattelite, show that it has such a perfectly thermal spectrum, that I think even Hoyle, will now abandon the Steady State theory.

Another attempt to avoid a beginning to time, was the suggestion, that maybe all the galaxies didn't meet up at a single point in the past. Although on average, the galaxies are moving apart from each other at a steady rate, they also have small additional velocities, relative to the uniform expansion. These so-called "peculiar velocities" of the galaxies, may be directed sideways to the main expansion. It was argued, that as you plotted the position of the galaxies back in time, the sideways peculiar velocities, would have meant that the galaxies didn't all meet up. Instead, there could have been a previous contracting phase of the universe, in which galaxies were moving towards each other. The sideways velocities could have meant that the galaxies didn't collide, but rushed past each other, and then started to move apart. There wouldn't have been any singularity of infinite density, or any breakdown of the laws of physics. Thus there would be no necessity for the universe, and time itself, to have a beginning. Indeed, one might suppose that the universe had oscillated, though that still wouldn't solve the problem with the Second Law of Thermodynamics: one would expect that the universe would become more disordered each oscillation. It is therefore difficult to see, how the universe could have been oscillating for an infinite time.

This possibility, that the galaxies would have missed each other, was supported by a paper by Lifshitz and Khalatnikov in 1963. They claimed that there would be no singularities, in a solution of the field equations of general relativity, that was fully general, in the sense that it didn't have any exact symmetry. However, their claim was proved wrong, by a number of theorems by Roger Penrose and myself. These showed that general relativity predicted singularities, whenever more than a certain amount of mass, was present in a region. The first theorems were designed to show that time came to an end, inside a black hole, formed by the collapse of a star. However, the expansion of the universe, is like the time reverse of the collapse of a star. I therefore want to show you, that observational evidence indicates the universe contains sufficient matter, that it is like the time reverse of a black hole.

In order to discuss observations in cosmology, it is helpful to draw a diagram of events in space and time, with time going upwards, and the space directions horizontal. To show this diagram properly, I would really need a four dimensional screen. However, this institution could manage to provide only a two dimensional screen. I shall therefore be able to show only one of the space directions.

As we look out at the universe, we are looking back in time, because light had to leave distant objects a long time ago, to reach us at the present time. This means that the events we observe, lie on a cone in the space-time diagram, called our past light cone. The point of the cone is at our position, at the present time. As one goes back in time on the diagram, the light cone spreads out to greater distances, and its area increases. However, if there is sufficient matter on our past light cone, it will bend the rays of light towards each other. This will mean that, as one goes back into the past, the area of our past light cone will reach a maximum, and then start to decrease. It is this focussing of our past light cone, by the gravitational effect of the matter in the universe, that is the signal that the universe is within its horizon, like the time reverse of a black hole. If one can determine that there is enough matter in the universe, to focus our past light cone, one can then apply the singularity theorems, to show that time must have a beginning.

How can we tell from the observations, whether there is enough matter on our past light cone, to focus it. We observe a number of galaxies, but we can not measure directly how much matter they contain. Nor can we be sure that every line of sight from us, will pass through a galaxy. So I will give a different argument, to show that the universe contains enough matter, to focus our past light cone. The spectrum of the microwave background radiation, is characteristic of radiation that has been in thermal equilibrium, with matter at the same temperature. To achieve such an equilibrium, it is necessary for the radiation to be scattered by matter many times. For example, the light that we receive from the Sun, has a characteristically thermal spectrum. This is not because the nuclear reactions that go on in the center of the Sun, produce radiation with a thermal spectrum. Rather, it is because the radiation has been scattered by the matter in the Sun, many times on its way from the center.

In the case of the universe, the fact that the microwave background has such an exactly thermal spectrum, indicates that it must have been scattered many times. The universe must therefore contain enough matter, to make it opaque in every direction we look, because the microwave background is the same, in every direction we look. Moreover, this opacity must occur a long way away from us, because we can see galaxies and quasars at great distances. The greatest opacity over a broad wave band, for a given density, comes from ionized hydrogen. It then follows, that if there is enough matter to make the universe opaque, there is also enough matter to focus our past light cone. One can then apply the theorem of Penrose and myself, to show that time must have a beginning, if the General Theory of Relativity is correct.

Please wait while I load the rest of my lecture.

The focussing of our past light cone, implied that time must have a beginning, if the General Theory of relativity is correct. But one might raise the question, of whether General Relativity is correct. It certainly agrees with all the observational tests, that have been carried out. However these test General Relativity, only over fairly large distances. We know that General Relativity can not be quite correct on very small distances, because it is a classical theory. This means, it doesn't take into account the Uncertainty Principle of Quantum Mechanics, which says that an object can not have both a well defined position, and a well defined speed: the more accurately one measures the position, the less accurately one can measure the speed, and vice versa. Therefore, to understand the very high density stage, when the universe was very small, one needs a quantum theory of gravity, that will combine General Relativity, with the Uncertainty Principle.

Many people hoped that quantum effects, would somehow smooth out the singularity of infinite density, and allow the universe to bounce, and continue back to a previous contracting phase. This would be rather like the earlier idea of galaxies missing each other, but the bounce would occur at a much higher density. However, I think that this is not what happens: quantum effects do not remove the singularity, and allow time to be continued back indefinitely. But it seems that quantum effects can remove the most objectionable feature, of singularities in classical General Relativity. This is that the classical theory, does not enable one to calculate what would come out of a singularity, because all the Laws of Physics would break down there. This would mean that science could not predict, how the universe would have begun. Instead, one would have to appeal to an agency outside the universe. This may be why the Catholic church, was so ready to accept the Big Bang, and the singularity theorems.

Quantum effects do not seem to remove the singularity, at the beginning of time. But quantum theory is based on complex numbers, in an essential way, in contrast to classical theory, which is based on real numbers. In particular, quantum theory introduces the idea of complex time. That is, time measured with complex numbers, instead of just the ordinary, real numbers.

One can draw a diagram, with the ordinary, real, direction of time horizontal, and the imaginary direction of time vertical. Points on this diagram, that are on the same horizontal line, represent events that are separated by real intervals of time. But points on a vertical line, are separated by an imaginary interval of time. Imaginary intervals of time, behave just like a fourth direction of space, at right angles to the three normal directions of space. So the three space directions, and imaginary time together, make up a spacetime that is like Euclidean space, in that all directions are on the same footing. On the other hand, the three space directions, and real time, make up a spacetime that is like Minkowski space, in which the time direction is different from the space directions.

In the case of real time, there are only two possible behaviors: either time continues back into the past indefinitely. Or time had a beginning, at a singularity. However, in the imaginary direction of time, there is a third possibility: because imaginary time behaves like another direction in space, it is possible for space and imaginary time, to form a spacetime that is finite in extent, but doesn't have a boundary or edge. It would be like the surface of the Earth, but with two more dimensions. The surface of the Earth is finite in extent, but it doesn't have any boundary or edge. I have been round the world, and I have not fallen off.

Jim Hartle and I, have suggested that space and imaginary time together, are indeed finite in extent, but without boundary. If this is the case, there wouldn't be any singularities in the imaginary time direction, at which the laws of physics would break down. And there wouldn't be any boundaries, to the imaginary time spacetime, just as there aren't any boundaries to the surface of the Earth. This absence of boundaries, means that the laws of physics, determine the state of the universe uniquely, in imaginary time. But if one knows the state of the universe in imaginary time, one can calculate the state of the universe in real time, by analytical continuation. One would still expect some sort of Big Bang singularity in real time. After all, an analytic function that is not constant, must have a singularity somewhere. So real time would still have a beginning. But one wouldn't have to appeal to something outside the universe, like God, to determine how the universe began. Instead, the way the universe started out at the Big Bang, would be determined by the state of the universe in imaginary time. Thus, the universe would be a completely self contained system. It would not require us to postulate the existence of anything, outside the physical universe that we observe.

The no boundary condition, is the statement that the laws of physics hold everywhere. Clearly, this is something that one would like to believe, but it is a hypothesis. One has to test it, by comparing the state of the universe that it would predict, with observations of what the universe is actually like. If the observations disagreed with the predictions of the no boundary hypothesis, we would have to conclude the hypothesis was false. There would have to be something outside the universe, to wind up the clockwork, and set the universe going. Of course, even if the observations do agree with the predictions, that does not prove that the no boundary proposal is correct. But one's confidence in it would be increased, particularly because there doesn't seem to be any other natural proposal, for the quantum state of the universe.

What does the no boundary proposal predict, for how the universe began. In a quantum theory, the universe doesn't have just a single history. Rather, it can be thought of as having every possible history. With each history, is associated a complex number, called the amplitude. This determines how probable that history is. It can be thought of as the size of a wave, and its phase, that is, its position in the cycle of crests and troffs. The amplitude of a history, depends on what happens in that history. For most histories, the phase of the amplitude varies very rapidly, if the history is changed slightly. The amplitudes of such histories, will be very nearly cancelled therefore, by the amplitudes of slightly different histories. It will be like adding two waves, with the crests of one wave, coinciding with the troffs of the other. The waves will cancel each other out, and leave nothing. However, there will be certain histories, for which the amplitude does not change, if the history is changed slightly. In these cases, the amplitudes of slightly different histories will all add up. It will be like adding two waves, with the crests of one coinciding with the crests of the other. The waves will re-inforce each other. This will mean that these histories, are by far the most probable.

These histories, for which the amplitude doesn't change when the history is changed slightly, are solutions of the field equations of General Relativity. But they are solutions in which both time and space, can be complex quantities, rather than just real quantities, as in classical General Relativity. Of course, the time and space that one observes, are always real. But the use of complex solutions of the field equations, enables one to calculate the probabilities, of measuring different real quantities.

The universe is the same at every point in space, and in every direction, to a first approximation. So I will consider closed Friedmann solutions, which contain a massive scalar field, phi. In order that the solutions satisfy the no boundary condition, they must close up in the imaginary time direction, like the surface of the Earth closes up, at the North pole. One can think of the size of the universe, as being like the length of an East West circle around the North pole, and the imaginary time direction, as being like the North South direction. On this analogy, the North pole of the Earth, would be like the beginning of the universe in imaginary time. The universe would start out with zero size, just as the length of an East West circle, is zero at the North pole. As imaginary time increases, the universe would expand in size, just as the length of an East West circle, increases when one moves South from the North pole. The universe would reach a maximum size in imaginary time, just as the length of an East West circle, reaches a maximum at the equator.

One can consider the behavior of the solutions, on a diagram in which real time is shown in the horizontal direction, and imaginary time in the vertical direction. One can choose to measure time from the point like the North pole, where the size of the universe is zero. One imposes the condition that this is a regular point of spacetime, just as the North pole is a regular point on the surface of the Earth. Then one can integrate the field equations, and determine the solution for all values of the complex time, given the value of the scalar field at the origin of time. In other words, there is a probable history for the universe, for each value of the scalar field at the north pole.

In the solutions, the size of the universe, will increase up the positive imaginary time axis, to a maximum value, and will then decrease to zero. This maximum value will be small, and will occur close to the origin, if the initial value of the scalar field is large. From the point on the imaginary axis where the size of the universe is a maximum, one can integrate the solution in the real direction of time. Work by one of my students, Glenn Lyons, has shown that the size of the universe, can remain almost real, as one follows the solution in this direction. For this to happen, the initial value of the scalar field, has to be chosen correctly. Of course, the size of the universe, and the value of the scalar field that one observes, are always real. But they are given by these solutions, to a good approximation. In a sense, one could say that the universe expanded from zero, to a certain size, in the imaginary direction of time, and then changed to expanding in real time. But this is just one way of looking at, what is really a sum over all complex histories.

One can't observe the absolute value of time, but only the time interval between events. Thus, along this horizontal line, from the maximum on the imaginary axis, time will appear real. The universe will expand at first in an inflationary, or exponential manner. This is like the behavior in the chaotic inflation model, proposed by Lindey. It would then go over to the normal hot Big Bang model.

In this simple model, the no boundary condition thus predicts inflation. However, one can also consider more general models, in which the universe is not exactly homogeneous and isotropic. If the departures from homogeneity are large enough, they will prevent the universe from inflating. However, preliminary calculations, indicate that the probabilities of very inhomogeneous universes, are low. Thus the no boundary proposal, predicts inflation with high probability. It also predicts the spectrum of small departures, that would be expected from an exact Friedmann model. This has almost the scale-free form, that is thought to be required for the standard model of galaxy formation. Thus the predictions of the no boundary proposal, seem to be consistent with observation. This does not prove that the no boundary hypothesis, is correct. But it gives one more confidence in it.

What does the no boundary proposal predict for the future of the universe? Because it requires that the universe is finite in space, as well as in imaginary time, it implies that the universe will recollapse eventually. However, it will not recollapse for a very long time, much longer than the 15 billion years it has already been expanding. So, there are one or two more immediate problems that we will have to deal with, before we need worry that the end of the universe is nigh. These solutions will not close up smoothly in imaginary time, like at the south pole. The reason is, that the solutions, do not represent the the whole history of the universe, but only the history up to a given size. A consequence of this, is that inhomogeneities in the universe, will continue to grow in the contracting phase. This means that the arrow of time, will not reverse in the contracting phase. A few years ago, I suggested that the Thermodynamic arrow of time, would point the other way, in the contracting phase. This would have meant that people living in the contracting phase, would have got younger, as the universe got smaller. However, I now realize I was wrong. People will continue to get older, even when the universe stops expanding, and starts to contract. So, I'm afraid it is no use waiting until the turn around in the universe, to return to your youth.

The conclusion is therefore, that the universe has not existed for ever. Rather, the universe, and time itself, had a beginning in the Big Bang, about 15 billion years ago. The beginning of real time, would have been a singularity, at which the laws of physics would have broken down. Nevertheless, the way the universe began, would have been determined by the laws of physics, if the universe satisfied the no boundary condition. This says that in the imaginary time direction, spacetime is finite in extent, but doesn't have any boundary or edge. The predictions of the no boundary proposal, seem to agree with observation. The no boundary hypothesis, also predicts that the universe will eventually collapse again. However, the contracting phase, will not have the opposite arrow of time, to the expanding phase. Because time is not going to go backwards, I think I better stop now. Thank you very much.

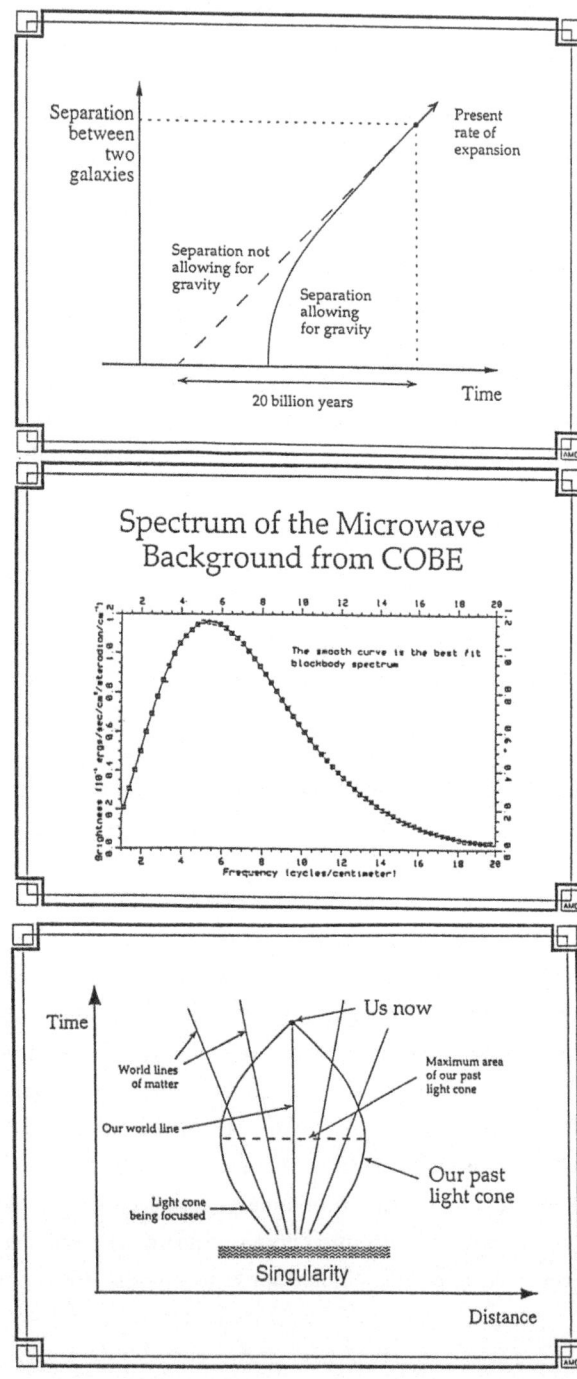

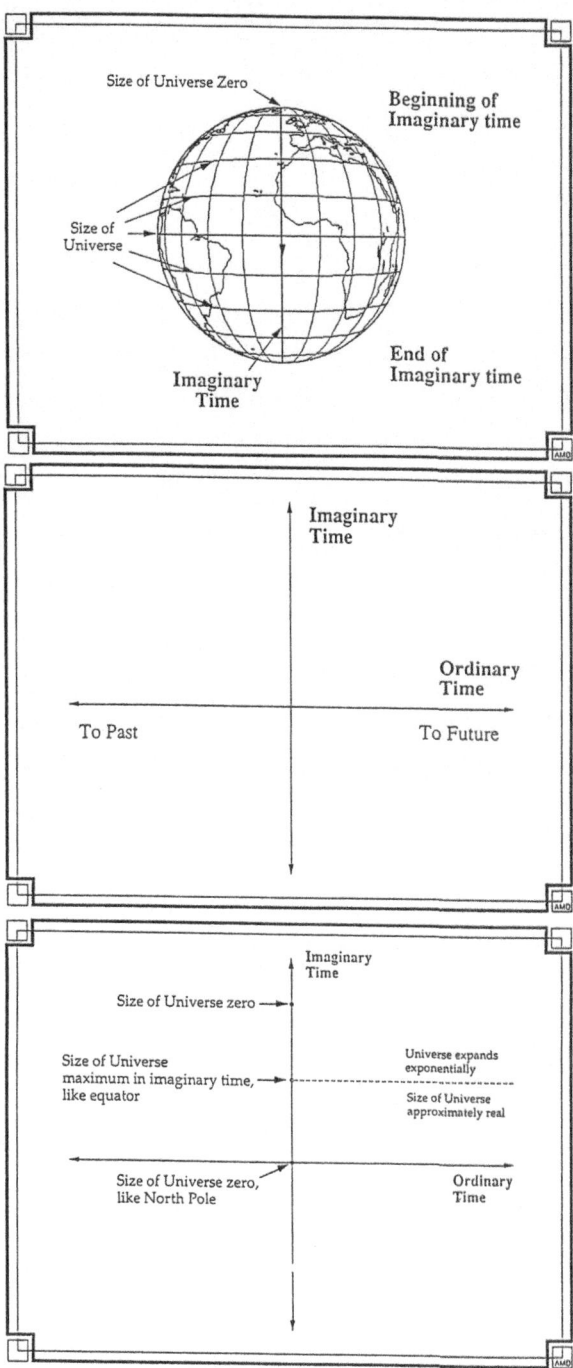

Size of Universe Zero

Beginning of
Imaginary time

Size of
Universe

Imaginary
Time

End of
Imaginary time

Imaginary
Time

Ordinary
Time

To Past

To Future

Imaginary
Time

Size of Universe zero →

Size of Universe
maximum in imaginary time, →
like equator

Universe expands
exponentially

Size of Universe
approximately real

Size of Universe zero,
like North Pole

Ordinary
Time

EXTENDED INFLATIONARY COSMOLOGY: A Primer

PAUL J. STEINHARDT
Department of Physics
University of Pennsylvania
Philadelphia, PA 19104

ABSTRACT. New approaches to inflationary cosmology have been developed which may ultimately solve the fine-tuning problem that plagues previous inflationary models. The distinctive feature is that the the inflationary phase transition is completed by bubble nucleation. The bubbles may lead to some interesting effects in the post-inflationary universe, including new seeds for galaxy formation and a unique gravitational wave signature.

1 Introduction

Inflationary cosmology, more than a decade old, is widely regarded as a simple and elegant solution to a number of cosmological problems that plague the standard big-bang picture.[1,2] Yet, it is also well known that all tenable models of inflation, whether one considers "new" inflation[3] or "chaotic" inflation[4] or variants[5] thereof, require a special choice of parameters and perhaps special initial conditions as well.[6] In most contexts, the special choice of parameters is obtained at the cost of an extreme and unnatural fine-tuning of microphysical coupling constants.

"Extended" inflation[7] is a qualitatively new approach to inflationary cosmology designed to redress the fine-tuning problem. Already, many interesting variants have been developed, including "generalized extended" inflation,[8] "hyperextended" inflation,[9] "double-field inflation,"[10] etc. At this point, it is debatable whether the fine-tuning problem is resolved or not; but, without question, extended inflation has greatly broadened the spectrum of tenable inflation models. Moreover, because extended inflation is qualitatively different from the conventional new inflation or chaotic inflation models, it leads to interesting, novel predictions for the post-inflationary universe.

Extended inflationary models harken back to Guth's original, "old" inflation concept.[2] Inflation is initiated when the universe is trapped in a false vacuum state by a large energy barrier during the course of a strongly first-order phase transition. The failure of old inflation, though, was that the universe could never escape the false vacuum state because the rate for tunneling through the barrier remains small compared to the inflationary expansion rate.[11] Extended inflation avoids the same failure by introducing mechanisms so that the tunneling rate eventually surpasses the expansion rate and, hence, the transition to the true vacuum can be completed.

The discovery of mechanisms for successfully completing first order phase transitions has two immediate and important implications. First, the cosmological ban on strongly first order phase transitions is lifted. Previously, the failure of old inflation was used to argue that there could never have been a strongly first-order phase transition since the big bang or else the universe would have been trapped forever in an inflationary false phase.[12] With the discovery of new mechanisms for successfully completing transitions by bubble nucleation, strongly first order phase transitions, whether inflationary or not, are permissible. Second, the fields driving the phase transition do not need to be weakly coupled as has been required by prior inflation models. Not only does this relieve the fine-tuning, but it eliminates any problems with reheating sufficiently after inflation.[6]

2 The Principles of Extended Inflation

To understand how extended inflation succeeds, it is useful to review how old inflation fails. Success or failure depends on the relation between the tunneling or bubble-nucleation rate, λ, to the expansion rate (Hubble parameter), H, which can be expressed in terms of a dimensionless bubble nucleation parameter, $\epsilon \equiv \lambda/H^4$. Guth and Weinberg[11] earlier showed that the false vacuum can be percolated by true vacuum bubbles only if ϵ exceeds a critical value, $\epsilon_{crit} \approx .02$. In old inflation, ϵ is time-independent since λ (which depends on the barrier shape) and H (which depends on the false vacuum energy) do not vary during inflation. Two dismal fates are possible: (a) $\epsilon < \epsilon_{crit}$, in which case the true vacuum bubbles never percolate and the universe inflates forever; or, (b) $\epsilon > \epsilon_{crit}$, in which case the true vacuum percolates, but so quickly that there is insufficient inflation to solve any cosmological problems.

One method of avoiding the dismal fates is to avoid bubble nucleation altogether. New inflation and chaotic inflation utilize this approach. For example, in new inflation, the energy barrier dissappears altogether as the universe supercools and the universe evolves slowly but continuously from the false to true vacuum phase.[3] However, experience over the past eight years has taught us that this approach requires special choices of parameters that make the models unattractive.[6,13]

Extended inflation models employ a different, almost obvious alternative approach: instead of allowing ϵ to remain time-independent, mechanisms are introduced so that ϵ is initially much less than ϵ_{crit} to achieve sufficient inflation, but then grows during inflation to a value $\epsilon > \epsilon_{crit}$ so that the phase transition can be completed. For typical (untuned) first order phase transitions, ϵ is exponentially small at the onset of inflation, $\mathcal{O}(e^{-40})$ or smaller. Hence, an increase in ϵ by 20 or more orders of magnitude is sought.

How can $\epsilon \equiv \lambda/H^4$ be made time-dependent? One possibility is that λ can be made to increase during the transition. For example, if σ, the order-parameter field for the transition, couples to another field, ϕ, which evolves during inflation in such a way that the energy barrier decreases during inflation (e.g., due to a coupling like $-\alpha\phi^2\sigma^2$ with increasing ϕ), λ dramatically increases during inflation. This concept has been attempted in "double-field inflation,"[10] which is closely related to an earlier scheme proposed by Linde and Kofman.[14] The disadvantage of this approach is that one has to introduce a special interaction and fine-tune the coupling, α, to achieve acceptable inflation.[10] Perhaps a more promising scheme is exemplified by "generalized extended" inflation,[8] in which ϕ rescales both the kinetic and potential energy for the σ field as it evolves, as might occur in particle physics models with scale invariant interactions. For appropriate rescalings, the net effect is to alter both λ and H during the phase transition, resulting in a net increase in ϵ. For simplicity, we will focus in this paper on models in which λ is constant, but H decreases during inflation, which is yet another approach for increasing ϵ.

Since $H = (8\pi G\rho_F/3)^{\frac{1}{2}}$ (where G is Newton's constant and ρ_F is the false vacuum energy density due to the σ field), decreasing H can be obtained by either decreasing G or decreasing ρ_F. In many cases, the two possibilities are equivalent under a Weyl transformation of the metric: a theory with constant G and decreasing ρ_F can be Weyl transformed into a theory with decreasing G and constant ρ_F.

To be more specific, consider the typical Lagrangian density for old inflation:

$$\mathcal{L}_{old} = \frac{1}{G}\mathcal{R} - \frac{1}{2}(\partial_\mu\sigma)^2 - V(\sigma) + \mathcal{L}_{other}, \tag{1}$$

where $\mathcal{R}$ is the scalar curvature, $V(\sigma)$ is the potential for the inflaton field σ which drives inflation (i.e., the false vacuum corresponds to $\sigma = 0$ or $V(\sigma = 0) = \rho_F$), and $\mathcal{L}_{other}$ is the contribution of all other matter fields. The analysis of old inflation concluded that a large energy barrier in $V(\sigma)$ separating $\sigma = 0$ from the true vacuum keeps the universe trapped forever in the false ($\sigma = 0$) phase.[11] The Lagrangian density may easily be modified so that G decreases:

$$\mathcal{L}_G = \frac{f(\phi)}{G}\mathcal{R} - \frac{1}{2}(\partial_\mu\phi)^2 - W(\phi) - \frac{1}{2}(\partial_\mu\sigma)^2 - V(\sigma) + \mathcal{L}_{other}, \tag{2}$$

where $f(\phi)$ and $W(\phi)$ are chosen so that $f(\phi)$ increases during inflation. Since the effective Newton's constant is $G_{eff} = G/f(\phi)$, increasing $f(\phi)$ results in decreasing G_{eff}, decreasing H, and increasing ϵ. Increasing the bubble nucleation parameter, ϵ, is what is desired to complete the inflationary phase transition. The equation of motion for ϕ is:

$$\ddot{\phi} + 3H\dot{\phi} + \frac{1}{2}\frac{(1+6f'')f'\dot{\phi}^2}{f+3f'^2} + \frac{fW'(\phi) - 2f'(W(\phi)+V(\sigma))}{f+3f'^2} = 0 \tag{3}$$

Typically, $W(\phi)$ is taken to be zero or at least negligible compared to ρ_F, and $V(\sigma) = \rho_F$ provides the driving force for ϕ during inflation. In some models, maintaining the condition on $W(\phi)$ as ϕ grows during inflation requires fine-tuning. However, the condition might be automatically enforced in models with scale-invariant interactions; for example, in supergravity and superstring models, the dilaton or moduli couple to $\mathcal{R}$ as desired, but $W(\phi)$ is expected to be exactly zero or at least negligibly small during inflation.[17]

There is great freedom in choosing $f(\phi)$, so long as it increases during inflation. In the first extended inflation model,[7] $f(\phi)$ was set equal to $\xi\phi^2$, resulting in a model that is mathematical equivalent to a Brans-Dicke theory[15] with Brans-Dicke parameter, $b = 1/8\xi$. For pure Brans-Dicke theories in which ϕ continues to evolve to the present epoch, radar-ranging experiments[16] imply $b > 500$ or $\xi < .00025$. However, pure Brans-Dicke theory is not natural from a particle physics point-of-view. For more plausible theories, such as scale-invariant models in which a dilaton acts as the Brans-Dicke field, $W(\phi)$ is expected to become non-zero when supersymmetry and/or conformal invariance is spontaneously broken at low energies — the dilaton develops a small mass.[17] In this case, $W(\phi)$ would pin the value of ϕ, but only after the symmetry is broken at energies which might be small compared to the inflation scale. After pinning, the theory reduces to conventional Einstein gravity in the late history of the universe and, hence, the radar-ranging experiment place no constraint on ξ. Work on "hyperextended" inflation models went further to show that virtually any $f(\phi)$ that increases monotonically with ϕ during inflation can be employed.[9] The more general $f(\phi)$ have technical advantages[9] some of which will be mentioned in later sections.

A Weyl transformation of the metric, $g_{\mu\nu} \to f^{-1}(\phi)g_{\mu\nu}$ converts a theory with decreasing G to conventional Einstein gravity with decreasing ρ_F:[18]

$$\mathcal{L}_G \to \mathcal{L}_{\rho_F} = \frac{1}{G}\mathcal{R} - \frac{1}{2f}(\partial_\mu\phi)^2 - \frac{1}{f^2}W(\phi) - \frac{1}{2f}(\partial_\mu\sigma)^2 - \frac{1}{f^2}V(\sigma) + \mathcal{L}'_{other}, \quad (4)$$

where $f(\phi)$ appears linearly (quadratically) in the kinetic (potential) terms of $\mathcal{L}'_{other}$. In this conformal frame, one might be tempted to conclude that the gravitational sector is of the conventional Einstein form, but the vacuum energy density, $V(\sigma) = \rho_F$, is rescaled by a time-varying factor, $1/f^2(\phi)$, during inflation. In reality, the theory is equivalent to the original; part of the gravitational interaction is now stored in the ϕ field which couples in a special way to all matter (in order for the equivalence principle to be obeyed). A truly different model can be obtained, though, by removing the coupling of ϕ to other fields, or altering the relative couplings to the kinetic and potential terms. This new theory would still allow the inflationary phase transition to be completed, but would not require a change in the gravitational interaction (and, hence, would not be constrained by experimental limits on varying G).

3 Constraints on Extended Inflation

Achieving sufficient inflation and completing the phase transition are achieved in the models outlined above for a very wide range of parameters. Three additional constraints must be satisfied to obtain a successful model:

3.1 SMALL AMPLITUDE DENSITY PERTURBATIONS

Obtaining small amplitude, adiabatic fluctuations represents a severe constraint in new and chaotic inflation models,[6] but is a qualitatively different kind of constraint in extended inflation. In all inflationary models, a nearly scale-invariant spectrum of gaussian, adiabatic fluctuations is generated due to de Sitter fluctuations in the fields evolving continuously during inflation. In new and chaotic inflation, the density fluctuation amplitude can be shown to be:

$$\frac{\delta\rho}{\rho} \sim \frac{H^2}{\dot{\sigma}}, \tag{5}$$

where σ is the scalar field which slowly rolls during the phase transition and the expression is to be evaluated during the last 60 e-foldings of inflation when fluctuations on observable distance scales were generated.[19] The amplitude is a ratio of two quantities that depend upon the same field, σ: H^2 depends upon the false vacuum energy density of the σ field and $\dot{\sigma}$ is the velocity of the same field during inflation. Hence, it is no surprise that the dimensionful parameters cancel from the ratio and the amplitude reduces to a ratio of dimensionless parameters. The isotropy of the microwave background constrains $\delta\rho/\rho < 10^{-5}$, which imposes a constraint that the dimensionless parameters be unnaturally small; for example, the quartic self-coupling of the σ field is constrained to be less than 10^{-12}.

In extended inflation, ϕ, rather than σ, evolves during inflation and determines when inflation ends. Hence, one can show that[7,20,21,22]

$$\frac{\delta\rho}{\rho} = \frac{H^2}{\dot{\phi}}. \tag{6}$$

In this case, H^2 depends upon the σ field vacuum energy, as before, but the denominator depends on ϕ, a completely independent field characterized by a different mass scale: $\phi \sim \mathcal{O}(M_{Planck})$. Hence, $\delta\rho/\rho$ reduces to a ratio of dimensionful quantities, $\delta\rho/\rho \propto M^2/M_{Planck}^2$, where M is the energy scale for the phase transition. By choosing $M < 10^{16}$ GeV or so, small amplitude fluctuations can be obtained without fine-tuning of dimensionless couplings. This relieves the usual fine-tuning problem; and, since the field can have strong couplings, reheating to high temperature after inflation can be easily achieved.

One might object that the tuning of dimensionless parameters has merely been replaced by a hierarchy of dimensionful mass scales. Indeed, at this stage, there is no rigorous argument to show which approach is superior. However, it can be said that we know that mass hierarchies exist in nature (and one of the goals of unified theories is to generate mass hierarchies from a fundamental theory with a single scale). By contrast, we have no evidence for the exponentially small dimensionless parameters required by new and chaotic inflation. It should also be emphasized that new and chaotic inflation require exponentially small dimensionless parameters, but not too small or there is a problem with reheating after inflation. In extended inflation, which depends on a mass hierarchy rather than small dimensionless coefficients, the reheating problem after inflation is avoided. Any small parameters in extended inflation (e.g., in $W(\phi)$) can be arbitrarily small (even zero) without adversely affecting the scenario.

3.2 ACCEPTABLE BUBBLE DISTRIBUTION

A key constraint in extended inflation that has no analogue in new or chaotic inflation models is that there be an acceptable bubble-size distribution.[20,23,24] Initially, almost all the energy in the bubbles is contained in the bubble-walls. When the walls of different bubbles collide, the coherent energy in the bubble walls is converted into radiation and matter. The radiation pressure drives the relativistic particles to stream inwards and homogenize the energy over the bubble. However, for very large bubbles, there has not been sufficient time since inflation for the radiation to reach the interior. These bubbles represent large amplitude ($\delta\rho/\rho \sim 1$) fluctuations with wavelength of order the bubble radius. One must ensure that the fractional volume occupied by bubbles of order the horizon size at decoupling, $F(R > R_D)$, not exceed 10^{-5} in order to avoid unacceptable distortions of the microwave background. Most bubbles are of order the horizon scale at the end of inflation, H_{end}^{-1}, where $H_{end}R_D \approx 10^{25}$ — the "dangerous," large bubbles lie at the far tail of the bubble distribution. Nevertheless, they provide a non-trivial constraint on inflationary models.

In old inflation, the bubble nucleation parameter, $\epsilon = \lambda/H^4$ is time-independent, so that the number of bubbles produced per Hubble volume is the same each Hubble time. Since the bubble size depends on the age of the bubble, the bubble-size distribution is scale-invariant: there are as many older bubbles of bigger radius as there are younger bubbles of smaller radius. In computing the fractional volume, $F(R > R_D)$, the bigger bubbles carry greater weight and $F(R > R_D)$ is divergent! As Guth and Weinberg concluded, the bubbles produced in old inflation produce an unacceptably inhomogeneous universe.[11]

In the original extended inflation model with non-minimal coupling, $f(\phi) = \xi\phi^2$, $H \propto 1/t$ or $\epsilon \propto t^4$. Hence, with increasing ϵ, more younger/smaller bubbles are produced compared to older/bigger bubbles. One can show that $F(R > R_D)$ is no longer divergent:[20,21,24] $F(R > R_D) = [H_{end}R_D]^{-32\xi}$. In order to satisfy $F(R > R_D) < 10^{-5}$, it is necessary that $\xi > .005$. For a pure Brans-Dicke theory, this means that the Brans-Dicke parameter ($= 1/8\xi$) must be less than 25, which conflicts with well known radar-ranging limits. However, as emphasized above, the radar-ranging constraints only apply for pure Brans-Dicke theory in which ϕ continues to evolve as a free field through the present epoch. If ϕ has a small potential, $W(\phi)$, which locks it to a constant value sometime after inflation, the radar-ranging limits are irrelevant and $\xi > .005$ is acceptable (and even a natural range).

In hyperextended inflation[9] and some generalized extended inflation[8] models, ϵ increases exponentially fast near the end of inflation. The ratio of younger/smaller bubbles to older/bigger bubbles is greater still. In this case, one finds $F(R > R_D) \approx$

$\mathcal{O}(1)/[\log{(H_{end}R_D)}]^3$, which generally satisfies the 10^{-5} bound without any special adjustment of parameters. Note, however, that the distribution is relatively flat so that there may be a non-negligible number of bubbles extending up to moderate scales (few orders of magnitude smaller than R_D).

3.3 $\dot{G}$ Approaches Zero by Nucleosynthesis

Another new constraint encountered in extended inflation is that the effective gravitational constant must not vary significantly after the nucleosynthesis epoch. To achieve the proper primordial abundances of helium, deuterium and lithium, it is necessary that G equal its present value to within 20 to 40 per cent, assuming three light neutrino species and a primordial helium abundance near 24 per cent.[26,27]

There are three basic ways of satisfying the $\dot{G}$ constraint. First, if ϕ has a small mass, m_ϕ, compared to the mass scale of the inflationary phase transition (i.e., in Eq. 3, $W(\phi)$ is non-zero and has a minimum), ϕ will oscillate around a fixed value beginning when the Hubble parameter H becomes of order m_ϕ, perhaps long after inflation ends. The model for the ϕ is basically a standard type of induced gravity model. The oscillations are damped by the Hubble expansion and ϕ settles at a fixed value, thereby fixing G_{eff}. The nucleosynthesis bound implies a constraint on the oscillation amplitude which appears to be easy to satisfy. Second, even if $W(\phi) = 0$, ϕ can driven to a constant value if $f(\phi)$ reaches a local maximum.[9] During inflation, we assumed that $f(\phi)$ is a monotonically increasing function of ϕ. If $f(\phi)$ has a maximum, though, and if that maximum is reached sometime after inflation ends, ϕ remains fixed at that value according to Eq. 3. The result is a somewhat novel type of induced gravity model.[9] As in the first approach, ϕ will oscillate around the fixed point and the nucleosynthesis bound results in a similar, mild constraint on the amplitude as before. A third approach is a model in which G never changes. As mentioned at the end of Sec. 2, it is conceivable for ϕ to couple like a Brans-Dicke field to σ, but not to other matter fields. In this case, ϕ affects the expansion rate during inflation when σ dominates the energy density, but ϕ does not affect G at all, so the nucleosynthesis bound is trivially satisfied.

4 New Features of the Post-Inflationary Universe

The qualitative differences between extended inflation and prior inflation models lead to some interesting, potentially different predictions for the post-inflationary universe:

4.1 New, Non-Gaussian, Isocurvature Component of Density Fluctuation Spectrum

The bubbles produced by the tunneling σ field provide a new component of density perturbation to be added to the gaussian spectrum of adiabatic fluctuations associated with the evolving (Brans-Dicke like) ϕ field. In the previous section, we discussed the constraints on the bubble-size distribution needed to ensure that the microwave background is not unacceptably distorted. Assuming that the constraints are satisfied, there will be no bubbles of radius $R_D \sim 1000$ Mpc or bigger within the observable universe. However, the spectrum of bubbles may still extend as high as 1-10 Mpc, and even much smaller bubbles may leave interesting traces. As noted previously, almost all of the energy gained in tunneling from the true to the false vacuum phase is contained within the bubble wall until the walls collide. At that point, the energy is converted to thermal radiation, massive particles, and black holes. The thermal pressure drives the radiation inward to fill the void in the bubble interior. Massive particles and objects, though, are left behind. The consequence is an isocurvature perturbation with wavelength of order the bubble radius with the topology of massive particles lying on a soap-bubble type network. Summing over the bubbles, one obtains a contribution to the density fluctuation spectrum that is **non-gaussian** (set by the bubble-wall distribution), **non-adiabatic** (isocurvature), and **non-scale invariant** (since the bubble distribution dies off at large wavelengths) contribution to the density perturbations — violating every conventional inflationary prediction! While isocurvature perturbations have been discussed in the context of new and chaotic inflation in the past,[25] they always required extraordinary complex and finely-tuned potentials. In extended inflation, the isocurvature contribution from bubbles is a generic feature. A key, open issue is to determine the evolution of this contribution to the density fluctuations and its ultimate effect on galaxy formation.

4.2 New Source of Gravitational Radiation

Bubble nucleation will produce a distinctive source of gravitational radiation.[28] Crudely, a fraction of the thermal radiation produced in bubble collisions will be in the form of gravitational radiation with wavelength of order the horizon size at the end of inflation redshifted to the present epoch. The ratio of gravitational radiation to photons will remain nearly constant until the present epoch, but the wavelength will be redshifted. Assuming that the critical temperature of the phase transition is of order 10^{14} GeV or so ($H \approx 10^9$ GeV), produces a sharp peak in the gravitational wave spectrum in the 10 m range with amplitude roughly five orders of magnitude higher than the spectrum from conventional inflationary models. The amplitude and wavelength are in a range that may be detectable in forthcoming experiments.

4.3 Incorporating Baryosynthesis, Cosmic String and Global Texture Scenarios

In conventional inflation models, the couplings of the inflaton field are fine-tuned to be so small that the reheating temperature is 10^9 GeV or less. The reheat temperature is so low that one would normally conclude that it is problematic to proceed with grand unification scale baryosynthesis, or to generate cosmic string or global texture defects that have been suggested as possible contributors to large-scale galaxy structure. In the past, additional fine-tuning is assumed so that the massive defects can be generated at such a low reheat temperature, and new methods for generating baryosynthesis are invoked. I am told[29] that Zeldovich once remarked that "if you buy inflation, sell cosmic strings" for this very reason. However, with extended inflation, it is much easier to generate massive defects.[30] The most natural way is to choose the inflationary phase transition to be the same one that produces the desired defect. Since the transition is completed by bubble nucleation, the usual Kibble mechanism[31] can be invoked to produce the defects. (In new or chaotic inflation, the observable universe lies in one coherent region or one bubble, so no defects are generated. In extended inflation, the observable universe is the union of many bubbles, so many defects lie within the observable universe.) Hence, it is plausible to buy both inflation and cosmic strings (or global textures).

Caveat Emptor: It is required that, if any defects are produced, they be defects which are useful or at least innocuous for big bang cosmology (like strings and global texture) and not the defects that are harmful (like massive domain walls or monopoles). Since most particle models predict a series of phase transitions, most of which produce harmless or useful defects, the qualitative condition does not appear to be difficult to satisfy.

4.4   OSCILLATIONS IN G:

It is unlikely that G rests at its present value just as inflation ends; if G is ultimately fixed by some potential, $W(\phi)$, or by an extremum of $f(\phi)$ (see discussion in Sec. 2), most likely ϕ lies near but not at the minimum as inflation ends. At reheating, the forces driving ϕ change radically. During inflation, the leading driving force in Eq. 3 is the false vacuum energy, $V(\sigma = 0)$. At reheating, $V(\sigma) \to 0$; the remaining $f(\phi)$ contribution is proportional to $\mathcal{R}$, which is also zero during a radiation dominated epoch. The only non-zero contribution during the radiation-dominated era is $W(\phi)$, which was required to be scaled to some mass scale much smaller than the inflationary scale. $\mathcal{R}$ becomes non-zero when the universe re-enters the dominated epoch, at which time the $f(\phi)$ contributions may become important. It is conceivable that these terms are only important late in the history of the universe, during nucleosynthesis or even much later when the universe re-enters the matter-dominated epoch near decoupling. G would begin to oscillate about the present value beginning at the time these terms become important. The effect at nucleosynthesis would be that G probably differs from its present value. At present, a deviation of G by 20-40 percent is consistent with nucleosynthesis computations provided there are three light neutrino species and the helium abundance is near 24 per cent, which represents an additional constraint on model-building. However, if the helium abundance is revised to a value of 23 per cent or lower, the same computations will be inconsistent with three light neutrino species. A possible alternative is to suppose that G was 20 per cent or more lower than its present value during nucleosynthesis.[26,27] It is also conceivable that G is does not begin oscillating until close to the present epoch and that it continues to oscillate today. In this case, there will be an induced oscillation in the Hubble constant resulting in an "optical illusion"[32,33] in the galaxy redshift distribution;

there will appear to be a periodically spaced spherical shells of high galaxy density as a function of redshift surrounding any observer. This effect might explain the recently reported periodicity[34] in pencil-beams surveys of the galaxy redshift-distribution by Broadhurst, et al. Both the nucleosynthesis and periodicity effects appear to dovetail with time-varying G suggested by the extended inflation scenario.

Hence, extended inflation may not only provide a more natural approach to inflationary cosmology, but may also produce interesting new effects in the post-inflationary universe. Some of these new effects may cause us to revise the traditionally stated inflationary predictions and/or may broaden the role that inflation played in the evolution of the universe.

I would like to express my thanks to Prof. K. Sato and Dr. J. Yokoyama for their gracious support and hospitality. In preparing this review, I have benefitted greatly from conversations with E. Kolb, D. Salopek and M. Turner, and from collaborations with D. La, E. Bertschinger and F. Accetta. This work was supported in part by U.S. DOE Grant No. DOE-EY-76-C-02-3071.

References

[1] For a recent review of conventional inflation models, see A. D. Linde, **Particle Physics and Inflationary Cosmology**, (Gordon and Breach, New York, 1990).

[2] A. H. Guth, *Phys. Rev.* D23, 347 (1981).

[3] A. D. Linde, *Phys. Lett.* **108B**, 389 (1982); A. Albrecht and P. J. Steinhardt, *Phys. Rev. Lett.* **48**, 1220 (1982).

[4] A. D. Linde, *Phys. Lett.* **129B**, 177 (1983).

[5] Examples of variants include inflation driven by quantum cosmology, J. B. Hartle, and S.W. Hawking, *Phys. Rev.* D28, 2960, (1983); or power-law inflation, L. F. Abbott and M. B. Wise, *Nucl. Phys.* B244, 541 (1984), and F. Lucchin and S. Mataresse, *Phys. Rev.* D32, 1316 (1985); or induced-gravity inflation, F. S. Accetta, D. J. Zoller, and M. S. Turner, *Phys. Rev.* D31, 3046 (1985), and F. Lucchin, S. Matarese, and M. D. Pollock, *Phys. Lett.* **167B**, 163 (1986).

[6] P. J. Steinhardt and M. S. Turner, *Phys. Rev.* D29, 2162 (1984).

[7] D. La and P. J. Steinhardt, *Phys. Rev. Lett.* **62**, 376 (1989).

[8] R. Holman, E. W. Kolb, S. L. Vadas, and Y. Wang, Fermilab Preprint FNAL-PUB-90/147-A (1990).

[9] P. J. Steinhardt and F. S. Accetta, *Phys. Rev. Lett.* **64**, 2740 (1990); see also J. D. Barrow and K. Maeda, Preprint WU-AP/04/89; and J. Garcia-Bellido and M. Quiros, CERN preprint TH.5674/90 (1990).

[10] F. C. Adams and K. Freese, MIT Preprint (1990).

[11] A. H. Guth and E. J. Weinberg, *Nucl. Phys.* **B212**, 321, (1983).

[12] A. D. Linde, *Phys. Lett.* **B70**, 306 (1977); A. H. Guth and E. J. Weinberg, *Phys. Rev. Lett.* **45**, 1131 (1980); P. J. Steinhardt, *Nucl. Phys.* **B179**, 492 (1981).

[13] F. C. Adams, K. Freese and A. H. Guth, MIT Preprint (1990).

[14] L. A. Kofman and A. D. Linde, *Nucl. Phys.* **B282**, 555 (1987); E. T. Vishniac, K. Olive and D. Seckel, *Nucl. Phys.* **289**, 717 (1987); and L. A. Kofman, A. D. Linde, and J. Einasto, *Nature* **326**, 48 (1987).

[15] C. Brans and R. H. Dicke, *Phys. Rev.* **24**, 924 (1961).

[16] R. D. Reasenberg, et al., *Astrophys. J.* **234**, L219 (1979).

[17] See, for example, M. B. Green, J. H. Schwarz, and E. Witten, **Superstring Theory: 2** (Cambirdge Univ. Press, Cambridge, 1987), pp. 326-330, 403-404.

[18] B. Whitt, *Phys. Lett.* **145B**, 176 (1984); K. Maeda, *Phys. Rev.* **D39**, 3159 (1989).

[19] J. Bardeen, P. J. Steinhardt and M. S. Turner, *Phys. Rev.* **D28**, 679 (1983); A. H. Guth and S.-Y. Pi, *Phys. Rev. Lett.* **49**, 1110 (1982); A. A. Starobinskii, *Phys. Lett.* **B117**, 175 (1982); S. W. Hawking, *Phys. Lett.* **B115**, 295 (1982).

[20] D. La, P. J. Steinhardt, and E. W. Bertschinger, *Phys. Lett.* **B231**, 231 (1989).

[21] E. W. Kolb, D. S. Salopek and M. S. Turner, Fermilab Preprint, FNAL-PUB-90/116-A (1990).

[22] J.-C. Hwang, U. of Texas Preprint (1990).

[23] D. La and P. J. Steinhardt, *Phys. Lett.* **B220**, 375, (1989).

[24] E. J. Weinberg, *Phys. Rev.* **D40**, 3950 (1989).

[25] See, for example, D. S. Salopek, J. R. Bond, and J. M. Bardeen, *Phys. Rev. D* **40**, 1953 (1989); and A. A. Starobinskii, *Pis'ma Zh. Eksp. Teor. Fiz.* **42**, 399 (1989) [JETP Lett. **42**, 152 (1985)].

[26] See, for example, discussions by J. Audouze and G. Steigman in these proceedings.

[27] F. S. Accetta, L. M. Krauss and P. Romanelli, Yale preprint YCTP-P1-90, to appear, *Phys. Lett.*, (1990).

[28] M. S. Turner and F. Wilczek, Fermilab Preprint (1990).

[29] E. W. Kolb, private communication.

[30] J. D. Barrow, E. J. Copeland, E. W. Kolb, and A. R. Liddle, FNAL-PUB-90/98-A (1990); E. J. Copeland, E. W. Kolb, and A. R. Liddle, FNAL-PUB-90/56-A (1990).

[31] T. J. Kibble, *J. Phys.* *A***9**, 1387 (1976).

[32] M. Morikawa, Univ. of British Columbia preprints 90-0208 and 90-0380 (1990). See also contribution in this proceedings.

[33] C. T. Hill, P. J. Steinhardt and M. S. Turner, Fermilab Preprint, FNAL-PUB-90/129-T (1990).

[34] T. J. Broadhurst, R. S. Ellis, D. C. Koo, and A. S. Szalay, *Nature* **343**, 726 (1990).

THE INFLATON SECTOR OF EXTENDED INFLATION

Edward W. Kolb

NASA/Fermilab Astrophysics Group
Fermi National Accelerator Laboratory
P.O. Box 500, Batavia, IL   60510   USA
     and
Department of Astronomy and Astrophysics & Enrico Fermi Institute
The University of Chicago
5640 South Ellis Ave., Chicago, IL   60637   USA

ABSTRACT. In extended inflation the inflationary era is brought to a close by the process of percolation of true vacuum bubbles produced in a first-order phase transition. In this paper I discuss several effects that might obtain if the Universe undergoes an inflationary first-order phase transition.

1. Baryogenesis [1]

One of the most important results in particle astrophysics is the development of a framework that provides a dynamical mechanism for the generation of the baryon asymmetry. The baryon number density is defined as the number density of baryons, minus the number density of antibaryons: $n_B \equiv n_b - n_{\bar{b}}$. Today, $n_B = n_b = 1.13 \times 10^{-5}(\Omega_B h^2)$ cm^{-3}. Of course, the baryon number density changes with expansion, so it is most useful to define a quantity B, called the *baryon number of the universe*, which is the ratio of the baryon number density to the entropy density s. Assuming three species of light neutrinos, the present entropy density is $s = 2970$ cm^{-3}, and the baryon number is $B = 3.81 \times 10^{-9}(\Omega_B h^2)$. Primordial nucleosynthesis provides the constraint $0.010 \leq \Omega_B h^2 \leq 0.017$,[2] which implies $B = (3.81 \text{ to } 6.48) \times 10^{-11}$.

A key feature of inflation is the creation of a large amount of entropy in a volume that was at one point in causal contact. The creation of entropy in inflation would dilute any pre-existing baryon asymmetry, so it is necessary to create the asymmetry after, or very near the end of, inflation. In order for the baryon number to arise after inflation in the usual picture, it is necessary for three criteria to be satisfied: baryon number (B) violating reactions must occur, C and CP invariance must be broken, and non-equilibrium conditions must obtain. There are two standard scenarios for baryogenesis:[3] In the first picture the baryon asymmetry is produced by the "out of equilibrium" B, C, and CP violating decays of some massive particle, while the second scenario involves the evaporation of black holes.

In the out of equilibrium decay scenario, the most likely candidate for the decaying particle is a massive boson that arises in Grand Unified Theories (GUTs). In the simplest models, the degree of C and CP violation is larger for Higgs scalars than for the gauge vector bosons, so we will assume that the relevant boson is a massive Higgs particle. This Higgs is also taken to be different from the inflaton. The Higgs of GUTs naturally violate B. The origin of the C and CP violation necessary for baryogenesis is uncertain. It is practical simply to parameterize the degree of C and CP violation in the decay of the particle. To illustrate such a parameterization, imagine that some Higgs scalar H has two possible decay channels, to final states f_1, with baryon number B_1, and f_2, with baryon number B_2. Consider the initial condition of an equal number of H and its antiparticle, $\bar{H}$. The H's decay to final states f_1 and f_2 with decay widths $\Gamma(H \to f_1)$ and $\Gamma(H \to f_2)$, while the $\bar{H}$'s decay to final states $\bar{f}_1$ and $\bar{f}_2$ with decay widths $\Gamma(\bar{H} \to \bar{f}_1)$ and $\Gamma(\bar{H} \to \bar{f}_2)$. The decays produce a net baryon asymmetry per H–$\bar{H}$ given by

$$\epsilon \equiv \sum_{i=1,2} B_i \frac{\Gamma(H \to f_i) - \Gamma(\bar{H} \to \bar{f}_i)}{\Gamma_H}, \tag{1}$$

where Γ_H is the total decay width. Of course ϵ can be calculated if one knows the masses and couplings of the relevant particles. Reasonable upper bounds for ϵ are in the range of 10^{-2} to 10^{-3}, but it could be much smaller. For more details, the reader is referred to Ref. (3).

The non-equilibrium condition is most easily realized if the particle interacts weakly enough so that by the time it decays when the age of the Universe is equal to its lifetime, the particle is nonrelativistic. Then the decay products will be rapidly thermalized, and the "back reactions" that would destroy the baryon asymmetry produced in the decay will be suppressed.

In most successful models of new inflation the reheat temperature is constrained to be rather low. This is due to the fact that new inflation requires flat scalar potentials in order for inflation to occur during the "slow roll" of the scalar field toward its minimum. In order to maintain the flatness of the potential, the inflaton field must be very weakly coupled to all fields so that one-loop corrections to the scalar potential do not interfere with the desired flatness of the potential. The feeble coupling of the inflaton to other fields means that the process of converting the energy stored in the scalar field to radiation ("re"heating) is inherently inefficient. Although it is possible to overcome this difficulty in several ways, it remains a concern for new inflation.

The thermalization process of bubble wall collision at the end of extended inflation provides a natural arena for baryogenesis in the early Universe, as it automatically creates conditions far from thermal equilibrium, exactly as required for B, C, and CP violating GUT processes to produce an asymmetry.

Our only assumption about first-order inflation is that the parameter that determines the efficiency of bubble nucleation, $\epsilon(t) = \Gamma(t)/H^4(t)$, where Γ is the nucleation rate per volume and H is the expansion rate of the Universe, has a time dependence that suppresses bubble nucleation early in inflation, then rapidly increases so inflation is brought to a successful conclusion in a burst of bubble nucleation.

In order to keep the discussion as general as possible, consider the salient features of the potential in terms of a few parameters that can be easily identified with any scalar potential that undergoes spontaneous symmetry breaking. The parameters of the potential are assumed to be: 1.) σ_0, the energy scale for SSB, i.e., the VEV of the scalar field. 2.) λ, a dimensionless coupling constant of the inflaton potential. We will assume that the potential is proportional to λ. 3.) ξ, a dimensionless number that measures the difference between the false and the true vacuum energy density via $\rho_V = \xi \lambda \sigma_0^4$. ξ must be less than unity for sufficient inflation to occur.

From these few parameters it is possible to find all the information required about the bubbles formed in the phase transition. For instance, an important parameter is the size of bubbles nucleated in the tunnelling to the true vacuum. In the thin-wall approximation, the size of a nucleated bubble is given by $R_C \sim 3(\xi\lambda^{1/2}\sigma_0)^{-1}$. Bubbles smaller than this critical size will not grow, and it is exponentially unlikely to nucleate bubbles larger than this critical size. We will assume that all the true-vacuum bubbles are initially created with size $R = R_C$.

Another interesting parameter is the thickness of the bubble wall separating the true-vacuum region inside from the false-vacuum region outside the bubble. For the potential described above, the bubble wall thickness is $\Delta \sim (\lambda^{1/2}\sigma_0)^{-1}$. Note that the ratio of the bubble thickness to its size is $\Delta/R_C \sim \xi$; as advertised, if $\xi \ll 1$, the thin-wall approximation is valid. We note here that the results are (probably) valid even in the absence of the thin-wall approximation. Finally, the energy per unit area of the bubble wall is $\eta \sim \lambda^{1/2}\sigma_0^3$.

It is necessary to have some idea of the size of bubbles at the end of inflation, when bubbles of true vacuum percolate, collide, and release the energy density tied up in the bubble walls. The bubbles of true vacuum are nucleated with size $R = R_C$. After nucleation the bubble will grow until it collides with other bubbles.

Bubbles nucleated at late time will have little growth in coordinate radius, and any increase in the physical size of such a bubble is due solely to the growth in the scale factor between the time the bubble is nucleated and the end of inflation.

The physical size of a bubble nucleated at time t_{NUC} is related to its coordinate size by $R(t_{NUC}) = r(t_{NUC})a(t_{NUC}) = R_C$. If there is negligible growth in the coordinate size of the bubble between the t_{NUC} and end of inflation t_{END}, then at the end of inflation the bubble will have a physical size $R(t_{END}) \equiv R = r(t_{NUC})a(t_{END}) = R_C[a(t_{END})/a(t_{NUC})]$. Assume that the burst of bubble nucleation at the end of inflation leads to bubbles all of the same size, $R = \alpha R_C$, where $\alpha \equiv a(t_{END})/a(t_{NUC})$.

Now we have the picture of the Universe at the end of extended inflation. To a good approximation the Universe is percolated by bubbles of true vacuum of size $R = \alpha R_C$, with all the energy density residing in the bubble walls. The next step is to examine how the release of energy from the bubble walls into radiation via bubble wall collisions takes place.

Now concentrate on a single bubble of radius $R = \alpha R_C$. The collisions of the bubble walls produce some spectrum of particles, which are subsequently thermalized. We need to estimate the typical energy of a particle produced in these collisions. When a bubble forms, the energy of the false vacuum has been entirely transformed into potential energy in the bubble walls, but as the bubbles expand, more and more of their energy becomes kinetic and the walls become highly relativistic. A simple calculation shows that if the bubble has expanded by a factor of α since nucleation, then only $1/\alpha$ of its energy remains as potential energy. The numerical simulations of bubble collisions by Hawking, Moss, and Stewart[4] demonstrate that during collisions the walls oscillate through each other, and it seems reasonable that the kinetic energy is dispersed at an energy related to the frequency of these oscillations (see their discussion of phase waves). The kinetic energy is presumably dispersed into lower energy particles, and does not participate in baryogenesis. We are more interested in the fate of the potential energy. The bubble walls can be imagined as a coherent state of inflaton particles, so that the typical energy of the products of their decays is simply the mass of the inflaton. This energy scale is just equal to the inverse thickness of the wall. Note that by the time the walls actually disperse, most of the kinetic energy has been radiated away,[4] so the walls are probably no longer highly relativistic.

The probable first step in the reheating process is converting this coherent state of Higgs into an incoherent state. The next step would be the conversion of the incoherent state of Higgs into other particles either through decay of the Higgs, or through inelastic scattering.

We are assuming that baryon-number violating bosons H will be produced in the process. The σ field is typically in the adjoint representation of the gauge group, while H is typically in the fundamental or some other representation. It is possible to envision some symmetry forbidding a direct σ–H coupling, or that the coupling is very small compared to other couplings. If this is the case, production of H relative to other particles will be suppressed by some power of the small coupling constant. However in the generic case where all couplings are of the same magnitude there will be no suppression. Of course the ultimate answer is model dependent but calculable.

As discussed earlier, bubbles do not grow substantially before percolation in our idealized extended inflation model. Hence α remains not too far from 1, although a growth by a factor of 1000 even will not necessarily rule out the model. The bubble wall collisions yield a significant amount of the original false-vacuum energy in the form of potential energy, giving rise to high energy particles. The potential energy in the bubble walls is given by $M_{\text{POT}} = 4\pi\eta R^2 \sim 4\pi\lambda^{1/2}\sigma_0^3 R^2$. Taking the mean energy of a particle produced in the collisions to be of the order of the inverse thickness of the wall, $\langle E \rangle \sim \Delta^{-1}$, the mean number of particles produced in the collisions from the wall's potential energy is $\langle N \rangle \simeq M_{\text{POT}}/\langle E \rangle \sim 4\pi\Delta\lambda^{1/2}\sigma_0^3 R^2$.

In general, the bubble collisions will produce all species of particles, at least all species with masses not too large compared to $\langle E \rangle$. In the following we will assume that this is the case for the baryon-number violating Higgs particles. If the Higgs mass exceeds Δ^{-1} by a significant amount, we can expect some suppression, presumably exponential, in the number of Higgs formed. This possibility will be discussed later. For now, we simply parameterize the fraction of the primary annihilation products that are supermassive Higgs by a fraction f_H, which in general will depend on the masses and couplings of a particular theory in question. The typical number of Higgs particles produced per bubble is $\langle N_H \rangle \sim f_H\langle N \rangle \sim 4\pi f_H\Delta\lambda^{1/2}\sigma_0^3 R^2$.

Now assume that the only source of the supermassive Higgs is from the primary particles produced in the bubble-wall collisions. This will be true if the reheat temperature, T_{RH}, is below the Higgs mass.

The Higgs particles produced in the wall collisions decay, producing a net baryon asymmetry ϵ per decay, where ϵ is given in Eq. (1). Hence, the excess of baryons over antibaryons produced from a single bubble, $N_B = N_b - N_{\bar{b}}$, is given by

$$N_B = \epsilon\langle N_H \rangle \sim 4\pi\epsilon f_H\sigma_0^2 R^2, \tag{2}$$

where we have substituted in for the bubble thickness. This results in a baryon number density of

$$n_B = N_B/(4\pi R^3/3) = 3\epsilon f_H\sigma_0^2 R^{-1}. \tag{3}$$

Now calculate the entropy generated in bubble-wall collisions. As stated above, the potential energy of a bubble is $M_{\text{POT}} = 4\pi\sigma_0^3\lambda^{1/2}R^2$. Including the (possibly dominant) kinetic energy contribution, the total mass of the bubble is $M_{\text{BUB}} = 4\pi\sigma_0^3\lambda^{1/2}R^2\alpha$. Thermalization of the mass in the bubble walls will redistribute this energy throughout the bubble, resulting in a radiation energy density

$$\rho_R \sim M/(4\pi R^3/3) \sim 3\lambda^{1/2}\sigma_0^3\alpha/R = \xi\lambda\sigma_0^4, \tag{4}$$

which is just the false vacuum energy. The reheat temperature is related to the radiation energy density via $\rho_R = (g_*\pi^2/30)T_{RH}^4$, where g_* is the effective number of degrees of freedom in all the species of particles which may be formed in the thermalization process. From this we obtain the entropy density, s, produced by the thermalization of the debris from bubble-wall collisions:

$$s = \frac{2\pi^2}{45} g_* T_{RH}^3 \sim g_*^{1/4} \xi^{3/4} \lambda^{3/4} \sigma_0^3. \tag{5}$$

Eqs. (3) and (5) give $B \equiv n_B/s = \epsilon f_H \alpha^{-1} g_*^{-1/4} \lambda^{-1/4} \xi^{1/4}$.

Provided the mass of the Higgs is less than T_{RH}, one might conjecture that f_H is given simply by g_H/g_*, where g_H is the number of Higgs degrees of freedom; that is, all suitably light particles are produced equally. In general the situation will be more complex, and the fraction of Higgs produced will depend on the various couplings in the theory. This introduces a model dependence into the picture, though in fact one can always regard ϵf_H as a single unknown parameter. For simplicity, we assume here that all particles are indeed produced equally. Substituting this gives the final result $B = \epsilon g_H \alpha^{-1} g_*^{-5/4} \lambda^{-1/4} \xi^{1/4}$. This allows us to make numerical estimates of B based on sample values of these parameters. Notice that the dependence on both λ and ξ, which are the two parameters on which the inflaton's potential depends, is very weak. The important contributions are the degree of asymmetry in CP violating Higgs decays, the number of particle species available for production in the wall collisions and the factor α by which bubbles expand before colliding.

It is also interesting to note the possibility of isothermal perturbations arising from the thermalization process. While we have assumed throughout this paper that at percolation all the true vacuum bubbles have the same size, the full picture is somewhat more complicated, as bubbles formed earlier in inflation will grow to larger sizes than those formed right at the end. While homogeneity of the microwave background requires large bubbles to be suppressed, one would still expect to see a range of sizes of small bubbles, and hence spatial variations in the ratio of baryon number density to entropy density from point to point.

In conclusion then, we have seen that the result of the first-order phase transition bringing extended inflation to an end is an environment well out of thermal equilibrium. In such conditions baryogenesis via the decay of baryon number violating Higgs particles can proceed, and we have demonstrated a means by which the baryon number can be estimated. The mechanism has further been shown to work for a large range of model parameters and to have the capability of predicting a baryon asymmetry of the required magnitude.

For more details on baryogenesis, the reader is referred to the original paper of Barrow, Copeland, Kolb, and Liddle.[1]

2. Black Holes [5]

There are three possible sources for the formation of small primordial black holes after extended inflation. Holes may form via the gravitational instability of inhomogeneities formed during the thermalization phase; there is the possibility of trapped regions of false vacuum (within their Schwarzschild radii) caught between bubbles of true vacuum;[6] and there is the possibility that black holes are formed in the collision process.[4]

Unfortunately, the technical details of even estimating the typical number density and mass of the black holes formed by these processes are quite difficult. Some progress in this direction was made by Hawking, et al.,[4] in the context of the original inflationary scenario, and more recently Hsu[7] has examined black hole production from false vacuum regions in extended inflation. In order to keep the discussion on a more general footing, for now simply assume that some fraction β of the energy after collisions is in black holes, while the remaining $1 - \beta$ is in radiation,[8] and later consider the various outcomes implied by the differing values of β.

The total energy density at the end of extended inflation is partitioned between the energy density of radiation, ρ_R, and black holes, ρ_{BH}:

$$\rho(t_{\rm END}) = \rho_R(t_{\rm END}) + \rho_{BH}(t_{\rm END})$$
$$\rho_R(t_{\rm END}) = (1 - \beta)\rho(t_{\rm END}) = \frac{\pi^2}{30}g_* T_{RH}^4$$
$$\rho_{BH}(t_{\rm END}) = \beta\rho(t_{\rm END}) = M_0 n_{BH}(t_{\rm END}), \tag{6}$$

where T_{RH} is the reheat temperature, M_0 is the initial mass of the black holes formed (for convenience we will assume that they all have the same mass), and n_{BH} is the number density of black holes. The time $t_{\rm END}$ can also be expressed in terms of $\rho(t_{\rm END})$:

$$t_{RH}^2 \equiv \left(\frac{3}{32\pi}\right)\frac{m_{Pl}^2}{\rho(t_{\rm END})}. \tag{7}$$

(For matter domination, the factor $3/32\pi$ is replaced by $1/6\pi$.) From $H_{\rm END}$ and ρ we also define a "horizon mass" at the end of inflation:

$$M_{\rm HOR} = \frac{4\pi}{3}\rho(t_{\rm END})(2t_{RH})^3 = \left(\frac{3}{32\pi}\right)^{1/2}\frac{m_{Pl}^3}{\rho^{1/2}(t_{\rm END})}. \tag{8}$$

(The right hand side is the same in the matter dominated case.) $M_{\rm HOR}$ represents the mass within the "physics horizon," at the end of inflation, and plays the same role as the mass within the horizon in the standard FRW model.

Once formed, the black holes evaporate at a rate given by

$$\dot{M}_{BH} = -\frac{g_*}{3}\frac{m_{Pl}^4}{M_{BH}^2}, \tag{9}$$

which leads to a time dependence of the black hole mass of

$$M_{BH}^3(t) = M_0^3 - g_* m_{Pl}^4(t - t_{\rm END}). \tag{10}$$

It is convenient to define a black hole lifetime,

$$\tau \equiv M_0^3/g_* m_{Pl}^4, \tag{11}$$

and the expression for the mass as a function of time becomes $M(t) = M_0[1 - (t - t_{\rm END})/\tau]^{1/3}$. The evaporation ends at time $t_{BH} = t_{RH} + \tau$.

Black holes radiate as blackbodies with temperature $T_{BH} = m_{Pl}^2/8\pi M_{BH}$. This allows us to calculate what is, for our purposes, the most important quantity—the number of particles emitted during the course of the evaporation. Let us first calculate the number of particles emitted while the black hole is between the temperatures T and $T + dT$. The change in mass of the black hole, dM, which is the amount of energy radiated as particles, is given by

$$dM = \frac{m_{Pl}^2}{8\pi} \left(\frac{1}{T} - \frac{1}{T + dT} \right). \tag{12}$$

Each emitted particle has energy $3T$ (the mean energy of a particle in a Maxwell-Boltzmann distribution at temperature T), so the number of particles emitted between those temperatures is just

$$dN = \frac{m_{Pl}^2}{24\pi T} \left(\frac{1}{T} - \frac{1}{T + dT} \right) = \frac{m_{Pl}^2}{24\pi T^3} dT. \tag{13}$$

Integrating this, we find that the number of particles emitted as the black hole temperature increases from its initial temperature T_0 to ∞ is

$$N = \frac{4\pi M_0^2}{3 m_{Pl}^2}. \tag{14}$$

Note that this gives the total number of particles emitted.

It is interesting to consider the possibility that amongst the particles radiated are Higgs bosons, again denoted as H, whose decay can lead to the baryon asymmetry. Again, B will depend upon the fraction of the particles emitted as H, denoted as f_H. To determine the appropriate form for f_H, the initial temperature of the black hole at formation may be important. If it is less than the mass of the Higgs boson, m_H, then the thermal spectrum in the initial phase of the evaporation will not include Higgs as the typical energy is not high enough to produce so massive a particle. Only when the black hole temperature has increased to m_H will the thermal radiation include a significant fraction of Higgs. This can lead to an overall suppression in the number of Higgs produced during the complete course of the evaporation. Once the temperature is high enough to radiate Higgs, we expect that the energy of radiated particles will be distributed evenly amongst all radiated species, so that f_H is a constant given by g_H/g_*.

Black hole evaporation affects the evolution of both components of the total mass density. Since the hole mass is decreased by evaporation, the evolution of the black hole energy density, which in the absence of evaporation would be that of nonrelativistic matter $(\rho_{NR} \propto a^{-3}$, where a is the scale factor), is altered. The production of radiation from the hole evaporation also modifies the evolution of radiation energy density, which normally scales as a^{-4}. Of course, the departure of the energy densities from the normal evolution is most pronounced around the time $t \sim t_{RH} + \tau$. An exact treatment of this effect is given in Ref. (5), where a network of equations is derived describing the evolution of the different components of the energy density and also the evolution of the baryon asymmetry. In order to understand the general results, let us for the moment ignore the complication resulting from the decrease of the hole mass.

Two different situations arise, depending on whether black holes or radiation dominate the energy density of the Universe at the time the holes evaporate.[8] If $\beta < 1/2$, then the evolution of the scale factor is that appropriate to a radiation-dominated Universe, i.e., $a(t) \sim t^{1/2}$, and the energy density of black holes goes as $a^{-3} \propto t^{-3/2}$, while that of radiation goes as $a^{-4} \propto t^{-2}$. Therefore, provided their lifetime is sufficiently long, black holes will come to dominate the Universe at a time $t_* = t_{END}(1 - \beta)^2/\beta^2$, and hence if $\tau > t_* - t_{END}$, they will come to dominate before their evaporation. If $\beta > 1/2$, black holes dominate even initially. If the black holes dominate before evaporation, then their evaporation produces not only the baryons, but also the entropy.

For the details of the calculations the reader is referred to Ref. (5). Here I shall simply summarize the results.

First consider the case where black hole evaporation occurs before domination. This corresponds to small β and initially light black holes. Since the black holes never dominate, the Universe expands like a radiation-dominated Universe, with $a \propto t^{1/2}$. If the black holes evaporate before domination, their radiation will not significantly change the background entropy density.

In this case the final baryon asymmetry is

$$B_A \equiv \frac{n_B}{s} = \frac{1}{2}\epsilon f_H \left(\frac{45\pi}{g_*}\right)^{1/4} \left(\frac{M_0}{m_{Pl}}\right)^{1/2} \left(\frac{M_0}{M_{\mathrm{HOR}}}\right)^{1/2} \frac{\beta}{(1-\beta)^{3/4}}, \qquad (15)$$

where we have used Eq. (8). Note that the penultimate factor gives the initial black hole mass as a fraction of the horizon mass.

Now consider the second possibility, that holes evaporate after they dominate the energy density. This divides into two further sub-cases; in the former, black holes come to dominate at time t_* as defined earlier, while in the latter black holes dominate immediately after formation.

In the first of these sub-cases, once $t > t_*$ the scale factor evolves as appropriate for a matter-dominated Universe, $a(t) \sim t^{2/3}$, and so $\rho_{BH}(t) = \rho_{BH}(t_*)(t_*/t)^2$ and $\rho_R(t) = \rho_R(t_*)(t_*/t)^{8/3}$, with the energy densities equal at t_*.

The evaporation of a single black hole gives a baryon number $n_B = \epsilon f_H N\, n_{BH}(t_{BH})$. This time, though, the entropy is also determined by the other black hole evaporation products, as they provide the dominant contribution. The result for the baryon number is

$$B_{B1} = \frac{1}{2}\epsilon f_H \left(\frac{45\pi}{g_*}\right)^{1/4} \left(\frac{M_0}{m_{Pl}}\right)^{1/2} \left(\frac{M_0}{M_{\mathrm{HOR}}}\right)^{1/2} (1-\beta)^{1/4} \left(1 + \frac{\tau}{t_{RH}}\right)^{-1/2}. \qquad (16)$$

This expression is very similar to that obtained in the "evaporation before domination" scenario; in particular the black hole mass appears in the same functional form, and the prefactors are all the same with the exception of the β term, which naturally has changed as we move to a different physical situation. The last factor demonstrates how a long black hole lifetime dilutes the baryon asymmetry obtained; if τ is very small this factor is just equal to one, while for $\tau \gg t_{RH}$ we get a reduction in the baryon asymmetry by a factor of about $\sqrt{M_0^3/M_{\mathrm{HOR}}m_{Pl}^2 g_*}$. Clearly, this factor can be important for long-lived (initially massive) black holes. These are also exactly the type of holes that one might expect to survive long enough to come to dominate even if β is originally substantially less than $1/2$.

We now examine the second sub-case of black hole domination—that in which the black holes dominate even initially. The black hole energy density is now given by $\rho_{BH}(t) = \rho_{BH}(t_{RH})(t_{RH}/t)^2$, and

$$B_{B2} = \frac{1}{2}\epsilon f_H \left(\frac{45\pi}{g_*}\right)^{1/4} \left(\frac{M_0}{m_{Pl}}\right)^{1/2} \left(\frac{M_0}{M_{\mathrm{HOR}}}\right)^{1/2} \beta^{1/4} \left(1 + \frac{\tau}{t_{RH}}\right)^{-1/2}, \qquad (17)$$

which is just Eq. (16) multiplied by $(\beta/(1-\beta))^{1/4}$. This factor represents the dilution of the black hole energy density up to domination. As expected, Eqs. (16) and (17) match in the case of marginal domination where $\beta = 1/2$. The β dependence in Eq. (17) simply reflects the fraction of the horizon mass contributed by black holes. It differs from Eq. (16) because here there is no evolution in the initial radiation-dominated phase, hence no era of dilution before domination. In the case of Eq. (16) an extra multiplier of $[(1 - \beta)/\beta]^{1/4}$ is needed to account for the evolution in the initial radiation-dominated phase.

This completes the set of results for the different regions of domination, and is summarized in Table I. Many more details are to be found in the paper of Barrow, Copeland, Kolb, and Liddle.[5]

Table I. Results for the baryon number produced by black hole evaporation depend upon β (the fraction of the energy of the Universe in black holes at $t = t_{\rm END}$, where $t_{\rm END}$ is taken to be the end of inflation), t_* (the time at which the black holes dominate the mass of the Universe), and $\tau = M_{BH}^3/g_* m_{Pl}^4$ (the lifetime of a black hole of mass M_{BH}).

β	τ	$B \equiv n_B/s$
$\beta < 1/2$	$\tau < t_* - t_{\rm END}$	Eq. (15)
$\beta < 1/2$	$\tau > t_* - t_{\rm END}$	Eq. (16)
$\beta > 1/2$	independent of τ	Eq. (17)

3. Topological Defects [9]

I have already discussed the generation of adiabatic density fluctuations during extended inflation. However there might very well be a different mechanism for the formation of structure after extended inflation, namely the formation of topological defects in the inflaton field formed as it passes through the phase transition. Calculations of the false-vacuum decay rate made so far consider the evolution from a false-vacuum state to a *unique* true-vacuum state. However, the inflaton is far more likely to have degenerate minima, especially if it is part of a grand-unified Higgs sector.

Recall the picture of defect formation in a smooth second-order phase transition.[10] At early times the universe was very hot and the fields describing interactions were in a highly symmetric phase. However as the universe expanded and cooled, symmetry breaking occurred, which may have left behind remnants of the old symmetric phase, possibly in the form of strings, domain walls or monopoles. Here, we concentrate on strings.

Strings appropriate to galaxy formation are required to have a line density of $G\mu \sim 10^{-6}$, where $\mu \sim \sigma_0^2$, corresponding to a breaking scale of 10^{-3} Planck masses. Unfortunately, generic new and chaotic inflationary scenarios occur at or below this energy scale, and hence the strings form before or early in the inflationary epoch and are rapidly inflated away. It has been demonstrated that the universe cannot be made to reheat after inflation to sufficiently high temperatures as to restore the symmetry of the string-forming field and allow a new phase of string formation after inflation.[11,12] This leads to the incompatibility of cosmic strings with new or chaotic inflation. These arguments apply whether the inflaton and the cosmic string fields are the same field or different ones (in chaotic inflation the inflaton field can never be identified with the cosmic string field as the symmetry is broken even initially). In the case where the cosmic string field is distinct from the inflaton field, models have been proposed which resolve the conflict. The model of Vishniac, Olive, and Seckel[12] couples the inflaton and the string field in a particular way, but the only motivation for doing this is to solve the strings-inflation problem, so their solution appears unnatural. More recently, Yokoyama[13] has suggested that a non-minimal coupling to gravity of the string field can hold it in its symmetric phase during inflation, and allow strings to form at the end of inflation.

Now consider the picture of string formation in extended inflation, where the fact that the transition is first order has crucial consequences. As the Universe cools from high temperatures, a complex scalar field is trapped in a false-vacuum state and the Universe enters a phase of rapid power-law expansion. Bubbles of true vacuum then begin to nucleate and grow at the speed of light. Due to the presence of event horizons in the inflating Universe they grow to a constant comoving volume which depends on their time of formation. The important ingredient to our scenario is that each bubble forms independently of the rest, and so there is no correlation between the choice of true vacuum made in each bubble from the selection of degenerate true vacua. Eventually the bubbles grow and collide, finally percolating the Universe and bringing the inflationary era to an end.

At the end of inflation, the collision of bubble walls (in which all the energy is held) produces particles and causes thermalization of the energy. However, because the scalar field is only correlated on the scale of a bubble, we can expect topological defects to be present. The usual arguments state that there is typically of order one cosmic string per correlation volume of the scalar field, and hence we expect roughly one string per mean bubble size at the end of inflation.

This model for the formation of strings allows for the existence of large voids, which would be a consequence of the rare large bubbles. Although the typical string separation at the end of inflation is ξ_{eff}, extended inflation allows for the possibility of rare large bubbles, formed by quantum tunnelling early in inflation. The true vacuum formed inside bubbles contains no matter (any matter originally in that volume is assumed to be inflated away while the scalar field dominates the energy density). All the energy of the Universe after inflation is contained in the walls of the expanding bubbles which collide to form matter and to cause thermalization of the energy density. After collisions, matter will flow back into the void, though as it cannot travel faster than light, we can calculate the minimum time the bubble will require to thermalize. A large bubble will have a coherent scalar field vacuum and hence no strings will be formed within it—we can thus expect the interior of the bubble to evolve into a large region void of strings. If cosmic strings are to provide the seeds for galaxy formation, then we can expect to see few or no galaxies within the void. The presence of voids is an additional property of this model which may help explain observed large-scale structure.

In fact, at the time of percolation the bubbles may have a range of sizes, which can lead to the formation of an initial string network differing from the usual one. As the correlation length is essentially just the bubble size, and because there would appear to be no *a priori* reason why bubbles everywhere should be *exactly* the same size (at small sizes the assumption of a scale-invariant bubble size distribution would seem more reasonable), the strings will be formed with a randomly spatially varying correlation length. This will presumably lead to higher densities of strings in some regions than others, which again may have implications for structure formation, depending on how much the effects of the initial string distribution might be wiped out by the future evolution and decay of strings. One desirable effect of a more dilute string network would be to avoid the uncomfortable bounds from gravitational wave production from small string loops.[14] The fact that the correlation length will generically be greater (and in some models perhaps much greater) than that of the Kibble mechanism may also have important implications, though perhaps not as great as one might naively suppose if the small strings rapidly disappear from the network once string evolution commences.

These formation arguments can be equally well applied to the cases of domain walls and monopoles, again giving rise to an estimate of order one defect per bubble at the time of bubble collision. In the case of domain walls this will give rise to an excessive number, and will be disallowed on cosmological grounds. Hence, any extended inflation model featuring a potential with domain wall solutions (i.e., a disconnected vacuum manifold) can be ruled out. The situation is less clear for monopoles, because the correlation length may well be substantially greater than that of the Kibble mechanism and hence proportionally fewer monopoles are expected. However, standard estimates of the cosmological monopole abundance[15] give values of perhaps twenty orders of magnitude in excess of the Parker limit,[16] so the correlation length would have to be increased by seven or eight orders of magnitude before being within experimental limits—such an increase seems very unlikely.

If we consider the unification to be part of a grand-unified theory, the problem of monopole overproduction must be addressed, as any breaking to the symmetry of the standard model must produce monopoles at some stage. The simplest method is to arrange for monopoles to be formed in a partial symmetry breaking and then later inflated away in a second transition.

For more details, the reader should see the original paper of Copeland, Kolb, and Liddle.[9]

4. Gravity Waves [17]

One of the most interesting new features of a completed first-order phase transition is the observation by Turner and Wilczek that a significant amount of gravity waves might be produced in the reheating process.[17]

The beauty of this observation is that it is largely independent of the details of the particular extended inflation model. In the picture of reheating I have been describing here, bubbles of size R and mass M_{BUB} collide. Furthermore, the bubbles are most likely relativistic, or at least semi-relativistic, when they collide. The luminosity in gravity waves emitted in such a close encounter can be estimated from the quadrupole formula:

$$L_{GW} \sim G_N \left(\frac{d^3 Q}{dt^3}\right)^2 \sim G_N \frac{M_{BUB}^2}{R^2}. \tag{18}$$

Thus a bubble collision releases an energy E_{GW} given by

$$E_{GW} \sim R L_{GW} \sim G_N \frac{M_{BUB}^2}{R}, \tag{19}$$

in the form of gravitational radiation with wavelength R.

Of course it is most useful to compare this energy with the total energy released in the bubble collision. Since the total mass of the bubble, M_{BUB}, is eventually released into radiation, then the ratio of the gravitational wave energy density to the radiation energy density is

$$\epsilon_{GW} \equiv \frac{E_{GW}}{M_{\text{BUB}}} \sim G_N \frac{M_{\text{BUB}}}{R}. \qquad (20)$$

If this is true after extended inflation, then the present ratio would be approximately $g_*(\text{today})/g_*(T_{RH}) \sim 0.01$ times this value. Since the contribution to Ω in radiation is today about $3 \times 10^{-5} h^{-2}$, and ρ_{GW} and ρ_R both decrease in expansion as a^{-4}, this implies that today $\Omega_{GW} h^2 \sim 10^{-5} \epsilon_{GW}$.

The wavelength of the gravitational radiation today would simply be the wavelength at creation, $\lambda(T_{RH}) \sim R$, redshifted by the expansion of the Universe:

$$\lambda(\text{today}) = R[a(\text{today})/a(T_{RH})] \sim RT_{RH}/2.7 \text{ K} \sim 4 \times 10^{26} R(M/10^{14}\text{GeV}), \qquad (21)$$

where again we have assumed that the re-heat temperature is comparable to the mass scale of symmetry breaking M.

Now the question is what to use for R. Turner and Wilczek make the reasonable assumption that the size of the bubble is the particle horizon at the end of extended inflation. If this is true, then $G_N M_{\text{BUB}}/R$ is about unity, $\epsilon_{GW} \sim 0.01$, and $\Omega_{GW} h^2 \sim 10^{-5}$. The fact that ϵ_{GW} is about unity in this case is easy to understand: masses the size of the horizon are moving about with velocities of about the velocity of light! This choice for R also predicts $R \sim H^{-1} \sim m_{Pl}/M^2 \sim 2 \times 10^{-19}(10^{14}\text{GeV}/M)^2\text{cm}$, which leads to a present wavelength for the gravity waves of $\lambda(\text{today}) = 8 \times 10^3(10^{14}\text{GeV}/M)$ cm. This is quite interesting because it is within the sensitivity and wavelength range of LIGO II and other large second-generation interferometric detectors.

However it might be equally possible that the bubbles are much smaller. The smallest they might be is R_C, their critical size. Let's take the pessimistic view that $R \sim M^{-1}$. If this is true, then $\epsilon_{GW} \sim G_N M_{\text{BUB}}/M = M_{\text{BUB}} M/m_{Pl}^2$. If the bubble size is M^{-1} and the false-vacuum energy is of order M^4, then $M_{\text{BUB}} \sim M$, and $\epsilon \sim M^2/m_{Pl}^2 \sim 10^{-10}(M/10^{14}\text{GeV})^2$. This would lead to a present value of $\Omega_{GW} h^2 \sim 2 \times 10^{-15}(M/10^{14}\text{GeV})^2$. Another price to be paid is that the present wavelength of the gravity waves would be much smaller: $\lambda(\text{today}) = 8 \times 10^{-2}$ cm. This is too small in magnitude and wavelength for interferometric detectors.

Clearly the correct answer is model dependent. The latter assumption is most likely far too pessimistic, while the former assumption may turn out to be somewhat optimistic.

ACKNOWLEDGEMENTS

I would like to acknowledge my collaborators on the work reported here: John Barrow, Ed Copeland, and Andrew Liddle. All of the work reported here is a result of collaborative efforts with them. This work has been supported by the Department of Energy and NASA (grant #NAGW-1340).

REFERENCES

1. J. D. Barrow, E. J. Copeland, E. W. Kolb, and A. R. Liddle, Fermilab Report FNAL-PUB-90/62A.

2. See, e.g., K. A. Olive, "Big Bang Nucleosynthesis: Theory and Observations", Univ. Minnesota preprint (1990); B. E. J. Pagel, in *Evolutionary Phenomena in Galaxies*, eds. J. E. Beckman and B. E. J. Pagel (Cambridge University Press, Cambridge, 1989).

3. See, e.g., E. W. Kolb and M. S. Turner, *The Early Universe* (Addison-Wesley, Redwood City, CA., 1990).

4. S. W. Hawking, I. G. Moss, and J. M. Stewart, *Phys. Rev. D* **26**, 2681 (1982). It should be noted that the collision of two bubbles of equal size may not at all approximate the process of reheating at the end of extended inflation.

5. J. D. Barrow, E. J. Copeland, E. W. Kolb, and A. R. Liddle, Fermilab Report FNAL–PUB–90/98A.

6. H. Kodama, M. Sasaki, and K. Sato, *Prog. Theor. Phys.* **68**, 1979 (1982), and references therein.

7. S. Hsu, "Black Holes from Extended Inflation," Lawrence Berkeley Laboratory preprint, unpublished.

8. J. D. Barrow, *Mon. Not. R. Astron. Soc.* **192**, 427 (1980); J. D. Barrow and G. G. Ross, *Nucl. Phys.* **B181**, 461 (1981); J. D. Barrow, *Fund. Cosmic. Phys.* **8**, 83 (1983).

9. E. J. Copeland, E. W. Kolb, and A. R. Liddle, *Phys. Rev. D* to be published.

10. T. W. B. Kibble, *J. Phys. A* **9**, 1387 (1976).

11. M. D. Pollock, *Phys. Lett.* **185B**, 34 (1987).

12. E. T. Vishniac, K. A. Olive and D. Seckel, *Nucl. Phys.* **B289**, 717 (1987).

13. J. Yokoyama, *Phys. Lett.* **212B**, 273 (1988); J. Yokoyama, *Phys. Rev. Lett.* **63**, 712 (1989).

14. A. Albrecht and N. Turok, *Phys. Rev. D* **40**. 973, (1989); F. R. Bouchet, D. P. Bennett, and A. Stebbins, *Nature* **335**, 410 (1988); F. S. Accetta and L. L. Krauss, *Phys. Lett.* **233B**, 93 (1989).

15. J. Preskill, *Phys. Rev. Lett.* **43**, 1365 (1979).

16. E. N. Parker, *Ap. J.* **160**, 383 (1970).

17. M. S. Turner and F. Wilczek, Fermilab Report FNAL–PUB–90/198A.

Inflation in Generalized Einstein Theories

K. MAEDA [1] , A.L. BERKIN [1], and J. YOKOYAMA[2]
(1) *Dept. of Physics, Waseda University, Tokyo 169-50, Japan*
(2) *Dept. of Physics, University of Tokyo, Tokyo 113, Japan*

ABSTRACT. We classify inflationary models into three types: (1) the Einstein gravity theory with inflaton scalar field, (2) modified gravity theories with ordinary matter fluid and (3) the hybrid of these two models. Extended inflation and soft inflation belong to the third class. We then analyze the details of soft inflationary models, which have two scalar fields: one the standard inflaton, whose potential is exponentially coupled to the other field. We consider satisfaction of the inflationary constraints. New inflation works well, with the coupling constant near the values allowed by grand unified theories. For chaotic inflation with a massive inflaton, we find successful inflation without fine-tuning of the coupling constant or initial data.

The inflationary cosmology purports to solve many longstanding cosmological problems, and has been the subject of much investigation during the previous decade. However, there is no fully satisfactory model for the source of inflation yet. In order to look for a more satisfactory model, we shall first classify the inflationary models proposed so far.

The action contains two parts; the gravity and matter sectors written as $S = \int d^4x \sqrt{-g}$ ($L_{\text{gravity}} + L_{\text{matter}}$). In the prototype old inflation, we assume Einstein's general relativity for L_{gravity} and a Higgs scalar field in GUTs with vacuum energy for L_{matter}, which leads to an inflationary expansion of the universe. Although such a model failed because the present universe is not realized, it is not the only model. We have several choices for L_{gravity} and L_{matter}. In Table 1, we place the proposed inflationary models into three classes.

class	L_{gravity}	L_{matter}	models
class I	Einstein gravity	vacuum energy (GUTs)	old inflation new inflation chaotic inflation power-law inflation
class II	generalized Einstein theories (GETs)	usual matter	R^2- inflation induced gravity inflation non-minimal scalar inflation Kaluza-Klein inflation
class III	generalized Einstein theories (GETs)	vacuum energy (GUTs)	extended inflation soft inflation

Table 1

The generalized Einstein theories (GETs) contain Jordan-Brans-Dicke (JBD) theory, the Fujii-Zee induced gravity model, R^2- theory and 4-dimensional effective theories from higher-dimensional theories such as Kaluza-Klein or superstring models.

The difficulty of the models in Class I and II is either we do not have a successful inflationary model in particle physics such as GUTs or some simple models need an extremely small coupling constant, which leads us to wonder whether such a model is natural (the so-called fine-tuning problem). We may also have a fine-tuning problem of initial data in some models. The question arises whether we can find a natural inflationary model in particle physics or with a simple potential without any fine-tuning in class III. Extended inflation was proposed to avoid the difficulties of old inflation[1]. We proposed soft inflation[2], where the potential of the inflaton is either new inflation or chaotic type. In this paper, we shall present a more detailed analysis of soft inflation[3].

The class III models are conformally equivalent to the class I model with two scalar fields; an inflaton field ψ and a scalar field ϕ coupled to gravity[4]. We then use as our action

$$ S = \int d^4x \sqrt{-g} \left[\frac{1}{2\kappa^2} R - \frac{1}{2}(\nabla\phi)^2 - \frac{1}{2}e^{-\gamma\kappa\phi}(\nabla\psi)^2 - e^{-\beta\kappa\phi}V(\psi) \right], \tag{1} $$

where $\kappa^2 = 8\pi G$, $V(\psi)$ is the inflaton potential and β, γ are dimensionless coupling constants, which are fixed by parameters in GETs. The constraints on these parameters give limits on β and γ (e.g. $\beta < 0.1$ in JBD theory, from the observational constraint $\omega > 500$). In Table 2, we list the values of β, γ for each theory.

theories	L_{gravity}	ϕ	β	γ
Jordan-Brans-Dicke theory	$\frac{1}{16\pi}[\Phi R - \frac{\omega}{\Phi}(\nabla\Phi)^2]$	$(\frac{2\omega+3}{2})^{1/2}\ln(\kappa^2\Phi/8\pi)$	$(\frac{8}{2\omega+3})^{1/2}$	$\beta/2$
induced gravity	$\frac{1}{2}[\epsilon\Phi^2 R - (\nabla\Phi)^2]$	$(\frac{1+6\epsilon}{\epsilon})^{1/2}\ln(\kappa\sqrt{\epsilon}\Phi)$	$(\frac{16\epsilon}{1+6\epsilon})^{1/2}$	$\beta/2$
higher-dimensional theories 1	$\frac{1}{2\kappa^2}[\Phi^2 R + \frac{4(D-1)}{D}(\nabla\Phi)^2]$	$[\frac{2(D+2)}{D}]^{1/2}\ln\Phi$	$(\frac{8D}{D+2})^{1/2}$	$\beta/2$
higher-dimensional theories 2	$\frac{1}{2\kappa^2}R^{(4+D)}$	$[\frac{2(D+2)}{D}]^{1/2}\ln\Phi$	$(\frac{2D}{D+2})^{1/2}$	0
superstring model	10-dim supergravity	ϕ_S	$2\sqrt{2}$	$\beta/2$
		ϕ_T	$-\sqrt{6}$	β

Table 2

We consider three conditions for successful inflation:

(a) horizon problem

$$a_f/a_0 > 10^{30} \tag{2}$$

where subscripts f and 0 denote the values at the end and beginning of inflation, respectively.

(b) density perturbations

$$\delta\rho/\rho < 10^{-4} \tag{3}$$

where the formula for density perturbations is given by

$$\delta\rho/\rho \sim H^2 \max\left[|\dot\phi|, e^{-\gamma\kappa\phi/2}|\dot\psi|\right] /(\dot\phi^2 + e^{-\gamma\kappa\phi}\dot\psi^2) \tag{4}$$

(c) reheating temperature

$$T_{RH} > 10^{10}\text{GeV} \tag{5}$$

for baryogenesis.

Since the details of the analysis are given in Refs. [2], [3] and a short description is also in this volume (Berkin *et. al.*), we show the results by figures. In the figures, H, D and RH give the constraints from the horizon problem, density perturbations and reheating temperature, respectively. The shaded region is permitted.

(1) new inflation model

$(V = V_0 - \frac{1}{4}\lambda\psi^4)$

(i) $\gamma = 0$

Fig.1(a) shows the allowed region in (λ, energy scale $V_0^{1/4}$), setting $\beta = 0.1$ and $\phi_0 = 10m_{PL}$, where λ is a coupling constant of GUTs. The value of simple SU(5) GUT is shown by $\times$. Fig.1(b) denotes the allowed region for initial data ϕ_0 in terms of β, assuming simple SU(5) GUT model. Although such simple GUT models are compatible with successful inflation, we need fine-tuning of initial data.

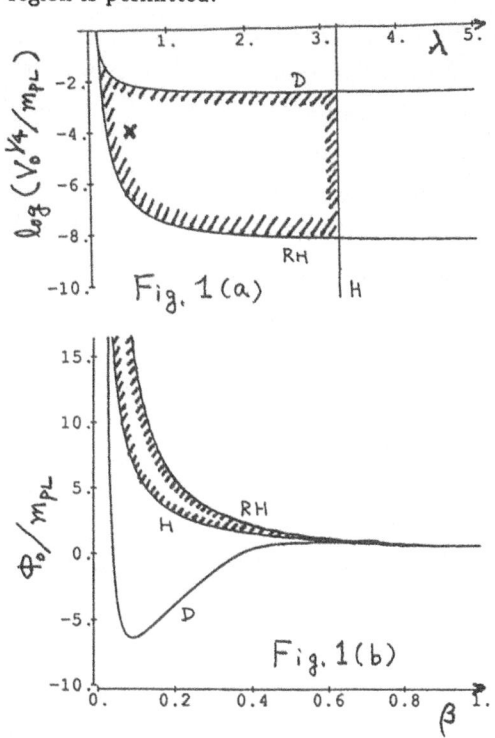

Fig. 1 (a)

Fig. 1 (b)

170

(ii) $\gamma = \beta/2$

Fig.2(a), (b) are similar to Fig.1(a),(b). Although the simple SU(5) GUT model does not give successful inflation, the critical value of the coupling constant is not extremely small ($\lambda < 0.02$), as in the conventional new inflation model ($\lambda < 10^{-12}$). We do not need fine-tuning of initial data. In Fig.2(b), we set $\lambda = 0.015$ in order to find successful model.

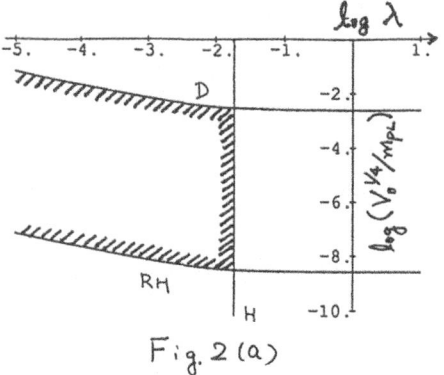

Fig. 2 (a)

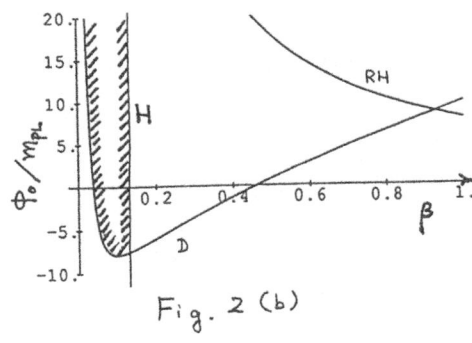

Fig. 2 (b)

(2) chaotic inflation model ($V = \frac{1}{4}\lambda\psi^4$)

(i) $\gamma = 0$

Fig.3 shows the allowed region of the coupling constant λ in terms of β, setting $\phi_0 = 10 m_{PL}$. Although the value of λ need not be small as in conventional chaotic inflation ($\lambda < 10^{-12}$), we need fine-tuning of the coupling constant for successful inflation.

(ii) $\gamma = \beta/2$

Fig.4 is the same as Fig.3 except for the value of γ. The constraints become worse than the conventional model. This result does not depend on the initial value of ϕ_0.

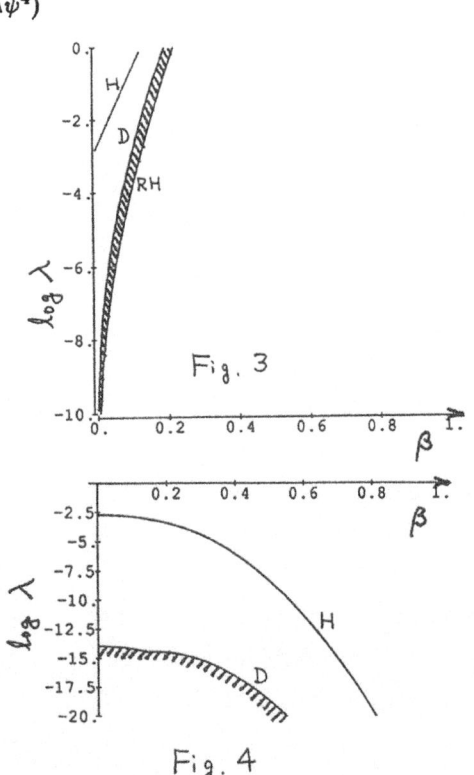

Fig. 3

Fig. 4

(3) chaotic inflation model ($V = \frac{1}{2}m^2\psi^2$)

(i) $\gamma = 0$

Fig.5 shows the allowed region in the mass parameter m in terms of β, setting $\phi_0 = 10m_{PL}$. The result is quite similar to the case **(2)-(i)**.

(ii) $\gamma = \beta/2$

Fig.6(a) is the same as Fig.5. The constraint becomes much better ($m < 10^{-(2\sim3)}m_{PL}$) than the convensional chaotic inflationary model ($m < 10^{-6}$ m_{PL}). Fig.6(b) shows the allowed region of initial data ϕ_0 in terms of β, setting $m = 10^{-4}m_{PL}$. The result is not so sensitive to the value of m. We also do not need fine-tuning of initial data.

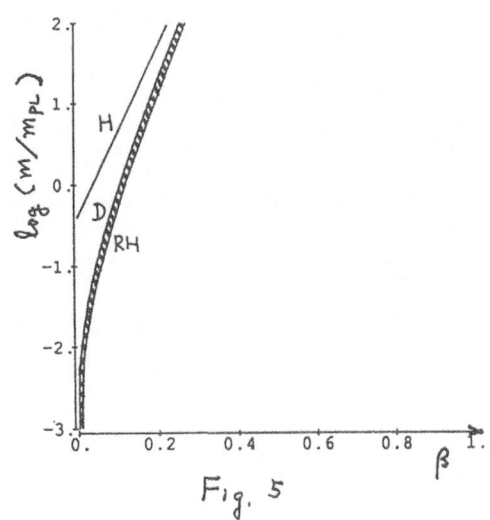

Fig. 5

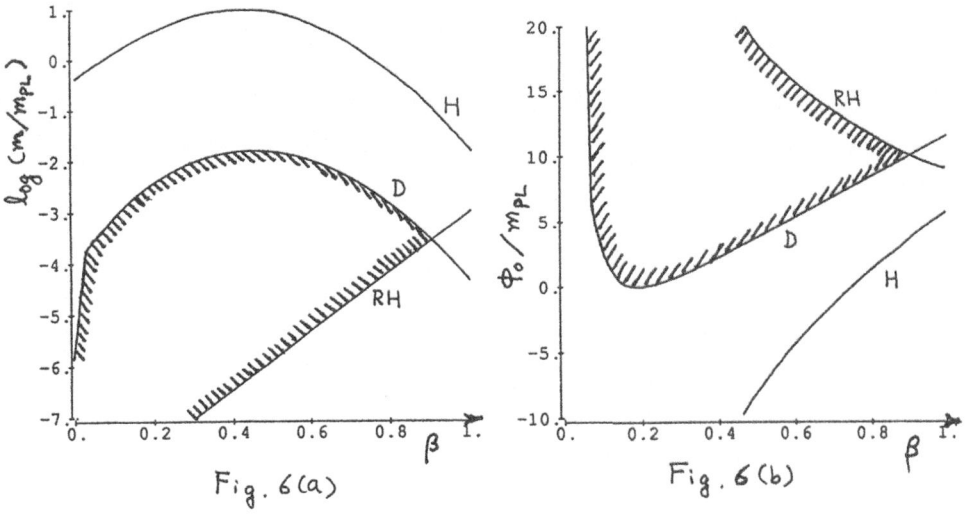

Fig. 6(a)   Fig. 6(b)

We summarize the results in Table 3.

potential	$\gamma = 0$	$\gamma = \beta/2$
new inflation $(V = V_0 - \frac{1}{4}\lambda\psi)$	GUT : compatible fine-tuning of initial data	simple SU(5) : excluded, but $\lambda \leq 0.02$ (not extremely small) no fine-tuning
chaotic inflation $(V = \frac{1}{4}\lambda\psi)$	$\lambda \sim O(1)$: compatible fine-tuning of λ for given β	$\lambda \leq 10^{-12}$
chaotic inflation $(V = \frac{1}{2}m^2\psi)$	$m \leq m_{PL}$: compatible fine-tuning of m for given β	$m \leq 10^{-(2\sim3)} m_{PL}$ no fine-tuning

Table 3

We may wonder what happens in the original frame, which is related by a conformal transformation to the Einstein frame presented here. Since the relations are given as[2],[3]:

$$(a) \quad \left(\frac{a_f}{a_0}\right)_{\text{original}} \sim \left(\frac{a_f}{a_0}\right)_{\text{conformal}}^{(1-\beta^2/4)}, \tag{6}$$

$$(b) \quad \left(\frac{\delta\rho}{\rho}\right)_{\text{original}} \sim \left(\frac{\delta\rho}{\rho}\right)_{\text{conformal}}, \tag{7}$$

$$(c) \quad (T_{RH})_{\text{original}} \sim (T_{RH})_{\text{conformal}} \ (\beta = 0) \ \geq \ (T_{RH})_{\text{conformal}}, \tag{8}$$

the conclusion does not change much, and the reheating constraint becomes loosened.

We should investigate the spectrum of density perturbations and the possible scenarios of large scale structure formation. The present model in the Einstein conformal frame may also provide a natural scenario of a decaying cosmological constant, although we need more detailed analysis. These will be discussed in future work.

REFERENCES:

[1] D. La and P.J. Steinhardt, Phys. Rev. Lett.62, 376 (1989).

[2] A.L. Berkin, K. Maeda, and J. Yokoyama, Phys. Rev. Lett. 65, 141 (1990).

[3] A.L. Berkin and K. Maeda, preprint WU-AP/10/91 (1991).

[4] K. Maeda, Phys. Rev. D39, 3159 (1989), and citations within.

Baryogenesis in the Universe

A.D. DOLGOV[*]
*Institute of Theoretical and Experimental Physics,
Moscow, 117259, USSR.*

ABSTRACT   A review of the models of baryogenesis is given. The content is the following: 1. Introduction and history. 2. Three principles of baryogenesis. 3. Heavy particle decays. 4. Baryogenesis with conserved baryonic charge. 5. Baryogenesis, inflation, and CP-violation. 6. Baryogenesis and large scale Universe structure. 7. Baryonic charge condensate. 8. Baryogenesis and CPT. 9. Baryosynthesis by topological defects. 10. Baryosynthesis on electroweak scale. 11. Conclusion

1. Introduction and history.

The dominance of matter over antimatter in the Universe is now well explained. The explanation maybe a little bit too good because we have a large number of theoretical models and do not know which is the right one. The purpose of this talk is to review different ideas in baryogenesis and to see whether it is possible to distinguish between them by astronomical observations. The data are not conclusive however. It is even not known whether all the visible part of the Universe consists of matter or there could be some spatially separated regions formed of antibaryons and positrons. What is definite that our Galaxy is build of matter only. As for other far-away galaxies they may consist either of matter or antimatter but not both. We have to conclude that either baryons dominate over antibaryons all over the Universe or they are separated by large scale

$$\ell_B \gg \ell_{\text{gal}} \tag{1.1}$$

where ℓ_B is a characteristic size of the region occupied by baryons or by antibaryons and $\ell_{\text{gal}} = 10 \div 100$kps is the size of galaxies. The challenge is clear either to find a model of charge asymmetry of the Universe or a mechanism of matter-antimatter separation.

The history of the problem goes up to the celebrated paper by Sakharov (1967) in which three basic principles of the baryogenesis have been formulated. They are:

---

[*] From October 1, 1990 to September 31, 1990, Yukawa Institute for Theoretical Physics Kyoto University, Kyoto 606, Japan.

a. Baryonic charge nonconservation.

b. Breaking of particle-antiparticle symmetry.

c. Deviation from thermal equilibrium.

Further development has shown that neither of them is obligatory and the observed picture can be explained without any one of them.

After the paper by Kuzmin (1970) who proposed a slightly different from Sakharov (1967) version of the model the topic was forgotten for almost a decade. With advent of grand unification models it was understood that baryons are not conserved and the papers by Ignatiev, Krasnikov, Kuzmin, and Tavkhelidse (1978) and especially by Yoshimura (1978) have stimulated "big-bang" of papers on the subject which is still active.

The excess of baryons over antibaryons averaged over the scale $\ell < \ell_B$ is given by the expression

$$\beta = \frac{N_B - N_{\overline{B}}}{N_\gamma} = 2.10^{-8} h_{100}^2 \frac{\Omega_B}{\Omega_{\text{tot}}} = 10^{-9} - 10^{-10} \tag{1.2}$$

where $N_{B,\overline{B},\gamma}$ is the number density of corresponding particles, $h_{100} = H_0/100\text{km}$ /sec/ Mpc, H_0 is the present-day value of the Hubble parameter, and Ω, is the ratio of the total (sub-tot) or baryonic (sub-B) energy density to the closure energy density. Note that if $N_B = N_{\overline{B}}$ down to the temperature $T \approx m_N/50 = 20\text{MeV}$ then the annihilation would diminish N_B/N_γ to the vanishingly small value of 10^{-19}. So already at these temperatures baryons and antibaryons should be spatially separated by the distance $\ell > 100\text{sec}$ which is much larger than the horizon $\ell_h \approx 4 \cdot 10^{-3}\text{sec}$.

2. Three principles of baryogenesis.

a. Baryonic charge nonconservation is in a very good shape from theoretical point of view. There are many theories predicting transformation of baryons into other particles. As for experiment, neither proton or nucleus decay nor $(n - \overline{n})$-oscillations have been observed up to now and the only "experimental" evidence in favor of B-nonconservation is presented by the baryon asymmetry of the Universe. Inflationary Universe model makes the cosmological arguments very strong. Indeed if B is conserved, sufficiently long inflation with $H_I t_I > 70$ could not be realized.
b. The position of charge symmetry breaking is opposite to that of B-nonconservation. The decay $K_L^0 \rightarrow 2\pi$ and the different branching ratios in $K_{\ell3}$-decays unambiguously show that particles and antiparticles are really different. As for the theory, there are several theoretical models of CP-violation but neither of them can be considered as established.
c. The necessity of statistical nonequilibrium for baryogenesis follows from the fact that the equilibrium distribution functions are determined by the particle energy E and its chemical potential μ only. If charge is not conserved the corresponding chemical potential vanishes in equilibrium. Masses of particles and antiparticles should be the same because of the CPT-theorem. So the equilibrium number densities of particles and anti-particles should be equal and no charge asymmetry arises.

In the following sections we see how these principles work to generate $< B > \neq 0$ in the Universe and also discuss the models which are effective even if one of conditions $a, b,$ or c is not fulfilled.

3. Heavy particle decays.

By heavy particles we mean here those which mass is large in comparison with temperature. This historically first scenario of baryogenesis is now slightly out of fashion but it is still workable and could produce the observed asymmetry in some specific models. Baryosynthesis in grand unification models has been reviewed by Kolb and Turner (1983) and the reader is addressed there for the long list of references to earlier papers. Grand unification models meet difficulties in creating the observed baryon asymmetry, first, because the reheating temperature after inflation is generically below the grand unification energy scale and, second because the numerical value of β_{GUT} is rather low. Decays of gauge bosons X of grand unification indeed do not conserve baryonic charge but the resulting baryon asymmetry is small. For the case of kinetic (in contrast to chemical) equilibrium:

$$\beta \simeq (10^{-1} \div 10^{-2}) \frac{\epsilon}{Q \ln Q} \tag{3.1}$$

where $\epsilon = (\Gamma_X - \Gamma_{\overline{X}})/\Gamma_X, Q \approx m_{pl}\Gamma_X/\sqrt{K}m_X^2 \approx 10^2$, and $K \approx 10^2$ is the number of bosonic and $\frac{7}{8}$ -fermionic degrees of freedom in the plasma. If kinetic equilibrium with respect to the decay products is not restored the value of β is $\sqrt{\ln Q}$ times larger (Dolgov, 1980a).

If the decaying particles are far from equilibrium it would increase the result for β by one or even two orders of magnitude. Such a case can be realized in the process of (re)heating after inflation if the inflaton decays or the decays of heavy particles produced by coherent oscillations of the inflaton field do not conserve B (Dolgov and Linde, 1982). The problem of low reheating temperature still persists.

Another model of fast equilibrium breaking has been considered by several groups of authors (Lazarides, Panagiotakopoulos, and Shafi, 1986; Yamamoto, 1987; Mohapatra and Valle, 1987). It is assumed that there is a light boson g decaying with B-nonconservation. This boson is assumed to be coupled to Higgs field ϕ by the term

$$\mathcal{L}_{\text{int}} = \lambda|g|^2|\phi|^2 \tag{3.2}$$

Spontaneous symmetry breaking resulting in vacuum condensate of ϕ gives rise to a large mass of $g, \delta m_g^2 = \lambda| < \phi > |^2$. If the breaking of the symmetry is fast then the number density of g which was originally about the photon number density, $N_g \simeq 0.1T^3$, remains the same despite the large mass of $g, m_g \gg T$. So all of a sudden g-bosons become out-of-equilibrium and the baryon asymmetry originating from their decays would be unsuppressed. Unfortunately the asymmetry is deluted by the entropy produced by g-decays or $g\overline{g}$-annihilation.

It has been discovered recently that the sign of baryon excess generated by heavy particle decays is not prearranged by the sign of CP-symmetry breaking but also determined by the kinetics of the processes (Yokoyama, Kodama, Sato and Sato, 1987; Yokoyama, Kodama and Sato, 1988).

In all these models the characteristic energy scale is very high, far beyond the access of existing accelerator facilities. To meet the demands of experimentalists the model of low temperature baryogenesis has been constructed (Dimopoulos and Hall, 1987). It has been assumed that there exist new light particles with B-nonconserving interactions. The proton decay is properly suppressed because leptonic charge is conserved. Baryonic excess in the model is generated in the GeV-MeV range of temperature. What makes the model interesting is that its predictions can be checked in direct particle physics experiments.

4. Baryogenesis with conserved baryonic charge.

At first sight the idea seems crazy but the discovery of black hole evaporation (Hawking, 1974) makes it sound. Indeed as it has been noted by Zel'dovich (1976) charge asymmetric and B-conserving decays of evaporated particles can give rise to an excess of baryons over antibaryons outside of the black hole while the opposite amount of baryonic charge is buried inside the disappearing black hole. In this model baryonic charge can be conserved in particle interactions and all the nonconservation arises due to topological nontriviality of space-time.

To be more precise let us consider the following toy model. Let A be a massive boson decaying into heavy baryon H and light antibaryon $\overline{L}$ or vice versa, $\overline{H}$ and L. Because of CP-violation the branching ratios of the decays should be different

$$\Delta = \frac{\Gamma(A \to L\overline{H}) - \Gamma(A \to \overline{L}H)}{\Gamma_{\text{tot}}^A} \tag{4.1}$$

The produced baryons propagating in gravitational field of the black hole can be captured by the latter. The probability of capture is large for heavier particles. This would result in a net flux of baryonic charge from the black hole. The resulting baryon asymmetry of the Universe is given by the expression (Dolgov, 1980b):

$$\beta \approx 0.1 K^{-3/4} \frac{\Gamma_{\text{tot}}^a}{m_A} \frac{\rho_{BH}}{\rho_{\text{tot}}} \left(\frac{m_{pl}}{m_{BH}}\right)^{1/2} \Delta \tag{4.2}$$

Here $(\rho_{BH}/\rho_{\text{tot}})$ is the relative energy density of black holes with mass $M_{BH}, m_{pl} \approx 10^{19}$ GeV is the Planck mass and M_A is the mass of the decaying meson.

5. Baryogenesis, inflation and CP-violation.

Inflationary Universe model has opened new possibilities for baryogenesis. One could expect that quantum fluctuations in the expanding Universe produce a nonzero density of baryonic charge or a nonzero scalar field at microscopically small scale. Usually these fluctuations remains unobservable being averaged over macroscopic scale. Inflation could however drastically change the situation exponentially fast transforming microscales into macroscales. Moreover the amplitude of scalar field fluctuations is known to rise (Vilenkin and Ford, 1982; Linde, 1982) so that astronomically large regions with nonzero baryonic charge density if B is not conserved (Affleck and Dine, 1985) or regions filled by classical complex scalar field ϕ which simulates $C(CP)$-breaking can be formed. When inflation is over ϕ goes to the equilibrium point and if the latter is at $\phi = 0$ no CP-violation connected with ϕ survives. In a sense this resembles spontaneous breaking of $C(CP)$- invariance (Lee, 1973) but without domain walls. Another difference is, that $C(CP)$ is broken only for a short time when ϕ is out of equilibrium that is when $\phi \neq 0$. If baryogenesis proceeds just at that period the resulting charge density could be determined by ϕ and would have nothing to do with low energy CP- violation observed in particle physics. In what follows such a breaking of CP- invariance is called stochastic CP-breaking.

In principle explicit breaking of CP-invariance could be distinguished from spontaneous or stochastic ones by observations. If CP is broken explicitly then N_B/N_γ=const all over the Universe at least in the standard model. It has been shown by Yokoyama, Kodama and Sato (1988) however that in two-inflaton model there could be both regions with $B > 0$ and $B < 0$. Still models with spontaneous or stochastic CP-breaking offer richer possibilities. From the point of view of observations the result depends upon the relation between the present-day horizon size $\ell_u \approx 10^{10} y$ and the characteristic scale of variation of baryonic charge density ℓ_B.

If $\ell_B > \ell_u$ there is no qualitative difference with the case of explicit CP-violation. If $\ell_{gal} < \ell_B < \ell_u$ (Sato, 1981) there would be a lot of antimatter in our part of the Universe probably separated by voids from the regions filled by matter so that there should be no contradictions with observations. A feeble annihilation in those almost empty regions could be the source of the observed γ background radiation. The big contrast in baryonic charge density or in other words large isocurvature fluctuations existing in this model might give rise to unacceptable angular fluctuations of microwave background radiation (Efstathiou and Bond, 1987).

If CP is spontaneously broken in the first order phase transition during inflationary stage one may expect the formation of three dimensional baryonic or antibaryonic islands washed by the ocean of dark matter (Dolgov and Kardashev, 1986; Dolgov, Illarionov, Kardashev and Novikov, 1987). The island size is not specified by the model and can be the order of ℓ_u but still inside the present-day horizon. If this is the case one would observe no luminous matter outside certain red-shift value. The model predicts low angular fluctuations of the microwave background radiation because the latter comes to us from baryon-empty space. It has been noted in the above mentioned papers that the distribution of baryonic matter inside the islands could rather naturally be spatially periodic. In fact the periodicity in matter distribution is a generic feature of models of baryogenesis with spontaneous or stochastic CP-breaking during inflationary stage (Chizhov and Dolgov, 1990). Since the result is rather surprising we discuss it in some details in the next section.

6. Baryogenesis and large scale Universe structure.

A complex scalar field ϕ could acquire a nonzero vacuum expectation value either as a result of a phase transition in the course of the Universe expansion or because of the rise of quantum fluctuations during inflationary stage. Anyway if not generic it is at least very natural that a classical scalar field is generated during inflation. It is also natural that ϕ depends on the space coordinates and that the characteristic scale of its variation, r_0, is inflated up to cosmologically large size. We assume that it is sufficient to expand $\phi(r, t)$ in r only up to the first term:

$$\phi(r) = \phi_0(1 + \frac{rn}{r_0}) \tag{6.1}$$

where n is a unit vector.

During inflationary stage ϕ remains almost constant because of the large Hubble friction. When inflation is over ϕ starts to oscillate around equilibrium with a decreasing amplitude. Since ϕ is a complex field it induces CP-violation. If $(\dot{\phi}/\phi)$ is slow in comparison with baryosynthesis rate then the number density of the produced baryons gives a snapshop of $\phi(r, t_B)$ where t_B is the moment of baryogenesis. A possible anharmonicity in the equation of motion of ϕ would give rise to oscillations of ϕ as a function of r despite that initially ϕ is a monotonic function of r. To see this let us consider the potential of ϕ of the form

$$u(\phi) = \lambda|\phi|^{2n}/n \tag{6.2}$$

Let us forget for the time being about the Universe expansion and possible variation of the phase of ϕ. The space derivative term in the equation of motion is not essential because it is exponentially damped by inflation. So the dependence on the space coordinates can be taken into account adiabatically. In this case the following conservation law is valid

$$|\dot{\phi}|^2 = (\lambda/n)(|\phi_i|^{2n} - |\phi|^{2n}) \tag{6.3}$$

We assume for simplicity that initially $\dot{\phi} = 0$ so that $\phi_i(r)$ is the initial value of ϕ. Making the substitutions $t = (n/\lambda)^{1/2}\phi_i^{1-n}(r)y$ and $\phi = \phi_i(r)z(y)$ we see that $z(y)$ satisfies the equation $(z')^2 = 1 - z^{2n}$ so it is a periodic function of y with the fixed period $2B(1/2, 1/2n)/n$. So if $\phi_i(r)$ is a linear function of r (6.1) the function $\phi(r, t)$ is an oscillating function of r with slowly changing amplitude and the period

$$\frac{\Delta r}{r_0} = \frac{2B(1/2, 1/2n)}{\lambda^{1/2}n^{1/2}(n-1)t\phi_0^{n-1}} \tag{6.4}$$

For details see Chizhov and Dolgov (1990).

Thus we came to the conclusion that models of early baryosynthesis rather naturally predict periodic distribution of luminous matter in the Universe. This is in concordance with the data presented on this Conference by Szalay (1990). A very interesting possibility is that of ϕ oscillating around zero, so that baryon and antibaryon layers alternate. If this is true the neighboring part of the Universe at the distance of about 100Mpc would consist of antimatter. Large angular fluctuations of microwave background radiation generated by the isocurvature density perturbations would be absent if the island Universe model is valid and we are situated near the center of the island. Another possibility is that ϕ has oscillated around nonzero complex value or in addition to CP-violation induced by ϕ there is a constant contribution into CP-odd amplitude coming from another mechanism. In this case the baryonic number density would oscillate around nonzero average value.

7. Baryonic charge condensate.

If a scalar field has nonzero baryonic charge and if the latter is nonconserved quantum fluctuations during inflation result in nonvanishing baryonic charge density,

$$< B^2 >^{1/2} \approx H_I^3 \qquad (7.1)$$

This fact has been used by Affleck and Dine (1985) for a baryogenesis scenario. In the original version of the model the generated baryon-to-entropy ratio was huge, $\beta \gtrsim 1$ but it seems that more natural is $\beta \ll 1$ (Dolgov and Kirilova, 1989).

Let us consider the following toy model which well reproduces the essential features of the realistic scenario. Let the scalar field possessing nonzero baryonic charge is described by the Lagrangian

$$\mathcal{L}(\chi) = |\partial \chi|^2 - [m^2|\chi|^2 + \frac{\lambda_1}{2}|\chi|^4 + \frac{\lambda_2}{4}\chi^4 + \frac{\lambda_2^*}{4}\chi^{*4}] \qquad (7.1)$$

Baryonic current is defined in the usual way

$$j_\mu^B = i\chi^* \overleftrightarrow{\partial}_\mu \chi \qquad (7.2)$$

and it is nonconserved if $\lambda_2 \neq 0$.

There exist the so-called flat directions in the potential in many supersymmetric GUT theories. In our model it corresponds to $m = 0$ and $\lambda_1 = |\lambda_2|$. In this case quantum fluctuations in De Sitter background push the field along the valley. The fluctuations of χ in the orthogonal direction give rise to nonzero $< B^2 >$ which is about H_I^3 by dimensional ground. When inflation ends the field starts to roll down to the origin due to a small mass term, $m \ll H_I^2$. When it moves along the valley, B is not conserved and in the average $< B >= 0$. For sufficiently small χ, $m^2|\chi|^2$ dominates in the potential energy and $< B >\neq 0$ is conserved. During this stage

χ decays into fermions generating the baryon asymmetry of the Universe. Thus the amount of the baryonic charge accumulated by χ is determined by quantum fluctuations on the inflationary stage and can be of either sign (clock-wise or anti-clock-wise rotation). If the damping of the phase variation of χ is caused only by the Hubble friction then χ quickly rotates around the origin and correspondingly the produced baryon asymmetry is very large.

This is not necessarily the case however. The coupling of χ to fermions

$$\mathcal{L}_{\text{int}} = g\chi\overline{\phi}_i\phi_j + \text{h.c.} \tag{7.3}$$

results in strong particle production when χ oscillates perpendicular to the valley and to the damping of the baryonic charge. The oscillation frequency is $\omega_{\text{osc}} = \sqrt{\lambda} < \chi >$ where $< \chi >$ is the field amplitude in the valley, while effective mass of fermions induced by $< \chi >$ is $g < \chi >$. The particle production rate is very sensitive to the relation between λ and g^2. With relatively small variation of (λ/g^2) it is possible to proceed from complete suppression of the phase fluctuations to complete unsuppression. This opens an interesting possibility of generation a small baryonic asymmetry together with a large leptonic one if $(B-L)$ is not conserved or different leptonic charges are not conserved separately (Dolgov and Kirilova, 1989). Cosmological model with large leptonic asymmetries was discussed by Kolb and Turner (1987). The electronic asymmetry of the order of

$$(L_e/N_\gamma) \approx 10^{-2} \tag{7.5}$$

would make primordial nucleosynthesis with large baryonic energy density, $\Omega_B \approx 1$, compatible with observations (Terasawa and Sato, 1988).

As it has been already noted the baryogenesis in this model could proceed without CP-violation with the sign of baryonic asymmetry determined by quantum fluctuations. The characteristic scale of the region with definite baryonic charge during inflation is evaluated as

$$\ell_B^{(I)} = H_I^{-1}\exp\{(\frac{2\pi < \chi >}{H_I})^2\} \simeq H_I^{-1}\exp\{\frac{O(1)}{\sqrt{\lambda}}\} \tag{7.6}$$

To get $\ell_B > \ell_{\text{gal}}$ at the present time one needs $\lambda < 10^{-4}$.

If $(g^2/\lambda)_L > (g^2/\lambda)_B$ then the leptonic asymmetry would be larger than the baryonic one since the damping of leptonic charge due to particle production would be weaker. If simultaneously $\lambda_L > \lambda_B$ then $\ell_L < \ell_B$ and if the scale of leptonic charge variation is smaller than the present-day horizon, say, $\ell_L \approx \ell_{\text{gal}}$ then there could be observable variations of the primordial abundances of light elements e.g., of ^{4}He in different regions of the sky. This gives a chance to confirm (but not to reject) the model.

8. Baryogenesis and CPT.

CPT is the only rigorously proven symmetry in the family of different combinations of C, P and T. In fact all others are known to be broken. There is no known way to describe consistently CPT-breaking but if it is nevertheless broken the masses of particle and antiparticles should be different and this gives rise to baryonic asymmetry in thermal equilibrium,

$$\beta = \frac{N_B}{N_\gamma} \approx \frac{1}{\pi} \frac{\delta m^2}{T_{\Delta B}^2} \tag{8.1}$$

where $\delta m^2 = \overline{m}^2 - m^2$ is the mass difference between antiparticles and particles and $T_{\Delta B}$ is the temperature when the processes with B-nonconservation are frozen.

CPT-violation can be also mimicked by a time-dependent external field. In this case it also is possible to generate baryon asymmetry in thermal equilibrium (Cohen and Kaplan, 1988) but now without breaking of any sacred principles. To this end a scalar field coupled to baryonic current has been introduced

$$\mathcal{L} = \frac{1}{f} \partial_\mu \phi \cdot j_B^\mu \tag{8.2}$$

where f is a constant with dimension of mass. For homogeneous but time-dependent field $\phi(t)$ this coupling induce nonzero baryonic charge chemical potential $\mu_B = \dot{\phi}/f$, and correspondingly nonzero baryonic charge density.

More or less homogeneous field ϕ could be generated during inflation and in the post-inflationary stage it would go down the potential in accordance with classical equation of motion. While B-nonconserving processes are in equilibrium the baryonic charge density is given by the expression

$$N_B = \frac{1}{6} q_B \mu_B T^3 (1 + \frac{\mu_B^2}{\pi^2 T^2}) \tag{8.3}$$

where q_B is the baryonic charge of the particles (quarks). When these processes freeze and j_B^μ becomes effectively conserving, expression (8.2) formally vanishes but just because of the breaking of equilibrium with respect to B-nonconservation baryonic asymmetry which existed in equilibrium do not disappear completely. The observed value of β can be obtained in this model if $f > 3 \cdot 10^{13}$GeV (Cohen and Kaplan, 1988). The model predicts isocurvature density perturbations (Turner, Cohen and Kaplan, 1989). In fact this is a generic feature not only of this model but also of all the models discussed in secs. 5 - 7 above.

9. Baryosynthesis by topological defects.

It is well known that topological defects formed by the phase transitions during Universe cooling down could give rise to B-nonconserving processes. An example of that is the baryon decay catalysis by magnetic monopoles of grand unification models (Rubakov, 1981, 1982; Callan, 1982). Unfortunately this process could not produce the observed baryon asymmetry of the Universe partly because of low monopole number density and partly because of little deviation from thermal equilibrium.

Cosmic strings could participate in creation of baryon asymmetry in two ways, first, by nonequilibrium production of heavy particles either from collapsing string loops (Bhattacharjee, Kibble and Turok, 1982) or from cusps or kinks (Kawasaki and Maeda, 1988) and, second, by the catalysis of B-nonconservation analogously to the Rubakov-Callan effect (Gregory, Davis and Brandenberger, 1988). A similar mechanism which is somewhat more effective is the nonabelian Aharonov-Bohm effect (Perivolaropoulos, Matheson, Davis and Brandenberger, 1990).

10. Baryosynthesis on electroweak scale.

All previously considered scenario of baryogenesis are based on physics beyond the standard $SU(3) \times SU(2) \times U(1)$ model. Baryon asymmetry however might be generated by the usual electroweak interactions with the known particles only.

Formally the Lagrangian of electroweak theory conserves baryonic charge but quantum corrections are known to break the conservation (t' Hooft, 1976a,b). The effect is connected with quantum tunneling in the functional space of gauge field with a change of topology and so it is exponentially suppressed in $1/\alpha$. It makes the effect practically unobservable. One might hope however that with rising temperature or energy the B-changing transition would proceed not via the barrier penetration but due to classical motion over the barrier. The rate of the B-nonconserving processes Γ_B has been calculated by Kuzmin, Rubakov and Shaposhnikov (1985) with exponential accuracy and by Arnold and McLerran (1987) who have evaluated the pre-exponential factor. The result is

$$\frac{\Gamma_B}{H} = 10^{24} (\frac{m_w}{T})^2 \exp(-120 C \frac{m_w}{T}) \tag{10.2}$$

where C is a numerical constant of the order of unity and H is the Hubble parameter. The result shows that $\Gamma_B/H \gg 1$ for T above a few hundred GeV's and then quickly drops to zero. This means that baryonic charge disappear at large temperatures because of thermal equilibrium and later on B-nonconserving processes goes out of equilibrium so fast that there is no time to generate baryon asymmetry. So any pre-existing baryon asymmetry would be washed out on electroweak temperature scale and new one could not be generated. This is true of course if initially $B - L = 0$, because electroweak interactions even with quantum anomaly conserve $B - L$.

There are several ways out of this difficulty. The simplest one is the model with low temperature baryogenesis, $T_{\Delta B} < T_{EW}$. Another possibility is initially nonzero $(B - L)$. This could be either because of high temperature baryogenesis with nonconserved $(B - L)$ or due to generation of leptonic asymmetry by Majorana mass term as has been proposed by Fukugita and Yanagita (1986).

Later on Shaposhnikov (1988) advocated the idea that the energy of the system is degenerate with respect to the topological charge of the gauge fields. In this case the spontaneous formation of gauge field configuration with nonzero topological charge should take place in the hot primeval plasma above the symmetry breaking point. When the electroweak symmetry is broken the degeneracy is destroyed and the system goes to the state with zero topological charge producing equal baryonic charge. This process could provide the observed baryon asymmetry for sufficiently low mass of the Higgs boson, $m_H < 60\text{GeV}$.

As a whole the model of electroweak baryogenesis is very attractive but there is a subtle point concerning the change of topology in the space of gauge fields. At high temperatures it can be realized by specific field configurations called sphalerons (Dashen, Hasslacher and Neveu, 1974; Klinkhammer and Manton, 1984; Forgacs and Horvath, 1984). Estimate (10.2) is valid if sphalerons are in thermal equilibrium with primeval plasma. The problem of formation of coherent field configuration with large size, $L = m_W^{-1} \gg T^{-1}$ is not clear however. One may speak about it if the thermal energy of particles inside the configuration $E_L = T^4 L^3 = T^4/m_W^3$ is smaller than the energy of this configuration $M_L = m_W/\alpha_W \approx 10\text{TeV}$. Otherwise thermal fluctuations should destroy it (if the latter is not a topologically stable object). The condition $E_L < M_L$ is realized in the temperature region for which $\Gamma_B/H = 10^{-2} \div 10^{-3}$ and the B-nonconserving processes are not effective. Another problem which deserves consideration in this connection is the production of classical field configuration in particle collision.

There is a recent activity concerning B-nonconservation in two-body collisions (Ringwald, 1988; McLerran, Vainshtein and Voloshin, 1990). The authors have argued that electroweak interactions become strong in TeV-energy-range and that the baryonic charge nonconservation in multiparticle production is not suppressed. This conjecture is subject to criticism (Zakharov, 1990) and theoretically the problem is not yet settled down but if it is confirmed by experiment in the next generation accelerators it would mean that baryon asymmetry of the Universe could be created in the standard $SU(3) \times SU(2) \times U(1)$ model.

11. Conclusion.

There are plenty of models in the market satisfying a variety of tastes. All of them could give $\beta = 10^{-9} - 10^{-10}$ with large proton life-time, $\tau_p > 10^{32}y$. Low temperature baryogenesis is now in fashion so that the constraints imposed on (re)heating temperature in inflationary models are satisfied. The models without B-nonconservation, or without C- and CP- violation, or without thermal equilibrium breaking are possible.

Inflation opens a new possibility of charge symmetric Universe as a whole with B and $\overline{B}$ separated by large distance. Early baryosynthesis models with spontaneous or stochastic C, CP-breaking at inflationary stage naturally predict periodic distribution of matter in the Universe. The characteristic scale is model dependent and the recently observed structure might be compatible with it. More natural and more exciting possibility is that of alternating baryonic and antibaryonic layers. If this is the case a care should be taken of angular variation of the microwave background temperature. The latter would be sufficiently small in spherically symmetric case if the Earth is in the centre of the world as the archaic people believed. An interesting question in this connection is whether it possible to distinguish antibaryonic matter from the baryonic one at the distance of about 100Mpc. In any case new models of baryosynthesis naturally give rise to isocurvature perturbations.

A testable by observations prediction indirectly connected with baryogenesis is the variable at galactic scale leptonic asymmetry which could give rise to spatial variations of the primordial abundances of light elements. Except for these striking possibilities there are seemingly no chances to test the discussed above models by astronomical observations. Particle physics might slightly help in this respect if baryonic charge nonconservation at TeV-energy scale is observed.

References

Affleck, I. and Dine, M. (1985). A New Mechanism for Baryogenesis. Nucl. Phys. **B249**, 361-380.

Arnold, P. and McLerran, L. (1987). Sphalerons, Small Fluctuations, and Baryon Number Violation in Electroweak Theory. Phys. Rev. **D36**, 581-595.

Bhattacharjee, R., Kibble, T.W., and Turok, N. (1982). Baryon Number Generation from Collapsing Cosmic Strings. Phys. Lett. **B119**, 95-96.

Callan, C,G. (1982). Disappearing Dyons. Phys. Rev. **D25**, 2141-2146.

Chizhov, M.V. and Dolgov, A.D. (1990). Baryogenesis and Large Scale Structure of the Universe. Submitted to ZhETF.

Cohen, A. and Kaplan, D. (1988). Spontaneous Baryogenesis. Nucl. Phys. **B308**, 913-928.

Dashen, R., Hasslacher, B., and Neveu, A. (1974). Nonperturbative Methods and Extended Hadron Models in Field Theory. III. Four-Dimensional Nonabelian Models. Phys. Rev. **D10**, 4138-4142.

Dimopoulos, S. and Hall, J. L. (1987). Baryogenesis at the MeV Era. Phys. Lett. **196B**, 135-141.

Dolgov, A.D. (1980a). Kinetics of Generation of Baryon Asymmetry of the Universe. Yadernaya Fizika **32**, 1606-1621.

Dolgov, A.D. (1980b). Quantum Evaporation of Black Holes and Baryon Asymmetry of the Universe. ZhETF, **79**, 337-349.

Dolgov, A.D., Illarionov, A.F., Kardashev, N.S., and Novikov, I.D. (1987). Cosmological Model of Baryonic Island. ZhETF. **94**, 1-14.

Dolgov, A.D. and Kardashev, N.S. (1986). Spontaneous Baryonic-Antibaryonic Symmetry Violation and the Island Universe Model. Space Reseacrh Int. Preprint-1190.

Dolgov, A.D. and Kirilova D.P. (1989). Condensate of Baryonic Charge and Baryogenesis. Preprint JINR P2-89-873.

Dolgov, A.D. and Linde, A.D. (1982). Baryon Asymmetry in the Inflationary Cosmology. Phys. Lett. **116B**, 329-334.

Efstathiou, G. and Bond, J.B. (1987). Microwave Anisotropy Constraints on Isocurvature Baryon Models. Mon. Not. Roy. Astr. Soc. **227**, 33p-38p.

Forgacs, P. and Horvath, Z. (1984). Topology and Saddle Points in Field Theories. Phys. Lett. **138B**, 397-401.

Fukugita, M. and Yanagita, T. (1986). Baryogenesis without Grand Unification. Phys. Lett. **174B**, 45-47.

Gregory, R., Davis, A.C., and Brandenberger, R. (1988). Cosmic String Catalysis of Skyrmion Decay. Nucl. Phys. **B323**, 187-208.

Hawking, S.W. (1974). Black Hole Explosions? Nature. **248**, 30-31.

Ignatiev, A,.Yu., Krasnikov, N.V., Kuzmin, V.A., and Tavkhelidze, A.N. (1978). Universal CP-Noninvariant Superweak Interaction and Baryon Asymmetry of the Universe. Phys. Lett. **76B**, 436-438.

Kawasaki, M. and Maeda, K. (1988). Baryon Number Generation from Cosmic String Loops. Phys. Lett. **208B**, 84-88.

Klinkhamer, F.R. and Manton, N.S. (1984). A Saddle Point Solution in the Weinberg Salam Theory. Phys. Rev. **D30**, 2212-2220.

Kolb, E.W. and Turner, M.S. (1983). Grand Unified Theories and the Origin of the Baryon Asymmetry. Ann. Rev. Nucl. Part. Sci. **33**, 645-696.

Kolb, E.W. and Turner, M.S. (1987). Electroweak Anomaly and Lepton Asymmetry. Mod. Phys. Lett. **A2**, 285-291. Kuzmin, V.A. (1970). CP-Noninvariance and Baryon Asymmetry of the Universe. Pis'ma ZhETF. **12**, 335-337.

Kuzmin, V.A., Rubakov, V.A., and Shaposhnikov, M.E. (1985). On Anomalous Electroweak Baryon Number Nonconservation in the Early Universe. Phys. Lett. **B155**, 36-42.

Lazarides, G., Panagiotakopoulos, C., and Shafi, Q. (1986). Baryogenesis and the Gravitino Problem in Superstring Models. Phys. Rev. Lett. **56**, 557-560.

Lee, T.D. (1973). A Theory of Spontaneous T-Violation. Phys. Rev. **D8**, 1226-1239.

Linde, A.D. (1982). Scalar Field Fluctuations in the Expanding Universe and the New Inflationary Universe Scenario. Phys. Lett. **116B**, 335-339.

McLerran, L., Vainshtein, A., and Voloshin, M. (1990). Electroweak Interactions Become Strong at Energy above 10TeV. Phys. Rev. **D42**, 171-179.

Mohapatra, C.N. and Valle, J.W.F. (1987). Late Baryogenesis in Superstring Models. Phys. Lett. **186B**, 303-308.

Perivolaropoulos, L., Matheson, A., Davis, A.C., and Brandenberger, R.H. (1990). Nonabelian Aharonov-Bohm Baryon Decay Catalysis. Phys. Lett. **B245**, 556-560.

Ringwald, A. (1988). Rate of Anomalous Electroweak Baryon and Lepton Number Violation at Finite Temperature. Phys. Lett. **B201**, 510-516.

Rubakov, V.A. (1981). Superheavy Magnetic Monopoles and Proton Decay. Pis'ma ZhETF. **33**, 658-660.

Rubakov, V.A. (1982). Adler-Bell-Jackiw Anomaly and Fermion Number Breaking in the Presence of a Magnetic Monopole. Nucl. Phys. **B203**, 311-348.

Sakharov, A.D. (1967). Breaking of CP-Invariance, C-Asymmetry, and Baryon Asymmetry of the Universe. Pis'ma ZhETF. **5**, 32-35.

Sato, K. (1981). Cosmological Baryon Number Domain Structure and the First Order Phase Transition of a Vacuum. Phys. Lett. **99B**, 66-70.

Shaposhnikov, M.E. (1987). Baryon Asymmetry of the Universe in Standard Electroweak Theory. Nucl. Phys. **B287**, 757-775.

Szalay, A. (1990). Proceedings of this Conference.

Turner, M.S., Cohen, A.G. and Kaplan, D.B., (1989). Isocurvature Baryon Number Fluctuations in Inflationary Universe. Phys. Lett. **B216**, 20-26.

Terasawa, N. and Sato, K. (1988). Lepton and Baryon Number Asymmetry of the Universe and Primordial Nucleosynthesis. Prog. Theor. Phys. **80**, 468-476.

t'Hooft, G. (1976a). Symmetry Breaking through Bell-Jackiw Anomalies. Phys. Rev. Lett. **37**, 8-11.

t'Hooft, G. (1976b). Computation of the Quantum Effects due to a Four-Dimensional Pseudoparticle. Phys. Rev. **D14**, 3432-3450.

Vilenkin, A. and Ford, L.H. (1982). Gravitational Effects upon Cosmological Phase Transition. Phys. Rev. **D26**, 1231-1241.

Yamamoto, K. (1987). A Model for Baryogenesis in Superstring Unification. Phys. Lett. **194B**, 390-396.

Yokoyama, J., Kodama, H., Sato, K., and Sato, N. (1987). Baryogenesis in the Inflationary Universe. The Instantaneous Reheating Model. Int. J. Mod. Phys. **A2**, 1808-1828.

Yokoyama, J., Kodama, H., and Sato, K. (1988). Baryogenesis through Nonequilibrium Processes in Inflationary Universe. Prog. Theor. Phys. **79**, 800-818.

Yoshimura, M. (1978). Unified Gauge Theory and the Baryon Number of the Universe. Phys. Rev. Lett. **41**, 281-284; 42, 746 (Erratum).

Zakharov, V. (1990). Classical Corrections to Instanton Induced Interactions. Preprint TPI-MINN-90/7-T.

Zel'dovich, Ya.B. (1976). Charge Asymmetry of the Universe as a Result of Black Hole Evaporation and the Asymmetry of Weak Interactions. Pis'ma ZhETF. **24**, 29-32.

FORMATION OF TOPOLOGICAL DEFECTS
IN THE INFLATIONARY UNIVERSE

Jun'ichi Yokoyama and Michiyasu Nagasawa
Department of Physics, Faculty of Sciences,
The University of Tokyo, Tokyo 113, Japan

ABSTRACT. It is argued that topological defects of typical grand unification scales are produced by the curvature-induced phase transition more likely than by the Kibble mechanism. Properties of the new phase transition mechanism is clarified and its cosmological consequences are discussed.

It was more than a decade ago when unified gauge theories of elementary interactions got united with cosmology of the early universe. One of the most important predictions of these unified theories is that broken symmetry today would be restored at high temperatures. If we apply it to the conventional big bang cosmology naively, the universe would have experienced various types of phase transitions at different energy scales.

Some of them may have left observable relics or topological defects. They are classified to several types, namely, domain walls, strings, monopoles, and textures. Which type of relics would appear is determined by the homotopy class of the vacuum manifold after phase transition. First domain walls appear in the phase transition $\mathcal{G} \longrightarrow \mathcal{H}$ if the resultant vacuum manifold $\mathcal{M} \equiv \mathcal{G}/\mathcal{H}$ has the property $\Pi_0(\mathcal{M}) \neq \mathbf{1}$. The simplest example is to break a discrete symmetry as $\mathcal{G} = Z_2$, $\mathcal{H} = \mathbf{1}$, $\mathcal{M} = Z_2$, and $\Pi_0(\mathcal{M}) = Z_2$ [1]. Second the condition for the existence of strings in symmetry breaking $\mathcal{G} \longrightarrow \mathcal{H}$ is that the vacuum manifold $\mathcal{M}$ is not simply connected, or $\Pi_1(\mathcal{M}) \neq \mathbf{1}$. Strings will be present when, for example, $U(1)$ symmetry is broken [2]. Third monopoles are produced if the manifold of degenerate vacuum states $\mathcal{M}$ contains unshrinkable surfaces, or it satisfies $\Pi_2(\mathcal{M}) \neq \mathbf{1}$. This condition is fulfilled if the vacuum manifold contains $U(1)$ symmetry. Thus grand unified theories, in which electromagnetic $U(1)$ symmetry is unified in a larger group, inevitably predicts the existence of magnetic monopoles [3]. Finally textures may appear if $\Pi_3(\mathcal{M}) \neq \mathbf{1}$. For example, by breaking a global $SU(2)$ simmetry one can produce global textures [4].

It has been described by the Kibble mechanism [5] how they may be produced in the cosmological phase transitions. To illustrate this, consider a simple Lagrangian,

$$\mathcal{L}_\chi = \frac{1}{2}(D_\mu \chi)^a (D^\mu \chi)^a - V_0[\chi], \qquad V_0[\chi] = \frac{\lambda}{4}(\chi^a \chi^a - v^2)^2, \qquad (1)$$

where $D_\mu = \partial_\mu - ieA_\mu$ or $D_\mu = \partial_\mu$ depending on whether the symmetry is gauged or not. The superscript a is the index of internal degree of freedom. At high temperature, the potential acquires finite temperature correction,

$$V_T[\chi] = V_0[\chi] + (c_1 e^2 + c_2 \lambda) T^2 \chi^a \chi^a + ..., \tag{2}$$

and the critical temperature of symmetry restoration is given by $T_c \approx \sqrt{\lambda/(c_1 e^2 + c_2 \lambda)} v$, where c_1 and c_2 are constants determined by the structure of the Lagrangian. The stability of the potential against radiative corrections requires $\lambda \lesssim \alpha^2$ where α is the gauge coupling strength. Hence we have $T_c \sim ev \sim v$, of order of the energy scale of the model.

Among the above four types of defects, strings or global textures of the typical grand unification scale with $v \approx 10^{16}$ GeV may provide the origin of density fluctuations responsible for the large-scale structure formation [6]. On the other hand, magnetic monopoles of this scale would cause a disaster because they are overproduced to drive the universe to recollapse soon after the phase transition, if it is properly described by the above Kibble mechanism [7].

There is, however, a crutial assumption in the Kibble's scenario that is not necessarily justified in the early universe, the assumption of thermal equilibrium at the grand unification era. In the Kibble mechanism, it has been taken for granted that the scalar field χ was already in thermal equilibrium well before $T = T_c$. However, in order to attain thermal equilibrium from an arbitrary initial state, it is necessary that particle interaction rate with χ, Γ, should exceed the cosmic expansion rate H [8]. That is,

$$\Gamma = \langle n\sigma c \rangle \approx \frac{N}{\pi^2} T^3 \frac{\alpha^2}{T^2} > H = \left(\frac{8\pi^3 g_* T^4}{90 M_{pl}^2}\right)^{\frac{1}{2}}, \tag{3}$$

where we have considered thermalization by massless (gauge) particles. Hence T should satisfy

$$T < T_{eq} \equiv 3 \times 10^{14} \left(\frac{\alpha}{0.02}\right)^2 \left(\frac{N}{10}\right) \left(\frac{g_*}{200}\right)^{-\frac{1}{2}} \text{GeV}, \tag{4}$$

where α, N and g_* are (gauge) coupling strength, number of relativistic modes interacting with χ, and total relativistic degrees of freedom, respectively. Though all these parameters are model dependent, it is not always true that T_{eq} is as large as T_c for cosmologically interesting topological defects.

Thus the universe may never have experienced the symmetric phaseof χ. This means that we cannot calculate the abundance and distribution of the defects from first principle but that they are determined by the initial configuration of the classical universe, unless it started evolution in thermal equilibrium state so that the Kibble mechanism could work.

It is more natural, however, to expect that our universe started its classical evolution out of a chaotic state goverened by quantum and thermal fluctuations and that it underwent inflation [9] or exponential expansion to be isotropic and globally homogeneous as observed today [10]. It is after the reheating epoch that the universe was first dominated by radiation in thermal equilibrium. Unfortunately, the maximum temperature it experienced, or the reheat temperature, cannot be so high as the grand unification scale in order to avoid too much gravitational waves or density fluctuations [11] and/or too many gravitinos [12] to be produced after inflation. Therefore in such a scenario thermal phase transitions of that energy scale cannot take place. Thus it is impossible to keep adequate density of, say, cosmic strings or global textures to form large-scale structures thereby.

In order to resolve this problem we must invent a mechanism of non-thermal phase transitions. Fortunately, if we realize the fact that cosmic expansion is so fast in the early universe that the finite-curvature effect may play an important role, we may obtain a natural mechanism as we show below [13].

In general, scalar field χ may couple with the spacetime curvature scalar $\mathcal{R}$. That is, χ's potential may have an additional term $V_c[\chi] = \frac{1}{2}\xi\mathcal{R}\chi^a\chi^a$, where ξ is the coupling strength between χ and $\mathcal{R}$. Such a coupling term is, of course, meaningless in the Minkowski spacetime where $\mathcal{R} = 0$, but this term may affect or even control the phase transition in the universe [14]. The values of ξ often considered in literatures are $\xi = 1/6$, the conformal coupling, and $\xi = 0$, the minimal coupling. Even in the latter case, radiative corrections induce nonzero ξ. For example if χ is gauged, we have $\xi = O(e^2) > 0$. Hence we consider the case ξ has some positive value. If $\xi\mathcal{R} > \lambda v^2$, the symmetry of χ is restored. In the inflationary stage, the scalar curvature is given by $\mathcal{R} \cong 12H^2$. In most of viable inflationary universe models such as chaotic inflation [10], extended inflation [15], or soft inflation [16] models, the Hubble parameter decreases gradually even in the inflationary phase. Hence it is possible that phase transition takes place during inflation due to the gradual decrease of $\mathcal{R}$. In such a scenario phase transition is triggered by quantum fluctuations intrinsic to the inflating spacetime rather than thermal fluctuations. Here we discuss the dynamics of phase transitions during inflation as well as their cosmological implications.

To do this we adopt the following simple Lagrangian for the moment.

$$\mathcal{L}_\chi = \frac{1}{2}(\partial\phi)^2 - \frac{1}{2}m^2\phi^2 + \frac{1}{2}(\partial\chi)^2 - V[\chi], \quad V[\chi] = \frac{\lambda}{4}(\chi^2 - v^2)^2 + \frac{1}{2}\xi\mathcal{R}\chi^2. \tag{5}$$

The minimally coupled scalar field ϕ with mass $m \sim 10^{13}$GeV, which is constrained by the magnitude of relic density fluctuation, induces the chaotic inflation and χ is responsible to domain wall formation. In the inflationary stage, evolution of $\phi(t)$, the scale factor $a(t)$, and $\mathcal{R}(t)$ is given by

$$\phi(t) = \phi_0 - \frac{mM_{pl}}{2\sqrt{3\pi}}(t - t_0), \quad a(t) = a_0\exp\left[\frac{2\pi}{M_{pl}^2}(\phi_0^2 - \phi(t)^2)\right], \quad \mathcal{R}(t) = \frac{16\pi m^2}{M_{pl}^2}\phi(t)^2 - 2m^2,$$
$$\tag{6}$$

respectively. Thus $\mathcal{R}$ decreases gradually as ϕ evolves towards $\phi = 0$.

The symmetric state $\chi = 0$ becomes classically unstable when $M_{\chi\text{eff}}^2 \equiv V''[\chi = 0] = -\lambda v^2 + \xi\mathcal{R} < 0$. However, while $|M_{\chi\text{eff}}^2| \ll H^2$, quantum fluctuations dominates its evolution. In this stage, evolution of the long-wavelength part of χ is described by the following Langevin equation [17].

$$\frac{d\chi}{du} = -\frac{V'[\chi]}{3H^2(u)} + \frac{f(u)}{H(u)}, \tag{7}$$

where we have used the e-folding number of cosmic expansion factor $u \equiv \ln a(t) - \ln a(t_0)$ as a new time variable and the Hubble parameter is given by $H^2(u) = H_0^2 - \frac{2}{3}m^2u$. Here

$f(u)$ represents quantum noise with the correlation $\langle f(u_1)f(u_2)\rangle \simeq \frac{H^4(u)}{4\pi^2}\delta(u_1 - u_2)$. From the above Langevin equation yields the following equation of motion of $\langle\chi^2(u)\rangle$.

$$\frac{d}{du}\langle\chi^2(u)\rangle \simeq \frac{2M^2_{\chi\text{eff}}}{3H^2(u)}\langle\chi^2(u)\rangle + \frac{H^4(u)}{4\pi^2}, \tag{8}$$

where the term proportional to $\langle\chi^4(u)\rangle$ has been neglected. The first term on the right-hand side represents classical potential force while the second term induces diffusion due to quantum fluctuations. Since the typical magnitude of $\chi(u)$ is $\sqrt{\langle\chi^2(u)\rangle}$, we may find when its evolution becomes governed by the potential force rather than quantum fluctuations by comparing the magnitudes of these two terms.

We can readily solve the above equation to get,

$$\langle\chi^2(u)\rangle = \left[\langle\chi^2(0)\rangle + \frac{H_0^2}{4\pi^2}\int_0^u du' e^{8\xi u'}\left(1 - \frac{2m^2u'}{3H_0^2}\right)^{\frac{\lambda v^2}{m^2}+2\xi+1}\right]e^{-8\xi u}\left(1 - \frac{2m^2u}{3H_0^2}\right)^{-\frac{\lambda v^2}{m^2}+2\xi}, \tag{9}$$

where we have defined t_0 the epoch $M^2_{\chi\text{eff}}$ vanishes. If $\xi \gtrsim 1/12$ and inflation lasts long enough by the time when $M^2_{\chi\text{eff}} = H^2$, $\langle\chi^2(0)\rangle$ turns out to be negligible and neglection of the term proportional to $\langle\chi^4(u)\rangle$ in (8) is also justified. Then we find that classical potential force surpasses quantum fluctuation around

$$u \simeq \frac{1}{12\xi}\sqrt{\frac{\lambda v^2}{m^2}} \equiv \frac{c}{12\xi} \equiv u_c, \tag{10}$$

when we may predict where topological defects will appear after completion of the phase transition.

Therefore we may identify the minimum size of the correlation length of χ with the Hubble length at this epoch, which is stretched up to

$$l = \frac{1}{H(u_c)}\exp\left[\frac{1}{8\xi}\left(c - \frac{2}{3}\sqrt{c}\right)\right] \equiv \frac{e^{n_{cf}}}{H(u_c)}, \tag{11}$$

at the end of inflation where the exponent n_{cf} is the e-folding number of inflation after $u = u_c$. If we require $n_{cf} \lesssim 65$ or

$$\left(\frac{\lambda}{10^{-3}}\right)\left(\frac{v}{10^{16}\text{GeV}}\right)^2\left(\frac{m}{10^{13.5}\text{GeV}}\right)^{-2} \lesssim 5\xi, \tag{12}$$

topological defects may exist within the current horizon. Their distribution at $u = u_c$ is determined by quantum fluctuations which is correlated beyond $H^{-1}(u_c)$ and whose spectrum in the epoch $|M^2_{\chi\text{eff}}| \ll H^2$ is given by $|\chi_k|^2 \simeq \frac{H^2}{2k^3}$. The resultant initial distribution of defects exhibits a fractal structure as seen in Fig. 1 which is very different from that expected in thermal phase transition (Fig. 2)[1].

---

[1]See Ref.[18] for more accurate treatments.

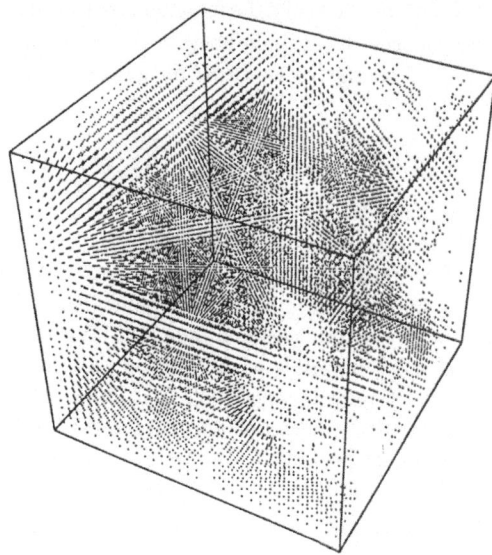

Fig. 1 Structure of domain walls predicted by the curvature-induced phase transition during inflation, which is shown in 32^3 lattices. The separation of each lattice points corresponds to the minimum size of the correlation length. Dotted points are in the state $\chi = v$, while empty points are in $\chi = -v$. Domain walls are present in their boundaries.

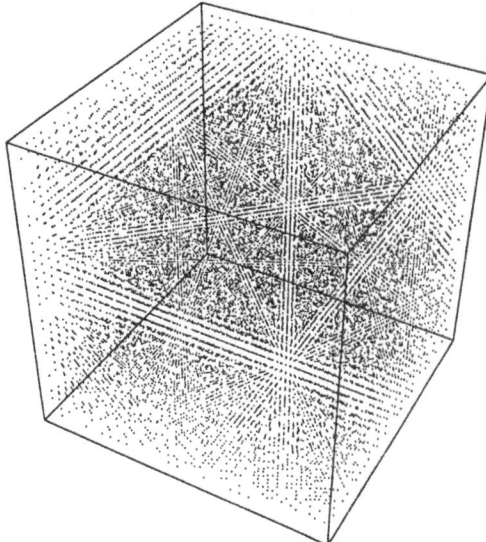

Fig. 2 Structure of domain walls predicted by the thermal phase transition. Typical separation of walls is simply given by the correlation length.

In order to ensure the existence of defects on a specific scale, we should have $M^2_{\chi\text{eff}}(u) \gtrsim H^2(u)$ when the scale leaves the Hubble horizon during inflation [19]. For example if we demand the existence of strings or textures on galactic scales, the following constraint should be satisfied.

$$\left(\frac{\lambda}{10^{-3}}\right)\left(\frac{v}{10^{16}\text{GeV}}\right)^2\left(\frac{m}{10^{13.5}\text{GeV}}\right)^{-2} \lesssim 4\xi - \frac{1}{3}, \tag{13}$$

which can easily be satisfied with natural values of λ.

In summary we have argued that the Kibble mechanism may not work to produce topological defects at typical grand unification scales unless the universe started its evolution in a very special state and that it is likely that phase transition is governed by evolution of $\mathcal{R}$ during inflation with non-minimal coupling [13]. With this mechanism cosmic strings and/or global textures are compatible with inflation and their distribution may be considerably different from the standard one [18,20]. It is also possible that magnetic monopoles with mass 10^{15-18}GeV are both harmless and detectable by future experiments [21].

This work was partly supported by the Japanese Grant in Aid for Science Research Fund of Ministry of Education, Science and Culture 02740189.

References

1. Ya.B. Zel'dovich, I.Yu. Kobzarev and L.B. Okun, ZhETF **67**(1974)3.
   [Sov. Phys. JETP **40**(1975)1.]
2. H.B. Nielsen and P. Olesen, Nucl. Phys. **B61**(1973)45.
3. G. 't Hooft, Nucl. Phys. **B79**(1974)276.
   A.M. Polyakov, ZhETF Pis'ma **20**(1974)430. [JETP Lett. **20**(1974)194.]
4. R.L. Davis, Phys. Rev. **D35**(1987)3705.
5. T.W.B. Kibble, J. Phys. **A9**(1976)1387.
6. A. Vilenkin, Phys. Rep. **121**(1985)265.
7. J.P. Preskill, Phys. Rev. Lett. **43**(1979)1365.
8. J. Ellis and G. Steigman, Phys. Lett. **B89**(1980)186.
   J. Yokoyama, Phys. Rev. Lett. **63**(1989)712.
9. K. Sato, Mon. Not. Roy. Astron. Soc. **195**(1981)467.
   A.H. Guth, Phys. Rev. **D23**(1981)347.
10. A.D. Linde, Phys. Lett. **B129**(1983)177.
11. M.D. Pollock, Phys. Lett. **B185**(1987)34.
12. M. Kawasaki and K. Sato, Phys. Lett. **B189**(1987)23.
13. J. Yokoyama, Phys. Lett. **B212**(1988)273.
14. G.M. Shore, Ann. of Phys. **128**(1980)376.
    B. Allen, Nucl. Phys. **B226**(1983)228.
15. D. La and P.J. Steinhardt, Phys. Rev. Lett. **62**(1989)376.
16. A.L. Berkin, K. Maeda, and J. Yokoyama, Phys. Rev. Lett. **65**(1990)141.
17. A.A. Starobinsky, Lecture Notes in Physics **246**(Springer, 1986)107.
18. M. Nagasawa et al., in preparation.
19. H.M. Hodges and J.R. Primack, preprint SCIPP 90/12.
20. E.T. Vishniac, K.A. Olive and D. Seckel, Nucl. Phys. **B289**(1987)717.
21. J. Yokoyama, Phys. Lett. **B231**(1989)49.

NON-ZEL'DOVICH FLUCTUATIONS FROM INFLATION

JOEL R. PRIMACK
Santa Cruz Institute for Particle Physics
University of California
Santa Cruz, CA 95064
U.S.A.

ABSTRACT. I review the recent work of the Santa Cruz group on the generation of non-Zel'dovich (including non-Gaussian) fluctuations in chaotic inflation. With a single inflaton having the most general quartic polynomial potential, most of the space of the parameters of the potential corresponds to fluctuations with an approximately Zel'dovich spectrum. To the extent that significant deviations from the Zel'dovich spectrum arise, the spectrum characteristically has a dip at a particular scale; in this case the usual upper limit on the quartic coefficient is relaxed and the reheat temperature is correspondingly increased. In the context of the cold dark matter model, such a dip spectrum may help explain both increased large scale structure and earlier galaxy formation. If we consider a general inflaton potential, it is possible to invert the slow-roll equations of motion and find the potential corresponding to almost any desired fluctuation spectrum. In models with multiple inflatons such as double inflation, de Sitter fluctuations in the second inflaton generated during the inflationary period controlled by the first inflaton make it impossible to preset the value of the second inflaton, and thus to get structure in the fluctuation spectrum on a cosmologically interesting scale. Regarding non-Gaussian fluctuations (nGf), we distinguish between local and non-local nGf. With a single inflaton, the inflationary nGf are negligible if the fluctuation amplitude lies below the upper limit from nonobservation of cosmic background radiation fluctuations. Several nonstandard possibilities arise in the generation of non-local nGf such as cosmic strings and texture, with the corresponding scalar fields either free or coupled to the inflaton or to curvature. In particular, such cosmic strings typically correspond to a strongly Type I superconductor, a case that should be investigated further.

1. Introduction

It is standard to assume that the primordial perturbations are Gaussian random fluctuations with a scale-free (flat, Harrison-Zel'dovich) spectrum. There are three basic reasons for studying non-standard — i.e. non-flat Gaussian or non-Gaussian — primordial fluctuations: (1) non-standard fluctuations may be needed to reconcile theories such as cold dark matter (CDM) with observations; (2) non-standard fluctuations may arise in theoretical models for the origin of fluctuations; and (3) an understanding of the consequences of non-standard fluctuations may help in analyzing observations. I will discuss each of these motivations briefly in turn.

193

Cold Dark Matter (CDM) is remarkably successful as a theory of galaxy and cluster formation, but it predicts less structure on very large scales than is apparently observed. Although there are several known physical mechanisms that can add additional large-scale power to the CDM fluctuation spectrum in order to improve the agreement with the observed galaxy and cluster angular correlations, bulk motion, and very large scale structures such as voids, filaments, and "Great Walls", these mechanisms generally increase the predicted amplitude of the microwave background anisotropies. For example, within the context of standard primordial fluctuations, increasing the baryonic fraction or lowering the total cosmological density Ω below unity each increase the power in the fluctuation spectrum at wavelengths of tens of Mpc —but at the cost of possible incompatibility with cosmic background radiation (CBR) $\Delta T/T$ constraints (Bardeen, Bond and Efstathiou (1987); Blumenthal, Dekel and Primack (1988), Holtzman (1989), Holtzman and Primack (1990)). Non-standard primordial fluctuations may help alleviate such problems.

To the extent that non-standard fluctuations are generated in rather generic situations, we may have them whether we like it or not. Any model, such as cosmic strings, in which the primordial fluctuations arise in connection with extended physical structures necessarily gives rise to non-Gaussian fluctuations (hereafter abbreviated nGf), since there must be phase correlations between different wave numbers. We propose to refer to these as *non-local* nGf. (We introduced this terminology in Kofman et al. (1990), to which we also refer the reader for an introduction to the generation of fluctuations in inflation and a review of much recent work.) Inflationary models without strings that produce non-flat perturbations may also lead to the generation of nGf. The non-Gaussian random field of perturbations is often simply a non-linear function of some other Gaussian random field. We propose to call such perturbations *local* nGf.

In any of the current models for the origin of structure in the universe, CBR fluctuations are expected at levels that should soon be observable. They may be non-Zeldovich and/or non-Gaussian, and analysis procedures must be developed with these possibilities in mind. A particular problem with nGf is that, while Gaussian fluctuations are fully specified merely by giving the power spectrum $P(k)$, much additional information is in principle needed to specify nGf fully. To the extent that we can specify the sorts of nGf to look for, the analysis may be simplified. Non-linear gravitational evolution in structure formation will also generate nGf, and the question arises how to distinguish these from primordial nGf.

In order to try to clarify the implications of non-standard perturbations for cosmology, my colleagues and I have addressed the following questions:

a) Are there certain characteristic sorts of non-standard fluctuations that arise in models of chaotic inflation, and how can these fluctuations be most usefully characterized?

b) How "likely" are such non-standard fluctuations? In particular, what are the constraints they must satisfy, what regions of parameter space correspond to them, and what dynamical issues arise?

This talk reviews our results so far. It is organized as follows. The next section concerns the sort of non-flat fluctuations that arise in a single-inflaton model with a polynomial potential, and §3 discusses more generally the connection between a given fluctuation spectrum and the single-inflaton potential that produces it. §4 discusses a dynamical problem with double inflation and similar schemes. §5 discusses nGf, both the extent to which local nGf can arise in single-inflaton chaotic inflation models, and some novel non-local nGf that can arise in models with additional scalars associated with cosmic strings or texture. Our conclusions are summarized in §6.

2. Non-flat $P(k)$ from a Single-Inflaton Model with a Polynomial Potential

The standard treatments of chaotic inflation (see, e.g., Linde (1990), Olive (1990)) assume the rather special case of a power-law inflaton potential: $V(\phi) = \lambda \phi^n$, with $n = 4$ or 2. Hodges, Blumenthal, Kofman, and Primack (1990) (hereafter HBKP) considered instead the potential

$$V(\phi) = A \left(\frac{\phi^4}{4} + \alpha \frac{\phi^3}{3} + \beta \frac{\phi^2}{2} \right) + V_0. \tag{1}$$

This is the most general renormalizable potential, and it is also the first few terms in a Taylor series expansion of any potential. (A linear term could always be eliminated by shifting ϕ.) We assume that V_0 is adjusted so that $V(\phi_0)$ vanishes at the global minimum ϕ_0. Since V is symmetric under $(\phi, \alpha) \rightarrow (-\phi, -\alpha)$, it suffices to consider only $\phi > \phi_0$. Also, we consider only values of α, β with no secondary minimum for $\phi > \phi_0$, i.e. with no false vacuum in the path of the inflaton; this eliminates the dotted regions of Figure 1(a). (b) Values of potential parameters A, α, and β corresponding to an inflationary scenario with a nearly scale-invariant spectrum with an appropriate amplitude.

We analyzed this model both analytically, using the slow-roll approximation, and also numerically. In Figure 1(b), we show the parameters of the potential corresponding to the fluctuation spectrum $\delta \equiv (\delta \rho / \rho)_\lambda$ such that $4 \times 10^{-6} < \delta < 1.3 \times 10^{-5}$ over length scales λ from $1 - 10^3$ Mpc. Most of α, β parameter space corresponds to a nearly flat spectrum. However, for α small and negative and β slightly larger than α^2, the inflaton encounters a near inflection point of the potential; and this can lead to a feature in the fluctuation spectrum: a rise, a fall, or more generally a dip. The depth of the dip is about a factor of 3 for $|\alpha| \sim 1$, and its location is given approximately by

$$\beta \approx \alpha^2 + \pi^2 \alpha^6 / (3 N_{LSS}^2), \tag{2}$$

where $N_{LSS} \approx 50$ is the number of e-folds from large-scale-structure scales to the end of inflation. Figure 2 shows a blow-up of regions of parameter space corresponding to a fairly deep dip spectrum. Note that the quartic coefficient A is considerably larger here than for most of the parameter space (cf. Figure 1 (b)), because of the near cancellation between the quartic and cubic terms. The larger curvature near the origin leads to a higher reheat temperature at the end of inflation, which might be desirable. Figure 3 shows typical dip spectra corresponding to two choices of the parameters, and Figure 4 shows the potential for the latter choice — note that $V(\phi)$ nearly has an inflection point.

Spectra with dips like those in Figure 3, if located around 5 Mpc (the scale on which the spectrum is normalized), may help reconcile CDM with observations. The increase in power (over a pure Zel'dovich spectrum) on small scales could lead to earlier galaxy formation, and the increase on larger scales would lead to more large scale structure. Holtzman and Primack (1990) have found that even the deep-dip spectrum (B) of Figure 3 is compatible with present CBR anisotropy limits, and that the corresponding prediction for the cluster autocorrelation function is consistent with the available data.

How "likely" are such dip spectra? It is obvious from Figure 3 that they arise in only a small region of (α, β) parameter space, but perhaps this is compensated by the corresponding increase in the allowed range for the quartic coefficient A. Since we have no idea what measure to use, we cannot calculate a corresponding a priori probability.

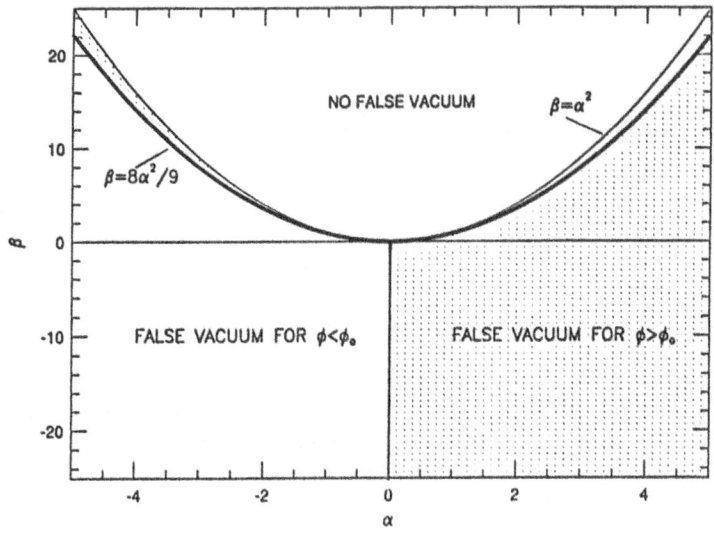

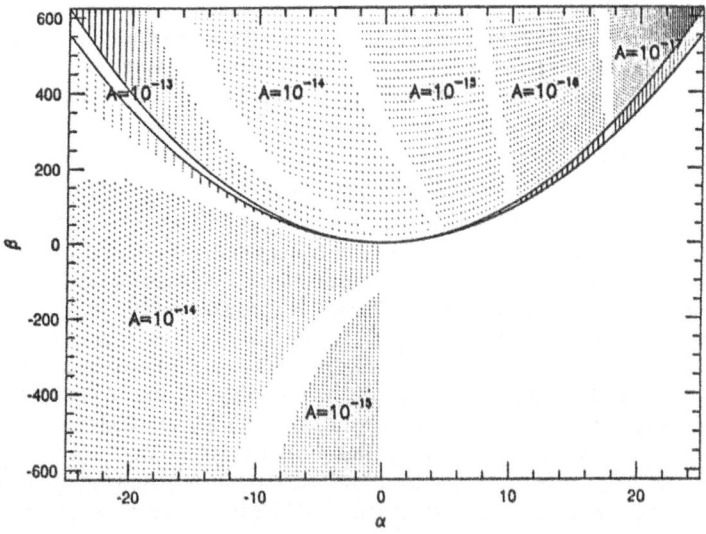

Figure 1. (a) Parameter space for the potential (1). For $\beta > 8\alpha^2/9$, the global minimum of the potential is at $\phi_0 = 0$, and is nonzero elsewhere. If $\beta \leq \alpha^2$ the potential has a false vacuum, unless the vacuum is degenerate (heavy-weight lines) or there is an inflection point (light-weight lines). The dotted regions correspond to a potential with a false vacuum to the right of the global minimum of the potential, which we do not consider.

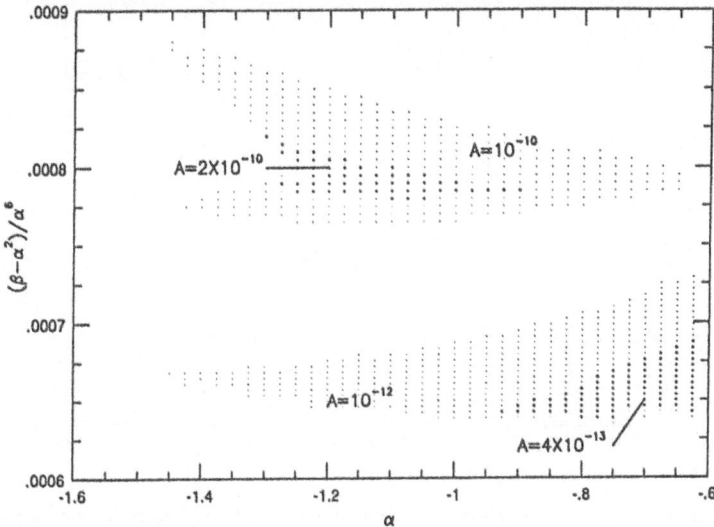

Figure 2. Regions of α, β parameter space, for a number of values of the quartic coupling A, that give spectra of sufficiently small amplitude over the range $1 - 10^3$ Mpc and also vary by a factor of at least 5 over this range.

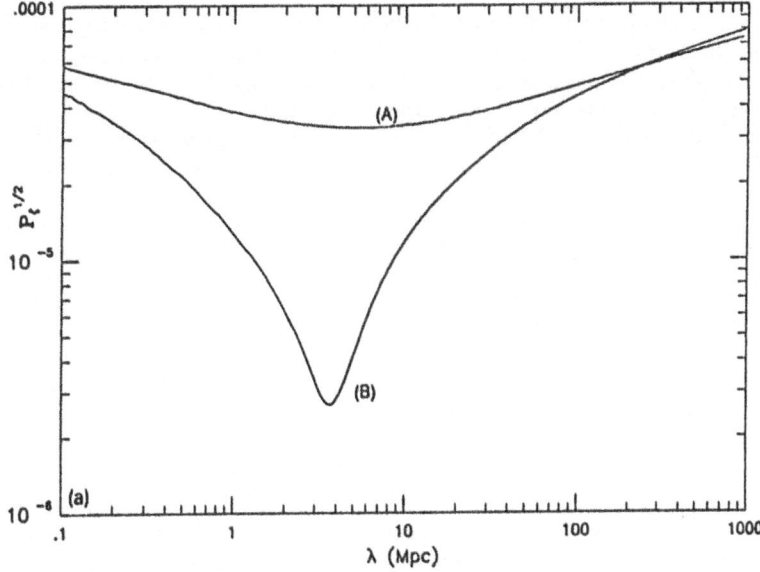

Figure 3. The power spectrum $P_\zeta^{1/2}(\approx 7.5\delta)$ as a function of length scale, for two sets of potential parameters: $(\alpha, \beta) =$ (A) (-2.3, 5.41), (B) (-1.3, 1.6938). In both cases, $A = 2 \times 10^{-11}$. The spectrum was obtained by numerically solving for the evolution of the metric perturbations.

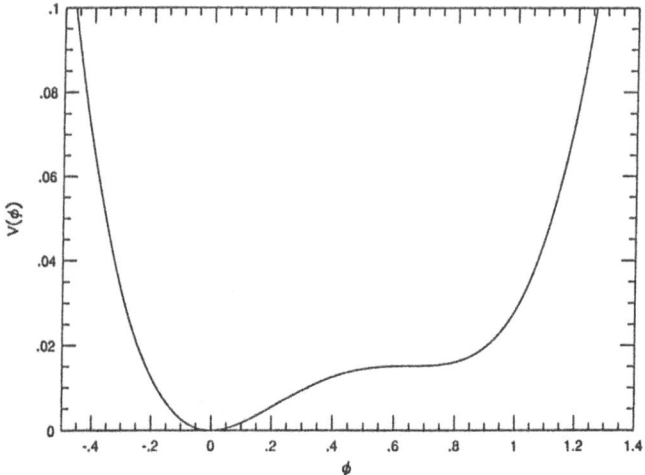

Figure 4. The potential $V(\phi)$ (in arbitrary units), for case (B) of Figure 3. The scalar field ϕ is in Planck mass units. (Figures 1 - 4 are from HBKP, which also includes a plot of the dip spectra over a much larger range of length scales.)

3. "Designer Inflation" with a Single Inflaton

Even with only a single inflaton, chaotic inflation is like Alice's restaurant: you can have (almost) any fluctuation spectrum you want. The price is that the potential $V(\phi)$ becomes fairly arbitrary, but it is perhaps misguided to demand that $V(\phi)$ be renormalizable, for example, since Planck-scale physics may provide a cutoff.

Hodges and Blumenthal (1990) have shown that it is possible to invert the slow roll equations in order to obtain the potential and ϕ implicitly in terms of the length scale λ:

$$V(\lambda) = \left[\frac{1}{V_0} - \frac{48}{m_{Pl}^3} \int_{\lambda_0}^{\lambda} \frac{d\lambda}{\lambda P(\lambda)} \right]^{-1}, \quad \phi(\lambda) = \phi_0 + \frac{1}{m_{Pl}} \sqrt{\frac{6}{\pi}} \int_{\lambda_0}^{\lambda} \frac{V^{1/2} d\lambda}{\lambda P^{1/2}}. \tag{3}$$

Figure 5(a) shows some examples of fluctuation spectra $P(\lambda)$, and 5(b) shows the corresponding potentials $V(\phi)$. Applying Eqs. (3) to the "mountain" spectrum represented by the thin solid line in 5(a) gives the potential represented by the same weight line (second from the bottom) in 5(b). The heavy solid line in 5(a) shows the result of integrating the equations numerically (rather than using the slow-roll approximation) to obtain the spectrum; its agreement with the initial spectrum shows that the slow-roll approximation is not bad even for this peaked spectrum. Note that the long-dash line in 5(a) represents a constant (i.e., exactly flat) spectrum; the corresponding potential in 5(b) is of the form $V_0[1 - a(\phi - \phi_0)]^{-2}$.

As a bonus, Hodges and Blumenthal (1990) were also able to deduce new constraints on reheating after inflation.

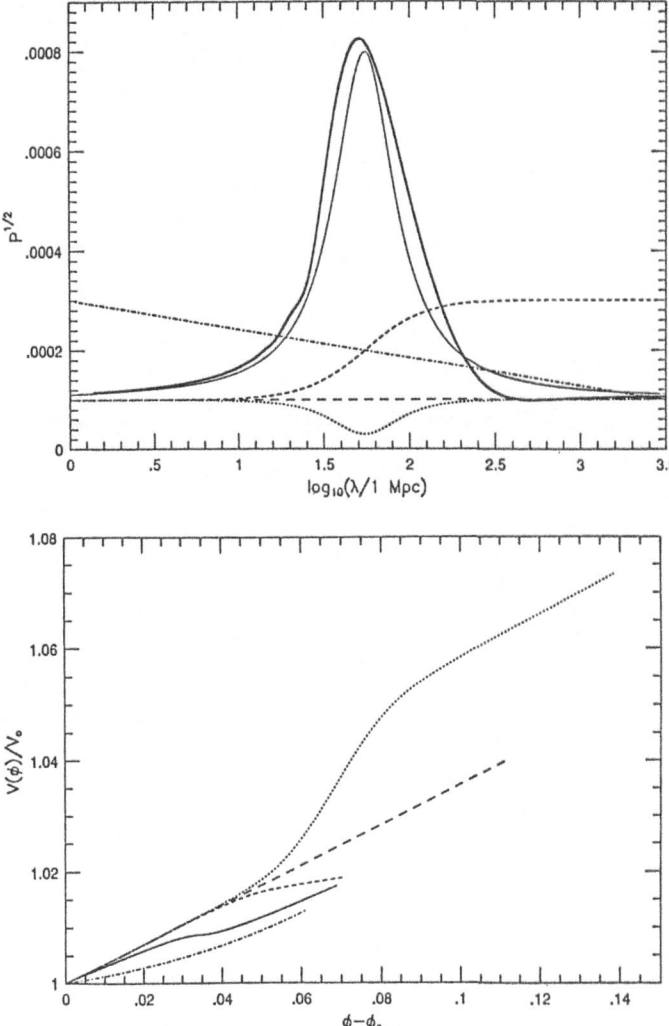

Figure 5. Plotted are (a) some sample fluctuation spectra, and (b) the corresponding potentials $V(\phi)$, with ϕ in Planck units. The initial value of the potential was taken to be $V_0 = V(1\text{Mpc}) = 10^{-12}m_{Pl}^4$, for each of these examples. Only the range of $P(\lambda)^{1/2}$ shown in (a) is converted to $V(\phi)$ in (b), to illustrate the mapping from λ-space to ϕ-space. (From Hodges and Blumenthal (1990).)

4. Trouble with Double Inflation

In the double inflation model (Starobinsky (1985), Silk and Turner (1987), Kofman and Linde (1987)), a second inflaton field must start at a precisely tuned value in order that a break in the power spectrum appears on astrophysical scales of interest. Hodges (1990) shows that it is unlikely that double inflation might break scale invariance and help produce

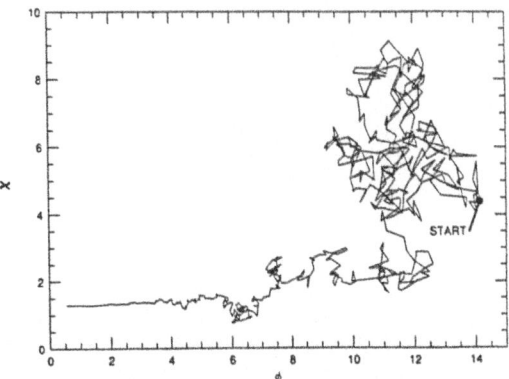

Figure 6. Evolution of the fields ϕ, χ (in Planck mass units). The initial value of ϕ corresponds to the Planck density, and the initial value of χ is chosen to optimize the likelihood of double inflation being important. In this (typical) case, the power spectrum does not contain a break on an interesting scale.

cosmological large scale structure on the relevant scale, if inflation starts near the Planck scale. The cumulative effects of quantum fluctuations in an auxiliary field χ, during the evolution of the first inflaton ϕ, lead to a large dispersion in the probability distribution of χ when the auxilliary field becomes responsible for inflation. It follows that double inflation, and similar mechanisms that require tuning of the initial values of auxiliary scalar fields, are unlikely to produce breaks on scales of interest. Hodges (1990) used a simple model to estimate the probability that the break in the spectrum from double inflation would occur in the region of large scale structure, and confirmed the estimate by evolving the Lengevin fluctuation equation. Figure 6 shows an example of the evolution of two uncoupled inflatons χ, ϕ with quartic couplings $\lambda_\chi << \lambda_\phi$. In the absence of de Sitter fluctuations, χ would initially have remained constant as ϕ decreased, thus realizing the necessary condition for the break in the spectrum to occur on a multi-Mpc scale. But the inflationary fluctuations make that unlikely.

Note that, unlike the polynomial potential model of §2 where we do not know how to estimate the probability of a given given set of polynomial parameters and the corresponding fluctuation spectrum, here the probability is calculalable in terms of dynamical effects — and it is small!

5. Non-Gaussian Fluctuations

5.1. NGf from a Single-Inflaton Model

If one chooses an initial $\phi(t_i)$ at initial time t_i, then one can solve for $\mathcal{P}(\phi, t_o)$, the distribution function today for an ensemble of universes of typical size H_o^{-1}. The distribution $\mathcal{P}$ will generally be very sensitive to the initial condition $\phi(t_i)$. The Fokker-Planck and Langevin equations (Starobinsky (1986) and these proceedings) solved analytically as well as numerically (Hodges (1989), Bond, Salopek, and Bardeen (1989)) show that for initial conditions corresponding to universes with the size of the present horizon, non-Gaussian effects in $\mathcal{P}(\phi, t)$ are negligible if the fluctuation amplitude does not exceed the usual constraints. This result was checked for a wide class of potentials $V(\phi)$ including chaotic, new- and power-law inflations. For initial conditions corresponding to an ensemble of universes with sizes much greater than our horizon, non-Gaussian effects could be significant (Matarrese, Ortolan and Lucchin (1989)). Physically, this means that for a large initial value $\phi(t_i)$, the duration of inflation is long, and the generated long-wave quantum fluctuations accumulate so that they significantly affect the classical rolling down of the scalar field. On the other hand, for $\phi(t_i)$ much less than the value at which fluctuations dominate the evolution of ϕ as happens in "eternal inflation" (Linde 1986), there is insufficient time to generate such large fluctuations. Then, non-Gaussian effects are negligible, even on scales well outside our present horizon. (A more detailed treatment of nGf in single-inflaton models is given in HBKP.)

In models having fluctuation spectra that are non-Zel'dovich, it is possible that the amplitude on very small scales is much larger than is allowed on large scale structure scales (*e.g.,* , from the upper limits on the anisotropy of the cosmic background radiation). In this case, it is important to understand the constraints on the underlying spectrum from the requirement that overproduction of small black holes does not lead to premature matter dominance of the universe. It follows from our earlier work HBKP that, even in models with a single inflaton, nGf are expected in this case. Blumenthal, Hodges, and I regard this as an additional motivation to analyze the primordial black hole question in detail. Is there a reasonable choice of renormalizable single-inflaton potential which can lead to a phenomenologically acceptable production of small black holes in the early universe, perhaps even enough to make the dark matter? We're working on it.

5.1. Non-local nGf with Auxilliary String or Texture Fields

Hodges and Primack (1990) examine mechanisms, several of which we propose, to generate structure formation through either gauged or global cosmic strings or global texture (Turok 1989), within the framework of chaotic inflation. The requirement that inflationary density fluctuations are not so large as to conflict with observations leads to a number of constraints on model parameters. We point out that inflaton decoherence at the end of inflation can lead to a scaling pattern of global strings/texture in certain inflationary models. If formed after inflation, gauged cosmic strings correspond to a Type I superconductor, and naturalness suggests that it is strongly Type I. Similarly, gauged strings formed during inflation via conformal coupling $\xi = 1/6$ to the cosmological curvature, in a model suggested by Yokoyama (1989) in order to evade the millisecond pulsar constraint on cosmic strings (Albrecht and Turok 1989, Accetta and Krauss 1989, Bouchet and Bennett 1990), are expected to be strongly Type I. Since the energy of strongly Type I strings is mainly in the scalar field, strings with winding number higher than one are stable; also the intercommutation properties of such strings may be novel. The possibility that cosmic strings correspond to this case has apparently not been considered. We have also improved upon the understanding of the formation process of strings and texture during inflation, in particular the alternative fractal string scenario put forth by Vishniac, Olive and Seckel (1987).

6. Conclusions

In terms of the two questions about non-standard fluctuations posed in the Introduction, we have made only partial progress. We have been able to characterize the the fluctuation spectra that arise in single-inflaton chaotic inflation with a polynomial potential, and to show that significant nGf do not arise in this case. But the likelihood and character of non-standard fluctuations in more general cases remains unclear.

I had hoped in embarking on this study of inflationary fluctuations that some characteristic type or types of local nGf would arise — this would give observers something particular to look for. Unfortunately, it did not turn out to be so simple, although my colleagues and I, the CITA group (cf. Salopek, these proceedings), and others have found many interesting things. Perhaps the theoretical situation will clarify after the observers have found primordial nGf — or perhaps the primordial fluctuations really are of the simple Zel'dovich form!

ACKNOWLEDGMENTS. I am very grateful to George Blumenthal, Hardy Hodges, and Lev Kofman, my collaborators in the work I have described here, for the insights they have shared with me. This work was supported in part by NSF grant PHY-8800801 and by a UCSC faculty research grant.

REFERENCES

Accetta, F. R., and Krauss, L. (1989) Nucl. Phys. B319, 747.

Albrecht, A., and Turok, N. (1989) Phys. Rev. D40, 973.

Bardeen, J., Bond, J., and Efstathiou, G. (1987) Ap. J. 321, 28.

Blumenthal, G., Dekel, A., and Primack, J. R. (1988) Ap. J. 326, 539.

Bond, J., Salopek, D., and Bardeen, J. (1989) Phys. Rev. D40, 1753.

Bouchet, F. R., and Bennett, D. P. (1990) Phys. Rev. D41, 720; Astrophys. J. Lett. 354, L41.

Hodges, H. (1989) Phys. Rev. D39, 3568.

Hodges, H. (1990) Phys. Rev. Lett., 64, 1080.

Hodges, H., and Blumenthal, G. (1990) Phys. Rev. D, in press.

Hodges, H., Blumenthal, G., Kofman, L. and Primack, J. R. (1990) Nucl. Phys. B335, 197. (HBKP)

Holtzman, J. (1989) Astrophys. J. Suppl. 71, 1-24.

Holtzman, J., and Primack, J. R. (1990), submitted to Ap. J.

Kofman, L., Blumenthal, G., Hodges, H., and Primack, J. R. (1990) in D.W. Latham and L.N. da Costa, (eds.), Proc. of Workshop on Large Scale Structure and Peculiar Motions in the Universe, Rio de Janeiro, Brazil, May 22-27, 1989. Publ. Astron. Soc. Pacific, in press.

Kofman, L. and Linde, A. (1987) Nucl. Phys. 282B, 555.

Linde, A. (1986) Phys. Lett. 175B, 395.

Linde, A. (1990) Particle Physics and Inflationary Cosmology, Harwood Academic Publishers, New York.

Matarrese, S., Ortolan, A. and Lucchin, F. (1989) Phys. Rev. D 40, 290.

Olive, K. (1990) Phys. Rep. 190, 307.

Silk, J., and Turner, M. (1987) Phys. Rev. D35, 419.

Starobinsky, A. (1985) JETP Lett. 42, 152.

Starobinsky, A. (1986) in. H. J. de Vega and N. Sanchez (eds.), Field Theory, Quantum Gravity and Strings Springer Lecture Notes in Physics, 246, p. 107.

Turok, N. (1989) Phys. Rev. Lett. 63, 2625.

Vishniac, E., Olive, K., and Seckel, D. (1987) Nucl. Phys. B289, 717.

Yokoyama, J. (1989) Phys. Rev. Lett. 63, 712.

MAGNETIC THEORY OF GRAVITATION

Y.M. CHO
Department of Physics
Seoul National University
Seoul 151-742, Korea

ABSTRACT. Recently a generalization of Einstein's theory of gravitation has been proposed which can accommodate the gravitational monopoles. The generalization is made in such a way that the equations of motion become exactly symmetric under the dual transformation of the curvature tensor. The theory is characterize by a fundamental scale κ which can be interpreted as a "magnetic" mass. In exact analogy with the Dirac's generalization of Maxwell's theory, the theory can be made sensible only if the energy E is quantized by the condition $\kappa E = 2\pi n$, where n is an integer. We review the recent development on the subject.

Long time ago Dirac[1,2] has taught us that the electrodynamics can be made consistent with the existence of the magnetic monopoles if the celebrated charge quantization condition is imposed. Inspired by an apparent analogy between the electrodynamics and the general relativity in the post-Newtonian limit, Zee[3] has recently made a very interesting conjecture that the Einstein's theory of gravitation could be generalized to include the "gravitational monopoles". Certainly a most interesting possibility in this generalization would be a quantization of energy, because the "charge" in general relativity is nothing but the energy. Recently it has been shown that[4] indeed the generalization is possible provided that one is willing to accept an energy quantization requirement. We review the recent progress on the theory of gravitational monopoles.

The generalization can be made in exact analogy with the Dirac's generalization of the electrodynamics. All we need is to enlarge the superspace (i.e., the space of the metric) to allow the string singularities of the metric, under the condition that the strings should be *physically* unobservable. Since the classical dynamics of gravitation is governed by the geodesic equation, a necessary condition for the invisibility of the string is that the (singular) metric does not create a physical singularity in the geodesic equation. This invisibility of the string at the classical level, however, is not sufficient to guarantee the invisibility of the string at the quantum level. To assure that the string remains invisible at the quantum level, we need the quantization of energy as we will see in the followings.

Let us first show that the general relativity admits a gravitational monopole[5,6], if the metric is allowed to have a string singularity. To keep the analogy between the Dirac's generalization and ours as far as possible, we start by writing the most general stationary metric $g_{\mu\nu}$ $(\mu, \nu = 1, 2, 3, 4)$ as

$$ds^2 = -\phi(dt + B_i dx^i)^2 + g_{ij}dx^i dx^j \tag{1}$$

where g_{ij} $(i, j = 1, 2, 3)$, B_i, and ϕ are functions of the spatial coordinates only. The fact that the most general stationary metric can always be put into this form can easily be understood if one regards the Einstein's theory as a $(3 + 1)$-dimensional Kaluza-Klein theory[7,8] and dimensionally reduces it to the 3-dimensional space, assuming that the metric admits a time-like Killing vector. Notice that in this Kaluza-Klein point of view the space-time can be viewed as a principal fibre bundle in which the time-axis becomes the "internal" fibre over the 3-dimensional space. Thus B_i may be interpreted as a "gauge potential" which has the following gauge degree of freedom

$$B_i \longrightarrow B_i' = B_i - \partial_i \Lambda. \tag{2}$$

This gauge degrees of freedom is guaranteed by the general invariance of Einstein's theory under the coordinate transformation

$$t \longrightarrow t' = t + \Lambda. \tag{3}$$

Now we choose the polar coordinates and let

$$\phi = \phi(r), \quad B_i = \frac{\kappa}{4\pi}(1 - \cos\theta)\partial_i\varphi$$

$$(ds^2)_3 = g_{ij}dx^i dx^j = f(r)dr^2 + r^2(d\theta^2 + \sin^2\theta d\varphi^2), \tag{4}$$

where we have introduced a scale parameter κ to keep B_i dimensionless. Then one can easily show that the vacuum Einstein's equation admits the following asymptotically flat solution

$$\phi(r) = 1 - 2\frac{Gm\left(r^2 - \frac{\kappa^2}{64\pi^2}\right)^{1/2} + \frac{\kappa^2}{64\pi^2}}{r^2}$$

$$f(r)\phi(r) = \left[1 - \frac{\kappa^2}{64\pi^2 r^2}\right]^{-1}. \tag{5}$$

Obviously the potential B_i describes a "monopole" to which a string singularity is attached, but now as a component of the metric. The solution is characterized by two parameters, the monopole strength κ and the inertial mass m, so that it actually describes a "gravitational dyon". Notice that one may call κ a "magnetic mass" since it has the dimension of a length.

One might notice that the above solution is not exactly new, because one can easily show that it is locally identical to the well-known Taub-NUT solution[5,6]. However, we emphasize that *there is one big difference between the Taub-NUT solution and ours, and that is in the global topology of the space-time*. The asymptotic topology of the Taub-NUT space-time at the spatial infinity is well-known to be isomorphic to S^3, which makes it totally unphysical because it has a periodic time which violates the causality and admits no reasonable space-like hypersurface. But here we require the asymptotic topology of our space-time to be $R^1 \times S^2$ (where R^1 represents the time axis). How can this be possible? Notice that the reason why the S^3 topology of Taub-NUT space-time is forced upon us is because this guarantees the metric to be smooth everywhere on the space-time manifold. But obviously we can escape from this pathological S^3 topology if we are willing to accept the string singularity[9]. So we choose to accept the string as it is to retain the physically desirable space-time topology, but will get rid of the string by making it physically unobservable.

What should we do to make the string invisible? At the classical level the string is by itself invisible[4] because it can easily be shown that *the string does not produce any gravitational effect which can be detected by a classical neutral test particle*. To see this notice that the geodesic equation of the metric is given by

$$\frac{d^2\tau}{ds^2} + \frac{\partial_i\phi}{\phi}\frac{dx^i}{ds}\frac{d\tau}{ds} = 0$$

$$\frac{d^2x^k}{ds^2} + \Gamma_{ij}{}^k\frac{dx^i}{ds}\frac{dx^j}{ds} = \phi G_{ij}g^{jk}\frac{dx^i}{ds}\frac{d\tau}{ds} - \frac{1}{2}(\partial_i\phi)g^{ik}(\frac{d\tau}{ds})^2 \tag{6}$$

where

$$\frac{d\tau}{ds} = \frac{dt}{ds} + \frac{\kappa}{4\pi}(1 - \cos\theta)\frac{d\varphi}{ds}$$

and s is the affine parameter, $\Gamma_{ij}{}^k$ is the 3-dimensional connection obtained with g_{ij}, and G_{ij} is the field strength of the potential B_i. Notice that the metric (4) admits four Killing vectors[5,10] corresponding to the time-translation and a space-time rotation. Consequently the geodesic equation allows four constants of motion, the total energy E and the total angular momentum $\vec{J}$ which can be expressed as a sum of the orbital angular momentum $\vec{L}$ and the spin angular momentum $\vec{S}$ (due to the gravitational field of the magnetic mass) which is always orthogonal to $\vec{L}$. Choosing $\vec{J}$-axis as the z-axis one can easily integrate (6) to obtain

$$\phi\frac{d\tau}{ds} = E$$

$$f\phi(\frac{dr}{ds})^2 + \epsilon_4 \phi \frac{J^2 \sin^2 \theta}{r^2} \phi = E^2$$

$$\frac{d\theta}{ds} = 0 \quad (\cos \theta = \frac{\kappa E}{4\pi J}) \tag{7}$$

$$\frac{d\varphi}{ds} = \frac{J}{r^2},$$

where ϵ_4 is the signature factor of the trajectory. The result shows that the geodesic trajectory which lies on a cone[10] does not recognize the existence of the string. This is exactly what one would have expected, because asymptotically the geodesic equation becomes very similar to the one which describes the motion of a charged particle around a Dirac monopole. This assures that at the classical level the string becomes completely invisible.

Notice that the Riemannian curvature of the metric (4) becomes spherically symmetric and does *not* contain any string singularity. This, together with the complete invisibility of the string at the classical level, suggests that the string is not a physical singularity but a simple coordinate singularity[4]. To show that this is indeed the case, notice that the gravitational monopole can be described by any metric which may be related to (4) by a general coordinate transformation. So consider two metrics $g_{\mu\nu}^{(\pm)}$ which have potential $B_i^{(\pm)}$ given by

$$B_i^{(\pm)} = \frac{\kappa}{4\pi}(\pm 1 - \cos\theta)\partial_i \varphi.$$

Clearly both $g_{\mu\nu}^{(+)}$ and $g_{\mu\nu}^{(-)}$ describe the same monopole since they are related by the following coordinate transformation

$$t \longrightarrow t' = t + \frac{\kappa}{2\pi}\varphi. \tag{8}$$

The coordinate transformation moves the string along the negative z-axis to the positive z-axis. So dividing the space-time into two cross sections[2] and sewing them together in the overlapping region with (8), one can remove the string completely from our space-time. This shows that *the string is not a physical singularity but indeed an artifact of the coordinates we have chosen, which is why it can not be detected by a classical test particle.*

To make the string invisible at the quantum level, however, we must impose the following quantization condition of energy E

$$\kappa E = 2\pi n \quad (n; \text{integer}). \tag{9}$$

We present three independent arguments for this:

A) Consider the quantum wave function Ψ of a particle with energy E rotating around the string, and let

$$\Psi(t, \vec{x}) = e^{-iEt} R(\vec{x}).$$

Then under the transformation (8) the wave function should transform as

$$\Psi(t, \vec{x}) \longrightarrow \Psi(t', \vec{x}) = e^{-i\frac{\kappa E}{2\pi}\varphi} \Psi(t, \vec{x}). \tag{10}$$

But since the wave function should remain single-valued under the coordinate transformation, one need to have the quantization condition. This of course is the argument given by Zee[3].

B) We have emphasized that the string is not a physical singularity. Nevertheless let us try to treat the string *as if* it is real. Then asymptotically the string of the metric (4) becomes nothing but the spinning string[11] which can be described in the cylindrical coordinates by

$$ds^2 = -(dt + \frac{\kappa}{2\pi} d\varphi)^2 + d\rho^2 + \rho^2 d\varphi^2 + dz^2. \tag{11}$$

In principle this string could be detected by a scattering of a test particle around it. The quantum scattering cross section of a scalar particle with energy E around the spinning string is known to be[12]

$$\frac{d^2\sigma}{d\varphi dz} = \frac{1}{2k\pi} \frac{\sin^2(\kappa E/2)}{\sin^2(\varphi/2)}, \tag{12}$$

where k is the momentum of the particle. Clearly the string becomes invisible under the quantum scattering if and only if the quantization condition is satisfied. The quantum scattering of a spin 1/2 particle around the string[13] gives the same conclusion. This argument is particularly enlightening because it tells us that, even if one tries to treat the string as physical, the quantization condition will forbid us to detect it with a quantum scattering of a test particle.

C) Remember that our space-time may be viewed as a principal fibre bundle in which the time-axis becomes the fibre over the 3-dimensional space. In this picture the string singularity manifests itself when the bundle is regarded trivial. But of course we can always make the string disappear by introducing two cross sections and making the fibre bundle topologically non-trivial[2,14], provided that the quantization condition holds true. Indeed the coordinate transformation (8) tells us how one can remove the string with a mixing of the time coordinate with the azimuthal coordinate, just as Wu and Yang[2] have told us how one can remove the Dirac string with a mixing of the internal coordinate (i.e., the fifth-coordinate) with the azimuthal coordinate. In fact, when the time is assumed to be periodic, this mixing is precisely what one need to obtain the S^3 topology of the Taub-NUT space-time. We emphasize, however, that our way of removing the string need not necessarily require the physical time to be periodic, in as much as the Dirac's theory does not necessarily require the existence of a 5-dimensional space-time. According to (8), the periodicity of the azimuthal coordinate requires us to identify (t', φ) with $(t' + \kappa n, \varphi + 2\pi n)$, but not with $(t' + \kappa n, \varphi)$. This assures us that our time must be helical[13], but not periodic.

In all the above arguments the striking similarity between the Dirac string and ours is unmistakable. Again *it should be made absolutely clear that the quantization of energy has nothing to do with the periodic time coordinate* as has been repeatedly suggested in the literature[5,6]. In retrospect the quantization condition (9) could easily have been understood from the following simple dimensional argument. The existence of a gravitational monopole necessarily requires the existence of a fundamental length scale κ which in turn implies the existence of a fundamental energy scale, and hence the quantization of energy.

We will now show how Einstein's equation should be generalized[4] in the presence of a gravitational monopole. To do this let us first define the dual curvature tensors as follows

$$R^*_{ABCD} = \frac{1}{2}\epsilon_{AB}{}^{MN}R_{MNCD}$$

$$R^*_{AB} = g^{CD}R^*_{ACBD} = -\frac{1}{2}\epsilon_A{}^{PQR}R_{PQRB} \tag{13}$$

$$R^* = g^{AB}R^*_{AB} = \frac{1}{2}\epsilon^{ABCD}R_{ABCD}.$$

With this the first and second Bianchi identities can be expressed as

$$R^*_{AB} = 0$$

$$\nabla^A R^*_{ABCD} = 0. \tag{14}$$

This should be contrasted with the following expressions of Einstein's equation

$$R_{AB} = -8\pi G(T_{AB} - \frac{1}{2}Tg_{AB})$$

$$\nabla^A R_{ABCD} = -8\pi G[\nabla_C(T_{DB} - \frac{1}{2}Tg_{DB}) - \nabla_D(T_{CB} - \frac{1}{2}Tg_{CB})], \tag{15}$$

where the second equation follows from the first one with the help of the Bianchi identities. This shows that Einstein's theory of gravitation is "maximally asymmetric" under the dual transformation. With this clarification it now becomes clear what we should do to accommodate the gravitational monopoles. *We need to have two ("magnetic" as well as "electric") energy-momentum tensors which must couple to the curvature tensor in such a way that the generalized theory becomes symmetric under the dual transformation.* But obviously this can be made possible only if the Bianchi identities are violated. So we only require the curvature to be metric-compatible,

$$R_{ABCD} = R_{[AB][CD]} \tag{16}$$

and replace the first Bianchi identity by

$$R_{[ABC]D} = -\frac{1}{3}\epsilon_{ABC}{}^{M}R^*_{MD}$$

$$R^*_{[ABC]D} = \frac{1}{3}\epsilon_{ABC}{}^{M}R_{MD}. \tag{17}$$

Notice that now the Ricci tensors are no longer symmetric,

$$R_{[AB]} = -\frac{1}{2}\epsilon_{AB}{}^{CD}R^*_{CD}$$

$$R^*_{[AB]} = \frac{1}{2}\epsilon_{AB}{}^{CD}R_{CD}. \tag{18}$$

With this we propose the following generalization of Einstein's equation

$$R_{(AB)} - \frac{1}{2}Rg_{AB} = -8\pi G T_{AB}$$

$$R^*_{(AB)} - \frac{1}{2}R^*g_{AB} = -8\pi G' S_{AB}, \tag{19}$$

where G' and S_{AB} are the gravitational constant and the energy-momentum tensor of the magnetic matter. To ensure the conservation of the two energy-momentum tensors, we now need to generalize the second Bianchi identity. To see how, let

$$\nabla^{A}R_{ABCD} = J_{BCD}$$

$$\nabla^{A}R^*_{ABCD} = K_{BCD}$$

and find

$$\nabla^{B}[R_{(AB)} - \frac{1}{2}Rg_{AB}] = \frac{1}{2}\epsilon_{A}{}^{BCD}(K_{BCD} + \nabla_{B}R^*_{CD})$$

$$\nabla^{B}[R^*_{(AB)} - \frac{1}{2}R^*g_{AB}] = -\frac{1}{2}\epsilon_{A}{}^{BCD}(J_{BCD} + \nabla_{B}R_{CD}). \tag{20}$$

This tells that the second Bianchi identity should be modified in such a way that the following equalities hold,

$$\nabla^{A}R_{A[BCD]} + \nabla_{[B}R_{CD]} = 0$$

$$\nabla^{A}R^*_{A[BCD]} + \nabla_{[B}R^*_{CD]} = 0. \tag{21}$$

This completes the desired generalization of Einstein's theory.

A slightly different generalization is possible[4] if one is willing to accept non-symmetric energy-momentum tensors. Notice that (18) with (20) can be written as

$$\nabla^B[R_{AB} - \frac{1}{2}Rg_{AB}] = \frac{1}{2}\epsilon_A{}^{BCD}K_{BCD}$$

$$\nabla^B[R^*_{AB} - \frac{1}{2}R^*g_{AB}] = -\frac{1}{2}\epsilon_A{}^{BCD}J_{BCD}.$$

So one may have

$$R_{AB} - \frac{1}{2}Rg_{AB} = -8\pi G T_{AB}$$

$$R^*_{AB} - \frac{1}{2}R^*g_{AB} = -8\pi G' S_{AB} \tag{22}$$

together with the following modification of the second Bianchi identity

$$\nabla^A R_{A[BCD]} = 0$$

$$\nabla^A R^*_{A[BCD]} = 0. \tag{23}$$

In this generalization T_{AB} and S_{AB} become non-symmetric (but conserved) energy-momentum tensors.

We have shown that Einstein's theory of gravitation could be generalized to accommodate the gravitational monopoles, exactly as Dirac has shown that the Maxwell's theory could be generalized to accommodate the magnetic monopoles. But obviously our generalization opens up much more questions, fundamental as well as phenomenological, than we have tried to answer in this talk. A more detailed discussion on these questions will be published elsewhere[15].

Acknowledgement

It is a great pleasure to thank S. O. Ahn for the encouragements. The work is supported in part by the Ministry of Education and by the Korea Science and Engineering Foundation.

References

1 P. A. M. Dirac, Proc. Roy. Soc. **A133**, 60(1931).
2 T. T. Wu and C. N. Yang, Phys. Rev. **D12**, 3845(1975).
3 A. Zee, Phys. Rev. Lett. **55**, 2379(1985).
4 Y. M. Cho, SNUTP 90-07, submitted to Phys. Rev. Lett.

5 E. T. Newman, L. Tamburino, and T. Unti, J. Math. Phys. **4**, 915(1963); C. W. Misner, J. Math. Phys. **4**, 924(1963); M. Demianski and E. T. Newman, Bull. Acad. Pol. Sci. **14**, 653(1966); J. S. Dowker, Gen. Rel. Grav. **5**, 603(1974).

6 S. Ramaswamy and A. Sen, J. Math. Phys. **22**, 2612(1981); P. O. Mazur, Phys. Rev. Lett. **57**, 929(1986).

7 Y. M. Cho, J. Math. Phys. **16**, 2029(1975); Y. M. Cho and P. G. O. Freund, Phys. Rev. **D12**, 1711(1975); Y. M. Cho and P. S. Jang, Phys. Rev. **D12**, 3789(1975).

8 Y. M. Cho, Phys. Lett. **186B**, 38(1987); Y. M. Cho, Phys. Lett. **199B**, 358(1987); Y. M. Cho and D. S. Kimm, J. Math. Phys. **30**, 1570(1989).

9 A. Ashtekar and A. Sen, J. Math. Phys. **23**, 2168(1982).

10 R. L. Zimmerman and B. Y. Shahir, Gen. Rel. Grav. **21**, 821(1989).

11 A. Staruszkiewicz, Acta. Phys. Pol. **24**, 734(1963); J. R. Gott III and M. Alpert, Gen. Rel. Grav. **16**, 243(1984); D. Deser, R. Jackiw, and G. 't Hooft, Ann. Phys. (NY) **152**, 220(1984).

12 G. 't Hooft, Comm. Math. Phys. **117**, 685(1988); S. Deser and R. Jackiw, Comm. Math. Phys. **118**, 495(1988); Y. Aharonov and D. Bohm, Phys. Rev. **115**, 485(1959).

13 P. de Sousa Gebert and R. Jackiw, Comm. Math. Phys. **124**, 229(1989); M. Alford and F. Wilczek, Phys. Rev. Lett. **62**, 1071(1989).

14 Y. M. Cho, Phys. Rev. Lett. **44**, 1115(1980); Y. M. Cho, Phys. Rev. **D23**, 2415(1981).

15 Y. M. Cho and D. H. Park, to be published.

CHAOTIC INFLATION AND THE OMEGA PROBLEM

G. J. MATHEWS, F. GRAZIANI, and H. KURKI-SUONIO
University of California
Lawrence Livermore National Laboratory
Livermore, Ca 94550

ABSTRACT. An overview of the omega problem is presented along with discussions of several ways to resolve this problem in the context of chaotic and stochastic inflation.

1. Introduction

One of the most compelling features of inflationary cosmology is that it provides a solution to the flatness problem, i.e. an explanation of the fact that the ratio, Ω, of the present mass density to the closure density is so close to unity. Indeed, inflationary models naturally lead to a present value of the closure parameter which is equal to unity to many significant figures. At the same time, however, there are at least some observational indications that the present value of the closure parameter may be significantly less than unity. At the very least it is probably safe to say that there is no convincing evidence that the present mass density of the universe is not less than half of the closure density. This is the omega problem. The purpose of this paper will be to take the possibility that there is an omega problem seriously and look within the context of inflationary scenarios for a natural explanation as to why the present value of the closure parameter might be so small.

2. The Omega Problem

Let us begin with a brief review of observational determinations of Ω. There are seven methods. Three of them seem to indicate that Ω is less than unity and the others are inconclusive. The most direct determination of the closure parameter is from mass-to-light ratios. These ratios increase with distance scale from the solar neighborhood (where $M/L \sim 2$ implies $\Omega \sim 0.002$) to the supercluster scale (where, for example, virgocentric infall of the Local Group implies (Davis and Peebles 1983) $M/L = 350, \Omega = 0.35 \pm .02$).

The largest distance scale for which one may determine the closure parameter is from the ratio of the streaming velocity to the Hubble velocity, $\frac{v}{v_H}$, of the local supercluster with respect to the cosmic microwave background. The streaming velocity is related to the closure parameter and the local density distribution via the relation (Peebles 1976),

$$\frac{v}{v_H} = \Omega^{0.6}\frac{\delta\rho}{\rho} .$$ (2.1)

If the IRAS survey of galaxies is used to infer the mass distribution, $\frac{\delta\rho}{\rho}$, then a value of Ω near unity results (Yahil, Walker, and Rowan-Robinson 1986). However, there is reason to believe (Fukugita 1990) that the IRAS survey is biased against the large mass concentrations associated with giant ellipticals. Hence, $\frac{\delta\rho}{\rho}$ is underestimated and Ω is

213

overestimated. Correcting for this bias reduces the upper limit from streaming motion to $\Omega \lesssim 0.4$.

Similarly, galaxy number counts as a function of redshift at first seemed to indicate a large value for Ω (Loh and Spillar 1986). However, it has been pointed out that there are large corrections to the luminousity function for faint galaxies (Caditz and Petrozian 1989) as well as corrections for galactic evolution and merging (Bahcall and Tremain 1988). These corrections introduce large uncertainties ($\Delta\Omega \sim \pm 1$) and can reduce the inferred value of the closure parameter to $\Omega \sim 0.2 \pm 1.0$. On the other hand, it has recently been shown (Fukugita, et al. 1990) that galaxy number counts as a function of magnitude are only consistent with $\Omega \sim 0.1$ independent of galactic evolution corrections. Redshift-magnitude (Yoshii and Takahara 1988) or redshift-angular-diameter studies (Sandage 1988) are presently too uncertain ($\Omega \sim 1 \pm 1$) to add any further significant constraint.

Perhaps, it is also worth pointing out that, as we have heard at this conference (Audouze 1990; Steigmann 1990), the standard big-bang upper limit to the baryonic contribution to the closure parameter based upon primordial nucleosynthesis is $\Omega_b \leq 0.06$. We have also heard (Sato 1990; Mathews, et al. 1990) that an inhomogeneous model with optimum baryon diffusion could increase this limit to $\Omega_b \lesssim 0.3$ for $Y_p \lesssim 0.24$. Although it is popular to propose dark-matter candidates to increase the total Ω above the baryonic contribution, none of these candidates have yet been detected. It is at least possible that there is only baryonic matter and $\Omega = \Omega_b \lesssim 0.3$.

Hence, at the present time there is no convincing observational evidence that the universe is closed. At the very least, one can say that a value of $\Omega \lesssim 0.5$ is not presently ruled out. It is worthwhile, therefore, to consider how inflationary scenarios might be naturally altered to allow for a present closure parameter less than unity.

The present value of the closure parameter is determined from whatever curvature, k, may have been present prior to inflation, and the amount of expansion, i.e.

$$\Omega = 1 + \frac{k}{H_0^2 R_0^2} \ , \tag{2.1}$$

where H_0 is the present Hubble parameter, and R_0 is the present scale factor. For the purposes of this discussion we take $\frac{k}{H_i^2 R_i^2} \sim -1$ just before inflation. The present value of Ω is then determined from the value of $H_f^2 R_f^2$ at the end of inflation times the factors by which these parameters have changed from the end of inflation through the present matter dominated epoch. A value of $\Omega \lesssim 0.5$ corresponds to $ln(\frac{R_f}{R_i}) \lesssim 60 + ln(\frac{T_b}{10^{15}GeV})$.

It should also be pointed out that there is another implicit assumption in any inflationary model which attempts to allow for a present value of the closure parameter less than unity. At the time when Ω begins to deviate from unity the observable horizon corresponds to the horizon before inflation. Fluctuations in the initial curvature would then lead to fluctuations in the present microwave background temperature. We must therefore assume that, although the initial curvature may have been large, it was smooth. We suggest at least one reason why this might be a reasonable assumption. A smooth curvature would minimize the Einstein-Hilbert action. Thus, if the birth of the universe from the Planck time can be described by a functional integral in Euclidean space-time, deviations from the minimum action would have been exponentially suppressed.

3. New Inflation

It is worthwhile to first review the recent demonstration by Steinhardt (1990) that the new inflationary model (Linde 1983) can not simultaneously satisfy $\Omega \lesssim 0.5$ and the constraint from the isotropy of the microwave background temperature, $\frac{\Delta T}{T} \lesssim 10^{-4}$. For a new inflationary potential of the form, $V(\phi) = V_0 - \lambda\phi^n$, in the slow rolling approximation, the number of expansion e-folds during inflation is given by;

$$\ln\left(\frac{R_f}{R_b}\right) = \int H\,dt = \frac{3H^2}{|V''|} = \frac{3H^2}{\lambda n(n-1)\phi_b^{n-2}} \tag{3.1}$$

whereas, the background fluctuations are given by (Guth and Pi 1986; Steinhart and Turner 1986),

$$\frac{\Delta\rho}{\rho} = \frac{4\Delta T}{T} = \frac{H^2}{\dot{\phi}_h} \tag{3.2}$$

where $\dot{\phi}_h$ is evaluated during the inflationary epoch at the point when the present horizon stretches beyond the apparent horizon, H^{-1}. For an inflation occuring at $T \sim 10^{15}GeV$, this occurs about 60 e-folds before the end of inflation (cf. Sec. 2). Thus,

$$60 \approx \frac{3H^2}{\lambda n(n-1)\phi_h^{n-2}} \, . \tag{3.3}$$

In the slow rolling approximation,

$$\dot{\phi}_h \approx \frac{-V'(\phi_h)}{3H} = \frac{\lambda n\phi_h^{n-1}}{3H} \, . \tag{3.4}$$

Thus,

$$\frac{\Delta\rho}{\rho} = \left(\frac{3H^2}{\lambda n(n-1)\phi_h^{n-2}}\right)\left(\frac{(n-1)H}{\dot{\phi}_h}\right) \, . \tag{3.5}$$

This implies that $\phi_h \gtrsim 6 \times 10^5(n-1)H$. Taking the ratio of eq. (3.1) to (3.3) and setting $\phi_b \approx H$ implies,

$$\frac{ln(R_f/R_b)}{60} \gtrsim [6 \times 10^5(n-1)]^{n-2}. \tag{3.6}$$

For any potential with $n > 2$ then one requires $ln(R_f/R_b) \gg 100$ and a present closure parameter equal to unity to many significant figures.

4. Chaotic Inflation

The constraint derived above does not, however, apply to chaotic inflationary scenarios (Linde 1983). Such models are characterized by the presence of domains of homogeneous inflaton field which appear shortly after the Planck time. For a simple $V(\phi) = \lambda\phi^n$ potential the constraint conditions can be derived as above except that now one can have $\phi_b > \phi_h > H$.

Figure 1 shows numerical calculations of the present closure parameter and microwave background energy-density fluctuations as a function of the initial value for the inflaton field, ϕ_0, for different values of the potential strength, λ, for a $V(\phi) = \frac{\lambda}{4}\phi^4$ potential. It can be seen that $0.1 \lesssim \Omega \lesssim 0.5$ implies an initial inflaton field, $\phi_0 \approx 4.5 \pm 0.1$, depending upon λ. Models with smaller λ require somewhat smaller ϕ_0. When λ is small it takes a longer time before the vacuum energy density exceeds the particle energy density. Thus, the inflation happens later, with a longer time scale, leading to a lower reheating temperature. After inflation the universe needs to expand by a smaller factor to reach the present temperature. Correspondingly, to have Ω deviate from unity by the present time requires less inflation. At the same time a value of $\lambda \lesssim 10^{-12}$ gives $\frac{\Delta\rho}{\rho} \leq 10^{-4}$. For small enough ϕ_0 that Ω presently deviates significantly from unity, this is satisfied with a somewhat smaller λ. (for $\Omega = 0.5$, $\lambda \leq 10^{-13}$; for $\Omega = 0.1, \lambda \leq 10^{-14}$). This is because fluctuations with a wavelength on the scale of the present microwave background surface of last scattering were in this case generated at the very beginning of inflation, when $\dot{\phi}$ had not yet accelerated to its "slow-rolling" value.

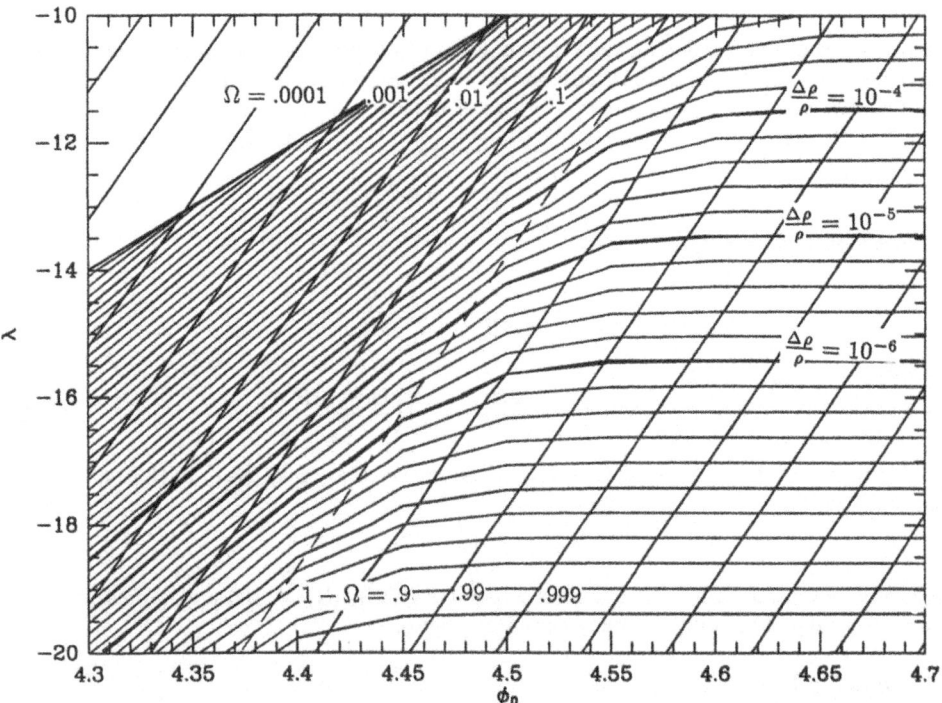

Figure 1. Contours of constant Ω and $\frac{\Delta\rho}{\rho}$ for different initial values of the potential strength, λ, and initial inflaton field value, ϕ_0, in a chaotic inflation model with $V(\phi) = \frac{\lambda}{4}\phi^4$. Values of Ω and $(1 - \Omega)$ are indicated on the diagonal lines. Contours in constant $\frac{\Delta\rho}{\rho}$ are labeled on the right.

Thus, the omega problem and flatness problem can be simultaneously satisfied in chaotic inflation (although only at the expense of a "smallness" problem for λ). It is intriguing, therefore, to ask what is the probability that the initial value for the inflaton field just happened to be that which would produce a present value of $\Omega \sim 0.1 - 0.5$.

For the chaotic inflation scenario to be describable by the classical evolution equations we require that $V(\phi) \lesssim M_{Pl}^4$. This limit implies an upper limit to the initial inflaton field amplitude. For example, $\lambda \sim 10^{-12}$, implies $\phi_0 \lesssim 1000 \, M_{Pl}$. Figure 2 shows the differential probability for a given value of Ω today under the assumption that any value of ϕ_0 from 0 to 1000 is equally probable. An integral of the region between $0.1 \lesssim \Omega \lesssim 0.5$ gives a probability $\sim 3 \times 10^{-5}$. This integrated probability is not altogether satisfactory. Although it is not vanishingly small, neither is it near unity. One would like to find a reason why our universe might have prefered such a small value for the initial inflaton field. One possibility is that (if the notion of thermal equilibrium is relevant near the Planck time) a Boltzmann factor, or a temperature-dependent, $m^2 T^2$, term in the effective potential might have skewed the probability toward smaller ϕ. The onset of inflation would quickly decrease the temperature to zero. Nevertheless, the memory of the early thermodynamic history might have favored a small value for ϕ_0. As another possible explanation we consider the stochastic inflation models.

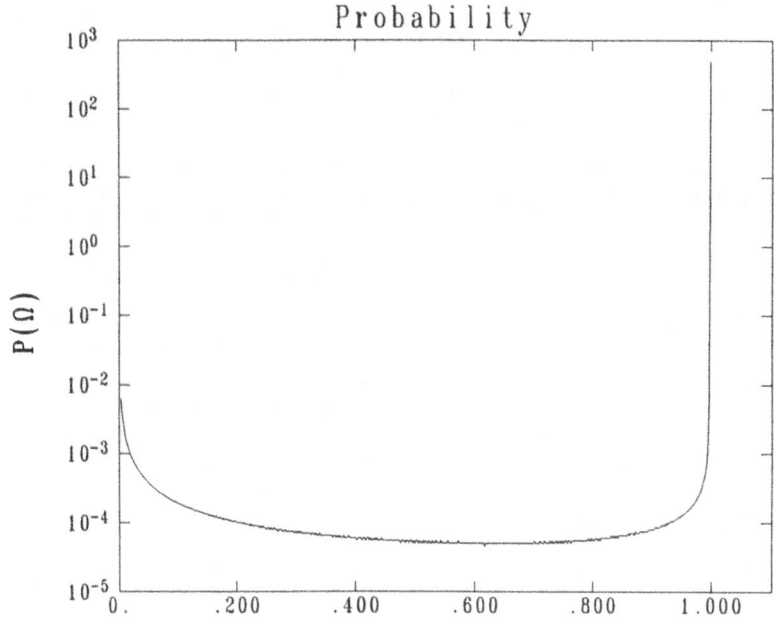

Figure 2. Differential probability for a particular present value of Ω for a chaotic inflationary model with equal probability for any initial value of the inflaton field from 0 to 1000 M_{Pl}.

5. Stochastic Inflation

In stochastic inflation (Starobinsky 1984; Graziani and Olynyk 1985; Mazenko 1986; Rey 1987; Goncharov, Linde, and Mukhanov 1987; Bardeen and Bublik 1987; Graziani 1988; Ortolan, Lucchin, and Matarrese 1988; Nambu, and Sasaki 1989) one assumes the existence of semiclassical long range quantum fluctuations in the inflaton field such that inflation occurs in an ensemble of subdomains each of which creates new subdomains as it inflates. The fact that there is at least some probability which can random walk up the potential, and the fact that higher field values are associated with larger inflation, together imply that the inflation process never ends, i.e. there is a continuous creation of new inflating subdomains (Starobinsky 1984). As Starobinsky (1984) and others (cf. Goncharov, Linde and Mukhanov; Nambu and Sasaki 1989) have shown, the slow-rolling stochastic evolution can be described in terms of a Fokker-Planck equation,

$$\frac{dP(\phi)}{dt} = \frac{\partial}{\partial \phi}[KP(\phi)] + \frac{\partial^2}{\partial \phi^2}[DP(\phi)] , \tag{5.1}$$

where the diffusion coefficients in the slow rolling de Sitter phase are given by,

$$K \approx \frac{V'}{3H} , \tag{5.2}$$

and

$$D \approx \frac{H^3}{8\pi^2} . \tag{5.3}$$

The long-time solution of this equation (Starobinsky 1984) then gives the desired probability distribution for different values of the inflaton field before classical roll over and thermalization.

$$P(\phi) \sim exp\Big[\frac{3M_{Pl}^4}{8}\Big(\frac{1}{V(\phi)} - \frac{1}{V(\phi_0)}\Big)\Big] . \tag{5.4}$$

Since this solution assumes slow rolling and only de Sitter evolution, it is only valid for $1 \lesssim \phi \lesssim 1000$. Nevertheless, it has the desired property of maximum probability for small ϕ. Unfortunately, however, it is too good. For $\lambda \sim 10^{-12}$, the probability is vanishingly small for any values of ϕ greater than unity (including the desired value of $\phi \sim 4.5$). Our idea has been to look at this distribution more carefully to see if the singularity near small ϕ can be avoided. Our reasoning is as follows. The physical probability distribution will be given by the field probability distribution in eq. (5.4) times a factor accounting for the inflated volume of the domain (Goncharov, Linde, and Mukhanov 1987).

$$P_p = P(\phi)e^{3 \int H dt} . \tag{5.5}$$

The volume factor is a steeply increasing function, $\phi \sim e^{\phi^2}$. Thus, if a way can be found to disperse the singularity near $\phi = 0$ such that the probability distribution appears gaussian, $P(\phi) \sim e^{\frac{-\phi^2}{\sigma^2}}$, then the physical probability distribution will have a peak at $\phi \sim 4.5$ for σ sufficiently large, $\sigma \sim 5M_{Pl}$.

Our approach has been to study the evolution of a more general Fokker-Planck equation which is valid through the transition to a Friedmann expansion and which does not rely upon the slow-rolling approximation which breaks down near the bottom of the potential. We do, however, neglect gravitational back reaction-effects. The present work should thus be thought of as a mean field approximation to the exact relativistic solution.

For example, our generalized Fokker-Planck equation for an effective potential of the form $V(\phi) = \frac{m^2}{2}\phi^2$ is,

$$\frac{dP(\phi)}{dt} = -\frac{\partial}{\partial\phi}\Big[\xi P\Big] + \frac{\partial}{\partial\xi}\Big[3H\xi P\Big] + \frac{\partial}{\partial\xi}\Big[m^2\phi P\Big]$$

$$+\frac{\partial^2}{\partial\phi^2}\Big[D_{\phi\phi}P\Big] + \frac{\partial^2}{\partial\phi\partial\xi}\Big[(D_{\phi\xi} + D_{\xi\phi})P\Big] + \frac{\partial^2}{\partial\xi^2}\Big[D_{\xi\xi}P\Big] , \qquad (5.6)$$

where ξ is the conjugate momentum to ϕ. The diffusion constants are defined in terms of the scale factor and Fourier amplitudes, $\psi_k(t)$, for the short-range fluctuating component of ϕ,

$$D_{\phi\phi} = \frac{\epsilon^3}{4\pi^2}|\ddot{R}| \; \dot{R}^2 \; |\psi_k(t)|^2 \qquad (5.7)$$

$$D_{\xi\xi} = \frac{\epsilon^3}{4\pi^2}|\ddot{R}| \; \dot{R}^2 \; |\dot{\psi}_k(t)|^2 \qquad (5.8)$$

$$D_{\phi\xi} + D_{\xi\phi} = \frac{\epsilon^3}{4\pi^2}|\ddot{R}| \; \dot{R}^2 \; \frac{\partial}{\partial t}|\psi_k(t)|^2 . \qquad (5.9)$$

The diffusion coefficients are evaluated at $k = \epsilon\dot{R}$ where ϵ is a small parameter, $\epsilon \ll 1$.

Our hope has been that by studying a more general Fokker-Planck solution the singularity in the long term solution can be avoided. Our studies of the behavior of this more general equation, however, have indicated that the quantum fluctuations damp out significantly by the time the distribution populates the bottom of the potential. Figure 3 demonstrates how neither classical oscillations of the field nor thermal fluctuations during reheating could produce a large enough dispersion in the long time solution to produce the desired peak.

Figure 3 shows the envelope of trajectories of the classical field at the bottom of the potential for different initial values of the inflaton field. This calculation is in the limit of weak reheating. Note that even without reheating the maximum amplitude of the oscillations is only $\sim 0.1 M_{Pl}$. Thus, classical oscillations of the field do not provide enough width to the distribution. Similarly, in the limit of strong reheating, the amount of thermal energy available is miniscule compared to the amount of vacuum energy required to produce a significant flattening of the distribution. So thermal fluctuations do not solve the problem either.

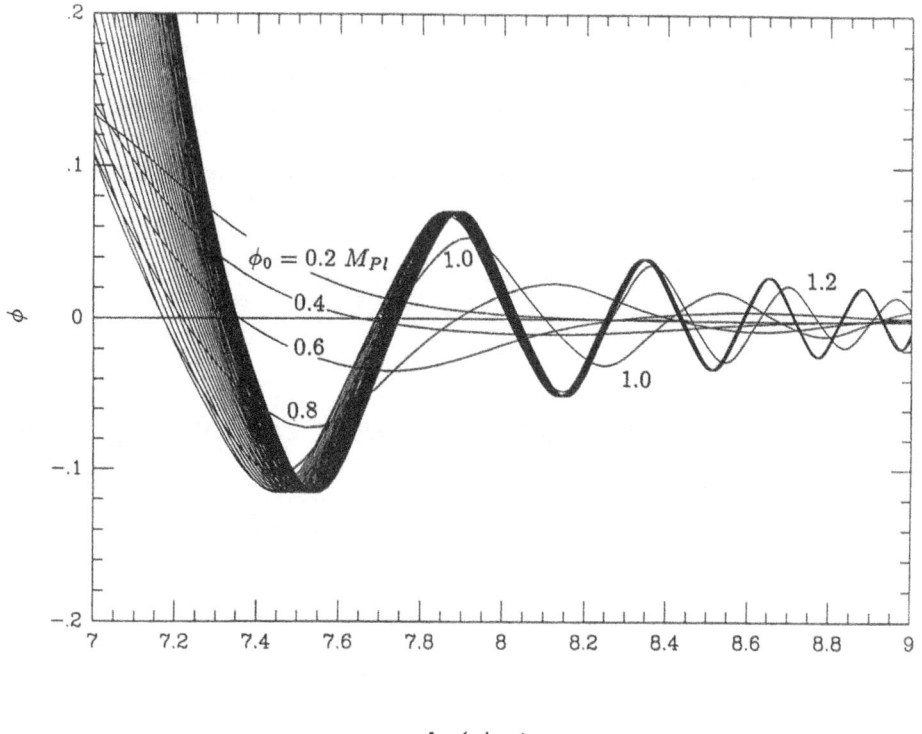

$$log(t/t_{Pl})$$

Figure 3. Envelope of the time evolution of the inflaton field starting from different initial values, ϕ_0.

We note that one possible solution to this problem is at least hinted at in the recent work of Nambu and Sasaki (1989). They showed that by explicitly including the volume term into the Fokker-Planck equation, i.e.

$$\frac{dP(\phi)}{dt} = \frac{\partial}{\partial \phi}[KP(\phi)] + \frac{\partial^2}{\partial \phi^2}[DP(\phi)] + 3(H - <H>)P(\phi) , \qquad (5.10)$$

that an analytical solution can be found which avoids the singularity for small ϕ. Indeed their solution even appears to allow for a peak in the distribution for finite ϕ. Figure 4 shows examples of the analytic solution from Nambu ans Sasaki (1989) for several different values of λ. It appears possible to have a peak in the distribution for $\phi \sim 5$. However, for this value of λ the microwave background fluctuations are unacceptably large, and the peak occurs at a value for the potential energy which is greater than the Planck scale.

The only way we have found to solve the Ω problem in the context of stochastic inflation is by utilizing a generalized effective potential of the form,

$$V(\phi) = \lambda[\frac{\alpha}{2}\phi^2 + \frac{\beta}{3}\phi^3 + \frac{1}{4}\phi^4] . \qquad (5.11)$$

The distribution shown in figure 5 was generated from such a potential with $\alpha = 2.32$, $\beta = -3.02$, and $\lambda = 10^{-12}$. A universe which rolls down from the peak in the distribution would have present fluctuations in the microwave background of $\frac{\Delta T}{T} \sim 5 \times 10^{-5}$ and $\Omega \sim 0.3$.

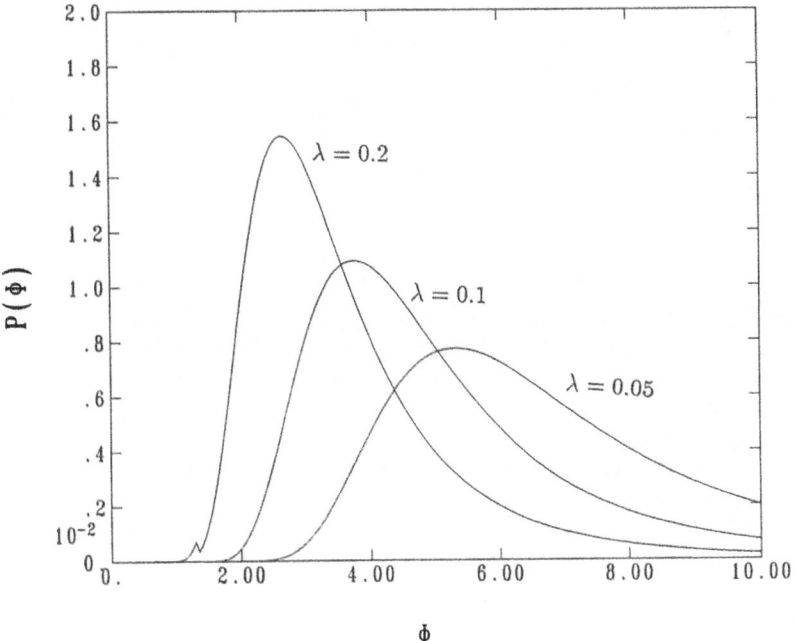

Figure 4. Stationary renormalizable analytic solution (Nambu and Sasaki 1989) for the probability distribution of ϕ for different potential strengths, λ.

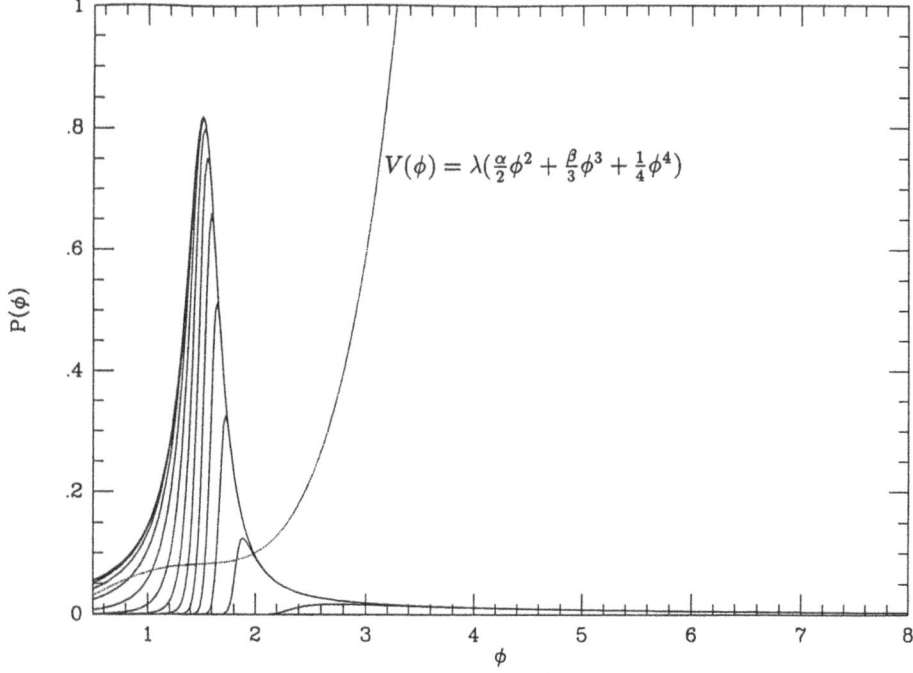

$$V(\phi) = \lambda(\tfrac{\alpha}{2}\phi^2 + \tfrac{\beta}{3}\phi^3 + \tfrac{1}{4}\phi^4)$$

Figure 5. Example of the development of a stationary solution for $P(\phi)$ from the stochastic evolution of a system with a generalized effective potential. For this calculation probability flows in from the right as new universes are born and exits on the left as universes roll down the potential and thermalize.

The authors acknowledge useful conversations with N. J. Snyderman, M. J. Rees, and J. R. Bond. Work performed under the auspices of the U. S. Department of Energy by the Lawrence Livermore National Laboratory under contract W-7405-ENG-48.

REFERENCES

Audouze, J. 1990 (this Conf. Proc.).

Bahcall, S. R. and Tremaine, S. 1988, *Ap. J. Lett.*, **326**, L1

Bardeen, J. and Bublik, G. J. 1987, *Class. Quant Grav.*, 4, 573.

Caditz, D. and Petrosian, V. 1989, *Ap. J. Lett.*, **337**, L65.

Davis, M. and Peebles, P. J. E. 1983, *Ann. Rev. Astr. Ap.*, **21**, 109.

Fukugita, M. 1990 (this Conf. Proc.); and "Proc. 3rd Nishinomiya Yukawa Memorial Symposium," Nishinomiya, Japan, 10-11 Nov. 1988.

Fukugita, M., Takahara, F., Yamashita, K., and Yoshii, Y. 1990, *Submitted to Ap. J.*

Goncharov, A. S., Linde, A. D., and Mukhanov, V. F. 1987, *Int. J. Mod. Phys.*, **A2**, 561.

Graziani, F. 1988, *Phys. Rev.*, **D38**, 1122, 1131, 1808.

La, D. and Steinhardt, P. J. 1989, *Phys. Rev. Lett.*, **62**, 376.

Linde, A. D. 1983, *Phys. Lett.*, **B129** 177.

Loh, E. and Spillar, E. J. 1986, *Ap. J. Lett.*, **307**, L1.

Mathews, G. J., Meyer, B. S., Alcock, C. R., and Fuller, G. M. 1990, *Ap. J.*, **358**, 36.

Mazenko, G. 1986, *Phys. Rev.*, **D 34**, 2223.

Nambu, Y. and Sasaki, M. 1989, *Phys. Lett.*, **B219**, 240.

Ortolan, A., Lucchin, F., and Matarrese, S. 1988, *Phys. Rev.*, **D 38**, 465.

Peebles, P. J. E. 1976, *Ap. J.*, **205**, 318.

Rey, S. J. 1987, *Nucl. Phys.*, **B284**, 706.

Sandage, A. 1988, *Ann. Rev. Astron. Ap*, **26**, 561.

Sato, K. 1990 (this Conf. Proc.).

Starobinsky, A. A. 1984, in *Fundamental Interactions*, V. N. Ponomarev, ed., (MPGI press, Moskow) p. 54.

Steigman, G. 1990 (this Conf. Proc.).

Steinhardt, P. J. 1990, *Nature*, **345**, 47.

Yahil, A., Walker, D., and Rowan-Robinson, M. 1986, *Ap. J. Lett.*, **301**, L1.

Yoshii, Y. and Takahara, F. 1988, *Ap. J.*, **326**, 1.

LATE-TIME COSMOLOGICAL PHASE TRANSITIONS

DAVID N. SCHRAMM
The University of Chicago
5640 S. Ellis Avenue, Chicago, IL 60637
and
NASA/Fermilab Astrophysics Center
Box 500, Batavia, IL 60510-0500

ABSTRACT. It is shown that the potential galaxy formation and large-scale structure problems of (1) objects existing at high redshifts ($Z \gtrsim 5$), (2) structures existing on scales of $100 Mpc$ as well as velocity flows on such scales, and (3) minimal microwave anisotropies $\frac{\Delta T}{T} \lesssim 10^{-5}$ can be solved if the seeds needed to generate structure form in a vacuum phase transition after decoupling. It is argued that the basic physics of such a phase transition is no more exotic than that utilized in the more traditional GUT scale phase transitions, and that, just as in the GUT case, siginificant random gaussian fluctuations and/or topological defects can form. Scale lengths of $\sim 100 Mpc$ for large-scale structure as well as $\sim 1 Mpc$ for galaxy formation occur naturally. Possible support for new physics that might be associated with such a late-time transition comes from the preliminary results of the SAGE solar neutrino experiment, implying neutrino flavor mixing with values similar to those required for a late-time transition. It is also noted that a see-saw model for the neutrino masses might also imply a tau neutrino mass that is an ideal hot dark matter candidate. However, in general either hot or cold dark matter can be consistent with a late-time transition.

Introduction

The purpose of this paper is to describe the current situation regarding late-time cosmological phase transitions as mechanisms for generating structure in the Universe[1]. This subject has received a tremendous boost by the combination of a variety of preliminary observations regarding large-scale structure and galaxy formation coupled with recent hints that the solar neutrino experiment may require new neutrino physics involving small masses and flavor changing with energy scales involved being appropriate to a late-time phase transition.

In this paper we will discuss plausible late-time phase transitions (LTPT) and compare their plausibility with other mechanisms for generating structure in the Universe. In particular, we will note that the physics of late-time transitions is really no different from the basic physics required for GUT scale phase transitions used in more traditional models. We will also note that the recent reports from the Soviet-American Gallium Experiment (SAGE) seem to imply that neutrino physics involves masses and flavor mixings that might be appropriate to a late-time transition. We will then discuss the range of possible structures that might be generated in an LTPT, noting that they can yield multiple great walls and velocity flows, objects at high redshift, and a variety of intricate and unusual patterns of the type that are beginning to be seen in redshift surveys. In particular, we will note that LTPT can generate small spherical objects such as "bags" or "balls" of wall or textures which can serve as seeds in hot dark matter models for galaxy formation and can possibly explain the seed spacings necessary for Vornoi tessalation models[2,3] of the Broadhurst *et al.*[4,5] data. We will discuss how LTPT can give a combination of both random gaussian fluctuations as well as topological defects in this manner similar to any other phase transition in the early Universe. We will also note that because they form after recombination, many of the topological defects do not have the same consequences as those that formed before recombination. In particular, domain wall models can be made to work if the walls decay, or come from multiple minima, or have friction so that they move slowly, or if the walls split off into bags or balls of wall. All such models can yield interesting and exciting structures. Only in the case where infinite, stable walls dominate does one run into the one-wall domination problems of Ref. [6,7,8].

We will also discuss the resulting microwave fluctuations from LTPT. We will note that for a given size structure, if it is generated by an LTPT, one will obtain the minimal $\frac{\Delta T}{T}$ for that structure. Since an LTPT does not require fluctuations on the surface of last scattering, all induced microwave anisotropies are due to propagations through transparent medium effects.

This paper will conclude with a discussion of future obserations and how they will help verify or rule out LTPT. The paper will also outline future calculations that will be important in establishing or eliminating LTPT as a model for large-scale structure information.

Structure Formation

Before specifically going into late-time phase transitions, let us review the basic framework of structure formation in the Universe. In particular, let us note that structure formation requires that density fluctuations grow. In order for this to occur, $\rho_{m(atter)}$ must be greater than $\rho_{r(adiation)}$. If we define T_{eq} as the temperature where $\rho_m = \rho_r$, then for an $\Omega = 1$ universe with h_0 (the Hubble constant in units of $100 km/sec/Mpc$) equal to 0.5, equality is approximately 10^4 times the present temperature T_o. The horizon mass at T_{eq} is $\sim 5 \times 10^{16} M_{\odot}$

which gives a present comoving scale of $\sim 60 Mpc$. The recombination epoch T_{rec} for an $\Omega = 1$ universe occurs slightly after matter domination. At $T \sim 1000 T_o$ baryon fluctuations begin to grow after recombination and the horizon mass at recombination is about $10^{18} M_\odot$ with a comoving scale of $200 Mpc$. We also know that the fluctuations in the microwave background temperature at the time of recombination are less than a few parts in 10^5[9]. Thus, in traditional models with primordial fluctuations existing prior to matter domination, growth begins at matter domination with the limits from $\frac{\delta T}{T}$ forcing $\frac{\delta \rho}{\rho}$ to be less than the order of 10^{-4} since

$$\frac{\delta \rho_m}{\rho} \lesssim 3\frac{\delta T}{T} \lesssim 10^{-4}.$$

Since small fluctuation $\delta \rho$ grows linearly with $1+z$, this would mean that fluctuations could reach the order of unity only at the present epoch. Non-linear growth, and thus true structure formation, does not begin until $\frac{\delta \rho}{\rho}$ has reached unity (see Figure 1). Thus, in the

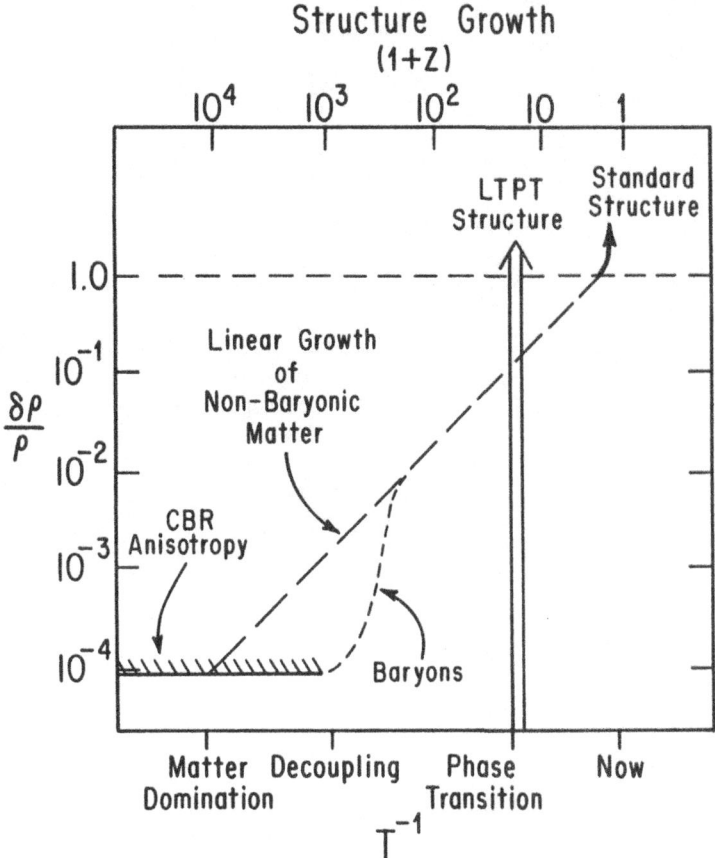

Figure 1. The growth of density fluctuations with the expansion of the universe. NOte that LTPT can yield non-linear growth and thus structure at epochs much earlier than standard primordial models.

standard model, the existence of objects at $z > 1$ (see for example Gunn, Schneider, and Schmidt[10]) requires that there be fluctuations far larger than the average in order that these objects currently exist. As Efstathiou and Rees[11] point out, the gaussian fluctuation model for primordial fluctuations would not allow a large number of quasar-like objects to form at $z \gtrsim 5$.

All models for structure formation require at least two basic ingredients for that structure:

(1) the matter,
(2) the seeds.

In traditional models, the seeds are random fluctuations in the density field generated at the end of the GUT phase transition, presumably accompanying inlflation[12,13].

The matter in any model of galaxy formation with $\Omega = 1$ consists of normal baryonic matter with Ω the order of 0.06 and some non-baryonic matter, either hot or cold, with Ω the order of 0.94.[14]. This is summarized in Table 1.

Table 1

MATTER

Baryonic $\Omega_b \sim 0.06$

VISIBLE $\Omega_{vis} \lesssim 0.01$

DARK

Halo
Jupiters
Brown Dwarfs
Stellar Black Holes

Intergalactic
Hot gas at $T \sim 10^5 K$
Stillborn Galaxies

Non Baryonic $\Omega_{nb} \sim 0.94$

HOT
$m_{\nu_r} \sim 25 eV$

COLD
Wimps/Inos $\sim 100 GeV$
Axions $\sim 10^{-5} eV$
Planetary Mass Black Holes

As was emphasized in reference [14], the robustness of the nucleosynthetic constraint telling us that $\Omega_{b(aryon)}$ is about 0.06 seems very solid. Since Ω associated with shining matter is less than 0.01, this tells us that the bulk of the baryons in the Universe are dark. Whether they are in condensed objects that would be in the halos of galaxies, such as brown dwarfs, jupiters or black holes, or whether they are in the form of some hot intergalactic gas at a temperature high enough to avoid the Gunn-Peterson Test, but low enough to avoid significant x-ray emission, or in the form of stillborn galaxies remains to be determined. In any case, it does seem clear that dark baryons must exist somewhere.

As to the non-baryonic dark matter the question remains as to whether it is in the form of matter that is slow moving at the time of galaxy formation, which has been dubbed "cold dark matter," or whether it is fast moving at the time of galaxy formation and dubbed "hot dark matter."

The seeds which clump the matter to form objects may be divided into two broad categories (see Table 2) which can further be subdivided. The two broad categories would be (1) random gaussian seeds, presumably induced by quantum fluctuations at the end of a phase transition, and (2) topological defects produced in a vacuum phase transition. For the random gaussian seeds, the traditional assumption has been that the phase transition is the one associated with inflation[12,13]. However, it has been shown that similar kinds of fluctuations can also be generated in late-time phase transitions[15,16]. Similarly, for the topological defects, they could be formed either at the end of a GUT phase transition ($\sim 10^{15} GeV$) or in some late-time transition[1,17,18]. In some sense this current division of random versus topological replaces the old division of adiabatic versus isothermal (or isocurvature). In fact, the current "random gaussian" are indeed "adiabatic" and the topological are isothermal and isocurvature. However, the latter have the new added feature of also being non-gaussian.

Table 2

SEEDS

I	RANDOM GAUSSIAN (Quantum)
	A. End of Inflation
	B. LTPT
II	TOPOLOGICAL DEFECTS
	A. GUT
	B. LTPT

Let's note that all models for galaxy formation require new fundamental physics beyond

$$SU_3 \times SU_2 \times U_1.$$

In particular, all non-baryonic dark matter, whether hot or cold, requires new physics, and similarly, all seeds, whether GUT scale or late-time and whether random gaussian or whether topological, require vacuum phase transitions. No model exists that does not invoke new physics. In fact, the existence of structure in the Universe is one of the most important clues to the existence of physics beyond the standard model.

We should also note that not all combinations of seeds and matter are possible. For example, if one uses random gaussian seeds, then the non-baryonic matter must be cold, whereas if one uses topological seeds, the non-baryonic matter can be either hot or cold. One should also note that baryonic halos would require hot dark matter and hence topological seeds. Thus, searches for the dark baryons will also help constrain the non-baryonic candidates.

All seed models require some form of vacuum phase transition. Thus, let us explore what possible phase transitions might occur (see Table 3). It should be noted in looking at Table 3 that of the three general classifications of cosmological phase transitions—the early, intermediate and late—the only ones that we absolutely know must have occurred are in the intermediate category when there is a horizon problem, namely that the horizon at the time of that transition is too small to generate galactic sized structure, and yet, the transition is not accompanied by significant inflation. The traditional early transitions have been used in the past because, while their horizon is small, inflation can amplify the effects to large scales. The other option, which we are advocating in this paper, is that of a late-time transition, where the universe waits until the horizon is sufficiently large that the physics of the phase transition directly yields the structures without having to use inflation to avoid the horizon problem.

Table 3

VACUUM PHASE TRANSITIONS

EARLY	(Small horizon but inflation)
	$\sim 10^{19} GeV$ - T.O.E.
	$\sim 10^{16} GeV$ - GUT
INTERMEDIATE	(Known to occur but horizon problem)
	$\sim 10^2 GeV$ - Electroweak
	$\sim 1 GeV$ - QCD
LATE	(Horizon large)
	$\sim 10^{-2} eV$ - Family symmetries, etc.

Potential Observations to be Explained

In the last couple of years there have been a number of observations affecting galaxy formation and large-scale structure that have been a potential problem for traditional models which invoked early random gaussian fluctuations. However, because each of these observations is new and has not stood the test of time, in this paper we refer to these as potential observations. In particular, many of the advocates of gaussian fluctuations and cold dark matter have tried to argue that these observations are statistical flukes that have yet to be established. Obviously, if these potential observations continue to hold up and are verified and are shown to be ubiquitous rather than statistical rarities, then the traditional models are in serious trouble. Table 4 summarizes these potential observations. Perhaps the most potentially damning would be observations of microwave anisotropies $\frac{\Delta T}{T}$ at levels significantly below 10^{-5}. However, at the present time, observations of small scale anisotropy are at the level of a couple times 10^{-5}. Observations on angular scales of degrees or more are also approaching a few 10^{-5}. As this paper is being written, the measurements have not yet reached the point of ruling out the model of random fluctuations. However, as noted by Smoot[9], within the not too distant future, COBE may be able to achieve limits as low as 3×10^{-6} on scales of a few degrees and larger, and antarctic studies may also push to similar levels on somewhat smaller scales, as might the baloon studies of Meyer at MIT.

Table 4

POTENTIAL OBSERVATIONS

1. $\frac{\delta T}{T} \lesssim 10^{-5}$

2. Structures $\gtrsim 100 Mpc$

3. Large coherent velocity flows

4. Objects existing at $z \gtrsim 5$

5. Large cluster - cluster correlations

The next observation that can be a potential problem for traditional models is the existence of structures with scales greater than the order of $100 Mpc$. In particular, the great wall observed by Geller and Huchra[19] shows that there is at least one such wall in the Universe. The observations of Broadhurst et al.[4,5] show evidence for a multiplicity of such great walls with the characteristic spacing comparable to the size of the Geller-Huchra wall itself. While much debate has been made about whether or not the multiple walls of Broadhurst et al. are periodic or quasi-periodic, it does seem clear from their observations, as well as the work reported by Szalay[20], that there is significant structure in the Universe on scales of $\sim 100 Mpc$. This is thoroughly supported by the large coherent velocity flows where the Seven Samurai[21] and others have found evidence for the existence of an object they call the "Great Attractor" towards which the Virgo cluster and the Hydro-Centaurus cluster all seem to be flowing with a velocity $\sim 600 km/sec$. This again seems to indicate evidence of structures on the scales of at least $60 Mpc$.

Perhaps most constraining of the traditional astronomical measurements is the existence of objects at very large redshifts. In particular, Gunn, Schneider, and Schmidt[10] have found a quasar with a redshift of 4.73. As Efstathiou and Rees[11] have noted, if such objects are ubiquitous, this would be fatal for primordial gaussian fluctuation models. Similarly, if one ever finds an a quasar-type object at much larger redshifts, that would also be fatal.

Another potentially fatal observation for gaussian fluctuation models comes from the work of Bahcall and Soneira[22], and Klypin and Khlopov[23] where they find that clusters of galaxies seem to be more strongly correlated with each other than galaxies are correlated with each other. While Primack and Dekel[24] have warned of the dangers of projection effects on such observations, it seems difficult to understand how projection effects would give the fractal-like behavior[25]. Furthermore, the southern hemisphere work of Huchra[26] also seems to support high cluster correlations. Most recently van den Bergh and West[27] have also found similar correlations for the CD galaxies observed at cluster centers. The CD's should not have the projection effect problems because redshifts are known. Even Primack and Dekel now acknowledge that there seems to be some excess in cluster correlations. If such large correlations turn out to be real, they too cannot be easily explained in the gaussian model, and, as Szalay and Schramm[25] note, they seem to be best fit by some sort of fractal-like pattern, as one might get from topological defects induced by a phase transition.

Late-Time Transitions

By late-time transition we will mean any non-linear growth occurring shortly after recombination. As mentioned above, such non-linear growth can be related either to a gaussian pattern or to a topological pattern such as walls, strings or textures. It is also possible that some normal random gaussian pattern from the very early universe could be triggered to undergo non-linear growth by some sort of phase transition or related phenomenon occurring after recombination. An example of this latter case would be the neutrino flypaper model of Fuller and Schramm[28].

In general we will see that these late-time transitions can give the smallest possible $\frac{\Delta T}{T}$ for a given size structure. They can produce non-gaussian structural patterns, fractal-like with large velocity flows. It might be noted that the co-moving horizon at the time of the transition is not too different than the scale associated with the largest structures observed. No model of primordial fluctuations naturally imbeds this horizon scale onto the structural pattern. If some non-linear growth is associated with the patterns, the horizon scale can be imposed on the structure.

Another very dramatic advantage of late-time transitions, illustrated in Fig. 1, is that it can produce structure with $\frac{\delta\rho}{\rho} \gtrsim 1$ at $z \gtrsim 10$. Thus, one could have significant structure and a significant number of objects at high redshift, which is a problem in any normal model with the seeds forming prior to recombination.

Let us now explore the possible physics that might give rise to a late-time transition, that is, a transition with a critical temperature between $0.001eV$ and $1eV$. It might be noted that in some sense it is a "hierarchy" rather than a "fine-tuning" problem to obtain a transition in this temperature range. We are trying to find a small mass scale somewhat analogous to how one would like to find the mass scale of the electron, or, for that matter, the Z^0 boson, when the natural mass scales to the problem are closer to $10^{19}GeV$, as in superstring models, or to 0. The hierarchy problem of trying to find the intermediate scale of the electroweak interaction of somewhere between the quark-lepton scale and the GUT or Planck scale has traditionally been approached with either a supersymmetric solution or a dynamical solution ("technicolor"). This supersymmetric solution, in some sense, is analogous to the model proposed in the appendix of Hill, Schramm and Fry[1], denoted as HSF, which is an adaptation of the Hill-Ross[29] mechanism. A dynamical solution which has been proposed by Dimopoulos[30] involves a shadow $SU3$. The scale of a physics that might be associated with an HSF mechanism was relating to the MSW mixing solution to the solar neutrino problem.

The MSW[31,32] mixing solution to the solar neutrino problem is achieved if the neutrino mass difference squared, δm^2, is of the order of 10^{-4} to $10^{-7}eV^2$, or, in other words, neutrino masses of the order of a fraction of an electron volt. If we assume, following HSF, that the neutrino masses are generated by a pseudo-Nambu-Goldstone boson mechanism with mass

$$m_\phi \sim \frac{m_\nu^2}{f}$$

and with a transition occurring at $T_{crit} \sim m_\nu$, and if we further assume that the coupling f is related to the GUT scale, since we want to imbed this in some sort of unified theory, then the Compton wavelength $\lambda_\phi \sim 1 Mpc$, in other words, a galactic scale. The density of the ϕ field at the time of the transition is the order of the cosmological density, in other words,

$$\frac{\rho_\phi}{\rho} \sim 1.$$

(Note that this is natural for phase transitions, whereas the requirement for primordial transitions to have small fluctuations, as inflation requires, is a fine tuning requirement.) Furthermore, the average spacing of the nucleation sights, L, can be estimated from Coleman's theory on spontaneous nucleation to yield spacings today that are $\sim 100 Mpc$:

$$\frac{R_H}{L} \sim Log(\frac{M_p}{T_{crit}})$$

$$L_{co} \equiv L(1 + z_{crit}) = \frac{6000}{Log(\frac{M_p}{t_{crit}})} \, (\frac{0.5}{h_0}) \, (1 + z_c)^{-1/2}$$

where $1 + z_{c(rit)} = \frac{T_{crit}}{T_0}$, R_H is the horizon radius at z_c and $M_p \sim 10^{19} GeV$. This yields for $T_{crit} \sim 10^{-2}eV \, to \, 10^{-3}eV \; L_{co(moving)} \sim 40$ to $120 Mpc$.

Recent impetus for new physics at this energy scale has come from the SAGE experiment which detects neutrinos from the PP chain in the sun. The previous solar neutrino experiments, the Chlorine and the Kamiokande experiments, are mainly sensitive to the rare ^8B branch of the solar energy generating reactions. It is well established that the ^8B experiments have seen fluxes at levels somewhat below theoretical predictions[33]. However, there has always been the worry that the ^8B channel may be supressed due to astrophysical effects since its yield is very temperature sensitive. However, the PP chain that produces the neutrinos to be detected by SAGE must work if the sun is burning by fussion. Thus, the report[34] of no significant counts above background after five months of running the gallium experiment when they expected nineteen counts for the standard model implies that something is happening to the neutrinos on their way between emission and arrival at earth. Of course, the present results are very preliminary. Questions with regard to estimates of background, counting efficiencies, systematics, statistics, etc., remain, but the tantalizing hint that the ν_e's mixed into some other species of neutrino on their way out of the sun is certainly exciting. The final state of this experiment will not be known for several years. In 1991 we will begin to have results from a similar gallium experiment operated by the GALLEX collaboration in the Grand Sasso Tunnel in Italy. Their chemistry is somewhat cleaner and we will have an independent check. Furthermore, both of these gallium experiments will be callibrated using ^{51}Cr sources of MeV neutrinos. Thus, one will have a true check of their counting efficiencies, etc., and both of these experiments will run for a long-enough time that the statistics will reach significant levels. If the neutrinos really are mixing on their way out of the sun, then the MSW solution is probably valid and we are in the realm discussed above.

It might also be noted that a simple application of the Gell-Mann–Ramond–Slansky see-saw model[35] for neutrino masses yields some interesting implications. If we assume that there is a mass hierarchy in the neutrinos with the electron neutrino having negligible mass, the μ the intermediate mass and the τ the heaviest, and we assume that the mixing of the ν_e in the sun goes to its nearest neighbor family, the ν_μ, then the ν_μ is carrying most of the mass of the MSW δ_m^2. The see-saw mechanism argues that

$$m_{\nu_i} \sim \frac{m_{f_i}^2}{M}$$

for a given family, or, in other words,

$$m_{\nu_\tau} \sim m_{\nu_\mu} (\frac{m_{f_3}}{m_{f_2}})^2.$$

If we use lepton masses for the fermion masses, this yields a ν_τ mass in the neighborhood of a few eV. However, if we use heavy quark masses, then, since the top quark mass is $\gtrsim 100$ times that of the charm quark, this yields ν_τ masses in the neighborhood of 10 to 100 eV, making it perfect hot dark matter. It might also be noted that the see-saw mass scale, M, in this picture, ends up being the order of 10^9 to $10^{12} GeV$, which happens to be the only window allowed for the DFS-axion[36] scale. It might further be noted that if the non-baryonic dark matter is indeed the τ neutrino, then one is required to dismiss primordial gaussian fluctuations.

Note that even if the MSW mixing is $\nu_e - \nu_\tau$, the LTPT possibility is still there, but then all neutrinos would be light and could not serve as HDM. It is interesting that in this latter case the see-saw M is the GUT scale.

Structure from LTPT

LTPT can produce vacuum fluctuations of the random gaussian character just as could be generated at the end of inflation[11], however, as emphasized in Ref. [15,16], these structures will have a quantum scale that is the order of a galaxy size, and the bosons associated with the fluctuations might even serve as the dark matter of the universe.

The other alternative for LTPT is to produce topological structures. Just as early universe phase transitions can produce strings and/or textures, LTPT can also produce such objects. Furthermore, LTPT can produce walls which are a problem for primordial phase transitions. However, there is a problem for some walls, depending on the nature of the interaction potential. LTPT that have a $\lambda\phi^4$ potential will end up with one wall dominating as was demonstrated in Ref. [6,7,8]. However, this problem of one wall dominating can be surmounted in a variety of ways which have varying degrees of attractiveness, depending on the eyes of the beholder. For example, in the HSF phase transition, the walls are sine-Gordon rather than $\lambda\phi^4$. As Widrow has shown[37], the sine-Gordon walls can yield "bags" of wall or "balls" of wall which survive several expansion times. These bags or balls can then serve as seeds in galaxy formation, and thus, it is their amplitude that becomes a deciding factor for $\frac{\Delta T}{T}$ limits as opposed to the energy scale of the infinite walls which can be made quite small. This latter point was emphasized by Hill, Schramm and Widrow[17]. Another way of avoiding single wall dominance is the decaying wall model of Kawano[38] where the walls serve as seeds and then decay away. It is also possible to escape one-wall domination with a large number of minima in the potential. Perhaps the most dramatic way of escaping one-wall domination, thus keeping a network of walls, as shown in Fig. 2A, is if the walls have friction with the ambient medium, whether it be neutrinos or the remaining baryonic and/or non-baryonic matter in the universe[39]. Alessandro Massarotti has shown that friction can in many reasonable cases slow the walls down sufficiently that they do not evolve to the one-ball domination situation. In this case, one retains a complex network with L for the wall being much less than the horizon size.

WALL NETWORK

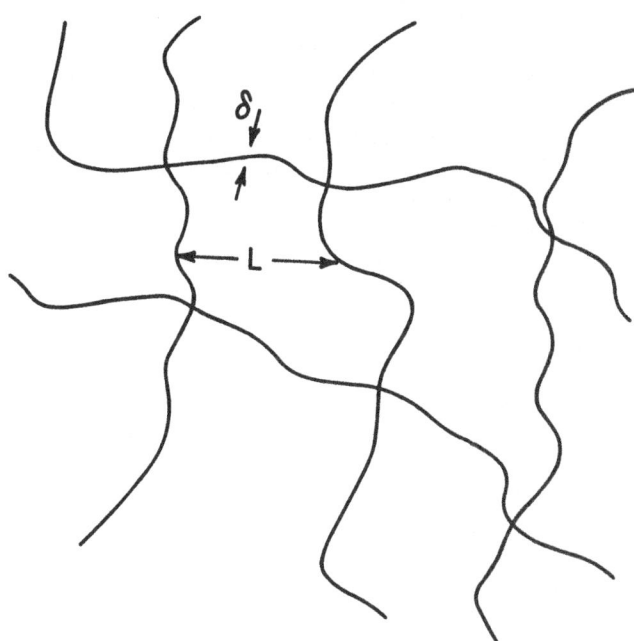

Figure 2a. A generic wall network defining the wall thickness δ and the characteristic spacing of structure L.

236

It might be noted that long walls gravitationally repel rather than attract[40,41], whereas balls of wall are attractive seeds, thus a combined network of balls and slowed-down long walls can yield a complex structure which may be even of a fractal character in agreement with the claims of Schramm and Szalay[25] from cluster correlations.

In addition to walls, LTPT can also produce textures[42] or non-topological solitons[43]. In these latter cases, or with the bags of wall dominating, one will have networks more closely ressembling Fig. 2B and Fig. 2A. It should be noted that the parameters L and δ and the nature of the structures generated are dependent on the model for the LTPT. It should also be noted that questions of the detailed physics of imbedding the LTPT into some larger GUT or TOE are dependent on the unification model. HSF have shown that a reasonable toy model can be constructed which can give a phase transition. These phase transitions in many ways are quite analogous to the axion-producing phase transition which has a coupling at a scale near to the order of $10^{11} GeV$, far above the QCD phase transition scale of the order of GeV. And like the axion, the particle involved in the LTPT of HSF has a pseudo-Nambu-Goldstone boson. However, instead of being related to the strong interaction and quarks, in the LTPT case it is related to the neutrinos and probably to family symmetry.

SEED NETWORK
(Bag, balls-of-wall, Textures, etc)

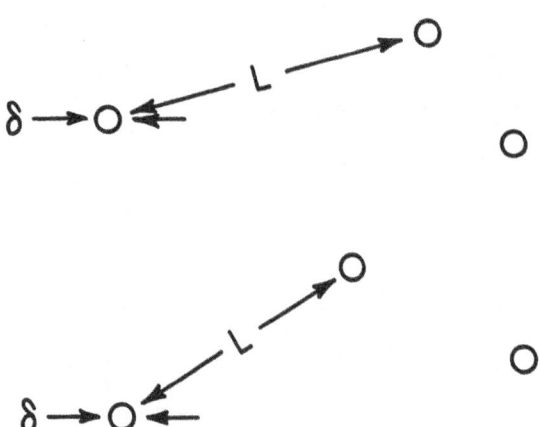

Figure 2b. A generic network for seed generation with seed size, δ and seed separation L.

Generating seeds at an LTPT might be advantageous for producing the multiple walls of Broadhurst et al.[4,5]. In particular, Icke and Weygaert[2], and Coles[3] have independently demonstrated that the phenomenological Vornoi tessalations of the intersection of expanding rarefaction shells give a very good fit to large scale structure if the nodes of these tessalations are fit to the Abell clusters. In particular, they note that one gets quasi-periodic walls at $\sim 130 Mpc$ with cluster correlation functions that are quite strong and follow the fractal behavior of Schramm and Szalay. However, the seed distribution required to give this tessalation causes a conflict with the microwave background radiation, if the seeds are generated prior to the decoupling. However, an LTPT could remedy that. Similarly, an LTPT can provide the seeds to enable hot dark matter to work as a galaxy formation model (see, for example, Ref. [44]). It might be noted that the typical bag of wall can easily yield a galaxy or a quasar-forming seed.

We can estimate the mass associated with a wall in the following way:

Let $\sigma =$ energy density per unit area, that is:

$$\sigma \equiv \rho_w \delta \sim \frac{4 \times 10^{-5} s \rho_0 \delta}{h_0^2} (1 + z_c)^4$$

where

$$\delta \equiv \text{thickness}$$

$$\rho_w = \text{density of the wall}$$

$$s = \frac{\rho_w}{\rho_r} \text{ at } z_c$$

$$\rho_0 = 3 \times 10^{11} h_0^2 M_\odot / Mpc^3$$

then

$$M_w \sim \pi \sigma L^2 \sim 3 \times 10^7 (1 + z_c)^4 (\frac{\delta}{Mpc})(\frac{L}{Mpc})^2 M_\odot.$$

and for stable walls,

$$\Omega_w(z) \sim \frac{3}{4} \frac{\sigma}{\rho_0 L (1 + z)^3}.$$

Note that Ω_w, at the present epoch, can be the order of unity. Wall domination can occur at the present epoch if

$$z_c \gtrsim 11 (\frac{L}{\delta})^{\frac{1}{4}} (\frac{h_0^2}{s})^{\frac{1}{4}} - 1$$

for stable walls. It might be noted that if wall domination occurs at the present epoch, as long as there are multiple walls, rather than just one wall dominating, one has the interesting situation where the expansion of the universe is no longer following the normal matter-dominated relationship, and, in particular, one can achieve ages greater than $\frac{1}{H_0}$. Such a situation may be a solution to the age-Hubble constant problem if h_o is ever shown to be greater than 0.7.

It might also be noted for topological structure generated by LTPT that the structure is relatively independent of whether the non-baryonic dark matter is hot or cold.

Microwave Anisotropies

Since LTPT provide no fluctuations on the surface of last scattering, all fluctuations from the microwave background must be due to the differential redshift-blueshift non-cancellation due to a changing potential in the transparent medium or due to scattering of the microwave photons off of moving objects. One can estimate the potential change due to the ϕ field itself generated in the phase transition and by the dynamic motion of the structures and the Doppler shift thereby produced. One can also do the classical Rees-Sciama and Sachs-Wolf calculations for the $\frac{\Delta T}{T}$ generated by existing objects[45,46]. We can estimate its effects roughly in the following way: The static effects will dimensionally go as

$$\frac{\delta T}{T} \sim \Omega_w (\frac{L}{R_H})^2 \sim G\sigma L$$

The time-changing effects can be estimated by multiplying the static effect by $\frac{V}{c}$. While different people remember different formulations of these things, one can show that because of the nature of walls and other topological systems, the effects can be reduced to the form $G\sigma L$ times $\frac{V}{c}$ or $\frac{V^2}{c^2}$. Since any walls or topological seeds we ever see must be moving with $V < c$, the dominant effect will in general go like $G\sigma L$, which can be shown to yield the result:

$$\frac{\delta T}{T} \sim 10^{-8} (\frac{1+z_c}{10})^4 \, s(\frac{\delta}{Mpc})(\frac{L}{Mpc})$$

$$\sim 10^{-6} \text{ for } L \sim 100 Mpc, \ \delta \sim 1 Mpc, \ z_c \sim 10.$$

Note that this yields $\frac{\delta T}{T} \sim 10^{-6}$ even for an L of $100 Mpc$. The distribution, however, of these fluctuations depends very much on the detailed topological nature of the structures produced. In particular, Turner, Watkins and Widrow[46] have shown that balls of wall tend to produce spikes very similar in nature to the spikes that textures produce. A general formalism showing the wide range of structures in non-gaussian microwave background fluctuations has been developed by Goetz and Noetzold[47].

In general, one can see that if structure of size L is generated by a late-time phase transition, and L is the maximum size of structure produced in that transition, then the late-time transition does give the minimum $\frac{\delta T}{T}$ for that structure. Of course the question is what is the characteristic size L of structure generated in a transition. For $\lambda\phi^4$ structures, L goes to the horizon size, in which case $\frac{\delta T}{T}$ gets larger than current observational limits. However, as mentioned above, many other possibilities can be generated in LTPT. With L at present being a somewhat freely adjustable parameter, depending on the model, the amount of friction, the decay of the walls, etc.

Future

Obviously, one of the key things that is needed for the future of LTPT is the development of more realistic particle models. While there are hints from the solar neutrino problem that interesting new flavor physics is involved at the low energy scale, truly consistent models embedded within a complete general framework remain to be worked out. Another important aspect of many of the topological models is the development of dynamical calculations with matter involved. LTPT take place when the universe is matter dominated. The evolution of the structure interacting with the matter, including friction effects, as pointed out by Massarotti, is critical. These dynamical calculations become quite complex in the same manner that the evolution of cosmic string networks became quite complex. As we've seen with cosmic string networks, it's taken many years for convergence to occur on the results. Similar complications may occur in the complex topological models generated in late-time phase transitions. This is an area that needs exploration.

Once one begins to have models for the structures generated in LTPT, one needs to generate full N-body calculations analogous to those that were carried out with cold dark matter, so that one can see whether or not realistic galaxies and structures that look like observations really do occur. The complexity of the topological structures, particularly since long walls can repel and balls of wall attract, could be quite interesting. Can one indeed generate Vornoi tessalation patterns? This remains to be seen.

Of course, from the observational point of view, the two critical things are the microwave background and the large-scale structure. In particular, if the microwave structures are indeed shown to be significantly less than 10^{-5}, whereby traditional primordial structure generating models are ruled out, then LTPT becomes attractive. Furthermore, since many LTPT involve distributions of fluctuations that are non-gaussian[47,48,49], finding such non-gaussian patterns becomes extremely important. Of course on the observational astronomy side, the actual determination of the three-dimensional large scale structure of the universe is the bottom line. Is this structure fractal in nature? Does it have patterns implying topological seeds? Can it be generated by the gravitational evolution of initial random gaussian fluctuation?

Acknowledgements

I would like to thank my collaborator on LTPT, Chris Hill, for inspiration on all aspects of this problem. I would also like to acknowledge useful discussions with my colleagues working on this exciting LTPT problem: Savas Dimopoulos, Jim Fry, George Fuller, Günter Götz, Dirk Nötzold, Bill Press, Grahm Ross, Barbara Ryden, David Spergel, Albert Stebbins, Glen Starkman, Michael Turner, Rick Watkins, Larry Widrow. This work was supported in part by NSF grant AST 88-22595 and by NASA grant NAGW 1321 at the University of Chicago and by the DoE and NASA grant NAGW 1340 at the NASA/Fermilab Astrophysics Center.

References

[1] Hill, C., Schramm, D.N., and Fry, J.N. (1990) "Cosmological structure formation from soft topological defects," Comments Nuclear and Particle Physics, 19, 25-39.

[2] Icke, V. and van de Weygaert, R. (1990) "Vornoi Cosmology," in press, Quarterly Journal of the Royal Astronomical Society.

[3] Coles, P. (1990) "Understanding recent observations of large-scale structure," submitted to Mon. Notices Royal Astronomical Society.

[4] Broadhurst, T., Ellis, R., Koo, D., and Szalay, A. (1990) "Large-scale distribution of galaxies at the Galactic poles," Nature, 343, 726.

[5] Koo, D.C. and Kron, R.G. (1988) "A deep redshift survey of field galaxies," in F.G. Kron and A. Renzini (eds.), Towards Understanding Galaxies at Large Redshifts, Kluwer Academic Publishers, Dordrecht, p.209.

[6] Press, W., Ryden, B., and Spergel, D. (1989) "Dynamical evolution of domain walls in an expanding universe," Astrophysical Journal, 397, 590.

[7] Hodges, H. (1988?) "Domain walls with bound Bose condensates," Physical Review D, 37, 3052-5.

[8] Stebbins, A.J. and Turner, M.S. (1989) "Is the great attractor really a great wall?" Astrophysical Journal, 339, L13.

[9] Smoot, G. (1990) "The Microwave background," Proceedings of the Tokyo meeting on Primordial Nucleosynthesis, September 1990.

[10] Schneider, D.P., Schmidt, M., and Gunn, J.E. (1989) "A Quasar at $z = 4.73$," Astron. Journal, 98, 1951-1958.

[11] Efstathiou, G. and Rees, M.J. (1988) "High red-shift quasars in the Cold Dark Matter cosmogony," Mon. Notices Royal Astronomical Society, 230, 5-11.

[12] Olive, K. (1990) "Inflation," Physics Reports, 190, 309-403.

[13] Kolb, E. and Turner, M. (1989) The Early Universe, Addison Wesley, San Francisco.

[14] Schramm, D.N. (1990) "Big bang nucleosynthesis: the standard model and alternatives," Proc. of the 1990 Nobel Symposium, Graäftåvalen, Sweden.

[15] Wasserman, I. (1986) "Late phase transitions and spontaneous generation of cosmological density perturbations," Physics Review Letters, 57, 2234.

[16] Press, W., Ryden, B., and Spergel, D. (1990) "Single mechanism for generating large-scale structure and providing dark missing matter," Physical Review Letters, 64, 1084.

[17] Hill, C.T., Schramm, D.N., Widrow, L.M. (1990) "Late-time phase transitions and large scale structure," Proc. of the XXVth Rencontres de Moriond at Les Arcs, Savoie, France, March 1990, in press, and Science, submitted.

[18] Frieman, J., Hill, C. and Watkins, R. (1990) "Late-time phase transitions," Fermilab Preprint, in press.

[19] Geller, M. and Huchra J. (1989) "Mapping the universe," Science, 246, 897.

[20] Szalay, A. (1990) "Correlations of galaxies over cosmic scales," Proc. IUPAP Conf. Primordial Nucleosynthesis and Evolution of Early Universe, Tokyo, September 1990.

[21] Faber, S. and Dressler, A. (1988) "The Great attractor," Proc. Texas Symposium on Rel. Astrophysics, Dallas.

[22] Bahcall, N. and Soneira, R. (1983) "The Spatial correlation function of rich clusters of galaxies," Astrophysical Journal, 270, 20-38.

[23] Klypin, A. and Kopylov, A. (1983) "Cluster correlations,' Soviet Astr. Lett., 9, 41-46.

[24] Primack, J. and Dekel, A. (1990) U. of California at Santa Cruz preprint.

[25] Szalay, A. and Schramm, D.N. (1985) "Are galaxies more strongly correlated than clusters?" Nature, 314, 718-719.

[26] Huchra, J. (1990) Harvard University preprint.

[27] van den Bergh, S. and West, J. (1990) "Correlation Functions of CD Galaxies," Dominion Astronomical Observatory preprint.

[28] Fuller, G. and Schramm, D.N. (1990) "Neutrino flypaper and the formation of structure in the universe," University of California at San Diego preprint.

[29] Hill, C. and Ross, G. (1988) "Pseudo-Goldstone bosons and new macroscopic forces," Physics Letters B, 203, 125.

[30] Dimopoulos, S. (1990) private communication.

[31] Mikheyev, S.P. and Smirnov, A.Yu.. (1986) "Resonant amplification of ν oscillations in matter and solar-neutrino spectroscopy," Nuovo Cimento, 9C, 17.

[32] Wolfenstein, L. (1978) "Neutrino oscillations in matter," Physical Review D, 17, 2369.

[33] Bahcall, J. (1989) Neutrino Astrophysics, Cambridge University Press, Cambridge, England and New York.

[34] SAGE colloboration (1990) Int. High energy Meeting, Singapore, 1990.

[35] Gell-Mann, M, Ramond, P. and Slansky, R., (1979) "Complex spinars in unified theories," in Supergravity, P. Van Nieuwenhuizen and D. Freedman (eds.), North Holland Pub. Co.

[36] Dine, M., Fischler, W., and Srednicki, M. (1981) "A Simple solution to the strong CP problem with a harmless axion," Physics Letters B, 104, 199.

[37] Widrow, L. (1990) "Wall interactions," CfA/Harvard preprint.

[38] Kawano, L. (1990) "Evolution of domain walls in the early Universe," Physical Review D, 41, 1013.

[39] Massarotti, A. (1990) "Evolution of light domain walls interacting with dark matter," Fermilab Pub-90/77-A, and Physical Review D, in press, 1990.

[40] Ipser, J. and Sikivie, P. (1984) "Gravitationally repulsive domain walls," Physical Review D, 30, 712.

[41] Götz, G. (1990) "Scalar field configurations with planar and cylindrical symmetry: Thick domain walls and strings," Fermilab PUB-90/35-A.

[42] Turok, N. (1989) "Textures," Physical Review Letters, 63, 2625.

[43] Lee, T.D. (1987) "Soliton stars and the critical masses of black holes," Physical Review D, 35, 3637.

[44] Bertschinger, E., Scherrer, R., Vilenson, J., and VanDalen, A. (1990) "Seeded hot dark matter," Ohio State preprint.

[45] Rees, M. and Sciama, D. (1968) "Cosmological background radiation," Nature, 217, 511.

[46] Sachs, R.K. and Wolf, A.M. (1967) "Perturbations of a cosmological model and angular variations of the microwave background," Astrophysical Journal, 147, 74.

[47] Turner, M., Watkins, R., and Widrow, L. (1990) "Microwave distortions from collapsing domain-wall bubbles," Astrophysical Journal Letters, submitted.

[48] Götz, G. and Nötzold, D. (1989) "On thick domain walls in General Relativity," Fermilab PUB-89/235-A and (1990) Physical Review D, submitted; and Götz, G. and Nötzold, D. (1989) "An exact solution for a thick domain wall in general relativity," Fermilab PUB-89/236-A and (1990) Physical Review D, submitted.

[49] Stebbins, A. (1990) "Microwave anisotropy signatures," CITA preprint.

False-Vacuum Decay in Generalized Extended Inflation

Richard Holman
Physics Department,
Carnegie Mellon University,
Pittsburgh, PA   15213

ABSTRACT. We study false-vacuum decay in context of generalized extended inflationary theories, and compute the bubble nucleation rates for these theories in the limit of $G_N \to 0$. We find that the time dependence of the nucleation rate can be *exponentially* strong through the time dependence of the Jordan–Brans–Dicke field. This can have a pronounced effect on whether extended inflation can be successfully implemented.

1. Introduction

Extended inflation [1] is the latest addition to the panoply of inflationary models. It is a revision of old inflation [2] that avoids the "graceful exit" problems associated with the original inflationary model. It does this by using a Jordan–Brans–Dicke (JBD) theory of gravity [3] which has *power law* solutions, instead of exponential ones, for a vacuum dominated universe.

As is well known, the "graceful exit" problem of old inflation concerned the percolation of the bubbles of true vacuum formed in the exponentially expanding sea of false vacuum. The relevant parameter is $\epsilon \equiv \lambda/H^4$ where λ is the rate at which bubbles of the true vacuum nucleate from the false vacuum and H is the Hubble parameter. During old inflation these are both constant. The problem that arises is that if ϵ is small enough to allow the "inflaton" field (whose potential drives inflation) to remain in the false vacuum long enough for suffcient inflation to occur, then the bubbles of true vacuum will *not* percolate to form a cluster large enough to fit the region that would later become the observable universe. On the other hand, if we make this parameter large enough for percolation to take place ($\epsilon \gtrsim \epsilon_{\text{crit}} \simeq 0.23$ [4]), insufficient inflation will take place.

Extended inflation evades this problem by making H time dependent so that ϵ grows with cosmic time as $\epsilon \propto t^4$. Thus, ϵ can start small enough so that the universe inflates by an amount sufficient to solve all the standard cosmological conundrums and then it can grow large enough for percolation to take place.

In this article, we review some work [5] concerning the details of the calculation of the nucleation rate λ that enters the percolation parameter ϵ. In all previous work, this rate was taken to be constant. What we will show here is that the *generic* circumstance is that λ will depend on the JBD field Φ. This possibility then enlarges the class of models that may have good extended inflationary behaviour.

2. Generalized Extended Inflation

Let σ denote the inflaton field whose potential $V(\sigma)$ has both a metastable and a stable ground state. We also define Φ to be the JBD field. The generic action we consider here is:

$$S = \int d^4x \sqrt{-g} \left[-\Phi R + \omega g^{\mu\nu} \frac{\partial_\mu \Phi \partial_\nu \Phi}{\Phi} \right.$$
$$\left. + F(\Phi) \frac{1}{2} g^{\mu\nu} \partial_\mu \sigma \partial_\nu \sigma - G(\Phi) V(\sigma) \right]. \tag{1}$$

The action has been written in the *Jordan Conformal Frame* where the effective Newton's constant is the JBD field G_N. In the original extended inflationary model, $F(\Phi) = G(\Phi) = 1$.

We now want to compute the bubble nucleation rate for σ to tunnel from the metastable state at $\sigma = \sigma_{FV}$ to the true ground state $\sigma = \sigma_{TV}$. The usual way to do this would be to use the standard euclidean bounce formalism developed by Callan and Coleman [6] (and later generalized to include gravity by Coleman and DeLuccia [7]). Unfortunately, this formalism is inapplicable, due to the fact that the JBD field is "rolling" during the bounce. Thus we have to solve the two field tunneling problem! At this time, this is beyond our capabilities, but there is way to generate a sensible approximation scheme which will allows us to compute the Φ dependence of the nucleation rate.

The first step is to perform a conformal transformation on the above action which renders the action identical to one in which the theory of gravity is standard GR (this conformal frame is known as the *Einstein Conformal Frame*. This is done as follows:

$$g_{\mu\nu} = \Omega^2 \overline{g}_{\mu\nu}$$
$$\sqrt{-g} = \Omega^4 \sqrt{-\overline{g}}$$
$$R = \Omega^{-2} \overline{R} - 6\Omega^{-3} \overline{\Box} \Omega, \tag{2}$$

where the overbars denote quantities in the Einstein conformal frame, and $\Omega^{-2} = 16\pi G_N \Phi$. We define a new field ψ via the relation

$$\psi \equiv \psi_0 \ln (16\pi G_N \Phi), \tag{3}$$

with $\psi_0^2 \equiv (3 + 2\omega)/16\pi G_N$. With these definitions, the action of Eq. (1) expressed in the Einstein frame becomes

$$\overline{S} = \int d^4x \sqrt{-\overline{g}} \left[-\frac{\overline{R}}{16\pi G_N} + \frac{1}{2} \overline{g}^{\mu\nu} \partial_\mu \psi \partial_\nu \psi \right.$$
$$\left. + f(\psi/\psi_0) \frac{1}{2} \overline{g}^{\mu\nu} \partial_\mu \sigma \partial_\nu \sigma - g(\psi/\psi_0) V(\sigma) \right]. \tag{4}$$

where

$$f(\psi/\psi_0) \equiv \exp(-\psi/\psi_0) F(\Phi), \qquad g(\psi/\psi_0) \equiv \exp(-2\psi/\psi_0) G(\Phi), \tag{5}$$

and $\Phi = \Phi(\psi/\psi_0)$ is understood.

From the action in the Einstein frame, we see that gravitational effects can be frozen out of the bounce calculation by taking the $G_N \rightarrow 0$ limit *and* freezing out the evolution of ψ during the bounce. Note that we are *not* freezing out ψ's evolution in *real time* however. As long as the effective Planck mass scale induced by the JBD field is much larger than the mass scales associated with σ, this approximation should be reliable.

We thus arrive at the following truncated Euclidean action for the inflaton σ:

$$\overline{S}_E = \int d^4x \left[f(\xi)\frac{1}{2}\partial^\mu\sigma\partial_\mu\sigma + g(\xi)V(\sigma) \right], \tag{6}$$

where $\xi \equiv \exp(\psi/\psi_0)$. To calculate the bubble nucleation rate per unit physical three-volume, we must compute the bounce action B and prefactor A that appear in the formula for the rate [6]:

$$\overline{\lambda} = A\exp(-B) \tag{7}$$

If we rescale the coordinates to

$$\hat{x}^\alpha = \sqrt{\frac{g(\xi)}{f(\xi)}}\, x^\alpha \tag{8}$$

we can rewrite the action of Eq. (6) as

$$\overline{S}_E = \frac{f^2(\xi)}{g(\xi)} \int d^4\hat{x} \left[\frac{1}{2}\hat{\partial}^\mu\sigma\hat{\partial}_\mu\sigma + V(\sigma) \right] = \frac{f^2(\xi)}{g(\xi)} S_0 \tag{9}$$

where S_0 is the Euclidean action of the standard theory (i.e., the action of Eq. (4) with $\xi = 1$). Clearly, this implies that the bounce configuration σ_B is related to the bounce of the theory containing $\hat{\sigma}_B$:

$$\sigma_B(x) = \hat{\sigma}_B(\sqrt{g(\xi)/f(\xi)}\,x). \tag{10}$$

The bounce action is

$$B(\xi) = \frac{f^2(\xi)}{g(\xi)}B_0 \qquad [\xi = \exp(\psi/\psi_0)], \tag{11}$$

where B_0 is the (ξ-independent) bounce action calculated for the theory with $\xi = 1$ ($\psi = 0$). The fact that the coupling of ψ into the action of Eq. (6) can be factored out by means of coordinate rescaling is essential in enabling us to carry out our calculation. The prefactor A from Eq. (2.7) is given by [6]

$$A = \left| \frac{\det'[S_E''(\sigma_B)]}{\det[S_E''(\sigma_{FV})]} \right|^{-1/2} \prod_\mu \left(\frac{C_\mu}{2\pi} \right)^{1/2}. \tag{12}$$

Here, σ_{FV} is the false-vaccum configuration, σ_B is the bounce solution, and $\det'$ indicates that the functional determinant is to be evaluated in the subspace orthogonal to the four translational zero modes. The C_μ are normalization factors of the zero modes of the operator $S_E''(\sigma_B)$.

Performing the functional variation of the Euclidean action yields

$$\begin{aligned} A_{DET} &\equiv \left| \frac{\det'[S_E''(\sigma_B)]}{\det[S_E''(\sigma_{FV})]} \right|^{-1/2} \\ &= \left| \frac{\det'[-f(\xi)\partial^2 + g(\xi)V''(\sigma_B)]}{\det[-f(\xi)\partial^2 + V''(\sigma_{FV})]} \right|^{-1/2}. \end{aligned} \tag{13}$$

To determine the ξ dependence of the above expression, we observe that if $\Psi_\theta(x)$ is the eigenfunction of the operator $-\hat{\partial}^2 + V''(\hat{\sigma})$ with eigenvalue θ, then

$$[-f(\xi)\partial^2 + g(\xi)V''(\sigma)]\Psi_\theta(\sqrt{g(\xi)/f(\xi)}\,x) = g(\xi)[-\hat{\partial}^2 + V''(\sigma)]\Psi_\theta(\hat{x})$$

$$= g(\xi)\,\theta\,\Psi_\theta(\sqrt{g(\xi)/f(\xi)}\,x), \qquad (14)$$

i.e., $\Psi_\theta(\sqrt{g(\xi)/f(\xi)}\,x)$ is the eigenfunction of the operator $-f(\xi)\partial^2 + g(\xi)V''(\sigma_B)$ with eigenvalue $g(\xi)\theta$. Since the primed determinant has four eigenvalues fewer than the unprimed one, we have

$$A_{DET} = \{[g(\xi)]^{-4}\}^{-1/2}\,\widehat{A}_{DET} = g^2(\xi)\widehat{A}_{DET}. \qquad (15)$$

The C_μ are defined so that the properly normalized modes are $C_\mu^{-1/2}\partial_\mu\sigma_B$ ($\mu = 1,\cdots,4$). Thus, $C_\mu = \int d^4x(\partial_\mu\sigma_B)^2$ (no sum over μ implied), and for an $O(4)$-symmetric bounce, the C_μ are all equal. The ξ dependence of C_μ can easily be found:

$$C_\mu = \int d^4x(\partial_\mu\sigma_B)^2 = f(\xi)/g(\xi)\widehat{C}_\mu. \qquad (16)$$

Hence, the nucleation rate in the Einstein frame is

$$\overline{\lambda}(\overline{t}) = f^2(\xi)\widehat{A}\exp(-B_0 f^2(\xi)/g(\xi)). \qquad (17)$$

Now we may find the nucleation rate in the Jordan frame. Recall that $\exp(\psi/\psi_0) = 16\pi G_N\Phi$, and that the nucleation rate in the Jordan frame is related to that in the Einstein frame by

$$\lambda \equiv \frac{dP}{d^4x\sqrt{-g}} = \frac{dP}{d^4x\sqrt{-\overline{g}}}\frac{\sqrt{-\overline{g}}}{\sqrt{-g}} = \Omega^{-4}\overline{\lambda}$$

$$= \xi^2 f^2(\xi)\widehat{A}\exp(-B_0 f^2(\xi)/g(\xi))$$

$$= \widehat{A}F^2(\Phi)\exp\left\{-B_0 F^2(\Phi)/G(\Phi)\right\}. \qquad (18)$$

$\widehat{A}$ and B_0 are Φ independent and depend only upon the inflaton potential. B_0 is dimensionless, while $\widehat{A}$ has mass dimension 4.

If we take simple power law forms for $F(\Phi)$ and $G(\Phi)$ i.e.

$$F(\Phi) = (16\pi G_N\Phi)^n, \quad G(\Phi) = (16\pi G_N\Phi)^m \qquad (19)$$

we find that the nucleation rate in the Jordan frame is given by:

$$\lambda = \widehat{A}(16\pi G_N\Phi)^{2n}\exp[-B_0(16\pi G_N\Phi)^{2n-m}] \qquad (20)$$

In the original extended inflation model $m = n = 0$. However, there are a variety of situations such as dimensionally reduced models and models in which the inflaton has different gravitational couplings to different sectors [8] where neither m nor n need vanish.

3. Conclusions

We see then that the generic situation will require that the nucleation rate be dependent on the JBD field. This can be used to great advantage in attempting to solve some of the problems encountered by the original extended inflation model [9].

This work of R.H. was supported in part by the Department of Energy grant #DE-AC02-76ER3066. I would like to thank my collaborators, Rocky Kolb, Yun Wang, Sharon Vadas and Erick Weinberg for making this project (and others!) so much fun.

References

1. D. La and P. J. Steinhardt, *Phys. Rev. Lett.* **62** (1989) 376; *Phys. Lett.* **220B** (1989) 375; See also P. Steinhardt's contribution in this volume.

2. A. H. Guth, *Phys. Rev. D* **23** (1981) 347.

3. P. Jordan, *Zeit. Phys.* **157** (1959) 112; C. Brans and C. H. Dicke, *Phys. Rev.* **24** (1961) 925.

4. A. H. Guth and E. W. Weinberg, *Nucl. Phys.* **B212** (1982) 321.

5. R. Holman, E. W. Kolb, S. L. Vadas, Y. Wang, Carnegie Mellon preprint CMU-HEP90-13, Fermilab preprint FNAL-PUB-90/147-A (to appear in Phys. Lett. B); R. Holman, E. W. Kolb, S. L. Vadas, Y. Wang, and E. J. Weinberg, *Phys. Lett.* **237B** (1990) 37.

6. S. Coleman, *Phys. Rev. D* **15** (1977) 2929; C. G. Callan and S. Coleman, *Phys. Rev. D* **16** (1977) 1762.

7. S. Coleman and F. De Luccia, *Phys. Rev. D* **21** (1980) 3305.

8. R. Holman, E. W. Kolb, S. Vadas, Y. Wang, Carnegie-Mellon University preprint CMU-HEP90-09, Fermilab preprint FNAL-PUB-90/99-A (1990) (to appear in *Phys. Rev. D*) R. Holman, E. W. Kolb, and Y. Wang, *Phys. Rev. Lett.* **65** (1990) 17.

9. E. J. Weinberg, *Phys. Rev. D* **40** (1989) 3950; D. La, P. J. Steinhardt and E. Bertschinger, *Phys. Lett.* **231B** (1989) 231.

RECONCILING A SMALL DENSITY
PARAMETER TO INFLATION

Yasunori FUJII and Tsuyoshi NISHIOKA

Institute of Physics
University of Tokyo-Komaba
Meguro-ku, Tokyo, 153 Japan

Abstract A theoretical model is proposed in which a decaying cosmological constant fills up the gap between $\Omega = 1$ as expected from the inflationary scenario and the observed small value $\Omega \lesssim 0.1$.

The inflationary scenario predicts the density parameter Ω which is very close to 1, apparently in contradiction with many observational results showing much smaller value, typically $\Omega \lesssim 0.1$. There has been a suggestion that the deficit can be filled up with a nonzero cosmological constant Λ.[1] This is in fact supported strongly by a recent detailed analysis of the number count of faint galaxies.[2] The data is fitted quite well by $\Omega \approx 0.1$ and $\Omega_\Lambda = \Lambda/(8\pi G \rho_{\text{critical}}) \approx 0.9$ with $k = 0$; even better than by the fit with $\Lambda = 0$ and $k = -1$.

In spite of the phenomenological success of this reconciliation scenario, one still faces a serious difficulty. The value of the required Λ is smaller by about 120 orders than the theoretically natural expectation $\sim m_{\text{Pl}}^2$; a well-known fine-tuning problem. The small value itself can be understood if Λ is not a true constant but decays like $\sim t^{-2}$. Substituting the present age of the universe $\sim 14\,\text{Gyr} \sim 10^{60} m_{\text{Pl}}^{-1}$ for t gives the right order of magnitude of the desired Λ; the present-day Λ is small only because our universe is old.[3]

This is still not enough to account for the behavior of the fit in [2]. We recall that the decaying cosmological constant scenario has been implemented often in terms of a scalar field $\phi(t)$. After a Weyl rescaling (conformal transformation) to make the gravitational constant to be a true constant, the effective cosmological constant is nothing but the scalar field energy which decays naturally like $\sim t^{-2}$ asymptotically, the same way as the ordinary matter density falls off. The most crucial element in the fit in [2] is, however, the presence of an energy density that behaves differently from the ordinary matter density, more specifically falling off more slowly hence giving the scale factor $a(t)$ an extra acceleration as compared with the standard behavior $a(t) \sim t^{2/3}$ in the spatially flat universe of $k = 0$.

We attempted to modify the scalar-tensor theory to have the scalar field energy which falls off like $\sim t^{-2}$ as an overall behavior but accompanied with some local deviations to simulate a nearly constant energy density for some duration of the cosmic time. We achieved this by introducing another scalar field $\Phi(t)$ together with a specific potential.[4]

At first we anticipated that a massive scalar field Φ would oscillate giving a periodic deviation of the scale factor from the overall smooth behavior $\sim t^{2/3}$; for some length of time $a(t)$ would be accelerated resulting in the behavior as required in the fit of [2]. As we found, however, Φ shows a damped oscillation as a function of $\log t$; the energy density thus falls off like $\sim t^{-3}$, too fast to affect the late-time cosmology. We then introduced, from a purely phenomenological point of view, a potential involving $\sim \Phi^2 \sin(\omega \ln \phi)$ where ω is a parameter. As a consequence Φ now shows a quasi-periodic oscillation, a flip-flop behavior versus $\log t$ (see Figs. 3-5 in [4]). This is a nonlinear effect known as relaxation oscillation due to a delicate competition between the restoring force and the cosmological "frictional force." The quasi-period is entirely different from any time scales prepared in the theory, essentially of the order of $m_{\rm Pl}^{-1}$. In this scenario the universe has likely experienced several number of such quasi-oscillations since the beginning if measured in $\log t$.

Every time when Φ suddenly drops or rises, the universe plunges into a transient period in which the scalar field energy, the effective cosmological constant, is dominant. Otherwise, in fact most of the logarithmic time, the universe is in a normal state in which the vacuum energy is negligibly small. During the "first-half" of each transient epoch, $a(t)$ shows a slight acceleration (followed by a deceleration) against the smooth background behavior. It is rather easy to adjust parameters such that one of such accelerations occurs around the present epoch in accordance with what is meant in the fit of [2]. We can do this keeping the success of the nucleosynthesis by the standard cosmology; the process must have occurred during one of the normal periods in which Φ was stuck to one of the nearly constant values thus leaving the scale factor evolving in the standard manner.

We emphasize that the scenario of a decaying cosmological constant allows a large enough Λ in the past to cause inflation. In this way we have now a coherent picture of the universe covering from primordial inflation to the present epoch, without appealing to unnatural fine-tuning. A number of details are still yet to be completed. In particular, the special potential mentioned above needs a derivation from a more fundamental theoretical basis, though $\ln \phi$ is a new field defined naturally after the Weyl rescaling.

Finally, the quasi-periodic behavior of this nature may result in large-scale structure somewhat similar to that reported recently.[5]

References

[1] See, for example, Peebles, P.J.E. Nature **321** (1986), 27; Fukugita, M. in Proc. The Third Nishinomiya-Yukawa Symposium, eds. H. Sato and H. Kodama, Springer Verlag, Berlin, Heidelberg, 1990.

[2] Fukugita, M., Takahara, F., Yamashita, K. and Yoshii, Y. Ap. J. to be published.

[3] See, Fujii, Y. and Nishioka, T. Phys. Rev. **D42** (1990), 361, and the papers cited therein.

[4] Fujii, Y. and Nishioka, T. Phys. Lett. **B**, to be published.

[5] Broadhurst, T.J., Ellis, R.S., Koo, D.C. and Szalay, A.S. Nature, **343** (1990), 726.

SOFT INFLATION: A MODEL FOR EASING CONSTRAINTS

ANDREW L. BERKIN,[1] KEI-ICHI MAEDA[1] and JUN'ICHI YOKOYAMA[2]
[1] *Physics Dept., Waseda Univ., Okubo 3-4-1, Shinjuku-ku, Tokyo 169, Japan*
[2] *Dept. of Physics, Faculty of Sciences, University of Tokyo, Tokyo 113, Japan*

ABSTRACT. The cosmology resulting from two coupled scalar fields, one which is either a new or chaotic type inflaton and the other which has an exponentially decaying potential, is studied. Such a potential arises in superstring theories and in the conformally transformed frame of generalized Einstein theories like the Jordan-Brans-Dicke theory. The constraints necessary for successful inflation are examined. Conventional GUT models such as SU(5) are compatible with new inflation, while restrictions on the self-coupling constant are significantly loosened for chaotic inflation when the exponential potential is fundamental, but no significant improvement for the latter occurs when in the conformal frame.

Although the inflationary universe solves many cosmological problems, as yet no fully satisfactory model exists. Suppression of density perturbations forces the self-coupling to excessively small values, killing new inflation and imposing a fine-tuning problem on chaotic inflation. In this work [1], the constraints are loosened via the action

$$S = \int d^4x \sqrt{-g} \left[\frac{1}{2\kappa^2} R - \frac{1}{2}(\nabla\phi)^2 - \frac{1}{2}(\nabla\psi)^2 - e^{-\beta\kappa\phi} V(\psi) \right], \tag{1}$$

where $\kappa^2 = 8\pi G$, $V(\psi)$ is the inflaton potential and β is a dimensionless coupling constant, which must be smaller than $\sqrt{2}$ to guarantee power-law inflation. Such potentials arise in superstring or supergravity models, as well as in the conformally transformed frame of generalized Einstein theories. As the inflaton rolls down the relatively flat $V(\psi)$, ϕ evolves along the exponential potential, resulting in power-law inflation. The exponential potential acts to produce a much smaller effective self-coupling of the inflaton field, thus "softening" the constraints so that new inflation is possible and chaotic inflation is less fine-tuned.

Neglecting $\ddot{\psi}$ and $\dot{\psi}^2/2$ in the field equations coming from (1) gives the attractor solution

$$\kappa\phi = \kappa\phi_c + \frac{2}{\beta}\ln(t/\kappa), \tag{2}$$

$$a = a_0(t/t_0)^{2/\beta^2}, \tag{3}$$

$$f(\psi) = f(\psi_0) - (1 - \beta^2/6)\ln(a/a_0), \tag{4}$$

where a is the scale factor, for general V we have

$$\exp[\beta\kappa\phi_c] \equiv \frac{\beta^4\kappa^4 V(\psi)}{12(1 - \beta^2/6)}, \quad f(\psi) \equiv \kappa^2 \int d\psi \frac{V}{V'}, \tag{5}$$

and the subscript 0 denotes the value at t_0 when the universe enters the inflationary phase.

The potential for new inflation may be written as $V(\psi) = V_0 - \lambda\psi^4/4$, where V_0 is the GUT scale. For chaotic inflation, the potential used is $V(\psi) = \lambda_n\psi^n/n$, with n an even integer. Inflation ends at t_f when the slow rolling approximation breaks down, equivalent to $\partial^2(\ln V)/\partial\psi^2 \approx 3\kappa^2(1 - \beta^2/6)^{-1}$. Then

$$f(\psi) = \kappa^2 V_0/2\lambda\psi^2 \quad \text{and} \quad \kappa\psi_f = [\kappa^4 V_0/\lambda(1 - \beta^2/6)]^{1/2} \quad \text{for new inflation} \tag{6}$$

$$f(\psi) = \kappa^2\psi^2/2n \quad \text{and} \quad \kappa\psi_f = [n(1 - \beta^2/6)/3]^{1/2} \quad \text{for chaotic inflation} \tag{7}$$

251

252

The horizon problem is solved if the amount of expansion is greater than 10^{30} times the ratio of reheat temperature to Planck mass. For reheating, a temperature greater than 10^{10} GeV is required for standard baryogenesis, although models with far lower temperatures exist. Density perturbations, given by

$$\frac{\delta\rho}{\rho} \sim H^2 \frac{\max\{|\dot{\phi}|, |\dot{\psi}|\}}{\dot{\phi}^2 + \dot{\psi}^2}\bigg|_h \; ; \quad t_h \ \text{s.t.} \ \frac{a_f}{a_h} = e^{70} \tag{8}$$

must be less than 10^{-4} in accordance with the microwave background.

For new inflation, these constraints yield

$$\psi_0 < \psi_{cr,H} \equiv 0.16 H_0 \lambda^{-1/2} e^{\beta\kappa\phi_0/2} \left[1 + 0.017 \ln\left(V_0^{1/4}/10^{15}\text{GeV}\right)\right]^{-1/2}$$

$$\psi_0 > \psi_{cr,RH} \equiv 0.19 H_0 \lambda^{-1/2} \beta e^{\beta\kappa\phi_0/2} \left[1 + 0.094 \ln\left(V_0^{1/4}/10^{15}\text{GeV}\right)\right]^{-1/2}$$

$$\psi_0 < \psi_{cr,D} \equiv 0.15 H_0 \lambda^{-1/2} e^{\beta\kappa\phi_0/2}$$
$$\times \left\{1 - 0.20\beta^{-2}\left[1 + 0.14\ln\beta - 0.29\ln\left(V_0^{1/4}/10^{15}\text{GeV}\right)\right]\right\}^{-1/2}$$

for the horizon problem, reheating, and density perturbations. Note that $|\dot{\phi}|_h > |\dot{\psi}|_h$ unless $\beta \lesssim 10^{-10}\lambda^{-1/2}(V_0^{1/4}/10^{15}\text{GeV})^2$. If the condition $\lambda \lesssim 0.02 e^{\beta\kappa\phi_0}$ is satisfied, the natural initial condition of $\psi_0 \gtrsim H_0$ is allowed. The allowed region of parameter space then includes the standard SU(5) GUT model [1].

For chaotic inflation, the constraints may be written as

$$\psi_0 > \psi_{cr,H} \equiv 2.3 m_{PL} n^{1/2}$$

$$\psi_0 < \psi_{cr,RH} \equiv 2.6 m_{PL} n^{1/2} \beta^{-1}$$

$$\psi_0 > \psi_{cr,D} \equiv \begin{cases} 1.5 m_{PL} n^{1/2} \beta^{-1}[1 + 2.5\beta^2]^{1/2} & \text{for } \beta < (n/140)^{1/2} \\ 1.4 m_{PL} n^{1/2} \beta^{-1}[1 + 2.9\beta^2]^{1/2} & \text{for } \beta > (n/140)^{1/2} \end{cases}$$

where $\beta \lessgtr (n/140)^{1/2}$ corresponds to $|\dot{\phi}|_h \lessgtr |\dot{\psi}|_h$, respectively, and the initial energy scale was taken to be the Planck energy. Much lower values of ψ_0 are possible compared to the standard case, and hence much larger values of λ. For example, $\lambda_4 < 5.8 \times 10^{-3}\beta^4 e^{\beta\kappa\phi_0}$ for the quartic case and $\lambda_2 < 0.15 m_{PL}^2 \beta^2 e^{\beta\kappa\phi_0}$ for the massive scalar field.

If the model is derived from a conformally transformed generalized Einstein theory, then a non-standard $e^{-\beta\kappa\phi/2}(\nabla\psi)^2/2$ kinetic term appears in (1). ψ now rolls faster than with canonical kinetic terms, therefore ϕ does not evolve as much and some of the benefit of the effective coupling is lost. For new inflation, a ϕ_0 twice as large restores the beneficial results discussed above. However, with chaotic inflation only an order of magnitude better results are possible compared to the standard one field case. When transforming back to the original frame, the amount of inflation and density perturbations are affected only slightly, while reheating becomes easier as there is no longer an exponential potential.

In conclusion, a fundamental exponential potential coupled to an inflaton potential produces a successful inflationary scenario for both new and chaotic potentials. When the exponential potential comes from conformally transformed generalized Einstein theories, the success is still retained with new inflation, but the fine tuning problem reappears with chaotic potentials.

REFERENCE: [1] A.L. Berkin, K. Maeda, and J. Yokoyama, Phys. Rev. Lett. 65, 141 (1990), and citations within.

STOCHASTIC INFLATION LATTICE SIMULATIONS:
ULTRA-LARGE SCALE STRUCTURE OF THE UNIVERSE

D.S. Salopek
NASA/ Fermilab Astrophysics Center
P.O. Box 500 MS-209
Batavia, Illinois, USA 60510

ABSTRACT

Non-Gaussian fluctuations for structure formation may arise in inflation from the nonlinear interaction of long wavelength gravitational and scalar fields. Long wavelength fields have spatial gradients $a^{-1}\nabla$ small compared to the Hubble radius, and they are described in terms of classical random fields that are fed by short wavelength quantum noise. Lattice Langevin calculations are given for a 'toy model' with a scalar field interacting with an exponential potential where one can obtain exact analytic solutions of the Fokker-Planck equation. For single scalar field models that are consistent with current microwave background fluctuations, the fluctuations are Gaussian. However, for scales much larger than our observable Universe, one expects large metric fluctuations that are non-Gaussian. This example illuminates non-Gaussian models involving multiple scalar fields which are consistent with current microwave background limits.

1. INTRODUCTION

There are a growing number of cosmological observations[1-5] that are in apparent conflict with the simplest model of structure formation, the Cold Dark Matter model. One possibility is that the Gaussian initial conditions described by a Zeldovich scale-invariant spectrum that typically arise from inflation are incorrect. The view that will be adopted here is that the inflation model is basically correct although one should refine the calculations to incorporate nonlinearities among multiple scalar fields. As the first step, nonlinearities will be incorporated at wavelengths larger than the Hubble radius. The initial conditions for structure formation are generated stochastically as quantum modes expand beyond the Hubble radius. One can then investigate whether non-Gaussian fluctuations for structure formation can arise from inflation. I report the results of a collaboration[6,7] with J.R. Bond of the Canadian Institute for Theoretical Astrophysics, Toronto, Canada.

There were two clues that indicated that the nonlinear evolution of long wavelength fields was tractable. (1) For a single scalar field, Bardeen, Steinhardt and Turner[8] demonstrated using linear perturbation theory that there was a remarkable constant of integration ζ when the physical wavelength of a comoving mode exceeded the Hubble radius. (2) When two spatial points are separated by more than the Hubble radius, they are no longer in causal contact; they essentially evolve as independent Universes. Hence, linear perturbation theory is not essential. In a significant improvement over homogeneous mini-superspace models, one can in fact generalize ζ to nonlinear multiple fields.[6]

One begins by decomposing all fields into long wavelength (denoted by a bar) and short wavelength components (denoted by δ),

$$\phi_j(T,x) = \bar{\phi}_j(T,x) + \delta\phi_j(T,x), \quad g_{\mu\nu}(T,x) = \bar{g}_{\mu\nu}(T,x) + \delta g_{\mu\nu}(T,x), \qquad (1.1)$$

where the Hubble radius, H^{-1}, marks the boundary between long and short. The metric will be assumed to have the diagonal form,

$$ds^2 = -N^2(T,x)dT^2 + e^{2\alpha(T,x)}\big((dx^1)^2 + (dx^2)^2 + (dx^3)^2\big),$$

described by a lapse function, $N(T,x)$, and an inhomogeneous scale factor, $e^{\alpha(T,x)}$. Hence, the dynamic effects of gravitational radiation are neglected which is typically an excellent approximation. The long wavelength scalar and gravitational field equations will be solved nonlinearly (Sec. 2) whereas the short wavelength modes will still (unfortunately) be treated in linear perturbation theory (Sec. 3). As the comoving short wavelength modes expand beyond the Hubble radius, one assumes that they become classical and they add a stochastic kick[9−12] to the long wavelength background.

2. EVOLUTION OF LONG WAVELENGTH FIELDS

The long wavelength ADM equations that will be required in this report are,

$$\bar{H}^2 = \frac{8\pi}{3m_P^2}\left(\frac{1}{2}\sum_l \bar{\Pi}^{\phi_l\,2} + V(\bar{\phi}_j)\right) \tag{2.1a}$$

$$\bar{H}_{,i} = -\frac{4\pi}{m_P^2}\sum_l \bar{\Pi}^{\phi_l}\bar{\phi}_{l,i} \tag{2.1b}$$

$$\Delta\bar{\phi}_j = \bar{\Pi}^{\phi_j}\bar{N}\Delta T + \Delta S_{\phi_j}. \tag{2.1c}$$

$\bar{H}$ is the Hubble parameter, whereas the $\bar{\Pi}^{\phi_j}$ are the scalar field momenta. Eqs. (2.1) follow from the standard scalar field equations and Einstein's equations with all second order spatial gradients neglected. First order spatial gradients are retained otherwise one is describing homogeneous mini-superspace which is too limited for the applications considered. The first two equations, the energy constraint and the momentum constraint, do not explicitly contain noise terms, whereas the third one, the evolution equations for $\bar{\phi}_j$ contain a contributions from short wavelength quantum noise, ΔS_{ϕ_j}, that have crossed the Hubble radius. The remaining evolution equations for $\bar{\Pi}^{\phi_j}$ and $\bar{H}$ also contain noise terms but they are not required if one neglects decaying modes. The lapse function will be specified when one chooses the time parameter (Sec. 3).

The crucial step in solving eqs.(2.1) is to integrate the momentum constraint. The Hubble parameter is a function of the scalar fields and possibly the time parameter, T:

$$\bar{H} \equiv \bar{H}(\bar{\phi}_k(x),T), \quad \text{where} \quad \bar{\Pi}^{\phi_k} = -\frac{m_P^2}{4\pi}\frac{\partial\bar{H}}{\partial\bar{\phi}_j}(\bar{\phi}(x),T). \tag{2.2}$$

(If noise term is not important, one can actually show that the Hubble parameter does not depend explicitly on time,[6] $\bar{H} \equiv \bar{H}(\bar{\phi}_k)$.) For example, the spatial derivative of the Hubble parameter in (2.1b) may be expanded leading to,

$$\frac{\partial\bar{H}}{\partial\bar{\phi}_j}\bar{\phi}_{j,i} = -\frac{4\pi}{m_P^2}\bar{\Pi}^{\phi_j}\bar{\phi}_{j,i},$$

which may satisfied if one identifies the scalar field momenta with the partial derivative of $\bar{H}$ as in (2.2). Substituting the momenta into the energy constraint leads to the separated Hamilton-Jacobi equation (SHJE),

$$\bar{H}^2 = \frac{m_p^2}{12\pi} \sum_l \left(\frac{\partial \bar{H}}{\partial \bar{\phi}_l}\right)^2 + \frac{8\pi}{3m_p^2} V(\bar{\phi}_j), \tag{2.3}$$

a partial differential equation which does not depend explicitly on the time coordinate nor on the spatial variables. In this sense, it is a completely covariant equation. The momentum constraint essentially patches together the various spatial points to make one Universe.

The separated Hamilton-Jacobi equation is self-contained. Its solution is not unique, but in many cases of physical interest there is typically an attractor solution to which almost all solutions approach. For example, for a single scalar field interacting through an exponential potential,[13]

$$V(\phi) = V(0) \exp\left[-\sqrt{\frac{16\pi}{p}} \frac{\phi}{m_p}\right], \tag{2.4}$$

the attractor solution[6] of (2.3) is

$$\bar{H}_{att}(\bar{\phi}) \equiv \bar{H}(0) \exp\left(-\sqrt{\frac{4\pi}{p}} \frac{\bar{\phi}}{m_p}\right), \quad \bar{H}(0) = \left[\frac{8\pi V(0)}{3m_p^2(1 - \frac{1}{3p})}\right]^{1/2}. \tag{2.5}$$

The parameter p describes the flatness of the potential; inflation occurs if $p > 1$. Of course, other solutions exist[6] but they describe the decaying mode that always appears in cosmological models. If one simply neglects the decaying mode, then the Hubble parameter of stochastic inflation may be identified with $\bar{H}_{att}(\bar{\phi}_k)$ and one may safely ignore the explicit time dependence appearing in (2.2). With this additional assumption, one need not consider the remaining evolution equations that were dropped in (2.1).

In fact, one of the big advantages of applying Hamilton-Jacobi theory to stochastic inflation is that growing and decaying modes are cleanly separated even when the slow-roll approximation is not valid. Another advantage is that gauge ambiguities are not as problematic as in linear perturbation theory because one does not introduce a fictitious homogeneous background. Since the constraints have been eliminated, only the evolution equations for the scalar fields, eq.(2.1c), remain to be solved. Given the attractor solution of the SHJE, $\bar{\Pi}^{\phi_k}$ is assumed to be a known function of $\bar{\phi}_k$ through (2.2). When the time parameter is specified in Sec. 3, the lapse function will also be a known function of the scalar fields.

The initial conditions for the long wavelength problem are generated by short wavelength quantum fluctuations whose wavelength exceeds the Hubble radius. By assuming that they become classical when they cross the horizon,[9,10] one circumvents the problem of quantization of the long wavelength gravitational field which appears to be inconsistent beyond the semi-classical approximation.[6,7] However, using an exact solution of the long wavelength Wheeler-DeWitt equation for single scalar field with an exponential potential (2.4), one can nonetheless estimate that quantum gravity corrections are of the order of

$$\text{Quantum Corrections} \sim 4\pi \frac{H^2}{m_p^2}, \tag{2.6}$$

where H is the value of the Hubble parameter when the comoving scale of interest (typically $\approx 3000h^{-1}$Mpc) crossed the horizon during inflation. For single scalar field models that are consistent with microwave background limits, $\bar{H} \approx 10^{-5}m_p$, the correction is approximately one in a billion. For many models, it is then an excellent approximation to treat the long wavelength fields classically, although a probabilistic description is essential.

3. QUANTUM NOISE FROM SHORT WAVELENGTH FIELDS

Using the stochastic long wavelength formalism, one may answer questions which could not be adequately addressed in homogeneous mini-superspace quantum cosmology: What is the time parameter? What is the initial choice of the probability distribution?

3.1 THE CHOICE OF TIME PARAMETER

In order to describe the evolution of short wavelength quantum fluctuations on an inhomogeneous long wavelength background, it proves convenient to choose conformal time τ as the time parameter so that the long wavelength metric has the form

$$ds^2 = e^{2\bar{\alpha}(\tau, x^j)}(-d\tau^2 + dx^2 + dy^2 + dz^2). \tag{3.1}$$

If the fields evolve, slowly then conformal time may shown to be given by

$$\tau \sim -\frac{1}{\bar{H}e^{\bar{\alpha}}}. \tag{3.2}$$

In practice, one employs

$$T = \ln(\bar{H}e^{\bar{\alpha}}) \tag{3.3}$$

as the time parameter because it is easier to apply in numerical analyses.

One can show that in linear perturbation theory about a long wavelength background, a Fourier mode solution for the scalar field is an excellent approximation,

$$\delta\phi_j(\tau, x) = e^{-\bar{\alpha}(\tau, x)}e^{ik \cdot x}e^{-ik\tau}/\sqrt{2k}, \tag{3.4}$$

when the physical wavelength $e^{\bar{\alpha}}k^{-1}$ is much shorter than the Hubble radius, H^{-1}. Eq.(3.4) corresponds to the positive energy solution that describes the quantum mechanical ground state. The effects of long wavelength inhomogeneities are contained in the spatially dependent background scale factor, $e^{-\bar{\alpha}(\tau, x)}$. At very short wavelengths, metric fluctuations beyond the long wavelength background are not important at least in linear perturbation theory. The amplitude is normalized in analogy to the analysis of linear quantum fluctuations on a homogeneous time dependent background[14] where one employs the equal time quantum commutator relations,

$$[\phi_j(\tau, x), \Pi^{\phi_k}(\tau, x')] = i \, e^{-2\bar{\alpha}} \delta^3(x - x') \delta_j^k.$$

When the physical wavelength of a mode approaches the Hubble radius the approximate solution (3.4) breaks down, but this is precisely when long wavelength evolution becomes important. At the start of the timestep, ΔT, one adds to the background a noise impulse, ΔS_{ϕ_j}, which consists of all those Fourier modes that will have crossed the Hubble radius during the time step,

$$\bar{\phi}_j(T, x) \rightarrow \bar{\phi}_j(T, x) + \Delta S_{\phi_j}, \quad \text{where} \quad \Delta S_{\phi_j} = \bar{H}\left[\bar{\phi}(T, x)\right] \sum_{e^T \le |k| < e^{T+\Delta T}} \frac{e^{ik \cdot x}}{\sqrt{2k^3}} a_j(k),$$

(3.5)

In (3.5), I have used the notation,

$$\sum_k \equiv \int \frac{d^3k}{(2\pi)^3},$$

and I have applied the horizon crossing expression $e^{-\bar{\alpha}(T,x)} = \bar{H}(\bar{\phi}_j(T, x))/k$. Here, $a_j(k)$ is a classical complex Gaussian random field,

$$a_j^*(k) = a_j(-k) \quad \text{and} \quad < a_j(k)a_{j'}(k') >= (2\pi)^3 \delta^3(k + k')\delta_{jj'},$$

(3.6)

which imitates quantum fluctuations which are Gaussian in linear perturbation theory. The noise term (3.5) differs from the usual perturbation calculation on a homogeneous background in that it is modulated by the local value of the background Hubble parameter $\bar{H}\left[\bar{\phi}(T, x)\right]$.

The choice of time parameter (3.3) leads to the following definition of the lapse function,

$$\bar{N}^{-1} = \dot{T}/\bar{N} = \dot{\bar{\alpha}}/\bar{N} + \frac{1}{\bar{H}}\dot{\bar{H}}/\bar{N} = \bar{H} - \frac{m_p^2}{4\pi}\frac{1}{\bar{H}}\sum_m\left(\frac{\partial \bar{H}}{\partial \bar{\phi}_m}\right)^2,$$

(3.7)

The formulation of the stochastic problem is now complete. One solves (2.1c) numerically on a lattice. $\bar{\Pi}^{\phi_k}$ and $\bar{N}$ are expressed in terms of the attractor solution through (2.2) and (3.7). The noise term is given in (3.5).

4. STOCHASTIC INFLATION LATTICE SIMULATIONS

4.1 INITIAL CONDITIONS AND THE FOKKER-PLANCK EQUATION

A lattice simulation of inflation begins homogeneously at $T = T_0$ with $\bar{\phi}_j = \bar{\phi}_{j0}$; inhomogeneities are produced only subsequently by quantum noise. If $P(\bar{\phi}_j|T; \bar{\phi}_{j0}, T_0)$ denotes the scalar field distribution on a uniform T surface, then the initial probability distribution for the scalar field is a δ function:

$$P(\bar{\phi}_j|T; \bar{\phi}_{j0}, T_0) = \delta^n(\bar{\phi}_j - \bar{\phi}_{j0}).$$

(4.1)

The probability function P is an example of a limited statistic that gives a partial understanding of a complicated lattice simulation. From now on, only long wavelength fields will be considered and thus the bar notation will be dropped.

For a single scalar field with an exponential potential (2.4), the Fokker-Planck equation that describes the evolution of the probability function with the full metric back reaction is then

$$\frac{\partial P}{\partial T} = -\frac{m_P}{\sqrt{4\pi p}}\frac{1}{1-1/p}\frac{\partial P}{\partial \phi} + \frac{H^2(0)}{8\pi^2}\frac{\partial^2}{\partial \phi^2}\left[\exp(-\sqrt{\frac{16\pi}{p}}\frac{\phi}{m_P})P\right] \tag{4.2}$$
$$+ \delta(T - T_0)\delta(\phi - \phi_0).$$

The first term on the right hand side describes the classical drift of the scalar field down the potential, whereas the second, $(\partial^2/\partial \phi^2)(H_{att}^2(\phi)P)/(8\pi^2)$, is the result of quantum noise that causes the diffusion of the probability function. The third term incorporates the initial conditions (4.1). There are factor ordering problems and other corrections associated with this equation,[7] but I will ignore them here. The solution of (4.2) will be referred to as the Green's function:[7]

$$P(\phi|T; T_0, \phi_0) = \sqrt{\frac{16\pi}{pm_P^2}}\, y^{-1}e^{-(1+z^2)/y}\, z^2\, I_0(2z/y). \tag{4.3a}$$

where I_0 is the modified Bessel function of order zero, $I_0(x) = J_0(ix)$. The functions $z(\phi, T)$ and $y(T)$ are given by

$$z(\phi, T) = \exp\left(\sqrt{\frac{4\pi}{p}}\frac{\phi - \phi_0}{m_P} - \frac{T - T_0}{p - 1}\right), \tag{4.3b}$$

$$y(T) = \frac{1}{\pi}(1 - 1/p)\frac{H^2(\phi_0)}{m_P^2}\left[1 - e^{-2(T-T_0)/(p-1)}\right]. \tag{4.3c}$$

(The coefficient $H(\phi_0)$ should not be confused with the $H(0)$ of (2.5), which it equals only for $\phi_0 = 0$). At late times, $T - T_0 \to \infty$, quantum diffusion is no longer important because the stochastic force is proportional to the Hubble parameter which decreases in time, and the probability distribution on the lattice evolves as a wave of fixed shape,

$$\lim_{T-T_0 \to \infty} P(\phi|T; T_0, \phi_0) = f_{\phi_0}\left(\phi - \phi_0 - \frac{m_P}{\sqrt{4\pi p}}\frac{1}{(1-1/p)}(T - T_0)\right).$$

The Fokker-Planck equation for an exponential potential was first analyzed by Ortolan, Matarrese and Lucchin[15,16] who applied both analytic approximations was well as numerical methods. Eq. (4.3) is the *exact* Green's function solution of the improved Fokker-Planck equation (4.2).

4.2 NUMERICAL SIMULATIONS

The late time results, $T - T_0 \to \infty$ of two 64^3 lattice simulations for a single scalar field interacting through an exponential potential with p=5 are shown in Figs. 1-3. In Fig. 1a, I show contour plots on a surface of constant T corresponding to -2, -1, 0, 1, 2 σ scalar field fluctuations from the mean for a lattice simulation that began homogeneously with $H(\phi_0) = 10^{-5}m_P$, consistent with microwave background anisotropy limits. In Fig. 2a, only -2σ fluctuations are shown for the benefit of the reader. If one included the effects of beam smearing, these figures would correspond to microwave background maps at angular scales greater than 2^0. One may actually perform a hypersurface transformation to

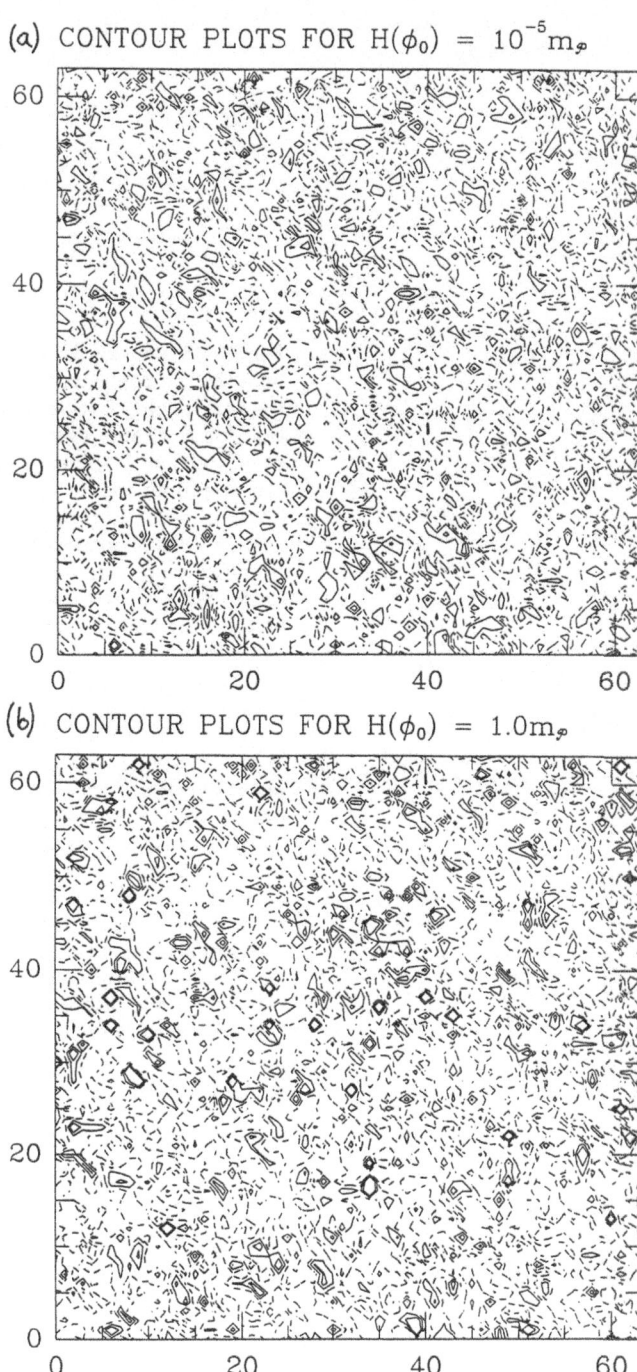

(a) CONTOUR PLOTS FOR H(ϕ_0) = 10^{-5}m$_{\varphi}$

(b) CONTOUR PLOTS FOR H(ϕ_0) = 1.0m$_{\varphi}$

Fig. 1 Contour maps of scalar field fluctuations for a two dimensional slice of a 64^3 lattice simulation for an exponential potential stochastic inflation model with $p = 5$. The initial configurations were homogeneous, with $H(\phi_0)/m_P = 10^{-5}$ for (a) and $H(\phi_0) = 1.0m_P$ for (b). The solid contours correspond to -2σ and -1σ deviations from the scalar field mean, (*i.e.,* high energy density regions) and the broken contours correspond to 0, 1, 2σ fluctuations. The mean has been subtracted out. The initial condition for (a) was chosen to yield scalar field fluctuations that lead to structure compatible with current microwave background anisotropy limits; the fluctuations are Gaussian-distributed to high accuracy. (b) is one of the simplest models where non-Gaussian statistics can arise in cosmology. The map is in initial co-moving position rather than final physical position and has a uniform value of $H(\phi)e^{\alpha} = e^T$, where T is the time at which the slice is viewed. Because fluctuations are much larger than allowed by present microwave background limits, the size of the lattice is much larger than our present horizon size, and as a result this map has no observable consequences.

260

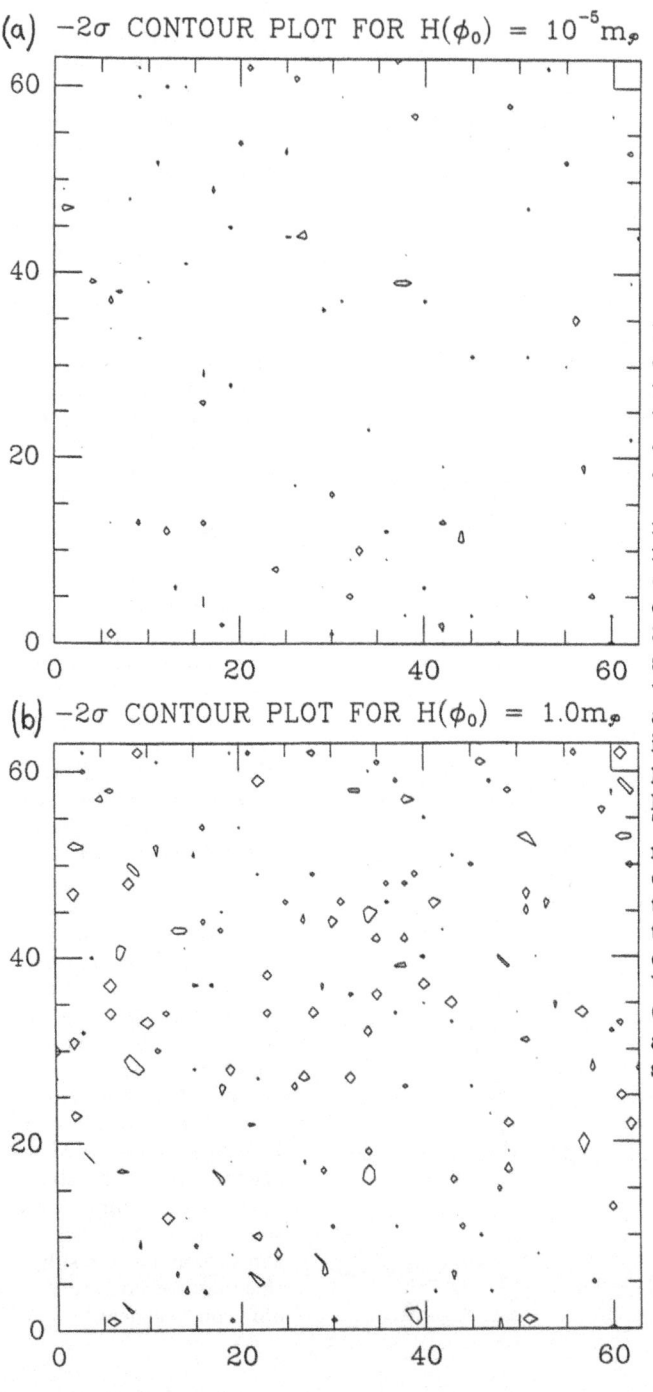

(a) -2σ CONTOUR PLOT FOR $H(\phi_0) = 10^{-5} m_\varphi$

(b) -2σ CONTOUR PLOT FOR $H(\phi_0) = 1.0 m_\varphi$

Fig. 2 This figure shows only the -2σ contours of the scalar field distribution for the same cases as in Fig.1. In (a), there are 91 points with $\phi < -2\sigma$, which is approximately what is predicted from Gaussian statistics (94). For (b), because of the long tail in Fig. 3, the high energy density areas with negative values of the scalar field are more numerous than in the Gaussian example of (a): there are 192 points with $\phi \leq -2\sigma$. The high energy density regions are more strongly clustered at a shorter distance scale. In alternative models which could describe our observable Universe, a significant non-Gaussian tail would have an important impact on structure formation.

show that $\Delta\phi$ on a uniform T slice is actually proportional to the microwave background anisotropy,[7]

$$\Delta T_{cmb}/T_{cmb} = \sqrt{4\pi p}\,(1 - 1/p)\left(\Delta\phi(T)/m_P\right)/5\,. \tag{4.4}$$

In Fig. 3a, for a constant value of T, I have plotted the distribution on the lattice of the scalar field variable

$$\chi = (\phi - \phi_{cl})/H(\phi(0)) \tag{4.5}$$

where ϕ_{cl} is the classical attractor trajectory,

$$\phi_{cl}(T) = \phi_0 + \frac{m_P}{\sqrt{4\pi p}}\frac{1}{1 - 1/p}(T - T_0)\,.$$

Agreement with the exact Green's function solution (4.3) (broken curve) is excellent, and the distribution is Gaussian to an excellent approximation.

Non-Gaussian fluctuations arise when the initial value of the Hubble parameter is comparable to the Planck scale. In Fig. 3b, I show for $H(\phi_0) = 1.0 m_P$ that the histogram of scalar field values from a lattice calculation agrees well with the exact solution. Once again, contour plots for -2, -1, 0, 1, 2 σ fluctuations from the mean are shown in Fig. 1b whereas only -2σ contours are given in Fig. 2b. The non-Gaussian contours look remarkably different from the Gaussian case, essentially because of the extended high energy density tail that appears in Fig. 1b. These points have been able to diffuse to relatively high energy densities because the stochastic force is proportional to the Hubble parameter. Figs. 1 and 2 are plotted using comoving spatial variables, x, whereas if one used physical space, then the high energy density regions would actually dominate the volume because they have almost eternally inflated.[12] In fact, a numerical instability occurs when $H(\phi_0) > m_P$ because some lattice points may actually diffuse to $\phi = -\infty$ in a finite period of time.[7] Here, I have simply removed from the lattice all those points that decrease below χ_{cut},

$$\chi_{cut} \equiv -5\sqrt{\frac{p-1}{8\pi^2}}\,. \tag{4.6}$$

This simulation with its large initial value of the Hubble parameter cannot describe our observable Universe otherwise microwave background limits which are determined by (4.4) would be violated. In fact, the lattice size in Figs. 1b, 2b is more than 10^{20} times larger than that in Figs. 1a, 2a because as the scalar field rolls down from $1.0 m_P$ to $10^{-5} m_P$, the Universe expands by a factor of

$$\text{Relative Size} = (10^5)^{p-1} \sim 10^{20}, \quad \text{for} \quad p = 5.$$

The relative size depends sensitively on the free parameter p, eq.(2.4), which also controls the slope of the primordial fluctuation spectrum in the Newtonian potential,[7]

$$\mathcal{P}_{\Phi_H} \propto k^{-2/(p-1)}.$$

$p = \infty$ is the scale-invariant Zeldovich spectrum. If one normalizes the Cold Dark Matter fluctuation spectrum at galaxy scales, then p cannot be much smaller than 5 otherwise large angle microwave background limits would be violated in Figs. 1a, 2a.[7]

Fig. 3 As a check of the numerical method, the final scalar field distributions from the lattice simulations of Fig.1a,b (solid histogram) are shown to agree with the exact solutions of eq.(4.3) (dashed curves). For (a), with $H(\phi_0) = 10^{-5} m_P$, this is a Gaussian distribution. In (b), of the initially $64^2 = 4086$ points, 22 paths wandered beyond χ_{cut}, eq.(4.6), and were discarded, as described in Sec. 4.2. Simulations with even larger $H(\phi_0)$ have a much larger loss of trajectories: *e.g.*, with $H(\phi_0) = 3m_P$, approximately 10% of the points are discarded. In single scalar field models, significant non-Gaussian distributions are generated only if the scalar field begins with a Hubble parameter $H(\phi_0) \approx m_P$, corresponding to a patch of the Universe $\approx 10^{20}$ times larger than our observable patch.

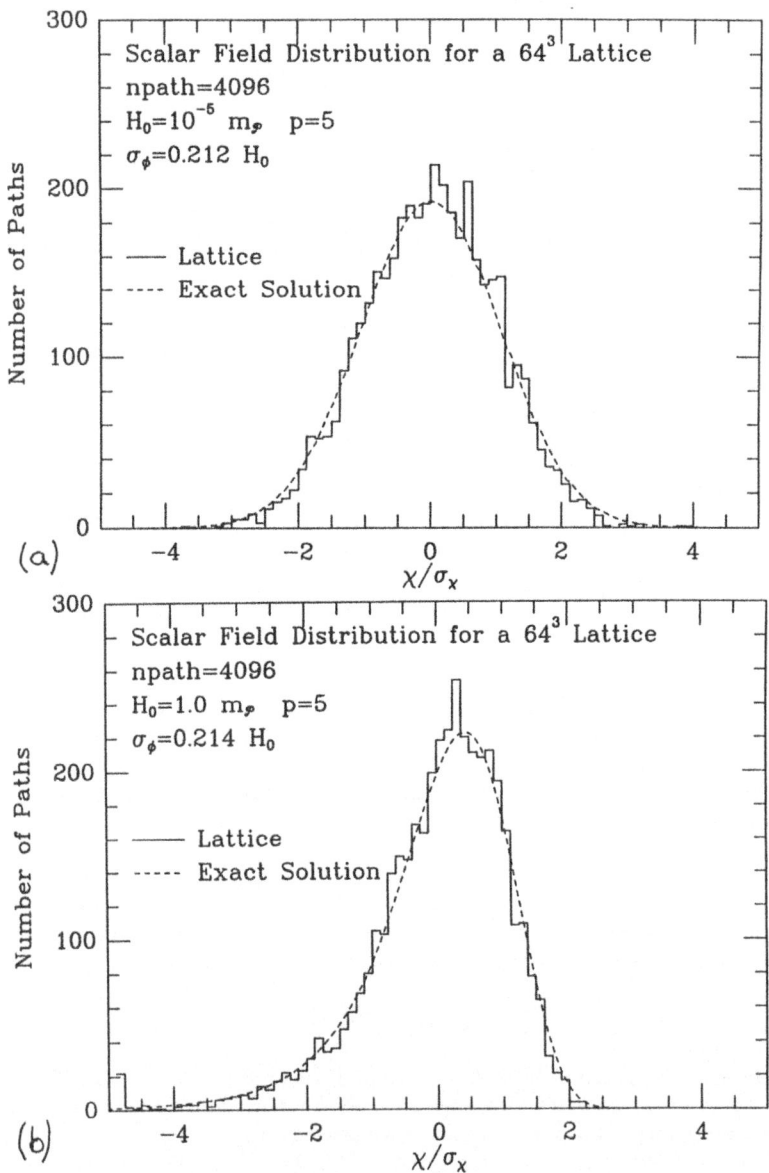

5.   DISCUSSION AND CONCLUSIONS

Theoretical large scale structure models have not kept pace with the growing number of cosmological observations. It is a challenge to the theorist to propose alternative scenarios which are consistent with the observations. Here, I have summarized the first steps towards producing models which produce calculable non-Gaussian fluctuations from inflation. Non-Gaussian fluctuations can arise from nonlinear long wavelength evolution whose signature may perhaps be observed in the near future from microwave background fluctuations.

Typically for single scalar field models, the nonlinear evolution of long wavelength fields does not produce significant non-Gaussian fluctuations in our observable Universe because microwave background anisotropy limits force the initial value of the Hubble parameter to be quite small. However, for scales much larger than our observable Universe, non-Gaussian fluctuations do arise from lattice points that almost eternally inflate. The stochastic formalism has been sufficiently developed that one can now consider multiple scalar field models that produce non-Gaussian fluctuations in our observable Universe.[17]

Exact solutions of Starobinski's[10] Fokker-Planck equation have been given for a $\lambda\phi^4/4$ potential by Yi, Vishniac and Mineshige (YVM).[18] In their case non-Gaussian fluctuations can be significant at moderate deviations from the mean $\sim 6\sigma$ even when $H(\phi_0) \sim 10^{-3}m_{\mathcal{P}}$. For the case of an exponential potential, a careful numerical analysis of eq.(4.3) shows that non-Gaussian fluctuations are significant at 6σ for $H(\phi_0) > 0.02m_{\mathcal{P}}$. (By significant, I mean more than a 100% change in the probability distribution from a Gaussian one.) This mild discrepancy may signal that a $\lambda\phi^4/4$ is intrinsically different than a exponential potential. It would be interesting to see whether one could generalize their method of solution to the improved Fokker-Planck equation which includes the full nonlinear metric back-reaction.[7] In any case, the YVM effect is typically small because microwave background limits require a much smaller value of the Hubble parameter, $H(\phi_0) \lesssim 10^{-5}m_{\mathcal{P}}$. Furthermore, 2-3$\sigma$ fluctuations are typically of greater interest for the formation of galaxies.[19]

In the inflation model, quantum fluctuations in the scalar field are converted into metric fluctuations. Hence, one ultimately, requires a quantum theory of the gravitational field if one wishes to extend the formalism describing inflation. However, attempts to model spatial variations have employed at most linear perturbations on a homogeneous background.[20,14] In addition, quantization of the gravitational field using the Wheeler-DeWitt equation has encountered numerous difficulties including interpretation of negative probabilities, choice of time parameter and initial probability distribution, operator ordering problems, over-simplified models, *etc.* (consult ref. 21 for a recent review). Unfortunately, an attempt to a construct a quantum theory of long wavelength fields appears to be inconsistent with the momentum constraint beyond the semi-classical approximation.[6] In any case, one can give arguments that suggest that quantum corrections are typically small at long wavelengths. For practical purposes, one can simply assume that short wavelength quantum fluctuations become classical when the wavelength exceeds the Hubble radius. In this context, the Fokker-Planck equation and the associated Langevin equation have proven to be more useful than the Wheeler-DeWitt equation because they are solvable and because they admit a simple interpretation in terms of initial conditions for structure formation.

I would like to thank J.R. Bond for a fruitful collaboration on these stochastic inflation lattice topics. This work was supported by the U.S. Department of Energy and NASA at Fermilab (Grant No. NAGW-1340).

6. REFERENCES

[1] Bahcall, N. and Soneira, R., Astrophys. J.**270**, 70 (1983).

[2] De Lapparent, V., Geller, M.J. and Huchra, J.P., Ap. J. **302**, L1 (1986).

[3] Dressler, A., Faber, S.M., Burstein, D., Davies, R.L., Lynden-Bell, D., Terlevich, R.J. and Wegner, G., Astrophys. J. Lett. **313**, L37 (1986).

[4] Maddox, S.J., Efstathiou, G., Sutherland, W.J. and Loveday, J., Mon. Not. R. Astr. Soc., **242**, 43P (1990).

[5] Broadhurst, T.J., Ellis, R.S., Koo, D.C. and Szalay, A.S., Nature **343**, 726 (1990).

[6] Salopek, D.S. and Bond, J.R., *Nonlinear Evolution of Long Wavelength Metric Fluctuations in Inflation Models*, Phys. Rev. **D42** (in press, 1990).

[7] Salopek, D.S. and Bond, J.R., *Stochastic Inflation and Nonlinear Gravity*, Phys. Rev. D (in press, 1990).

[8] Bardeen, J.M., Steinhardt, P.J. and Turner, M.S., Phys. Rev. **D28**, 670 (1983).

[9] Vilenkin, A., Phys. Rev. **D27**, 2848 (1983).

[10] Starobinski, A.A., in *Current Topics in Field Theory, Quantum Gravity, and Strings*, Proc. Meudon and Paris VI, ed. H.T. de Vega and N. Sanchez **246** 107 (Springer-Verlag, 1986)

[11] Bardeen, J.M. and Bublik, G.J., Class. Quant. Grav. **5**, L113 (1988).

[12] Goncharov, A.S., Linde, A.D. and Mukhanov, V.F., Intern. Jour. Mod. Phys. **A2**, 561 (1987).

[13] Lucchin, F. and Matarrese, S., Phys. Rev. **D32**, 1316 (1985).

[14] Salopek, D.S, Bond, J.R. and Bardeen, J.M., Phys. Rev. **D40**, 1753 (1989).

[15] Ortolan, A., Lucchin, F. and Matarrese, S., Phys. Rev. **D38**, 465 (1988).

[16] Matarrese, S., Ortolan, A. and Lucchin, F. Phys. Rev. **D40**, 290 (1989).

[17] Salopek, D.S., in progress, (1991).

[18] Yi, I., Vishniac, E.T. and Mineshige, S., University of Texas Preprint (1990).

[19] Bardeen, J.M., Bond, J.R., Kaiser, N. and Szalay, A.S., Ap.J. **304**, 15 (1986).

[20] Halliwell, J.J. and Hawking, S., Phys. Rev. **D31**, 1777 (1985).

[21] Halliwell, J.J., Int. J. Mod. Phys. **A5**, 2473 (1990).

PURELY QUANTUM DERIVATION OF DENSITY FLUCTUATIONS IN THE INFLATIONARY UNIVERSE

Yasusada NAMBU and Misao SASAKI
Department of Physics, Kyoto University
Kyoto 606, JAPAN

It is commonly believed that cosmological objects such as galaxies were formed through gravitational instability of initially small density fluctuations. One of the great successes of the inflationary universe scenario is that it has a possibility to explain the origin of these initial fluctuations naturally by quantum fluctuations of the scalar field which drives inflation.

During the de Sitter phase, the wavelength of vacuum fluctuations of the scalar field is stretched rapidly by the exponential expansion of the universe. When the wavelength exceeds the Hubble horizon size, the fluctuation stops oscillating and is frozen to a constant value. What has been done commonly in the literature to evaluate the density fluctuation is that one first decomposes the scalar field into the homogeneous classical part and the quantum fluctuation part, solves each part separately, regards the quantum part as classical fluctuations when the wavelength exceeds the horizon size, and solves the classical evolution of the fluctuations which are consequently interpreted as density fluctuations that give rise to galaxies. However, the decomposition into classical and quantum parts is rather artificial. In particular, recent investigations by means of the stochastic approach to inflation indicate that the scalar field behaves highly stochastic initially and the notion of a homogeneous classical scalar field appears only as a result of the stochastic process. It is hence very much desirable to evaluate the density fluctuation amplitude based purely on quantum theoretical arguments, and see if the result is in accordance with the one obtained previously.

We calculated the amplitude of density fluctuations without introducing any concept of the classical scalar field. We shall only use purely quantum correlation functions of the scalar field and evaluate the two-point correlation function of the gauge invariant variable Δ which corresponds to energy density of the scalar field. This variable satisfies the following equation:

$$\overset{(3)}{\Delta}(\rho\Delta) = \overset{(3)}{\Delta}(-T^0_0) + 3Ha\partial^j(-T^0_j),$$

where T^a_b is the energy momentum tensor of the minimally coupled scalar field with negative mass square $m^2 < 0$. We assume background space-time is deSitter. Evaluating the two point correlation function, we get

$$\langle \rho\Delta(x)\rho\Delta(x')\rangle \approx -\frac{H^4}{8\pi^2}\left(\frac{2\eta^2}{r^2}\right)^{2+c}\langle\dot{\phi}^2(t)\rangle,$$

where $c \approx m^2/H^2 < 0$,

$$\left\langle \dot{\phi}^2(t) \right\rangle = \frac{-cH^4}{8\pi^2} \left(\frac{2\eta^2}{r_0^2} \right)^c .$$

with r_0 being an infrared cutoff which corresponds to the beginning size of the inflationary universe. We can get the power spectrum of the density fluctuation by Fourier transforming the above expression. The result becomes

$$\left\langle (\delta\rho/\rho)^2 \right\rangle_k \approx \frac{H^4}{(2\pi)^2 \left\langle \dot{\phi}^2(t_k) \right\rangle} (-k\eta)^c ,$$

where t_k is approximately the time at which the wavelength exceeds the horizon size in de Sitter phase. Since $|c| \ll 1$, this is a Zeldovich spectrum as expected. This result is in complete agreement with the one obtained previously where the existence of a homogeneous classical scalar field was assumed, provided that one interprets the expectation value $\langle \phi^2 \rangle$ as the square of the classical field φ^2.

The same result can be obtained by the stochastic approach to inflation, in which the modes with proper wavenumbers k/a greater than H are treated as a noise source. This implies that one can obtain the correct fluctuation amplitude also in the stochastic approach. In other words, one can apply the same argument as given here to the energy-momentum tensor of the coarse-grained scalar field defined on scales greater than the horizon size. An advantage of the stochastic approach is that the systematic understanding of the dynamics of inflation can be rather easily obtained. In particular, there is no need to assume the existence of an *a priori* classical field. In stead, the development of the homogeneous classical background out of the quantum field can be easily visualized and the interpretation of $\langle \phi^2 \rangle$ as φ^2 arises quite naturally.

References

Nambu, Y., and Sasaki, M.(1988),'Purely Quantum Derivation of Density Fluctuations in the Inflationary Universe', *Prog. Theor. Phys.* **83**, 37-50.

Nakao, K., Nambu, Y. and Sasaki, M.(1988),'Stochastic Dynamics of New Inflation', *Prog. Theor. Phys.* **80**, 1041-1068.

CONSTRAINTS ON THE COUPLING OF WEAKLY-INTERACTING PARTICLES TO MATTER FROM STELLAR EVOLUTION[*]

Naoki Iwamoto
Department of Physics and Astronomy
The University of Toledo
Toledo, Ohio 43606, U.S.A.

Axions couple to nucleons in both the KSVZ and DFSZ models.[1] The matrix elements for nucleon-nucleon bremsstrahlung (nn → nna, pp → ppa, and np → npa) are calculated and the energy loss rates from these processes have been obtained in the case where the nucleons are degenerate with different neutron and proton Fermi momenta.[2] The neutron-proton bremsstrahlung rate thus obtained differs from the previous results[3,4]: The phase space integrals are carried out explicitly for different neutron and proton Fermi momenta as compared with the approximate expression in Ref. 3. Furthermore, the rate vanishes when the proton concentration approaches zero, which is to be contrasted with the results in Ref. 4. The neutron-proton rate is found to be larger than the neutron-neutron rate by a factor 2-5 for typical values of the proton concentration in neutron star matter. Requiring the energy loss rate due to axion emission to be less than that of neutrino emission, one obtains the upper bound on the axion-nucleon coupling. This upper bound is slightly lower than the previous bound[5] due to the enhanced axion energy loss rate, which results from the inclusion of the neutron-proton process.

A similar analysis may be carried out for light supersymmetric particles. For higgsinos[6], the energy loss rates due to electron bremsstrahlung $e(Z,A) \rightarrow e(Z,A)\tilde{h}_0\tilde{h}_0$ and neutron-neutron bremsstrahlung nn → $nn\tilde{h}_0\tilde{h}_0$ are similar to the respective neutrino rates except for an extra factor $\cos^2(2b)$, where $b=\arctan(v_2/v_1)$ with v_1 and v_2 the vacuum expectation values of the neutral Higgs bosons. In the long-term cooling of neutron stars, the higgsinos and neutrinos escape freely from the star, so that the higgsino contribution is at most on the order of another neutrino species. Therefore, this gives much less restrictive constraints on the parameter b than in the supernova case,[7] where the freely escaping higgsinos can coexist and compete with the trapped neutrinos and b is constrained to be much closer to $\pi/4$. There might have been an interesting possibility that light sneutrinos[8] could be effective cooling agents for stars since their pair production through the wino channel could dominate over the neutrino pair production with the current lower bound on the wino mass[9] $m_{\tilde{W}} > 45.9$ GeV. However, the accelerator experiments on the width of the Z have ruled out light sneutrinos in the mass range $m_{\tilde{\nu}} < 30$ GeV within the past several months in 1990.[10,11]

In obtaining constraints on the coupling of new particles to matter, the neutrino energy loss is often used as a reference. In applying such a method, it is important to examine how much the new types of energy loss must exceed the neutrino energy loss in order to have an observable effect. In the context of thermal evolution of neutron stars, the standard (neutrino) scenario involves such uncertainties as the neutron star matter equation of state, the mass, and nucleon superfluidity. A specific value for the new particle-ordinary matter coupling may be said to be observationally excluded if the theoretically predicted x-ray flux with error bars (due to theoretical uncertainties) falls below the observed x-ray flux with observational error bars. This procedure necessarily involves the inclusion of the detector characteristics of future x-ray space missions, the estimates of the expected x-ray flux from and the background estimates for neutron stars inside historical supernova remnants. We are currently looking into this question by adding an extra energy loss rate (with specific temperature and density dependences) to the standard neutron star evolutionary code.[12]

* Supported in part by the U.S. National Science Foundation under Grant No. PHY90-08475.

1. R. D. Peccei, in CP Violation, edited by C. Jarlskog (World Scientific, Singapore, 1989).
2. N. Iwamoto, to be published.
3. R. Mayle et al., Phys. Lett. B 203, 188 (1988); 219, 515 (1989).
4. R. P. Brinkmann and M. S. Turner, Phys. Rev. D 38, 2338 (1988).
5. N. Iwamoto, Phys. Rev. Lett. 53, 1198 (1984); Phys. Rev. D 39, 2120 (1989).
6. H. E. Haber, preprint: SLAC-PUB-3834 (1985).
7. J. Ellis et al., Phys. Lett. B 215, 404 (1988).
8. L. E. Ibáñez, Phys. Lett. B 137, 160 (1984); J. S. Hagelin, G. L. Kane, and S. Raby, Nucl. Phys. B 241, 638 (1984); J. A. Grifols, M. Martinez, and J. Sola, Nucl. Phys. B 268, 151 (1986).
9. T. Barklow et al. (Mark II Collaboration), Phys. Rev. Lett. 64, 2984 (1990).
10. G. S. Abrams et al. (Mark II Collaboration), Phys. Rev. Lett. 63, 2173 (1989); D. Decamp et al. (ALEPH Collaboration), Phys. Lett. B 235, 399 (1990); M. Z. Akrawy et al. (OPAL Collaboration), Phys. Lett. B 240, 497 (1990); P. Abreu et al. (DELPHI Collaboration), Phys. Lett. B 241 493 (1990); B. Adeva et al. (L3 Collaboration), Phys. Lett. B 249 341 (1990).
11. L. M. Krauss, Phys. Rev. Lett. 64, 999 (1990); preprint: YCTP-P9-90 (1990).
12. H. Umeda, K. Nomoto, S. Tsuruta, and N. Iwamoto, work in progress.

FORMATION AND EVOLUTION OF
DOMAIN-WALL-NETWORKS

H. Kubotani, H. Ishihara and Y. Nambu

Department of Physics
Kyoto University
Kyoto 606, Japan

ABSTRACT. We investigate dynamical evolution of global strings and domain walls produced by the phase transition of a scalar field. We found that the system evolves in a quasi-equilibrium state.

Topological defects, produced by phase transitions, might be seeds of the structure of the universe. Recently, an idea of structure formation by extremely low energy (soft) domain walls is proposed by Hill, Schramm and Fry.[1]

By use of 2-dimensional numerical simulations, we investigate the dynamical evolution of the topological defects[2] described by the equation of a complex scalar field:

$$\partial_\eta^2 \phi + \frac{2}{\eta} \partial_\eta \phi - \nabla^2 \phi = -\frac{\partial V}{\partial \phi^*}.$$

The second term mimics the Hubble dumping. We consider two models specified by the potential. The potential of the first model is $V = \frac{\lambda}{4}(\phi^*\phi - \phi_0^2)^2$ and it produces global strings (pure string model). The second one has an additional potential $\delta V = \epsilon(\phi^{*3} + \phi^3)$ which breaks U(1) symmetry explicitly and it produces domain walls attached on the strings since the potential has shallow isolated minima (string-wall model).

The numerical simulations were run on a 400 × 400 mesh with the periodic boundary condition. The initial value of ϕ at each mesh is randomly distributed in a range $\phi^*\phi \leq \phi_0^2$ and the initial value of $\partial_\eta \phi$ is chosen to zero.

We see the time evolution of three components of the energy: kinetic energy ($E_{kin} \equiv \sum |\partial_\eta \phi|^2$), spatial derivative energy ($E_{sd} \equiv \sum |\nabla\phi|^2$), and potential energy ($E_{pot} \equiv \sum V$) (see Fig.1).

We find that in both systems the ratios of these components become constant in course of time. We call this stage a quasi-equilibrium stage. In the case of pure string model, E_{sd} dominates other components. This comes from spatial derivative of the angular component of ϕ. On the other hand, in the case of string-wall model, E_{sd} and E_{pot} share main part of the energy. It means the energy of the walls dominates the system. Since the ratio of E_{kin} is about 10% in both cases, typical velocity of the strings and walls is about 30 % of light velocity. The kinetic energy E_{kin} is supplied by pair annihilations of the strings.

270

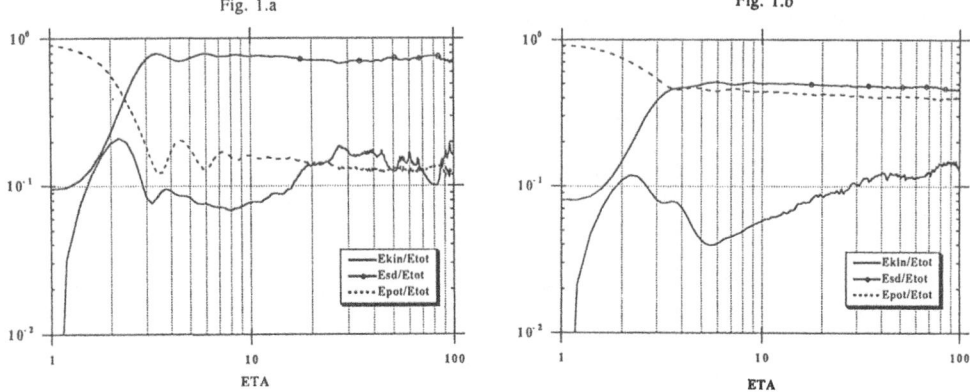

Fig.1 Evolution of components of energy. a) pure string model b) string-wall model.

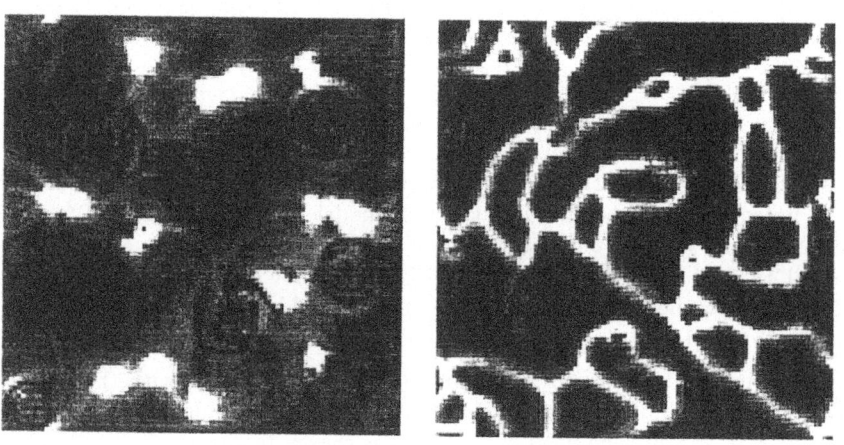

Fig.2 Snap shot of the system. Brightness means the energy density. a) pure string model b) string-wall model.

REFERENCES

1. C.T.Hill, D.N.Schramm and J.N.Fry, Com. Nucl. Part. Phys. **19** (1989) 25

2. W.H.Press, B.S.Ryden and D.N.Spergel, Ap.J. bf 347 (1989) 590,
   B.S.Ryden, W.H.Press and D.N.Spergel, Ap.J. bf 357 (1990) 293,
   L.Kawano, Phys. Rev. **D41** (1990) 1013.

CATASTROPHE OF SPACETIME IN THE EARLY UNIVERSE[#]

CAO SHENGLIN
Department of Astronomy
Beijing Normal University
Beijing
People's Republic of China

ABSTRACT. According to the some interesting subjects in the process of evolution of the universe, it is discussed that the catastrophe nature of the Finsler spacetime and its cosmological meaning. It is shown that the nature of evolution of the universe could be attributed to the geometric feature of the Finsler spacetime.

1. Finsler Metric

If we assume that the spacetime metric is defined by the form of the Finsler metric

$$ds^2 = (g_{ijkf}dx^i dx^j dx^k dx^f)^{\frac{1}{2}}, \quad i,j,k,f = 0,1,2,3. \tag{1}$$

According to cosmological symmetries, we may assume that

$$g_{ijkf} = \begin{cases} 1 & i = j = k = f, \\ 2h & i = j \neq k = f, \\ 0 & \text{for other cases.} \end{cases} \tag{2}$$

Where h is a control parameter depending on the opacity of the universe. h = 0, in the very early universe and h = ±1, in the present flat universe, generally speaking, $0 \leq |h| \leq 1$, during the evolutionary process of the universe.

According to the catastrophe theory, the functions $y = x^2$ and $y = x^4$ are topologically equivalent. But, the germ $y = x^2$ is topologically (and even differentiably) stable at zero. The germ $y = x^4$ is differentiably (and even topologically) unstable at zero. So, there is a great difference between the theories of geometry on the metric ds^2 and ds^4.

2. The Type of the Double-Cusp Catastrophe $T^4 + R^4$ and the Creation of Spacetime

For convenience, let us consider only 2-dimension case, and let

[#]
The project supported by National Natural Science Foundation of China

$$f(T,R) = T^4 + R^4 \tag{3}$$

it is a type of the double cusp catastrophe, so, it will have evolution, but, the equation $T^4 + R^4 = 0$ has 0 real root, so, it will have notion for observable.

As the catastrophe theory, spacetime manifold $M(T,R)$ is divided into four parts by the different value of the stability matrix

$$H(T,R) = \begin{pmatrix} 12T^2 & 0 \\ 0 & 12R^2 \end{pmatrix} = 144T^2R^2 \tag{4}$$

and

$$
\begin{aligned}
T^2R^2 &> 0 \quad \text{seed of time} \\
T^2R^2 &< 0 \quad \text{seed of space} \\
T^2R^2 &= 0 \quad \text{catastrophe set} \\
T = R &= 0 \quad \text{the origin.}
\end{aligned} \tag{5}
$$

3. The Type of the Double–Cusp Catastrophe $T^4 + R^4 - 2hT^2R^2$ and the Inflation

Let

$$f(T,R,h) = T^4 + R^4 - 2hT^2R^2, \quad -1 < h < 1 \tag{6}$$

it is a type of the double cusp catastrphe too. It will describe the inflation of the universe.

According to the four real roots of the stability matrix $H(T,R,h)$ the spacetime manifold $M(T,R)$ could be divided into 8 parts.

4. The Degenerate Form of the Double–Cusp Catastrophe $T^4 + R^4 \pm 2T^2R^2$ and Hot Big Bang

Where $h = 1$, the metric (6) will become degenerate form. The 4 real roots will become 2 double roots. So, the spacetime manifold could only be divided into 4 parts.

If the metric has to the form

$$f(T,R) = T^4 + R^4 - 2T^2R^2 = (T^2 - R^2)^2 \tag{7}$$

it just is Minkowskian space. If the metric has to the form

$$f(Y,Z) = Y^4 + Z^4 + 2Y^2Z^2 = (Y^2 + Z^2)^2 \tag{8}$$

it just is Euclidean space.

REFERENCES

Arnold V.I., Vavohnkc A.N., Cusein–Zade S.M., Singularities of differentiable maps, Vol. 1, Birkhauser 1985.
Arnold V.I., Catastrophe theory, Springer–Verlag 1986.
Asanov C.S., Finsler geometry, Relativity and Gauge theories, D. Reidel Publishing Company 1985.
Rindler W., Essential relativity, Springer–Verlag 1977.
Zeeman E.C., Catastrophe theory, Selected Papers 1972–1977, Addison–Wesley 1977.

(2+1)-DIMENSIONAL QUANTUM GRAVITY

Akio HOSOYA
Department of Physics,Tokyo Institute of Technology
Ohokayama, Meguroku, Tokyo 152
Japan

ABSTRACT. We consider the (2+1)-dimensional quantum gravity as a toy model. It is shown that the torus universe in the (2+1)-dimensional quantum gravity is a quantum chaos in a rigorous sense.

I would like to present a toy model of (2+1)-dimensional quantum universe which exhibits a feature of quantum chaos. I focus my presentation on the case of torus universe for the spatial manifold. In a sense, the (2+1)-dimensional pure Einstein gravity is an ideal toy model to see the global structure of space-time. The (2+1)-dimensional Einstein gravity contains no gravitational waves which are only local deformation modes in space-time and therefore irrelevant to the more interesting global motion of the spatial manifold. Actually, the (2+1)-dimensional Einstein space is locally flat. Therefore our 2-surface sweeps a part of the full Minkowski space. The motion of the 2-surface is not at all trivial, however, if its topology is nontrivial, *i.e.* its genus is non zero.[2] In that model, there remain only a finite number of degrees of freedom. In the case of torus for the 2-surface, in particular, the Teichmüller deformations induce a change from a fat torus to a slim torus and also a twist.

In the case of torus universe, the unimodular part of the metric, $\tilde{h}_{ij} = h_{ij}/\sqrt{h}$ is parametrized as

$$\tilde{h}_{ij} = \begin{pmatrix} (x^2 + y^2)/y, & x/y \\ x/y, & 1/y \end{pmatrix}$$

We substitute this parametrization for the spatial metric in to the standard Einstein-Hilbert action to get a reduced action in terms of the Teichmüller parameters x, y and the total volume of the universe v. It turns out that the super-metric of the superspace for the torus universe is given by the Poincaré metric:

$$ds^2 = \frac{1}{y^2}(dx^2 + dy^2).$$

The geodesic of the Lobachevski geometry defined by this metric is given by a semi-circle in the upper $x - y$ plane with its center on the x-axis, as is well known.

273

The Hamiltonian constraint becomes

$$y^2(P_x^2 + P_y^2) - v^2 P_v^2 = 0,$$

where P_x, P_y and P_v are conjugate momenta to x, y, and the volume of the universe v. The Hamiltonian constraint equation is very similar to the one for an relativistic point particle in curved space endowed with the Poincaré metric . The volume of the universe plays a role of time coodinate.

Following Dirac's prescription for the constraint equations in quantization, we replace the momenta by differential operators. The Hamiltonian constraint is interpreted as a wave equation to the state vector. The equation above is translated as

$$\left[\frac{\partial^2}{\partial s^2} - y^2(\partial_x^2 + \partial_y^2) \right] \psi(x, y, s) = 0$$

with $s = \log v$. The wave function $\psi(x, y, s)$ has also to be a global $SL(2, \mathbf{Z})$ diffeomorphism invariant in order to be well defined in the superspace Therefore, for the torus universe, the superspace is nothing but the fundamental region. The region $y \to \infty$ corresponds to the limit of the fattest torus and also of the thinest one. Actually they are identified by an element of $SL(2, \mathbf{Z})$, the inside-out operation of the torus.

We obtain

$$U_\nu(s, x, y) = \sum_{n \neq 0} \rho_\nu(n) \sqrt{y} K_{i\nu}(2\pi |n| y) e^{2\pi i n x} e^{-iEs} .$$

Here K is the so-called modified K Bessel function which approaches zero exponentially when its argument goes to infinity. A remarkable fact is that the coefficients $\rho_\nu(n)$ exhibit a kind of randomness. This is interesting because the classical geodesic motion in the fundamental region is known to be chaotic. The discrete eigenvalues ν are known only numerically. (The smallest one is $13,7797513 \cdots$).

The function U is called the Maass form, which is regular everywhere. That is, the singular universe has no chance to appear. The idea behind this property is similar to Hartle and Hawking's. In a sense the singularity of the space-time is circumvented in quantum cosmology of the torus universe. Although it is not unique, its variety is only discrete rather than continuous.

REFERENCE

A. Hosoya and K. Nakao Prog.Theor.Phys. 84 (90)739.

A STRINGY UNIVERSE SCENARIO

M. MASHINO AND H. MINAKATA
Department of Physics
Tokyo Metropolitan University
Setagaya, Tokyo 158
Japan

ABSTRACT. A scenario of the evolution of the early universe based on the fundamental string theory is described.

From where did our universe come? Of what is it ultimately composed? They are the deep questions asked by many people throughout the histry. The string theory seems to offer an answer that it ultimately came from strings. We attempt to construct a scenario of how the universe began with strings and how it have evolved to as it is now. The following is a short account of the work done in Ref.1. For details and for references see Ref.1.

The only guide available now, which we believe the best suited one, is the string thermodynamics. Its microcanonical formulation strongly suggests that highly ($\sim m_p$) energetic state of the matter takes the form of single long string. Then, it is natural to assume that the initial state of the universe is dominated by the single long string. Since it is the highly excited string state, it should allow the classical description.

This picture of the initial universe strongly resembles with that of the cosmic-string theory based on grand unification. We construct a definite picture of the evolution of the stringy universe by appealing to this similarity. In particular, what is of crucial importantceis the recognition that the network of the cosmic string inevitably reaches to the scaling solution. The long string, the loops which were chopped off the long string, and the radiations emitted from the loops are in peaceful coexistence in this scaling era. Namely, their energy densities, ρ_L, ρ_l, and ρ_r, behave as

$$\rho_i = C_i \frac{\mu}{R_H^2},$$
(1)

where R_H denotes the Hubble radius and i runs over L, l, and r. The point is that all the C_i are calculable.

One of the most important features of our stringy universe scenario is that their relative importance are determined self-consistently with the structure of the space-time. The latter is controlled by a key parameter, the stringy fraction s defined by

275

$$s = \frac{\rho_L(s) + \rho_1(s)}{\rho_L(s) + \rho_1(s) + \rho_r(s)} \ .$$

(2)

This equation determine the stringy fraction through the motion of the string in the background space-time, as represented in its right-hand-side. We suspect that the feature might be a low-energy manifestation of the fact that string theory unifies the matter with the space-time.

We present in Fig. 1 the computed results of the scaling densities of the long string, the loops, and the radiations.

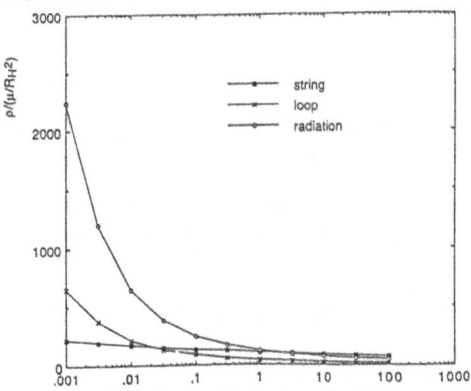

Figure 1. The energy densities ρ_L (solid square), ρ_1 (cross), and ρ_r (open circle) are plotted in units of μ / R_H^2 as functions of $\Gamma_{eff} = (n/2)\Gamma G\mu$ where $\Gamma \sim 50$ and n is the number of zero-mode radiations off loops. The scaling density parameter of Benett and Bouchet is adopted.

We are now dealing with the fundamental string theory with $G\mu \sim 10^{-3}$ rather than the GUT cosmic string with $G\mu \sim 10^{-6}$. Therefore, the region of interest of the value of Γ_{eff} is $1 \lesssim \Gamma_{eff} \lesssim 10$. We realize that the long string, the loops and the radiations have the energy densities of, roughly speaking, the same order of magnitude.

Since $G\mu \sim 10^{-3}$ one expects drastic violation of the bounds derived on various astrophysical setting. This trouble, however, can be avoided by the fact discovered by Witten that the superstring becomes axionic at the QCD phase transition. An axionic superstring is accompanied by domain wall and abruptly ($\sim 10^{-4}$ sec) disappears at the transition.

Our scenario so far described leaves the fate of the long strings open. It is tempting to speculate that they might play a role in the large-scale structure formation.

1. Mashino, M. and Minakata, H. (1990) "An attempt toward stringy cosmology" , Tokyo Metropolitan University preprint TMUP-HEL-9009.

The Constant-Mean-Curvature Slicing of
the Schwarzschild-de Sitter Space-Time

Ken-ichi Nakao

Uji Research Center, Yukawa Institute

for Theoretical Physics, Kyoto University, Uji 611, Japan

Numerical simulations are important not only for gravitational collapse and formation of neutron stars and black holes but also for the purpose of investigating the highly inhomogeneous stage of our universe. In particular, the 'cosmic no hair conjecture' connected with the inflationary scenario is one of the issues which should be examined by numerical simulations. For the numerical simulations, the coordinate condition is crucial to investigate the physically important region of a space-time. It is well known that the constant-mean-curvature(CMC) time slicing coordinate is useful for the asymptotically flat space-time. However, in order to investigate cosmological problems, we would like to know the applicability of that coordinate for the asymptotically non-flat space-time. As for inhomogeneous space-time with an asymptotically de Sitter space, the Schwarzschild-de Sitter solution is known. We investigate how the CMC hypersurfaces foliate the Schwarzschild-de Sitter space-time.

The metric of the Schwarzschild-de Sitter space-time is written in the ordinary Schwarzschild coordinate as

$$ds^2 = -C(r)dt^2 + C^{-1}(r)dr^2 + r^2(d\theta^2 + \sin^2\theta d\varphi^2), \tag{1}$$

where

$$C(r) \equiv 1 - \frac{2M_0}{r} - H_0^2 r^2, \tag{2}$$

M_0 is the gravitational mass and $H_0 \equiv \sqrt{\Lambda/3}$. For the case of $0 < 27M_0^2 H_0^2 < 1$, we have two event horizons: one is the cosmological event horizon at $r = r_c$ and another is that of the black hole (or white hole) at $r = r_h$.

We present the motion of the CMC hypersurface in the Kruscal diagram of the Schwarzschild-de Sitter space-time of $27M_0^2 H_0^2 < 1$ in Fig.1(a) and (b). There are

277

three limit radii on which the CMC hypersurfaces cease its time evolution as

$$r_{lim} = r_1 \equiv -\frac{1}{H_0\sqrt{3}}\left(1 - \sqrt{1 + \frac{3\sqrt{3}M_0}{H_0}}\right), \tag{3}$$

and

$$r_{lim} = r_{2,\pm} \equiv \frac{1}{H_0\sqrt{3}}\left(1 \pm \sqrt{1 - \frac{3\sqrt{3}M_0}{H_0}}\right), \tag{4}$$

with the following relation

$$r_1 < r_{2,-} < r_h < r_c < r_{2,+}. \tag{5}$$

(a) depicts hypersurfaces in the region within the cosmological event horizon $r \leq r_c$. The solid lines are the limit radii $r = r_1$ in the white hole (the lower half plane) and $r = r_{2,-}$ in the black hole (the upper half plane). On the other hand, (b) depicts the region out of the black hole and white hole event horizon $r \geq r_h$. As seen those figure, the CMC time slicing condition is appropriate for the numerical simulations for space-times with asymptotically de Sitter spacelike section.[1]

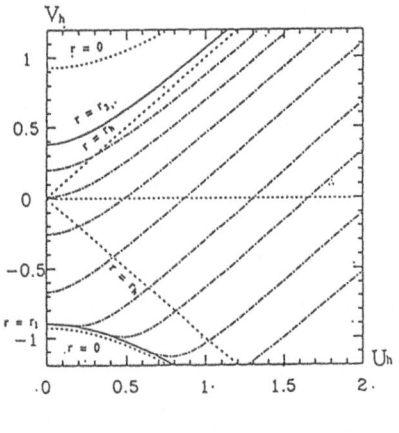

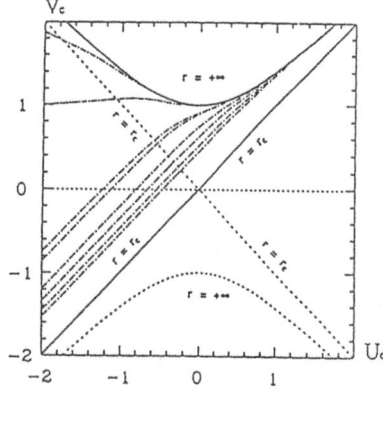

Fig.1 (a)   Fig.1 (b)

1. K. Nakao, K. Maeda, T. Nakamura and K. Oohara, YITP Preprint YITP/U-26 and Waseda University Preprint WU-AP/08/90 (1990)

SCHWARZSCHILD-DE SITTER TYPE WORMHOLE AND COSMOLOGICAL CONSTANT

Sung-Won KIM[1], Sung Ku KIM[2] and Jongmann YANG[2]
[1]*Department of Science Education*
[2]*Department of Physics*
Ewha Womans University
Seoul, 120-750, Korea

ABSTRACT. The two Schwarzschild-de Sitter spaces connected by a wormhole following the method of Visser is considered in order to find the effect of the cosmological constant on the Minkowskian wormhole. The dynamics of the wormhole is also compared with those of the Euclidean wormhole.

The metric element of the Schwarzschild-de Sitter (SdS) spacetime is given by

$$ds^2 = - (1-2M/r-r^2/R^2)dt^2 + (1-2M/r-r^2/R^2)^{-1}dr^2 + r^2d\Omega^2, \qquad (1)$$

where $R^2=3/8\pi\Lambda$ is the radius of the de Sitter universe without any mass. Here only $M/R<1/\sqrt{27}$ case is considered so that two horizons can exist: the larger radius, the cosmological horizon, r_C, and the smaller one (the size of the throat of the wormhole), the event horizon, r_H.

Similar to the method by Visser we make the SdS type wormhole model. Take two copies of SdS manifold, and remove from them the four-dimensional regions described by $\Omega_{1,2}=\{r_{1,2}<a|a>r_H\}$ with boundaries given by the timelike hypersurfaces $\partial\Omega_{1,2}=\{r_{1,2}=a|a>r_H\}$. By identifying these two hypersurfaces (i.e. $\partial\Omega_1=\partial\Omega_2$), the resulting spacetime is geodesically complete and possesses two asymptotically flat regions connected by a wormhole. Since the stress-enrgy tensor is delta function type at $\partial\Omega$, the Ricci tensor at the junction is calculated in terms of the discontinuity in the second fundamental forms. For the dynamic case (the radius a is not constant) by using the differences of the second fundamental forms over $\partial\Omega$ the field equation become

$$\sigma = - \frac{(1-2M/a-a^2/R^2+\dot{a}^2)^{1/2}}{2\pi a}, \quad \tau = - \frac{1}{4\pi a}\frac{1-M/a-2a^2/R^2+\dot{a}^2+a\ddot{a}}{(1-2M/a-a^2/R^2+\dot{a}^2)^{1/2}}, \qquad (2a,b)$$

where σ is the surface energy density and τ is the surface tension. Since this is built to avoid the event horizon, σ, $\tau<0$. These Eq.(2a,b) satisfy the same conservation law of stress-energy as the Schwarzschild case. Eq.(2a) means the dynamical equation

$$\dot{a}^2 - 2M/a - a^2/R^2 - (2\pi\sigma)^2 a^2 = -1 \tag{3}$$

By comparing with the Schwarzschild case when $\sigma=0$ (classical membrane) or $m=4\pi\sigma a^2$ (classical constraint), it is shown that the cosmological constant gives the possibility to blow up to infinity.

There is the need for some instanton field to build the Euclidean wormhole. The usual Giddings-Strominger type uses the three index antisymmetric tensor field as the axionic instanton field. Lee, Hong and Kim calculated the wormhole solution in Euclidean de Sitter spacetime. Its action with the cosmological constant Λ is

$$\int d^4x\sqrt{g} \; [\Lambda - R/16\pi G + f^2 H^{\lambda\mu\nu} H_{\lambda\mu\nu}], \tag{4}$$

where f is the Peccei-Quinn scale coupling. With the Euclidean metric of $S^3 R$ and the instanton field

$$ds^2 = dt^2 + a^2(t)d\Omega_3^2, \quad H_{\lambda\mu\nu} = h(t)\epsilon_{\lambda\mu\nu}, \tag{5}$$

where $d\Omega_3^2$ is the line element on the three-sphere and $\epsilon_{\lambda\mu\nu}$ is the volume three-form normalized, the H equation $dH=d^*H=0$ is obeyed by setting $h(t)=n/f^2a^3$, where n is the axion charge. The Einstein equation is then

$$\dot{a}^2 - 1 = - r_o^4/a^4 - a^2/R^2, \tag{6}$$

where $r_o^4=16\pi Gn^2/f^2$ is the radius of the wormhole when there is no cosmological constant. There are two radii: the radius of the parent universe determined by Λ and the one of the baby universe by n, the axion. Compared with the horizons of SdS, the instanton field replaces the matter of object for SdS while the horizons by Λ coincide.

Since Eq.(6) is calculated over the Euclidean metric, the transformation $a^2 \rightarrow -a^2$ makes Eq.(6) become the Minkowskian wormhole containing the instanton field instead of the matter field. The radius a is dynamically stable and is restricted to some finite value. The R-term, the cosmological constant, plays the role of avoidance of blowing up into the infinity, while Λ plays the reversed role in Minkowskian case. It says that two de Sitter universes are stably connected by a wormhole in Euclidean case.

This work was supported in part by KOSEF and Research Institute for Basic Sciences of Ewha Womans University.

Giddings, S. and Strominger, A. (1988) 'Axion-induced topology change in quantum gravity and string theory', Nucl. Phys. B306, 890-907.

Hong, J. and Kim, J. E. (1990) 'Drainage mechanism of positive cosmological constant', J. Kor. Phys. Soc. 23, 91-102.

Lee, K. (1988) 'Wormholes and Goldstone bosons', Phys. Rev. Lett. 61, 263-266.

Visser, M. (1989) 'Traversable wormholes from surgically modified Schwarzschild spacetimes', Nucl. Phys. B328, 203-212.

COBE: NEW SKY MAPS OF THE EARLY UNIVERSE

G. F. Smoot
Lawrence Berkeley Laboratory and Space Sciences Laboratory,
University of California, Berkeley,

ABSTRACT

This paper presents early results obtained from the first six months of measurements of the Cosmic Microwave Background (CMB) by instruments aboard NASA's Cosmic Background Explorer (*COBE*)[*] satellite and discusses the implications for cosmology. The three instruments: FIRAS, DMR, and DIRBE have operated well and produced significant new results. The FIRAS measurement of the CMB spectrum supports the standard Big Bang nucleosynthesis model. The maps made from the DMR instrument measurements show a surprisingly smooth early universe. The measurements are sufficiently precise that we must pay careful attention to potential systematic errors. The maps of galactic and local emission produced by the DIRBE instrument will be needed to identify foregrounds from extragalactic emission and thus to interpret the results in terms of events in the early universe.

INTRODUCTION

In the Big Bang scenario the light elements and the cosmic microwave background (CMB) are the most accessible relics from the eras during which the universe was a relatively structureless plasma prior to its evolving to the highly-ordered state observed today. In standard models of cosmology, CMB photons have travelled unhindered from the surface of last scattering in the early universe to the present era; as such, the CMB maps the large scale structure of space-time in the early universe. Despite a quarter-century of effort, no intrinsic anisotropy in the CMB has been detected. Likewise, there has been a fruitless search for distortions of the spectrum from a featureless blackbody.

COBE was launched on November 18, 1989 into a 900-km circular, near-polar orbit (inclination 99°). The Earth's gravitational quadrupole moment precesses the orbit to follow the terminator, allowing the instruments to point away from the Earth and perpendicular to the Sun to avoid both solar and terrestrial radiation. Throughout most of the year, the satellite orbit provides an exceptionally stable environment. During the two months surrounding summer solstice, the satellite is unable to shield the instruments and dewar from the Earth and Sun simultaneously, and the Earth limb becomes visible over the shielding surrounding the instrument aperture plane as the satellite passes over the North Pole. During the same period, the satellite enters the Earth's shadow as the orbit crosses over the South Pole.

FIRAS INSTRUMENT DESCRIPTION AND CMB SPECTRUM RESULTS

The purpose of the FIRAS is to compare the spectrum of the CMB with that of a precise blackbody, enabling the measurement of very small deviations from a Planckian spectrum. The FIRAS instrument covers two frequency ranges, a low frequency channel from 1 to 20 cm^{-1} and a high frequency channel

---

[*] The National Aeronautics and Space Administration/Goddard Space Flight Center is responsible for the design, development, and operation of the Cosmic Background Explorer. GSFC is also responsible for the software development through to the final processing of the space data. The COBE program is supported by the Astrophysics division of NASA's Office of Space Science and Applications.

from 20 to 100 cm^{-1}. It has a 7° diameter beam width, established by a non-imaging parabolic concentrator, which has a flared aperture to reduce diffractive sidelobe responses. The instrument is calibrated by a full beam, temperature-controlled external blackbody, which can be moved into the beam on command. The FIRAS is the first instrument to measure the background radiation and compare it with such an accurate external full-beam calibrator in flight. The spectral resolution is obtained with a polarizing Michelson interferometer, with separated input and output beams to permit fully symmetrical differential operation. One input beam views the sky or the full aperture calibrator, while the second input beam views an internal temperature controlled reference blackbody, with its own parabolic concentrator. Both input concentrators and both calibrators are temperature controlled and can be set by command to any temperature between 2 and 25 K. In standard operating condition the two concentrators and the internal reference body are commanded to match the sky temperature, thereby yielding a nearly nulled interferogram and reducing almost all instrumental gain errors to negligible values.

The external calibrator determines the accuracy of the instrument for broad band sources like the CMBR. It is a re-entrant cone shaped like a trumpet mute, made of Eccosorb CR-110 iron-loaded epoxy. The angles at the point and groove are 25°, so that a ray reaching the detector has undergone 7 specular reflections from the calibrator. The calculated reflectance for this design, including diffraction and surface imperfections, is less than 10^{-4} from 2 to 20 cm^{-1}. Measurements of an identical calibrator in an identical antenna using coherent radiation at 1 cm^{-1} and 3 cm^{-1} frequencies confirm this calculation. The instrument is calibrated by measuring spectra with the calibrator in the sky horn while operating all other controllable sources within the instrument at a sequence of different temperatures.

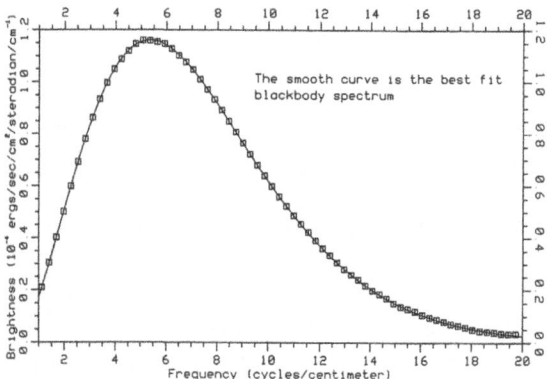

FIG. 1.—Preliminary spectrum of the cosmic microwave background from the FIRAS instrument at the north Galactic pole, compared to a blackbody. Boxes are measured points and show size of assumed 1% error band. The units for the vertical axis are 10^{-4} ergs s^{-1} cm^{-2} sr^{-1} cm.

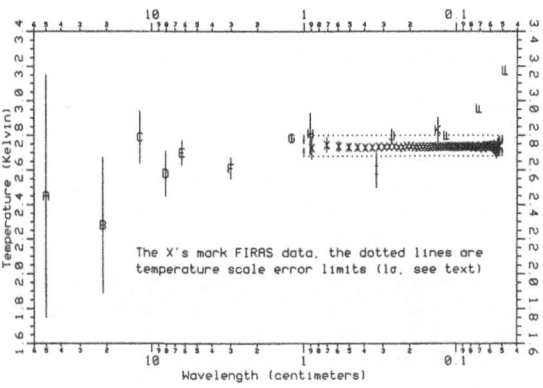

The first results of the FIRAS /1,2/ may be summarized as follows: The intensity of the background sky radiation is consistent with a blackbody at 2.735 ± 0.06 K. Deviations from this blackbody at the spectral resolution of the instrument are less than 1% of the peak brightness. The quoted error is primarily due to an uncertainty in the thermometer calibration; we expect to reduce this uncertainty by additional tests. The measured spectrum is shown in Figure 1, and is converted to temperature units in Figure 2, where it is compared to previous measurements. The deviations can be fitted to the Sunyaev-Zel'dovich form for Comptonization /3/, giving a limit on the y parameter of $|y| < 10^{-3}$ (3 σ). A fit to a Bose-Einstein distribution with a chemical potential gives a limit of $|\mu| < 10^{-2}$ (3 σ). This results in a strong limit on the existence of a smooth hot intergalactic medium: it can contribute less than 3% of the X-ray background radiation even at a reheating time as recent as z=2. There is no evidence of a distortion of the spectrum, such as that reported by Matsumoto et al. /4/, and the measured temperature is consistent with previous reports and the recent rocket result of Gush et al. /5/.

The variation of the spectrum with position in the sky as measured by the FIRAS is dominated by the dipole anisotropy of the CMB, plus a variation in the interstellar dust emission. A preliminary dipole anisotropy spectrum was determined by calculating the average spectra in two large circular regions of the sky each of angular diameter 60°, one centered at $(\alpha, \delta) = (11.1h, -6.3°)$ and the other at $(23.1h, 6.3°)$, which lie along opposite ends of the dipole axis as determined by the DMR. The difference between these spectra is fit extremely well by the difference of two blackbodies, and is consistent with a peak dipole amplitude of 3.3 ± 0.3 mK and the assumed dipole direction.

These precise limits to potential distortions covering the major portion of the CMB photons provide support for a key element in the models of Big Bang nucleosynthesis - the baryon to photon ratio and the interpretation of that number relative to the critical density. The FIRAS measurement actually counts the vast majority of photons and the precision of the measurement gives us the density constraint

$$\rho_\gamma = 4.722 \ (T/1K)^4 e^{4y} \ eV/cm^3 \sim 10^{-33} \ (To/2.7K)^4 e^{4y} \ gm/cm^3$$

$$n_\gamma = 20.286 \ (T/1K)^3 e^{3y} \ photons/cm^3 = 415. \ (To/2.735K)^3 e^{3y} \ photons/cm^3 \qquad (1)$$

The FIRAS measurements decrease the error on n_γ from greater than 20% to less than 5%.

DMR INSTRUMENT DESCRIPTION AND OPERATION

The *COBE* Differential Microwave Radiometers (DMR) instrument is intended to provide precise maps of the microwave sky on large angular scales. It consists of six differential microwave radiometers, two independent radiometers at each of three frequencies: 31.5, 53, and 90 GHz (wavelengths 9.5, 5.7, and 3.3 mm). At these frequencies the CMB dominates foreground galactic emission by at least a factor of roughly 1000. The multiple frequencies allow subtraction of galactic emission using its spectral signature, yielding maps of the CMB and thus the distribution of matter and energy in the early universe. Each radiometer measures the difference in microwave power between two regions of the sky separated by 60°. The combined motions of spacecraft spin (75 s period), orbit (103 minute period), and orbital precession (~1 degree per day) allow each sky position to be compared to all others through a massively redundant set of all possible difference measurements spaced 60° apart.

Each radiometer consists of a superheterodyne receiver switched at 100 Hz between two identical corrugated-horn antennas. The compact low-sidelobe antennas' main beam profile is well described by a Gaussian of 7° FWHM and point 60° apart, 30° to either side of the spacecraft spin axis /6/. The two channels at 31.5 GHz share a single antenna pair with an orthomode transducer splitting the input into opposite circular polarizations. Both channels share a common enclosure and thermal regulation system. The 53 and 90 GHz radiometers are similar but have two antenna pairs at each frequency, each with identical linear polarization response. A detailed description of the DMR instrument may be found in Smoot *et al.* /7/.

Three independent techniques are used to determine the radiometer calibration. Solid-state noise sources provide in-flight calibration by injecting broad-band microwave power into the front end of each radiometer at regular intervals (every two hours). Prior to launch the noise source signals were calibrated by comparing to the signal produced by targets of known, dissimilar temperatures (approximately 300 K and 77 K) covering the antenna apertures. The DMR observes the Moon for a fraction of an orbit during two weeks of every month providing an independent determination of the calibration factor. The Earth's ~30 km s^{-1} motion about the solar system barycenter produces a Doppler-shift dipole of known magnitude (~0.3 mK) and direction. The modulation in amplitude and direction is apparent, but at low signal to noise. Given sufficient observing time (>one year), this method may produce the most accurate determination of the absolute calibration of the DMR instrument. A forthcoming paper /8/ discusses the DMR calibration more fully.

DMR DATA REDUCTION AND ANALYSIS

A software analysis system receives data telemetered from the satellite, determines the instrument calibration, and inverts the difference measurements to map the microwave sky in each channel. Although the experiment has been designed to minimize or avoid sources of systematic uncertainty, both the instrument and the software can potentially introduce systematic effects correlated with antenna pointing, which would create or mask features in the final sky maps. Further details of the data processing algorithms may be found in Torres *et al. /9/*.

Table 1 summarizes limits to potential systematics at the time of the conference. The results in many cases are limited by sky coverage, signal to noise, and available analysis software. We anticipate increasingly tighter limits to potential systematics as coverage, integration time, and software improve.

TABLE 1 95% C.L. Upper Limits to Potential Systematic Effects

Foreground Emission	Peak Magnitude	Magnitude in Maps
COBE shield and dewar	<0.04 mK	<0.002 mK
Earth	<0.3 mK	<0.07
Moon (>25 degrees)	<0.04 mK	<0.02
Sun	<0.03 mK	<0.02
Planets	<0.26 mK	<0.26 a
Galaxy	<0.3 mK	<0.13 b
Extragalactic	<0.02 mK	<0.01
Orbit Environment	**Peak Magnitude**	**Magnitude in Maps**
Magnetic	<0.3 mK/G	<0.17
Thermal	<20 mK/K	<0.01
Voltage	<20 mK/V	<0.01
Cross-talk	<2 mK	<0.0001
Software Artifacts	**Peak Magnitude**	**Magnitude in Maps**
Solution Stability	<10^{-6}	<0.0001
Baseline Residuals	<2 mK	<0.03
Calibration Residuals	<2 %	<0.06
Absolute Calibration	<5 %	<0.17 c
Antenna Pointing	<1°	<0.03
Total Systematics		**<0.24 mK**

a Effect limited to one pixel
b Excluding data within 10° of galactic plane
c Dipole term only

The sky maps may in principle contain contamination from local sources or artifacts from the data reduction process itself. The data reduction process must distinguish cosmological signals from a variety of potential systematic effects. The most obvious source of non-cosmological signals is the presence in the sky of foreground microwave sources. These include thermal emission from the *COBE* spacecraft itself, from the Earth, Moon, and Sun, and from other celestial objects. Non-thermal radio-frequency interference (RFI) must also be considered, both from ground stations and from geosynchronous satellites. Although the DMR instrument is largely shielded from such sources, their residual or intermittent effect must be considered. A second class of potential systematics is the effect of the changing orbital environment on the instrument. Various instrument components have slightly different performance with changes in temperature, voltage, and local magnetic field, each of which can be modulated by the *COBE* orbit. Longer-term drifts can also affect the data. Finally, the data reduction process itself may introduce or mask features in the data. The DMR data are differential; the sparse matrix algorithm is subject to concerns of both coverage (closure) and solution stability. Other features of the data reduction process, particularly the calibration and baseline subtraction, are also a source of potential artifacts. All potential sources of systematic error must be identified and their effects measured or limited before maps with reliable uncertainties can be produced.

DMR RESULTS

Figure 2 shows the microwave sky at 3 mm on a linear scale. The most noticeable effect is the extreme uniformity of the CMB. Figure 3 shows preliminary maps of the microwave sky for each of the six DMR channels. The independent maps at each frequency enable celestial signals to be distinguished from noise or spurious features: a celestial source will appear at identical amplitude in both maps. The three frequencies allow separation of cosmological signals from galactic foregrounds based on spectral signatures. The maps are corrected to solar-system barycenter and do not include data with the Moon within 25° of an antenna; no other systematic corrections have been made. All six maps clearly show the dipole anisotropy and galactic emission. The dipole appears at similar amplitudes in all maps while galactic emission decreases sharply at higher frequencies, in accord with the expected spectral behavior.

An observer moving with velocity $\beta = v/c$ relative to an isotropic radiation field of temperature T_0 observes a Doppler-shifted temperature

$$T = T_0 \frac{(1 - \beta^2)^{1/2}}{1 - \beta\cos(\theta)} = T_0 [1 + \beta\cos(\theta) + \frac{1}{2}\beta^2 \cos(2\theta) + O(\beta^3)] \qquad (2)$$

The first term is the monopole CMB temperature without a Doppler shift. The second term, proportional to β, is a dipole distribution, varying as the cosine of the angle between the velocity and the direction of observation. The term proportional to β^2 is a quadrupole with amplitude reduced by $1/2 \ \beta$ from the dipole amplitude. The DMR maps clearly show a dipole distribution consistent with a Doppler-shifted thermal spectrum, implying a velocity for the solar system barycenter of $\beta=0.00123 \pm 0.00003$ (68% CL), or $v = 370\pm10$ km s^{-1} toward $(\ell,b) = (264°\pm2,49°\pm2°)$, where we assume a value $T_0=2.735$ K. The solar system velocity with respect to the local standard of rest is estimated at 20 km s^{-1} toward (57°,23°), while galactic rotation moves the the local standard of rest at 220 km s^{-1} toward (90°,0°) /10,11/. The DMR results thus imply a peculiar velocity for the Galaxy of $v_g = 550 \pm 10$ km s^{-1} in the direction (266° ± 2°, 30° ± 2°). This is in rough agreement with independent determinations of the velocity of the local group, $v_{lg} = 507 \pm 10$ km s^{-1} toward (264 °± 2°, 31° ± 2°) /12/.

Figure 4 shows the DMR maps with this dipole removed from the data. The only large-scale feature remaining is galactic emission, confined to the plane of the galaxy. This emission is present at roughly the level expected before flight and is consistent with emission from electrons (synchrotron and HII) and dust within the galaxy. The ratio of the dipole anisotropy (the largest cosmological feature in the maps) to the Galactic foreground reaches a maximum in the frequency range 60—90 GHz. There is no evidence of any other emission features.

286

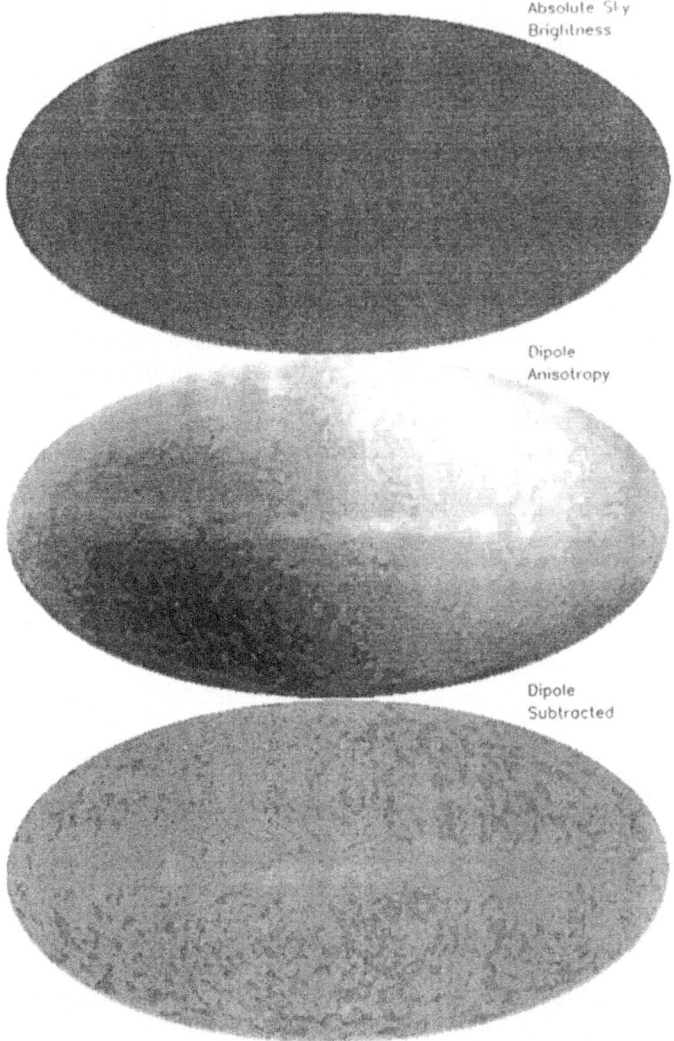

Figure 2 *COBE* DMR full sky maps of the temperature of the sky at 53 GHz (6mm wavelength). The maps are in galactic coordinates and have been corrected to solar system barycenter.

2.A Absolute Sky Brightness on a scale from 0 to 3.6 K

2.B Relative sky brightness with mean removed and scale to about -4 to +4 mK. The dipole anisotropy (0.1%) and galactic plane are clearly visible.

2.C Relative sky brightness with monpole and dipole removed. The galactic plane is only significant visible signal. The other structure evident is consistent with instrument observing noise.

90.0 GHz

53.0 GHz

31.5 GHz

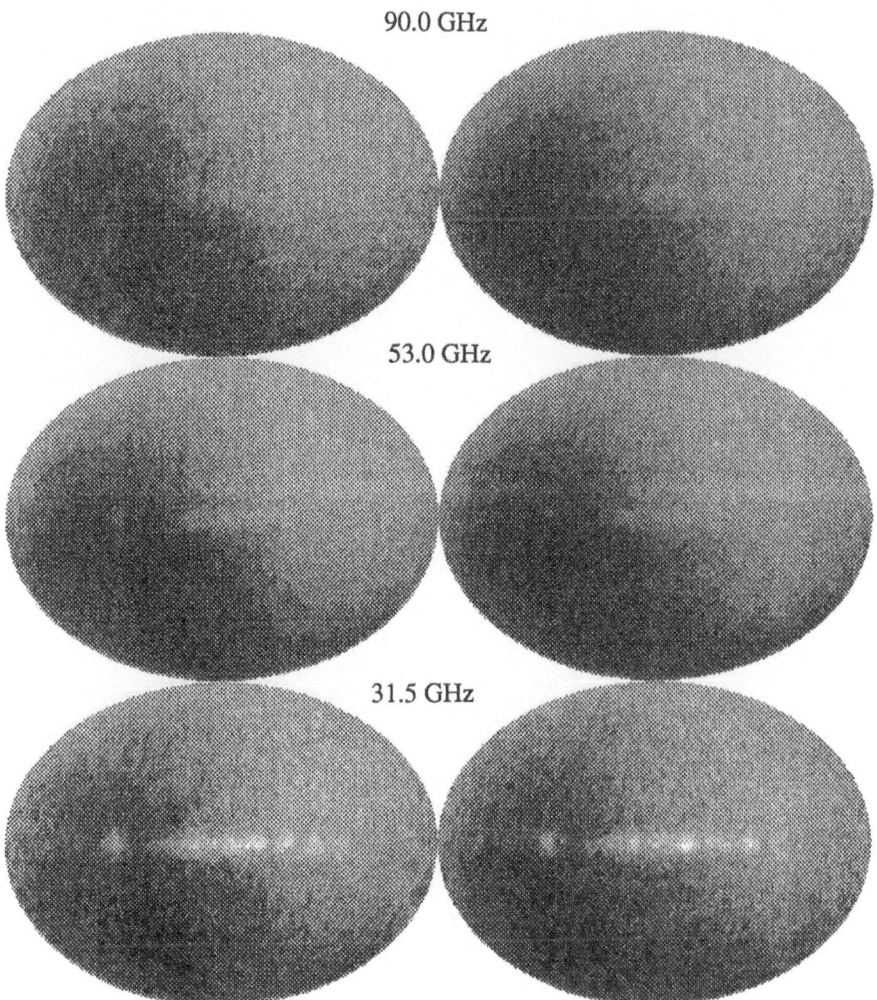

Figure 3. *COBE* DMR full sky maps of the relative temperature of the sky at frequencies 31.5, 53.0, and 90.0 GHz. The maps are in galactic coordinates and have been corrected to solar system barycenter. The maps show variations in received power and are insensitive to the mean temperature of about 2.735 K.

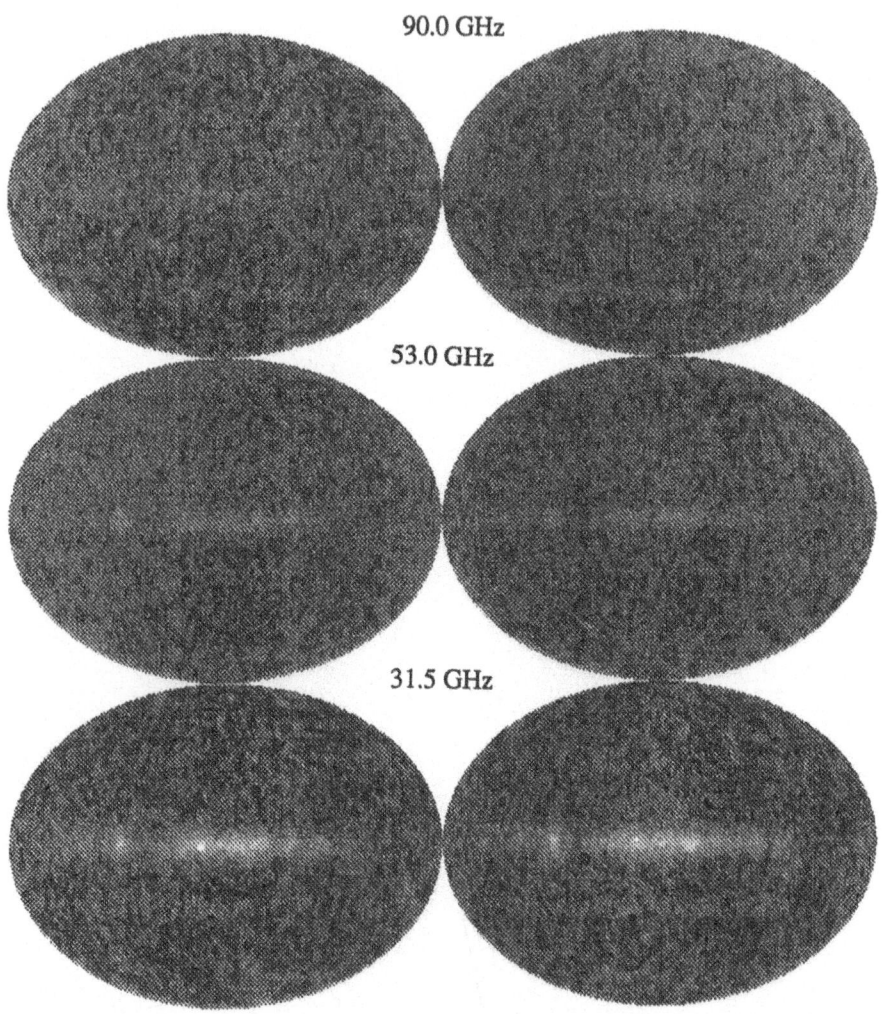

90.0 GHz

53.0 GHz

31.5 GHz

Figure 4. *COBE* DMR full sky maps at frequencies 31.5, 53.0, and 90.0 GHz. A dipole anisotropy corresponding to a motion of 370 km/s through a 2.735 K blackbody has been subtracted from the maps shown in Figure 3.

We have made a series of spherical harmonic fits to the data, excluding data within several ranges of galactic latitude. The only large-scale anisotropy detected to date is the dipole. Quadrupole and higher-order terms are limited to amplitude $\Delta T/T < 10^{-4}$. Similarly, a search for Gaussian or non-Gaussian fluctuations on the sky showed no features to limit $\Delta T/T < 10^{-4}$. The results are insensitive to the precise cut in galactic latitude and are consistent with the expected Gaussian instrument noise. The reported uncertainties are 95% confidence level unless otherwise stated, and include the effects of systematics as listed in Table 1.

DMR DISCUSSION

The DMR limits to CMB anisotropies provide significant new limits to the dynamics and physical processes in the early universe. The dipole anisotropy provides a precise measure of the Earth's peculiar velocity with respect to the co-moving frame. Limits to higher-order anisotropies limit global shear and vorticity in the early universe. If the universe were rotating (in violation of Mach's Principle), the resultant metric causes null geodesics to spiral; in a flat universe the resultant anisotropy is dominated by a quadrupole term /13,14/. The limit $\Delta T/T < 10^{-4}$ for quadrupole and higher spherical harmonics limits the rotation rate of universe to $\Omega < 3 \times 10^{-24}$ s^{-1}, or less than one ten-thousandth of a turn in the last ten billion years.

If the expansion of the universe were not uniform, the expansion anisotropy would lead to a temperature anisotropy in the CMB of similar magnitude. The large-scale isotropy of the DMR results indicate that the Hubble expansion is uniform to one part in 10^4. This provides additional evidence for hot big bang models of cosmology, and indicates that the currently observed expansion of the universe can be traced back at least to the radiation-dominated era.

Inhomogeneities in the density in the early universe also lead to temperature anisotropies in the CMB as the CMB photons climb out of varying gravitational potential wells /15,16/

$$\Delta T/T \sim \frac{1}{2} \frac{\delta \rho}{\rho} \left(\frac{H_0 L}{c} \right)^2 \tag{3}$$

The DMR results imply that the universe at the surface of last scattering was isotropic and homogeneous to the 10^{-4} level. The results have implications for structures beyond the present Hubble radius: on large scales the universe is isotropic and homogeneous. The large scale geometry of the universe is thus well-described by a Robertson-Walker metric with only local perturbations.

One such potential local perturbation is gravitational radiation. Long-wavelength gravitational waves propagating though this region of the universe distort the metric and produce a quadrupole distortion in the CMB /17/. Those on the surface of last scattering will chaotic fluctuations. The DMR is sensitive primarily to gravitational waves with scale sizes > 7° at the surface of last scattering, or ~200 Mpc today. The limits $\Delta T/T < 10^{-4}$ for quadrupole and higher spherical harmonics limit the energy density of single plane waves or a chaotic superposition to

$$\Omega_{GW} < 4 \times 10^{-3} \left(\frac{\lambda_{GW}}{10 \text{ Mpc}} \right)^{-2} h^{-2} \tag{4}$$

where Ω_{GW} is the energy density of the radiation relative to the critical density, λ_{GW} is the wavelength at the current epoch, and h is the Hubble constant in units 100 km s^{-1} Mpc^{-1}.

Cosmic strings provide another mechanism for local perturbations in the metric. They are nearly one-dimensional topological defects predicted by many particle physics gauge theories and are characterized by a large mass per unit length, μ /18/. The large mass and relativistic velocity produce CMB anisotropies through the relativistic boost and the Sachs-Wolfe effect (gravitational lensing alone does not produce anisotropy in an otherwise isotropic background). Many authors have calculated the anisotropy produced by various configurations of cosmic strings , with typical values /19,20/

$$\Delta T/T \sim 8\pi\beta\gamma \frac{G\mu}{c^2} \qquad (5)$$

The DMR experiment limits the existence of large-scale cosmic strings to $G\mu/c^2 < 10^{-4}$. There is no evidence for higher-order topological defects such as domain walls. The large scale geometry of the universe appears to be uniform and without defects.

The observed isotropy of the universe on large angular scales presents a major problem for cosmology. At the time of primordial nucleosynthesis the presently observable universe was divided into about 10^{25} causally independent regions. The observed uniformity of light elements implies a baryon density uniformity of $\delta\rho/\rho < 3$ during sysnthesis. /21/ At the surface of last scattering the horizon size was ~100 kpc (~2'). Regions separated by more than 2' were not in causal contact; consequently, DMR measures some 10^4 causally disconnected regions of the sky. Standard models of cosmology fail to explain why causally unconnected regions are the same to order unity, much less the 10^{-4} isotropy implied by the DMR observations. Inflationary scenarios provide one solution. In these models, the universe undergoes a spontaneous phase transition ~10^{-32} seconds after the Big Bang, causing a period of exponential growth in which the scale size increases by 30 to 40 orders of magnitude. The entire observed universe would then originate from a small pre-inflationary volume in causal contact with itself, eliminating the problem. In the simplest inflationary models, the pre-inflationary matter and radiation fields are diluted to zero along with any pre-existing anisotropies. The process of inflation, however, generates scale-free anisotropies with a Harrison- Zel'dovich spectrum which result in small but detectable CMB anisotropies in the present universe /22,23/. Although current DMR limits are an order of magnitude above the predicted spectrum, we anticipate achieving sufficient sensitivity over the planned mission to test the predictions of inflationary models.

A second major problem in cosmology is the growth of structure in the universe. The largest structures in the current universe (walls and voids) are observed to have density fluctuations $\delta\rho/\rho$ of order unity on scale sizes ~50 Mpc. Structures of this size are at the horizon scale at the surface of last scattering; consequently, the primordial density fluctuations are small and most of the growth is in the linear regime. The assumption of linear growth requires peculiar velocities ~ 0.01c in order to move the matter the required 10^8 light years of co-moving distance in the ~10^{10} years estimated to have elapsed since the surface of last scattering, an order of magnitude greater than the peculiar velocity inferred from dipole anisotropy. To explain the observed structure without violating limits on CMB anisotropy, and to generate the critical density required by inflationary models, many astrophysicists have turned to cosmological models in which most of the matter in the universe (> 90%) is composed of weakly interacting massive particles (WIMPs). The dynamical properties of this "dark matter" allow it to clump faster than the baryonic matter, which later falls into the WIMP gravitational potential wells to form the structures observed today. The gravitational potential and motion of these particles produce CMB anisotropy whose amplitude depends on the angular scale size. For scale size ~ 10' most reasonable models predict $\Delta T/T \sim 1—3 \times 10^{-5}$, depending on the average density of the universe /24/. Although current observations do not provide significant limits to these models, we anticipate that the DMR will provide a stringent test of such models as it continues to accumulate data.

DIRBE

The Diffuse Infrared Background Experiment (DIRBE) is designed to conduct a sensitive search for an isotropic cosmic infrared background (CIB) radiation over the spectral range from 1 to 300 micrometers. The cumulative emissions of pregalactic, protogalactic, and evolving galactic systems are expected to be recorded in this background. Since both the cosmic red-shift and reprocessing of short-wavelength radiation to longer wavelengths by dust act to shift the short-wavelength emissions of cosmic sources toward or into the infrared, the spectral range from 1 to 1000 micrometers is expected to contain much of the energy released since the formation of luminous objects, and could potentially contain a total radiant energy density comparable to that of the CMB. The discovery and measurement of the CIB would provide new insight into the cosmic 'dark ages' following the decoupling of matter and the CMB.

Observing the CIB is a formidable task. Bright foregrounds from the atmosphere of the Earth, from interplanetary dust scattering of sunlight and emission of absorbed sunlight, and from stellar and interstellar emissions of our own Galaxy dominate the diffuse sky brightness in the infrared. Even when measurements are made from space with cryogenically-cooled instruments, the local astrophysical foregrounds strongly constrain our ability to measure and discriminate an extragalactic infrared background. Furthermore, since the absolute brightness of the CIB is of paramount interest for cosmology, such measurements must be done relative to a well-established absolute flux reference, with instruments which strongly exclude or permit discrimination of all stray sources of radiation or offset signals which could mimic a cosmic signal.

The DIRBE instrument, a 10-spectral band absolute photometer with an 0.7° field of view, maps the full sky with high redundancy at solar elongation angles ranging from 64° to 124° to facilitate separation of interplanetary, Galactic, and extragalactic sources of emission. The approach is to obtain absolute brightness maps of the full sky in 10 photometric bands (J[1.2], K[2.3], L[3.4], and M[4.9]; the four IRAS bands at 12, 25, 60, and 100 micrometers; and 120-200 and 200-300 micrometer bands). In order to facilitate discrimination of the bright foreground contribution from interplanetary dust, linear polarization is also measured in the J, K, and L bands, and all celestial directions are observed hundreds of times at all accessible solar elongation angles (depending upon ecliptic latitude) in the range 64° to 124°. The instrument is designed to achieve a sensitivity for each field of view of $\lambda I(\lambda) = 10^{-13}$ W cm^{-2} sr^{-1} (1 σ, 1 year). This level is well below estimated CIBR contributions. Extensive modelling of the foregrounds, just beginning, will be required to isolate the extragalactic component.

The DIRBE instrument is an absolute radiometer, utilizing an off-axis folded Gregorian telescope with a 19-cm diameter primary mirror. The optical configuration is carefully designed for strong rejection of stray light from the Sun, Earth limb, Moon or other off-axis celestial radiation, or parts of the *COBE* payload. Stray light rejection features include both a secondary field stop and a Lyot stop, super-polished primary and secondary mirrors, a reflective forebaffle, extensive internal baffling, and a complete light-tight enclosure of the instrument within the *COBE* dewar. Additional protection is provided by the Sun and Earth shade surrounding the *COBE* dewar, which prevents direct illumination of the dewar aperture by these strong local sources. The DIRBE instrument, which is maintained at a temperature below 2 K within the dewar, measures absolute brightness by chopping between the sky signal and a zero-flux internal reference at 32 Hz using a tuning fork chopper. Instrumental offsets are measured by closing a cold shutter located at the prime focus. All spectral bands view the same instantaneous field-of-view, 0.7° x 0.7°, oriented at 30° from the spacecraft spin axis. This allows the DIRBE to modulate solar elongation angles by 60° during each rotation, and to sample fully 50% of the celestial sphere each day. Four highly-reproducible internal radiative reference sources can be used to stimulate all detectors when the shutter is closed to monitor the stability and linearity of the instrument response. The highly redundant sky sampling and frequent response checks provide precise photometric closure over the sky for the duration of the mission. Calibration of the photometric scale is obtained from observations of isolated bright celestial sources. Careful measurements of the beam shape in pre-flight system testing and during the mission using scans across bright point sources allow conversion of point-source calibrations to surface brightness calibrations.

Qualitatively, the initial DIRBE sky maps show the expected character of the infrared sky. For example, at 1.2 micrometers stellar emission from the galactic plane and from isolated high latitude stars is prominent. Zodiacal scattered light from interplanetary dust is also prominent. At fixed ecliptic latitude the zodiacal light decreases strongly with increasing solar elongation angle, and at fixed elongation angle it decreases with increasing ecliptic latitude. These two components continue to dominate out to 3.4 microns, though both become fainter as wavelength increases. The foreground emissions have relative minima at wavelengths of 3.4 microns and longward of 100 microns. A composite of the 1.2, 2.3, and 3.4 microns images is shown in Figure 5a. Because extinction at these wavelengths is far less than in visible light, the disk and bulge stellar populations of the Milky Way are dramatically apparent in this image. A composite of the 25, 60, and 100 micron images is shown in Figure 5b. At 12 and 25 micrometers, emission from the interplanetary dust dominates the sky brightness, again strongly dependent upon ecliptic latitude and elongation angle. At wavelengths of 60 micrometers and longer, emission from the interstellar medium dominates the galactic brightness, and the interplanetary dust emission becomes progressively less apparent. The patchy infrared cirrus noted in IRAS data is evident at all of these wavelengths. The DIRBE data will clearly be a valuable new resource for studies of the interplanetary medium and Galaxy as well as the search for the CIB.

IRAS measurements toward the south ecliptic pole have been compared with DIRBE results. The two experiments are seen to agree reasonably well at 12 and 25 micrometers, but the DIRBE results are substantially fainter at 60 and 100 micrometers, reaching a factor of 2.6 at 100 micrometers. We have made a more detailed comparison with IRAS data at several points on the sky. By choosing points of different brightness, we can distinguish zero-point offsets and gain differences. We find evidence for zero-point differences, evidently not constant in time, of a few MJy/sr at all IRAS wavelengths. These are most significant (as a fraction of total brightness) at the longest wavelengths. We also find that the IRAS dc gain at 60 and 100 micrometers is substantially too high, a result which can be understood in terms of the detector characteristics and the data reduction procedures used for IRAS image-format data (see the IRAS Explanatory Supplement (1988), pp. IV-9 and IV-10 for a discussion of the IRAS dc response). The combined effect of these discrepancies is particularly large at 100 micrometers toward faint sky regions. Since IRAS detector gain depends on source angular scale (a scan-modulated instrument) and source brightness, no simple conversion between the total brightness data from the two experiments is universally applicable. It should be noted that this discussion applies only to surface brightness measurements; point source flux calibration is not subject to similar problems.

CONCLUSIONS

Nine months after launch, the *COBE* instruments are working well and continue to collect data. FIRAS has measured the CMB spectrum quite precisely and future improvements and maps are expected. The FIRAS and DMR data show the expected dipole anisotropy, consistent with a Doppler-shifted thermal spectrum. Galactic emission is present at levels close to those expected prior to launch, and is largely confined to the plane of the galaxy. The results are currently limited by instrument noise and upper limits to potential sources of systematic error. There is no evidence for any other large-scale feature in the DMR maps. The DMR results limit CMB anisotropies on all angular scales >7° to $\Delta T/T < 10^{-4}$. The results are consistent with a universe described by a Robertson-Walker metric and show no evidence of anisotropic expansion, rotation, or defects (strings, walls, texture). As DMR sky coverage improves and the instrument noise per field of view decreases, we anticipate improved calibration, better estimates of potential systematics, and increasingly sensitive limits to potential CMB anisotropies. In principle, the DMR is capable of testing predictions of both inflationary and dark-matter cosmological models. DIRBE has produced exceptionally good and complete maps of the 1 to 300 micron sky. Data from all three instruments are fully consistent and support standard Big Bang nucleosynthesis.

ACKNOWLEDGEMENTS

Nine months after launch, the *COBE* instruments are working well and continue to collect datadue to the excellent work by the staff and managment of the *COBE* Project. It is a pleasure to thank and acknowledge the many people who worked on the project and contributed to its great success.

REFERENCES

1. Mather *et al.* 1990a, *Ap. J.*, **354**, L37-L41.
2. Cheng *et al.* 1990, *Bull. APS*, **35**, 937.
3. Zel'dovich Ya. B. and Sunyaev R. A. 1969, *Ap. Space Sciences*, **4**, 301.
4. Matsumoto *et al.* (1988), *Ap. J.*, **329**, 567-571.
5. Gush, H.P., Halpern, M. & Wishnow, E. (1990) *Phys. Rev. Lett.*, **35**, 937.
6. Toral, M.A., *et al.*, *IEEE Transactions on Antennas and Propagation*, **37**, 171 (1989).
7. Smoot, G.F., *et al.*, *Ap. J.*, **360**, 685 (1990).
8. Bennett, C.L., *et al.*, in preparation.
9. Torres, S., *et al.*, *Data Analysis in Astronomy*, ed. Di Gesu *et al.*, Plenum Press (1990).
10. Kerr, F.J., and Lyndon-Bell, D., *MNRAS*, **221**, 1023 (1990).
11. Fich, M., Blitz, L., and Stark, A., *Ap. J.*, **342**, 272 (1989).
12. Yahil, A., Tamman, A., and Sandage, A., *Ap. J.*, **217**, 903 (1977).
13. Collins, C.B., and Hawking, S.W., *MNRAS*, **162**, 307 (1973).
14. Barrow, J.D., Juskiewicz, R., and Sonoda, D.H., *MNRAS*, **213**, 917 (1985).
15. Sachs, R.K., and Wolfe, A.M., *Ap. J.*, **147**, 73 (1967).
16. Grischuk, L.P., and Zel'dovich, Ya. B., *Sov. Astron.*, **22**, 125 (1978).
17. Burke, W.L., *Ap. J.*, **196**, 329 (1975).
18. Vilenkin, A., *Physics Reports*, **121**, 263 (1985).
19. Stebbins, A., *Ap. J.*, **327**, 584 (1988).
20. Stebbins, A., *et al.*, *Ap. J.*, **322**, 1 (1987).
21. Yang, J., *et al.*, *Ap. J.*, **281**, 493 (1984).
22. Gorski, K., *Ap. J. Lett.* (submitted 1990).
23. Abbott, L.F., and Wise, M.B., *Ap. J. Lett.*, **282**, L47 (1984).
24. Bond, J.R. and Efstathiou, G, *Ap. J. Lett.*, **285**, L45 (1984).

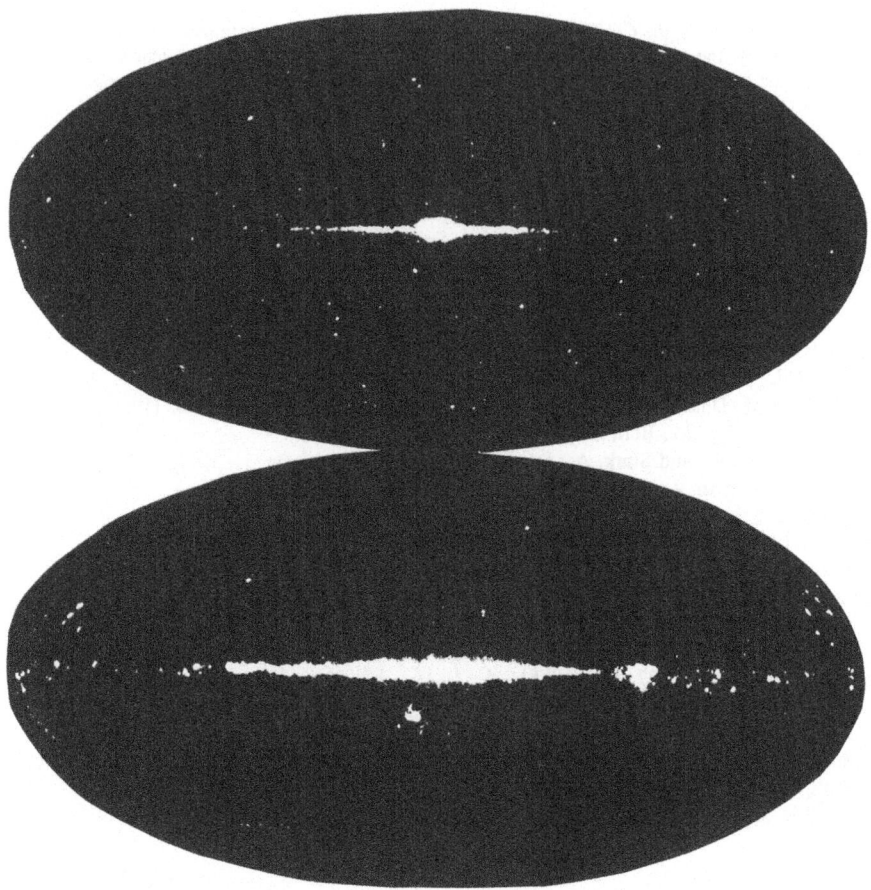

Figure 5. *COBE* DIRBE full sky maps in galactic coordinates. 5A. Combined 1.2, 2.2, and 3.4 micron map in respective blue, green, and red colors. The image shows both the thin disk and central bulge of the galaxy. 5B. Combined 25, 60, and 100 micron map in respective blue, green, and red colors. Both maps are preliminary results of a full sky scan by the DIRBE instrument. The discontinuties apparent are due primarily to the change in zodiacal light as the spacecraft and earth move around in the zodiacal dust but some part may be due to not haveing final instrument calibration.

LARGE SCALE COSMIC INSTABILITY

Craig J. Hogan
Yukawa Institute for Theoretical Physics, Kyoto University
Astronomy and Physics Departments, FM-20, University of Washington, Seattle, Washington 98195, USA

ABSTRACT. Spectral distortions in cosmic background radiation can create instabilities in matter, whose possible role in generating large scale structure is outlined here. A promising physical mechanism is sketched, based on Lyα resonance scattering, which exponentially amplifies perturbations in neutral hydrogen density on large scales after recombination.

1. Introduction

The gravitational growth of small amplitude perturbations and their collapse into bound astronomical systems successfully account in broad terms for the formation of galaxies and many of the statistical properties of their clustering and motions on large scales[1,2]. There is however no compelling observational reason to believe that gravity has always been the dominant agent for structuring the universe on large scales. Indeed, problems arise if the purely gravitational evolution of perturbations is traced back, as it often is, to the earliest moments of the big bang. For example, the presence of fluctuations at the epoch of cosmic recombination, when the cosmic microwave background radiation last interacted with matter, tends to generate anisotropy in the radiation which has not been observed. This difficulty is usually cited as evidence for a universe dominated by exotic primordial cold dark matter, but it could also be interpreted as a sign that the extrapolation of gravitational instability to very early times is incorrect.

Another reason to be unhappy with tracing the fluctuations back to the very beginning is simple skepticm. For example, conventional wisdom often attributes the binding energy of galaxy clusters (a conserved quantity in a purely gravitational theory) to a quantum instability within the first picoseconds of the big bang. This is a dramatic claim if true, but there is no evidence that things actually happened this way; it is usually supported by citing the lack of alternatives. The success of the "cold dark matter" matter model in interpreting relatively recent events need not be construed as a triumph for a Grand Scenario extending back to the Planck time. Something other than gravity may have intervened during the history of the universe to generate the fluctuations which were precursors of the observed structure, and it is prudent to think of possible alternatives.

In particular, thinking would be very different if there were known to be a true instability in ordinary matter at very low density which could spontaneously amplify primordial perturbations by such a large factor that it would effectively generate perturbations from scratch after cosmic recombination. If such an instability were to occur, it would play the central role in the early development of cosmic structure. It could circumvent the problem of excessive microwave background anisotropy even in open universes and for a wide variety of forms of dark matter, including hot massive neutrinos or ordinary baryons as well as cold dark matter. The properties of the largest astrophysical systems would have an astrophysical, rather than quantum, derivation. It would be ideal if the instability produced linear fluctuations with the right amplitude and spectrum to agree with that required empirically for the gravitational formation of large scale structure. But are there any remotely plausible physical mechanisms which can accomplish all of this?

The mass densities associated with astronomical systems are so low that such instabilities must involve long-range interactions. Aside from gravity however, the only *known* long-range force is electromagnetism. One can postulate new forces of nature or new active scalar fields in various guises (strings, textures, etc.[3,4,5]) which generate fluctuations in ordinary matter via gravitational or electromagnetic interactions. In this paper we concentrate instead on mechanisms by which ordinary electromagnetic radiation can trigger a macroscopic instability by itself; that is, under appropriate conditions, thermodynamic disequilibrium in the *spectrum* of homogeneous and isotropic background radiation field creates a dynamical instability in nearly homogeneous ordinary matter. This paper outlines some specific mechanisms for such instabilities and their possible applications to generating large scale structure.

2. Mock Gravity

There are two distinct types of instability to be considered. The first is known as the "mock gravity" instability, which occurs when absorbing particles are immersed in isotropic radiation; their mutual shadowing leads to an effective inverse square law attractive force[6,7,8]. The idea of making a long-range force with shadowing has a long history, including models for ordinary gravity dating back to Newton's contemporaries. The instability has some attractive features for manufacturing galaxies and large scale structure[9,10,11], especially the tendency to become "self-organizing" in the sense that the final distribution of material after the instability saturates is very insensitive to the initial perturbations[12,13]. However, there are implausible features with such schemes if one tries to create structure on really large scales, both in the amount of energy required and in the source of appropriate opacity in real astrophysical matter[14]. Moreover, an ironclad prediction of models which used dust opacity to make large scale structure was a significant ($\simeq 10\%$) distortion in the submillimeter part of the microwave background. The Berkeley-Nagoya detection of just such a background distortion was not confirmed by the COBE satellite[15], and detailed spectral modeling shows that no model parameters can produce the structure without exceeding the COBE limits on submillimeter distortions[16]. The mock gravity model with dust is dead; it is not clear whether it can be resurrected with other plausible sources of opacity.

3. Resonance Scattering Instability

The main focus here is on another more recent idea which operates on entirely different principles[17]. The minimal ingredients required here are just a single neutral atom of anything, and an isotropic background radiation field which has a photon occupation number increasing with frequency in the vicinity of the ground state resonance. Any peculiar velocity of the atom gets amplified by resonance scattering of the radiation, an effect which translates into a dynamical instability of matter in the limit of small amplitude perturbations. The structures produced in the end by the instability depend on the nonlinear feedback of matter motions on the radiation spectrum, and several approximate descriptions of this process are laid out below.

This instability could play an important role in structure formation if some new physics provides a suitable source of free energy in the radiation field. Pure primordial blackbody radiation has a monotonically decreasing photon occupation number $(e^{h\nu/kT} - 1)^{-1}$ and can never cause this instability. To make it happen, some new source of photons, such as a decaying particle, needs to be added to the standard big bang model to produce the energy in the spectral feature. The feature itself can arise from atomic transitions [18-21] or from continuum processes[22]. Thus far, spectra have been computed for only a small range of particle properties; spectra resulting from decaying massive $\simeq 50$ keV neutrinos (e.g. figure (4) of ref. 22) provide concrete examples of radiation fields which have both the intensity and the shape required to trigger an instability.

Suppose that some such process generates a steep spectral feature in the background radiation field which ends up at energy $h\nu > 10.2$ eV after recombination. When this feature redshifts to Lyα at 10.2eV, it triggers an instability in the neutral gas whereby some of the photon free energy is converted into the motions of matter. If the new radiation is sufficiently intense, the instability is able to create perturbations of the type required to generate galaxies and large scale structure. Because the instability amplifies perturbations exponentially, the original fluctuations can have a negligible amplitude. Although it is not a prediction of the bare-bones standard big bang model, and it is not clear whether there is a plausible decaying particle candidate for the energy source, the radiation intensity required to have all of these interesting consequences is surprisingly small by cosmic standards, and indeed may even be discouragingly difficult to detect with current techniques.

3.1 VELOCITY AMPLIFICATION

Consider an atom bathed in isotropic radiation with spectral energy density $i(\nu)$. Assume that at frequency ν the spectrum has logarithmic slope $\gamma \equiv d\ln i(\nu)/d\ln \nu$; that is, $i(\nu) \propto \nu^\gamma$. Now suppose the atom is moving through the background at velocity $\beta \equiv v/c$. To first order in β, the spectrum of the radiation in this moving frame, observed at an angle θ from the direction of motion, is[23]

$$i'(\nu, \beta, \theta) = i(\nu)[1 + \beta(3 - \gamma)cos(\theta)]. \tag{1}$$

For $\gamma > 3$, the monochromatic radiation brightness is actually smallest in the forward direction. This results because for $\gamma > 3$, the photon occupation number in phase space $N(\nu) = i(\nu)c^2/2h\nu^3$, a Lorentz invariant scalar, increases with frequency. In the moving frame, photons of a particular frequency coming from the forward direction have been boosted from lower energy where the occupation number is lower, so there are fewer of them than from the backwards direction. This "Compton-Getting effect" is a familiar one; it has been proposed, for example, as an observational tool in searching for line features in the background radiation[24].

A runaway increase in the atom's velocity results if a spectral feature with $\gamma > 3$ occurs near an atomic resonance. We focus here on interactions of atoms with radiation via coherent resonance scattering[25,26]. The classical cross section for unpolarized scattering from an electronic oscillator with frequency ν_0 is given by

$$\sigma(\nu) = \sigma_T[\nu^4/(\nu_0^2 - \nu^2)^2] \tag{2}$$

where $\sigma_T \equiv 8\pi r_e^2/3$ is the Thomson cross section. Expanding this expression about the line center ν_0 yields the classical damping profile, $\sigma = \sigma_T/4\delta^2 = (2\pi/3)r_e^2/\delta^2$, where the classical electron radius $r_e \equiv e^2/m_e c^2 = 2.8 \times 10^{-13}$cm, and $\delta = (\nu - \nu_0)/\nu_0$ is the fractional displacement of the photon from line center. In each resonance scattering event a neutral atom coherently scatters a photon into a random direction, the directional probability distribution having a $1 + cos^2(\theta)$ dependence for unpolarized radiation. Because of the symmetry of the scattering, the scattered radiation carries on average no momentum, so that the atom absorbs a mean momentum $h\nu/c$ in the original photon direction (even though in the center of mass frame the photon energy is always exactly unchanged by the scattering.)

For simplicity we consider here only this phase-coherent elastic interaction in the damping wings of the atomic line. Other quantum absorption and re-emission processes do not preserve phase coherence or photon number, and hence are inelastic or dissipative, requiring a more comprehensive analysis. Since these effects are only important in the narrow Doppler core of the line, they do not affect the bulk of the photons driving the instability in the regime of interest, so the essential features are well described by this simplified model.

Computing the force on a single moving hydrogen atom is then similar to the calculation of Thomson drag between free electrons and the cosmic background radiation[27]— but with a γ-dependent sign. The momentum transfer a moving atom receives from scattering the background radiation is obtained by integrating the absorbed momentum flux from all directions. We assume that γ is constant over some narrow range of frequencies in the vicinity of the Lyman α resonance at frequency ν_α. If we consider only photons farther than δ to one side of line center we find that for $\delta < \gamma^{-1}$

$$\dot{\beta} = -\int d\Omega \int d\nu \sigma(\nu) i'(\nu, \beta, \theta) cos(\theta)/mc^2 = (8\pi^2/9)\beta r_e^2 \nu_\alpha i(\nu_\alpha)(\gamma - 3)/mc^2 \delta, \tag{3}$$

yielding an exponentially growing velocity $\beta \propto e^{\omega t}$ with growth rate

$$\omega(\delta) = \omega_0/\delta, \tag{4a}$$

where

$$\omega_0 \equiv (8\pi^2/9)r_e^2 \nu_\alpha i(\nu_\alpha)(\gamma - 3)/mc^2 \tag{4b}$$

and m denotes the atomic mass. Thus, free energy in the form of a $\gamma > 3$ spectral distortion near Lyα tends to get channeled at this rate, for photons at fractional offset of order δ, into amplifying motions of matter.

3.2 SATURATION

If we turn our attention to the development of the instability in an extended medium (as opposed to the acceleration of a single atom) then it becomes important to keep track of the backreaction on the radiation field as the matter moves around. The spectral feature is modified by the motions of the matter. We do not attempt here to keep track of the detailed spatial statistical properties of the motions and the radiation reaction, but will describe the approximate behavior of the integral quantities—the average radiation spectrum and resulting rms matter velocities—with an argument based on energy balance.

The complement of the above instability in the matter is that photons change their energy slightly when they scatter off of atoms, leading to a steady degradation of the spectral feature. Consider a photon scattering off of an atom with velocity $\vec{\beta}$. If $\nu\vec{\Omega}$ and $\nu'\vec{\Omega}'$ are the pre-and post-scattering photon vectors, the energy change in a scattering is[28]

$$\frac{\nu'}{\nu} = \frac{1 - \vec{\beta}\vec{\Omega}}{1 - \vec{\beta}\vec{\Omega}' + (h\nu/\gamma m_e c^2)(1 - \vec{\Omega}\vec{\Omega}')} \tag{5}$$

where $\gamma = (1-\beta^2)^{-1/2}$. For very small $\beta < h\nu/mc^2$ the photon losses are dominated by Compton recoil. However, for larger β we have

$$\Delta\nu/\nu \simeq \vec{\beta}(\vec{\Omega}' - \vec{\Omega}), \tag{6}$$

the same energy shift one would find off of a moving mirror. Because of the symmetry of the scattering direction, to first order in β a scattered photon has on average the same energy as the incident one. However, individual scattering events can lead to energy gain or loss, depending on the direction, opening a channel for photons to change energy. For $\gamma > 3$, most of the scatterings at a given ν in the center of mass frame involve loss because of the preponderance of higher-energy photons coming from behind. Photon energy goes into matter motions until the spectrum is flattened to $\gamma = 3$ and the occupation number is the same for all photons.

The velocities resulting from the instability can be estimated from the available photon energy, as the acceleration continues according to equation (3) until the photon spectrum is changed. For photons with given offset δ, the photon energy losses must balance the kinetic energy gains of the matter due to these same photons. Let $\langle\beta^2\rangle$ denote the mean square velocity of the atoms; then the rate of change of kinetic energy density of the matter $(1/2)nm\langle\beta^2\rangle$ (where n denotes the number density of atoms) approximately equals the energy extracted from the photons. Using equations (3) and (4) we find that the spectrum at offset δ changes at a rate

$$\frac{d(\nu i(\nu))}{dt} \simeq -nm\langle\beta^2\rangle\omega_0/\delta^2 \tag{7}$$

Thus, eventually the velocities grow to the point where the instability saturates because of the photon reaction. Photons with the smallest δ, although they initially contribute the fastest growth rate, saturate more quickly at smaller velocities, which is why they can be neglected in the final outcome.

In an initially static gas, the instability continues until all the photons are exhausted. For a spectral feature of fractional width ϵ, the fraction extracted from each photon is of the order of ϵ so the final saturated velocity is

$$\langle \beta^2 \rangle \simeq 4\pi\epsilon^2 \nu i(\nu)/nmc^3. \tag{8}$$

In an expanding universe the instability saturates differently from the static case because adiabatic cooling caused by the expansion competes for the photon energy. The photons are redshifted by the cosmic expansion at the same time that they lose their energy to the motion of matter via the instability, so the efficiency is reduced. In the case where $\omega_0 > H$, where H is the expansion rate, the expansion can be ignored altogether, reducing to the static limit. In general however, losses via the expansion dominate a photon's energy loss so only a small portion of the energy is available for creating peculiar motions. Within a narrow spectral band, $|\delta| \lesssim \delta_{crit} \equiv \omega_0/H$, the growth rate of the instability exceeds H, so these photons create a nearly exponential instability as they do in the static case. $\langle \beta^2 \rangle$ grows rapidly until the feedback (7) is important, at which point energy loss for $|\delta| < \delta_{crit}$ is dominated by the instability. Each photon ends up losing a fraction $\simeq \min[\epsilon, \delta_{crit}]$ of its energy to the instability, the rest of it redshifting away in the expansion. If $\epsilon > \delta_{crit}$ then final matter velocity is now instead of equation (8),

$$\langle \beta^2 \rangle \simeq 4\pi\delta_{crit}\epsilon\nu i(\nu)/nmc^3 = 4\pi(8\pi^2(\gamma-3)/9)\epsilon[\nu i(\nu)]^2 r_e^2/Hnm^2c^5. \tag{9}$$

Note that a narrower feature is more efficient; if $\epsilon < \delta_{crit}$ it is as efficient as the static case.

3.3 KOMPANEETS EQUATION

It is worth pursuing the photon feedback more precisely than this rough energy argument allows. If successive photon scatterings occur off of atoms with uncorrelated velocities, we can use the Boltzmann equation to describe the evolution of $N(\nu, t)$. It takes a form similar to the familiar Kompaneets equation for Compton scattering in an electron gas at temperature T[26-28],

$$\frac{\partial N}{\partial t} = \frac{\sigma_T n_e h}{m_e c \nu^2} \frac{\partial}{\partial \nu} \left\{ \nu^4 [N(N+1) + \frac{kT}{h}\frac{\partial N}{\partial \nu}] \right\} + H\nu\frac{\partial N}{\partial \nu}. \tag{10}$$

This equation describes the diffusion of photons in energy, and can be carried over more or less directly to the resonant scattering model with appropriate substitutions $\sigma_T, n_e, m_e \to \sigma(\nu), n_H, m$, and recognizing that the diffusion is determined only by the variance of the three-dimensional velocity distribution of the scatterers, $kT \to mc^2\langle\beta^2\rangle/3$. We also add a term in the standard way to incorporate the cosmic expansion, and drop the first term (describing recoil) which is negligible for scattering off of atoms since $\langle\beta^2\rangle >> h\nu/mc^2$, to obtain

$$\frac{\partial N}{\partial t} = \frac{n_H\sigma(\nu)c\langle\beta^2\rangle}{3\nu^2} \frac{\partial}{\partial \nu}\left[\nu^4\frac{\partial N}{\partial \nu}\right] + H\nu\frac{\partial N}{\partial \nu}. \tag{11}$$

Integration of equation (11) leads to a self-consistent solution for the evolution of the photon distribution function in the presence of randomly moving matter.

By also integrating over N (equation (3)) we get the equation of motion for the matter, and a complete solution is obtained. The energy argument we used above assumed that before N has changed very much, we can impose our initial power law $N(\nu) \propto \nu^{\gamma-3}$; plugging this into equation (11) we obtain

$$\frac{\partial \ln N}{\partial t} \simeq \left[\frac{n_H \sigma c}{3}\langle\beta^2\rangle\gamma + H\right](\gamma - 3), \qquad (12)$$

which we justified in equation (7) directly from energy balance. In the opposite limit of the saturated instability, which will apply very close to the resonance, we expect $N(\nu, t)$ to achieve a steady state with $\partial N/\partial t = 0$.

This way of handling the cosmic expansion makes statistical presumptions about the matter velocities; they are split into a microscopic, random part and a a correlated, smooth macroscopic expansion. The competition between the two effects is clearly delineated in the two terms. For it to be a good approximation, the peculiar motions $\langle\beta^2\rangle$ must be greatest on scales smaller than a photon mean free path. This is not the case here. The atomic velocities encountered in successive scatterings are not actually uncorrelated, since the mean free path of these photons is small even at δ_{crit}. The traditional splitting off of the correlated motions into a separate *uniform* expansion term is perilous and seemingly arbitrary, since we can have large-scale perturbations in the expansion which are optically thick enough to trap their photons but which are nevertheless much smaller than the horizon.

In the opposite limit of spatial trapping we can view the photons as a gas. They form an unusual gas however since as photons are redshifted they become more numerous; that is, if we look at only the Hubble term in equation (11) we find that in terms of the scale factor a characterizing the physical size of a volume,

$$\frac{\partial \ln N}{\partial \ln a} = \frac{\partial \ln N}{\partial \ln \nu}. \qquad (13)$$

For $\gamma > 3$ the effective photon pressure increases as a region expands. Photons contribute a negative compressibility to the atomic gas, with a growth rate corresponding to the (imaginary) sound frequency $i\omega = (\partial p_\gamma/\partial\rho)^{1/2}/\lambda$, so that $\omega(\lambda) \simeq ((\gamma - 3)\delta(\lambda)\nu i(\nu)/nmc)^{1/2}/\lambda$. This classical form of thermodynamic instability is how the instability manifests itself in the thick limit. For this limit to apply, the diffusion rate $t_d^{-1} = c/\lambda^2 n\sigma(\delta)$ for photons to escape from a region of size λ must be less than the growth rate for the instability on that scale, which depends on the spectrum of the trapped photons. The diffusion distance traveled in a Hubble time by photons at offset δ is about $\lambda_{diff} \simeq \delta^{3/2}(H/r_e^2 nc)^{1/2}(c/H)$, which is small enough that this limit is actually a reasonable approximation for describing the initial development of the instability. However, it sheds no light on the saturation.

Thus, although we know there is an instability, and that it can greatly amplify initial perturbations, we do not have a precise way to compute the maximum amplitude or the spatial power spectrum of the fluctuations. The Boltzmann/Kompaneets approximations are not accurate for solving the large scale back-reaction problem, and the fluid approximation offers no way to compute the effects on the photon spectrum because it excludes the crucial element of spatial diffusion. What is required, but not yet available, is a computation which incorporates a simultaneous description of the spatial and spectral properties of the radiation distribution function, together with the spatial statistics of the matter velocities. This approach will also be necessary to compute the spatial power spectrum of fluctuations from the saturated instability.

3.4 LARGE SCALE VELOCITIES

We can use the results of the simple energy argument given above to relate the spectral intensity of the radiation to the amplitude of induced matter velocities and hence to the creation of structure. For a minimum estimate of the radiation intensity needed to generate the observed structure, we will assume that the bulk of kinetic energy density from the instability occurs on the same comoving scales which dominate rms galaxy velocities today. We can then use linear perturbation theory to relate the initial gas velocities to the current galaxy velocities, skirting for the moment the theoretically and observationally contentious question of what the coherence scale of the velocities actually is.

Assume for definiteness a flat ($\Omega = 1$) universe with a fractional baryon density Ω_b, and with the rest of the density collisionless, pressureless, and invisible. In linear theory, an initial velocity β_i in the baryon component introduced as a linear perturbation at redshift z_i leads asymptotically (for $z_i >> \Omega_b^{-1}$) to gravitational perturbations with peculiar velocity in the same component today[31]

$$\beta_0 = (2/5)(1 + z_i)^{1/2}\Omega_b\beta_i. \tag{14}$$

(For typical parameters, the initial baryon velocity is comparable to the final baryon velocity.)

Combining this with equation (9) and scaling appropriately from z_i to the present, $[\epsilon\nu i(\nu)]_i = (1 + z_i)^4[\epsilon\nu i(\nu)]_0$, $n_i = (1 + z_i)^3 n_0$, and $H_i = (1 + z_i)^{3/2}H_0$, where $H_0 = 100h\text{km sec}^{-1}\text{Mpc}^{-1}$ and $8\pi Gnm/3 = \Omega_b H^2$, we find that the energy density of the present day spectral feature must be at least

$$\left[\frac{4\pi\epsilon\nu i(\nu)}{c}\right]_0$$
$$= 7 \times 10^{-4} \text{ eVcm}^{-3} \left(\frac{\epsilon_i h^5}{\gamma - 3}\right)^{1/2} \left(\frac{\Omega_b h^2}{0.025}\right)^{-1/2} \left(\frac{\beta_0}{2 \times 10^{-3}}\right) \left[\frac{(1 + z_i)}{1000}\right]^{-9/4} \tag{15}$$

in order to generate present day peculiar motions of the order of β_0. We have adopted for normalization the value of Ω_b preferred from standard nucleosynthesis. The assumed velocity $\beta_0 c = 600\text{km sec}^{-1}$ is typical of observational estimates of rms velocities from data on large scale flows, galaxy reshift catalogs, and the peculiar motion of our own galaxy[32]. For comparison in these units, the microwave background energy density is $aT_{bb}^4 = 0.25\text{eVcm}^{-3}$ and the critical density is $10^4 h^2\text{eVcm}^{-3}$. One can verify that for a field of an intensity sufficient to generate final velocities β as large as this, δ_{crit} exceeds the Doppler width of the line core by a large factor, so our approximations, in particular our use of the classical damping profile throughout, are self-consistent. Roughly, $\delta_{crit} \simeq \beta(r_e^2 nc/H\epsilon)^{1/2}$, where the expression in parentheses is of the order of unity after recombination.

One should note that the same effects can in principle occur before hydrogen recombination, when the same instability can affect neutral or singly-ionized helium. In this case Thomson drag of free electrons on the blackbody radiation becomes another competing effect. Although this is not necessarily a fatal problem (as long as $aT_{bb}^4 \lesssim \nu i(\nu)/\delta_{crit}c$), in this paper we will not consider this possibility at length.

3.5 BACKGROUND RADIATION DISTORTIONS

Neutral atomic gas after recombination has almost no coupling to the microwave background; in the low frequency limit of Rayleigh scattering, $\sigma \propto (\nu/\nu_0)^4$ and after recombination $kT_{cbr} \lesssim h\nu_\alpha/40$. This justifies our neglect of radiation drag in the above analysis. The minimal interaction of neutral matter with blackbody photons allows fluctuations to be generated with unusually small microwave spectral distortions and anisotropy.

In the absence of dust obscuration, the background spectral feature which caused the instability should appear today in the background radiation spectrum. For interactions with the neutral hydrogen Lyman resonance after recombination, the feature with $\gamma > 3$ should appear redshifted in the mid to far infrared, shortwards of $0.12(1 + z)\mu m \approx 120\mu m$. If the instability is responsible for galaxy formation and clustering, the minimum radiation energy density is given in terms of model parameters by equation (15), and is generally of the order of $4\pi[\epsilon\nu i(\nu)]_0/c \simeq 10^{-3}aT_{bb}^4$.

It is possible that such a feature in cosmic background radiation could be detected by the DIRBE experiment on the Cosmic Background Explorer (COBE) satellite[33], which has an instrumental sensitivity of the order of $4\pi[\nu i(\nu)]_0/c \simeq 10^{-3}aT_{bb}^4$. Although this is a spectral region where features with $4\pi\nu i(\nu)/c \lesssim 0.1aT_{bb}^4$ may be very difficult to detect behind the Galactic and zodiacal emission, the isotropy and unusual spectral properties of this background may make it detectable for some choices of the model parameters. It is also possible that early star formation would create enough dust to absorb this radiation before it redshifts out of the ultraviolet, which would degrade the energy to what is now the submillimeter portion of the spectrum. In this case its spectral signature, though possibly detectable by the FIRAS instrument on COBE[15], would be indistinguishable from many other possible sources, including the stars themselves.

Although the resonance scattering instability utilizes some of the same ingredients as line radiation driven instabilities in stellar winds[34-36], it works entirely differently. For example, in the cosmic system we have a statistically homogeneous and isotropic system of gas and radiation, rather than a spherically symmetric stellar wind radiating net energy flux into empty space, and a finite fixed supply of resonant line photons, without constant replenishment from a central star.

It also offers an interesting contrast to mock gravity. Although both instabilities are powered by radiation pressure, the resonance coupling harnesses the energy of the photons much more efficiently; in mock gravity photons are discarded after one interaction, so only their momentum is absorbed. As a result, the intensity of the associated background distortions are much smaller for scattering. Dust continuum opacity is now not needed, since neutral atoms provide the key coupling by their resonance scattering. This results in a much shorter wavelength distortion than in mock gravity. Finally, the resonance scattering instability efficiently amplifies large-scale, small-amplitude modes, and saturates at a certain amplitude determined by the photon spectrum; thus it naturally produces perturbations similar to those which have been used for years as building blocks of the standard empirically motivated galaxy clustering models[37]. This feature is unusual in natural instabilities, which generally tend to saturate only upon reaching a highly clustered, nonlinear spatial structure.

I am grateful for useful discussions with M. Fukugita, H. Sato, and M. J. Rees. This work was supported by NASA grant NAGW-1703 and by an Alfred P. Sloan Foundation Fellowship.

4. References

1. Peebles, P. J. E. and Silk, J., *Nature* **346**, 233 (1990)
2. Frenk, C.S., in *Nobel Symposium No. 79: The Birth and Early Evolution of the Universe*, ed. B. Skagerstam (in press) (1991)
3. Ryden, B.S., Press, W. H. and Spergel, D. N., *Astrophys. J.* **357**, 293 (1990)
4. Turok, N., *Phys. Rev. Lett.* **63**, 2625 (1989)
5. Ostriker, J. P., Thompson, C., and Witten, E., *Physics Letters* **B180**, 231 (1986)
6. Spitzer, L., *Astrophys. J.* **94**, 232 (1941)
7. Gamow, G., *Rev. Mod. Phys.* **21**, 367 (1949)
8. Field, G. B., *Astrophys. J.* **165**, 29 (1971)
9. Hogan, C. J., and White, S. D. M., *Nature* **321**, 575 (1986)
10. Hogan, C. J., *Astrophys. J.* **340**, 1 (1989)
11. Hogan, C. J., *Nature* **338**, 132 (1989)
12. Hogan, C. J., and Woods, J., *Phys. Rev. Lett.*, submitted (1990)
13. Hogan, C. J., *Astrophys. J.*, in press (1990)
14. Wang, B., and Field, G. B., *Astrophys. J.* **346**, 3. (1989)
15. Mather, J. *et al.*, *Astrophys. J.* **354**, L37 (1990)
16. Bond, J. R., Carr, B. J. and Hogan, C. J., *Astrophys. J.*, in press (1991)
17. Hogan, C. J., *Nature*, submitted (1991)
18. Peebles, P. J. E., *Astrophys. J.* **153**, 1 (1968)
19. Krolik, J., *Astrophys. J.* **338**, 594 (1989)
20. Krolik, J., *Astrophys. J.* **353**, 21 (1990)
21. Lyubarsky, Y. E. and Sunyaev, R. A., *Astron. Astrophys.* **123**, 171 (1983)
22. Fukugita, M. and Kawasaki, M., *Astrophys. J.* **353**, 384 (1990)
23. Peebles, P.J.E. and Wilkinson, D.T., *Phys. Rev.* **174**, 2168 (1968)
24. de Bernardis, P., Masi, S., Melchiorri, F. and Melchiorri, F., *Astrophys. J.* **357**, 8 (1990)
25. Heitler, W., *The Quantum Theory of Radiation*, Oxford (1954)
26. Jackson, J. D. , *Classical Electrodynamics*, Wiley (1975)
27. Peebles, P. J. E., *Physical Cosmology*, Princeton (1971)
28. Sunyaev, R.A. and Zeldovich, Ya. B., *Ann. Rev. Astr. Ap.* **18**, 537 (1980)
29. Zeldovich, Ya. B. and Novikov, I.D., *Relativistic Astrophysics: Volume 2, The Structure and Evolution of the Universe*, (1983)University of Chicago Press
30. Rybicki, G. B. and Lightman, A. P., *Radiative Processes in Astrophysics*, Wiley (1979)
31. Hogan, C. J. and Kaiser, N., *Mon. Not. R. astr. Soc.* **237**, 31P-38P (1989)
32. Rubin, V.C. and Coyne, G.V.,eds., *Large Scale Motions in the Universe: A Vatican Study Week*, Princeton (1988)
33. Mather, J. *et al.*, *Astrophys. J.*, in press (1990)
34. Owocki, S. P., and Rybicki, G. B., *Astrophys. J.* **284**, 337-350 (1984)
35. Lucy, L. B., *Astrophys. J.* **284**, 351-356 (1984)
36. Owocki, S. P., and Rybicki, G. B., *Astrophys. J.* **299**, 265-276 (1985)
37. Peebles, P.J.E., *The Large Scale Structure of the Universe*, Princeton (1980)

GAS-INDUCED PRIMARY AND SECONDARY CMB ANISOTROPIES

J. Richard Bond and Steven T. Myers
CIAR Cosmology Program
Canadian Institute for Theoretical Astrophysics
University of Toronto, ON M5S 1A1, Canada

ABSTRACT *On scales less than a few degrees the primary CMB anisotropies (from linear processes occurring at photon decoupling) and the secondary CMB anisotropies (from nonlinear processes occurring later) arise mainly from gas-driven mechanisms such as Thomson scattering, inhomogeneous Compton cooling, and radiant emission by dust. We describe some of the current experiments which probe these anisotropies and the constraints on models that can be derived from them. We pay special attention to hierarchical Gaussian models of structure formation such as the $\Omega = 1$, $H_0 = 50$, adiabatic CDM model. For these theories, observations currently restrict the biasing factor defining the amplitude of the perturbations to be $b_\rho \gtrsim 0.7$ (for $\Omega_B \gtrsim 0.03$) assuming standard recombination. If there is early reionization, the best limits currently come from all-sky surveys probing the Sachs-Wolfe effect, giving $b_\rho \gtrsim 0.3$. The currently preferred value of the biasing factor (apart from the 'extra large scale power conundrum' of deep redshift surveys) is $b_\rho \sim 1 - 1.5$, so modest experimental sensitivity improvements will test these theories. We also present our methods for generating large catalogues of galaxy groups and clusters using a hierarchical Gaussian peaks formulation. These are used to generate simulated maps of secondary CMB anisotropies due to the Sunyaev-Zeldovich effect and the associated soft X-ray emission viewable with ROSAT. The non-Gaussian nature of the resulting image statistics and the ability to superpose various backgrounds and foregrounds are essential for optimally designing observational tests of these models. Both aspects are well-handled by the map synthesis technique. The prospects for ambient SZ observation depend strongly on the primordial amplitude of the perturbations but radio source confusion will make it difficult at Rayleigh-Jeans wavelengths. We also sketch how the hierarchical peaks method can be applied to the construction of primeval galaxy catalogues and display illustrative far infrared maps of dust-laden star-bursting galaxies at $z \sim 5$ that may be detectable with ground-based sub-mm telescopes.*

1. INTRODUCTION

Background radiations in all wavebands ranging from the X-ray to the radio provide invaluable windows on the universe at moderate to high redshift. The deepest probe is provided by the primordial cosmic microwave background (CMB). To avoid excessive distortion of the $T = 2.74K$ blackbody spectrum, most of the energy in it must have arisen prior to a redshift $z_P \sim 10^{6.4}$, when the energy injected could have been thermalized by free-free absorption and double Compton scattering. Energy released between z_P and $z_{BE} \sim 10^{4.6}$ gives rise to a Bose-Einstein distribution characterized by a nonzero photon chemical potential. From COBE's FIRAS and the University of British Columbia's COBRA experiments, we now know that at most 1% of the CMB energy could have been injected in this 'Bose-Einstein' era. Since the medium is still extremely optically thick to scattering at that time, anisotropies would be wiped out up to the time of photon decoupling. If energy is injected later than z_{BE}, the gas will Compton cool, yielding the characteristic Kompaneets distortion shape. FIRAS and COBRA limit this amount of energy to be 0.4% of that in the CMB. As well, photon energy injected at shorter wavelengths than the bulk of the CMB will be redshifted into a secondary background. If it is starlight from the first generation of stars, the redshifted waveband will be the near infrared, unless there is a sufficient dust cover between the sources and us that it gets absorbed and reradiated into a redshifted waveband in the far infrared to submillimetre. In addition, optical, UV and even higher energy backgrounds are expected from starlight and accretion-light generated at lower redshifts and from the gravitational binding energy released in the formation of cosmic entities. The sequence of energy releases will be characterized not only by monopole quantities such as the globally-averaged spectrum, but also by telltale angular patterns that might be used to probe the sort of objects that give rise to the backgrounds and their clustering properties. These backgrounds may be the only window we have to the Universe at moderate redshift.

1.1 Overview of Primary and Secondary Anisotropies

In this paper, we adopt the somewhat arbitrary split between primary and secondary CMB anisotropies given by Bond (1988) to describe these angular patterns. *Primary* anisotropies are those that were generated when the Universe was (predominantly) in the linear regime. They are invaluable as a direct probe of the amplitude and statistics of primordial perturbations without the complications engendered by the nonlinear and dissipative processes that characterize *secondary* anisotropies, which are all the rest. In particular, secondary ones arise once objects collapse and liberate energy. In many models of structure formation, the distinction between primary and secondary is quite blurred. For example, short distance scale structure can go nonlinear shortly after (or even before) photon decoupling and lead directly to anisotropies. This might be expected in cosmic string theories and in scenarios with a large amount of power on small scales such as isocurvature baryon models (§4.1).

When the photon transport equations are linearized, the terms which act as sources driving the development of primary anisotropies are: (1) the Sachs–Wolfe source, most important on large angles, (2) the photon 'bunching' source, important on intermediate angles and, for isocurvature perturbations, on large angles as well, and (3) the electron velocity source due to the Doppler effect in the Thomson scattering terms, an effect which is significantly diminished by the fuzziness of the last scattering surface (through destructive interference of opposing contributions to $\Delta T/T$ from the troughs and crests that straddle the photon decoupling surface). This damping is important for waves whose size is smaller than the width of the last scattering surface. There is a corresponding 'coherence' angle θ_c, which ranges from $\sim 2' - 10'$ (for normal recombination) to a few degrees (for models with early reionization). This scale θ_c gives the characteristic smoothness of the hills and valleys of the primary $\Delta T/T$ pattern. However, each of the three sources contribute to a rich anisotropy pattern that appears on all higher angular scales (Fig.2.1). The relative magnitude at any given scale depends upon the shape and statistical distribution of the primordial spectrum, upon the global geometry and upon the baryon and dark matter abundances.

One may hope that the statistics of the initial perturbations can be inferred from those of the primary anisotropies. For example, initially Gaussian density perturbations, the form most commonly assumed, breed Gaussian-distributed $\Delta T/T$ patterns. In many respects the most interesting anisotropies are those that arise only from the inhomogeneous rippling of our past light cone associated with metric perturbations. Above the angle subtended by the horizon at photon decoupling, which is a few degrees if recombination is normal (so decoupling occurs at $z \sim 1000$), we expect these Sachs-Wolfe anisotropies to dominate and give us a clean window through which to view the primordial geometry. Whether there is anything to see at large angles, however, depends very much upon the shape of the primordial power spectrum. If we can pin the amplitude of primordial perturbations to the structure we see, and if the fluctuations are initially scale-invariant as inflation theories most often predict, then these large-angle anisotropies should be visible with COBE's DMR. On the other hand, it is also possible that these very large scales, which are unrelated to any of the collapsed structure we see in redshift surveys of any sort, have no significant fluctuations at all, in which case we must be content with what we can learn from the scales over which gas dynamical processes can operate.

Although primary small-scale anisotropies are severely damped if there is early reionization (the photon-decoupling surface is up to Gpc thick rather than 10's of Mpc), quadratic nonlinearities in the Thomson scattering do not suffer dramatic destructive interference, since different wave-modes are coupled (Vishniac 1987). This effect leads to significant anisotropies in some theories such as the isocurvature baryon models (Efstathiou and Bond 1987). However, if the gas moves relative to the primordial density perturbations (nonlocal biasing), then the quadratic nonlinearity assumption has to be abandoned in favour of a more powerful treatment of Thomson scattering from the nonlinear clumps of ionized flowing gas.

Once matter begins to collapse, the gas can heat up through shocks thermalizing the released gravitational energy or through other heat sources such as thermonuclear or accretion energy. At high redshifts ($z \gtrsim 10$) the main cooling mechanism is Compton cooling. If the Universe is optically thick to Compton scattering at that time then we do not expect to see anisotropies. It is only after the Thomson optical depth falls below unity that a nonuniform pattern emerges to tell us of the state of the heated gas. In explosion and other models with significant energy release beyond the obligatory gravitational binding energy, this can occur early enough that the Sunyaev-Zeldovich (SZ) anisotropies that accompany Compton cooling distortions offer a window to the Universe at $z \sim 10$. In hierarchical models in which only binding energy release heats the gas to X-ray emitting temperatures, SZ anisotropies may tell us only about the $z \lesssim 1$ universe when pancakes, filaments, and clusters form in abundance (§4).

Within the collapsing gas, stars will also liberate potentially observable energy, and blanketing dust may shift the energy into a distortion of the CMB. This dust could be generated in the first stars that form at $z \sim 300 - 10$, or might not appear in significant abundance until galaxies form at $z \sim 20 - 3$ (the respective redshifts being very model-dependent). The anisotropy pattern would be quite different in the two cases. We also address this in §4.

A secondary background is characterized by a monopole, which gives the angle-averaged energy spectrum, by a dipole, largely induced by the motion of the earth relative to the rest frame of the radiation, which therefore should nearly coincide with the CMB dipole, and by higher multipoles which provide a rich storehouse of information on the origin of the background in question. For low multipoles, there will be a Sachs-Wolfe effect similar to that for primary anisotropies, but, given the current COBE constraints on the amplitude of the background, it is unlikely to be competitive with the primary multipoles. Where anisotropies generated by gas or dust processes are most interesting is for multipoles above about 50; *i.e.*, on angular scales below a few degrees.

1.2 Gaussian Fields Are Maximally Random

The initial conditions most often invoked for cosmic structure formation are linear perturbations distributed as a homogeneous and isotropic Gaussian random field. All that is required to fully specify the statistics is a single function, the (linear) density fluctuation spectrum, $d\sigma_\rho^2/d\ln k \equiv \frac{k^3}{2\pi^2} \langle |(\delta\rho/\rho)(k)|^2 \rangle$, giving the contribution of (statistically independent) modes of comoving wavenumber $\vec{k}$ to the *rms* linear density fluctuations σ_ρ. Although Gaussian perturbations generated by quantum noise during inflation has served as the standard model of the past decade, non-Gaussian initial perturbations may arise in inflation models with more than one scalar field, or in theories with phase transitions of various sorts.

For Gaussian models, the shape of $d\sigma_\rho^2/d\ln k$ prior to the onset of nonlinearity is determined by the linear evolution of the post-inflation spectrum. The overall normalization amplitude of the linear spectrum is parameterized by a biasing factor b_ρ (Bardeen *et al.* 1986), which is one if mass traces light and greater than one if galaxies are more clustered than the mass distribution.

Since one of the aims of this paper is to emphasize the difference in observing strategies that one would adopt to test non-Gaussian rather than Gaussian anisotropies, we first describe the fundamental characteristic of Gaussian fluctuations. Consider a general statistical distribution functional $\mathcal{P}[F(\vec{x})]\mathcal{D}F(\vec{x})$ giving the probability of a field configuration $F(\vec{x})$ of a random field F. Define the entropy of this probability to be

$$\text{Entropy}[\mathcal{P}] = - \int \mathcal{P}[F(\vec{x})] \ln \left(\mathcal{P}[F(\vec{x})] \right) \mathcal{D}F(\vec{x}).$$

Among all of the distributions with a specified spectrum $d\sigma_F^2/d\ln k$, the Gaussian one is the one which maximizes the entropy. This lemma is the basis of a picturesque way of thinking of non-Gaussian fluctuations. Fig. 2.1(b) displays typical 2D angular power spectra for primary anisotropies (for the CDM model of §3) and secondary anisotropies (for the SZ and dust models of §4). The statistics of the primary ones are Gaussian, and so the maps are maximally random distributions of the power available. The best observing strategy is then to concentrate the observing time on just a dozen or so patches of the sky because you are bound to hit something (see Readhead *et al.* 1989, Appendix A, for a quantification of this point). The secondary ones are non-Gaussian and the power in the maps (*e.g.*, Fig. 4.1, 4.3) is more concentrated around the 'hot' or 'cold' spots than in the Gaussian case. In that case, a better observing strategy is to sample many patches at lower sensitivity to look for the regions of high power concentration.

§2 reviews the experimental status, §3 reviews the current constraints on the scale invariant models, in particular the CDM model, from primary anisotropies, and §4 presents new maps of the X-ray background and SZ anisotropies simulated using a hierarchical peaks approach to identify groups and clusters of galaxies. Application to far-infrared emission from bursting primeval galaxies is also sketched. The non-Gaussian maps of secondary anisotropies for the SZ and dust models shown in Figures 4.1 and 4.3 should be compared with the Gaussian maps of primary CDM and primary isocurvature baryon anisotropies displayed in Bond and Efstathiou (1987) and Bond (1990). We also show the extent to which the OVRO small angle experiments are likely to suffer contamination from secondary anisotropies and find that faint radio sources completely dominate the secondary signal.

In this paper, we let $h \equiv H_0/100 \, \mathrm{km} \, \mathrm{s}^{-1} \, \mathrm{Mpc}^{-1}$ and Ω_{nr} denote the energy density relative to closure in (stable) non-relativistic particles (e.g., CDM and baryons). Inflation implies that Ω is nearly unity, while the Hubble parameter must be in the 40 to 50 range to avoid a time crisis for globular clusters and nuclear cosmochronology, and Ω_B must be $\lesssim 0.06(2h)^{-2}$ to maintain the successes of primordial nucleosynthesis. We use the minimal CDM model ($\Omega = \Omega_{nr} = 1$, h=0.5, adiabatic scale invariant perturbations) to provide our standard illustration of structure in the CMB. The main free parameter is b_ρ. Currently it is thought to lie in the range 1-2.6, with the 1-1.5 range now preferred over the higher ~ 2 value. However, for many of the calculations in this paper, the defining features depend only upon a relatively limited band of k-space. For example, for the group and cluster catalogues of §4, the important waveband is from $\sim 2 \, h^{-1} \mathrm{Mpc}$ to $\sim 8 \, h^{-1} \mathrm{Mpc}$. We can think of b_ρ as a way of characterizing the primordial amplitude of the perturbations in this waveband. The role of the wavelengths longward of this band is to set the large scale clustering of the objects. To be sure, these correlations may be observable in the background radiations we discuss, but it is the nature and abundance of the objects themselves which in the first instance characterize the radiation pattern. We therefore believe that, although we used the CDM model in §4, our results are robust for a more general class of Gaussian hierarchical clustering theories.

2. ANISOTROPY EXPERIMENTS

It is customary to categorize observations as very small ($< 1'$), small ($1'$-$30'$), intermediate ($30'$-$2°$), or large scale ($> 2°$). Large angle measurements do not probe gas dynamical processes very well, but are superb for probing primordial metric fluctuations, thereby providing a direct window on the early universe. They can be either space-based, such as RELICT 1 (Klypin et al. 1987) and COBE's DMR (Smoot et al. 1991), balloon-borne, such as those of Boughn et al. (1990) and Meyer, Cheng and Page (1990), or ground-based, such as the Tenerife experiment (Davies et al. 1987) and the interferometric observations of Timbie and Wilkinson (1988).

Currently, the highest sensitivity intermediate angle experiment is that of Meinhold and Lubin (1990), who took advantage of the superior observing conditions at the South Pole. The data was taken at a wavelength of $\lambda = 3.3$mm in a region at $\alpha = 21^h.5$, $\delta = -73°$ where galactic dust emission as seen by IRAS at $\lambda = 100\mu$ was found to be low, with an expected contribution of 18-36 μK rms. Nine fields were linked together in a strip from $\alpha = 20^h.52$-$22^h.48$. Each measurement consisted of a single difference of half cycles of a $1°4$ sinusoidal chop with a 30' beam (FWHM). A mean level and gradient was removed from each observing cycle through the fields. A total of 43 hours of useful integration time was obtained. For uncorrelated Gaussian fluctuations, the 95% (Bayesian) credible limit is $\delta T/T < 3.4 \times 10^{-5}$. There are a number of other intermediate angle experiments now underway, using balloons, the South Pole, and other extremely cold places such as Saskatchewan, Canada.

Currently, the most sensitive small angle experiment is that of Readhead et al.(1989), who used the 40-m radiotelescope at the Owens Valley Radio Observatory (OVRO) at a frequency of 20 GHz. A "double-differencing" switching scheme was used with Gaussian beams (1.8 FWHM) separated by 7.15, for a set of 8 fields near the North Celestial Pole (NCP). The 8 fields were observed for a total of 398 hours; one field was discarded from the subsequent analysis due to the presence of a faint but significant nonthermal extragalactic radio source in the beam. A 95% credible limit of $\delta T/T < 1.7 \times 10^{-5}$ between single uncorrelated beams was derived for the 7 remaining fields.

The Readhead *et al.* experiment pushed for maximum sensitivity to Gaussian fluctuations by observing a relatively small number of fields with low individual noise levels (31 μK 1σ). However, if the anisotropies are non-Gaussian, then the impressive limits we find in §3 for the OVRO 8-field and South Pole 9-field experiments do not apply. To constrain such models, a large statistical sample of fields is required. However, the penalty is that each field will have a higher noise level for the same amount of observing time. An example is the RING experiment, a survey of 96 fields undertaken with the OVRO 20 GHz system (Myers 1990, Myers, Readhead, and Lawrence 1991). The 96 fields are arranged in a ring around the NCP at a declination of $88°10'42''$. Double switching is performed as in the 8-field experiment, and the geometry is such that the reference fields in the switching cycle are the main fields for adjacent observations — thus the RING is interlocked, which provides extra checks on possible systematic errors in the radiometry. With an average 'error bar' of 113μK (1σ), the expected 95% upper limit for 96 fields is $\delta T/T < 1.8 \times 10^{-5}$, nearly as good as was seen in the 8-field experiment; however, an increased variance in the data points is found above the level expected for the noise alone and an upper limit of $\delta T/T < 5.4 \times 10^{-5}$ between uncorrelated single beams is found. Unfortunately, the enhancement of the signal from non-Gaussian fluctuations afforded by the enlarged area coverage also increases the contamination from discrete extragalactic radio sources. We discuss the effect of this 'secondary background' on the RING and 8-field experiments in §4.3. A new Owens Valley experiment with a 7.5 FWHM beam and 24.5 separation is also underway, using a dedicated 5-metre diameter radiotelescope at 32 GHz, which is expected to probe fluctuations below the 10^{-5} level.

On even smaller angular scales, at or below 1', the VLA has been used to probe fluctuations at 5 GHz (Martin and Partridge 1988 and Fomalont *et al.* 1988) and at 15 GHz (Hogan and Partridge 1989). Because of the lower observing frequency, contamination from discrete extragalactic radio sources within the imaged field has plagued the 5 GHz efforts. After careful subtraction of the contribution of these objects, 95% limits of $\delta T/T < 6 \times 10^{-5}$ on the scale of 1' and $\delta T/T < 1.2 \times 10^{-4}$ on 18'' have been reached. The 15 GHz limits are somewhat worse due to poorer system performance of the VLA at this frequency. New observations at 8.5 GHz using new low noise receivers have produced limits competitive with the filled-aperture experiments: preliminary 95% limits quoted by Partridge (1991) give $\delta T/T < (1-2) \times 10^{-5}$ on the scale of 1' and $\delta T/T < 6 \times 10^{-5}$ on 10''.

There has also been recent interest in higher frequency experiments at very small scales using sub-millimetre telescopes. Kreysa and Chini (1988) used the IRAM Pic de Valeta dish with a resolution of 11'' FWHM in a double-differencing three-beam experiment with a throw angle of 30'' to search, at $\lambda = 1300\mu$, for emission from quasars that had been detected by IRAS at 100μ. Most of their fields gave no quasar signal, and could be treated as blank sky. The 95% credible limit for 21 of their 25 fields is $\delta T/T < 3.2 \times 10^{-4}$. They argued that 4 fields could be rejected since it was assumed the quasars were the sources. The JCMT experiment of Church *et al.* (1990) was a three beam experiment at 800μ, with an 18'' resolution and a throw angle of 40'', which gave a 95% credible limit of 1.4×10^{-3}. In spite of the poorer sensitivity, this JCMT result gives competitive limits on dust-generated anisotropy because the wavelength is shorter than IRAM (§4.4). SCUBA (Cunningham and Gear 1990) is an array instrument to be used on the 15m JCMT in a few years which will give diffraction-limited performance for each pixel at 855μ (14'' FWHM, with 37 pixels), and

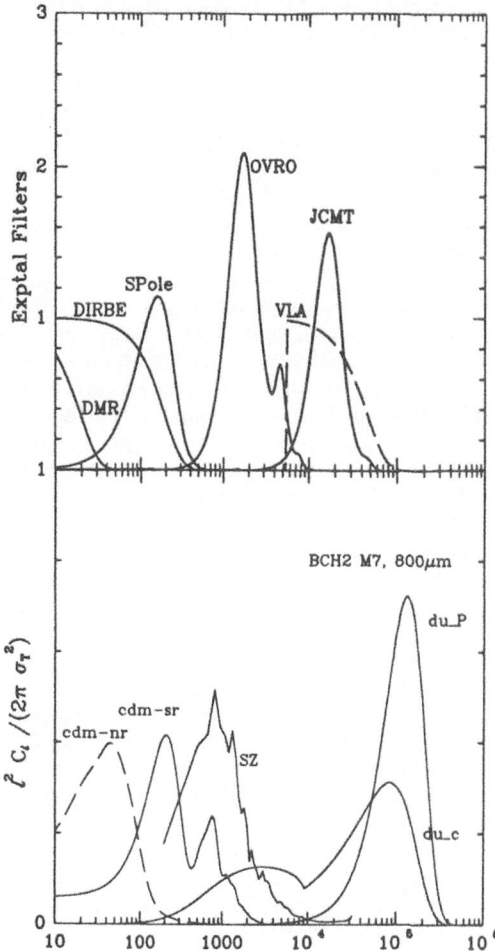

FIGURE 2.1 (a) *Experimental filters as a function of the multipole ℓ. When they are multiplied by the angular power spectra shown in (b) and integrated over* ln ℓ, *they give the variance in the anisotropies expected in each field of the relevant experiment. We can roughly consider* ℓ *to be in inverse radians, so the* ℓ*-pole probes angles around* 3438/ℓ *arcminutes. The COBE FIRAS and DMR filters and the RELICT filter are the same,* 7°. *The DIRBE filter is a single beam one with a* 42' *beam. OVRO is Owens Valley with a* 1.8' *beam and* 7.15' *throw. The RING experiment has the same filter. JCMT is the Church et al. (1990) experiment, good for probing dust emission anisotropies. VLA roughly corresponds to the VLA in D-mode at* 8.5 *GHz. The South Pole filter is that for the Meinhold and Lubin (1990) experiment. (b) Angular power spectra* ℓ²Cℓ/(2π) *are plotted for primary CDM anisotropies with standard recombination (cdm-sr) and with early reionization (cdm-nr), for the SZ effect from groups and clusters (as derived from a* 4° × 4° *map similar to Fig.4.1a), and for a dust model of primeval galaxies emitting in the sub-mm at* z ∼ 5 *(showing separately the Poissonian piece, du_P, and the continuous clustering piece, du_c). The normalizations* σ_T *are given in the text.*

438 μ (7″ FWHM, with 91 pixels). Although it is similar to the Church *et al.* experiment in resolution and configuration, it promises a vastly improved sensitivity per unit integration time.

The ability of these experiments to probe various theories of anisotropy can be illustrated by comparing the multipole band that the experiments are sensitive to with the angular power spectra of the theories, C_ℓ, as we do in Figure 2.1. If we expand the microwave background pattern in multipole moments, $\Delta T/T(\hat{q}) = \sum_{\ell m} a_{\ell m} Y_{\ell m}(\hat{q})$, where $\hat{q}$ is the angular direction of the incoming photons, then $C_\ell = \langle |a_{\ell m}|^2 \rangle$. For primary anisotropies and Gaussian initial perturbations, only C_ℓ is needed to fully specify a realization of the anisotropy pattern. For secondary anisotropies, which are generically non-Gaussian, an infinite number of higher order (reduced) correlation functions of $a_{\ell m}$ is needed to realize the patterns. The multipole band can be characterized by a filter function W_ℓ such that the *rms* fluctuations in a given experimental field is $(\Delta T/T)^2_{rms} = \sum_\ell (2\ell+1) W_\ell C_\ell / (4\pi)$.

The angular spectra shown in Fig.2.1b are for models described in §3 and §4. We refer the reader to these sections for details. We first make some general comments. The South Pole experiment has a multipole range ideal for optimizing the signal from power spectra like cdm-sr but a filter with a beam about 3 or 4 times larger is better for probing cdm-nr. OVRO filters are good for SZ, but smaller scale ones are not unless one is looking at specific clusters, and primeval galaxy backgrounds from dust are best probed with arcsecond resolution: the 18″ is almost too large (§4.4).

The quantities actually plotted in Fig.2.1b are $\ell^2 C_\ell / (2\pi \sigma_T^2)$, the power per log of wavenumber, normalized to have unit area under the curves. The normalization factor σ_T gives the total *rms* fluctuations for the model in question. The 'cdm-sr' primary anisotropies (for the $\Omega = \Omega_{nr} = 1$, $\Omega_B = 0.1$, h=0.5 CDM model with standard recombination) have $b_\rho \sigma_T = 3.5 \times 10^{-5}$; with no recombination (or very early reionization), $b_\rho \sigma_T = 2.5 \times 10^{-5}$. For the ambient SZ anisotropies, the power spectrum was computed by Fourier transforming a $4° \times 4°$ map similar to Fig. 4.1a except unfiltered. For the waves within the box, $\sigma_T = 6 \times 10^{-6}$ (at radio frequencies), significantly smaller than the primary ones.

The angular power from dust-emission in the pregalactic and protogalactic medium is concentrated at sub-arcsecond scales, corresponding to scales just above the size of the sources. The discreteness of the objects leads to a Poissonian 'white noise' power spectrum (du_P) just above these scales. To illustrate this we have chosen a specific model that we use extensively in §4.4 which has primeval galaxies radiating at $z \sim 5$ with comoving number density n_{G*} that are clustered as well. At 800μm, the normalization in $\Delta T/T$ is $\sigma_T = 8.5 \times 10^{-3} (0.02(\text{h}^{-1}\text{Mpc})^{-3}/n_{G*})^{1/2}$, which scales with the abundance in the classic $1/\sqrt{N}$ Poissonian fashion. Thus, if the source density is about that of bright galaxies now, $\sim 0.02(\text{h}^{-1}\text{Mpc})^{-3}$, then the fluctuation power is huge, and this remains so even when one filters with the JCMT beam. On the other hand, if the objects are dwarf galaxies with a density 100 times greater, seeing this discreteness noise becomes difficult. The dust power extending to smaller ℓ is the contribution from galaxy clustering. For the case shown here, with nonlinear clustering following a $\xi \sim r^{-1.8}$ correlation function attached to a larger scale linear clustering and biasing contribution (the two parts of du_c), the normalization is $\sigma_T = 8.8 \times 10^{-4}$ at 800μm. This would dominate the anisotropy signal in COBE's DIRBE experiment for example.

3. CONSTRAINTS FROM PRIMARY ANISOTROPIES

In this section, we review the current state of the limits on theories of structure formation assuming Gaussian initial conditions such as the CDM model. At the moment the constraints from large angle experiments such as COBE's DMR and the earlier Soviet

RELICT 1 satellite experiment and from recent balloon experiments (Boughn et al. 1990, Meyer et al. 1990) do not restrict inflation theories with normal recombination as much as intermediate and small angle experiments do. For scale invariant $\Omega = \Omega_{nr} = 1$ theories, the large-angle 95% confidence limit on the amplitude parameter is $b_\rho \gtrsim 0.2 - 0.4$, with the range depending upon the specific observation and upon such parameters as Ω_B, but not very much upon whether there was early reionization or not.

Intermediate and small angle constraints on scale invariant theories were derived by Bond, Efstathiou, Lubin and Meinhold (1991, hereafter BELM) who compared the South Pole and OVRO data with the theoretical predictions calculated in linear perturbation theory using the methods of Bond and Efstathiou (1987) and Efstathiou and Bond (1986, 1987). The Bayesian analysis procedure used by BELM has three great advantages: Firstly, for Gaussian anisotropies the integral over the theoretical amplitudes can be performed analytically, which allows a large number of theories to be constrained with very little computer time. This contrasts with a frequentist-based Monte Carlo approach used by Vittorio et al. (1991) who also analyzed the Meinhold and Lubin data (and obtained results in quantitative agreement with BELM). Secondly, one can incorporate properly modelled systematic effects such as residual gradients from atmospheric effects, which had to be removed from the Meinhold and Lubin signal. Thirdly, one can combine data sets to place more stringent constraints than the individual experiments give by themselves.

In the following table, we list the 95% credible limits derived by BELM on b_ρ for $\Omega = \Omega_{nr} = 1$, h=0.5, CDM models as a function of Ω_B, for OVRO data alone, the Meinhold and Lubin South Pole data alone, and for both combined. Here, SR denotes standard recombination and NR denotes no recombination, which is a limiting case of early reionization.

	Ω_B	SPole+OVRO	SPole	OVRO
SR	0.01	0.52	0.40	0.28
SR	0.03	0.66	0.45	0.43
SR	0.1	0.95	0.63	0.63
NR	0.1	0.27	0.27	0.06
SR	0.2	1.3	0.93	0.75

We therefore conclude that the experiments are very near to the predicted levels of the SR models, but are still far off the NR models, for which the large angle limits are currently better. This situation will change once the experiments with 1° to 2° beams begin to deliver their data. When we analyzed the 96-field RING data using the Bayesian approach, we found b_ρ limits significantly lower than those obtained for the South Pole and OVRO 7-field observations, due to the larger error bars and residual signal. As we discuss in §4.3, the central issue of source contamination must be well-modelled for the RING to improve the limit.

The angular CMB power spectrum for a massive neutrino-dominated universe is quite similar in shape to that of the CDM model but the overall amplitude must be higher to ensure that enough structure forms. Indeed, the relation between the redshift z_{nl} at which linear theory predicts that the rms fluctuations in the density first reach unity is related to b_ρ by $1 + z_{nl} \approx 1.1/b_\rho$. Thus, we must have $z_{nl} \lesssim 0.7$, which makes it very unlikely that these hot dark matter models will work without some fix such as an isocurvature component from strings, small mass seeds, or whatever.

There are indications that the CDM model does not have enough large scale power in the density fluctuations to explain the observed clustering. Bardeen, Bond and Efstathiou

(1987) found that modifying the contents of the universe so that linear evolution of an initially scale invariant power spectrum would lead to more large scale power than the CDM model has has a price: it also invariably leads to higher amplitude small and intermediate angle anisotropies, and therefore even tighter constraints on b_ρ than those in the Table. Breaking initial scale invariance to get more large scale power also gives larger anisotropies and tighter b_ρ limits. The paradox of strong large scale clustering in a universe with low CMB anisotropy is growing. Fixes such as early reionization will only delay the day of reckoning a bit since experiments are being undertaken which optimally probe NR models. Other fixes such as non-local biasing of light compared with the mass distribution may solve the clustering problem, but cannot give CMB isotropy.

4. SECONDARY ANISOTROPIES FROM NONLINEAR STRUCTURE

Current constraints on secondary radiation backgrounds, even those from COBE, still allow reasonably large energy densities Ω_{RT} (in units of the critical energy density). Expected thermonuclear, accretion and other sources must give $\Omega_{RT} \gtrsim 10^{-8} h^{-2}$ and would quite plausibly give Ω_{RT} around $10^{-6} h^{-2}$. (See Bond, Carr and Hogan (1986, 1991), hereafter BCH1 and BCH2.) For comparison, the total CMB energy is $\Omega_{cmb} \approx 25 \times 10^{-6} h^{-2}$. The most popular models for spectral distortions of the CMB are emission from primeval dust, which FIRAS constrains to be $\Omega_{RT} h^2 \lesssim 0.6 \times 10^{-6}$ over the band $500 - 10^4 \mu$, and Compton cooling of hot gas, which COBE and COBRA constrain to be $\Omega_{RT} h^2 \lesssim 10^{-7}$ (scaling as $\bar{y}/10^{-3}$). COBE could achieve limits on $\Omega_{RT} h^2$ as low as $\sim 3 \times 10^{-8}$ for general distortions, and as low as 3×10^{-9} for Compton distortions. DIRBE could detect a background below 10^{-7} from the far to near infrared, if the large foregrounds can be properly subtracted. Thus, although we should expect that an observable background will be found, to unravel its nature the correlations in the background will be fundamental, and these will definitely be non-Gaussian.

The minimal extension of the linear theory of primary anisotropies of §3 to get a non-Gaussian background is obtained by going to second order perturbation theory. These quadratic nonlinearities can play a very important role in constraining models with early reionization such as the classic isocurvature baryon models of the 1970s (Peebles 1987, Efstathiou and Bond 1987). Although there is no well motivated particle physics model for the generation of these, open universes with initially Gaussian fluctuations in the baryon density with a power law spectrum $d\sigma^2_{n_B}/d\ln k \propto k^{3+n_s}$ are often assumed, with n_s treated as a parameter describing the local spectral shape. Large scale clustering data motivate values in the range $n_s = 0$ (Poisson seed model) to $n_s = -1$. The combination of the Meinhold–Lubin and OVRO data can be used to rule out much of Ω_B–n_s space for $n_s \lesssim 0$ (BELM); and $n_s = -3$, the scale invariant slope, is very strongly ruled out. However, there is little reason to suppose that isocurvature perturbations would be Gaussian and power law. Even if they are, the second order perturbation theory calculations may not describe the anisotropies from the heated matter, although for the constraints derived from the OVRO experiment to be invalidated fairly large scale gas transport at these early epochs is required. A general treatment of Thomson scattering by nonlinear bulk-flow currents would be required to address this, a subject not yet well explored.

The obvious way to solve nonlinear gas problems is with large scale hydrodynamical simulations of the pregalactic and intergalactic medium. Even with the greatly expanded computational power that is being applied to 3D hydrodynamic/N-body cosmology codes, we are still some distance from generating realistic radiation patterns because of resolution limitations. We expect that the future will bring truly large scale simulations many Gpc in depth which will be solved at all of the interesting sites in the volume by spatially adaptive hydro techniques. Until that day it is worthwhile to explore the parameter space using a variety of approximate methods.

4.1 Constructing Catalogues for Secondary Anisotropies

The most often used method for estimating average backgrounds utilizes the Press-Schechter (PS) mass function for Gaussian hierarchical theories. In spite of the fact that this has recently been put on a firmer theoretical foundation and shown to accord reasonably well with the mass function of groups found in N-body studies (*e.g.*, Bond, Cole, Efstathiou and Kaiser 1991), the PS mass function does not give information on fluctuations. What is required is an assignment scheme for a point process. One attempt within the basic PS framework is the 'brick' model used by Cole and Kaiser (1988) for calculating SZ from clusters. Schaeffer and Silk (1988) used the Schaeffer ansatz for the form of the entire nonlinear hierarchy of n-point functions of galaxies to estimate SZ anisotropies for clusters.

Bond (1988) advocated shot noise models consisting of a catalogue of objects (the shots) with assumed profiles for the gas and dust surrounding the shots. For a Gaussian hierarchical model such as the CDM scenario, the shots were taken to be density peaks of the initial Gaussian random field filtered on a variety of scales. In shot noise models, anisotropies are a result of two effects: the Poisson noise associated with the finite number of sources and the continuous clustering of the shots. For SZ anisotropies, groups and clusters associated with various filtering scales were used. Variants of the same method were used to estimate SZ anisotropies in explosion models as well, with the shots being defined by the centres of non-overlapping voids. Bond (1988) and BCH2 used shot noise models to estimate angular power spectra and *rms* anisotropies from dust emission.

These calculations can give the average distortion and the *rms* anisotropies as a function of beam size semi-analytically, but to get the complete statistical distribution in practice requires Monte Carlo simulations. The limited hydrodynamical (Ryu *et al.* 1990, Cen *et al.* 1990) and sticky-particle (*e.g.*, Thomas and Carlberg 1990, Klypin and Kates 1991) studies that have been performed to date have given maps from which the full statistical information could be drawn, except that the single 3D boxes used in the simulations were of relatively small volume. So far one cannot satisfy at the same time the competing needs to treat large scale clustering, to have high resolution in the collapsed objects, and to deal with the rare events that give the largest excursions in the anisotropy signals.

We now describe the current status of a program that we have undertaken to make realistic secondary background maps, applying a significant improvement in the Gaussian peaks approach to create simulated catalogues of objects for hierarchical Gaussian models of structure formation such as the CDM model.

The major obstacle in the Bond (1988) approach was the cloud-in-cloud problem of overcounting objects of smaller mass that would have merged into higher mass ones. We avoid this in the following: We find the (linear) density peaks on a hierarchy of filter scales $R_{f1}, ..., R_{fn}$ that are above some threshold f_v associated with complete collapse of at least the central neighbourhood of the peak. From each peak we then go out in sequential radial shells and determine the average overdensity within that radius. When this value has fallen below f_v, we let that define the radius r_{pk} of the peak. The threshold we chose, $f_v = 1.686$, is that value for which the shell of matter at that radius will have collapsed to a point in a spherical model. The mass at larger radii will still be infalling, while that at smaller radii will have passed through the centre, and so, presumably, will have virialized with the rest of the interior mass. Although there could be a sudden increase in overdensity at some shell at a radius $r > r_{pk}$, we expect this to be picked up as part of a peak on a larger filtering scale. The locally sphericalized profile can also be used to determine the binding energy per mass within the shell and thus the final virial velocity and temperature associated with the collapsed entity. The cloud-in-cloud problem is solved by having peaks of larger extent annihilate peaks of smaller extent.

Some progress can be made analytically. One can evaluate the average number of peaks per unit filtering radius which cross through the critical f_v contour (Bond 1989, Bond, Cole, Efstathiou and Kaiser 1991). Analytic determination of the peak mass and the peak annihilation in the way we do it here is not feasible since it depends on non-local properties of the field. Monte Carlo simulations are required, and, fortunately, these are feasible.

The physics defining the profiles about those peaks that survive in the catalogue is currently put in by hand, motivated by observed profiles and by the gross parameters such as binding energy found in the spherical model. Evolution in the peak population with redshift translates into evolution of the shots, their clustering and their profiles. Additional dynamics is included in the simulations by moving the peaks according to the Zeldovich approximation. To avoid the problems using this when there is too much nonlinear dynamics (caustics wash out), we use the displacement field smoothed over the Gaussian scale $R_f = H_0^{-1}\sigma_v/\sqrt{3}$, where σ_v is the current 3D *rms* velocity dispersion determined from linear theory; this choice is motivated by the Zeldovich filtering scale found by Bond and Couchman (1987).

A reasonable next step is to make better use of the initial conditions around each peak by doing local hydrodynamic collapses. This is not feasible for the number of peaks we have to deal with, but what is reasonable, but for the future, is to simulate generic local collapses and derive the peaks' properties from these, along the lines of Evrard's (1990) 3D hydro simulations of clusters.

4.2 SZ and X-ray Maps from Group and Cluster Catalogues

To illustrate our approach, we describe our construction of group and cluster catalogues for the CDM model with biasing factors $b_\rho = 1$, 1.4 and 1.7. We choose some maximum angle over which we wish to construct maps, here 5°. We then lay down an inverted pyramid of boxes which contains within it a 5° cone. To cover the region out to redshift 1.5, beyond which there are very few groups, eleven layers of boxes of size $200\,h^{-1}$Mpc are needed. For each box, we construct a linear density F_k in Fourier space. For each of 15 Gaussian filtering scales R_{fj}, we FFT the filtered $F_k \exp[-(kR_{fj})^2/2]$, forming $F(\vec{r}; R_{fj})$. If there is significant power beyond the fundamental mode in the box, we add by brute force a contribution to F from very long wavelengths. For the CDM model in boxes of this size, this was not necessary. We find the peaks of each $F(\vec{r}; R_{fj})$ realization. Upon completion of the peak-finding for all 15 filters, we then determine the radii r_{pk} subtended by the peaks in Lagrangian (initial condition) space. The radius r_{pk} is defined to be that r at which the (unfiltered) volume-averaged value $\bar{F}(< r)$ first drops below f_v. The mass associated with the peak is $M = (4\pi/3)\rho_{cr}\Omega_{nr}r_{pk}^3$, where ρ_{cr} is the comoving critical density. Our filters cover a mass range from $10^{13.5}$ up to 10^{16} M$_\odot$.

Assuming a sphericalized profile, we can also determine the binding energy per mass within r_{pk} (assuming as well that $\Omega = \Omega_{nr} = 1$, in which case E/M is independent of redshift, unless there is mixing (shell crossing through r_{pk}) or radiation loss or gain):

$$[E/M](< r_{pk}) = -\frac{1}{2}(Har_{pk})^2 S(r), \quad S(r) = r_{pk}^{-5}\int_0^{r_{pk}} \bar{F}(< r)\,dr^5; \qquad (4.1a)$$

$$\bar{n}_B = 4.2 \times 10^{-4} \Omega_B h^2 [f_v(1+z)]^3 \, \text{cm}^{-3}, \quad T_V = \frac{m_N(-2[E/M](<r_{pk}))}{3Y_T}. \quad (4.1b)$$

The average density of baryons within r_{pk} is given by eq.(4.1b), as is the temperature of the uncooled baryons T_V, assuming the dark matter within r_{pk} is relaxed to a hydrostatic equilibrium with a uniform virial velocity which the uncooled gas shares. Here, m_N is the nucleon mass and Y_T is the number of particles per baryon in the gas (1.69 if it is fully ionized.

We then determine the displacement vectors $s_i(\vec{r})$ by FFT of $s_i(\vec{k}) = -ik^{-1}\hat{k}F_k$ $\exp[-(kH_0^{-1}\sigma_v/\sqrt{3})^2/2]$, which we use to give the peak dynamics in the Zeldovich approximation.

We then move on to the next box.

After all boxes are completed, we trim the catalogue by having the peaks found using large filter radii annihilate those found using smaller filter radii that overlap with them. Among overlapping peaks of the same filter radius, the bigger ones annihilate the littler ones.

For the profile of the gas, we assume a β-model with $\beta = 2/3$ as observed from the X-ray profiles of clusters (e.g., Mushotsky 1988). This corresponds to a roughly isothermal equation of state. That is, we keep T_V uniform throughout the cluster, but we redistribute the uncooled baryons into the density profile $n_B = n_{Bc}/(1 + r^2/r_{core}^2)$ which we truncate at r_{pk}. For simplicity, at this stage in our program, we have taken a fixed core radius of 300 kpc, motivated by X-ray cluster observations. Hydrodynamical models are needed to find the true dependence of r_{core} on cluster mass and evolution.

We are just now in the process of calibrating our results by detailed comparison with group catalogues constructed from N-body simulations. Thus, although our approach seems to be physically reasonable, it may require further refinement. Another crucial issue is whether the theoretical cluster catalogues agree with observed cluster and group catalogues. Although it has been known since the beginnings of the biased CDM model that high biasing factors yield anemic clusters, and that this could serve as a powerful test of the model, it has been difficult to quantify because of uncertainties in whether peak and PS models are accurate, the small number of clusters found in N-body studies, and the somewhat sketchy observational data on clusters. In particular, there has been a strong suspicion that the optically determined velocity dispersions of clusters were often overestimated. The recent work of Edge et al. (1990) using X-ray catalogues seems to confirm this. Our current $b_\rho = 1$ and 1.4 catalogues give temperature distribution functions which bracket that of Edge et al.

Because clusters are rare events defined by the tail of a Gaussian probability distribution, it is not surprising that the mass function will be highly sensitive to even the modest changes in the biasing factor used here. The most dramatic way to illustrate this is to show what various experiments would see as a function of biasing factor for fixed values of the other physical parameters such as core radius and Ω_B. Figure 4.1a,b are $4° \times 4°$ SZ contour maps smeared with the 1.8' OVRO beam. The contours are negative, dropping by factors of two from the starting value of -5×10^{-6}. The value $\Omega_B = 0.05$ chosen for this map is the preferred number from primordial nucleosynthesis, but the amplitude scales with Ω_B. However, the appropriate Ω_B to use is an effective value which excludes those baryons in the groups and clusters that are locked into stars, and therefore it may vary with group mass.

For the $b_\rho = 1$ map, the average y-parameter is $\bar{y} = 1.2 \times 10^{-6}$. (Recall that $\overline{\Delta T/T} = -2\bar{y}$ in the Rayleigh Jeans regime of the CMB spectrum. Hereafter, we shall assume Rayleigh Jeans wavelengths in the results we quote.) The rms is $\Delta T/T_{rms} = 2.7 \times 10^{-6}$, a discouragingly small number when compared with the sensitivity of current experiments. However, the minimum value is -4.6×10^{-5}, and there are other large excursions from the

318

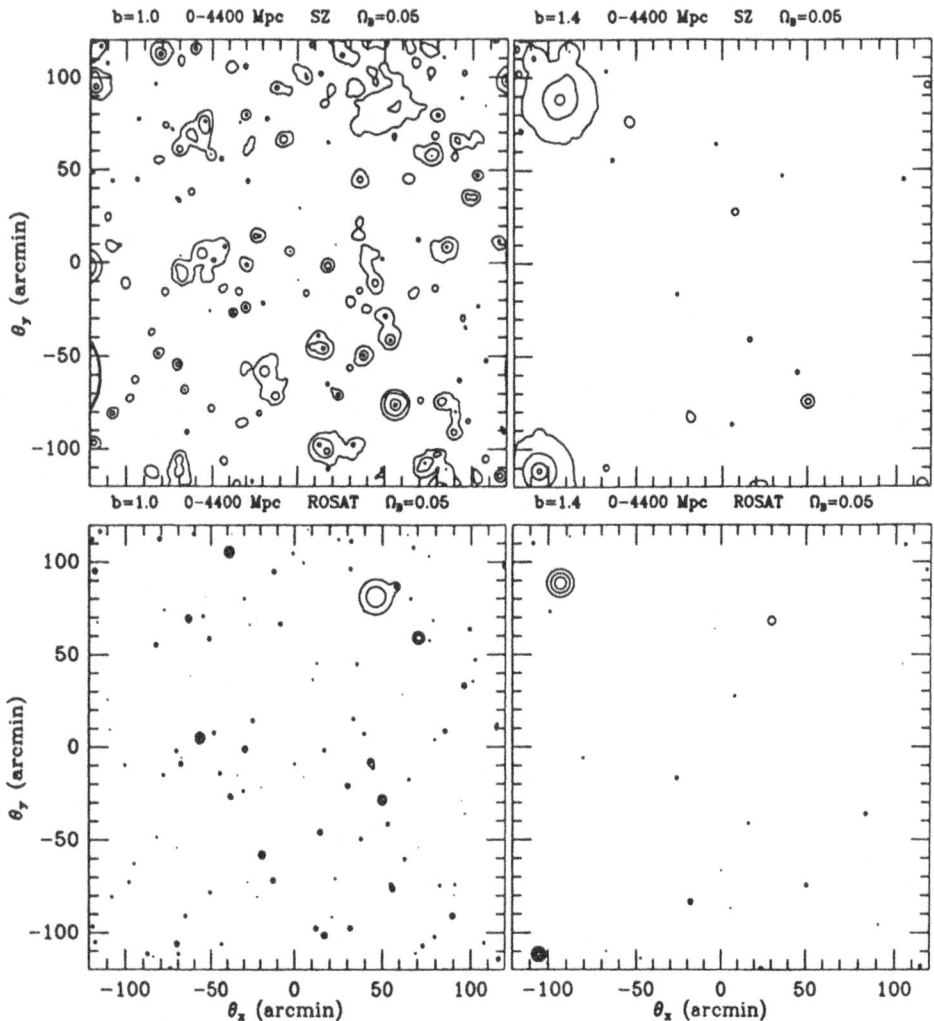

FIGURE 4.1 $4° \times 4°$ *contour maps of SZ sources smeared with a 1.8' beam from the (a)* $b_\rho = 1$ *and (b)* $b_\rho = 1.4$ *catalogues. The contour levels decrease by factors of two from* $-5 \times 10^{-6}(\Omega_B/0.05)$. *(c) and (d) give the corresponding* $b_\rho = 1$ *and 1.4 maps of X-ray emission from groups and clusters as would be seen by the ROSAT PSPC instrument. The contours increase by factors of two from a flux* $2.5 \times 10^{-15}(\Omega_B/0.05)^2$ *erg* cm^{-2} s^{-1}. *ROSAT's sensitivity in pointing mode is* $\sim 10^{-14}$.

rms, as expected in such a non-Gaussian map. For the $b_\rho = 1.4$ map, $\bar{y} = 5 \times 10^{-7}$ and the *rms* is $\Delta T/T_{rms} = 2.2 \times 10^{-6}$, with minimum value -4.6×10^{-5}. Although these numbers appear not to be very sensitive to b_ρ, there is a dramatic visual difference: the $b_\rho = 1.4$ map is almost clean except for a large cluster which happens to be in the patch, and which completely determines the *rms*. This cluster is a more powerful radiator in the $b_\rho = 1$ case, but the Zeldovich dynamics has moved it outside of the map. (It appears in the 5° map of Fig. 4.2.) The $b_\rho = 1.7$ map has only one small bit popping up through 10^{-5}. These maps graphically illustrate the unreliability of *rms* estimates as a guide to experimental design.

Using the cluster catalogue, we can also construct the X-ray maps which correspond to the SZ maps. We passed the redshifted bremsstrahlung spectrum for each cluster through an approximate ROSAT filter (0.1 − 2 kev) to get the sky surface brightness of these sources. In Fig. 4.1c and d we give the $b_\rho = 1$ and 1.4 maps of the flux predicted for ROSAT's PSPC from these sources. The minimum contour is at $2.5 \times 10^{-15}(\Omega_B/0.05)^2$ erg cm^{-2} s^{-1}, and subsequent contours increase by a factor of 2 over previous ones. The 5σ sensitivity for the ROSAT all-sky survey was estimated to be $\sim 2 \times 10^{-13}$ before launch, and the highly sampled ecliptic pole region 5σ sensitivity was estimated to be $\sim 2 \times 10^{-14}$. Post-launch results indicate a slightly worse all-sky sensitivity, but a minimum flux sensitivity a factor of two better for the pole. Thus, deep observations in pointed mode look marginally promising for $b_\rho = 1$ and not very promising for $b_\rho = 1.4$. The $b_\rho = 1.7$ map is dismal. We have not applied any filter to the maps, although, in pointed mode, there is a $\sim 30''$ resolution. The ROSAT field-of-view is 2°, although only for the inner 1° is the resolution undegraded.

As in the SZ maps, these ROSAT maps indicate that it will be difficult to detect the influence of hot gas at $z \sim 0.5$ unless there is extra energy injected beyond the gravitational binding energy of groups and clusters.

4.3 The Influence of Radio and SZ Sources on OVRO

A severe problem in looking for the 'ambient' SZ effect at Rayleigh Jeans wavelengths is the confusion from discrete non-thermal extragalactic radio sources, which themselves generate a non-Gaussian radiation background. As a reasonable approximation, the spatial distribution can be taken as almost Poissonian, with a nearly power-law number-flux count relation. At frequencies above 15 GHz, where the attenuation of the galactic non-thermal background allows CMB experiments to be effective, the number-flux slope at faint flux density levels is sufficiently flat ($N \propto S^{-2.1}$) that the expected contribution to the *rms* of an experiment increases as the area surveyed is enlarged. This foreground must be superposed on the anisotropy pattern produced by the primary and other secondary backgrounds, and in some instances may in fact dominate. The statistical signature of these confusing sources, which are positive fluctuations, is different than the signal from the SZ effect, which produces negative fluctuations, and from the primary anisotropies, which tend toward a zero-mean. A sufficiently large low-noise dataset may allow statistical separation of the different components. A more direct and effective decontamination would involve coincident (and preferably concurrent) observations at varying resolutions and wavelengths to allow size and spectral discrimination.

The best available datasets on intermediate angular scales or less, those of OVRO, VLA and the South Pole experiments, suffer in varying degrees from this contamination. The South Pole experiment is at a sufficiently high frequency that the discrete-source contamination is likely to be negligible; however, interference by dust is certain to be a problem as the sensitivity level is increased. The single-dish measurements of the OVRO 8-field and 96-field are especially susceptible to the effects of radio sources, due to the relatively low frequencies and the high gain of the antenna. One of the 8 fields in the former experiment

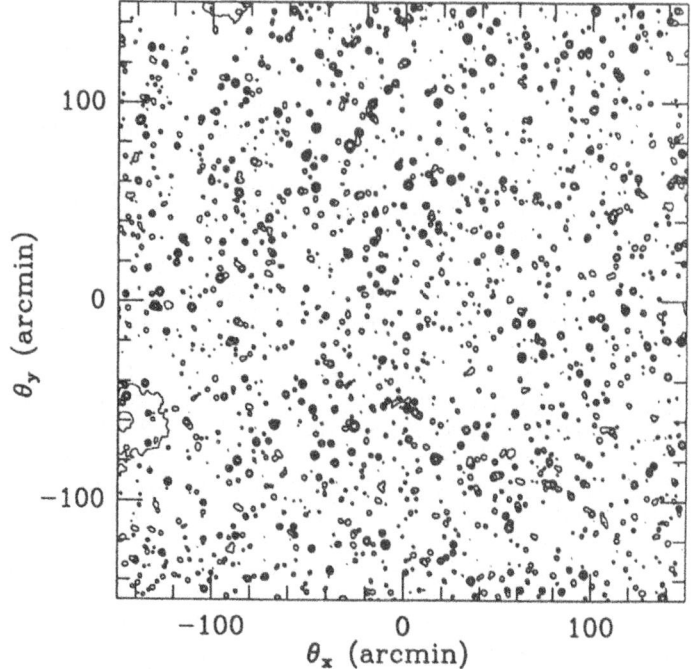

FIGURE 4.2 *Contour map of negative SZ sources (dotted) from the $b_\rho = 1$ catalogue with $\Omega_B = 0.05$ and positive radio sources (solid) drawn from faint counts extrapolated from the 5 GHz VLA data. The map is smoothed with the OVRO beam. The radio sources completely dominate the anisotropy signal. The lowest contour is 2×10^{-5} and the rest increase by factors of 4.*

was deleted from the sample because of a 20 mJy source in the beam, and the 96 fields of the RING survey exhibit signs of significant infection. Although interferometric observations from the VLA are also subject to this problem, because of the information automatically provided over a range of angular scales they can be partially decontaminated.

The observations of Fomalont *et al.* (1988) with the VLA at 5 GHz, in addition to producing the limits discussed in §2, predict that the differential source counts follow $dN/dS = 22S^{-2.1}$ sr^{-1} at the OVRO frequency of 20 GHz (assuming an average source spectral index of −0.7). The composite $5° \times 5°$ image of the $b_\rho = 1$ SZ model (see above) with a Poisson source model using this extrapolation of the VLA counts is shown in Figure 4.2. The only strong SZ sources are at the lower left edge of the map and at the top (just outside of the 4° map of Fig.4.1a).

For this map, the OVRO 8-field experiment would have a 37% chance of encountering at least one field contaminated at the 3σ level or worse (one was indeed found), as compared with only 0.4% for the $b_\rho = 1$ SZ model. The RING has a 68% chance of contamination compared with less than 0.01% for detection of SZ. Recent VLA data at 8.5 GHz have been reported by Windhorst to indicate that a population of flat spectrum radio sources may dominate at higher frequencies. Using an extrapolation of his 8.5 GHz counts, we find a map which is even more packed with radio sources. The 3σ encounter probability rises to

64% and 94% for the 8-field and the RING, respectively.

Separate observations using the VLA at 8.5 GHz (Myers 1990) of 20 RING fields, 7 of which had measured temperatures above 3σ, show that 4 contain milliJansky-level radio sources, 3 of which are sufficiently bright to account for the observed fluctuations with extrapolated spectra. Three others remain unaccounted for by VLA sources, and follow-up observations at higher frequencies are underway.

We conclude that, unless the foreground can be removed accurately, a successful detection of the ambient SZ effect from unidentified groups and clusters is unlikely at these low frequencies. Experiments at millimetre and submillimetre wavelengths are likely to provide better pros[ects for observation of the ambient SZ effect.

4.4 Dust Anisotropies from Primeval Galaxies

BCH1 showed that if galaxies (even dwarfs) exist at $z \sim 10$ then they cover the sky, and if they are dust-laden then all energy from the near IR to the X gets absorbed and re-emitted in the far IR, with a peak wavelength $\lambda \sim 500 - 700\mu$ which is relatively model insensitive. BCH1 and Bond (1988) showed that the intensity fluctuations in the sub-mm range would typically be $\Delta I_\nu/I_\nu \sim 0.01 - 0.1$ for a wide range of clustering models. BCH2 considerably extended the treatment of the emission and the expected anisotropy levels, using linear perturbation theory and shot noise models.

To focus the discussion, we consider BCH2 Models 7, 8 and 14 as typical examples of primeval galaxy dust models. For these, a burst of radiation generated at $z \gtrsim 5$ is absorbed by dust with a mass density Ω_{dust} that is located in primeval galaxies formed at $z = 5$ whose comoving number density is n_{G*}. The dust is assumed to be similar to typical Galactic dust. If n_{G*} is large, the depth across each individual galaxy is small, and the BCH2 linear perturbation treatment is appropriate. With $\Omega_{dust} = 10^{-5}$, which corresponds to Population I abundances in these galaxies, 86% of the energy is absorbed and re-radiated into a submillimetre background, while the remaining 14% is unabsorbed, contributing to a near infrared background; with $\Omega_{dust} = 10^{-6}$, only 18% is in the sub-mm and the rest is in the near IR. With a burst energy $\Omega_{RT}h^2 = 10^{-7}$, the average dust temperature is $T_d = 22$K, with the peak in $d\Omega_R/d\ln\lambda$ occurring at $\lambda \approx 500\mu$m. With the larger energy $\Omega_{RT}h^2 = 10^{-6}$, the dust is hotter, with average temperature 35K, and a spectral peak at $\lambda \approx 300\mu$m. However, since this peak emission band is shortward of where the current 6×10^{-7} FIRAS constraint comes from, this model is not yet ruled out. For this model, the IRAM and JCMT experiments, assuming their 95% upper limits are for rms anisotropies (non-Gaussian problems here, see below), give a source density limit of $n_{G*} \gtrsim 0.06(h^{-1}Mpc)^{-3}$, a factor of 3 above the bright galaxy abundance. If we assume that the primeval galaxies clustered with a $\gamma = 1.8$ power law correlation function, like galaxies that we see now do, then the limit on their clustering length (when they emit) is $r_o \lesssim 4 h^{-1}Mpc$, comparable to the current observed length. With the smaller energy, the constraints are a factor of 30 worse on source abundance and 7 worse on clustering length; $i.e.$, uninteresting.

For this same model with a $\Omega_{RT}h^2 = 10^{-6}$ energy release, we can predict rms levels in W cm^{-2} sr^{-1} for various experiments, assuming a source abundance of $0.02(h^{-1}Mpc)^{-3}$, self-similar nonlinear clustering and biased linear clustering:

DIRBE (200μm)	SIRTF (200μm)	JCMT (800μm)	JCMT (500μm)
0.5×10^{-13}	4×10^{-13}	6×10^{-13}	30×10^{-13}

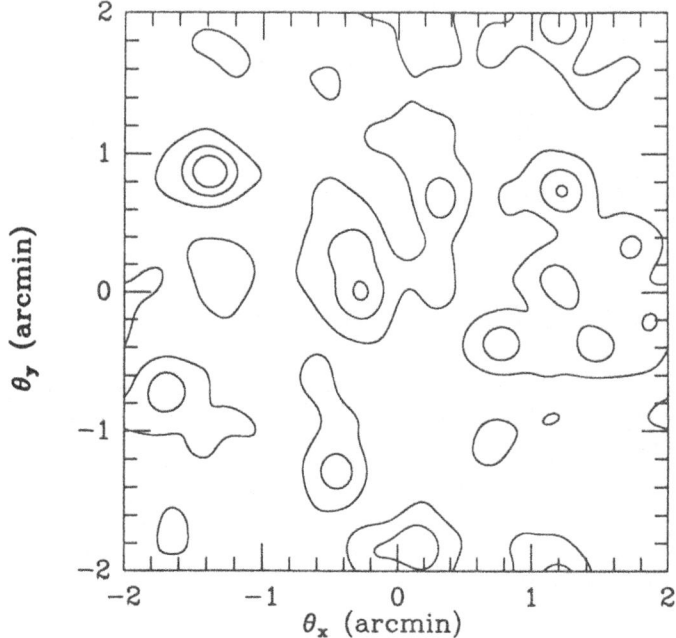

FIGURE 4.3 A 4' × 4' contour map for dust-emission from primeval galaxies at $z \sim 5$ convolved with an 18" beam appropriate for the Church et al. (1990) 800µm JCMT experiment. The lowest contour is 10^{-13} W cm^{-2} sr^{-1} and the rest increase linearly in equal steps.

These estimates are obtained by multiplying the du_P and du_c power spectra in Figure 2.1b by the appropriate filters in Fig.2.1a (SIRTF's is somewhat similar to OVRO's), and integrating over $d\ln\ell$. The predicted levels are achievable with all of these instruments. However, the foregrounds will make background anisotropy determinations at 200µm and 500µm decidedly difficult, even though these are the bands where the largest signals are expected. Although the non-Gaussian aspects of the patterns make the use of *rms* values suspect, the relatively large beams of sub-mm experiments smooth the primeval galaxy images sufficiently that the pattern approaches Gaussian and there are few blank patches of sky (see Fig. 4.3). It is easy to invent models with smaller anisotropies. Indeed, for a given Ω_{RT}, the JCMT estimates in the Table scale as $n_{G*}^{-1/2}$, and thus increasing the number of sources produces smaller relative anisotropies. As well, over the relatively short intergalactic distances gas dynamical processes are likely to be effective, and these could equally well wash out as enhance anisotropies. Another unresolved issue is whether gravitational lensing will effectively smear the anisotropies over small angular scales.

If the number n_{G*} is small, then the galaxies may be optically thick and perturbation calculations will not be appropriate. To adequately deal with such sources, physical modelling of the emission from each source is required. Here we give an example using the hierarchical peaks method for identifying primeval galaxies in a $b_\rho = 1.4$ CDM model. For illustration, we used just two filter scales, to identify primeval dwarf galaxies and primeval 'bright' galaxies. The objects had masses $6 \times 10^{10} \Omega_B \, M_\odot$ and $8 \times 10^{11} \Omega_B \, M_\odot$ in baryons and star-burst level luminosities $\sim 10^{44}$ erg/s and $\sim 10^{45}$ erg/s, respectively. The dust temperature was taken to be a constant 30K. Normal dust with a dust-to-baryon fraction in the galaxies of 0.2% was assumed. The scaling of the luminosity and of our Fig.4.3 map contours goes as $(T_d/30K)^5$ and linearly with the dust-to-baryon ratio. We assume these primevals were bursting between $z = 6$, before which few exist anyway, and an arbitrary cutoff redshift we chose to be $z = 4$. With these parameters, the energy in the background from these objects would be $\Omega_{RT} h^2 = 2.5 \times 10^{-7}$, 0.1% of the total energy in the CMB.

The $4' \times 4'$ in Fig.4.3 is for the flux integrated over the entire dust emission spectrum. In that case, the lowest contour is 10^{-13} W cm^{-2} sr^{-1}, with the scaling given above. The radiation pattern is smoothed with an 18$''$ beam, used in the chop and wobble experiment of Church *et al.* (1990), which had a throw of 40$''$. The *rms* relative intensity fluctuations in the map are 0.5 and the maximum value is 4.1 times the average intensity distortion. We look forward to JCMT's SCUBA, which should be able to probe intensity levels like these of busting primeval galaxies. If so, then the non-Gaussian aspects of the radiation pattern would provide a powerful probe of the high redshift protogalactic medium including how galaxies are clustered at that time.

Acknowledgements: No attempt at complete references was made in this sightseeing tour. JRB would like to thank his collaborators Shaun Cole, George Efstathiou, Bernard Carr, Craig Hogan and Nick Kaiser. STM would like to thank Tony Readhead and Charles Lawrence. We would also like to thank Marguerite Pierre for enlightening us about ROSAT's performance and Rogier Windhorst for worrying us even more about faint radio source contamination. Support of a Canadian Institute for Advanced Research Fellowship, a Steacie Fellowship and the NSERC of Canada is gratefully acknowledged by JRB.

6. REFERENCES

Bardeen, J.M., Bond, J.R., Kaiser, N. and Szalay, A.S. 1986, *Ap. J.*, **304**, 15.
Bardeen, J.M., Bond, J.R. and Efstathiou, G. 1987, *Ap. J.*, **321**, 28.
Bond, J.R., Carr, B.J. and Hogan, C.J. 1986, *Ap. J.* **306**, 428 [BCH1]; 1991, *Ap. J.* **367**, 420 [BCH2].
Bond, J.R. and Efstathiou, G. 1987, *M.N.R.A.S.* **226**, 655.
Bond, J.R. 1988, in *The Early Universe*, Proc. NATO Summer School, Vancouver Is., Aug. 1986, ed. Unruh, W.G. (Dordrecht:Reidel); 1989, in *Frontiers of Physics— From Colliders to Cosmology*, ed. Campbell, B. and Khanna, F. (Singapore: World Scientific); 1990, in *The Cosmic Microwave Background: 25 Years Later*, p. 45-65. ed. Mandolesi, N. and Vittorio, N. (Dordrecht: Kluwer).
Bond, J.R. and Couchman, H.M.P. 1987, in *Proceedings of the Second Canadian Conference on General Relativity and Relativistic Astrophysics*, eds. C. Dyer, A. Coley, (Singapore: World Scientific).
Bond, J.R., Efstathiou, G., Lubin, P.M. and Meinhold, P.R. 1991, *Phys. Rev. Lett.*, in press.
Bond, J.R., Cole, S., Efstathiou, G. and Kaiser, N. 1991, *Ap. J.*, in press.
Boughn, S.P., Cheng, E.S., Cottingham, D.A. and Fixsen, D.J. 1990, preprint.
Cen, R.Y., Jameson, A., Liu, F. and Ostriker, J.P. 1990, *Ap. J. Lett.*, **362**, L41.
Church, S.E., Lasenby, A.N. and Hills, R.E. 1990, preprint.
Cole, S. and Kaiser, N. 1989, *M.N.R.A.S.*, **233**, 637.

COSMIC X-RAY BACKGROUND

H. INOUE
Institute of Space and Astronautical Science
3-1-1, Yoshinodai, Sagamihara, Kanagawa 229, Japan.

ABSTRACT. Recent progress in observational study of the cosmic X-ray background (CXB) is reviewed mainly based on results from the large area counter on board Ginga. According to the Ginga fluctuation analysis of the sky background (Hayashida 1990), the $\log N - \log S$ relation of the form $N \propto S^{-3/2}$ in the 2–10 keV band seems to hold down to a flux level as low as 10^{-13} erg cm^{-2} s^{-1}, and the integrated flux of all discrete sources above 10^{-13} erg cm^{-2} s^{-1} amounts to about 30% of the CXB. The angular auto-correlation function of the CXB obtained with Ginga (Kondo 1991) shows that the contribution of the sources having a correlation length of 5 $(H/100$ km s^{-1} Mpc$^{-1})^{-1}$ Mpc is at most 50% of the CXB (95% confidence limit). This upper limit for the correlation length suggests that more than 50% of the CXB do not come from members of clusters of galaxies within $z < 1$. Some possibilities for the CXB origin are discussed.

1   Introduction

A significant amount of X-rays was discovered to come from all directions of the sky in the first rocket experiment that detected an extra-solar X- ray source (Giacconi et al. 1962). Since then, a great deal of effort has been made to understand the origin of the X-ray background (XRB), however it has not fully been solved yet. For early work see a review by Schwartz and Gursky (1974), and for more recent reviews see e.g. Bolt (1987) and Giacconi and Zamorani (1987).

The characteristics of the XRB are very different below and above 2 keV. In the soft X-ray band strong anisotropy over the sky has been observed (see e.g. Tanaka and Bleeker 1977; Marshall and Clark 1984). The XRB in the soft X-ray band is regarded as thermal emission from hot plasmas in our Galaxy.

On the other hand, above 2 keV, the XRB is highly uniform except for an excess component along the galactic plane. The excess along the plane is considered to be associated with our Galaxy (Koyama 1989, and references therin), whereas the rest of the emission is believed to be of extragalactic origin. In this review, only the XRB above 2 keV at high galactic latitude will be discussed. Hereafter, we call the XRB of extragalactic X-ray background (CXB) to distinguish it from the Galactic origin.

2 CXB spectrum and anisotropy

Marshall et al. (1980) examined the energy spectrum of the CXB obtained with HEAO–1 A2 and found that it can be represented well by a thermal bremsstrahlung spectrum with temperature of about 40 keV in the energy range from 3 to 50 keV. The 40 keV thermal emission is valid for the CXB spectrum up to about 100 keV, but the CXB spectrum has bump near a few MeV (Rothschild et al. 1983). Figure 1 shows the CXB spectrum.

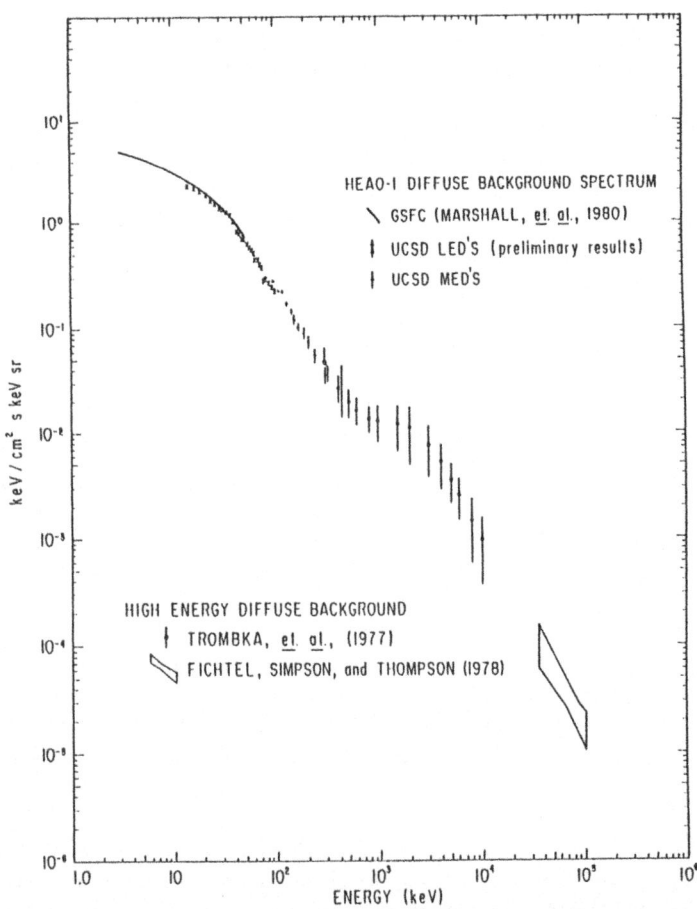

Fig.1 Spectrum of the cosmic X-ray background (from Rothschild et al. 1983).

Large scale structure has been searched for in X-ray data from various satellites. Protheroe et al. (1980) and Shafer (1983) reported on a dipole anisotropy in the CXB. Besides this, no significant anisotropy on a scale of 3–27 degrees was found in the HEAO–1 data (Shafer 1983; Persic et al. 1989) nor on scales of several arc-minutes in the Einstein experiment (Barcons and Fabian 1989). Recent Ginga observations (Carrera et al. 1990; Kondo 1991) of the CXB also show no significant anisotropy on angular scale of 2–10 degrees.

3   Diffuse origin of the CXB

Optically thin thermal emission from hot gas filling the universe has been considered to be a possible origin of the CXB; a bremsstrahlung spectrum fits the data in the energy range 3–50 keV very well (Marshall et al. 1980). However, this possibility has been ruled out by a lack of any high frequency distortion in the microwave background spectrum observed with COBE (Mather et al. 1990); the contribution of hot gas in the universe to the CXB flux could be at most 3%.

4   Discrete source origin of the CXB

4.1 $\log N - \log S$ relation for X-ray sources

Another possibility for the origin of the CXB is that the observed CXB flux is supplied by a collection of unresolved, weak discrete sources. The contribution to the CXB (< 3keV) from sources with flux levels above 4×10^{-12} erg cm^{-2} s^{-1} in the 0.3–3.5 keV band was estimated to be about 30% (Giacconi et al. 1979). However, most of the CXB flux is emitted in the range 3–50 keV. A more direct estimation of the discrete source contribution to the CXB in the 2–10 keV band has been carried out through the analysis of the sky background fluctuation observed with HEAO–1 A2 (Shafer 1983) and with Ginga (Hayashida 1990; Warwick and Stewart 1989).

Figure 2 shows the distribution of the Ginga sky background rate around the average value for (a) different sky directions and for (b) a particular direction. By comparing Fig.2-a with 2-b, we see that the background rate is different from directions in the sky, however no large scale structure was found. This indicates that the fluctuations are due to spatial statitstical fluctuations of unresolved sources. We can thus derive a constraint on the log $N - \log S$ relation for sources which contribute to the source confusion noise. The result is plotted in the log $N - \log S$ plane in Fig.3 (Hayashida 1990). This result is consistent with the $N \propto S^{-3/2}$ relation extrapolated from the Ginga high-latitude survey (Kondo 1991) and HEAO–1 A2 (Piccinotti et al. 1982) results, and reveals that the discrete-source-contribution to the CXB in the energy range 2 – 10 keV amounts to at least 30%.

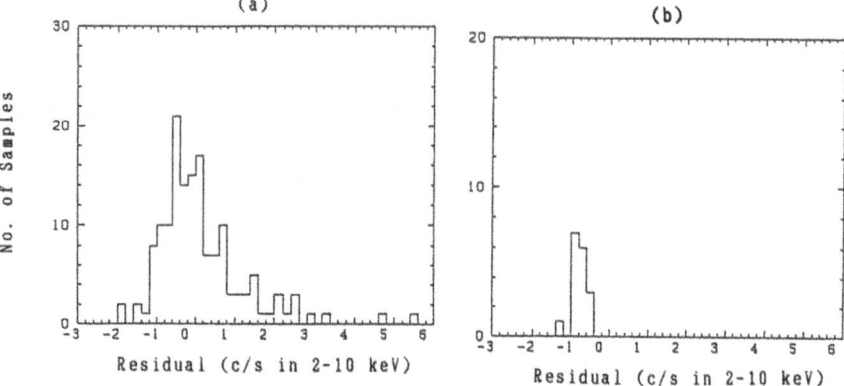

Fig.2 Distribution of the sky background around the average, observed with Ginga (Hayashida et al. 1989), for (a) different sky directions, and (b) a fixed sky direction.

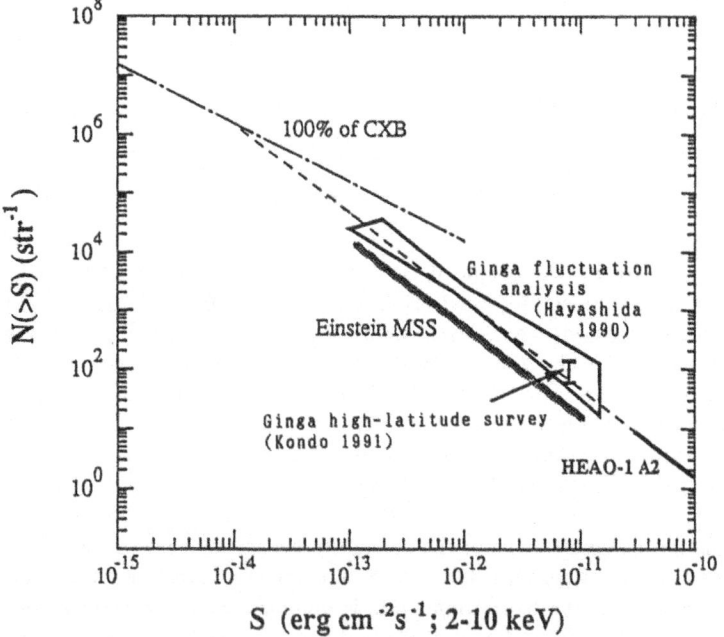

Fig.3 Log N – log S relations obtained from a Ginga high-latitude survey (Kondo 1991) and from a Ginga fluctuation analysis (Hayashida 1990). Results of the Einstein medium sensitivity survey (MSS), and of HEAO–1 A2 are also shown schematically. If a $N \propto S^{-3/2}$ relation (consistent with the Ginga and HEAO-1 A2 results) extends down to a flux level of 10^{-14} erg cm^{-2} s^{-1}, all of the CXB flux is explained.

4.2 Extrapolation of log N – log S relation of known types of sources

Clusters of galaxies (CG) are the majority of extragalactic X-ray sources in the flux range above 10^{-12} erg cm^{-2} s^{-1}, but active galactic nuclei (AGN) dominate in the flux range below 10^{-12} erg cm^{-2} s^{-1} (Gioia et al. 1984). Recent analyses show that the present volume density of high-luminosity CGs is greater than it was in the past ($z > 0.2$) (Gioia et al. 1990; Edge et al. 1990). Furthermore, the temperatures of hot plasmas in CGs range from 2 to 10 keV (Hatsukade 1989), and are too low to fit the CXB spectrum. The contribution of CGs to the CXB is at most 10% (e.g. Schmidt and Green 1986).

AGNs are the largest contributor to the CXB among sources with flux levels $10^{-13} \sim 10^{-12}$ erg cm^{-2} s^{-1}. Recent optical surveys of faint AGNs show that the surface density of optically selected, known types of AGNs seems to saturate at a value of several hundred per square degrees in the magnitude range below 24m (e.g. Boyle et al. 1987). On the other hand, the surface density of AGNs necessary to explain the entire CXB is several hundred per square degrees if the $N \propto S^{-3/2}$ relation holds down to flux levels of 10^{-14} erg cm^{-2} s^{-1} (see Fig.3; 1 steradian = 3.3 × 10^3 square degree). This means that known types of AGNs can explain the CXB flux only when the $N \propto S^{-3/2}$ relation extends down to 10^{-14} erg cm^{-2} s^{-1}. However, Hamilton and Helfand (1987) and Barcons and Fabian (1990) found that the log N – log S relation must flatten at a flux level around 10^{-14} erg cm^{-2} s^{-1}. The contribution of known types of extragalctic sources to the CXB is estimated to be 60% at most (e.g. Schmidt and Green 1986; Giacconi and Zamorani 1986). This is supported by the difference in the spectral slope between AGNs and CXB. Figure 4 shows the spectral energy-indeces of 28 AGNs against the cosmological redshift z observed with Ginga. As seen from this figure (see also Rothschild et al. 1983; Mushotzky 1984), the mean spectral index of QSOs and Seyferts is 0.7 $\sim$ 0.8 and is significantly higher than 0.4 of the CXB.

The determination of the mean spectral index of AGNs below the Ginga detection limit can also be addressed by analyzing the source confusion noise in the sky background (see section 4.1). The mean spectral index of the component which fluctuates for different parts of sky has been obtained (Hayashida 1990). The spectral slope is consistent with the average slope of the brighter AGNs measured one by one, and is significantly steeper than that of the CXB in the flux range around 10^{-12} erg cm^{-2} s^{-1}.

The above arguments strongly suggest that the majority of the sources responsible for the CXB are not known types of extragalactic sources with flux above about 10^{-13} erg cm^{-2} s^{-1} in the range 2–10 keV.

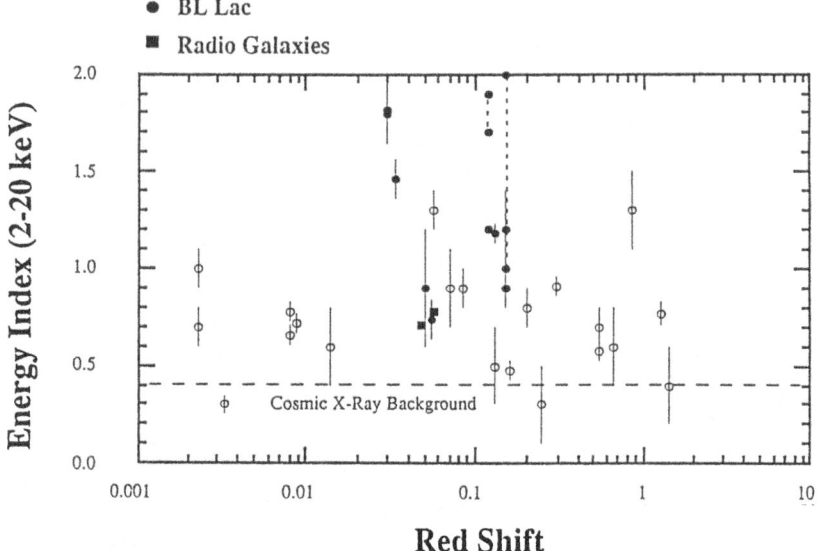

Fig.4 The distribution of the spectral index of AGNs (Ginga data) as a function of redshift factor, z. The spectral index of the CXB in the 2–10 keV range is also indicated.

4.3   Necessary conditions for main contributor to the CXB

Now, let us consider the necessary conditions for those unknown sources.

(1) Optical magnitude

As shown in the previous section, optically selected AGNs with the B-magnitude below 24m are insufficient in number to explain all of the CXB flux. In other words, the sources responsible for the CXB are optically very faint and the B-magnitude should be larger than 24m. If they are closer than $z = 3 \sim 4$, the absolute B-magnitude should be > -21M.

(2) Surface density and the angular correlation

If, as the Einstein fluctuation analysis (Hamilton and Helfand 1987; Barcons and Fabian 1990) predicts, the slope of the log N – log S curve is flatter than 1.5 at a flux level around 10^{-14} erg cm^{-2} s^{-1}, the surface density of the sources responsible for the CXB should be larger than a few thousands per square degrees.

The angular correlation of the sources is also given by the analysis of the auto-correlation function (ACF) of the CXB with HEAO–1 A2 (Persic et al. 1989; De Zotti et al. 1990) and with Ginga (Carrela et al.1990; Kondo 1991). The Ginga observations do not show any angular structure and require that the sources which have a correlation length of $5h^{-1}$ Mpc (h is the Hubble constant in unit of 100 km s^{-1} Mpc^{-1}) contribute less than 50% (95% confidence limit) to the CXB. This correlation length is a typical value of galaxies (Totsuji and Kihara 1969; Davis and Peebles 1983) and AGNs (Iovino and Shaver 1988), and represents the source clustering on a size of clusters of galaxies. Hence, if any types of sources in clusters of galaxies significantly contribute to the CXB, the correlation length of $5\ h^{-1}$ Mpc must appear in the auto correlation function. Since all known types of nearby sources are constituents of clusters of galaxies and clusters of galaxies are known to exist at $z < 1$, more than 50% of the CXB is considered to come from distant sources at $z > 1$.

(3) Deficiency of soft X-rays below 2 keV

As seen in Fig.3, the constraint on the log N – log S relation obtained with Ginga is inconsistent with that obtained with the Einstein Medium Sensitivity Survey (MSS; Maccacaro et al. 1982); the number from Ginga is significantly larger by a factor $2 \sim 3$ for the same flux range. This apparent discrepancy is probably due to the difference in the observation energy range between them. In the case of Einstein, the 2–10 keV flux was converted from the 0.3–3.5 keV flux on the assumption that sources had a power law spectrum with energy index of 0.7. The relatively small number of sources in the Einstein data indicates that the conversion factor from the 0.3–3.5 keV flux to the 2–10 keV flux should have been larger than that for the typical AGN spectrum and that the flux below $2 \sim 3$ keV is relatively small. This deficiency of the soft X-rays can be explained by absorption.

(4) Spectrum above 2 keV

The spectral slope of the CXB is 0.4 in the 2–10 keV range, while the AGNs with the average slope of $0.7 \sim 0.8$ contribute at least 30% to the CXB. Hence, the spectrum of the main contributor to the remaining CXB after subtracting the AGN contribution should have an almost flat spectrum with a slope close to zero in the 2–10 keV range.

The CXB slope becomes steep as the X-ray energy goes up above 10 keV and is about 1.4 in the 20–100 keV range. On the other hand, the power law spectrum of AGNs extends up to at least several 10 keV. Hence, the spectrum of the main contributor to the CXB should sharply fall as the X- ray energy increases up to several 10 keV.

The CXB spectrum becomes flat again as the X-ray energy increases close to 1 MeV. If the AGN spectra extends up to this energy range with a slope of about 0.7, the AGN contribution of $20 \sim 30\%$ to the CXB in the 2-10 keV range is sufficient to explain the flux in the "MeV bump". At present, however, we have little information about the AGN spectra in the energy range above 100 keV.

4.4   Some possibilities

(1) Normal galaxies

Griffiths and Padovani (1990) propose that star forming galaxies are a major component of the CXB in the 2–10 keV range. They estimated the relationship between X-ray and infrared luminosities for starburst/peculiar galaxies, and derived an X-ray luminosity function from the IRAS galaxy luminosity function. Using this luminosity function, the contribution of those galaxies to the CXB is estimated to be 20 ~ 30% with moderate evolution, rising to at least 50% with evolution similar to that of AGNs.

A major argument against this proposal is the spectrum. Griffiths and Padovani (1990) argue that star forming galaxies may be largely powered by massive X-ray binaries and that their hard X-ray spectra match the overall spectrum of the CXB in the 2–30 keV range. However, three starburst galaxies observed with Ginga commonly reveal rather soft spectra reproduced by thermal bremsstrahlung models with temperatures of 6 ~ 7 keV, without a siginificant amount of absorption (Ohashi et al. 1990a, 1990b).

(2) Distant Seyfert-1-like galaxies

Ginga observations of Seyfert 1 galaxies have revealed that their X-ray spectra are not described by a single power law, but have an emission line feature at 6 ~ 7 keV as well as an shallow and broad absorption-like feature at an energy of several to 15 keV and flatten at higher energies (Pounds et al. 1990; Matsuoka et al. 1990). Morisawa et al. (1990) discuss that if the flat spectra around 10–30 keV are redshifted by a factor of 3 ~ 4, a flat spectrum will be seen in the 2–10 keV range as is tha case in the CXB spectrum. However, if we want to explain the CXB origin in terms of highly redshifted AGNs, as Morisawa et al. (1990) propose, we have to introduce soft X-ray and optical light deficiencies for these sources unlike for nearby Seyfert 1 galaxies.

(3) Reflection-dominated hard X-ray sources

The emission line feature at 6 ~ 7 keV, the shallow and broad absorption feature and an excess above 15 keV commonly seen in the Seyfert 1 spectra can be reproduced very well by superposition of a component reflected by cool Thomson-thick matter on a single power law (see e.g. Inoue 1989). Fabian et al. (1990) found that the break at about 30 keV in the CXB spectrum fit to the spectrum of reflected X-rays well. Even though incident X-rays have a power law spectrum extending up to several 100 keV, Compton scattering decreases the energy of incident photons by a significant amount above 50 keV leading to a low apparent albedo at high energies. Since the X-rays below 10 keV are predominantly absorbed, the soft X-ray deficiency can also be explained.

The dominant appearence of the reflection spectrum can be realized when the central engine of the AGN is highly obscured by thick matter. The ratio between the reflection and the direct fluxes from the Seyfert 1 galaxies indicates that cool, thick matter around the central engine has a covering factor as large as or larger than 50%. This suggests that the thick material responsible for the reflection hides the central engine from us with a fairly large probability. Lack of optical broad lines of the Seyfert 2 galaxies is understood by introducing obscuration of the central part of the AGN, and, in fact, X-ray spectra of those galaxies observed with Ginga show clear evidence for high obscuration (Koyama et al. 1989; Awaki et al. 1990). Highly redshifted, Seyfert-2-like AGNs are a possible candidate for the origin of the CXB.

Heisler and Ostriker (1988) propose that very-high-redshifted QSOs ($z > 3 \sim 4$) exist but are obscured by matter in intervening galaxies. If the optical depth of the obscuring matter for Thomson scattering is much larther than unity, the X-ray spectrum will be similar to the reflection one. Hence, this is also another intresting possibility for the CXB.

ASTRO–D is the fourth X-ray astronomy mission of Japan, scheduled for launch in early 1993. It will carry nested thin-foil X-ray mirrors providing a large effective area over a wide energy range below 1 keV up to 10 keV. A set of CCD cameras and imaging gas scintillation proportional counters will be placed on the focal plane. With these instruments, ASTRO–D will be the first X-ray astronomy satellite capable of imaging the X-ray sky in a wide energy range up to 10 keV. The high sensitivity of ASTRO–D will enable us to observe very faint sources with flux down to 10^{-14} erg cm^{-2} s^{-1}. The detection limit will be able to resolve a larger fraction of the CXB into a number of faint sources than previously possible. We are looking forward to finding sources which are responsible for most part of the CXB in the deep survey images to be obtained with ASTRO–D.

The author would like to thank Prof. W.H.G. Lewin for his careful reading of the manuscript.

References

Awaki, H. et al. 1990, *Nature*, **346**, 544.

Barcons, X. and Fabian, A.C. 1989, *Mon. Not. Roy. Astron. Soc.*, **237**, 119.

Barcons, X. and Fabian, A.C. 1990, *Mon. Not. Roy. AStron. Soc.*, **243**, 366.

Bolt, E. 1987, *Physics Reports*, **146**, 215.

Boyle, B.J. et al. 1987, *Mon. Not. Roy. Astron. Soc.*, **227**, 717.

Carrera, F.J. et al. 1990, *preprint.*

Davis, M. and Peebles, P.J.E. 1983, *Astrophys. J.*, **267**, 465.

De Zotti, G. et al. 1990, *Astrophys. J.*, **351**, 22.

Edge, A.C. et al. 1990, *Mon. Not. Roy. Astron. Soc.*, **245**, 559.

Fabian, A.C. et al. 1990, *Mon. Not. Roy. Astron. Soc.*, **242**, 14.

Giacconi, R. et al. 1962, *Phys. Rev. Letters*, **9**, 439.

Giacconi, R. et al. 1979, *Astrophys. J. (Letters)*, **234**, L1.

Giacconi, R. and Zamorani, G. 1987, *Astrophys. J.*, **313**, 20.

Gioia, I.M. et al. 1984, *Astrophys. J.*, **283**, 495.

Gioia, I.M. et al. 1990, *Astrophys. J. (Letters)*, **356**, L35.

Griffiths, R.E. and Padovani, P. 1990, *Astrophys. J.*, **360**, 483.

Hamilton, T.T. and Helfand, D.J. 1987, *Astrophys. J.*, **313**, 93.

Hatsukade, I. 1989, *Ph.D.Thesis*, Osaka University.

Hayashida, K. 1990, *Ph.D.Thesis*, University of Tokyo.

Heisler, J. and Ostriker, J.P. 1988, *Astrophys. J.*, **332**, 543.

Inoue, H. 1989, in *Proc. 23rd ESLAB Symp. on Two-Topics in X-ray Astronomy*, (ESA, Noordwijk), p.783.

Iovino, A. and Shaver, P.A. 1988, *Astrophys. J. (Letters)*, **330**, L13.

Kondo, H. 1991, *Ph.D.Thesis*, Univ. of Tokyo.

Koyama, K. 1989, *Publ. Astron. Soc. Japan*, **41**, 665.

Koyama, K. et al. 1989, *Publ. Astron. Soc. Japan*, **41**, 731.

Maccacaro, T. et al. 1982, *Astrophys. J.*, **253**, 504.

Marshall, F.E. et al. 1980, *Astrophys. J.*, **235**, 4.

Marshall, F.J. and Clark, G.W. 1984, *Astrophys. J.*, **287**, 633.

Mather, J.C. et al. 1990, *Astrophys. J. (Letters)*, **354**, L37.

Matsuoka, M. et al. 1990, *Astrophys. J.*, **361**, 440.

Morisawa, K. et al. 1990, *Astron. Astrophys.*, **236**, 299.

Mushotzky, R.F. 1984, *Adv. Space Res.*, **3**, 10.

Ohashi, T. et al. 1990a, in *Windows on Galaxies*, ed G.Fabbiano, J.S.Gallagher and A.Renzini (Kluwer, Dordrecht), p.243.

Ohashi, T. et al. 1990b, *Astrophys. J.*, **365**, 180.

Persic, M. et al. 1989, *Astrophys. J. (Letters)*, **336**, L47.

Piccinotti, G. et al. 1982, *Astrophys. J.*, **253**, 485.

Pounds, K.A. et al. 1990, *Nature*, **344**, 132.

Protheroe, R.J., et al. 1980, *Mon. Not. Roy. Astron. Soc.*, **192**, 445.

Rothschild, R.E. et al. 1983, *Astrophys. J.*, **269**, 423.

Schmidt, M. and Green, R.F. 1986, *Astrophys. J.*, **305**, 68.

Schwartz, D.A. and Gursky, H. 1974, in *X-ray Astronomy*, ed. R. Giacconi and H. Gursky (Reidel, Dordrecht), p.359.

Shafer, R.A. 1983, *Ph.D.Thesis*, Univ. of Maryland.

Tanaka, Y. and Bleeker, J.A.M. 1977, *Space Sci. Rev.*, **20**, 815.

Totsuji, H. and Kihara, T. 1969, *Publ. Astron. Soc. Japan*, **21**, 221.

Warwick, R.S. and Stewart, G.C. 1989, in *Proc. 23rd ESLAB Symp. on Two-Topics in X-ray Astronomy*, (ESA, Noordwijk), p.727.

Large Scale Anisotropy of the CMB in an Open Universe and Constraints on the Models of Galaxy Formation

N.Gouda, N.Sugiyama and M.Sasaki*

Department of Physics, Kyoto University, Kyoto 606, Japan

* *Uji Research Center, Yukawa Institute for Theoretical Physics*

Kyoto University, Uji 611, Japan

ABSTRACT. We derived the formula by which we can calculate any multipole moment of cosmic microwave background anisotropy in an open universe. we also calculate by using this formula the rms quadrupole moment expected in some cosmological models and we restrict the models by comparing the calculated quadrupole moments with the observed upper limit of the quadrupole moment of the cosmic microwave background anisotropy.

1. Overview

1.1 LARGE SCALE STRUCTURES AND ANISOTROPY OF THE CMB

Large scale structures of the universe such as galaxies and clusters of galaxies are generally believed to be developed from small density inhomogeneities on large scales in the very early universe. Since such large scale density inhomogeneities inevitably give rise to temperature inhomogeneities, residual anisotropies of the cosmic microwave background(CMB) contain very important information about primordial inhomogeneities of the Universe. As a result, comparison of expected CMB anisotropies in a theoretical model for the formation of large scale structures with observed CMB anisotropies would provide stringent constraints on the model.

1.2. THE CMB ANISOTROPY ON SMALL ANGULAR SCALES

Much work has been done on the evaluation of CMB anisotropies in various scenarios. As for anisotropies on small angular scales, many papers have been published on expected CMB anisotropies in the three representative cosmological models;in baryon dominated universe model(BDM), in cold dark matter dominated models(CDM) and in hot dark matter dominated models(HDM). Those studies have already excluded models in a considerable range of cosmic parameters and initial conditions by comparing predicted anisotropies with observed upper limits of anisotropies on small angular scales. However, one cannot completely deny a possibility that small angle anisotropies were affected by yet unknown process after decoupling, besides Peebles' reionization model. In such a case, conclusions drawn in the previous studies may be drastically altered.

1.3 THE CMB ANISOTROPY ON LARGE ANGULAR SCALE

On the other hand, CMB anisotropies on large angular scales would have never been influenced by local processes such as reionization or gravitational lensing. In particular, the quadrupole moment of the CMB anisotropy, which is independent of our peculiar motion, serves as a very direct probe of primordial inhomogeneities of the Universe. The quadrupole moment has been calculated by many authors in various cosmological models. But all of these were unsatisfactory in the sense that their considerations were restricted either to the spatially flat background or to an incomplete evaluation of the quadrupole moment. That is, so far there has not appeared any work which considers an open universe model and includes both the generalized Sachs-Wolfe effect(Wilson 1983) and intrinsic photon fluctuations completely in evaluating the quadrupole moment. This unsatisfactory situation is mainly due to a technical difficulty in estimating the present quadrupole and/or higher multipole moments of the CMB anisotropy in an open universe. Now we have succeeded in deriving a formula which relates any multipole moment at a given time with multipole moments at a different time and with which one can calculate, up to numerical precision, *exactly* any multipole moment of the CMB anisotropy in an open universe(Gouda, Sugiyama and Sasaki 1990(a)). The resulting formula, of course, takes account of contributions from both intrinsic photon fluctuations and generalized Sachs-Wolfe effect.

2. Results

By using the formula mentioned above, we calculate the values of the rms quadrupole moments expected in some models and also give constraints on the models by comparing the results with the observation of the upper limit of the quadrupole moment $Q < 3 \times 10^{-5}$(Klypin *et al.* 1987). For adiabatic models, we find that all BDMs, low Ω_0 adiabatic CDMs and HDMs, all with $n \leq 1$, are excluded (Gouda, Sugiyama and Sasaki 1990(b)). This conclusion is similar to the one drawn from the small angle CMB anisotropies. As for isocurvature models, we find all models with $n \leq 1$ are excluded. This is more stringent than in the case of small angle anisotropies. As for Peebles' model, we find models with $n \geq 3$ are allowed.

3. Summary

We have calculated the rms quadrupole moment in a variety of open universe models. Comparing the results with the observed upper limit of the quadrupole moment, we have found that except for high Ω_0 adiabatic CDMs and HDMs, all conceivable models with $n \leq 1$ are excluded. Although the present conclusion is similar to the one obtained from the upper limit of small angle CMB anisotropies, it must be taken more seriously. The reason is that the quadrupole anisotropy is a faithful carrier of primordial features of density fluctuations, while small angle CMB anisotropies could be affected by phenomena which might have occurred after decoupling, such as reionization or gravitational lensing.

References

Gouda, N., Sugiyama, N., and Sasaki, M. 1990(a), *Large Angle Anisotropy of the Cosmic Microwave Background in an Open Universe*, preprint.

Gouda, N., Sugiyama, N., and Sasaki, M. 1990(b), *Constraints on Open Universe Models from Quadrupole Anisotropy of the Cosmic Microwave Background*, preprint.

Klypin, A.A., Sazhin, M.V., Strukov, I.A., and Skulachev, D.P. 1987, *Sov.Astron.Lett.*, **13**, 104.

Wilson, M.L. 1983, *Ap.J.*, **273**, 2.

THE BEST-FIT UNIVERSE

MICHAEL S. TURNER
NASA/Fermilab Astrophysics Center
Fermi National Accelerator Laboratory
Batavia, IL 60510-0500   USA
and
Departments of Physics and Astronomy & Astrophysics
Enrico Fermi Institute
The University of Chicago
Chicago, IL 60637-1433   USA

ABSTRACT. Inflation provides very strong motivation for a flat Universe, Harrison-Zel'dovich (constant-curvature) density perturbations, and cold dark matter. However, there are a number of cosmological observations that conflict with the predictions of the simplest such model—one with zero cosmological constant. They include the age of the Universe, dynamical determinations of Ω, galaxy-number counts, and the apparent abundance of large-scale structure in the Universe. While the discrepancies are not yet serious enough to rule out the simplest and "most well motivated" model, the current data point to a "best-fit model" with the following parameters: $\Omega_B \simeq 0.03$, $\Omega_{\mathrm{CDM}} \simeq 0.17$, $\Omega_\Lambda \simeq 0.8$, and $H_0 \simeq 70\,\mathrm{km\,sec^{-1}\,Mpc^{-1}}$, which improves significantly the concordance with observations. While there is no good reason to expect such a value for the cosmological constant, there is no physical principle that would rule such out.

1. Introduction

Over the past decade the infusion of ideas from particle physics with implications for the earliest history of the Universe and a growing body of cosmological data that can test these implications have led to a renaissance in cosmology. Several key cosmological parameters that seemed beyond the realm of explanation or prediction, can now be "predicted" by very well motivated theories of the early Universe. Among them are the baryon asymmetry of the Universe, the curvature radius of the Universe, the spectrum of primeval density perturbations, and the quantity and composition of matter in the Universe. Knowledge of these parameters is crucial to understanding how the Universe evolved to its present state—especially how structure formed.

Foremost of the attractive scenarios of the early Universe is inflation.[1] It provides a comprehensive scenario for the earliest history of the Universe and makes a number of robust predictions: (i) spatially-flat Universe;[#1] (ii) Harrison–Zel'dovich spectrum of scale-

---

[1]The spatially-flat Einstein–de Sitter model is favored for other reasons as well: (i) temporal Copernican principle—if $\Omega \neq 1$ the deviation of Ω from unity grows as a power of the scale factor, begging one to ask

invariant curvature perturbations;[2] and (iii) a spectrum of relic gravitational waves.[3]

(Some might dispute the "robustness" of these predictions. For example, it is not impossible for the density perturbations to be nonscale invariant,[4] and isocurvature perturbations can also arise.[5] It is possible to have just enough inflation so that Ω today is less than unity, although such a model begs the question of why the curvature radius is just today becoming comparable to the Hubble radius and would likely be in conflict with the observed large-scale isotropy and homogeneity as our inflationary region would be comparable in size to the present Hubble radius (see Silk and Turner[4]). The three inflationary predictions mentioned above are about as robust as theoretical predictions come! For further discussion of the "inflationary paradigm" see Ref. 6.)

The first two of these predictions have very important implications for structure formation, as they provide the initial data: the nature of the density perturbations, and the quantity and composition of matter in the Universe. Taking the simplest flat Universe model—one with zero cosmological constant—flatness ($\Omega_{TOT} = 1$) together with the primordial nucleosynthesis constraint to the baryon density[7]—$0.011h^{-2} \lesssim \Omega_B \lesssim 0.019h^{-2}$—implies that most of the matter in the Universe is nonbaryonic.[#2] (The present Hubble parameter $H_0 = 100h\,\mathrm{km\,sec^{-1}\,Mpc^{-1}}$ and $\Omega_i = \rho_i/\rho_{CRIT}$ is the fraction of critical density contributed by species i.) There are a number of early Universe relics that are promising candidates for the nonbaryonic component of the mass density: an axion of mass $10^{-6}\,\mathrm{eV}$ to $10^{-4}\,\mathrm{eV}$; a neutralino of mass from about $10\,\mathrm{GeV}$ to about $3\,\mathrm{TeV}$; and a light neutrino of mass $90h^2\,\mathrm{eV}$. (For a discussion of particle dark-matter candidates see Refs. 11.)

The neutrino is referred to as hot dark matter: Relic neutrinos have velocities close to the speed of light around the time the Universe became matter dominated and perturbations on scales less than about $40\,\mathrm{Mpc}/(m_\nu/30\,\mathrm{eV})$ are damped by neutrino free streaming. Structure forms from the "top down:" large objects (superclusters) form and then fragment into galaxies. Inflation-produced density perturbations and hot dark matter seem to be ruled out because galaxies form too late.[13]

The axion and neutralino behave as cold dark matter: Around the time the Universe becomes matter-dominated they have very small velocities, free streaming is unimportant, and perturbations on small scales survive unscathed.[#3] Inflation-produced density perturbations and cold dark matter is a far more promising scenario. Indeed, some (including this author) have called it the most well motivated, most detailed, and most successful model for structure formation yet proposed![15]

---

why Ω is just now beginning to differ from unity; and (ii) structure formation—in spatially-open models there is less time for the growth of density perturbations and larger initial perturbations are required; in fact, $\Omega \lesssim 0.3$ models with adiabatic density perturbations are inconsistent with the isotropy of the CMBR.

[2]Several means for avoiding the nucleosynthesis constraint have been suggested;[8,9] the one that has attracted the most interest involves inhomogeneities in the baryon-to-photon ratio, produced by a strongly first-order quark-hadron transition.[9] While this scenario initially looked promising, it is now clear that the light-element abundances predicted severely conflict with the observed abundances.[10]

[3]There is an intermediate possibility—referred to as warm dark matter—where the damping scale is about $1\,\mathrm{Mpc}$ (scale of galaxies); this case arises for a relic of mass about $1\,\mathrm{keV}$ with abundance about one-tenth that of a neutrino species.[14]

2. Successes of Cold Dark Matter (CDM)

Specifically, the cold dark matter scenario is: a flat Universe whose composition is $\Omega_B \sim 0.1 \ll \Omega_{CDM} \sim 0.9$ with $h \sim 0.5$ (to have a sufficiently old Universe) and inflation-produced Harrison–Zel'dovich curvature perturbations whose spectrum after the epoch of matter–radiation equality is[15]

$$|\delta_k|^2 = \frac{A k}{(1 + \beta k + \omega k^{1.5} + \gamma k^2)^2}. \tag{1}$$

Here δ_k is the amplitude of the Fourier component of comoving wavenumber k $(\equiv 2\pi/\lambda)$, A is an overall normalization constant, $\beta = 1.7(\Omega h^2)^{-1}$ Mpc, $\omega = 9.0(\Omega h^2)^{-1.5}$ Mpc$^{1.5}$, and $\gamma = 1.0(\Omega h^2)^{-2}$ Mpc^2. The epoch of matter–radiation equality, $t_{EQ} = 4.4 \times 10^{10} (\Omega h^2)^{-2}$ sec and $T_{EQ} = 5.5 (\Omega h^2)$ eV, is when subhorizon-sized perturbations can begin to grow. The *rms* mass fluctuation on scale λ, $(\delta M/M)_\lambda$, is related to δ_k by: $\delta M/M \simeq k^{3/2}|\delta_k|/\sqrt{2}\pi$. The spectrum given by Eq. (1) is characterized by $\delta M/M \to \lambda^{-2}$ for $\lambda \gg \lambda_{EQ} = 13(\Omega h^2)^{-1}$ Mpc and $\delta M/M \to$ const for $\lambda \ll \lambda_{EQ}$ (more precisely $\to \ln \lambda$). While the horizon-crossing amplitude of the inflation-produced curvature perturbations is scale-invariant, the spectrum of perturbations after matter–radiation equality has a scale: $\lambda_{EQ} = 13(\Omega h^2)^{-2}$ Mpc. That scale arises because subhorizon-sized density perturbations remain roughly constant in amplitude until the Universe becomes matter-dominated, and λ_{EQ} is the scale that crosses the horizon at matter–radiation equality. Note that the spectrum is a function of λ/λ_{EQ} only, and thus "shifts" right or left as $(\Omega h^2)^{-1}$. (For reasons that will soon become clear I have retained the Ω dependence throughout.)

Since the spectrum of perturbations decreases with increasing scale, small structures form first and larger structures form later[#4] ("bottom up" or hierarchical structure formation). Typical galaxies form relatively recently, red shifts $z \sim 1$ to 2, although "rare" objects such as QSOs and large radio galaxies can form earlier.[16] The formation of a galaxy begins with the dark matter halo; baryons within the extended halo then dissipate energy, collapse, and form the luminous disk. Rich clusters too should have formed relatively recently. The prediction of relatively recent galaxy and cluster formation is consistent with "deep" CCD exposures that reveal few high red-shift objects.[17] The successes of cold dark matter are many; they include:[18]

- Provides a detailed and comprehensive scenario

- Correctly accounts for many properties of galaxies including

    1. Number densities of galaxies of different types
    2. Internal properties of halos (flat rotation curves, rotation velocities, and mass densities)

- Accounts for observed galaxy clustering

- Predicts correct number density of clusters

- Accounts for clustering of clusters

[#4]Mergers also probably play an important role in the formation of larger objects.

- Predicts anisotropies of the cosmic microwave background radiation (CMBR) that are consistent with current limits and accessible in near-term experiments

In short, CDM is the most detailed and successful scenario of structure formation yet developed. The CDM Paradigm has served to focus and sharpen the questions that we ask about the formation of structure. At the very least CDM has served—and served well—as a foil for the growing database of cosmological observations.

3. Shortcomings of Cold Dark Matter

Cold dark matter is not without its shortcomings—perhaps serious enough to lead to its demise. For the most part its successful predictions involve the Universe on small scales— say less than about $20h^{-1}$ Mpc—where the observational data are relatively well established; its shortcomings involve observations on larger scales—where the data and their interpretations are less certain. The shortcomings of cold dark matter include:

- Predicts cluster–cluster correlation function amplitude that is about a factor of three too small

- Seems to predict less clustering on scales $\gtrsim 20h^{-1}$ Mpc than is indicated by recent determinations of the angular correlation function for the APM catalogue[19]

- May not be able to account for the large voids and the distribution of galaxies on thin surfaces surrounding voids seen in the CfA slices and in other surveys[20]

- May not be able to account for coherent structures as large and as thin as the so-called Great Wall[21]

- May not be able to account for the large bulk motion (about $700\,\mathrm{km\,s^{-1}}$) of the local $50h^{-1}$ Mpc neighborhood[22]

- May not be able to account for the "regularity" in red shifts seen in the recent pencil-beam survey of Broadhurst et al.[23]

These problems involve measurements that are on less firm ground and/or whose interpretations are less quantitative. For example, several authors have emphasized that the amplitude of the cluster–cluster correlation function[24] may have been overestimated due to selection effects associated with the way in which the Abell catalogue was constructed.[25] At present, there is no quantitative measure of the large-scale structure seen in the surveys mentioned, and to some eyes, numerical simulations of cold dark matter produce voids, Great Walls, and even red shift periodicity.[26]

The peculiar velocity field is a very powerful probe of the density field: Inhomogeneities in the matter distribution lead to peculiar velocities, and in linear theory $(\delta v/c)_\lambda \sim \Omega^{0.6}(\lambda/H_0^{-1})(\delta\rho/\rho)_\lambda$. The peculiar velocity field is almost unique in its ability to probe the density field; most other observations, e.g., red shift surveys, only determine the distribution of bright galaxies. However, peculiar velocities are difficult to measure because an accurate, independent measure of the distance to a galaxy is required. Moreover, the interpretation of the data is subtle. *If*, as the bulk motion data seem to indicate, a Great Attractor of mass $10^{16} M_\odot$ at a distance of about $40h^{-1}$ Mpc exists, this poses a real difficulty for CDM.

All of the above observations suggest that the cold dark matter scenario is deficient in large-scale power. There are other worrisome cosmological data:

(1) Age problems. The present age of a matter-dominated Einstein–de Sitter model $t_0 = 2/3H_0 \simeq 6.5h^{-1}$ Gyr. If the Hubble constant is greater than $65\,\mathrm{km\,s^{-1}\,Mpc^{-1}}$, then the age of the Universe is less than 10 Gyr, an age that is at best marginally consistent with other independent determinations. Conventional CDM all but requires $h = 0.5$. Likewise, the age of the Universe at a given epoch, $t(z) = 2H_0^{-1}/3(1+z)^{3/2}$, scales as H_0^{-1}; for a larger value of H_0 there is less time for an object at a given red shift to have evolved to its observed state. This may already be a problem for some high red-shift objects that appear highly evolved.[17]

(2) The Ω problem. Measurements of the mass density clearly indicate that the luminous component of the mass density is very small: $\Omega_{\mathrm{LUM}} \lesssim 0.01$. Determinations of the mass density "associated with bright galaxies" indicate that $\Omega_{\mathrm{ABG}} \simeq 0.1 - 0.3$, far greater than Ω_{LUM}, but significantly less than the predicted value of unity. There are ways of accommodating this disappointing fact. Since no rotation curve of a spiral galaxy has been seen to "turn over," the dark halos associated with spiral galaxies could be considerably larger than present estimates, perhaps even large enough to provide $\Omega = 1$. Likewise, it is possible that clusters are much larger than the distribution of galaxies indicate (e.g., if dynamical friction has caused galaxies to sink deep into the cluster potential). There is also the possibility that there is considerable mass density in unseen, low-luminosity galaxies that are more smoothly distributed—so-called biased galaxy formation.

(It should be mentioned that some determinations of Ω do give values close to unity; e.g., the reconstruction of the local peculiar velocity field using the distribution of matter as determined by the IRAS catalogue of infrared-selected galaxies provides a preliminary determination: $\Omega^{0.6}/b = 1.0 \pm 0.3$, where $1 \lesssim b \lesssim 3$ is the biasing factor.[27] Loh and Spillar[28] have used the galaxy count–red shift test with a sample of about 1000 field galaxies—red shifts out to 0.75—to infer $\Omega = 0.9^{+0.7}_{-0.5}$.[#5])

(4) Galaxy counts. The number of galaxies observed in a given solid angle $d\omega$ and given red shift interval dz depends upon the number density of galaxies $n_{\mathrm{GAL}}(z)$ and the cosmological model:

$$\frac{dN_{\mathrm{GAL}}}{d\omega dz} = \frac{n_{\mathrm{GAL}}(z)[zq_0 + (q_0 - 1)(\sqrt{2q_0 z + 1} - 1)]^2}{H_0^3(1+z)^3 q_0^4[1 - 2q_0 + 2q_0(1+z)]^{1/2}},$$

$$\simeq z^2 n_{\mathrm{GAL}}(z)[1 - 2(q_0 + 1)z + \cdots]/H_0^3, \tag{2}$$

where $q_0 = \Omega/2$ (for $\Lambda = 0$), and the second expression is valid for small z. For fixed number density of galaxies, the galaxy count increases with decreasing Ω (or q_0) because of the increase in spatial volume. The test has great cosmological leverage. Recent deep galaxies counts indicate an excess of galaxies at higher red shifts—indicative of a low value

---

[5]Their result has drawn much criticism; in part because their red shifts are not spectroscopically determined (they are determined by multi-band photometry) and because their results are sensitive to the assumptions that they make about galactic evolution.[28]

of Ω.[29] (If galaxy mergers are important—as they may well be in CDM—the number density of galaxies at higher red shifts would be expected to be larger.)

To summarize, the shortcomings of cold dark matter are deficient large-scale structure, deficient galaxy counts, the age problems, and the Ω problem.[#6] No one of these problems is sufficiently troublesome to falsify the cold dark matter paradigm—yet—but taken together they are worrisome. As we shall see, the addition of a cosmological constant simultaneously addresses all of these problems.

4. A Relic Cosmological Constant[#7]

The basic idea is simple; retain the flat Universe model, but add a cosmological constant. The model I discuss here is: (i) Hubble constant of around $70\,\mathrm{km\,s^{-1}\,Mpc^{-1}}$ ($h = 0.7$)—a nice compromise value; (ii) $\Omega_B \sim 0.03$—near the central value implied by nucleosynthesis; (iii) $\Omega_{\mathrm{CDM}} \sim 0.17$—sufficiently greater than the baryonic component so that the mass density is dominated by that of the cold dark matter; (iv) $\Omega_\Lambda = 0.8$—cosmological constant corresponding to an energy density $\rho_\Lambda \equiv \Omega_\Lambda \rho_{\mathrm{CRIT}} \simeq 3.2 \times 10^{-47}\,\mathrm{GeV}^4 = (2.4 \times 10^{-3}\,\mathrm{eV})^4$. I am not wed to these particular values and use them for definiteness. I will leave the question of motivation for the end. (If the ratio of Ω_{CDM} to Ω_B is somewhat smaller, decoupling can have an effect on the spectrum of density perturbations, which is to further boost power on large scales.[32] If the "best-fit model" is still deficient in large-scale power, this effect could help.)

For this model the total matter contribution $\Omega_{\mathrm{NR}} = \Omega_{\mathrm{CDM}} + \Omega_B = 0.2$, and today the vacuum energy density dominates the matter energy density by a factor of four. In general the ratio $\rho_{\mathrm{NR}}/\rho_\Lambda = 0.25(1 + z)^3$. At red shifts greater than about $z_\Lambda \simeq 0.59$ the matter energy density dominates, and the model behaves just a flat, CDM model.[#8] To determine when this model becomes matter dominated one simply sets $\Omega h^2 = \Omega_{\mathrm{NR}} h^2 \simeq 0.098$: $T_{\mathrm{EQ}} = 0.54\,\mathrm{eV}$; $t_{\mathrm{EQ}} \simeq 4.5 \times 10^{12}\,\mathrm{sec}$; and $z_{\mathrm{EQ}} \simeq 2300$. Once the radiation energy density is negligible ($z \ll z_{\mathrm{EQ}}$), the scale factor evolves as

$$R(t) = \left(\frac{\Omega_{\mathrm{NR}}}{\Omega_\Lambda}\right)^{1/3} \sinh^{2/3}\left(3\sqrt{\Omega_\Lambda} H_0 t/2\right), \qquad (3)$$

where the value of the scale factor today is taken to be one.

THE Ω PROBLEM

A cosmological constant behaves just like a uniform mass density (with equation of state $p = -\rho$). As such, it would not affect determinations of Ω based upon dynamics (e.g.,

---

[6]It is interesting to note that a neutrino-dominated Universe could also help with the Ω problem and the deficiency of large-scale structure. Because of their high velocities neutrinos tend to remain more smoothly distributed, and because structure forms from the "top down" there is more power on large scales.

[7]Cosmologists dating back to Einstein have "resorted" to a cosmological constant to solve their problems—Einstein to obtain static solutions, Hoyle, and Bondi and Gold to resolve the age crisis when the Hubble time was only 2 Gyr, and more recently Turner, Steigman, and Krauss[30] and Peebles[31] to solve the Ω problem.

[8]The very recent transition to vacuum energy domination occurs because the ratio of matter energy density to vacuum energy density varies rapidly, as R^{-3}.

galactic halos and cluster virial masses). These measurements of the masses of tightly bound systems are insensitive to the contribution of a uniform background energy density because the average density in these objects is much greater than the average density of the Universe. Likewise, determinations of Ω based upon the peculiar velocities induced by the lumpy matter distribution would only reveal the lumpy, matter component. Thus, all current dynamical determinations that indicate $\Omega \simeq 0.1 - 0.3$, would be consistent with a flat Universe ($\Omega = 1$) with $\Omega_{NR} = 0.2$.

THE AGE PROBLEMS

As is well appreciated the addition of a cosmological constant *increases* the age of a flat Universe. The age of a Λ model is

$$t(z) = \frac{2H_0^{-1}}{3\sqrt{\Omega_\Lambda}} \sinh^{-1}\left[\sqrt{\Omega_\Lambda/\Omega_{NR}}/(1+z)^{3/2}\right]; \tag{4a}$$

$$t_0 \equiv t(z=0) = \frac{2H_0^{-1}}{3\sqrt{\Omega_\Lambda}} \sinh^{-1}\left[\sqrt{\Omega_\Lambda/\Omega_{NR}}\right] = \frac{2H_0^{-1}}{3\sqrt{\Omega_\Lambda}} \ln\left[\frac{1+\sqrt{\Omega_\Lambda}}{\sqrt{\Omega_{NR}}}\right]. \tag{4b}$$

The present age of a Λ-model is always greater than $2H_0^{-1}/3$ and for $\Omega_\Lambda = 0.8$, $t_0 = 1.1H_0^{-1} \simeq 15.5\,\text{Gyr}$, an age which is comfortably consistent with the age as determined from the radioactive elements, the oldest globular clusters, and white-dwarf cooling (e.g., see Ch. 1 of Ref. 6 and references therein). Moreover, a Λ model is older than its matter-dominated counterpart at any given epoch, so that objects at a given red shift have had more time to evolve. For $z \gg z_\Lambda$, $t(z) \to 2H_0^{-1}/3\sqrt{\Omega_{NR}}(1+z)^{3/2}$, which is a factor of $\Omega_{NR}^{-1/2}$ older than a flat, matter-dominated model; at these early epochs the "best-fit model" is a factor of 1.6 older than the conventional CDM model.

LARGE-SCALE STRUCTURE

The spectrum of density perturbations at matter-radiation equality, $(\delta M/M) \propto k^{3/2}|\delta_k|$, decreases monotonically with λ and its wavelength scale is determined by the value of Ωh^2. The spectrum "shifts" to larger length scales as Ωh^2 is decreased. Using the conventional normalization, $(\delta M/M)_{\lambda \simeq 8h^{-1}\,\text{Mpc}} = 1$, decreasing Ωh^2 increases power on all scales greater than $8h^{-1}\,\text{Mpc}$. Said another way, the ratio of the characteristic scale in the spectrum, $\lambda_{EQ} = 13(\Omega h^2)^{-1}\,\text{Mpc}$, to the scale of nonlinearity in the Universe, $\lambda_{NL} \simeq 8h^{-1}\,\text{Mpc}$, is given by $\lambda_{EQ}/\lambda_{NL} \simeq 1.6/\Omega h$, and in the "best-fit model" this ratio is a factor of 3.5 greater than in conventional cold dark matter ($\Omega = 1$ and $h = 0.5$), implying more power on large scales.

To be specific, normalizing spectrum (1) according to $(\delta M/M)_{\lambda=8h^{-1}\,\text{Mpc}} = 1$,[#9] I find that: $A = 4.4 \times 10^6\,\text{Mpc}^4$ for $\Omega = 1$ and $h = 0.5$ (conventional CDM) and $A = 2.5 \times 10^7\,\text{Mpc}^4$ for $\Omega_{NR} = 0.2$ and $h = 0.7$ ("best-fit model"). On large scales ($\lambda \gg \lambda_{EQ}$) $\delta M/M \propto \sqrt{A}/\lambda^2$; it follows that $\delta M/M$ for the "best-fit model" is a factor of 4.7 bigger on large scales.

---

[9] I have used the "top hat" window function [$W(r) = 1$ for $r \leq r_0$ and $= 0$ for $r \geq r_0$] to define M, so that $(\delta M/M)^2 = (9/2\pi^2) \int_0^\infty k^2|\delta_k|^2 [\sin(kr_0)/k^3r_0^3 - \cos(kr_0)/k^2r_0^2]^2\,dk$, where $r_0 = 8h^{-1}\,\text{Mpc}$.

GROWTH OF DENSITY PERTURBATIONS

Subhorizon-sized, linear density perturbations grow as the scale factor during the matter-dominated regime ($z \lesssim z_{EQ} \simeq 23000\Omega h^2$), and remain roughly constant in amplitude when the Universe is radiation dominated, curvature dominated ($z \lesssim z_{CURV} \simeq \Omega^{-1} - 2$; $z_{CURV} \simeq 3$ for $\Omega = 0.2$), or vacuum-energy dominated ($z_\Lambda \simeq [\Omega_\Lambda^{-1} - 1]^{1/3} - 1 \simeq 0.59$). For a nonflat, $\Omega = 0.2$ model the reduction in the growth of perturbations relative to a flat model is very significant: about a factor of 20. By contrast, in flat-Λ models perturbations grow almost unhindered until the present,[31,33] and in the "best-fit model" the growth factor is only a factor of 0.8 less than z_{EQ}, or about 1800. For conventional CDM the growth factor $z_{EQ} \simeq 5800$, only about a factor of three more growth.

MICROWAVE ANISOTROPIES

For conventional CDM the predicted CMBR temperature anisotropies are about a factor of three or so below the current level of observed isotropy (depending upon the angular scale and biasing factor b).[34] One might worry that because the "best-fit model" has more power on large scales and the growth factor for perturbations is smaller the predicted CMBR anisotropies might violate current bounds. That is not the case. The reason involves the angular size on the sky θ of a given scale λ at epoch z:

$$\theta(\lambda, z) = \lambda/r(z); \qquad (5a)$$

$$r(z) = \int_{t(z)}^{t_0} \frac{dt}{R(t)} = \frac{2H_0^{-1}}{3\Omega_\Lambda^{1/6}\Omega_{NR}^{1/3}} \int_{\sinh^{-1}\left[\sqrt{\Omega_\Lambda/(1+z)^3\Omega_{NR}}\right]}^{\sinh^{-1}\left[\sqrt{\Omega_\Lambda/\Omega_{NR}}\right]} \frac{du}{\sinh^{2/3} u}, \qquad (5b)$$

where $r(z)$ is the coordinate distance to an object at red shift z. In a flat, matter-dominated model $r(z) = 2H_0^{-1}\left[1 - 1/\sqrt{1+z}\right] \to 2H_0^{-1}$ for $z \gg 1$, and $\theta(\lambda, z \gg 1) \simeq 34.4''\,(\lambda/h^{-1}\,\mathrm{Mpc})$. For the "best-fit model" $r(z \gg 1) \simeq 3.9H_0^{-1}$ and $\theta(\lambda, z \gg 1) \simeq 17.7''\,(\lambda/h^{-1}\,\mathrm{Mpc})$.

In a flat Λ-model the horizon is further away and a given length scale has a smaller angular size. Since the temperature fluctuations on a given angular scale are related to the density perturbations on the length scale that subtends that angle at decoupling, in the "best-fit model" temperature fluctuations on a given angular scale are related to density perturbations on a *larger* scale λ. While the "best-fit model" has more power on a *fixed* (large) length scale, a fixed angle θ corresponds to a *larger* length scale, where the amplitude of perturbations is smaller because $\delta M/M$ decreases with λ.

Consider the temperature fluctuations on large-angular scales ($\theta \gg 1°$); they arise due to the Sachs–Wolfe effect and $(\delta T/T)_\theta \simeq (\delta\rho/\rho)_{HOR}/2$ on the scale $\lambda(\theta)$ when that scale crossed inside the horizon. For the Harrison–Zel'dovich spectrum the horizon-crossing amplitude is constant, so that $\delta T/T$ is independent of angular scale (for $\theta \gg 1°$). The CMBR quadrupole anisotropy is related to the amplitude of the perturbation that is just now crossing inside the horizon: $\lambda_{HOR} \sim 2H_0^{-1} \sim 12000\,\mathrm{Mpc}$ (conventional CDM) and $\lambda_{HOR} \sim 3.9H_0^{-1} \sim 16700\,\mathrm{Mpc}$ ("best-fit model"). Evaluating the normalized spectra on these scales it follows that the large-angle temperature fluctuations in the "best-fit model" are only a factor of 1.2 larger than for conventional CDM, in spite of the fact that the "best-fit model" has significantly more power on large scales.

The amplitude of the temperature fluctuations on small angular scales ($\theta \ll 1°$) is proportional to the amplitude of the density perturbations at the time of decoupling ($z_{\text{DEC}} \sim 1000$), on the scale $\lambda(\theta)$. In the "best-fit model" perturbations have grown by a factor of about $0.8z_{\text{DEC}}$ since decoupling, while those in the "most well motivated model" have grown by a factor of z_{EQ}. On the other hand the length scale corresponding to the angular scale θ is larger for the "best-fit model." The net result is that the temperature fluctuations on an angular scale of $1°$ are also only about a factor of 1.2 larger.

GALAXY COUNTS—AND OTHER KINEMATIC TESTS

Because the coordinate distance to an object of given red shift is greater in a flat Λ model, there is greater volume per red shift interval per solid angle, which increases the number of galaxies in $dzd\omega$. (The galaxy count test is discussed in more detail in Refs. 31 and 33.) To see roughly how this goes, consider the deceleration parameter of Sandage[35]

$$q_0 \equiv -\frac{\ddot{R}}{R_0 H_0^2} = \Omega(1 + 3p/\rho)/2 = (1 - 3\Omega_\Lambda)/2 \simeq -1.2, \tag{6}$$

where Ω is the total energy density ρ divided by the critical energy density and p is the total pressure. From Eq. (2) one can see that the galaxy-number count is significantly increased by the addition of a cosmological constant, $dN_{\text{GAL}}/dz = z^2 n_{\text{GAL}}[1 + 0.4z + \cdots]$ compared to $z^2 n_{\text{GAL}}[1 - 3z + \cdots]$.

There are other "kinematic tests" that could prove useful; they include the red shift–luminosity test (Hubble diagram), angle–red shift test, and look-back time–red shift test (for further discussion see Ref. 33). For small red shift different models can be parameterized in terms of Sandage's q_0. In this regard, the "best-fit model" is characterized by $q_0 = -1.2$ (for comparison, a conventional low-Ω, negatively curved model has $q_0 = \Omega/2$). To date, none of these tests have proved definitive, though some put great stock in the potential of the infrared ($2.2\mu m$ or K band) version of the Hubble diagram.[36]

LARGE-SCALE MOTIONS

The *rms* peculiar velocity of a volume defined by the "window function" $W(r)$, averaged over all such volumes in the Universe, is

$$\langle v^2 \rangle = \frac{1}{2\pi^2} \int_0^\infty k^2 |v_k|^2 |W(k)|^2 dk, \tag{7}$$

where in the linear perturbation regime the Fourier component of the peculiar velocity field $v_k(t) = -ikR(t)\dot{\delta}_k(t)/|k|^2$. For a flat, matter-dominated Universe, $|v_k| = H_0|\delta_k|/k$; while for a flat model with a cosmological constant $|v_k| = \Omega_{\text{NR}}^{0.57} H_0|\delta_k|/k$ (see Refs. 31 and 37). Using a gaussian window function $[W_{r_0}(r) = \exp(-r^2/2r_0^2)]$ and normalizing the spectrum as above, the *rms* peculiar velocity expected on the scale $r_0 = 50h^{-1}$ Mpc is[37]

$$v_{50} \simeq 83h^{-0.9} \, \text{km s}^{-1} \simeq 160 \, \text{km s}^{-1} \qquad (\Omega = 1, \ h = 0.5);$$

$$v_{50} \simeq 83\Omega_{\rm NR}^{-0.33} h^{-0.9}\,{\rm km\,s^{-1}} \simeq 200\,{\rm km\,s^{-1}} \qquad (\Omega_{\rm NR} = 0.2,\ h = 0.7).$$

While the *rms* peculiar velocity on the scale of 50 Mpc is still far short of $700\,{\rm km\,s^{-1}}$, it is larger, owing to fact that there is more power on large scales.[#10]

5. Concluding Remarks

Introducing a cosmological constant helps to resolve all the shortcomings of conventional cold dark matter ($\Omega_{\rm CDM} = 0.9$, $\Omega_B = 0.1$, and $h = 0.5$). In particular it eases the age problems, resolves the Ω problem, increases the number of galaxies expected in a given red shift interval, and leads to more power on large scales. At the same time, the predicted CMBR temperature anisotropies are only a factor of 1.2 larger than for conventional CDM.

The model can be tested in a number of ways, although the usual dynamical means of inferring Ω are not sensitive to Ω_Λ. Given its large deceleration parameter, $q_0 = 0.5(1 - 3\Omega_\Lambda) \simeq -1.2$, several of the classic kinematic tests—angle–red shift, galaxy count–red shift, lookback-time–red shift, and red shift-luminosity—may prove useful. It may well be that new tests that key on the the "hallmarks" of the Λ model—larger volume, older Universe, and more distance to the horizon—can be found.[#11]

The "best-fit" model has implications for dark matter. The abundance of a relic particle species $\Omega_X h^2$ is related to its fundamental properties (mass, couplings, etc.). In the "best-fit model" the value of $\Omega_X h^2$ is a factor of almost three smaller than in the conventional CDM model. For a thermal relic such as a neutralino, the relic abundance $\Omega_X h^2$ is proportional to the inverse of the annihilation cross section, implying that the annihilation cross section is about a factor of three larger. This fact has implications for dark matter searches: (i) The rates for indirect detection schemes that rely upon the annihilation products, e.g., high-energy neutrinos from annihilations in the sun, or high-energy positrons from annihilations in the halo, are increased by a factor of about three; (ii) The rates for direct detection, e.g., in bolometric or ionization detectors, which depend upon the cross section for elastic scattering with matter, are increased by the same factor because the elastic scattering cross section is related to the annihilation cross section by crossing symmetry.

One might wonder if in a Λ model one could revive hot dark matter, or do away with exotic dark matter all together. The revival of a neutrino-dominated Universe does not seem likely. The required neutrino mass is about 8 eV, implying a neutrino-damping scale of about 150 Mpc, which would further exacerbate the problems of a neutrino-dominated Universe. In a baryon-dominated Λ model perturbations are damped (by photon diffusion) on scales smaller than $\lambda_{\rm SILK} \simeq 1\,(\Omega_B h^2)^{-3/4}\,{\rm Mpc} \simeq 20\,{\rm Mpc}$, which results in the original Zel'dovich pancake scenario with supercluster-sized baryon pancakes. Because perturbations continue to grow until almost the present this scenario is better than a nonflat, low-Ω baryon-dominated model; however, this scenario is likely to have problems with CMBR anisotropies (re-ionizing the Universe might relieve this problem on angular scales smaller

---

[10]The comparison of theoretical expectations to the peculiar-velocity data of the Seven Samurai[22] is far more complicated than just computing $\langle v^2 \rangle$ for a gaussian window function.[38] The point I wish to make here is that adding a cosmological constant increases peculiar velocities.

[11]Along these lines, E.L. Turner has recently used the frequency of multiple image lensing of QSOs to argue against a large value of Λ, perhaps even precluding $\Omega_\Lambda = 0.8$.[39]

than about $7°$).[#12]

Finally, let me address the motivation for the "best-fit model." As its name suggests, it is a model motivated by observational data (and their current interpretation) and not aesthetics: Conventional cold dark matter is clearly better motivated. While the conventional CDM model has one question to answer—why the ratio of the baryon density to that of cold dark matter is of order unity[42]—in the "best-fit model" one must also address "why now?"—why is the cosmological constant just now becoming dynamical important? (This problem is similar to the flatness problem, where the question is, why is the curvature radius just now becoming comparable to the Hubble radius?) Moreover, there is the issue of the cosmological constant itself: At present there is every reason to expect a cosmological constant $\rho_\Lambda = \Lambda/8\pi G \sim m_{Pl}{}^4$ that is some 122 orders of magnitude larger than observations permit.[#13] (Supersymmetry *might* be able to help in this regard, reducing the estimate to $\rho_\Lambda \sim G_F^{-2}$, which is only 56 orders of magnitude too large!) The strongest statement that one can make in defense of a relic cosmological constant of the desired size is that no good argument exists for *excluding* it!

The additional of a cosmological constant to the cold dark matter paradigm resolves a number of its apparent shortcomings—and is the sole motivation for introducing it. Cold dark matter *sans* cosmological constant is still the most well motivated model. One should keep in mind that the observations or their interpretations could change, and cold dark matter could once again become both the most well motivated model and the best-fit model.

Acknowledgments

This work was supported in part by the DOE (at Chicago).

References

1. A. Guth, *Phys. Rev. D* **23**, 347 (1981); A.D. Linde, *Phys. Lett. B* **108**, 389 (1982); A. Albrecht and P.J. Steinhardt, *Phys. Rev. Lett.* **48**, 1220 (1982); D. La and P.J. Steinhardt, *Phys. Rev. Lett.* **62**, 376 (1989).

2. J.M. Bardeen, P.J. Steinhardt, and M.S. Turner, *Phys. Rev. D* **28**, 679 (1983); A. H. Guth and S.-Y. Pi, *Phys. Rev. Lett.* **49**, 1110 (1982); A. A. Starobinskii, *Phys. Lett. B* **117**, 175 (1982); S. W. Hawking, *Phys. Lett. B* **115**, 295 (1982).

3. V.A. Rubakov, M. Sazhin, and A. Veryaskin, *Phys. Lett. B* **115**, 189 (1982); R. Fabbri and M. Pollock, *Phys. Lett. B* **125**, 445 (1983); L. Abbott and M. Wise, *Nucl. Phys. B* **244**, 541 (1984); B. Allen, *Phys. Rev. D* **37**, 2078 (1988); M.T. Ressell and

---

[12] Peebles[40] has recently discussed this possibility. While re-ionization might be able to erase the primary CMBR fluctuations, the secondary fluctuations may be problematic.[41]

[13] There is one interesting explanation of why the cosmological constant is "probably" zero: Coleman and others[43] have argued that due to wormhole effects the wavefunction of the Universe is very sharply peaked at zero cosmological constant.

M.S. Turner, preprint (1989); M.S. Turner and F. Wilczek, *Phys. Rev. Lett.*, in press (1990).

4. See, e.g., J. Silk and M.S. Turner, *Phys. Rev. D* **35**, 419 (1987); D. Salopek, J.R. Bond, and J.M. Bardeen, *Phys. Rev. D* **40**, 1753 (1989).

5. D. Seckel and M.S. Turner, *Phys. Rev. D* **32**, 3178 (1985); A.D. Linde, *Phys. Lett. B* **158**, 375 (1985); M. Axenides, R. Brandenberger, and M.S. Turner, *Phys. Lett. B* **128**, 178 (1983); A. D. Linde, *Phys. Lett. B* **158**, 375 (1985); M. S. Turner, A. Cohen, and D. Kaplan, *Phys. Lett. B* **216**, 20 (1989).

6. E.W. Kolb and M.S. Turner, *The Early Universe* (Addison–Wesley, Redwood City, CA 1990), Ch. 8.

7. J. Yang, M.S. Turner, G. Steigman, D.N. Schramm, and K.A. Olive, *Astrophys. J.* **281**, 493 (1984); K.A. Olive, D.N. Schramm, G. Steigman, and T.P. Walker, *Phys. Lett. B* **236**, 454 (1990).

8. See, e.g., S. Dimopoulos, R. Esmailzadeh, L.J. Hall, and G.D. Starkman, *Phys. Rev. Lett.* **60**, 7 (1988); J.G. Bartlett and L.J. Hall, *Phys. Rev. Lett.*, in press (1990).

9. J. Applegate, C.J. Hogan, and R.J. Scherrer, *Phys. Rev. D* **35**, 1151 (1987); C. Alcock, G. Fuller, and G. Mathews, *Astrophys. J.* **320**, 439 (1987); R.A. Malaney and W.A. Fowler, *Astrophys. J.* **333**, 14 (1988).

10. See e.g., H. Kurkio-Sunio et al., *Phys. Rev. D* **38**, 1091 (1988), or the review by H. Kurkio-Sunio et al., *Astrophys. J.* **353**, 406 (1990); K. Sato and N. Terasawa, *Phys. Rev. D* **39**, 2893 (1989).

11. See, e.g., M. S. Turner, in *Dark Matter in the Universe*, eds. J. Kormendy and G. Knapp (Reidel, Dordrecht, 1989) or in *Proc. of Nobel Symposium No. 79: The Birth and Early Evolution of Our Universe*, eds. B. Gustafsson, Y. Nilsson, and B. Skagerstam (WSPC, Singapore, 1991); V. Trimble, *Ann. Rev. Astron. Astrophys.* **25**, 425 (1987); J. Primack et al., *Ann. Rev. Nucl. Part. Sci.* **38**, 751 (1989).

12. J.R. Bond and A. Szalay, *Astrophys. J.* **276**, 443 (1983).

13. S.D.M. White, C. Frenk, and M. Davis, *Astrophys. J.* **274**, L1 (1983); *ibid* **287**, 1 (1983); J. Centrella and A. Melott, *Nature* **305**, 196 (1982).

14. J.R. Bond, A.S. Szalay, and M.S. Turner, *Phys. Rev. Lett.* **48**, 1636 (1982). A keV axino is an example of a well motivated warm dark matter candidate; see, K. Rajagopal, M.S. Turner, and F. Wilczek, *Nucl. Phys. B*, in press (1991).

15. For a discussion of the cold dark matter paradigm for structure formation see Ch. 9 of Ref. 6; C.D.M. Frenk, in *Proc. of Nobel Symposium No. 79: The Birth and Early Evolution of Our Universe*, eds. B. Gustafsson, Y. Nilsson, and B. Skagerstam (WSPC, Singapore, 1991); or G. Efstathiou, in *Physics of the Early Universe*, eds. J.A. Peacock, A.F. Heavens, and A.T. Davies (Adam Higler, NY, 1990).

16. G. Efstathiou and M.J. Rees, *Mon. Not. R. astr. Soc.* **230**, 5p (1988).

17. L. Cowie, in *Proc. of Nobel Symposium No. 79: The Birth and Early Evolution of Our Universe*, eds. B. Gustafsson, Y. Nilsson, and B. Skagerstam (WSPC, Singapore, 1991); or in these proceedings.

18. G. Blumenthal, S.M. Faber, J.R. Primack, and M.J. Rees, *Nature* **311**, 517 (1984).

19. S.J. Maddox, G. Efstathiou, W.J. Sutherland, and J. Loveday, *Mon. Not. R. astr. Soc.* **242**, 43p (1990).

20. R.P. Kirshner, A. Oemler, P. Schechter, and S. Shectman, *Astrophys. J.* **248**, L57 (1981); V. de Lapparent, M.J. Geller, and J. Huchra, *Astrophys. J.*, **302**, L1 (1986).

21. M.J. Geller and J. Huchra, *Science* **246**, 897 (1989).

22. A. Dressler, D. Lynden-Bell, D. Burstein, R. Davies, S. Faber, R. Terlevich, and G. Wegner, *Astrophys. J.* **313**, L37 (1987).

23. T.J. Broadhurst, R.S. Ellis, D.C. Koo, and A.S. Szalay, *Nature* **343**, 726 (1990).

24. N.A. Bahcall, *Ann. Rev. Astron. Astrophys.* **26**, 631 (1988); N.A. Bahcall and R.M. Soneira, *Astrophys. J.* **270**, 20 (1983); M. Hauser and P.J.E. Peebles, *Astrophys. J.* **185**, 757 (1973).

25. W. Sutherland, *Mon. Not. R. astr. Soc.* **234**, 159 (1988); A. Dekel, G.R. Blumenthal, J.R. Primack, and S. Olivier, *Astrophys. J.* **338**, L5 (1989); S. Olivier, G.R. Blumenthal, A. Dekel, J.R. Primack, and D. Stanhill, *Astrophys. J.* **356**, 1 (1990).

26. S.D.M. White, C.D.M. Frenk, M. Davis, and G. Efstathiou, *Astrophys. J.* **313**, 505 (1987); D. Weinberg and J. Gunn, *Astrophys. J.* **352**, L25 (1990); C. Park, *Mon. Not. R. astr. Soc.* **242**, 59p (1990); C. Park and J.R. Gott, *Mon. Not. R. astr. Soc.*, in press (1990); J. Villumsen, Ohio State University preprint (1990); E. Bertschinger and J. Gelb, in preparation (1990).

27. E. Bertschinger, A. Dekel, and A. Yahil, in preparation (1990); also, M. Rowan-Robinson et al., *Mon. Not. R. astr. Soc.*, in press (1990); M.A. Strauss and M. Davis, in *Large-scale Motions in The Universe: A Vatican Study Week*, eds. V.C. Rubin and G.V. Coyne (Princeton Univ. Press, Princeton, 1988).

28. E. Loh and E.J. Spillar, *Astrophys. J.* **307**, L1 (1986); also see S. Bahcall and S. Tremaine, *Astrophys. J.* **326**, L1 (1988) for criticism of the Loh-Spillar work.

29. L. Cowie, in *Proc. of Nobel Symposium No. 79: The Birth and Early Evolution of Our Universe*, eds. B. Gustafsson, Y. Nilsson, and B. Skagerstam (WSPC, Singapore, 1991), or in these proceedings; D.C. Koo and A.S. Szalay, *Astrophys. J.* **282**, 390 (1984); T. Shanks, P.R.F. Stevenson, R. Fong, and H.T. MacGillivray, *Mon. Not. R. astr. Soc.* **206**, 767 (1984); D. Koo, in *The Epoch of Galaxy Formation*, eds. T. Shanks, A.F. Heavens, and J.A. Peacock (Kluwer, Dordrecht, 1989).

30. M.S. Turner, G. Steigman, and L. Krauss, *Phys. Rev. Lett.* **52**, 2090 (1984).

31. P.J.E. Peebles, *Astrophys. J.* **284**, 439 (1984).

32. G.R. Blumenthal, A. Dekel, and J.R. Primack, *Astrophys. J.* **326**, 539 (1988).

33. J. Charlton and M.S. Turner, *Astrophys. J.* **313**, 495 (1987).

34. N. Vittorio and J. Silk, *Astrophys. J.* **285**, L39 (1984); N. Vittorio et al., *Astrophys. J.* **341**, 163 (1989); N. Vittorio et al., *Astrophys. J. (Lett.)*, in press (1990); A.C.S. Readhead et al., *Astrophys. J.* **346**, 566 (1989); J.R. Bond and G. Efstathiou, *Astrophys. J.* **285**, L44 (1984); *Mon. Not. R. astr. Soc.* **226**, 655 (1987); J.R. Bond, in *Frontiers in Physics: From Colliders to Cosmology*, eds. B. Campbell and F. Khanna (WSPC, Singapore, 1990); J.R. Bond, G. Efstathiou, P. Lubin, and P. Meinhold, *Phys. Rev. Lett.*, in press (1991); or Ch. 9 of Ref. 6.

35. A. Sandage, *Astrophys. J.* **133**, 355 (1961).

36. H. Spinrad and S. Djorgovski, in *Observational Cosmology*, eds. A. Hewitt, G. Burbidge, and Li-Zhi Fang (Reidel, Dordrecht, 1987), p. 129.

37. N. Vittorio and M.S. Turner, *Astrophys. J.* **316**, 475 (1987).

38. See e.g., N. Kaiser and O. Lahav, *Mon. Not. R. astr. Soc.* **237**, 129 (1989); A. Dekel, E. Bertschinger, and S. Faber, *Astrophys. J.*, in press (1990); E. Bertschinger, "Large-scale Motions in the Universe: A Review," to published in the *Proceedings of the 1990 Moriond Workshop on Astrophysics*, in press (1990); *Large-scale Motions in The Universe: A Vatican Study Week*, eds. V.C. Rubin and G.V. Coyne (Princeton Univ. Press, Princeton, 1988).

39. E.L. Turner, *Astrophys. J. (Lett.)*, in press (1990).

40. P.J.E. Peebles, in preparation (1990).

41. E.T. Vishniac, *Astrophys. J.* **322**, 597 (1987); and work in progress (1990).

42. M.S. Turner and B.J. Carr, *Mod. Phys. Lett. A* **2**, 1 (1987).

43. S. Coleman, *Nucl. Phys* **B310**, 643 (1988); also see, S. Hawking, *Phys. Lett. B* **134**, 403 (1984); E. Baum, *Phys. Lett. B* **133**, 185 (1983).

LEP PHYSICS AND THE EARLY UNIVERSE

LAWRENCE M. KRAUSS[a]
Center For Theoretical Physics And Dept. Of Astronomy
Sloane Laboratory, Yale University, New Haven, CT 06511

ABSTRACT : I review the implications of the Z decay measurements at LEP (and SLC) for the early universe: (a) The Z width measurements, when combined with non-accelerator data rule out GeV range Dirac neutrinos, Majorana neutrinos, and sneutrinos as WIMP dark matter, rule out most explicit "cosmions", and strongly constrain neutralinos. When combined with data from other experiments, this implies that any WIMP is likely to be heavier than 10-20 GeV, and could easily be in the O(100 GeV-1 TeV) region; (b) The limit on N_V, when combined with big-bang nucleosynthesis (BBN) estimates, is consistent with both baryonic and non-baryonic dark matter and allows one to probe the consistency of BBN itself; (c) direct searches for WIMPs will probably require new detectors, sensitive to spin dependent interactions on possibly heavy targets, with event rates which could be 3-4 orders of magnitude smaller than those expected at present detectors; (d) limits on the Higgs and the top quark also have an impact on possible new processes in the early universe.

1. Introduction: The Z Width Reviewed:

There was a great deal of discussion at this meeting about the beautiful COBE results on the cosmic microwave background. The measured black-body spectrum provides incredibly strong constraints on possible energy release mechanisms in the pre- and post-recombination universe, and also on exotic scenarios for the formation of large scale structure. An equally significant experimental curve has recently been produced in particle physics. Like the CMB spectrum, it is also one whose shape has been predicted for over 20 years, waiting for experimentalists to measure it. Also, like the CMB spectrum, a single determination of its overall shape has provided very powerful new limits on exotica beyond the standard model. I refer of course to the measurement at LEP (and SLC) of the shape of the Z boson resonance. In less than a year of running, these measurements of Z decay parameters have significantly constrained our picture of the world. It is remarkable how even a small amount of unambiguous data about so fundamental a system can so strongly direct theory. In particular, the new limits on Z width constrain not only the number of light neutrino species in nature, but also the existence of new particles of mass up to 45 GeV which might be produced in Z decay. Among these are included many of the leading Weakly Interacting Massive Particle (WIMP) candidates for dark matter [1]. Those which remain viable will be far more challenging for experimenters to detect directly.

Before discussing the detailed constraints, and because it is important for everything I will have to say here, I will first review in detail how one infers a limit on the number of neutrinos

---

[a] Research supported in part by an NSF Presidential Young Investigator Award and by the DOE

from the Z decay measurements made at SLC and LEP. Such an analysis is at the very least pedagogically useful, and it also demonstrates some non-intuitive features of the measurement.

When Z bosons are produced at e+e- colliders, the shape of the Z resonance is most easily determined by observing the hadronic decays of the Z, which have by far the largest branching ratio, and so lead to the best statistics. By counting the number of decays as a function of energy, and carefully measuring the luminosity of the machine at each energy, the cross section for Z production can be mapped out, and compared to the standard Breit-Wigner form:

$$\sigma_h = \sigma_h^o \frac{s\,\Gamma_z^2}{\left(s - M_z^2\right)^2 + s^2 \Gamma_z^2 / M_z^2} \left[1 + \delta_{rad}(s)\right] \tag{1}$$

Here, $\sqrt{s}$ is the center of mass energy, $[1+\delta(s)]$ accounts for initial state radiation, and the Breit-Wigner resonance form is then described by 3 parameters: M_z, Γ_z, and σ_h^o, the mass, the total width, and the height at the peak respectively. These are displayed in the figure below.

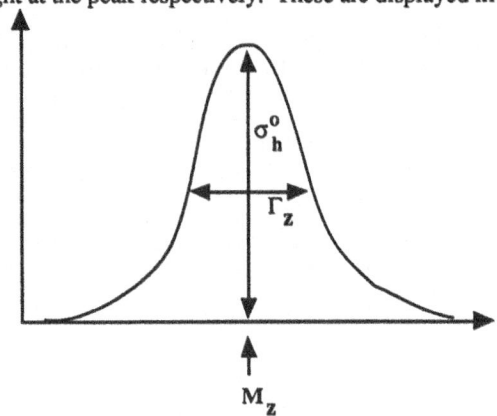

Not all these parameters are independent, however. In particular, in order that the integral under the curve give the total Z production cross section, the peak height σ^o and the width Γ are related, so that:

$$\sigma_h^o = \frac{12\pi\Gamma_e\Gamma_h}{M_z^2\Gamma_z^2} \tag{2}$$

where Γ_e and Γ_h are the Z partial decay widths into electrons and hadrons respectively.

The total Z width is *defined* to be: $\Gamma_z \equiv N_\nu\Gamma_\nu + 3\Gamma_e + \Gamma_h$, where Γ_ν is the partial width into massless neutrinos in the standard model, and lepton universality is assumed [1]. (Otherwise, the sum of partial widths into charged lepton pairs may be used in the definition.) This defines what is meant by the "number" of neutrinos, i.e.

$$N_\nu \equiv \frac{1}{\Gamma_\nu}\left(\Gamma_z - 3\Gamma_e - \Gamma_h\right) \tag{3}$$

One's naive expectation might be that to extract a limit on the number of neutrinos one would first fit the measurements to the Breit-Wigner form (1), determine a best fit value for Γ_z and then use (3). However, since each new neutrino species contributes about 6% to the width, an uncertainty in the width determination of 6% corresponds to an uncertainty of one extra neutrino type. Instead, since σ_h^o depends *quadratically* on Γ_z (because the integrated cross section for Z production is fixed), an extra neutrino species contributes about 12% to the peak cross section value. Thus, assuming the uncertainty in the peak cross section determination is no worse than that in the width, one can get *twice* as good a limit on the number of neutrinos by using the fit for σ_h^o and then using (2) to limit the number of neutrinos than one can get by using Γ_z alone. Moreover, σ_h^o is defined in (2) in terms of *ratios* of widths, and is thus less sensitive to uncertainties in the widths from the standard model, such as from the top quark mass, or the uncertainty in $\sin^2\theta_w$. Thus, using (2), we can write

$$N_\nu \equiv \frac{\Gamma_e}{\Gamma_\nu}\left(\sqrt{\frac{12\pi}{M_z^2 \sigma_h^o}}\sqrt{\frac{\Gamma_h}{\Gamma_e}} - 3\,\frac{\Gamma_h}{\Gamma_e}\right), \tag{4}$$

which expresses N_ν in terms of σ_h^o and in terms of the ratio Γ_h/Γ_e, which can either be extracted from the standard model, or more important, can be directly measured. In the latter case, the ratio Γ_e/Γ_ν must be taken from theory. (There is no such thing as completely model independent bound on N_ν.) The first limits on N_ν essentially used (2), (or (4))[2]. Next, the LEP groups began quoting limits from (3) and (4)[3]. Most recently, (4) has been used, with Γ_h/Γ_μ being directly measured. It is important, when comparing the results of different experiments, to make sure that the limits which are quoted are derived the same way, and also to attempt to use the firmest bounds. For example, the errors on the limits which were originally obtained using Γ_z are about twice as big as those obtained using σ_h.[4] Also a direct Γ_z bound is more susceptible to systematic errors in partial rates. For this reason a limit obtained by combining (3) and (4) is essentially dominated by the σ_h determination. (Also note that for this reason, the bound on extra neutrinos from (4) has an inherent insensitivity to unstable particles. Subsequent neutrino decays could effectively contribute to the measured hadronic partial width, and thus contribute to both the numerator and denominator of (2). The sensitivity to such an extra neutrino would then be reduced by a factor of 2. Other information from the experiments can be used however to directly rule out unstable neutrinos.)

Caveats aside, what is the current combined limit on N_ν from all the experiments? Below, I display the compilation I performed based on the initial LEP preprints[2], and the second round of preprints in March of this year, and finally the limits quoted at the International High Energy meeting in July, being as careful as possible to quote the limits derived from the same analysis in each experiment.

Experiment	N_ν Limit (Oct 89)	N_ν Limit (Mar 90)	N_ν Limit (July 90)
Mark II	2.8 ± 0.6	same	same
ALEPH	3.27 ± 0.30	3.03 ± 0.15	2.92 ± 0.12 (stat) ± 0.03 (thry)
DELPHI	$2.4 \pm 0.4 \pm 0.5$	3.05 ± 0.28	2.80 ± 0.16 (stat) ± 0.03 (thry)
L3	3.42 ± 0.48	3.11 ± 0.17[4]	2.95 ± 0.18 (stat) ± 0.03 (thry)
OPAL	3.12 ± 0.42	3.09 ± 0.19	2.84 ± 0.14 (stat) ± 0.03 (thry)
Average	3.13 ± 0.20	3.06 ± 0.09	(a) 2.89 ± 0.10 (stat +sys)*
			(b) 2.99 ± 0.09 (stat + sys)*

Finally, after having gone to all the trouble to explain these results, what is the point, i.e. what new information about dark matter may we glean? First, as I have stressed, N_ν is *defined* by (3) or (4) above. As such, it can represent more than just new light or massless neutrino species. In particular, any new undetected particle which couples to the Z and is lighter than half its mass can contribute to N_ν. This includes most WIMP candidates.

---

* The numbers quoted in this column were taken using experimental measured value of Γ_h/Γ_μ. If the value of Γ_h/Γ_μ is taken from the standard model, the mean value increases, and is given in (b).

2. New Constraints On Wimps, Etc:

We can determine the contribution of any new particle to the Z width as follows. Fermions (mass M and coupling $L = [g/2 \cos \theta_w](G_v \bar{\psi}\gamma^\mu\psi + G_a \bar{\psi}\gamma^\mu\gamma^5\psi)$ to the Z) contribute to the Z width as follows:

$$\Gamma = \frac{G_F M_Z^3}{6\sqrt{2}\pi}\left[\sqrt{1-4M^2/M_Z^2}\right]\left\{(G_v^2 + G_a^2)[1-M^2/M_Z^2] + 3(G_v^2 - G_a^2)M^2/M_Z^2\right\} \quad (5)$$

Complex scalars with mass M and $L = [ig K/2 \cos \theta_w](\varphi^* \partial^\mu\varphi - (\partial^\mu\varphi^*)\varphi)$ yield

$$\Gamma = \frac{G_F K^2 M_Z^3}{12\sqrt{2}\pi}\left[1-4M^2/M_Z^2\right]^{3/2} . \quad (6)$$

One can determine what fraction of a neutrino these might mimic, by comparing these widths to that for a massless v.

$$\Gamma(Z^0 \to \bar{v}v) = G_F M_Z^3/12\sqrt{2}\pi \approx 165\, MeV. \quad (7)$$

Then, by comparing the predictions to the 2 σ upper bound on N_v of 3.53 (Oct.), 3.25 (March), or 3.18 (July limit (b)), one can derive limits on new particle masses and couplings of new particles so that this bound is not exceeded. *For particles which couple with the same strength as neutrinos*, one finds[1]:

Particle Type	$N_v < 3.53$ (2σ)	$N_v < 3.25$ (2σ)	$N_v < 3.18$ (2σ)	
fermion	m ≥ O(30) GeV	m≥O(40) GeV	m≥O(42) GeV	(8)
scalar	m ≥ 0 GeV	m ≥ O(30) GeV	m≥O(35) GeV	

We shall see that most of the new constraints are insensitive to the precise bound on N_v used.

(a) **Dirac Neutrinos:** The prototypical WIMPs, heavy Dirac neutrinos were the first elementary particles whose non-thermal remnant abundance was calculated in a manner identical to that used previously to calculate successfully the remnant abundance of light elements such as Helium produced in the big bang. The idea is simple. Heavy neutrinos maintain an equilibrium density as long as their annihilation rate exceeds the expansion rate (assuming no particle antiparticle asymmetry). After this time, the ratio of neutrinos to thermal photons in the universe is frozen in at the Boltzmann factor appropriate to the temperature at freeze-out:

$$\frac{n_v}{n_\gamma} \approx exp-\left[\frac{m_v}{T_{freeze\,ou.}}\right] \quad (9)$$

$T_{freeze\,out}$ is fixed by the annihilation cross section. Since $<\sigma v> \approx G_F^2 m^2$ for Dirac neutrinos, it was shown in 1977[5] that if $m_v \approx 2$ GeV, (9) would imply a closure density of neutrinos in the universe today. Moreover, since the cross section increases quadratically with mass, *heavier* neutrinos will annihilate more efficiently, and so will have a *smaller* remnant mass density today. It is this dynamical coupling between weak interaction strength and GeV mass scale that makes WIMPs natural candidates for dark matter. Of course, if there is a particle-antiparticle asymmetry, this coupling is removed, and heavy neutrinos of any mass greater than 2 GeV could have a closure density today.

The LEP limits require the mass of a new stable Dirac neutrino to be greater than about 42 GeV. When this is incorporated in the calculations described above one finds that the remnant abundance of such particles, in the absence of an asymmetry, is far too small to make up all of even the galactic halo dark matter for masses between 40 GeV and about 2 TeV. Thus, the LEP results alone, in the absence of an asymmetry, rule out WIMP scale Dirac neutrinos as dark matter.

Fortunately, however, the LEP results can be supplemented by non-accelerator experiments, which can directly probe for a flux of dark matter WIMPS at the earth's surface. Low background Ge detectors are sensitive to ionization caused by energy deposited in elastic scattering off of nuclei for masses in excess of about 12 GeV, and the nonexistence of a signal above background[6] rules out Dirac neutrinos as galactic halo dark matter for 12 GeV $\leq$ m $\leq$ 2 TeV. This bound, which would require m $\leq$ 12 GeV, is exactly complementary to the LEP bound, which requires m $\geq$ O(40) GeV. **The combination completely rules out Dirac neutrinos as WIMPs.** (Note: neutrinos with mass in excess of 2 TeV, which would be super-weakly interacting with normal matter, and thus are unWIMP-like, remain viable, if they have a remnant asymmetry.)

(b) Majorana neutrinos: Majorana neutrinos are identical to their antiparticles and therefore cannot have a primordial asymmetry. Thus, their relic abundance is entirely determined by annihilations in the early universe. For Majorana neutrinos however, $<\sigma v> \approx G_F^2 \ p^2$, so annihilation is suppressed for non-relativistic particles, leading to a slightly higher value for the mass resulting in a closure density, of about 5 GeV[7]. The relic density falls above this roughly as $m^{-5/3}$. Thus, the newest LEP limits imply that the fraction of closure density (Ω) in Majorana neutrinos must be less than about .01, or about equal to the visible mass density in the universe today. **Thus, the result (8) by itself rules out Majorana neutrinos as WIMPs.**

(c) SUSY WIMPs: Low energy supersymmetry (SUSY) provides the most compelling WIMP dark matter candidates. If the SUSY breaking scale M is tied to the weak scale, then the lightest SUSY partner of ordinary matter (LSP) can get a mass of order $\alpha M \approx O(GeV)$. Moreover, since the other SUSY partners of ordinary matter which can mediate in scattering and annihilation processes have masses of order $M \approx M_w$, the LSP can have an interaction strength which is comparable to that of an ordinary massive neutrino. This combination makes it a natural WIMP![8] Which particle is lightest is model-dependent. Those often discussed include the sneutrino---the partner of the neutrino, and a "neutralino" --- the fermion partner of a linear combination of the photon, the Z, and the two Higgses present in SUSY models. I discuss these separately below:

(1) Scalar Neutrino: "Sneutrinos" couple to the Z with the same strength as neutrinos, and hence, from (2) will contribute $\approx 1/2$ as much to the Z width, for the same mass. Hence, while the initial LEP results did not rule out such a particle accessible in Z decay, the newest results imply m$\geq$ O(35) GeV. Again, non-accelerator data provides a strong complimentary upper bound on its mass. The WIMP direct detection experiments which limit neutrinos also limit sneutrinos to be lighter than O(12) GeV. This limit is supplemented by data from indirect detection using proton decay detectors. Sneutrinos are not only efficiently captured by the sun and earth[9], but they annihilate into light neutrinos, yielding a signal in proton decay detectors which is not seen, implying m $\leq$ O(3-5) GeV [10]. **Combining limits, sneutrinos are ruled out as WIMPs.**

(2) Neutralinos: The "neutralino" has sufficient flexibility so that it is not entirely ruled out on the basis of the Z decay limits alone. In low energy SUSY there are four neutral majorana fermion "partners" of ordinary matter--the photino, the zino, and two Higgsinos--- which are expected to be among the lightest states[8]. The states will mix in general, so that the mass eigenstates will be linear combinations of the weak eigenstates. In the minimal model the masses and couplings of the inos depend on four parameters: M_1 and M_2, the gauge fermion mass terms, the higgsino mass parameter μ, and the ratio of the two Higgs doublet expectation values v_2/v_1. If the model is unified at some scale then M_1 and M_2 are related by $M_1 = 5/3 \ \tan^2\theta_w M_2$, leaving three free parameters. Because of this larger freedom in model building, the constraints derivable from the Z decay width are less easily stated. For example, the pure neutralino states tend to decouple from Z decay. A pure photino or a pure Zino have no diagonal Z couplings,

since their boson partners have none. It also turns out that a light pure Higgsino tends to be the linear combination which has vanishing Z coupling.

Nevertheless, some non-trivial constraints on the parameter space are derivable. Even in the cases where the LSP decouples from Z decay, the other neutralinos can contribute to the Z width[11]. Moreover, "charginos", the fermionic partners of the charged gauge bosons and Higgses, can give even larger contributions to the Z width[12]. These particles can have triplet weak isospin quantum numbers, so that for a charged gaugino, the Z decay width can be larger than that for 4 neutrinos[12]. It is conventional to present constraints as curves in M-μ space, for a fixed value of v_1/v_2. This ratio reflects the origin of electroweak symmetry breaking. If the top quark mass term drives spontaneous symmetry breaking in the higgs sector, then v_1/v_2 must be greater than unity (in the limit $v_1/v_2=1$ the LSP tends to be pure photino or Higgsino and thus decouples from the Z), given that the current top quark mass lower limit is 89 GeV. As v_1/v_2 increases, the Z width constraints become more powerful.

To derive constraints on the Z width, one diagonalizes the neutralino and chargino mass matrices, finds the weak couplings of the mass eigenstates, and plugs them into (5)[11,12,13]. The largest decay branching ratios come from the charginos (see above), so that the lightest chargino is constrained to have a m$\geq$O($M_Z/2$) before its contribution to the Z width is below the present upper limit. The range in M-μ space ruled out by the requirement that m > O(40) GeV[12], is shown schematically below for two values of β. Also shown is the region of this parameter space which is ruled out by the requirement that the total measured Z decay width into neutralinos be less than .6 $\Gamma_Z(v)$[11]. Finally, I display estimates[14] of the cosmological mass density in the neutralino LSP, with the solid line for Ω=1 and the dashed curve for Ω=0.1. Squark and slepton and top masses $\approx$ O(100) GeV were assumed. Presumably the mass density of light neutralino WIMPs should lie between these two values. As these figures show, a significant region of the M-μ plane is constrained by the Z width limit.

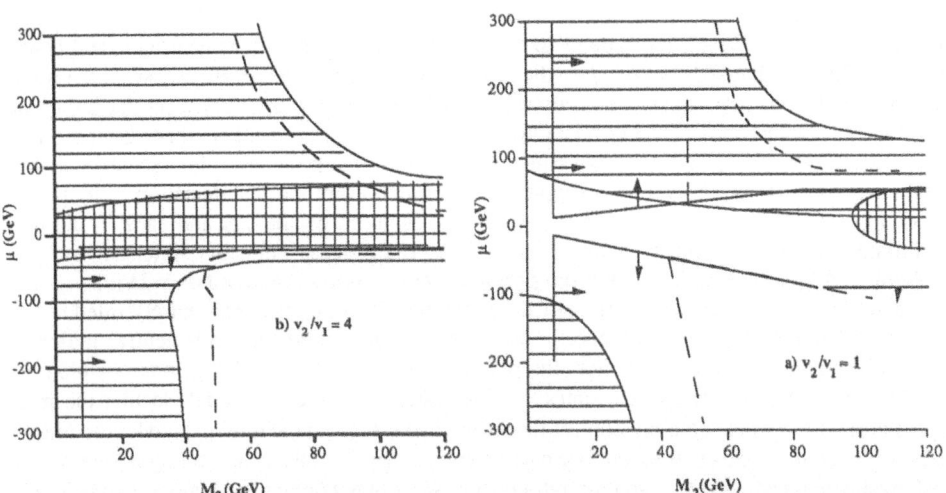

For v_2/v_1 > 1, much of the region M, $|\mu|$ < O(30) GeV is ruled out, and most of the remaining cosmologically interesting range involves large positive M and μ values. In fact[15], a large parameter range exists for which a cosmologically relevant LSP can be heavier than the Z and is nearly pure bino or Higgsino.

This brief foray into the SUSY phase space implies that the Z data suggests that the LSP is probably heavier than O(20) GeV, and is not likely to be a light pure photino, but rather a mix of neutralinos, whose couplings with ordinary matter might have significant scalar contributions. A recent detailed analysis of the SUSY phase space [16] confirms these estimates, and finds that unless the Higgs mass is light, $M_{LSP} > 15$-20 GeV, if it is to provide significant contribution to Ω. If one then incorporates the new LEP limits on a light Higgs (m>O(30) Gev) in this analysis and uses limits from CDF on other SUSY particles, we then obtain a lower bound on a cosmologically significant neutralino of 15-20 GeV, making it, in general, substantially heavier than initial estimates suggested. These are important results because: (a) a search of the available model space then suggests that if such a WIMP is to have a sufficient remnant abundance to make up the dark matter, it is likely to be as heavy or heavier than the W and Z particles; (b) this mass range is in the opposite direction of previous efforts to extend the range in which direct detectors were sensitive, and moreover, the properties of such a WIMP are such as to make it much more difficult to detect directly in underground detectors (see below).   Direct searches, which I believe are of vital importance, are now more challenging in light of LEP.

(d) **Exotica/cosmions**: Other than these WIMP candidates from particle physics, exotic objects have been proposed for purely astrophysical reasons. One of these is a "cosmion", a fixed abundance of which inside the sun might, for a mass of 4-10 GeV, lower the core temperature and reduce the solar neutrino flux. [17] This caused interest when it was suggested [18] that WIMPs might in principle be captured in the Sun over cosmological time to produce the required abundance, thus potentially solving both the dark matter and solar neutrino problems. However none of the standard WIMPs fit the severe requirements [19] which may make this solution appear contrived. Scattering cross sections 20-100 times larger than weak are required, as is the absence of annihilation inside the sun. However, in times of need ugliness is no obstacle, and theorists with time on their hands have produced exotic cosmion models. I describe here how the Z decay data can rule out many explicit models which previously appeared viable:

(a) Magnino, and Neutrino-Higgs: Raby and West proposed several cosmion models [20]. Each of these involved a Dirac neutrino with mass 5-10 GeV as a cosmion (with an additional interaction mediated either by a large magnetic moment, or by light Higgs exchange). The Z limit on extra massive Dirac neutrinos (and on a light Higgs) rules out both these possibilities.

(b) SUSY cosmions [21] : This model requires SUSY, and also a 1-2 Higgs. The only viable model proposed involved $v_2/v_1 \approx 1$, M=80-105, and μ=130-150 and a lightest chargino mass ≈ 30-40 GeV. These are ruled out by the Z decay data as discussed above, as is a light Higgs particle.

(c) E_6 cosmions [22]: This model proposed new v's in a 27 of E_6 with no Z couplings, but couplings to a new Z' particle. While such v's would not be produced in Z decay, there also exists one new doublet v per family. One might naively expect their mass to be about equal to that of the 5-10 GeV singlet cosmion. In this case this model would be ruled out by the Z width limit. Also, Z limits on light Z' particles are appearing which might also rule out such models.

(d) Colored Scalars: These models [23] are the least explicit, and hence difficult to rule out. But, all involve new colored scalars and heavy fermions. If they have standard weak quantum numbers, the Z data would then rule them out.

These results do not imply that all "cosmion" models, present or future are ruled out. They just make such a particle even more implausible. Moreover, a more model independent limit on cosmions from direct detection experiments [24], of m< 4GeV, now limits the available parameter space to a negligible region.

(e) **Generic WIMPs:** One can leave the realm of specific candidates, and examine general LEP constraints on arbitrary massive particles whose abundance in the universe today is determined by their annihilations (via Z exchange) in the early universe[16,25]. If one defines [16] a factor sin ϕ_Z which gives the suppression in the relative Z coupling compared to neutrinos, one finds that only WIMPs with sin ϕ_Z <0.3, and M>10 GeV for Majorana fermions and sin ϕ_Z <.003 and M> 6 GeV for Dirac fermions remain viable. If one combines the constraints imposed by Ge detection experiments, these limits are raised, and with small improvements, one might set "generic" fermion mass limits of O(20) GeV for WIMPs which annihilate through the Z.

3. The Z Width And Baryonic Dark Matter

It is by now very well known that the number of light neutrinos strongly impacts upon cosmology via primordial big bang nucleosynthesis. Traditionally, cosmologists have used the quantitative agreement between the predicted and the observationally inferred primordial abundances of light elements to limit the number of light neutrinos[26]. Now that this number has been well established experimentally via the Z width, one can hope to use it to further constrain other aspects of big bang cosmology. In particular, with the number of neutrinos known (*assuming* no other light states which can affect the expansion rate during helium production, but which do not couple to the Z), the uncertainty in the predicted abundance of ^{4}He is the smallest of all the light elements. Below I display the BBN predictions for the fraction by mass of ^{4}He (Y_p), for 3,4, and 5 neutrinos as a function of the baryon to photon ratio η[27], including the 2σ uncertainty due to our uncertainty in the BBN reaction rates (dominated by the uncertainty in the neutron half life). Also shown for 3 neutrinos is the reduced 2σ uncertainty with the most recent neutron half life measurement included.

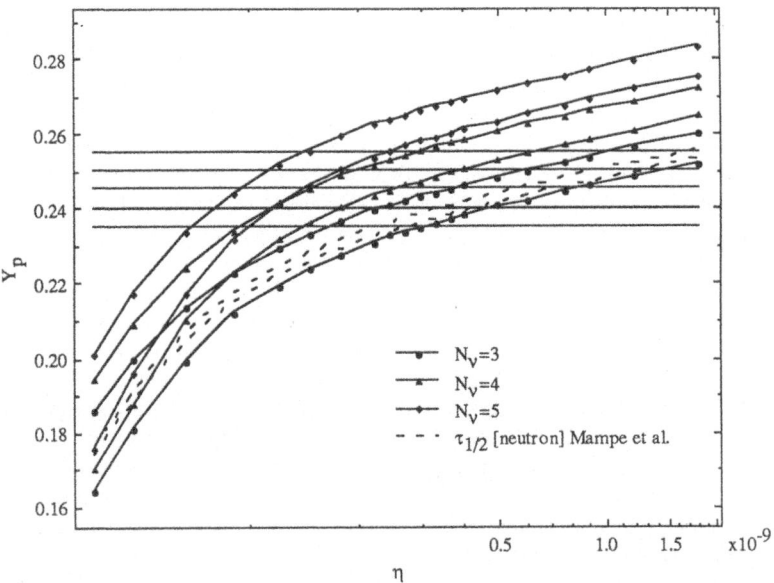

It is clear from the figure that if we can pin down the actual primordial abundance of ^{4}He we could place strong limits *from He alone* on the density of baryons in the universe today. The fraction of closure density in baryons (Ω_B) is determined from η as

$$\Omega_B = .0036 h_o^{-2} (T / 2.74 \text{ K})^3 (10^{10}\eta) \qquad (10)$$

where T is the present microwave background temperature, and h_o is limited by measurements of the Hubble constant today to be between .4 and 1. For example, if we were to limit the primordial abundance to be between 23.5 and 24% for example, then, *assuming at most 3 light neutrino equivalents*, η would be constrained to be between 2.5-4.8 x 10^{-10}. This in turn would limit Ω_B to be between 0.01 and 0.1. The lower bound (obtained for the extreme value $h_o = 1$, which leads to conflicts with limits on the age of the universe) is already suggestive that some dark matter must be baryonic. The upper limit is only marginally in agreement with the possibility that all dark matter inferred by virial estimates is baryonic. While such a narrow range for the primordial He abundance cannot be inferred from the present data[28], the Z width data, in combination with more careful analyses of big bang nucleosynthesis are bringing us closer definitively limiting the amount of baryonic dark matter in the universe. Indeed, it has been argued[28] that a reasonable upper limit on primordial ^{4}He is 24%, and a "best fit" value is 23%. In the former case, ^{4}He already gives the best upper limit on η, and in the latter case, the limit is so strong as to make BBN with 3 neutrinos inconsistent! Further developments are eagerly awaited.

4. Implications For Direct Wimp Detection

The Z width data effectively rules out all known WIMP candidates which interact coherently with total nuclear charge in low energy scattering processes. In addition, it suggests that the LSP, if it exists, is probably heavier than 10-20 GeV, and is not likely to be a pure photino. These have important implications for direct detection schemes, because they suggest that a viable WIMP is likely to be orthogonal to present detection schemes. Much of the thrust of ongoing WIMP detection experiments has been to develop sensitivity to probe for coherently interacting particles with mass less than 5-10 GeV. For Dirac neutrinos and sneutrinos, the LEP results now remove the need. Existing detectors have already ruled out this range for cosmions, so that motivation for continuing to probe the light mass range is also reduced. None of the present detectors is sensitive to the small rates which would result from the spin-dependent interactions, typical of neutralinos. While in the most optimistic case, the detection rates can increase linearly with neutralino mass (for fixed squark masses), this is only true for a target material heavier than the neutralino. Trying to maximize the recoil energy also suggests going to heavier nuclear targets. Recent work[29] suggests that the rates in nuclear targets is suppressed compared to original estimates for spin dependent scattering. The challenge to direct detection has increased! It may be that experimenters will have to wait first for a SUSY signal at LEP, the Tevatron, or the SSC, in order to optimize their detectors for maximize sensitivity to the resulting WIMP dark matter.

5. Other Early Universe Implications of LEP:

LEP has limited the standard electroweak model in two other ways which can impact upon cosmology. First and foremost, sensitive new limits on the parameters of the standard model, limiting isospin breaking and isospin conserving contributions from new heavy particles, severely constrain various possibilities for exotica beyond the standard model. The top quark is now expected to be in a mass range around 140 GeV. This has implications for any scheme of electroweak symmetry breaking, and in turn limits possible characteristics of the electroweak phase transition in the early universe, as well as possible models which might yield dark exotic dark matter. The limits obtained by combining LEP with other precision electroweak probes also appear to limit the possible existence of large numbers of new particles in the TeV range, as one might expect from technicolor models[30]. Next, LEP has apparently provided a new lower bound on a Higgs mass of about 30 GeV. This not only impacts in model building which is relevant for WIMP dark matter, as described above, but also strongly constrains the nature of the electroweak phase transition. For example, it rules out very low Higgs masses, which are associated with the possibility that our present vacuum is not stable, but merely metastable on cosmological timescales. Limits on the Higgs, and associated light particles may also be important for constraining various mechanisms for electroweak baryogenesis scenarios.

6. Conclusion:

It is remarkable to have data which conclusively limits an area in which there has been so much speculation over the last decade. The first results from physics at the Z have provided powerful new limits, not just on the number of light families, but on the nature of dark matter, and on possible new processes in the early universe. We can hope that the coming years may not only yield further constraints, but that a clear signal to guide us might emerge. At the very least, the present results not only reinforce the strong connection between particle physics and cosmology, they also demonstrate the importance of having actual data to constrain theory.

REFERENCES

1. most of the implications of Z decay for WIMPs discussed here is from L.M. Krauss, Phys. Rev. Lett. **64**, 999 (1990). See this for further references.

2. L3 , Phys. Lett **B231**, 509 (1989); ALEPH, Phys. Lett **B231**, 519 (1989) ; OPAL, Phys. Lett **B231**, 530 (1989); DELPHI, Phys. Lett **B231**, 539 (1989); G.S Abrams et al. Phys. Rev. Lett. 63 2173 (1989)

3. i.e. see L3, OPAL, ALEPH, and DELPHI reports, Moriond 90 Proceedings

4. LEP experimental results reported at the Singapore Internation High Energy meeting obtained from P. Colas, private communication, and from transcripts of lectures at Snowmass, July 1990.

5. B.W. Lee and S. Weinberg, Phys. Rev. Lett. **39** 165 (1977); D. Dicus et al, Phys. Rev. Lett.**39**, 168 (1977)

6. D. Caldwell et al, Phys. Rev. Lett. **61**, 510 (1988); S. Ahlen et al, Phys. Lett. **B195**, 603 (1987) For a description of a more recent experiment with Si, which is now reporting limits in the range of 5 GeV, see B. Sadoulet et al, Astrophys. J. Lett. **324**, 75 (1988)

7. L. M. Krauss, Phys. Lett 128B, 37 (1983); E. Kolb, K. A. Olive, Phys. Rev D33, 1202 (1986)

8. for a review: J. Polchinski in *Inner Space, Outer Space,* ed. E. Kolb et al, University of Chicago Press (Chicago 1986) or see J. Ellis et al, Nucl. Phys. **B238**, 453 (1984)

9. L. M. Krauss, Harvard preprint HUTP-85/A008a (1985); J. Silk, K. Olive and M. Srednicki, Phys. Rev. Lett. **55**, 257 (1985); L. M. Krauss et al, Astrophys. J. **299**, 1001 (1985) : A. Gould, Astrophys. J.**321**, 560 (1987); ; L. M. Krauss, M. Srednicki and F. Wilczek, Phys. Rev. D33, 2079 (1986)

10. i.e. see K. Olive and M. Srednicki, Phys. Lett. **B205**, 553 (1988)

11. i.e. R. Barbieri et al, Phys. Lett. **195B**, 500 (1987); V. Barger et al, Phys. Rev. **D28**, 2912 (1983)

12. A. Bartl et al, Wien pub. HEPHY-PUB 522/89 and refs therein

13. I. Adachi et al. Phys. Lett. **B218**, 105 (1989)

14. i.e. see J. Ellis et al, in ref 8.

15. M. Srednicki and K. Olive, Minnesota pub. UMN-TH-801/89

16. J. Ellis et al, Phys. Lett. **B289**

17. J. Faulkner, R. Gilliland, Astrophys. J. **299**, 994 (1985) ; D. Spergel, W. Press, Astrophys. J. **294**, 663 (1985)

18. L. M. Krauss, Harvard preprint HUTP-85/A008a (1985); W. Press, D. Spergel, Astrophys. J. **296**, 673 (1985)

19. L. M. Krauss et al, Astrophys. J. **299**, 1001 (1985)

20. S. Raby and G. B.West, Nucl. Phys. **B292**, 793 (1987); Phys. Lett. **194B**, 557 (1987); Phys. Lett. **202B**, 47 (1988)

21. G.F. Guidice, E. Roulet, Phys. Lett. **219**, 309 (1988)

22. G.G. Ross, G. C. Segre, Phys. Lett. **197**, 45 (1987)

23. G. Gelmini, L. Hall, M.J. Lin, Nucl. Phys. **B281**, 726 (1987)

24. D. Caldwell et al., Phys. Rev. Lett. **65**, 1305 (1990)

25. D.N. Schramm, Proceedings "La Tuile Workshop", 1990, Fermilab preprint 90/120-A

26. i.e. G. Steigman, D.N. Schramm, and J. E. Gunn, Phys. Lett. **B66**, 202 (1977)

27. This curve comes from L. M. Krauss, P. Romanelli, Astrophys. J. **358**, 47 (1990) .

28. i.e. see , B.E.J. Pagel, E.A. Simonson, Rev. Mexicana Aston. Astrof. **18**, 153 (1989).

29. A.F. Pacheco D. Strottman, Phys. Rev. **D40**, 2131 (1989); J. Engel, P. Vogel, Phys. Rev. **D40**, 3132 (1989); and most recently, F. Iachello, L.M. Krauss, and G. Maino, YCTP-P14-89 (April 1990), Phys. Lett. B, to appear

30. see for example W. J. Marciano, J. L. Rosner, Phys. Rev. Lett. **65**, 2963 (1990)

A SEARCH FOR DARK MATTERS IN THE KAMIOKANDE II

Y. SUZUKI

Institute for Cosmic Ray Research.
University of Tokyo
3-2-1, Midorichyo, Tanashi
Tokyo, 188 Japan

ABSTRACT. A search has been made for high energy neutrino events from the sun in the Kamiokande II detector as annihilation products of galactic cold dark matters. We found no excess neutrino events from the sun and set limits on the mass and the density of the dark matter candidates.

1. Introduction

Certain cold dark matter candidates have some possibility of being trapped in the SUN after losing their energy through elastic scattering[1][2]. The capture rate $\dot{N}_{cap}$ is determined by the mass m_x, the elastic cross section σ_{eff}, the density ρ_x and the velocity dispersion $\bar{v}$[2];

$$\dot{N}_{cap} = \left\{ \left[\sqrt{\frac{2\pi}{3}}\, \sigma_{eff} \left(\frac{\rho_x}{m_x}\right) \bar{v} \right] \times \left[\frac{M_{sun}}{m_H}\right] \times \left[\frac{3v_{esc}^2}{2\bar{v}^2}\right] \times 0.89 \right\} \times 2.31 \times \xi \times C_x(m_x),$$

where v_{esc} is the escape velocity of the sun. The factors $2.31 \times \xi \times C_x(m_x)$ are the corrections introduced by Gould[3], where $\xi = 0.75$ is a correction for a moving observer relative to the frame with isotopic velocity dispersion and $C_x(m_x)$ is the suppression factor from mass mismatching. For example, the dark matter with a mass of 10 GeV and with a typical weak interaction cross section, has a capture rate of $\sim 4 \times 10^{26}$/sec. If the mass is below ~ 3 GeV, dark matters escape out of the sun and not to be accumulated.

These dark matters accumulate in the solar core, and annihilate yielding high energy neutrinos from the sun[4][5][6]. The annihilation rate is

$$\dot{N}_a = \left(\frac{1}{2}\right) \dot{N}_{cap} \left(\frac{N_x}{N_{eq}}\right)^2,$$

where N_x is the present number of the dark matter being captured in the sun. The ratio $N_x/N_{eq} = 1$ for an equilibrium[4]. The equilibrium time depends on the capture rate and the annihilation coefficient. It is, for example, 2.7 million years for 10 GeV Dirac neutrinos.

Total neutrino flux on the surface of the earth is

$$f(E_{v_i}) = \frac{\dot{N}_a}{4\pi R^2} \left(\frac{dn_V}{dE_V}\right) P_f(E_{v_i}),$$

where R is the sun-earth distance, dn_V/dE_V is the neutrino spectrum and $P_f(E_{v_i})$ is the absorption probability of neutrinos within the sun's media. The absolute flux is determined by the mass, the elastic cross section, the density and the velocity. The spectrum is determined by the branching ratio of each annihilation process. Neutrinos from the annihilation of heavy Dirac neutrinos have almost monochromatic energy since the pair production of neutrinos dominate. The annihilation of Majorana particles through Z^0 dominantly produces heavy quarks yielding a broad energy spectrum. The production of the massless neutrinos is suppressed due to Fermi statistics of Majorana particles. Photinos annihilate through scalar fermion exchange yielding mainly τ leptons producing v_τ. Therefore the detection of heavy Dirac neutrinos and scalar neutrinos which also produce monochromatic neutrinos through zino exchange is relatively easier than the detection of the Majorana particles.

Large underground experiments like the Kamiokande II[7] are primarily sensitive to most of those dark matter candidates with the mass greater than 3 GeV. However, recent results from LEP rejected many of these dark matter candidates. Dirac neutrinos with the mass less than 43 GeV, Majorana neutrinos with the mass less than 35 GeV are excluded[8][9]. These dark matter candidates are now supposed to be heavier than previously favoured.

2. Flux limits of high energy neutrinos from the sun

A search has been made for high energy neutrino events from the sun in the Kamiokande II detector[9]. Three data sets were used in the analysis; (1) events originating in the fiducial volume and contained within the detector (fully contained events: 4.4 kt·yr), (2) events originating in the fiducial volume but not fully contained in the detector (vertex contained events: 1.6 kt·yr), and (3) upward-going muons produced in the surrounding rocks by v_μ interactions (165m²·yr). The directions of incident neutrinos are well defined since only the events with visible energy of larger than ~ 1 GeV are selected and the upward-going muons are supposed to be created by neutrinos above 10 GeV. No evidence of an enhancement toward the direction of the sun was found.

The neutrino flux limits from the sun for each kind of neutrinos were obtained after corrected for the detection efficiency and the acceptance. The integrated neutrino flux ($\phi_{v_i}+\phi_{\bar{v}_i}$) limits are shown in Fig. 1. In this calculation the flux of neutrinos and antineutrinos is assumed to be same, which is suitable for most of the dark matter candidates.

3. A Dark Matter Search

For a search for a particular dark matter candidate, we did not simply use the integrated flux limits shown in Fig. 1 which was obtained with general loose selection criteria. Monte Carlo calculations were performed to obtain the neutrino spectrum and the detection efficiency and the expected number of events with optimized selection criteria for each dark matter candidate to get better signal to background ratio. In these calculations we have assumed $\rho_x \sim 0.3 \text{GeV/cm}^3$, $v_x \sim 300 \text{km/sec}$ and that the numbers of particles and anti-particles are the same. For lightest super symmetric particles ($\tilde{v}, \tilde{\gamma}, \tilde{h}$), parameters are adjusted to get $\Omega_x h_{1/2}^2 = 1$. The details of the parameter selections can be found elsewhere[10].

The observed and the expected number of events for ν_D, ν_M, $\tilde{\gamma}$, $\tilde{h}$, $\tilde{\nu}_e$, $\tilde{\nu}_\tau$ using the contained events are shown in Fig. 2. Those for ν_D, ν_M, $\tilde{\nu}_\mu$ using the upward-going muons are shown in Fig. 3. The expected event rate and the muon fluxes of the Majorana particles are calculated for two assumptions on the spin couplings of the elastic cross sections. The expected event rate including the EMC spin effect[11] give higher than ones from the naive quark model for the Majorana neutrinos and give less for $\tilde{\gamma}$ and $\tilde{h}$. Taking these differences as a systematic uncertainty, no stringent limits can be set for $\nu_M, \tilde{\gamma}, \tilde{h}$. The mass limits are obtained for Dirac and scalar neutrinos;

$$3\,\text{GeV} < m_{\nu D} \lesssim 100\,\text{GeV},$$

$$3\,\text{GeV} < m_{\tilde{\nu}_e} \lesssim 90\,\text{GeV},$$

$$3\,\text{GeV} < m_{\tilde{\nu}_\mu} \lesssim 90\,\text{GeV, and}$$

$$4\,\text{GeV} < m_{\tilde{\nu}_\tau} \lesssim 90\,\text{GeV}.$$

The upper bound of the mass for the Dirac neutrinos was limited at the threshold of the W^+W^- production which was not taken into account in the present analysis. The dominant annihilation process for the scalar neutrinos is the Majorana part of the zino exchange. The annihilation process to virtual Z_0 is insignificant in the sun since the process is taken through p-wave and is proportional to the velocity of the particles. If the scalar neutrinos exist in the sun and mass of scalar neutrinos must be less than $m_{\tilde{z}}$, therefore the region $m_{\tilde{\nu}} < m_z$ can be exploited since we assume conservatively $m_{\tilde{z}} = m_z$.

If the coupling of Dirac fermions to Z_0 are suppressed through mixing with an inert state by a factor $\sin\phi_z$[9], then the excluded region can be calculated for parameters, m_ν vs $\sin\phi_z$. The results by the Kamiokande II together with the ones obtained by LEP[9] and Ge-experiment[12] are shown in Fig. 5.

References

[1] G.Steigman et al., Astron. J. **83**, 1050 (1978); D.N.Spergel and W.H.Press, Astrophys. J. **294**, 663 (1985); L.M.Krauss et al., Astrophys. J. **299**, 1001 (1985); J.Faulkner and G.Gilliland, Astrophys. J. **299**, 994 (1985).

[2] Press and Spergel, Astrophysical Journal, **296**, 679 (1985).

[3] A. Gould, Astrophysical Journal, **321**, 571 (1987).

[4] T.K.Gaisser, G.Steigman, and S.Tilav, Phys. Rev. **D34**, 2206 (1986).

[5] J.S.Haglin et al., Phys. Lett. **B180**, 375 (1986).

[6] M. Srednicki et al., Nucl. Phys. **B279**, 804 (1987).

[7] The members of the Kamiokande II are K.S.Hirata, K.Inoue, T.Kajita, T.Kifune, K.Kihara, M.Nakahata, K.Nakamura, S.Ohara, N.Sato, Y.Suzuki, Y.Totsuka and Y.Yaginuma (ICRR,Univ. of Tokyo), M.Mori, Y.Oyama, A.Suzuki, K.Takahashi and M.Yamada (KEK), M.Koshiba and K.Nishijima (Tokai Univ.), T.Suda and T.Tajima (Kobe Univ.), K.Miyano, H.Miyata and H.Takei (Niigata Univ.), Y.Fukuda, E. Kodera, Y.Nagashima and M. Takita (Osaka Univ.), K.Kaneyuki and

T. Tanimori (Tokyo Teck), and W.E.Beier, L.R.Feldscher, E.D.Frank, W.Frati, S.B.Kim, A.K.Mann, F.M.Newcomer, R.Van Berg and W.Zhang (Univ. of Penn)

[8] L.M.Krauss, Phys. Rev. Lett. **64**, 999 (1990); F.Dydak, Talk at the 25th International Conference on High Energy Physics, Singapore, August 1990.

[9] J.Ellis et al., Phys. Lett. **B245**, 251 (1990).

[10] N. Sato et al., to be published in Phys. Rev. D.

[11] J.Ashman et al. Phys. Lett. **B206**, 364 (1988); J.Ellis et al., Phys. Lett. **B198**, 393 (1987); J. Ellis and R.A.Flores, Nucl. Phys. **B307**, 883 (1988).

[12] J.Ellis et al., Phys. Lett. **245**, 251 (1990); D.Caldwell et al., Phys. Rev. Lett. **61**, 510 (1988).

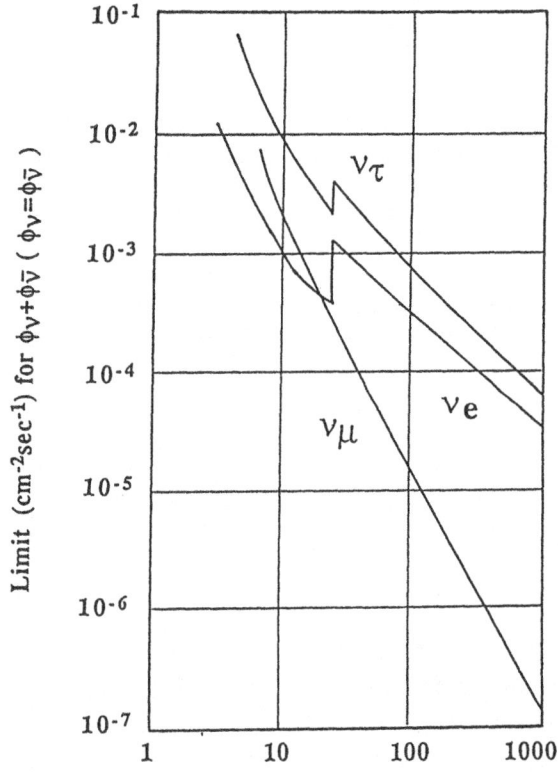

Threshold neutrino energy (GeV)

Figure 1. The integrated flux limits on the high energy neutrinos from the sun. The limits are obtained for each neutrino species and assumed that the neutrino flux and antineutrino flux are the same. The assumption is suitable for most of the dark matter candidates. The results below 25 GeV for v_e and v_μ were obtained by the analysis of the contained events, and those above 25 GeV by that os the upward-going muons.

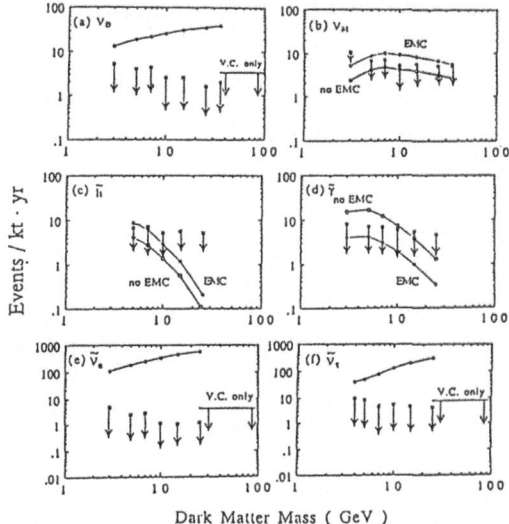

Figure 2. The background subtracted upper limits (shown by arrows) on the event rates of the contained neutrino events from the sun for (a) v_D, (b) v_M, (c) $\tilde{h}$ (d)$\tilde{\gamma}$, (e) $\tilde{v}_e$ and (f) $\tilde{v}_\tau$ at 90% CL. The closed and open circles connected by the strait lines show expected event rate for the dark matter annihilation in the sun. In (b), (c) and (d), the closed circles are those obtained with the EMC spin effect. In (a), (c) and (f), the limits indicated by 'V.C. only' are those obtained only from the vertex contained events.

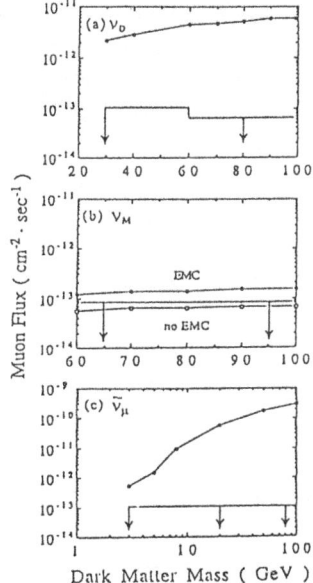

Figure 3. The background subtracted upper limits on the muon flux. The optimized selection criteria were used for each dark matter candidate; (a) v_D, (b) v_M,(c) $\tilde{h}$ (d) $\tilde{v}_\mu$ at 90% CL. The closed and open circles show the expected muon flux due to the interaction of the high energy neutrinos from the sun. The EMC spin effect is adopted in the results shown by the closed circle.

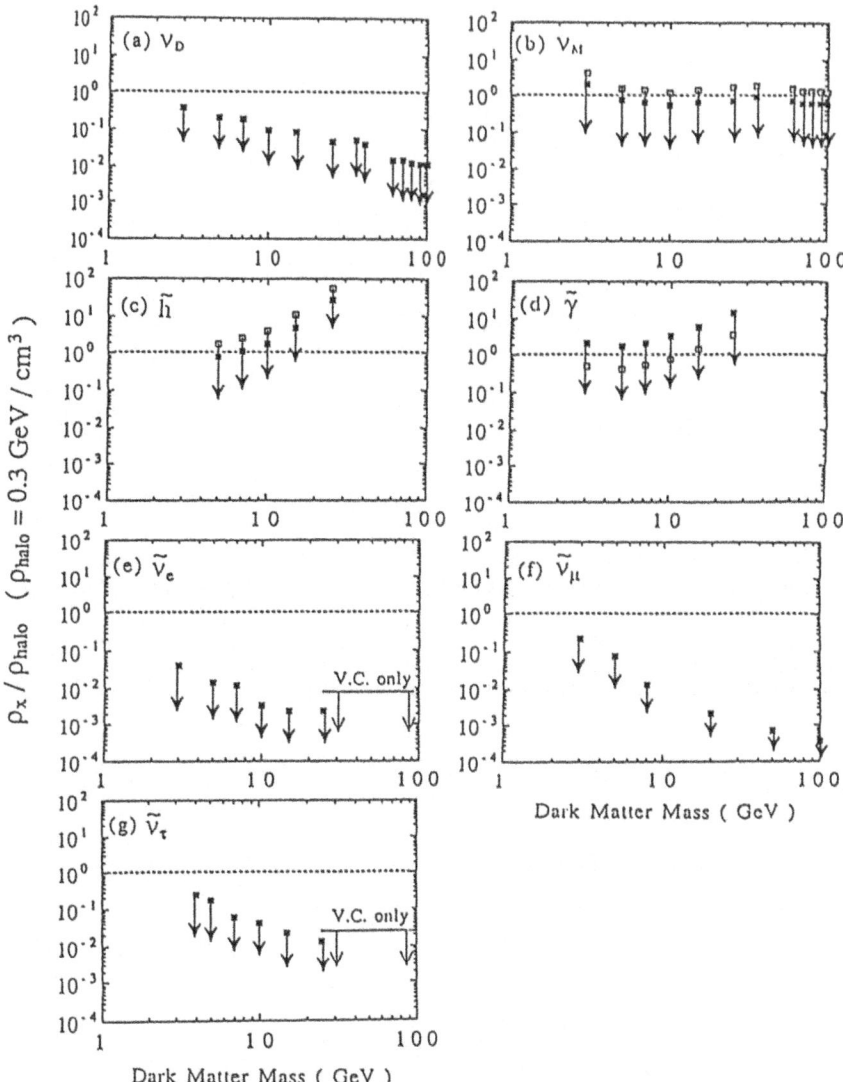

Figure 4. The upper limits on the dark matter densities, ρ_x / ρ_{halo} for (a) ν_D, (b) ν_M, (c) $\tilde{h}$ (d)$\tilde{\gamma}$, (e) $\tilde{\nu}_e$, (f) $\tilde{\nu}_\mu$ and (g) $\tilde{\nu}_\tau$ at 90% CL with $\rho_{halo} = 0.3$ GeV/cm^3. In (b), (c) and (d), the closed squares are obtained from the elastic cross sections with the EMC spin effect.

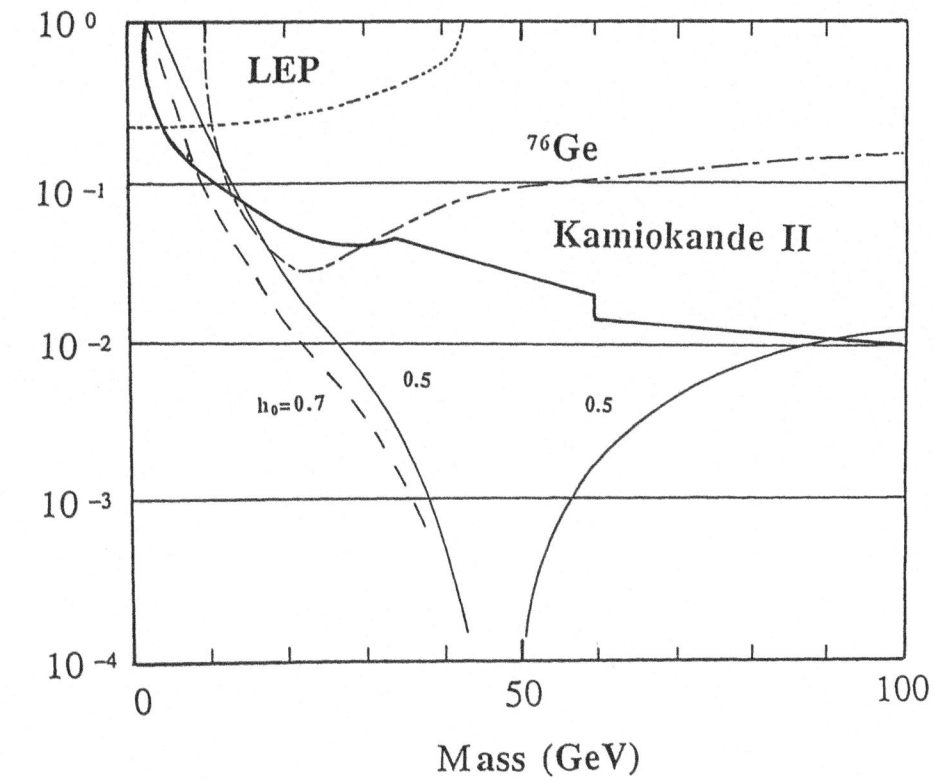

Figure 5. The excluded region of the mass and the coupling strength of the Dirac fermion obtained by the present experiment, the result from LEP[9] and the germanium[12] experiment. The little region[9] around 10 GeV and $\sin^2\phi_z \sim 0.1$ was excluded solely by the Kamioka experiment.

BARYONIC DARK MATTER

Joseph Silk

Mt. Stromlo and Siding Spring Observatories, The Australian National University,
and
Departments of Astronomy and Physics, Center for Particle Astrophysics,
University of California, Berkeley

ABSTRACT.   Both canonical primordial nucleosynthesis constraints and large-scale structure measurements, as well as observations of the fundamental cosmological parameters, appear to be consistent with the hypothesis that the universe predominantly consists of baryonic dark matter. I review the arguments for BDM to consist of compact objects that are either stellar relics or substellar objects. I describe several techniques for searching for halo BDM.

I. INTRODUCTION

A common theme to the best physical theories is their ultimate simplicity. It is precisely this aspect that makes inflationary cosmology so appealing, with its predictions that the universe be isotropic, homogeneous, and spatially flat, with a gaussian, scale-invariant spectrum of primordial density fluctuations that eventually seeds large-scale structure. Unfortunately, apart from the homogeneity and isotropy that are seen in the deep galaxy counts and, to an unprecedented level ($\frac{\delta T}{T}$ now approaching an upper limit of $\sim 10^{-5}$ on angular scales from 90 degrees to one arc-minute), in the cosmic microwave background, nature has not been very co-operative. Mounting astronomical evidence points to a breakdown in scale invariance, with indications of excessive large-scale power in the galaxy distribution. More seriously perhaps, despite two decades of intensive searching for dark matter, the case for a critical density remains unacceptably weak.

A low Ω universe, with $\Omega \sim 0.1$, thus retains considerable appeal: here Ω is the ratio of mean cosmological density to the Einstein-de Sitter value. It is not only consistent with all dynamical measurements, but directly implied by a large body of astronomical data. These include M/L determinations on the scale of galaxy clusters and superclusters, as well as time-scale arguments, and, most recently, the deep galaxy counts whose continuing increase as $\sim 10^{0.4m}$ to $\gtrsim$ 27th B magnitude can most simply be accounted for with the large volume element available at $z \sim 1$ in a low Ω cosmology. Scale invariance is naturally broken in a low density cosmology at the onset of the de Sitter-like phase of the expansion. This may have relevance for the relic spectrum of density fluctuations that generated large-scale structure. More importantly, perhaps, for large-scale structure is the requirement from CMB isotropy constraints of primordial entropy fluctuations : an adiabatic fluctuation spectrum would generate excessive $\frac{\delta T}{T}$ if $\Omega \sim 0.1$ because of the diminished period available for fluctuation growth.

The most tantalizing support for low Ω refers only to the baryonic component and comes from the success of primordial nucleosynthesis, which, with LEP now requiring precisely 3 neutrino flavors from Z-boson decays and the neutron lifetime accurately measured, has remarkably little freedom in the canonical Big Bang model to account for the abundances of He^4, He^3, H^2, and Li^7. The only free parameter is the baryonic density, which is constrained to be $\Omega_b h^2 \approx 0.015$ with an uncertainty of at most a factor of 2 (Steigman, these proceedings). Hence if a low Hubble constant, $h = H_0/100 \text{km s}^{-1} \text{Mpc}^{-1} \approx 0.5$ is adopted, the possibility arises that all of the matter in the universe could by baryonic in nature, with most of the baryons being dark regardless of

371

whether there is a non-baryonic component or not. A minimal hypothesis is therefore that dark halos consist predominantly of baryonic dark matter, on the logical grounds that the most likely forms of baryonic dark matter, to be described below, would have formed in a similar manner to, but preceded, the luminous regions of galaxies. The baryonic dark matter provided a deep gravitational potential well containing the gas that eventually condensed into stars.

If a significant component of non-baryonic dark matter exists, in the form of weakly interacting relic particles, for example, or of massive (10 - 100 eV) neutrinos, its most appealing role would be to provide an almost uniform, and hence difficult to measure, substratum that contributes to the mean value of Ω. Perhaps the non-baryonic dark matter dominates galaxy clusters, although there is no direct evidence for this despite the fact that the mass-to-light ratios measured for clusters are several times greater than the largest values determined from galaxy rotation curves. Clusters contain only a small fraction of all galaxies and make only a small contribution to the overall mass budget of the universe. It is more likely that over the ~ 10 Mpc scale of superclusters, where the highest Ω values are determined, recourse must be had to non-baryonic dark matter.

This offers the intriguing possibility that it is the primordial fluctuations in the non-baryonic dark matter distribution that are responsible for generating large-scale structure via graviational instability, including superclusters, voids and large-scale flows, whereas it is the baryonic component, containing its own independently generated primordial fluctuations on smaller scales, that feeds galaxy formation. Such a bifurcation in the power distribution of fluctuations has been anticipated in hybrid models that appeal to hot and cold dark matter, the cold component being now identified with baryonic dark matter and the hot component with massive neutrinos. A similar situation might also arise in multiple inflation models, or if a late phase transition were responsible for large-scale structure. One might even generate both primordial entropy and curvature models for the fluctuation spectrum, as would be expected in some extended inflation models.

In this review, I will attempt to narrow the range of BDM candidates by appealing to astrophysically motivated plausibility arguments. Section II presents the case for primordial stars as the source of BDM in the form of compact stellar remnants. I review the objections to this hypothesis in Section III, and find that none are wholly convincing. In Section IV, I present the arguments for and against the possibility that BDM consists of sub-stellar mass objects, specifically brown dwarfs. A final section is devoted to describing the experimental searches that are presently underway for BDM.

II. PRIMORDIAL STARS AS BDM

Given a theory of present-day star formation, one might hope to be able to apply this in the context of galaxy formation to explore the nature of dark matter halos. Unfortunately, there is no compelling theory of star formation, and resort must be made to phenomenological arguments to probe the nature of primordial stars, if indeed there were such objects. Primordial stars are attractive as a source of BDM for the following reason. The only forms of dark matter which are known to exist in a significant amount are compact stellar remnants, specifically white dwarfs and neutron stars. These form a substantial fraction (at least 10 per cent, and most likely as much as 50 per cent) of the mass measured in the luminous regions of galaxies. The solar neighbourhood is the best studied example, but there is good reason to believe that in old stellar systems, including globular clusters and elliptical galaxies, these remnants contribute up to 50 per cent of the mass within the half-light radius, where substantial spectroscopic data and mass modelling is available. While one can always invent more exotic forms of BDM, and suggestions span the range from intergalactic snowballs and planets to massive black holes, the lack of evidence for such objects existing anywhere in the universe where they contribute substantially to the local mean mass density will provide sufficient reason to avoid any further mention of them in this review.

It is first necessary to establish that primordial stars existed. By primordial stars, I mean either stars that formed in an extremely metal-poor environment, characteristic either of the pregalactic era or the formation phase of galaxies. Successive generations of stars polluted the interstellar medium as the chemical abundances gradually evolved to a level approaching the nucleosynthetic yield. Note first that there is a direct existence proof : stars did indeed form out of an essentially pristine, pregalactic or protogalactic environment. There are two arguments to this effect. There are a few halo stars with metallicity much less than solar : the most metal poor star known has [Fe/H] $\approx$ -5. A star with such a low abundance is indistinguishable from the perspective of star formation theory from a star with zero metallicity. Moreover, in addition to this direct evidence for low mass, almost primordial stars, one can indirectly establish that the precursor heavy element factories of the most metal-poor, and hence oldest, stars in the galaxy were stars in the conventional mass range of 10 to $100\,M_\odot$. The reasoning relies on the observed abundance pattern of heavy elements in metal-poor halo main-sequence stars. These reflect the neutron capture odd-even signature characteristic of the r- and s-process nucleosynthesis that is believed to occur in stars.

Whether the initial mass function (IMF) of these stars differed from the current IMF is controversial. The excess of oxygen relative to iron observed in metal-poor stars ($[Fe/H] \lesssim -1$) may plateau at $[O/Fe] \approx 0.7$, as observed for halo giants (Barbuy 1988), or may continue to rise towards lower metallicities, as found for halo dwarfs (Abia and Rebolo 1989). If the former observation reflects the true situation, the simple interpretation is that of a time delay between the massive O-producing stars and the lower mass stars that fabricate the Fe, with a solar neighbourhood IMF. The latter observation, however, if substantiated, suggests that the primordial IMF was top-heavy, with the yield enhanced from massive stars relative to the standard nucleosynthetic yields.

Additional evidence that the primordial IMF was weighted towards more massive stars comes from the following, admittedly weak, arguments. The observed gas content of many spirals is inconsistent with that of a simple closed system of gas and stars evolving for a Hubble time with the canonical IMF. The lock-up rate of gas into compact stellar remnants, specifically white dwarfs, amounts to about 70 per cent. Consequently, the gas supply is exhausted after several billion years at the observed rate of star formation ($\sim 4\,M_\odot\,yr^{-1}$) for typical Sb/c disks, with a gas reservoir of $\sim 4 \times 10^9\,M_\odot$. The favoured resolutions to this problem are either gas infall, at a rate of at least $1\,M_\odot\,yr^{-1}$, a possibility suggested by theory for a cold dark matter-dominated universe but not supported by the observational evidence for circumgalactic high velocity gas flows at the present epoch, or else a primordial IMF that was deficient in low mass stars. Truncation of the IMF below, say, $2\,M_\odot$, would reduce the lock-up fraction to ~ 10 per cent, and extend the gas longevity to a Hubble time (Sandage 1986).

A second argument that is indicative of primordial IMF truncation relies on even more circumstantial evidence. The argument proceeds by analogy with starburst systems, galaxies which are undergoing unusually elevated levels of star formation. Extreme starbursts, characterized by a ratio of far infrared to blue luminosities that exceeds ~ 100, are almost invariably associated with tidally interacting galaxies. Moreover, when sufficient diagnostics are observed to study the star formation rate, duration and gas content, it is generally found that a low lock-up fraction is required to avoid premature exhaustion of the gas supply. This is accomplished if the IMF is truncated below 1 to 2 $M_\odot$. An IMF weighted towards massive stars may be a more physical realization of such a stellar mass function, and may indeed be observationally preferred for starburst galaxies (Scalo 1990). During galaxy formation, especially for the spheroidal components of galaxies, the star formation rate is also inferred to have been high. The time-scale over which the bulk of the spheroid stars formed was 10^8 to 10^9 yr, similar to the analogous time-scale, defined as *mass of available gas per unit star formation rate* in a starburst. In addition, galaxy mergers are believed to have been frequent at earlier epochs of the universe, and are indeed a common theme in most galaxy formation theories. Numerical simulations suggest that the near-resonant interaction of gas-rich galaxies should be highly dissipative, and provide a plausible trigger for a vigorous and efficient episode of star formation.

There are also indications that in some circumstances the IMF may be truncated at the high mass end. Cooling flows in galaxy clusters are the most extreme example of this phenomenon, where the purely circumstantial argument is based on the flows of $\sim 100 \, M_\odot yr^{-1}$ characteristically inferred for the $\sim 10^8 K$ intracluster gas without any known heat input that compensates for the cooling nor any observed sink for of order 90 per cent of the cooling gas. Low mass star formation is an attractive way to account for the fate of the cooling gas, although there is no direct evidence, for example from galaxy colours, to support this hypothesis.

Exclusively low mass star formation is known to occur in cold molecular clouds, the sites of T-associations, although the upper mass limit of ~ 1 to $2 \, M_\odot$ is larger than the value that would be consistent with, for example, the cooling flow observed in M87. Globular clusters provide an example of a primordial site where low mass star formation may have predominated. There are two arguments in favour of this view. Direct measurements of the initial mass function in some globular star clusters find that it continues to rise at least to $\sim 0.2 \, M_\odot$, the observational cut-off. Moreover, there are indications that the IMF slope is positively correlated with the cluster core relaxation time-scale (Richer 1990). The steepest slope measured, for ω Cen, may therefore be most representative of the primordial IMF, and it appears to be steeper than the solar neighbourhood IMF. This implies that the primordial IMF, if representative of that in globular clusters, would have been completely dominated, in terms of integrated density, by the lowest mass stars of $\sim 0.1 \, M_\odot$. That there was a primordial paucity of massive stars in globular clusters is suggested by the unmeasurably low ($\lesssim 0.1$ dex) spread in metallicity that is observed for many of these systems. Self-enrichment in the gas-rich protoglobular cluster would have been avoided if the primordial IMF were truncated above $\sim 10 \, M_\odot$ and supernovae consequently were not produced. There are, of course, alternative explanations of the low metallicity dispersion in globulars. Indeed, in two of the most massive globulars, a finite dispersion is found that actually is characteristic of self-enrichment by supernovae, and increasing numbers of neutron stars are also being found in globular clusters such as M15 and ω Cen: 5 and 10, respectively, as of this date, in the form of recycled pulsars (Manchester 1990).

III. COMPACT STELLAR REMNANTS AS BDM?

A plausible scenario, in the context of conventional cosmogonical schemes, can readily be constructed for generation of a considerable amount of dark baryonic matter in the form of compact stellar remnants. Consider the following scheme.

In a BDM-dominated universe ($\Omega_b = 0.1$), commence with a spectrum of primordial isocurvature fluctuations $|\delta_k|^2 \propto k^n$, with n observationally constrained to satisfy $0 \gtrsim n \gtrsim -1$ from large-scale power constraints, or $0 \gtrsim n \gtrsim -1/2$ from microwave background constraints. The associated density fluctuations are $\delta\rho/\rho \propto M^{-(n+3)/6} \propto M^{0.4}$ to $M^{-0.5}$, and, when appropriately normalized to the observed large-scale structure, provide an acceptable fit to the various constraints. The fluctuation amplitude grows as $(1 + z)^{-1}$ over the redshift range $z_{dec} \gtrsim z \gtrsim \Omega_b^{-1}$, and with mass-scale $\sim 10^{13} h^{-1} \, M_\odot$ just being non-linear at present, we infer that mass-scales containing 10^7 ($n = -1$) to 10^9 ($n = 0$) $M_\odot$ went non-linear at decoupling of matter and radiation,

$z_{dec} \approx 1000$. The Jeans mass is $\sim 10^6 \, M_\odot$ immediately after decoupling, so these scales are gravitationally unstable. I shall assume that a large fraction (f_{bdm}) of the mass fragments into stars, before the photoionization of diffuse gas is effective and Compton drag halts further collapse. Only at $z \lesssim 100$ can the gas collapse freely, when Compton drag is no longer effective (Hogan 1978). A rough estimate of the BDM fraction is given by requiring that the universe be photoionized by the BDM precursors, so that

$$f_{bdm}\epsilon_{nuc}f_i\rho_b = n_iI_h = \rho_b m_p^{-1}tt_{rec}^{-1}I_h,$$

where f_i is the fraction of the stellar luminosity in ionizing photons of number density n_i, ϵ_{nuc} is the nuclear burning efficiency, I_h is the hydrogen ionization potential, and t_{rec} is the hydrogen recombination time. I infer that $f_{bdm} = 0.08(1 + z)(0.007/\epsilon_{nuc})(0.2/f_i)$. Star formation occurs over $1000 \gtrsim z \gtrsim 100$, with reionization of the intergalactic medium occurring as early as $z \sim 500$. Only after $z \lesssim 100$ does Compton drag relax and allow diffuse clouds to develop and galaxy formation to be initiated, with large-scale structure being generated by $z \sim 10$. Because of Compton cooling within the star forming clouds, early star formation should be efficient. HII regions and hot supernova remnant shell interiors, whose overpressure and eventual overlap is responsible for driving the galactic winds that deplete the gas reservoirs of dwarf galaxies, are quenched by this process at $z \gtrsim 100$. Any supernova-driven turbulence associated with merging of cooling shells is highly supersonic, and most of this energy is dissipated before any coherent outflow can be established.

The objections to such a scheme for producing primeval compact remnants as BDM, and raised to counter earlier versions of this hypothesis, are twofold: too much diffuse background light is generated and the intergalactic medium is overenriched (Carr et al. 1984). Neither of these arguments is insuperable. Consider first the issue of the diffuse extragalactic radiation field. The high redshift inferred for the bulk of the primordial star formation results in a relatively modest energy excess: $\Omega_{rad} = 7 \times 10^{-6}(\Omega_b/0.1)(\epsilon_{nuc}/0.007)[100/(1 + z)]$. Relative to the CMB, this amounts to $\Omega_{rad}/\Omega_{CMB} = 0.3h^2$. Excess CMB energy of this magnitude was actually reported in an experiment that subsequently was retracted : the current COBE limit on distortions near the peak of the CMB is $\Delta\rho_\gamma/\rho_\gamma \lesssim 0.01$. The expected wavelength range of the redshifted starlight is unlikely to be near the CMB peak at millimeter wavelengths, unless there is a substantial opacity in intergalactic or circumgalactic dust. One would expect the excess radiation background to appear either between 10 and 100μm, if the star forming environment is transparent at 1000A, or to be degraded to 300 - 3000μm if the star forming clouds produce copious quantities of dust. Only in the latter case is the COBE limit an important constraint, since observations at far-infrared wavelengths do not yet have sufficient sensitivity to limit $\Omega_{rad}/\Omega_{CMB}$ to less than 10 per cent.

Overenrichment is a more serious problem, but there are severe theoretical uncertainties in making this a quantitative issue. Nucleosynthetic yields are poorly known for current generation stars, let alone for primordial stars. While supernovae are expected to eject considerable amounts of heavy elements regardless of initial abundances, the same may not be true for intermediate mass stars which are the primary sources of stellar production of helium, carbon and nitrogen. For example, helium shell flashes are suppressed at $[CNO] \lesssim -4$ (Fujimoto et al. 1984). This means that convection is suppressed, and dredge-up of heavier elements produced in He-burning will not contaminate the hydrogen layer of AGB stars during the period of prolific mass-loss that characterizes this evolutionary phase. Depending on the precise form of the primordial IMF, yields may be greatly reduced. Even with conventional yields, one can fine-tune the IMF, if all stars are in the 2 to 5 $M_\odot$ mass range, so that relatively little He, C, N overproduction occurs (Ryu et al. 1990). A modest overenrichment can be cured by appeal to uncertainties in diffusion and dispersal of the heavy elements. Indeed, some pre-enrichment may even be required, as for example appears to be the case for the galactic disk, to account for the paucity of metal-poor G-dwarfs.

IV. BROWN DWARFS AS BDM

Brown dwarfs, defined to be objects below the hydrogen burning limit that are undergoing gravitational contraction, are much less vulnerable to disproof as BDM than are compact stellar remnants, since their observable signatures are more elusive. I have already described the phenomenological reasoning, based on inferences from galaxy cluster cooling flows and the IMF and metallicity dispersion within globular clusters, which (weakly) supports the thesis that primordial star formation was weighted towards brown dwarfs. Further theoretical support comes from the following, equally weak, argument. Stars form via fragmentation of massive gas clouds. There is no reason to believe that the fragmentation process has any advance knowledge of the fact that the minimum hydrogen burning mass for a star is between $0.08\,M_\odot$ at solar metallicity and $0.1\,M_\odot$ at primordial abundance levels. Opacity-limited fragmentation of a self-gravitating cloud results in a minimum fragment mass of $\sim 10^{-3}\,M_\odot$ under highly idealized assumptions, such as spherical symmetry, neglect of non-linear effects including fragment interactions that might lead to coalescence or disruption, neglect of accretion, and radiation transfer in a uniform medium that ignores the reality of cooling in a highly inhomogeneous cloud. Most of these effects tend to increase the minimum fragment mass attained after an initial free-fall time, for which a best guess might be $\sim 0.01\,M_\odot$ (Boss 1988).

A direct test of this is not possible with the available dynamic range of numerical collapse simulations. However, a seed mass of this order also appears to be required in theories of low mass star formation that appeal to spherical accretion onto the non-homologous core of a cold molecular cloud, and successfully account for the first appearance of gravitationally-powered stars in the Hertzsprung-Russell diagram, on the T-Tauri star birth line (Stahler 1988).

There are two principal difficulties with the brown dwarf hypothesis. The efficiency of brown dwarf formation seems arbitrary, since one cannot utilize the hypothesis of self-regulation via energy input from forming or dying stars to ensure that *some* diffuse gas remains to later form galaxies.. It is these processes that determine star formation efficiency as observed in molecular clouds and as outlined in the preceding discussion of compact stellar remnant formation. The solution would seem to require postulating that some massive stars form coevally with the brown dwarfs, to supply the necessary energy input for self-regulation to limit the mass fraction that condenses into compact objects. It clearly would be disastrous for the model if too much or too little of the primordial gas was converted into brown dwarfs.

Another objection focusses on observational searches for brown dwarfs in the solar vicinity. To date, not a single brown dwarf detection has been confirmed. The searches utilize astrometric or velocity variations of nearby (within 5 pc) late-type stars. Several companions have been detected with masses down to $0.08\,M_\odot$, but no candidates with masses in the range 0.02 to $0.08\,M_\odot$ have been discovered. The mass function of companions cuts off below $\sim 0.08\,M_\odot$. Only a limited range of mass ratios is probed, and this result does not of course tell us anything about free-floating brown dwarfs. For the brown dwarf hypothesis to survive, one would have to argue that isolated brown dwarfs form differently from those in binary systems, or that primordial metallicities and physical conditions strongly enhanced the formation of brown dwarfs.

V. THE SEARCH FOR BDM

MACHO candidates for BDM are observable. A number of experimental searches are presently underway. I divide the ensuing discussion into consideration of white dwarfs, brown dwarfs, neutron stars, and a generic approach to BDM.

A. White dwarfs

The compact stellar remnants that are most easily reconciled with diffuse background light limits and primordial enrichment constraints are white dwarfs. Dark halos which contain an appreciable fraction of their mass in the form of white dwarfs may be detectable via the likely presence of white dwarf binaries, which result in the formation of Type Ia supernovae. Up to 20 per cent of binaries in the initial mass range 6 to 18 $M_{\odot}$ eventually merge as degenerate dwarfs and explode as supernovae within $\sim$ 17 Gyr. Observations of supernovae in elliptical galaxies and in galaxy clusters suggest that the binary fraction in white dwarf BDM must be an order of magnitude less than found for a similar mass precursor population in the solar vicinity (Smecker and Wyse 1990). This assumes that one is looking at populations with similar ages and spatial distributions when comparing observed and predicted supernova rates. However, the double degenerate merger model for Type Ia supernovae has several uncertain parameters which make the asymptotic prediction of supernova rates, after a Hubble time has elapsed, correspondingly uncertain.

A second approach to searching for halo white dwarfs relies on direct detection. Discovery of an abrupt decline in the luminosity function of old disk white dwarfs fainter than $L_V = 10^{-4.5} L_{\odot}$ is attributed to formation of the galactic disk about 9 Gyr ago. White dwarfs are still visible for a further 6 Gyr, although their visual luminosity declines by an order of magnitude over this time. The existing surveys for high proper motion halo white dwarfs only probe down to 18th magnitude. It is quite feasible to develop a CCD survey that will search for the predicted frequency ($\sim$ 1 /sq.deg.) of high velocity halo white dwarfs expected, out to $\sim$ 100 pc distance (Tamanaha et al. 1990), provided the limiting survey magnitude is extended to 22 in the I ($\sim$ 9000A) band.

B. Brown dwarfs

The cumulative effect of a halo's worth of brown dwarfs results in a diffuse far infrared glow, which peaks at a wavelength and with an intensity that depends on the adopted IMF. Over the mass range suggested here of 0.01 to 0.1 Ms for putative brown dwarfs, the resulting flux should be detectable between $2\mu m$ and $20\mu m$ for nearby edge-on spiral galaxies by the future generation of space-infrared telescopes (Adams and Walker 1990). Direct detection of the nearest halo brown dwarfs may also be feasible in the mass range immediately below the hydrogen-burning limit. These brown dwarfs have very long cooling time-scales, and the nearest of them may be detectable as high proper motion objects in deep large area near infrared surveys.

C. Neutron stars

The abundances of neutron stars, as inferred from the frequencies both of low mass x-ray binaries and of recycled millisecond pulsars, in globular clusters appear to be enhanced with respect to the disk and the bulge populations. Tidal capture of a neutron star by a main sequence star followed by Roche lobe overflow, as the hydrogen burning star exhausts its fuel supply and evolves into a red giant, provides an important mechanism for generating both classes of objects, although neutron star capture by primordial main sequence binaries is probably a more significant contributor in globular clusters (Hut et al. 1990). Even if both processes occur with substantial probability in globular clusters, an appreciable fraction of the mass of a globular (about 1 percent for a Salpeter mass function, but possibly as high as 10 percent) is likely to be in the form of neutron stars. This does provide some weak justification to the search for neutron stars in BDM, which reflects a stellar population that either slightly predates or is coeval with that of the globular clusters.

More specifically, if the BDM objects formed in compact clusters, one would have had occasional captures by the few main-sequence stars that could plausibly have formed at the same time. Even if the ratio of low mass stars to neutron stars at formation is only one per cent, a potentially observable number of x-ray binaries is eventually produced as a consequence of tidal captures. By analogy with the inferred low mass x-ray binary and recycled pulsar population enhancement in the stellar content of globular clusters, one would expect the number of halo x-ray binaries to scale with spheroid luminosity. One thereby obtains an observable signature of BDM that is associated in particular with luminous spheroids and may contribute towards the x-ray emission observed from elliptical galaxies.

D. Generic BDM

Generic MACHOs, from sub-Jupiter masses up to several solar masses, may be detectable in a gravitational microlensing experiment. The probability of a gravitational lensing event due to passage of a lensing object in our halo in front of a background star in the LMC is extremely low (Paczynski 1986);

$$\tau = \pi n_x R_{crit}^2 a \approx 5 \times 10^{-7},$$

where n_x is the number density of MACHOs, a is the halo core radius, and significant amplification of the background star occurs if a halo object of mass m_x passes within a distance R_{crit} of the line of sight; $R_{crit} = \left(\frac{4Gm_x d}{c^2}\right)^2$, and $d \approx a$. The duration of an event is $\Delta t = R_{crit}/v_{rot}$ and the frequency of events is $\tau/\Delta t$, from which we infer that to observe an event of duration $\Delta t = 0.2(m_x/M_\odot)^{1/2}$yr with unit probability, it is necessary to monitor $\sim 2 \times 10^6$ LMC stars. This task may seem formidable, but two groups are preparing independent experiments to achieve a microlensing survey of the LMC with the goal of probing the mass range $10^{-8}\,M_\odot \lesssim m_x \lesssim 10\,M_\odot$. The sought-for signature is symmetric in time, and achromatic: hopefully these features will help distinguish microlensing from intrinsic stellar variability.

VI. CONCLUSIONS

Baryonic dark matter is a major constitutent of the universe. It exists in addition to any disk dark matter: the locally measured M/L ratio for the total mass surface density in the vicinity of the sun does not exceed 3, or in cosmological terms, amounts to $\Omega_{disks} \lesssim 0.002h^{-1}$. Disks contain most stars in the universe: averaging both over more evolved (higher M/L) disks of early type galaxies and spheroids suggests $\Omega_{stars} \approx 0.004^{-1}$. From nucleosynthesis constraints, one has $\Omega_b = 0.015h^{-2}$, whence the luminous baryon fraction is inferred to be $\sim 0.3h$ of all baryons. Inclusion of known gas reservoirs in galaxy groups and clusters increases this by at most fifty per cent. Thus, at least half of the baryons in the universe are dark baryons, not in intracluster gas or luminous stars.

One could easily hide these baryons in the form of a hot intergalactic medium that would be potentially detectable via diffuse emission (at x-ray or ultraviolet frequencies) or absorption towards high redshift quasars. However an equally compelling source of BDM is in the form of compact objects, which, unlike gas, also provide a possible explanation of halo dark matter. I have presented the case here that BDM MACHOs are compact stellar remnants, either white dwarfs or neutron stars, or brown dwarfs. Massive black holes, remnants of supermassive stars, provide a fourth option that I have not discussed simply because I give them lesser weight in the inevitable heirarchy of speculative hypotheses that must be invoked to develop a plausible astrophysics scheme for BDM formation. Several observational searches are underway for halo BDM: only success will justify these theoretical arguments.

I am grateful to Professor A. Rodgers for providing a hospitable and stimulating environment at Mt. Stromlo during the writing of this paper. My research has been supported at Berkeley in part by grants from NASA and NSF.

References

Abia, C. and Rebolo, R. 1989, *Astrophys. J.*, **347**, 186.

Adams,F. and Walker, T. 1990, *Astrophys. J.*, **359**, 57.

Barbuy, B. 1988, *Astrophys. J.*, **191**, 121.

Boss, A. 1988, *Astrophys. J.*, **331**, 370.

Carr, B.J., Bond, J. R. and Arnett, W. D. 1984, *Astrophys. J.*, **277**, 445.

Fujimoto, M., Iben, I., Chieffi, A. and Tornambe, A. 1984, *Astrophys. J.*, **287**, 749

Hut, P., Murphy, B. and Verbunt, F. 1990, *Astron. Ap.* (in press).

Manchester, R. 1990, private communication.

Paczynski, B. 1986, *Astrophys. J.*, **304**, 1.

Richer, H. 1990, preprint.

Ryu, D., Olive, K. A., and Silk, J. 1990,*Astrophys. J.*,**353**, 81.

Sandage, A. 1986, *Astrophys. J.*, **161**, 89.

Scalo, J. 1990, in *Windows On Galaxies*, A. Renzini, G. Fabbiano and J. S. Gallagher, eds., (Dordrecht:Kluwer) (in press).

Smecker, T. and Wyse, R. F. G. 1990, *Astrophys. J.* (in press).

Stahler, S. 1988, *Astrophys. J.*, **332**, 804.

Tamanaha, C., Silk, J., Wood, M. and Winget, D. 1990, *Astrophys. J.*, **358**, 164. .

Phenomenological Dark Matter Detection Rate-*from WIMP to SIMP-*

Humitaka SATO
Department of Physics
Kyoto University
Kyoto606
Japan

1. Two kinds of Dark matter

Freeze-out of interaction in the primordial fireball is one of the fundamental processes in the Big bang theory.[1,2] Relics of this process will provide us with the dark matter(DM). Unknown interactios in the energy region above the current accelerator energy might have brought us some new relic matter in the universe.

However the experimental result on Z width by LEP/SLC has ruined away an interesting mass range of the weakly interacting massive particle(WIMP) completely.[3] This result implied an exclusion of any new components besides the socalled standard theory of particle physics in the weak interaction. The WIMP scenario had predicted various facinating effects which might be detectable in near future.[1] And, then, this result would be quite disappointing, particularly for those who are planning the DM detection experiment.

The WIMP's were interesting objects because they might be a dominant component of mass in the cosmological large scale and they might be also detected directly by the present technique in a laboratory and/or indirectly by observing neutrinos and gamma- rays generated via annihilation and decay. The dominant mass and the detectability, however, will be two independent factors and it was rather an accidental situation that the WIMP in some mass range posseses these two properties. Therefore, these two properties should be considered separately. Hidden actors in these two phenomena would be different.

Aim of this paper is to search for loopholes of the DM particles which exhibit detectable evidences such as listed in the above. By taking the mass density and the direct detection as two separate observable constraints, we depict an "interesting" allowed regions for physical parameters of the DM in a purely phenomenological manner.

2. Crossing Symmetry

First, we note on the detection of some freez-out "symmetric" relic particle X in the halo: the detection rate through the recoil of the nuclei in detector is given as below[5];

$$r_x = < \sigma_x v >_s n_x \frac{\rho_H}{\rho_0} \frac{M_d}{m_N}$$

, where $< \sigma_x v >_s$ is the elastic scattering rate between the relic particle and the detctor nuclei, n_x is the average density of X and $\frac{\rho_H}{\rho_0}$ is an enhanced ratio in the halo. On the other hand, the abundance of the relic in a ratio to the photon relics is estimated roughly, neglecting all numerical coefficients of the order of unity[1,2], as

$$\beta_x = \frac{n_X}{n_\gamma} \sim \frac{1}{< \sigma_X v >_a m_x m_{Pl}}$$

and it is inversely propotional to $< \sigma_x v >_a$, m_{Pl} being the Planck mass.

Due to the crossing symmetry between annihilation and scattering, $< \sigma_x v >_s$ and $< \sigma_x v >_a$ are related each other.[4] Therefore, the above detection rate is independent of a coupling strength of the interaction, though it will depend on the type of interactions. This teaches us that an order of magnitude of the DM detection rate is determined not by a specific model of particle interaction but by the property of the expanding universe. We shall call this detection rate a phenomenological detection rate.

The above argument will be true if the annihilation of X into quarks is a dominant channel at the freezing epock. If this condition is not satisfied, the annihilation will proceed more completely. Then, we may conclude that the phenomenological detection rate gives the maximum rate.

If the ratio $< \sigma_x v >_s / < \sigma_x v >_a$ is roughly constant of the order of unity, the rate is solely dependent on the mass,i.e. inversely propotional to m_x. Therefore the canonical DM detection rate such as 1 event/day·kg computed for the WIMP is not specific to a special model of interaction but just determined by the mass of DM particle. What was specific to the WIMP scenario was that the dominant component of mass is deduced from the weakly interacting particles with mass of 2Gev.

In order to find the loophole of DM which is a dominant component of mass and/or is detectable by the WIMP detector, we have to introduce a larger annihilation cross section and an asymmetric abundance. To be consistent with the accelerator experiment, the mass of such particle must be sufficiently large. What we search for is "Strongly Interacting Massive Particle(SIMP)" rather than WIMP.

3. "Mass-dominant" and "Detectable" Dark Matter

We introduce three physical parameters, the mass m_x, the annihilation rate $< \sigma_x v >_a$ and the degree of asymmetry α_x. Since we have been used to the situation for the WIMP, let's employ here "WIMP unit" such as

$$M = \frac{m_x}{m_w}, A = \frac{< \sigma_x v >_a}{< \sigma_w v >_a}, S = \frac{< \sigma_w v >_s}{< \sigma_x v >_s}, C = \frac{A}{S}.$$

The crossing symmetry claims C is a number of order unity and C is independent on the velocity in the halo . As the degree of asymmetry α_x, we normalize it by the symmetric freeze-out abundance β_x such as

$$\alpha = \frac{\alpha_x}{\beta_x}$$

Now, the mass density, the detection rate and the abundance are written in the ratio to those for the WIMP as follows;

$$\rho \equiv \frac{\rho_x}{\rho_w} = \frac{\alpha}{A}$$

and

$$r \equiv \frac{r_x}{r_w} = \frac{\alpha}{MC}.$$

If we take the massive neutrino as the WIMP,$\Omega = 1$ for $m_w = 2Gev$ and one of the observational constraints is $\rho < 1$. If we had already completed the detector which can detect this massive neutrino, another constraint is $r < 1$. Furthermore, $\alpha > 1$.

We can depict the allowed region in the parameter space of M, A and α. The "interesting" region is the boundary along "mass dominant" of $\rho = 1$ or "detectable" of $r = 1$. Figures (i),(ii) and (iii) are the cross sections in the parameter space (α, A, M) for a given M, A and α respectively. The hatched region is an allowed region. (Here, we have assumed as $C \sim O(1)$.) The border (a)-(b) represents the "mass-dominant" or $\rho = 1$ and the border (b)-(c) does the "detectable" or $r = 1$. In outside of the border (a)-(b), the DM is over "mass-dominant" and, in outside of the border (b)-(c), over "detectable". In this parameter space, it is an accidential fine tuning that the both conditins of $\rho = 1$ and $r = 1$ are satisfied. So it will be more natural that "mass dominant" DM and "detectable" DM is different. Historically, DM was introduced as "mass dominant" one in astronomy. But, what the big bang introduced would be different from it and should not be coupled so strongly to it. Even if the relic DM is not "mass dominant", the "detectable" mass has its own right and the effort of detection in the laboratory should not be stopped. DM for astronomers and for laboratory experimentalists could be different. They had dreamed different dreams in the same bed.

Reference

1. E.W.Kolb and M.S.Turner,"The Early Universe", Addison-Wesly Pub Comp. 1990.
2. H.Sato, T.Matsuda and H.Takeda, Prog.Theor.Phys.Supplement No 49(1971),11.
3. K.Griest and J.Silk,Nature,333(1990),26.
4. K.Griest and B.Sadoulet, in "Proc. Second Particle Astrophysics School on Dark Matter", Erice,1988. preprint Fermilab-Conf-89/57-A
5. M.W.Goodman and E.Witten, Phys. Rev. D31(1986),3059.

385

Figure

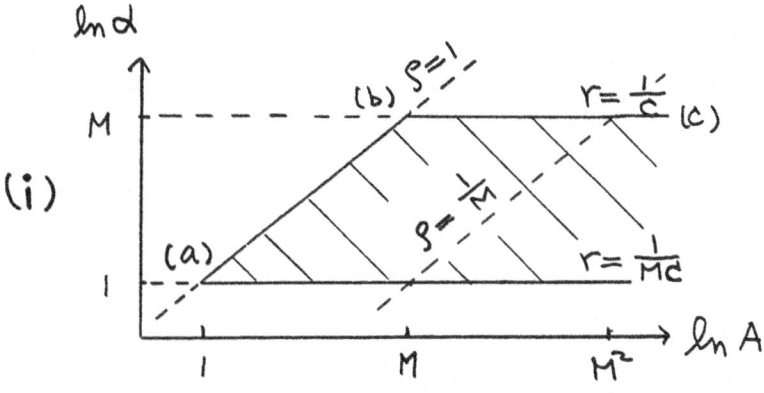

(i)

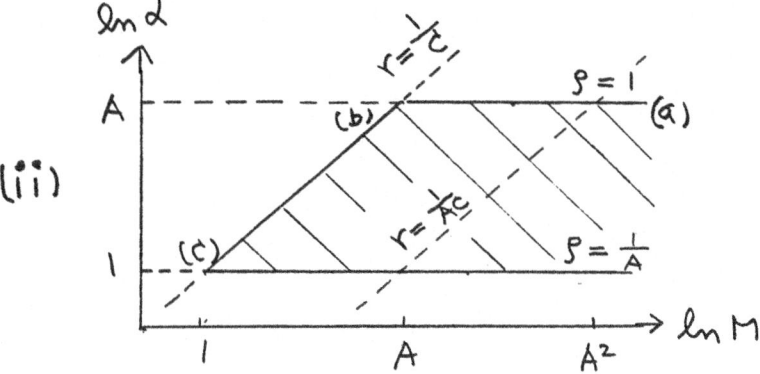

(ii)

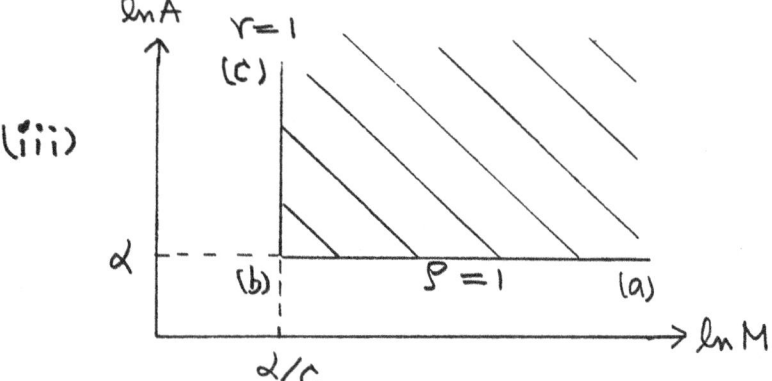

(iii)

X-Ray Iron Line of Cluster of Galaxies

Katsuji KOYAMA,
Department of Astrophysics
School of Science
Nagoya University
Furo-cho, Chikusa-ku
Nagoya, 464-01

ABSTRACT.   The evolution and structure of the intracluster gas are discussed based on the observation of the Ginga satellite.   We discovered large scale X-ray emission extending about 12 ° from the Virgo cluster.   The scan profile and the temperature structure suggest that the intra-cluster gas of Virgo cluster has two components around the bright elliptical galaxies, M87 and M49. The two-subclustering indicates that the Virgo cluster is now under strong evolution. We demonstrate that the simple evolutionary scenario can roughly explain the relation between X-ray luminosity and gas temperature from many clusters of galaxies observed with the Ginga satellite.

The iron abundance of the outer part of the Virgo cluster is only about 0.1-0.2 of cosmic value. We therefore interpret this to mean that the major components of intracluster gas are primordial. However the iron abundance of the Virgo cluster shows central concentration near M87, which indicates that the central part was contaminated by the gas ejected from each galaxy.   The iron abundance from individual clusters show a trend of anti-correlation with the gas temperature.   One possible reason for this anti-correlation is that the galaxy formation rate is smaller in clusters of hotter intra-cluster gas.

1.   Introduction

The discovery of line emissions from highly ionized iron in the X-ray spectrum of the cluster of galaxies provided us with a critical information for the origin of the X-ray emissions.

Fig.1

The X-ray spectrum
of the cluster obtained
with the Temma satellite
(Okumura *et al.* 1988).

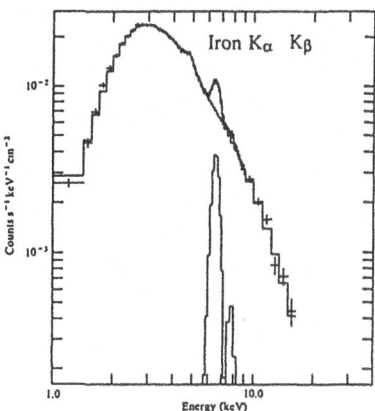

The best X-ray spectrum of the cluster above 2keV energy range has been obtained with the *Tenma* satellite. Figure 1 shows the X-ray spectrum of the Perseus cluster. We find clear peaks at energies of 6-8 keV. With the model fitting, Okumura et al. (1988) found that the center energies of these peaks are 6.7 keV and 8 keV with no significant broadning of the lines. Since these energies are fully consistent with the $K\alpha$ and $K\beta$ lines from Helium-like iron atom , and since the intrinsic line width is not broadened, we can conclude that the X-ray emission comes from a thin hot plasma with a temperature of 10^7-10^8 K. In fact, overall spectrum can be fitted with the thin thermal model of about 10 keV temperature.

Extensive imaging observations of the X-ray emissions from clusters of galaxies have been carried out with the Einstein Observatory (e.g. Forman and Jones). From the X-ray luminosity (L_X), size of plasma (D) and Temperature (T), we can estimate the total mass of hot gas and virial mass. Rothenflug and Arnaud (1985) have compiled the results and showed that the virial mass is about ten times larger than that of the gas mass. David *et al.* (1990) gave a relation between gas temperature and the ratio of gas mass and stellar mass. As temperature goes up, the gas-to-stellar mass ratio is also increased. In an extreme case , the gas mass is about ten times larger than the stellar mass. Therefore, the hot gas should play an essential role for the structure and evolution of the cluster of galaxies.

The *Ginga* satellite has a high sensitivity in the energy band of relevant cluster gas temperature, in particular in the iron -line energy. Here we report on the new X-ray results of clusters of galaxies made with the *Ginga* satellite.

2. Evolution of Clusters of Galaxies

We will start from the observation of the Virgo cluster. The Virgo cluster is the nearest rich cluster and has a large optical extension of about 12° along the major axis of the cluster. However the previous X-ray instrument had obtained X-ray emission within the small region less than 2 deg radius from M87.

We have conducted X-ray scanning observations along the cluster major axis which runs over M87 and M49 and two parallel paths about 3 degree away. To the scan profiles, we tried model fitting of surface brightness with the standard model;

$$S(r)= S(0) \{1+(r/a)^2\}^{-3\beta+0.5} \quad (r<r_{cut})$$

This model fitted well to the large scale scan profile with the same parameters of the *Einstein* IPC observation. Thus we find that the large scale extended emission is naturally extrapolated from the *Einstein* IPC result which gives the surface brightness within the small region near M87. Furthermore, we found extra component in the southern part of the Virgo cluster. The temperature distribution of the Virgo cluster along the major axis shows two sub-clustering of temperature; one is 2-keV component and the other is 3-keV component.

This sub clustering indicates that the Virgo cluster is not relaxed yet. With the numerical simulation of the evolution of the cluster, Many author reported that the same kind of sub-clustering is appeared during the evolutional stage. Therefore we see that the Virgo cluster is undergoing strong evolution at the present time. If the Virgo cluster is more evolved, the two sub-cluster merge with each other. By this evolution, the size of the cluster become small and the gas density become large. In this case, the gas temperature is proportional to the inverse of the radius;

$$T \propto R^{-1}$$

The X-ray luminosity is given with this equation;

$$L_x \propto n^2 R^3 T^{0.5} \propto T^{0.5} R^{-3}$$

Then we can get the luminosity and temperature evolution with the function;

$$L_x \propto T^{3.5}$$

The relation of the temperature and luminosity of many clusters observed with the Ginga satellite gave the best-fit relation of $L_x \propto T^{3.2}$, which is roughly consistent with the relation derived by the simple estimation assuming that the cluster is evolving to condense.

The luminosity evolution of the cluster of galaxies is also found in the logN-logS relation. Edge et al.(1990) found a clear deviation from the Euclidean distribution even at a small cosmological red-shift about 0.1-0.2. This indicates that the X-ray luminous clusters are undergoing strong luminosity evolution even at this moment.

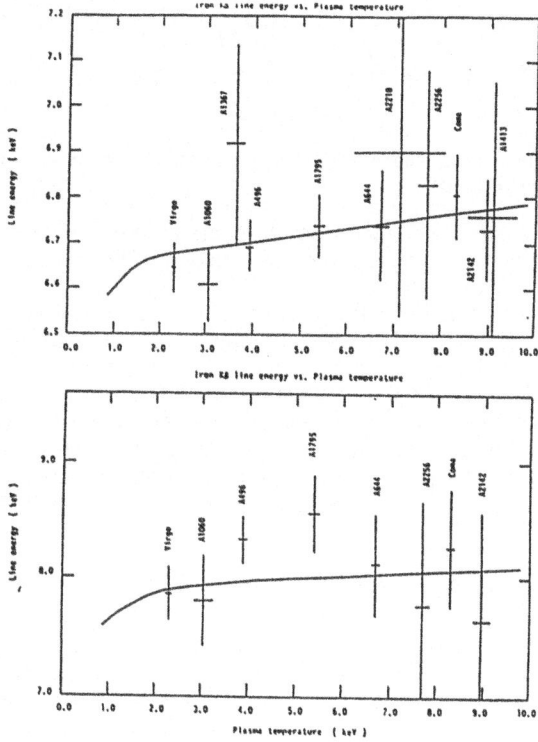

Fig.2   The relation between the gas temperature and iron line K_α (upper panel) and K_β (lower panel) energ (Hatsukade 1989).

3. Iron Line from Clusters of Galaxies

3-1 Iron distribution in the Virgo cluster

Since heavy elements such as iron are not primordial but originate in stars, the iron line is a powerful tool to study the origin of the intracluster gas.   Figure 2 gives a relation between the gas temperature and iron line K_α and K_β energies. The data fits well to these theoretical lines in which we assume the ionization equilibrium. Then, we can easily estimate iron abundance from the observed iron line intensity .

Koyama et al. (1991) obtained   a   distribution of iron abundance along the major axis of the Virgo cluster and found a clear peak near the cluster center at M87 (fig. 3).

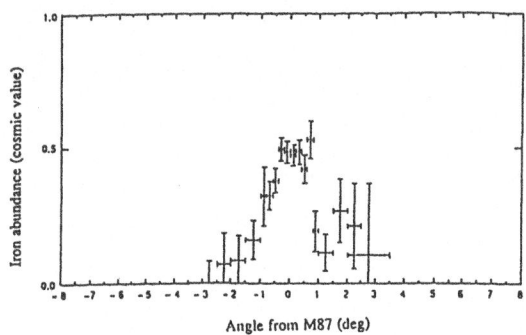

Fig.3   The distribution of iron abundance along the major axis of the Virgo cluster
(Koyama *et al.* 1990)

They concluded that the iron abundance from the central region (within 1° from M87) is about 0.5 cosmic value, while that at the outer region of the cluster (larger than 1° from M87) is 0.1~ 0.2 cosmic value.

Since the iron abundance of the outer part of the Virgo cluster is only 0.1-0.2 of cosmic value, we can assume   that the intracluster gas of Virgo should mainly originated from the primordial gas.   The outer region of the Virgo cluster has not yet been contaminated by the gas ejected from each galaxy.   By contrast, the central region of the cluster has been contaminated by a more efficient mass-loss from galaxies due to a denser galaxy distribution than the outer region of the cluster. Therefore the iron line distribution of the Virgo cluster gives us good example of tracing the hot gas evolution in the cluster. Another example was found in the recent study of the Perseus cluster by Ponman et al. (1990).

3-2 Iron Abundance and Gas Temperature

The Ginga field of view is too large to investigate the spatial structure other than the Virgo cluster.   Therefore we will discuss the mean iron abundance and the gas temperature of the cluster of galaxies.   Hatsukade (1989) summarized the iron abundance and the temperature from many clusters of galaxies. He found that the iron abundance is not constant but shows a decrease as the gas temperature increases.

We will assume that the intracluster gas consists of two components: the primordial gas which includes no iron, and the gas ejected from the galaxy with the cosmic abundance. Then the iron abundance in the intracluster gas is a simple function of the ratio of the primordial gas and the ejected gas.   From the observation, we found when the gas temperature is high, iron abundance is small. If we simply assume that the temperature is proportional to the depth of gravitational potential of the cluster. Then a cluster of deeper potential tend to have a smaller amount of ejected gas. This may be interpreted that the formation rate of the galaxy is small in a cluster of deep gravitational potential.

References

David, L. P. *et al*. 1990, *Astrophys. J.*, **356**, 32.
Edge,A.C, *et al.* (1990), *M.N.R.A.S.*, **245**, 559.
Forman, W. and Jones, C. 1982, *Ann. Rev. Astrophys*, **20**, 547.
Hatukade, I. (1989), Ph .D. Thesis, Osaka University
Koyama, K. *et al.* (1990), submitted to *Nature*
Okumura, Y. *et al.* (1988), *Publ. Astron. Soc. Japan,* **40**, 639.
Ponman, T. J., *et al.* (1990), *Nature* , **347**, 450.
Rothenflug,R., and Arnaud,M. (1985) , *Astr. Astrophys.* 144, 431
Takano, S. *et al.* (1989), *Nature* , **340**, 289.

DYNAMICAL EVOLUTION OF COMPACT GROUPS OF GALAXIES

K.KODAIRA
National Astronomical Observatory
Mitaka, Tokyo, 181
Japan
S.OKUMURA, J.MAKINO, T.EBISUZAKI, and D.SUGIMOTO
Department of Space and Earth Sciences
Faculty of Art, Univ.Tokyo
Komaba, Tokyo, 153
Japan

ABSTRACT. Observations about dynamical aspects of compact groups
of galaxies are reviewed, and N-body numerical simulations are
applied in order to lend theoretical interpretations to the
observational results. In connection with dynamical evolution time
scale the possible importance of an extended massive dark matter
envelope is pointed out in which a compact galaxy group is
embedded.

1. INTRODUCTION

Both of recent observational and theoretical studies seem to support
the view that gravitational interactions such as collisions and
merging continue to reshape galaxies and aggregations of galaxies
even to the present day long after the epoch of the initial galaxy
formation (cf.Schweizer 1989). This current view, however, is
mostly based upon the short time scale of dynamical evolution of
galaxy groups indicated by N-body simulations. In this paper we
will present a brief summary of observational investigations about
compact groups of galaxies, and report the preliminary results of
N-body simulations which have been carried out by the present
authors. Compact groups of galaxies most popularly can be
described as aggregates of galaxies whose projected separations are
typically on the order of the diameters of the galaxies themselves
(Rose 1977).

2. COMPACT GROUPS IN GENERAL

Rose (1977) listed about 30 compact groups of quartet galaxies
(n=4). He investigated in details two of them and concluded that
non-detection of any enhanced background light might negate

dynamical evolution of the group (Rose 1979). His spectroscopic observation led to mass-to-light ratios of 4 and 24 for the two groups, which were in apparent conflict with the massive halo hypothesis (Rose and Graham 1979). Turner and Gott (1976) published a catalog of groups of galaxies, and 10 dense groups among the entry and 8 from other sources were examined by Hickson, Richstone, and Turner (1977), to reveal that the size of the largest galaxy in each group is correlated with the mean separation among member galaxies, indicating stripping interaction. Heiligman and Turner (1980) investigated about 10 compact groups of $n=4$-8, and found that the luminosity function of the galaxies showed deficiency in the low luminosity part ($M_B \gtrsim -20$; $H_0=50$) relative to the Schechter's (1976) function. Merging of galaxies was suggested by them as one of possible causes of this anomaly.

Hickson (1982) defined compact groups of galaxies by applying restrictions on the number of member galaxies ($n \gtrsim 4$), on the separation to the nearest non-group galaxy ($\theta_N \geq 3\theta_G$), and on the average surface brightness of a group ($\bar{\mu}_G < 26.0$). Thus he listed about 100 groups and compared the properties of member galaxies with those of field galaxies and galaxies in rich clusters. They found fewer spirals in compact groups, in accordance to Postman and Geller (1984). Hickson (1982) also pointed out a possible deficiency of faint galaxies in compact groups, in particular, in groups with elliptical first-ranked galaxies. Hickson, Kindl, and Huchra (1988b) revealed high morphological concordance among member galaxies in Hickson's (1982) groups. They found, however, no strong correlation between dominant morphological type and galaxy space density, but significant correlation of morphology with group velocity dispersion and weak one with group optical luiminosity; more ellipticals for higher velocity dispersion, and for luminous groups.

3. SHAKHBAZYAN'S COMPACT GROUPS

Shakhbazyan (1957) and his colleagues published a series of lists of Compact Groups of Compact Galaxies which included about 400 groups (Shakhbazyan 1973; see references in Kodaira *et al.*1988). The SCGG list seems to be more comprehensive than others as far as sample numbers and the number range of member galaxies ($n=6$~20) are concerned, but is restricted to compact groups of compact galaxies. Our surface photometric studies using CCD imaging led to the conclusion that the Shakhbazyan's compact galaxies are luminous, but not extraordinary, ellipticals and S0's in a distance range of $z \approx 0.03$-0.11 (Kodaira *et al.*1988, 1990a, 1990b).

Our investigation suggests that a few objects are occasionally in the fore or background of the group but that majority of group members in the original lists have accordant redshift (Kodaira *et al.*1990a, 1990b; Hickson, Kindl, and Huchra, 1988a). There remains the question whether the accordant members are seen by chance on a line of sight or physically bound. The probability of chance coincidence was found to be $<10^{-4}$ for one of typical groups, SCGG 202

(n=17). This group turned out to be located on the far side of the
intracluster bridge between Coma Cluster and A1367 in the periphery
of a void (Kodaira et al.1988) supporting a physical nature of
SCGG's. Mamon (1986) argued based upon his numerical simulation
that roughly half of the compact groups in Hickson's catalog were
simply chance alignments of galaxies within loose groups. Based
upon the results of Monte Carlo numerical simulation, however,
Hickson and Rood (1988) argued that the probability of chance
coinsidence must be by about two orders of magnitude lower than
Mamon (1986) estimated. Under the assumption that SCGG's are
physically bound systems, we obtained mass-to-luminosity ratio
typically of M/L~100 and deficiency of faint galaxies (M_V>-20;
H_0=75) for 5 SCGG's (Kodaira et al.1990a).

4. DISCUSSION ON OBSERVATIONAL RESULTS

Previous investigations were concerned with compact groups of
rather small number of member galaxies (n=4~6), while SCGG covered
a wider range of n up to 20 or more. In figure 1 we have plotted
the number of member galaxies, n, versus the geometrical diameter
of a compact group, D, in units of 10^2Kpc for H_0=100Kms^{-1}Mpc^{-1}. The
plotted samples are 12 SCGG's whose redshifts are available and
those of Heiligman and Turner (1980). The two samples of Rose
(1979b) fall over the domain occupied by the latter. Hickson's
(1982; Hickson, Kindl, Huchra 1988b) samples also overlap with
Heiligman-Turner samples but scatter further to larger diameter
(see figure 2). However, when only the groups with no spiral
members are plotted, they distribute well in the same domain as
shown in figure 1.
      We suggest that compact groups, especially those mainly
composed of luminous non-spiral galaxies, may form a sequence
indicated by lines in figure 1. Although the number of SCGG's with
known redshift is only 12, these seem well to represent whole SCGG
samples (Kodaira et al.1990b). SCGG 1 is an extreme sample which
was detected by Shakhbazyan (1957) in the earliest phase of his
survey.
      In figure 2 we have plotted geometrical diameter, D, versus
group velocity dispersion σ for Hickson's (1982) groups according to
Hickson, Kindl, and Huchra (1988b), and for 5 SCGG's according to
Kodaira et al.(1990a). We notice that those of Hickson's groups
with non-spiral members are not only small in diameter but also
large in group velocity dispersion. Typical member number of these
Hickson's non-spiral groups is n=4-5, while n is typically ~10 for
SCGG's in figure 2. By defining indicative group mass as $m=\sigma^2 D/2G$,
we find m of the latter to be roughly one order of magnitude larger
than m of the former. This suggests either that galaxies in the
latter are substantially more massive than in the former, or that
the latter contain more dark matter than the former.
      We estimate the time scale of dynamical evolution of a compact
group by means of indicative crossing time, $t_c=D/\sigma$. This time scale
is as short as t_c~10^8yr for small non-spiral groups, and t_c~10^9yr

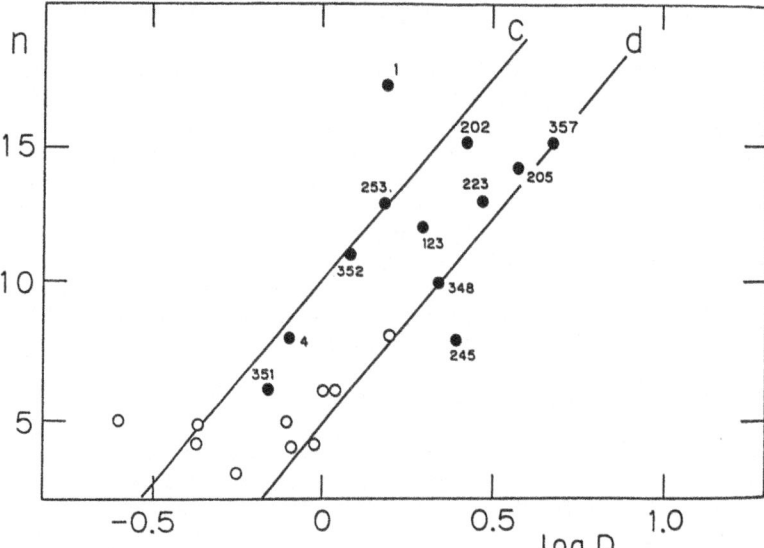

Fig.1. Number-Diameter Relation of SCGG's (filled circle) and Heilligman-Turner's groups (open circle)

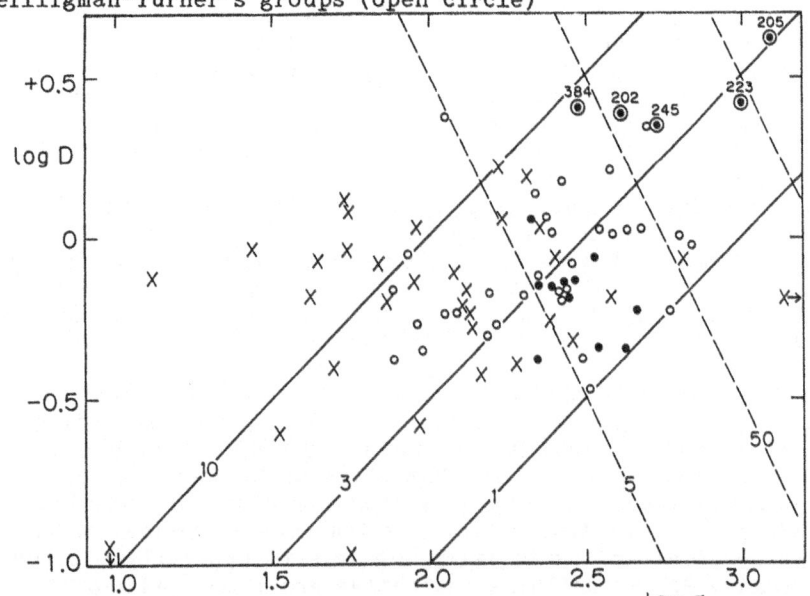

Fig.2. Diameter-Velocity Dispersion Plots for SCGG's (double circle), for spiral dominated (cross), elliptical dominated (circle), and non-spiral (filled circle) Hickson's groups. The lines for constant mass ($10^{12}m_\odot$; broken) and for constant crossing time (10^8yr) are indicated.

even for SCGG's. The short time scale as $t_c{\sim}10^8$yr, however, raises
the well-known question, how these groups can survive long enough
to be amply obseved at the present epoch, or how they can be
continuously produced from looser groups.

5. PREVIOUS NUMERICAL SIMULATION

A self-consistent N-body simulation of compact groups was carried
out by Ishizawa *et al.*(1983) with a dynamically unstable initial
state, while by Carnevali, Cavaliere, and Santangelo (1981) with a
virialized initial condition. The number of particles consisting a
galaxy was 20-100 and the number of group members was n=10-20, both
without dark matter. Ishizawa *et al.*(1983) found that a compact
group of galaxies evolved into one smooth merger remnant having few
surviving cores in it within a time scale of 10^{10}yr. The merging
rate was found to be at least by a factor 3 lower for the virialized
initial condition. Mergers tended to be formed during the initial
collapse phase when the velocity dispersion of galaxies was still
smaller than the internal velocity dispersion of galaxies.

Recently Barnes (1989) performed a self-consistent simulation
of 65,356 particles in 6 spiral galaxies of various masses, with a
massive dark halo surrounding each galaxy, starting from the
initial state of multiple system which instantaneously satisfied
the virial theorem. Due to the increased collisional cross-section
by the dark halo, a large merger was formed at $t{\approx}12$ in the standard
time units of $M_{gal}{=}1$ and $G{=}1$; $t{=}12$ corresponded to $2.5{\times}10^9$yr for a
unit mass of $1.2{\times}10^{10}m_\odot$ and a unit length of 30Kpc (see also Ishizawa
1986). With the increased number of particles, Barnes's (1989)
simulation could demonstrate that spiral galaxies merged into a
luminous elliptical.

Instead of treating galaxies in a group with individual
galaxies being composed of numerous particles, one can treat each
galaxy as a single particle in group dynamics, by incorporating
collisional effects which were evaluated with the self consistent
N-body simulations of two galaxy encounters (see references in
Ishizawa *et al.*1983, and Mamon 1987). As extension of the
single-particle approximation technique, Mamon (1987) performed a
comprehensive set of simulation allowing dark matter either in a
halo or in the diffuse background. The diffuse background of dark
matter, when it was included, was assumed to extend just over a
galaxy group, to have velocity dispersion of 0.8 times of that of
the galaxy group and mass of the level corresponding to a
mass-to-luminosity ratio of a group $M/L{\sim}100$. He simulated
Hickson's (1982) compact groups in virialized initial conditions, to
find that the groups could not satisfy the Hickson's (1982)
selection criteria any more typically after 6 or 30 half-mass
crossing time (1/30 or 1/8 H_0^{-1}) depending on whether or not the dark
matter was included. According to Mamon's (1987) results, no
compact group of n=8 survive past $0.5H_0^{-1}$ except for special ones,
which nevertheless all unstable within H_0^{-1}. The instability time
increases generally with larger group membership, and larger group

mass-to-luminosity ratio.

6. GRAPE SIMULATION

Suginmoto *et al.*(1989; see also Ito *et al.*1990 and Makino *et al.*1990)
developed a backend processor connected to a workstation. It is a
pipeline for calculating the gravitational force exerted on each
particle in an N-body system. This backend processor (called as
GRAPE - GRAavity PipE) facilitates N-body calculations equivalent
to 240 Mflops machine at this stage of development. It has been
applied to explore the dynamical interactions of elliptical galaxies
(Okumura *et al.*1990).

In connection to SCGG's, we are using the GRAPE system to
investigate the effects of dark matter background in which a
compact group of galaxies is embedded. The Plummer models of
potential and mass distribution are assumed as the initial
condition both for individual galaxies and for a group of 10 same
galaxies. The half-mass radii are taken as $r_{h,gal}$=0.77 for a galaxy
and $r_{h,gr}$=7.7 for the group of galaxies as well as for the dark
matter. Here, the standard units of M_{gal}=1, G=1, and E_{gal}=-1/4 are
used. The crossing time within a galaxy is $t_{cr,gal}$=2$\sqrt{2}$.

In the first case we simulate a model without any dark matter.
Each galaxy is represented by N_{gal}=1024 equal-mass particles. The
crossing time of galaxies through the group is $t_{cr,gr}$=20$\sqrt{2}$. Figures
3a and b show the initial configuration at t=0 and the final at
t=105=3.7$t_{cr,gr}$, respectively. The compact group evolves into two
merger galaxies, and one galaxy is escaping out from the frame.
Since the observed crossing time is typically $t_{cr,gr}$=10^9yr, such
evolution proceeds appreciably faster than H_0^{-1}.

Secondly, we simulate the case where 10 galaxies are embedded
in a diffuse dark matter envelope. We represent the dark matter by
N_{dm}=23040 particles in Virial equilibrium. Each galaxy is
represented by N_{gal}=256 particles of the same mass. The 10 galaxies
are in Virial equilibrium together with the dark matter. The
crossing time is $t_{cr,gr}$=2$\sqrt{20}$. The dynamical friction on galaxies due
to the dark matter is automatically taken into account. We find
that the group evolves toward a single giant merger. At
t=30=3.4$t_{cr,gr}$ (figure 3c), the identities of galaxies are still seen,
but at t=70=7.8$t_{cr,gr}$ they are dissolved into the merger (figure 3d).
The dark matter envelope gains energy from galaxies and expands
silightly. In this model the time scale of merging is somewhat
extended, yet not long enough compared to H_0^{-1}.

We simulate also the third case; It is the same as the second
case except that the dark matter is not represented by the
particles but merely by a fixed gravitational potential. Each
galaxy is represented by N_{gal}=1024 particles. In this case, there is
no effect of dynamical friction, which could correspond to the case
of very hot dark matter. At t=15017$t_{cr,gr}$, cores of 6 galaxies still
survive, though the particles expelled from galaxies form an
extended halo. The system looks like a compact group. This result
suggests that a compact group may survive longer than H_0^{-1} if the

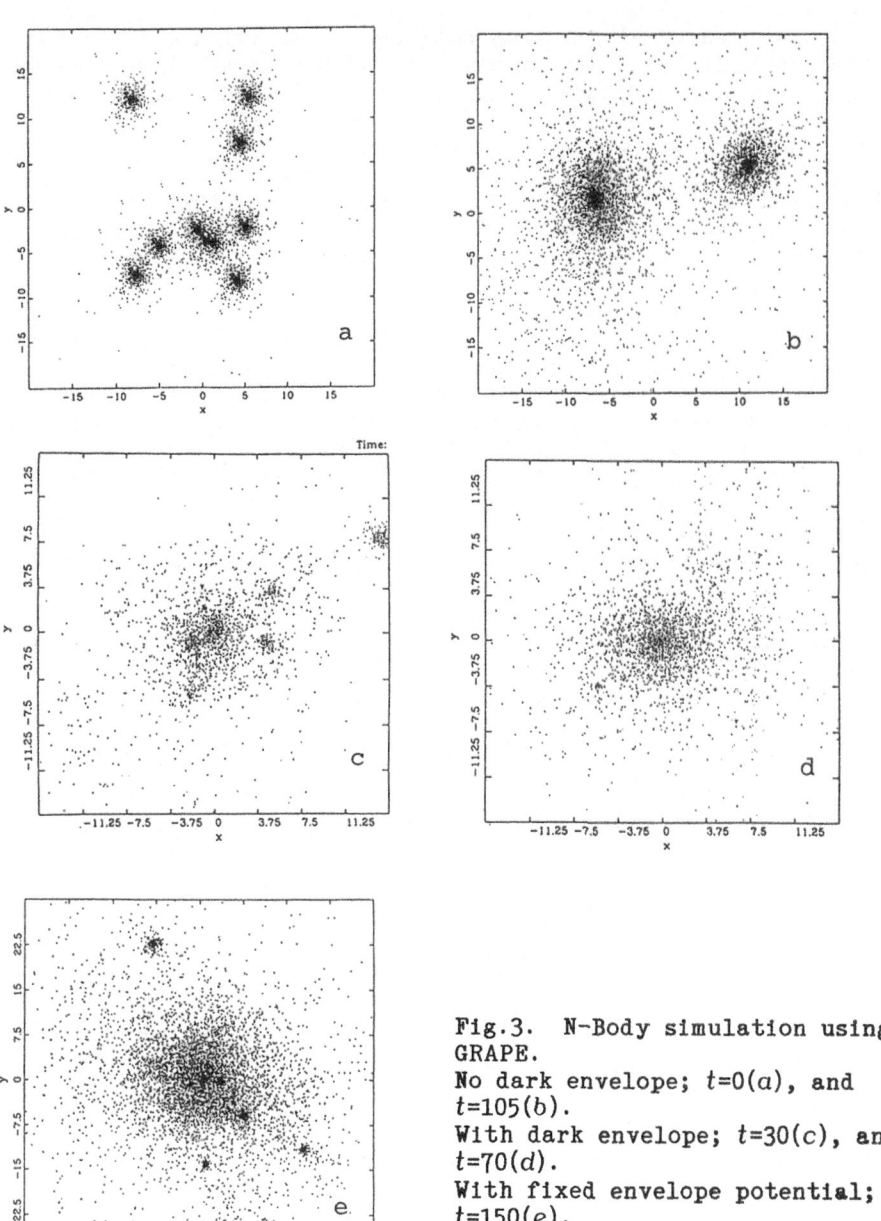

Fig.3. N-Body simulation using GRAPE.
No dark envelope; $t=0(a)$, and $t=105(b)$.
With dark envelope; $t=30(c)$, and $t=70(d)$.
With fixed envelope potential; $t=150(e)$.

effect of the dynamical friction between the galaxies and the dark matter is weak enough. In our simulation as well as those by Mamon (1987), the initial velocity dispersion of diffuse dark particles was assumed to be equal to or less than that of galaxy groups. When the velocity dispersion of dark particles exceeds that of galaxy groups, but with unchanged local density, the dynamical friction may substantially decrease. Such a model inevitably involves large extent and mass of the dark matter envelope, and its consequences shall be further investigated by the GRAPE team.

7. SUMMARY AND CONCLUSION

There are compact groups of galaxies preferably composed of luminous early-type galaxies with members of $n=4-20$. So far as they are not embedded in a massive dark envelope, compact groups of $n \lesssim 5$ have evolutionary time scale of $t_{evol} \sim 10^9 \mathrm{yr}$, to form final mergers. The small compact group of $n \lesssim 5$ are then quickly consumed and must be fed from larger groups. The larger groups of $n \gtrsim 10$ might be still evolving at present.

If a compact group of galaxies is embedded in a massive dark envelope, this tends to slow down the dynamical evolution of a compact group, but the degree of the deceleration depends upon the parameters of dark envelope, namely, total mass, particle mass, and velocity dispersion, relative to those of galaxy groups and of individual galaxies. We feel that some parameter sets may exist to allow compact groups to survive over H_0^{-1}.

References

Barnes, J.E.(1989) 'Evolution of Compact Groups and the Formation of Elliptical Galaxies' Nature, 338, 123-126.
Carnevali, P., Cavaliere, A., and Santangelo, P.(1981) 'Merging Instability in Groups of Galaxies' Astrophys.J., 249, 449-461.
Heiligman, G.M., and Turner, E.L.(1980) 'The Anomalous Luminosity Function of Galaxies in Compact Groups' Astrophys.J., 236, 745-749.
Hickson, P.(1982) 'Systematic Properties of Compact Groups of Galaxies' Astrophys.J., 255, 382-391.
Hickson, P., Kindle, E., and Huchra, J.P.(1988a) 'Discordant Redshifts in Compact Groups of Galaxies' Astrophys.J.(Letters), 329, L65-L67.
Hickson, P., Kindle, E., and Huchra, J.P.(1988b) 'Morphology of Galaxies in Compact Groups' Astrophys. J., 331, 64-70.
Hickson, P., Richstone, D.O., and Turner, E.L.(1977) 'Galaxy Collisions in Dense Groups' Astrophys.J., 213, 323-326.
Hickson, P., and Rood, H.J.(1988) 'The Nature of Compact Groups of Galaxies' Astrophys.J.(Letters), 331, L69-72.
Ishizawa, T., Matsumoto, R., Tajima, T., Kageyama, H., and Saki, H.(1983) 'Dynamical Evolution of a Compact Group of Galaxies' Publ.Astron.Soc.Japan, 35, 61-76.

Ishizawa, T.(1986) 'Simulation of Compact Groups of Galaxies'
    Astrophys.Space Science, 119, 221-225.
Ito, T., Makino, J., Ebisuzaki, T., and Sugimoto, D.(1990) 'A
    Special-purpose N-body Machine GRAPE-1' Computer Phys.Com.,
    60, 187-194.
Kodaira, K., Doi, M., Ichikawa, S., and Okamura, S.(1990a) 'An
    Observational Study of Shakhbazyan's Compact Groups of
    Galaxies II.SCGG202, 205, 223, 245, and 348' Publ.National
    Astron.Obs.Japan, 1, 283-295.
Kodaira, K., Iye, M., Okamura, S., and Stockton, A.(1988) 'An
    Observational Study of Shakhbazyan's Compact Group of
    Galaxies 202' Publ.Astron.Soc.Japan, 40, 533-545.
Kodaira, K., Sekiguchi, M., Sugai, H., and Doi, M.(1990b)
    'Redshift Observation of Shakhbazyan's Compact Groups of
    Galaxies and Their Number-Diameter Relation' submitted to
    Publ.Astron.Soc.Japan.
Makino, J., Ito, T., and Ebisuzaki, T.(1990) 'Error Analysis of
    the GRAPE-1 Special-Purpose N-Body Machine' submitted to
    Publ.Astron.Soc.Japan.
Mamon, G.A.(1986) 'Are Compact Groups of Galaxies Physically
    Dense?' Astrophys.J., 307, 426-430.
Mamon, G.A.(1987) 'The Dynamics of Small Groups of Galaxies I
    Virialized Groups' Astrophys.J., 321, 622-664.
Okumura, S., Ebisuzaki, T., Makino, J., Sugimoto, D., Ito, T.,
    and Kodaira, K.(1990) 'A Special Purpose Computer for
    Gravitational Many-Body Problem and Merger Simulation'
    Proc.IAU Symp.No.146, in press.
Postman, M., and Geller, M.J.(1984) 'The Morphology-Density
    Relation: The Group Connection' Astrophys.J., 281, 95-99.
Rose, J.A.(1977) 'A Survey of Compact Groups of Galaxies'
    Astrophys.J., 211, 311-318.
Rose, J.A.(1979) 'The Dynamical Nature of Compact Groups of
    Galaxies' Astrophys.J., 231, 10-22.
Rose, J.A., and Graham, J.A.(1979) 'Mass-to-Light Ratios of Two
    Compact Groups of Galaxies' Astrophys.J., 231, 320-326.
Schechter, P.L.(1976) 'An Analytic Expression for the
    Luminosity Function for Galaxies' Astrophys.J., 203, 297-306.
Schweizer, F.(1989) 'Merging Groups of Galaxies' Nature, 338,
    119-120.
Shakhbazyan, R.K.(1957) 'On a Star Cluster in Ursa Major'
    Astron.Tsirk SSSR, No.177, 11-12.
Shakhbazyan, R.K.(1973) 'Compact Group of Compact Galaxies'
    Astrofizika, 9, 495-498.
Sugimoto, D., Chikada, Y., Makino, J., Ito, T., Ebisuzaki, T.,
    and Umemura, M.(1990) 'A Special Purpose Computer for
    Gravitational Many-Body Problems' Nature, 345, 33-35.
Turner, E.L., and Gott, J.R.III.(1976) 'Groups of Galaxies.I.A
    Catalog' Astrophys.J.Suppl.,

CORRELATIONS OF SPIN ANGULAR MOMENTA
OF GALAXIES

Masanori IYE

National Astronomical Observatory, Mitaka, Tokyo 181 Japan

Abstract

The distributions of galaxy spin in the Local Super Cluster, groups of galaxies, and pairs of galaxies were studied by using a data base of spiral winding sense of galaxies with an assumption that all the spirals are trailing. The author argues that there is a possibility that the actual distribution of galaxy spin shows a dipole or a quadrupole component depending on the scenarios of galaxy formation. No significant correlation, however, was identified in any ensemble. Implications of the present study on the origin of the spin angular momenta of galaxies are discussed.

1. Introduction

The origin of spin angular momenta of disk galaxies is one of the most important issue that any successful theory of galaxy formation should explain consistently. Scenarios introducing concrete mechanisms to generate spin angular momentum in disk galaxies are 1) the primordial vorticity scheme (von Weizsaecker 1951), 2) the tidal torque scheme (Peebles 1969), 3) the pan-cake shock scheme (Doroshkevich 1973) and 4) the explosive formation scheme (Ostriker and Cowie 1981, Ikeuchi 1981). These schemes predict different distributions of galaxy spins in clusters of galaxies. It is therefore important to study the distribution of galaxy spins observationally. Observed correlation, if any, will provide a powerful key to check dynamical models of galaxy formation and evolution.

2. Spin Catalog

We propose here to use the projected spiral winding direction of galaxies to diagnose the distribution of spin angular momentum of galaxies. The winding direction of the spiral structure of disk galaxies, as we see on the sky, can be denoted either S-wise or Z-wise. Since every available evidence (*cf.* de Vaucouleurs 1958, Pasha 1985) indicates that spiral sturctures are predominantly trailing rather than leading, by assuming that all the spirals are trailing, we can distinguish the sign of the line-of-sight component of the spin vector just by the spiral winding sense. We identify that the spin vectors of S-wise spiral galaxies point towards us, while those of Z-wise spirals point away from us, when we take a right-handed system to define the spin vector (Fig.1).

I'd like to point out a simple but definite fact which strongly supports the interpretation that spirals are trailing. In order to determine whether the spiral structure of a particular galaxy is trailing or leading, one needs to know which side of the galactic disk is nearer to us. The current common understanding is that the side where the dark lane is more conspicuous is the near side. There was a big debate in 1950s as to the way to identify the near side of the galactic plane of a tilted galaxy. The arguments of each side, however, have never been very persuasive. Iye and Richter (1985) showed a definite way to identify the near side of the disk by measuring the differential reddening of globular clusters and applied this method to M31 (Fig.2). The result for M31 clearly supports the current

404

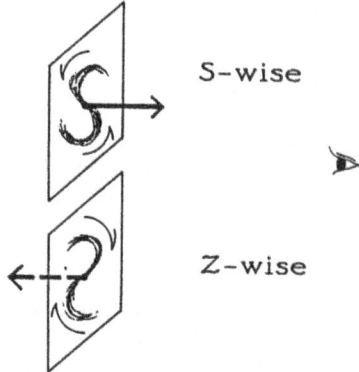

Fig.1   Spin vectors of an S-wise spiral and a Z-wise spiral

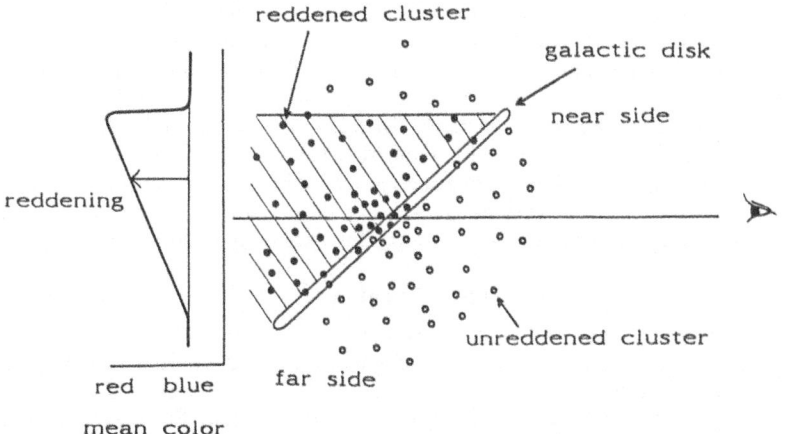

Fig.2   Geometry to show the presense of differential reddening of globular clusters in a galaxy with a tilted disk

interpretation that the side where we see more dark lanes is nearer to us and hence the dominance of trailing spirals.

We have compiled recently a new catalog for studying the distribution of the spin angular momenta of disk galaxies (Iye and Sugai 1990). It presents the sign of the line-of-sight

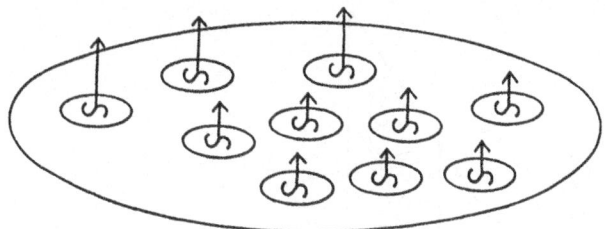

Fig.3 Expected coherence of galaxy spins in the primordial vorticity scheme

component of the spin angular momentum for a total of 8287 disk galaxies of types S0-Sm in the southern hemisphere selected from the ESO/Uppsala Survey of the ESO (B) Atlas (Laubert 1982). A similar catalogue was compiled for 1650 spirals in the northern hemisphere by Yamagata, Hamabe, and Iye (1981). We have therefore now a total of about 10000 galaxies in our S/Z data base. The probability of misidentifying the spiral winding sense is reduced to a level less than 0.5% by using a blind double checking scheme. The total numbers of S galaxies and Z galaxies agree remarkably well both in the northern and southern hemispheres. The number of galaxies in which S-wise and Z-wise spirals coexist is less than 10 out of 10000, implying that the occurrence of leading spirals is actually very rare.

3. Theoretical Predictions

3-1. Primordial vorticity

The primordial vorticity scheme argues that a large scale primordial eddy decayed into smaller eddies, which finally became individual galaxies. The spin angular momentum vectors of galaxies, therefore, would share a common orientation at the time of their formation. For an edge-on cluster, one should then be able to find a trend for the alignment of the major axis position angle of member galaxies to the major axis of the cluster. When we look at such a cluster from above, we would see predominantly S-wise galaxies rather than Z-wise galaxies. As for the Local Super Cluster, which we observe from inside, we would find a **dipole** bias of spin distribution such that S-wise spirals dominate in one hemisphere while Z-wise spirals dominate in the other hemisphere (Figure 3).

3-2. Pan-cake theory and explosive galaxy formation

In this scheme, the spin vectors lie parallel to the plane of the cluster and tangentially rather than radially (Doroshkevich 1973). The so-called pan-cake shock scenario predicts the occurrence of oblique shock at the cluster plane where the gas infalling from both sides of the plane collide. I expect further that the resulting radial inflow would tend to bounce away from the plane and generate spin angular momenta of galaxies. The major axis of galaxies tend to point toward the cluster center for face-on clusters. The spin vectors in the upper hemisphere are pointing counter clockwise when we view the cluster from above, whereas those in the lower hemisphere point clockwise (Fig.4). As for edge-on clusters,

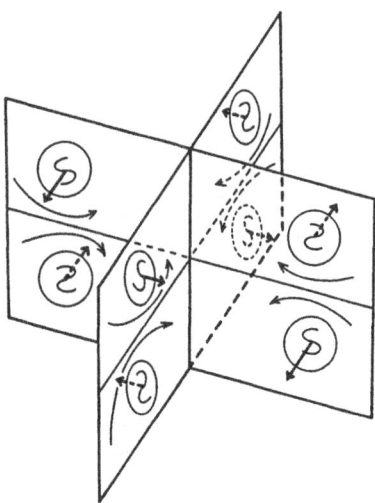

Fig.4 Expected distribution of galaxy spins in the pan-cake scheme

this scheme predicts the presence of a quadratic distribution of S and Z galaxies. As for the Local Super Cluster, since we are looking from inside but not generally from the right center, we shall see a **quadrupole** distribution of S-wise and Z-wise galaxies.

Since it is natural to expect the occurrence of oblique collision of shock fronts also in the explosive galaxy formation scheme, a cluster formed by explosive shock would show a similar spin distribution of galaxies as for the pan-cake shock scheme.

3-3. Tidal torque theory

If the tidal torque between individual galaxy and the whole cluster is responsible for generation of spins, the spin axes of galaxies tend to lie tangentially in clusters. If the tidal torques between a pair of galaxies is more important the spin vectors wouldbe parallel to the plane perpendicular to the line joining the pair of galaxies. The tidal torque scheme would show no strong bias in the distribution of spin vectors except for a possible correlation of spin vectors for the nearest neighbour galaxies.

3-4. Dynamical mixing of spin

Merging processes and close encounters of galaxies would have destroyed any original bias formed in the distribution of spin angular momenta of galaxies. Quantitative evaluation of the change of spin vectors due to the spin-orbital coupling during the close encounters is highly desired. It is important to emphasize that observationally derived bias, if any, in the distribution of spin vectors would place a strong constraint on theories of galaxy formation.

4. Statistical Analysis

We have compiled velocity information from the CfA redshift survey catalogue and from other sources into the present data base.

4-1. Local Supercluster

The distribution of spin vectors viewed from inside the cluster was studied. The dipole

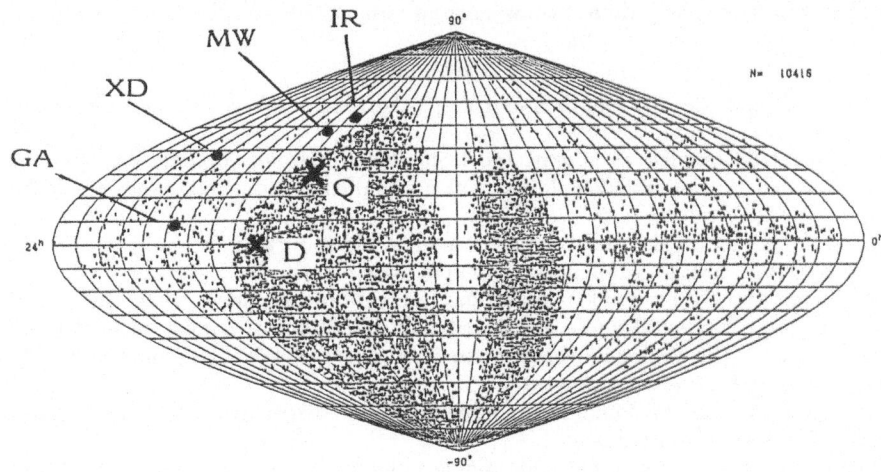

Fig.5 Distribution of galaxies compiled in the present spin data base is shown in the supergalactic coordinates. The observed directions of the dipole and the quadrupole components of galaxy spin distribution are shown together with other special directions of interest. D: observed dipole, Q: observed quadrupole, GA: great attractor (Lynden-Bell 1988), MW: microwave dipole (Strukov *et al.* 1988), IR: IRAS dipole (Strauss and Davis), XD: X-ray dipole (Miyaji and Boldt 1990).

and the quadrupole components in the distribution of the entire ensemble of galaxies in the present data base were evaluated and were compared with those obtained from Monte-Carlo simulations. The observed dipole and quadrupole intensities turned out to be nothing more than those expected from random Monte-Carlo simulations. I should say therefore that no statistically significant bias have been identified.

Although the intensities of observed dipole and quadrupole components are not at all strong, plotting the directions of their poles on the supergalactic coordinates might be of interest for comparison with other various poles (Fig.5).

4-2. Clusters, groups, and pairs of galaxies

An analysis for the Virgo cluster is under way. Distributions of the spiral winding sense were examined for 115 groups of galaxies. A χ^2 test shows that the observed distribution in 115 groups is completely consistent with a binomial distribution. Distributions of the spiral winding sense were studied for 5186 nearest pairs of galaxies in the present data base. A weak excess for coherent spin pairs over the opposite spin pairs was found at 1.4σ level.

4-3. Type segregation

S/Z number ratio were studied for each morphological type since the analyses for the northern galaxies showed rather strong bias especially for Sb and Sbc type (Thompson 1973 and Yamagata *et al.* 1981), which is very hard to explain. The present new data base for the southern hemisphere do not show such an unfavorable tendency.

5. SUMMARY

(1) Only 0.1% galaxies have both S and Z spirals. This implies that the occurrence of leading spirals is very rare.

(2) Coherent spin model, as is expected from the primordial vorticity scheme, would show a dipole distribution of S/Z galaxies, but no significant dipole component was found in the present data.

(3) Pan-cake model for the Local Super Cluster would show a quadrupole component in the S/Z distribtion, but no significant quadrupole was identified.

(4) Although the Monte-Carlo simulations for calibration shows that the observed strengths of dipole and quadrupole are statistically insignificant, it may be worth noting that the direction of the weak quadrupole component observed is not far from the directions of other dipoles and the great attractor.

(5) No significant bias was found in the number ratio of S-wise and Z-wise spiral galaxies for a total of 115 groups of galaxies.

(6) There is nothing but a weak predominance of coherent spin pairs over the opposite spin pairs at 1.4σ level.

In conclusion, coherent vorticity model does not comply with the present results. Tidal torque model, pan-cake model, and explosive formation model seem to comply with the present results.

REFERENCES

de Vaucouleurs,G. 1958, *Ap. J.*, **127**, 487.

Doroshkevich,A.G. 1973, *Astr. Ap.*, *Lett*, **14**, 11

Ikeuchi,S. 1981, *Publ. Astr. Soc. Japan*, **33**, 211

Iye,M. and Richter,O.-G. 1985, *Astr. Ap.*, **144**, 471.

Iye,M. and Sugai,H. 1990, submitted to *Ap. J.*

Laubert,A. 1982, *The ESO Uppsala Survey of the ESO(B) Atlas*, ESO

Lynden-Bell,D. 1988, *Ap. J.*, **320**, 19.

Miyaji, T. and Boldt,E. 1990, *Ap. J.*, **353**, L3.

Ostriker,J.P. and Cowie,L.L. 1981, *Ap. J.*, *Let*, **243**, L127.

Pasha,I.I. 1985, *Soviet Astr.Lett.*, **11**, 1.

Peeble,P.J.E. 1969, *Ap. J.*, **155**, 393.

Strauss,M.A. and Davis,M. 1988, in *Large Scale Structure of the Universe, IAU Symp* **130**, 191.

Strukov,I.A., Skulachev,D.P., and Klypin,A.A. 1988, ibid, 27.

Thompson,L.A. 1973, *Pub. A. S. P.*, **85**, 528.

von Weizsacker 1951, *Ap. J.*, **114**, 165.

Yamagata,T.,Hamabe,M.,and Iye,M. 1981, *Annals Tokyo Astron.Obs.*, *2nd Series*, Vol.18, No.3, 164

FORMATION OF BIPOLAR RADIO JETS AND LOBES
FROM ACCRETION DISK AROUND FORMING BLACKHOLE
AT THE CENTER OF PROTOGALAXIES

Y. UCHIDA[1], R. MATSUMOTO[2], S. HIROSE[1], and K. SHIBATA[3]

[1] *Department of Astronomy, University of Tokyo*
[2] *College of Arts and Sciences, Chiba University*
[3] *Department of Earth Science, Aichi University of Education*

ABSTRACT: We propose that radio jets and lobes from QSO's are "magnetic bipolar jets from forming blackholes", physically analogous to those of star-formation bipolar flows, but with very much greater energy due to very much greater depth in gravitational potential. We perform 2.5D MHD simulations for the situation in which the condensing mass of the accretion disk associated to the blackhole brought the magnetic flux with it, deforming the magnetic field into an hourglass shape. The differential rotation of the disk rotating at its neck continuously produce magnetic twists and send them out in the form of non-linear torsional Alfven waves to the bipolar directions. The gas of the disk atmosphere and the halo is accelerated helically when this non-linear torsional Alfven waves propagate through them. These NTAW's, at the same time, dynamically pinch the initially hourglass-shaped field into a collimated rod-shaped structure, and in some cases cause helical instability to make it into a winding structure. Jets from nearby AGN's may be due to a rejuvenated blackhole accretion by a similar process, with the dynamo-generated central poloidal field of the galaxy interacting with the stirred up gas, eg., in the merging of the galaxies.

The back-reaction of the process of the formation of the jet carrying away angular momentum exerts braking on the disk rotation, and allows an enhanced accretion of mass to the central blackhole, explaining at the same time the enhanced liberation of gravitational energy at the center.

1. Introduction

Active galactic nuclei (AGN's) and quasars (QSO's) often accompany radio jets and lobes. Some models have been proposed for the jet production in which the jets are attributed to outflows of gas and/or high energy particles somehow blown out from the active central source itself, and the "funnels", which are thought to be opened up along the axis of the accretion disk surrounding the central body collimate them into bipolar beams (Blandford and Rees 1974, review by Begelman et al.1984). The detail of the mechanisms for the energy liberation at the central object, for how to accelerate the wind from the central object, or for how to maintain the funnel in a favorable shape are, however, still open questions.

In the present paper, we discuss a new picture about the behavior of the disk, based on the gravo-magnetodynamical process of the jet formation proposed by the authors (Uchida 1989, Uchida and Hamatake 1989, Uchida 1990). The bipolar radio jets and lobes from quasars are thought to be produced in the maximum phase of mass accretion to the blackhole sitting at the center of a protogalaxy in our model proposed here, through a magnetohydrodynamical interaction of the rotating disk with large scale primordial magnetic field brought into it in the process of gravitational contraction. [Jets of nearby radio galaxies or AGN's may be due to the rejuvenated blackhole accretion at the center with restarted supply of gas due to stirring up of the gas, eg., in the merging galaxies.]

This working hypothesis came from a physical analogy of these, though there exists difference by many orders of magnitude in mass and spatial scales involved, to the case of star formation, in which also hypersonic bipolar jets are created in the final and maximum phase of mass accretion to the central forming star. The case of the forming star is better known to its detail compared with those of AGN's and QSO's, and evidence for the spinning of the bipolar flows has been obtained (Uchida et al.1987). This, together with their relation to the large scale magnetic field (Vrba et al.1988, Uchida et al. 1991), strongly supports our proposed gravo-magnetodynamic model (Uchida and Shibata 1985, Shibata and Uchida 1986) that the formation of these bipolar jets is intrinsically related to the action of squeezed (originally-interstellar) magnetic field. In this picture, the bipolar jets, spinning around their common axis in the same direction, come from the inner part of the disk, and not from the central object itself. The rest of the disk mass, together with the magnetic field, executes a spiral infall to the central forming star due to the loss of angular momentum in this process, explaining the enhanced release of gravitational energy [Bfrom the forming stars.

Our working hypothesis here, accordingly, implies that, though quite different in mass, energy, and spatial extension, the radio jets and lobes from AGN's and QSO's may be due to some corresponding mechanism taking place at the very center of them: We assume that this is associated with the formation of a massive blackhole, or more specifically, with the accretion of a massive disk to the blackhole.

In the following, we briefly describe the model situation, and discuss the obtained results of 2.5D MHD simulations for the formation of the jets and lobes, together with some conspicuous behavior of the disk (enhanced accretion, and magnetic reconnection which is likely to produce non-thermal phenomena) in response to the jet formation. Finally, we discuss the formation of an umbrella-like structure (radio lobes) and the wiggled structure of the jet part at large distances by the propagating non-linear torsional Alfven waves.

2. Model Situation and Evolutionary Picture

We first describe the physical situation of our model. We think of the formation of a massive blackhole in a dense central condensation of gas in a protogalaxy which, for example, may have drifted down towards the central region through some angular momentum loss mechanisms such as the magnetic braking due to a much milder production of torsional Alfven waves and slow helical flows from the condensing gas even in a larger scale (Uchida and Hamatake 1989), or as the passage of the gas through the spiral shocks produced in a barred potential as more widely discussed. We assume that the condensation of the gas to the blackhole + accretion disk system takes place in a large scale poloidal magnetic field which may be a part of the large scale primordial magnetic field squeezed during the formation process of the protogalaxy itself. In the case of the nearby galaxies, rejuvenated blackhole accretion in the central part of the galaxy may take place in the dynamo-generated global magnetic field, which is likely to be poloidal near the axis, in the event of merging of the galaxies. Magnetic field in both of these situations can be regarded as a poloidal field with large enough scale in terms of the blackhole and the disk associated to it at the center under consideration.

The gas forming the blackhole + accretion disk system is thought to have brought with it a part of this large scale poloidal magnetic flux in the process of condensation, and deforms the poloidal magnetic field into an hourglass-shape. The disk differentially rotating at its neck interacts with this *bunched* poloidal magnetic field of considerable strength, and produces magnetic twist about its axis continuously. The magnetic twist is accumulated and then released into the bipolar directions as non-linear torsional Alfven wave packets, and the mass in the atmospheric part of the disk and halo surrounding it will form a spinning flow into the bipolar directions accelerated by the relaxing non-linear torsional Alfven wave packets, since the free-fall velocity of the gas is smaller than the non-linearly enhanced Alfven velocity with which the twists relax.

The propagation of the non-linear torsional Alfven waves also causes a strong pinch effect which bunches the otherwise fanning out hourglass-shaped magnetic field into a long rod-shaped configuration. An extended umbrella-shaped region (=lobe part of the radio source) will develop at the front of this progressing pinch (=jet part of the radio source) which sweeps up and thrusts into the surrounding low-Alfven velocity region (especially at the edge of a sphere with $r \sim 10^{5.5}$pc; the mass inside this sphere have contracted into the central protogalaxy), tucking up the weak magnetic field. The jet part may suffer helical instability by its nature (because the jet part has an internal field with helical structure) when the ratio of toroidal to poloidal field exceeds certain value, for example, through the piling up of B_φ due to the decrease of propagation velocity towards the edge of this sphere.

The production of the torsional Alfven waves exerts magnetic braking to the disk material as its back-reaction, and enhances the final infall of the mass of the inner-most part of the disk, and certain fraction of a half of the liberated gravitational energy in this process is supplied to the production of the jets and lobes.

If we assume that the mass and the magnetic field in the formation of the protogalaxy come from an initial protogalactic medium inside a sphere of $R_P \sim 10^{5.5}$pc, having $n_P \sim 10^{-4 \sim -5}$ cm^{-3}, and a large scale magnetic field of $B_P \sim 10^{-9}$G, then $\sim 10^9$ M$_\odot$ of the mass condensed into the system consisted of a blackhole + disk (with, say, one percent of the blackhole mass. Also, the space surrounding them is supposed to be filled with a high temperature, low density halo), then, an estimate by assuming simple geometrical considerations leads to the mean values of n and B in the disk of,

$$n_{BD} = (40/3)\alpha_{PC}\alpha_{CB}\alpha_{BD}n_P(R_P/R_D)^3 \sim 10^{11-12} \text{ cm}^{-3} \tag{1}$$

$$B_{BD} = \alpha_{PC}\alpha_{CB}\alpha_{BD}B_P(R_P/R_{BD})^2 \sim 10^{0-1} \text{ G} \tag{2}$$

where, α_{PC}, α_{CB}, and α_{BD} are the fractions of mass collected into the central condensation from the primordial medium, and that into the blackhole from the condensation, and the fraction of the disk mass to the mass of the blackhole, respectively. These are arbitrarily assumed to be of order of 0.1, 0.1, and 0.01, respectively. R_D is the radius of the blackhole disk, $\sim 10^{-1}$pc. T_{BD}, the temperature of the disk, will depend on heating and radiation cooling etc., but is assumed to be $\sim 10^{4 \sim 5}$K for simplicity here.

3. Simulations of the Phenomena Occurring at the Center

Equation System : Non-dimensionalized equations governing our magnetohydrodynamic medium in a cylindrical coordinate (r, φ, z) are;

$$\frac{\partial \rho}{\partial t} + \nabla(\rho \mathbf{v}) = 0 \tag{3}$$

$$\frac{\partial}{\partial t}(\rho v_r) + \nabla(\rho v_r \mathbf{v}) - \frac{\rho v_\varphi^2}{r} + A_1 \frac{\partial p}{\partial r} - \frac{A_2}{c}((\mathbf{v} \times \mathbf{B}) \times \mathbf{B})_r - A_3 \rho g_r = 0 \tag{4}$$

$$\frac{\partial}{\partial t}(\rho v_z) + \nabla(\rho v_z \mathbf{v}) + A_1 \frac{\partial p}{\partial z} - \frac{A_2}{c}((\mathbf{v} \times \mathbf{B}) \times \mathbf{B})_z - A_3 \rho g_z = 0 \tag{5}$$

$$\frac{\partial}{\partial t}(r\rho v_\varphi) + \nabla(r\rho v_\varphi \mathbf{v}) - \frac{A_2}{c}\nabla(rB_\varphi \mathbf{B}) = 0 \tag{6}$$

$$\left(\frac{\partial}{\partial t} + \nabla \cdot \mathbf{v}\right)(p\,\rho^{-\gamma}) = 0 \tag{7}$$

$$\frac{\partial \mathbf{B}}{\partial t} - \nabla \times (\mathbf{v} \times \mathbf{B}) = 0 \tag{8}$$

where all the quantities are normalized by their values at $r = r_0$ in a scaled coordinate $r = R/R_s$ where R_s is the typical size of the system, and,

$$\mathbf{g} = -\nabla \psi \;, \quad \psi = -GM/r$$

$$A_1 \equiv v_{s0}^2/\gamma v_{\varphi 0}^2, \quad A_2 = V_{A0}^2/v_{\varphi 0}^2, \quad A_3 = v_{K0}^2/v_{\varphi 0}^2 \tag{9}$$

$$v_{s0}^2 = \gamma \Re T_0, \quad V_{A0}^2 = B_0^2/4\pi\rho_0, \quad v_{K0}^2 = GM/r_0.$$

where the quantities with subscript 0 are those at $r = r_0$ in the inner part of the disk.

Similarity of the Solutions : We now see the ground for the similarity of our present situation of the forming blackhole to that of the forming stars mentioned in section 2.

If we assume two situations in which initial distributions of ρ etc. and forces in the scaled coordinates are the same, *and if A_i's take common values*, respectively, then these two systems will behave in exactly a similar way, however different their scales may be, because then the equations are exactly the same in the scaled coordinate (Uchida and Shibata 1986). Only difference is the time scale.

In actuality, however, no rigorous similarity holds, but it is possible that some conceptual similarity may hold if these non-dimensional coefficients take values in similar ranges, respectively. In our present case, $A_1 \sim 10^{-4.6}, A_2 \sim 10^{-3.6}, A_3 \sim 1 + \delta$ for the values of B, ρ, etc. of section 2. A_1, and A_2 are smaller than those in the star formation case, but the same relations which are in the range categorically similar to those of a typical star formation case $(A_1 \sim 1.2 \times 10^{-3}, A_2 \sim 4.5 \times 10^{-3}, A_3 \sim 1 + \delta)$. Therefore,

$$A_1, \; A_2 \ll 1 \leq A_3 \tag{10}$$

in the disk, and,

$$A_2 \gg A_1, \; A_3 \tag{11}$$

at certain distance outside the disk hold in our present case as in the star formation case.

We therefore expect that the magnetic field, though pretty strong, is passively wound up by the rotation of the dense disk, but as soon as the effect of the magnetic twists emerge outside the disk, the unwinding twists, or the propagating non-linear torsional Alfven wave packets, begin to dominate the gas frozen to the magnetic field, but the axial field is strengthened more and provides the axis of the pinched configuration.

Since the non-dimensional coefficients, representing the ratios of forces, satisfy (10) and (11) in our present case as in the star formation case, we can expect conceptually similar behavior of the solutions in these two cases. We expect that the jets are much stronger in the present case since the driver, the rotating disk, is much more energetic and interact with a much tougher magnetic field which transmits the magnetic twists and the jets.

Initial Density and Pressure : We take the non-relativistic version of Abramowicz's disk model (Abramowicz 1978) as our initial model in which gas of a constant temperature distributes in a potential field including the centrifugal effect :

$$\psi = -\frac{GM}{r} + L_0^2 \varpi^{2(\alpha - 1)} + \frac{c_s^2}{(\gamma - 1)} = const \tag{12}$$

where $L = L_0 \varpi^\alpha$, and $P = K\rho^\gamma$. We further assume a high temperature hydrostatic halo concentrically distributing around the central blackhole,

$$\rho = \rho_0 \exp\{R_p(1/r - 1/r_0)\} \tag{13}$$

where $R_p = \gamma GM/\Re T_0$. We tailor the skirt of the atmosphere of the Abramowicz disk, so that the boundary is in pressure equilibrium with this halo. It is noted here that the relativity effects, both general and special, are not essential, and neglected for simplicity, in the range of radius relevant to our discussion.

Initial Magnetic Field : We assume that the classical Ferraro's law of isorotation holds in the dense enough deep interior of the disk,

$$\mathbf{A} \parallel \Omega \quad \text{for the region where} \quad \beta_{rot} \equiv \frac{\rho v_\varphi^2}{B^2/8\pi} \gg 1 \tag{14}$$

with $\mathbf{B} = rot\mathbf{A}$, and give $j_{\varphi 0}(r)$ in the surface part of the initial model so that $\int j_{\varphi 0} dS$ gives the "hourglass-shaped" field bunched by the disk, but the Lorentz force due to $j_{\varphi 0}$ on B is much smaller than the gravitational and centrifugal forces in the dense enough deep interior of the disk.

Numerical Method and Boundary Conditions : We solve the equation system numerically by using a modified Lax-Wendroff code which Shibata and Uchida developed in conjunction with the star formation problem. The boundary conditions on the boundaries of our $r - z$ plane are the same as those in the star formation cases.

Results of Simulations

We show a result of 2.5D MHD simulations in Fig.1. It is seen that the toroidal component of the magnetic field is created as the disk rotates, and as it accumulates it starts relaxing from both surfaces, and it swirls out the atmospheric gas into a spinning jet in the bipolar directions as expected from the analogy to the case of the forming stars. It is noted that a large part of the mass in the surface layers of the disk executes, by the loss of angular momentum through magnetic effect, a sliding-down motion towards the central object like an avalanche, releasing gravitational potential energy. The rotational kinetic energy increases in the infall, and the same process produces increasingly stronger nonlinear torsional Alfven waves continuously.

414

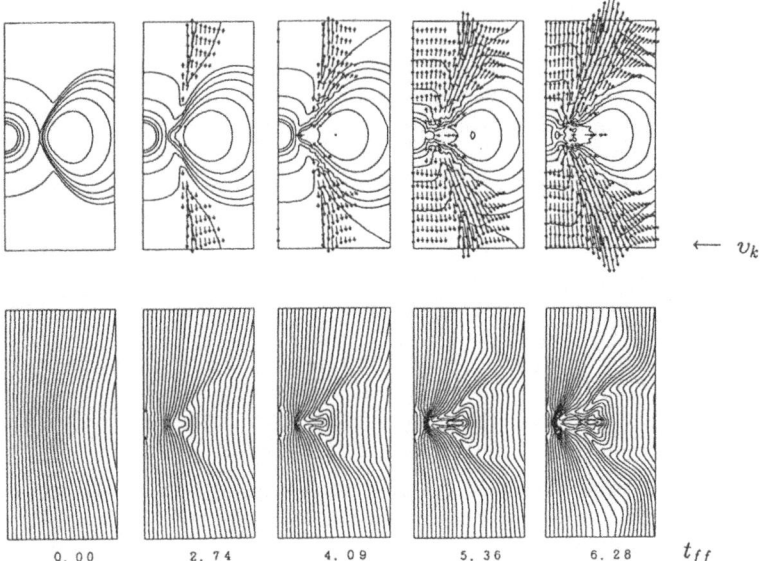

$\leftarrow \; v_k$

t_{ff}

Fig.1 : Results of 2.5D MHD simulations: ρ, $\mathbf{v}_p$, and $\mathbf{B}_p$. Magnetic braking causes the surface layers of the disk slide down, pulling magnetic field. B_φ is created and relaxes out swirling the atmospheric mass into cylindrical jets, and these contribute to a lartger loss of angular momentum.

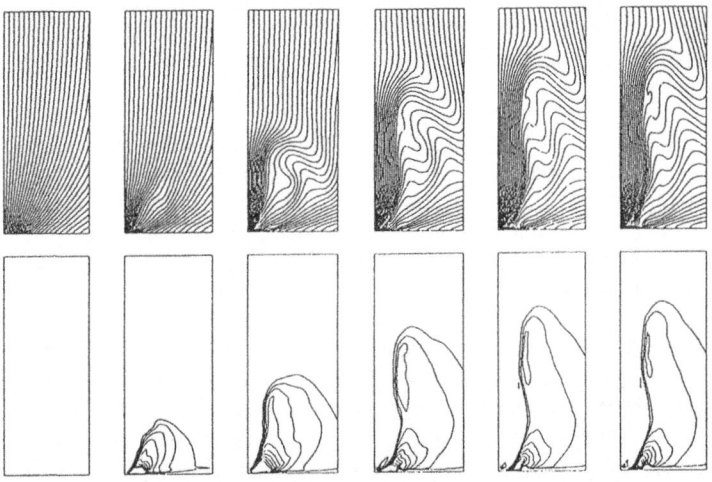

Fig.2 : Results of a large scale simulation: $\mathbf{B}_p$, and B_φ. The emitted B_φ packet pinches the initially fanning out field into a rod-like configuration (radio jet part). The pinching front tucks up the weak field into an umbrella-shape (radio lobe part).

Magnetic Reconnection and Non-thermal Activity : A remarkable point first revealed by our present simulations is the magnetic reconnection at the inner-most edge of the disk between the tips of the "avalanches" coming down from the upper and lower surfaces. This will prevent B_φ from increasing indefinitely, and converts the energy of this magnetic field into thermal (heating) and non-thermal (particle acceleration) phenomena (cf, Uchida 1986). This may be a natural explanation why there occur energetic non-thermal phenomena in the central part of AGN's and QSO's. The occurrence of the magnetic reconnection at the inner edge of the disk modifies the driving of the jets through its irregular bursty character, corresponding to the observed intermittency of the jets and the radio emission from the central source of AGN's and QSO's.

It is interesting to see that a part of the mass released in the reconnection is circulated to the equatorial part of the disk, whereas most of the rest is released to the blackhole, but seems to be buffered by the magnetic flux left in the central region, explaining some of the Ginga's recent observation (Inoue 1990) that there seems to exist something buffering the accreted mass, instead of allowing the mass to be simply swallowed by the central blackhole. The mass released by reconnection may fall by Rayleigh-Taylor type instability towards the central blackhole with magnetic interaction, whereas the particles accelerated in the magnetic reconnection process may escape along the reconnected magnetic field towards the jets and lobes with a re-acceleration process as mentioned below.

4. Propagation of the Torsional Alfven Waves at Large Distances

We briefly mention the propagation of the thus-generated torsional Alfven waves at large distances. This is more straightforwardly dealt with in the hourglass-shaped field geometry in large scale. (A deformed-multipole type poloidal field which may exist in the case of AGN's due to galactic dynamo process will also do the same, because it has a very large scale compared with the compact disk surrounding the blackhole, and once a packet of strong toroidal field is propagated along the fanning out poloidal field, the toroidal field and its pinched guiding poloidal field get very much bunched and get stronger than the global field in the bulge part of the parent galaxy, and the large scale weak field may easily be "punched through" by the "sweeping pinch" packet, and stretched out with it very easily. Thus the result will not be very different between the initially open field case and the initially closed field case, except for the presence of weak return field lines in the initially closed field case mimicing the observation better. We here show the results of the open field case for simplicity.)

Results of Simulations in Large Scale : Figure 2 shows the calculation in a large scale. It should be noted that a simulation with a realistic geometrical scale ratios can not be made due to the limited available dynamic ranges in scales in the numerical calculation even by the largest supercomputers today. This is due to the difficulty of using high enough mesh resolution. The real ratio of the sizes of the lobes to that of the "engine" may amount to 10^8, but the dynamic range in the spatial scales in the simulation can only be 10^3 at the most. So, we have to be contented by having a "toy-model" in which the essential physical behavior is seen with this practical restriction in the size ratios of the simulations.

With this in mind, we can see that the dynamically sweeping pinch effect squeezes the poloidal field towards the axis in a nonlinear way, as the large amplitude torsional Alfven waves (which are produced in the differentially rotating disk and released along the bipolar directions) progress into the medium surrounding the (proto)galaxy. The weak field is tucked up into an "umbrella" shape which mimics the lobes of the radio sources as first pointed out by Uchida and Hamatake (1989). The umbrella shape may be more realistic in the case of initially closed galactic dynamo field, by stretching out the initially-closed field to the bipolar directions as seen in Cygnus-A case (Dreher et al.1989).

It should be noted that the strongest pinch effect is seen at the "hub" of the "umbrella"-shaped lobe structure, corresponding to the "hot spots" in the observation of the synchrotron radiation.

5. Discussion

It is an advantage of our magnetic model that the large scale helical structure like that in 3C449, which is plane-symmetric with respect to the plane perpendicular to the jet axis, rather than point-symmetric with respect to the central source, may be explicable by magnetic helical instability if calculated in 3D, rather than a precessional-source hypothesis which should result in a point-symmetrical structure.

Also, acceleration of particles is very naturally explained in our magnetic model. The necessary re-acceleration of the primary high-energy electrons produced in the magnetic reconnection may be executed in our model by Fermi-I acceleration mechanism between successive torsional Alfven wave packets sent out by the sporadic winding up process affected by the bursty reconnection. Two successively propagating packets provide us with the Fermi-I process, because the earlier-going one gets slower than the later-going one due to the decrease in the Alfven velocity with the distance in our present model.

References
Abramowicz,M., Jaroszynski,M., and Sikora,M., 1978, *Astron. Astrophys.*, **63**, 221.
Begelman,M.C., Blandford,R.D., and Rees,M., 1984, *Rev.Mod.Phys.*, A**56**, 255.
Blandford,R.D., and Rees,M.J., 1974, *Mon.Not.R. Astron. Soc.*, **169**, 395.
Dreher,J.W., Carilli,C.L., and Perley,R.A., 1987, *Astrophys.J.*, **316**, 611.
Inoue,H., 1990, *Adv. Space Res.*, **10**, (2)-153.
Matsumoto,R., Uchida,Y., Hirose,S., Ferrari,A., and Shibata,K.,1991, in preparation.
Shibata,K., and Uchida,Y., 1986, *Publ.Astron. Soc.Japan*, **38**, 631.
Uchida,Y., and Shibata,K., 1985, *Publ.Astron. Soc.Japan*, **37**, 515.
Uchida,Y., 1986, *Astrophys. Space Sci.*, **118**, 443.
Uchida,Y., and Shibata,K., 1986, *Can.J.Phys.*, **64**, 507.
Uchida,Y., Kaifu,N., Shibata,K., Hayashi,S.-S., Hasegawa,T., and Hamatake,H., 1987, *Publ.Astron. Soc.Japan*, **39**, 907.
Uchida,Y., 1989, *Adv. Space Res.*, **10**, (9)-31.
Uchida,Y., and Hamatake,H., 1989, in *Accretion Disk and Magnetic Fields in Astrophysics*, ed. Belvedere,G. (Reidel), p233.
Uchida,Y., 1990, in *Galactic and Extragalactic Magnetic Fields*, eds. Beck,R., Kronberg,P., and Wielebinski,R., p425.
Uchida,Y., Fukui,Y., Minoshima,Y., Mizuno,A., Iwata,T., and Takaba,H., 1991, *Nature*, in press.
Vrba,F.J., Strom, S.E., and Strom, K.M., 1988, *Astron. J.*, **96**, 680.

AN EVOLUTIONARY UNIFIED SCHEME
FOR RADIO-LOUD QUASARS AND BLAZARS

F. VAGNETTI[1], E. GIALLONGO[2], and A. CAVALIERE[1]

[1] *Astrophysics, Dept. of Physics, II University of Rome, I-00173 Rome*
[2] *Astronomical Observatory of Rome, I-00040 Monteporzio*

The extragalactic sources called "blazars" (Angel and Stockman 1980) are characterized by flat spectrum radio emission, strongly variable optical-IR power-law continuum, and strong ($p \gtrsim 3\%$) and variable optical polarization. The set of blazars includes the BL Lacertae objects and many (virtually all, after Fugmann 1988) radio-loud flat-spectrum quasars. The two subsets differ mainly for the strength of the emission lines and for the apparent distribution in cosmological depth. In fact, at variance with the quasars, BL Lac objects are classified as "lineless" active galactic nuclei and are mostly selected at low redshifts on the basis of the absorption lines of the underlying elliptical galaxies (Burbidge and Hewitt 1987). But several *transitional* objects show considerable variability in the equivalent width (EW) of the emission lines and are classified sometimes as (high redshift) BL Lac objects and sometimes as quasars (Antonucci et al. 1987).

Within the *unified schemes*, the blazars are objects relativistically beamed toward the observer, while misaligned objects would be observed as radio galaxies and/or steep-spectrum quasars (Barthel 1989). We try a connection between high and low redshift blazars in terms of an *evolutionary unified scheme* that adds the time dimension to the geometrical scheme (Vagnetti, Giallongo and Cavaliere 1991, VGC). As for the EW of the emission lines, we propose that in cosmological time strong-lined change into weak-lined objects, based on the following consideration. In the framework of luminosity evolution, two components of different behavior are envisaged for the optical band: the isotropic "thermal" component L_i typical of radio-quiet and steep-spectrum radio-loud quasars, that mainly excites the lines; and a beamed component L_b which dims more slowly to remain dominant at low redshifts, possibly swamping some broad emission lines in BL Lac Objects.

To disentangle the evolution of the beamed component, we first test the dependence of the optical to radio ratio as a function of the radio power L_R and of the look-back time $T(z)$, separately for flat-spectrum and steep-spectrum quasars. We perform a linear regression analysis, trying the simple form: $\log(L_{opt}/L_R) = A_L \log L_R + A_z T + A$. The data base and the results of the analysis are fully reported in VGC. The confidence regions of the two interesting parameters A_L and A_z are shown in Fig. 1 for a particular case. The distributions of $\langle L_{opt}/L_R \rangle$ of flat-spectrum and steep-spectrum quasars differ: the flat set is shifted toward slower evolution while the steep one is shifted toward faster evolution, with the best fits differing by $\gtrsim 2\sigma$. We use these different correlations for the two radio spectral sets to deduce the evolutionary properties of the two luminosity functions in the optical band. Assuming pure luminosity evolutions both in the radio and optical bands, i.e. $L_R \propto \exp(k_R T)$ and $L_{opt} \propto \exp(k_{opt} T)$, the evolutionary parameters are related by: $k_{opt} = k_R(A_L+1) + A_z \ln 10$; in turn, k_R is estimated through a maximum likelihood analysis of the radio luminosity function. We find $k_R = 5.2$ and $k_{opt} = 3.9$ for the flat-spectrum quasars, $k_R = 6.3$ and $k_{opt} = 6.2$ for the steep-spectrum quasars.

The slower evolution of flat-spectrum quasars in the optical can be understood within the *evolutionary unified scheme*, in terms of a slower evolution of the beamed component compared to the isotropic one. The observed evolution may be caused by the secular change

418

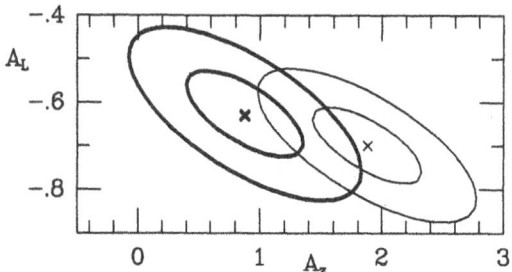

Fig. 1. Confidence regions at 1 and 2 σ levels for each of the two interesting correlation parameters A_L and A_z, with luminosities in units of 10^{33} erg s^{-1} Hz^{-1} at 2.7 GHz and at 5550Å and $T(z)$ in units of 19.6 Gyr; $\Omega_o = 0.2$. Thin contours: steep-spectrum quasars ($\alpha \geq 0.2$); thick contours: flat-spectrum quasars ($\alpha < 0.2$). The two best fits are represented by crosses. The separation is as sharp as could be expected considering the basic continuity of the two spectral sets (cf. VGC).

of the intrinsic jet power L_j and/or of the Lorentz factor Γ: $L_b(z) \propto L_j(z)\Gamma^{2+\alpha}(z)$ (see e.g. Urry and Shafer 1984). Since the evolution in the radio band is stronger than in the optical, a spectral evolution is implied. This may be explained within the inhomogeneous synchrotron model of Marscher (1980), whose spectrum is broken in the IR at a frequency $\nu_2 \propto \Gamma$. An increase of Γ by a factor $\simeq 1.5$ from $z \simeq 2$ to $z \simeq 0$ causes a drift of the spectral break toward higher frequencies, accounting for the difference of k_R and k_{opt} given above. Circumstantial evidence for such increase of Γ in the cosmic time is discussed in VGC.

It is now possible to understand the peculiar position of BL Lac Objects. The observed EW of the emission lines ought to be reduced by a factor $L_i/[L_i + L_b]$. Considering the evolution of the flat-spectrum quasars as an upper limit to the evolution of the beamed component L_b, the evolutions of the two components of the optical luminosity ought to differ by an amount $\Delta k \gtrsim 2.3$. This implies a reduction in EW by a factor $\sim 1/5$ from $z = 2$ to $z = 0$, enough to explain the absence of emission lines of intermediate strength such as OIII ($\lambda 5007$) and Hβ. The stronger Hα is more difficult to swamp, and in fact is often observed even in local BL Lac objects (VGC). Moreover, since the average $\langle L_{opt}/L_R \rangle$ and the density of local BL Lac objects agree with the extrapolations at the appropriate L_R and z found for flat-spectrum quasars from the analysis above, typical BL Lac objects can be considered as the low z end of the spectral and statistical evolution of flat-spectrum quasars.

The evolutionary unified scheme implies also a connection between the parent populations, namely between the radiogalaxies of different powers and morphologies. This might be understood in terms of an evolution towards an increasing richness of the environment (Prestage and Peacock 1988), and agrees with the observations of weak nuclear activity in some low redshift radio galaxies (De Robertis and Yee 1990).

References

Angel, J. R. P., and Stockman H. S. 1980, *Ann. Rev. Astr. Ap.* **8**, 321.
Antonucci, R. R. J., Hickson, P., Miller, J. S., and Olszewski, E. W. 1987, *A. J.* **93**, 785.
Barthel, P. D. 1989, *Ap. J.* **336**, 606.
Burbidge J., and Hewitt, A. 1987, *A. J.* **92**, 1.
De Robertis, M.M., and Yee, H.K.C. 1990, *A. J.* **100**, 84.
Fugmann, W. 1988, *Astron. Astrophys.* **205**, 86.
Marscher, A. P. 1980 , *Ap. J.* **235**, 386.
Prestage, R. M., and Peacock, J. A. 1988, *M.N.R.A.S.* **230**, 131.
Urry, C.M., and Shafer, R.A. 1984, *Ap. J.* **280**, 569.
Vagnetti, F., Giallongo, E., and Cavaliere, A. 1991, *Ap. J.* **368**, in press (VGC).

MAGNETOHYDRODYNAMICAL ENERGY EXTRACTION FROM A KERR BLACK HOLE

M. YOKOSAWA
Dept. of Physics
Ibaraki Unuversity
Mito 310
Japan

ABSTRACT. We solve the freezing equation of the magnetic field and give the exact nonstationary solution for the variation of magnetic field with a given axially symmetric accretion. We study the dynamical evolution of the energy density distribution of the magnetic field and discuss the energy transfer through the surface of a stretched horizon.

1. Introduction

X-ray emission is observed in many active galactic nuclei (AGN). Observations have showed the X-ray spectra of a substantial sample of Seyferts to be remarkably uniform, being well described by a simple power-law model. If the bulk of X-ray luminosity were synchrotron or inverse Compton radiation, it would require the structue of AGN in which the nonthermal emission is far superior to thermal one. The magnetic field could play a principal role in the required model of AGN because, for instance, the magnetic fields interacting with a black hole bring about the electromagnetic power or the magnetic field interacting with infalling "cold clouds" would supply a large bulk of energetic particles in the same way as in solar flares. Thus, we report here the structures of the magnetic fields formed near the black hole and the energy extraction from a black hole by magnetohydrodynamical process.

2. Nonstationary Solution for Variation of Magnetic field with Accretion on to a Rotating Black Hole

We obtaine the exact nonstationary solution for the variation of magnetic field with a given axially symmetric accretion. The magnetic field is considered as frozen-in the matter and homogeneous at the initial moment. The magnetic field lines near a rotating black hole are twisted by the frame-dragging effect and the strongly winding field is formed around the event horizon. The horizon is then seen as wrapped by a bundle of magnetic field lines .

3. Rotating Energy of a Black Hole Stored in Magnetosphere

Using the nonstationary solution of the magnetic field, we discuss the dynamical evolution of the energy density distribution of the field and also discuss the energy transfer through the surface of a stretched horizon. The radially infalling gas enhancese both the radial componet of the magnetic field $\tilde{B}_{\hat{r}}$ and the meridian one $\tilde{B}_{\hat{\theta}}$. The hole's rotation strengthens the toroidal component $\tilde{B}_{\hat{\phi}}$. Therefore the gravitational binding energy and the hole's rotation energy are stored in the magnetosphere formed around the black hole. The stored energy in a region should be evaluated by " the energy-at-infinity" E_{∞}. Initially the energy density $\epsilon_{\infty}(x)$ is negative over the most of the area around the horizon. After the delayed time, $t > t_d$, the positive peak in $\epsilon_{\infty}(x)$ is formed at the stretched horizon and then the electromagnetic energy density at any position in the magnetosphere increases with time. The large amount of the electromagnetic energy caused by the hole's rotation is stored in the region whose distance from the center of the black hole is several times as long as the horizon radius. We investigate the transfer of the electomagnetic energy through the surface of the stretched horizon. When $(\tilde{B}_{\hat{\phi}}\tilde{B}_{\hat{r}})_{at\ r_{st}} < 0$, this part acts as the extraction of the energy from the black hole, that is, the rotational energy of the hole is tranferred to the outer space by means of the magnetic field. When $(\tilde{B}_{\hat{\phi}}\tilde{B}_{\hat{r}})_{at\ r_{st}} > 0$, conversely the hole gains the rotational energy from its enviromental magnetic field. We separate the flux S_{∞} in two parts, $S_{\infty} = S_{PF} + S_T$, where S_{PF} is due to the Poynting flux and S_T is due to the work rate by the torque. Since initially the absolute value of S_T is larger than S_{PF}, the total flux S_{∞} is negative. This stage may be explained by the process that the negative energy of the electromagnetic field falls into the black hole and then the energy of the black hole is transferred to the outer space. However this qualitative explanation is not precisely true since one of the turning time from the negative value to the positive one is not consistent with the other. The infalling matter generates the radial field $B_{\hat{r}}$ and then the turning time of the energy flux is not consistent with that of the energy density at any time.

After a delayed time, $t > t_d$, the absolute values of S_T and S_{PF} are almost same and therefore the positive net flux S_{∞} is extremely small in comparison with S_{PF} or $|S_T|$. In this stage, the work rate S_T due to the electromagnetic stress $T_{\hat{j}\hat{k}}$ is comparable to the Poynting flux measured in LNRF.

The black hole loses its rotational energy by the electromagnetic field and at the same time may gain the rotational energy by infalling gas acted by the electromagnetic force. The net variation of the hole's rotating energy should be evaluated by the study of the magnetohydrodynamics with the reaction of the electromagnetic force on the gas. Nevertheless it is worthwhile for the energetics of AGN to estimate the extraction rate of the hole's rotating energy by the electromagnetic field. The transfer rate S_T is obtained in order of magnitude by setting $\tilde{B}^{\hat{\phi}}\tilde{B}^{\hat{r}} \sim -a\tilde{B}^2, \varpi_{st} \sim r_h/\sqrt{2}, \omega_{st} \sim a/2Mr_h, A_{st} \sim 4\pi(r_h^2 + a^2)$:

$$|S_T| \sim (\frac{a}{M})^2 \tilde{B}^2 r_h^2 \sim 5 \cdot 10^{44} erg/sec (\frac{a}{M})^2 (\frac{M}{10^8 M_{\odot}})^2 (\frac{\tilde{B}}{10^4 G})^2.$$

Thus, reasonable astrophysical parameters may lead to power outputs of the magnitude $\sim 10^{44} erg/sec$ seen in AGN.

Reference

Yokosawa, M., Ishizuka, T., and Yabuki, Y. (1990) "Extraction of rotating energy from a black hole by frozen-in magnetic field in an accretion flow", *Publ. Astron. Soc. Japan*, submitted.

SPHERICAL SYMMETRIC MODEL FOR CALCULATING LARGE PECULIAR VELOCITIES OF GALAXIES

Xiang Shouping(2) Cheng Fuzhen(1,2) Liu Jianmin(2)
1. Center of Astronomy and Astronomy and Astrophysics, CCAST(World Lab.)
2. Meter for Astrophysics, University of Science and Technology of China, Hefei, Anhui, P. R. China

I. Theory

Assuming Gaussian primodial mass perturbation and spherical symmetric model. We present a different way to calculate the distribution of large scale peculiar velocities. Our method averts from using thelinear perturbation approximation and window function which are widely used in the fluid models, so that some uncertainties such as the real distribution of the number density of galaxies could be avoided.

Consider a spherical shell with a physical radius r_i at initial time t_i. The evolution of the radius of the thin shell is given in the parametized form (peebles 1980). Further, we obtain the peculiar velocity U_p of the shell relative to the Hubble flow.

We assume that the probability density of the initial perturbation δ_i is Gaussian

$$P(\delta_i) = \frac{1}{\sqrt{2\pi}\,\sigma(r_i)} e^{-\frac{\delta_i}{2\sigma^2(r_i)}} \qquad (1)$$

where $\sigma(r_i)$ is the expected value δ_i. The expectation of U_p on the whole space for a fixed r can be obtained by

$$V_p^2(r) = \int V_p^2 P(\delta_i)d\delta_i \Big/ \int P(\delta_i)d\delta_i \qquad (2)$$

where the integrations should be performed along the constant r curve.

II. Discussion and Conclusion

The results given by our model are approximately consistent with that given by the fluid (Vittorio et al., 1987) and the conmic string model (Brandenberger et al., 1987).

421

422

It is shown in Fig.1 that the peculiar velocities decrease
with increasing of distance scale in the case of initial
Gaussian perturbation spectrum, and it is still hardly to
explain the large peculiar velocities reported by lynden-
bell et. al. (1988).

A reasonable explanation for the disagreement between
observation and theories is that there may be a huge mass-
concentration region (Great Attractor) out of the distance
of 60h^{-1}Mpc away from the Local Group. But it is still uncertain that whether or not the Great Attractor exists due
to the largeerrors in distance measurements. On other hand
some non-standard model or non-gravitational processes
might be needed for explanation of the formation of the
Great Attractor.

References

Brandenberger, K. et al., Phys. Rev., D36
Lynden-Bell, D. et al., Ap. J., 326(1988), 19.
Peebles, P. J. E., Laarg-Scale Structure of
    Universe (1980), Princeton.
Vittorio, N., Turner, M.S., Ap. J., 316(1987), 457.

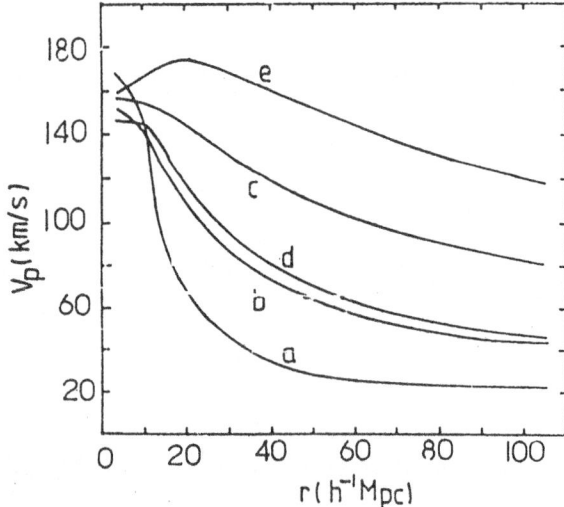

Fig.1. The distribution of peculiar velocities with
respect scales. a) Baryonic universe, h=1; b) CMD,
adiabatic perturbation, h=1; c) CDM, isocurvature
perturbation, h=1; d) HDM, H=1; e) CDM, isocurvature
perturbation, h=0.5.

ON THE ORIGIN OF COSMOLOGICAL MAGNETIC FIELDS

T. TAJIMA, S. CABLE, K. SHIBATA
Department of Physics and Institute for Fusion Studies
The University of Texas at Austin
Austin, Texas 78712

ABSTRACT. It is shown that a plasma with temperature T sustains fluctuations of electromagnetic fields and particle density even if it is assumed to be in a thermal equilibrium. The level of fluctuations in the plasma for a given wavelength and frequency of electromagnetic fields is rigorously computed by the fluctuation-dissipation theorem. A large zero frequency peak of electromagnetic fluctuations is discovered. We show that the energy contained in this peak is complementary to the energy "lost" by the plasma cutoff effect. The level of the zero (or nearly zero) frequency magnetic fields is computed as $\langle B^2 \rangle^0 / 8\pi = \frac{1}{2\pi^3} T(\omega_p/c)^3$, where T and ω_p are the temperature and plasma frequency. This is the theoretical minimum magnetic field strength, as no turbulence is assumed. The size of the fluctuations is $\lambda \sim (c/\omega_p)(\eta\tau)^{1/2}$, where η and τ are the collision frequency and the lifetime of magnetic fields fluctuations. The level of magnetic fields is significant at the early radiation epoch of the Universe: Magnetic fields as great as 10^{16} Gauss is generated at $t = 10^{-2}$ sec.

The Fluctuation-Dissipation Theorem in the Radiative Plasma

In the epoch from 10^{-2} sec after the big bang and till 10^{13} sec the constituent matter was electrons (and positrons), photons, and protons (and other light ions), which make up a hot plasma. The most important interaction in this period is the coupling between photons and leptons, collective or individual. We shall show that the presence of plasma plays an important role in shaping the radiation spectrum.

In or near thermal equilibrium the plasma has thermal fluctuations, whose level is related to the medium's dissipative characteristics and the temperature T, as formulated in the fluctuation-dissipation theorem (Kubo, 1957). We find an expression for the fluctuation spectrum of the magnetic field in an equilibrium plasma as a function of frequency. This is accomplished by deriving the magnetic fluctuations in wavenumber and frequency space $\langle B^2 \rangle_{k\omega} / 8\pi$ from the fluctuation-dissipation theory, then integrating over wavenumber. $\langle B^2 \rangle_\omega / 8\pi$ is nearly a black-body spectrum at high frequencies, but, when plasma collisionality is taken into account, it has a high, narrow peak at frequency $\omega = 0$.

We look at waves in a homogeneous isotropic nonmagnetized equilibrium plasma. The strength of magnetic field fluctuations is

$$\frac{\langle B_{tot}^2 \rangle_{k\omega}}{8\pi} = \frac{i}{2} \frac{\hbar}{e^{\hbar\omega/T} - 1} \frac{c^2 k^2}{\omega^2} \left\{ \Lambda_{22}^{-1} + \Lambda_{33}^{-1} - \Lambda_{22}^{-1*} - \Lambda_{33}^{-1*} \right\} , \tag{1}$$

where

$$\Lambda_{ij}(\omega, \mathbf{k}) = \frac{c^2 k^2}{\omega^2} \left(\frac{k_i k_j}{k^2} - \delta_{ij} \right) + \epsilon_{ij}(\omega, \mathbf{k}) , \tag{2}$$

423

where $\epsilon_{ij}(\omega, \mathbf{k})$ being the dielectric tensor of the plasma. Combining Eqs. (1) and (2) after some algebra, we obtain

$$\frac{\langle B^2 \rangle_{\mathbf{k}\omega}}{8\pi} = \frac{2\hbar\omega}{e^{\hbar\omega/T} - 1} \eta\omega_p^2 \frac{k^2 c^2}{\omega^2} \frac{1}{\left[\omega^2 - k^2 c^2 - \omega_p^2\right]^2 + \eta^2 \left[\omega - k^2 c^2/\omega\right]^2} . \tag{3}$$

In order to obtain the frequency spectrum of magnetic fluctuations $\langle B_\omega^2 \rangle /8\pi = \int d\mathbf{k} \langle B_{\mathbf{k}\omega}^2 \rangle /8\pi$, integration of Eq. (3) is executed. It involves a special technique (Tajima $et\ al.$ 1990). Through this algebra and integration, we obtain the frequency spectrum of magnetic fluctuations:

$$\frac{\langle B^2 \rangle_\omega}{8\pi} = \frac{1}{\pi^2} \frac{\hbar\omega'}{e^{(\hbar\omega_{pe}/T)\omega'} - 1} 2\eta' \left(\frac{\omega_{pe}}{c}\right)^3 \int_0^{x_{cut}} dx \frac{x^4}{(\omega'^2 + \eta'^2)x^4 + \cdots}$$

$$+ \frac{\hbar(\omega'^2 - \omega_p'^2)^{3/2}}{2\pi(e^{(\hbar\omega_{pe}/T)\omega'} - 1)} \left(\frac{\omega_{pe}}{c}\right)^3 \Theta \left(\omega - \sqrt{c^2 k_{cut}^2 + \omega_p^2}\right) , \tag{4}$$

where $x_{cut} = k_{cut}\, c/\omega_{pe} \sim 1$, $\omega' = \omega/\omega_{pe}$ and Θ is the Heaviside step function. The corresponding wavenumber spectrums through the integral of Eq. (1) over ω gives the expression

$$\frac{\langle B^2 \rangle_{\mathbf{k}}}{8\pi} = \frac{\hbar\, k^2\, c^2}{\left(e^{\hbar/T (\omega_p^2 + k^2 c^2)^{1/2}} - 1\right)(\omega_p^2 + c^2 k^2)^{1/2}} + T \frac{\omega_p^2}{\omega_p^2 + c^2 k^2} . \tag{5}$$

The first term in (5) corresponds to the radiation energy, while the second corresponds to the $\omega \sim 0$ fluctuation modes. In the classical limit the sum of the first and second terms yield T in Eq. (5), the equipartition law.

In Table I we summarize our results. We survey physical quantities of importance that characterize the radiation epoch (or the plasma epoch).

Table I — The zero frequency magnetic fluctuations in early Universe. ($t = 10^{-2}, 1$, and 10^{13} sec after the big bang). The temperature T, density of the plasma electrons, the zero frequency magnetic fluctuations B, and the plasma beta β.

		$t = 10^{-2}$	$t = 1$	$t = 10^{13}$	$t = 3 \times 10^{17}$ sec
T	eV	10^7	10^6	0.4	$T_\gamma = 0.0003$
n	cm^{-3}	5×10^{34}	4×10^{31}	10^3	10^{-6}
B	Gauss	10^{16}	10^{13}	10^{-12*}	
β		1	$10 - 10^2$	10^{15*}	
*: no dynamo effects included					

References

Kubo, R., 1957, J. Phys. Soc. Jpn. **12**, 570.

Tajima, T., K. Shibata, S. Cable, and R. Kulsrud, 1990, submitted to Ap. J.

The Hawaii Deep Survey — Implications for Cosmology and Galaxy Formation

Lennox L. Cowie
Institute for Astronomy, University of Hawaii

ABSTRACT. The Hawaii deep survey has now established 2.2 micron galaxy counts to a K magnitude of 23. The low number density of these counts at the faint end favors a $q_0 = 0.5$, $\Lambda = 0$ geometry which minimizes the available cosmological volume. However these models provide no explanation for the blue galaxies which are responsible for the large number of faint blue counts. Combining spectroscopic data with K magnitudes for a blue selected sample shows that the large number of galaxies at $B = 23 - 24$ appears to be a population of blue dwarf galaxies lying at typical redshifts of 0.3 to 0.4. There are no obvious descendants of these galaxies at the present epoch. The dwarf galaxies contain as much K light as the normal galaxy population and may dominate the baryonic and dark matter content of the universe.

1. Introduction

In principle, number counts of faint galaxies should be a powerful probe of the cosmological geometry. However, in practice optical and radio galaxy counts contain much more information about the evolution of the galaxy or its nucleus then about the cosmology. The simple reason for this is that the nuclear activity responsible for the radio emission and the massive stars responsible for the blue light are transient events by cosmological standards. It is therefore possible for the galaxies' properties to change enormously with redshift and correspondingly hard to compare the current galaxy population with its progenitor population to interpret the number counts.

The situation is quite different in the near infrared. In a 2.2 micron (K band) galaxy sample we are seeing light from near-solar mass stars whose lifetimes are comparable to the age of the galaxy. This means that evolutionary corrections are much smoother aand can be more securely modelled. There is also the advantage that out to substantial redshifts the K band samples the relatively well understood optical portion of the galaxy spectrum so other uncertainties in the modelling are

425

removed or minimized. Finally because the optical and near IR spectrum is generally much flatter than the UV portion the dimming with redshift in the near IR is generally much smaller than in the optical and we sample to large redshift at much brighter magnitudes. This last effect is important because galaxy counts that reach faint enough limits to break through and see the whole galaxy occupied cosmological volume contain information about the geometry and the redshift of galaxy formation just as star counts in our own galaxy contain information about its shape and size.

Motivated by these considerations the Hawaii deep survey was designed to obtain deep infrared galaxy samples using the new infrared arrays that have recently become available. The survey was designed to establish the K-band galaxy counts to $K = 23$ and to provide a well-defined large sample of K-selected galaxies, with corresponding optical colors, which can be used as a basis for spectroscopic studies. The follow up spectroscopic studies are nearly complete for a $B < 24$ selected sample but are only just begining for the K-selected sample. The optical data and the spectrosopy of the blue selected sample are described in Lilly,Cowie and Gardner (1991;LCG) while the K band counts and the properties of the K band sample are given in Cowie et al. (1991).

2. Optical and Infrared Galaxy Counts

The current status of the differential galaxy counts in B (4500 Å) and K is given in Figure 1. Consider first the B band counts, which I am going to compare with number-count models, both with and without galaxy evolution, that I have adapted from those of Yoshii and Takahara (1988). There are numerous assumptions in these models about the present population mix and galaxy luminosity function and the nature of galaxy evolution (that is, how its luminosity and spectral type changed with time, younger galaxies generally being more luminous and bluer); these are well described by Ellis (1983). However, there are some very basic features that are more or less model independent. First, it is possible to roughly model the counts to $B = 21$ (even with a model with no galaxy luminosity evolution, but see Maddox et al. 1990). Beyond this point actual counts tend to rise more rapidly than any of the non-evolving predictions (the dashed lines), even including the most extreme effects of geometry: an open universe has more volume at high redshift than a closed one and a zero-curvature universe with $\Omega_0 = 0.1$ and positive cosmological constant has even more; these models therefore predict more faint-end counts but this effect is small compared with the count discrepancy.

Luminosity evolution has a much larger effect, as can be seen from Figure 1 where the evolving models are shown as solid lines. With the luminosity evolution included, one sees galaxies to much larger distances and consequently the models predict many more counts, but of course they predict that the excess counts should be produced by extended tails or secondary peaks *at high redshift*. Prior to the advent of spectroscopic data, this seemed a good explanation for the excess counts but as the faint-object spectra have become available in the last couple of years it has become clear that it must be ruled out (cf. LCG). To demonstrate this I show

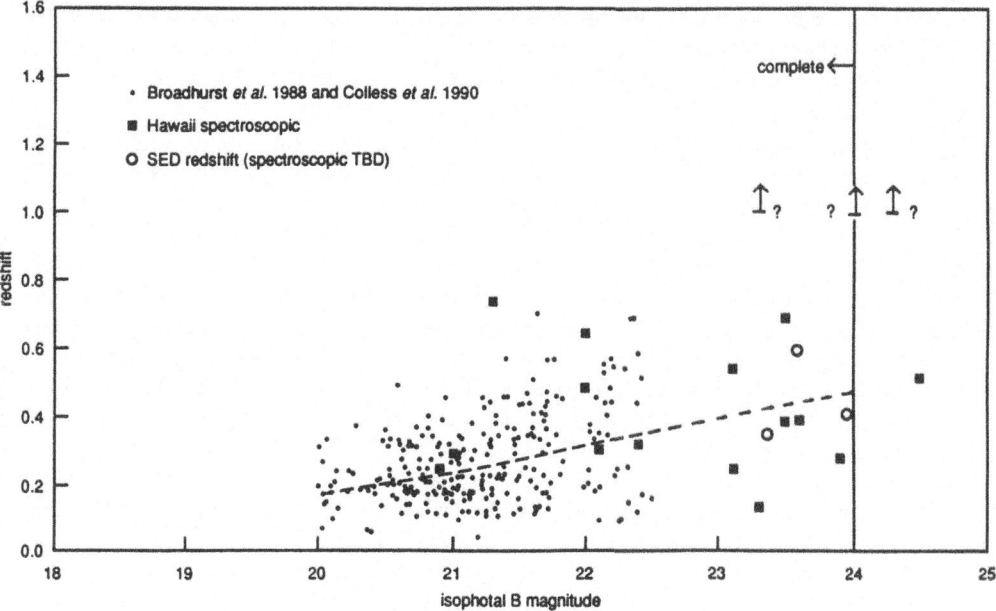

Figure 2 Spectroscopic samples for B magnitude limited galaxies. The dots show the $B < 21$ and $B < 22.5$ samples of Broadhurst et al. (1988) and Colless et al (1990). The latter is about 80% complete. The sold squares show the Hawaii spectroscopic data (e.g. Lilly et al. 1990) which is nearly complete to $B = 24$. (Open circles are three objects that remain to be observed spectroscopically shown at redshifts estimated from their colors.) Only one object at $B < 24$ defied spectroscopic identification and could be at high z (upward-pointing arrow); nearly all objects are at $z < 0.6$. The dashed line shows a predicted mean redshift for a model with no galaxy evolution; it provides a remarkably good fit.

cosmological and one astrophysical, with no clearcut way to decide between them. That is, either there is a major population of blue dwarf galaxies that was present at $z \sim 0.3 - 0.4$ but that was not included in the assumed present-day luminosity functions (this is discussed in much more detail below) or alternatively there must be much more volume at low redshift than even a $q_0 = 0.02$ Friedman model would predict. The latter could be the case, for example, in a zero curvature $\Omega = 0.1$ universe with a cosmological constant, a possibility that has many other attractions such as solving the cosmological timescale problem and easing the difficulties of large scale structure. As is shown in Figure 1, such a model (with some galaxy evolution included) can fit the number counts (e.g. Lilly, Cowie and Gardner 1991, Fukugita et al. 1990) and yet not violate the spectroscopic constraints.

However, we now have the K band number counts that, as we discussed in the introduction, are much more sensitive to cosmological geometry than the blue counts are. This effort has produced the result that can be seen in Figure 1. Basically, at around $K = 19$ the counts stop rising rapidly, they move to a much shallower slope. What is suggested here is that the fainter magnitudes no longer sample to larger distances; we must have reached at last the magnitude at which we have broken through to the end of the galaxy-occupied cosmological volume. At this point the flat or slowly-rising counts represent the faint-end shape of the galaxy luminosity function averaged through this volume.

If this is the case we are in a much stronger position to decide what the cosmological geometry is than when we only had the ambiguous still-rising B band counts to work with. Comparing the model predictions with the data, the absolute K number density at the faint end favors a $q_0 = 0.5$ universe (Figure 1) and this is a relatively robust result. To see why, suppose the galaxy population completely occupies a volume to a redshift z_f where galaxy formation occurred. Then the number density of galaxies in a given observed magnitude interval per comoving volume is $f(M) = N(M)/V(z_f)$, where $N(M)$ is the observed number of counts (just under 10^5 deg^{-2} mag^{-1} from $K = 20$ to $= 23$) and $V(z_f)$ is the comoving volume to redshift z_f corresponding to the observed solid angle. When the counts are flat, not rising with magnitude, we must be sampling all populations at the faint end of the luminosity function and so we can approximate $f(M)$ with the local density of galaxies at several magnitudes below standard luminosity, a number that is quite well determined at about 2×10^{-3} $(H_0/50$ km s^{-1} Mpc$^{-1})^3$ Mpc^{-3} (e.g. Efstathiou et al. 1988) and depends only weakly on luminosity at these magnitudes. The local number density of counts that would be inferred from the observed counts at $K = 21.5$ and 22.5 is shown in Figure 3 as a function of cosmological geometry and z_f. What can be seen here is that in a $q_0 = 0.5$ geometry we get reasonably good agreement for a wide range of z_f while for an open model we require $z_f \sim 2$ and for zero curvature cosmological constant models $z_f \sim 1.5$. Given the amount of evidence we have that galaxy formation began beyond these epochs the data strongly favors a $q_0 = 0.5$ model.

One might reasonably wonder whether luminosity evolution or some other astrophysical effect could change the number density of galaxies to explain this result, but the problem is that we have here a *deficiency* in the number of objects observed compared to local density, not an excess as we have in the optical, and this is much harder to find excuses for. If we add populations, include luminosity evolution or have galaxies in fragments at higher z we predict *more* counts, not fewer, and are forced even further toward a low-volume geometry. Even worse, the stars that currently give the K band light were there for the whole cosmological lifetime of the galaxy so there is no way to 'dim' the galaxy relative to its current luminosity.

3. Dwarf Galaxies and the Faint Blue Number Counts

For any model which roughly fits the K band counts we can see from Figure 1 that we now have the problem of explaining a huge excess of faint blue galaxies. Another way of expressing the relative overabundance of faint blue galaxies and underabundance of faint IR galaxies is to note that the K-band galaxy sample shows a rapid trend to bluer colors at fainter magnitudes and that there are relatively few

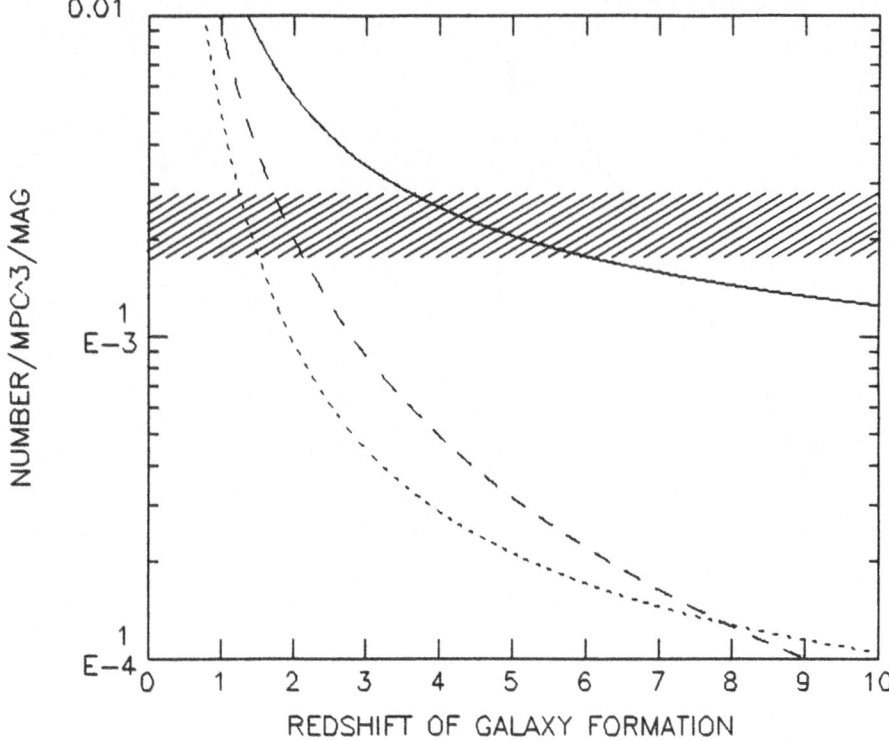

Figure 3 The average number density of galaxies per observed magnitude is shown for the $K = 21.5$ counts as a function of the epoch of formation and the cosmological geometry: $q_0 = 0.5$ (solid lines); $q_0 = 0.02$ (dashed lines); zero-curvature, $\Omega_0 = 0.1$, cosmological constant (dotted lines). The shaded region shows the local density of galaxies at 2.5 to 3.5 magnitudes below the Schechter luminosity, based on the summary of Efstathiou et al. (1988). The closed case can be roughly consistent over a wide range of formation epoch, but models with larger volume elements fail radically unless the bulk of galaxies formed very recently.

faint objects that are red in $(B-K)$. This is shown in Figure 4 which is a histogram of the $(B-K)$ color in various magnitude ranges. Objects which are not detected at the $1\ \sigma$ level in B are shown as open regions on the histogram and could lie anywhere to the red. Compared to the observations the models we discussed above predict redder distributuions. The most extreme are the non evolving models (we show the $q_0 = 0.02$ case) but even a $q_0 = 0.5$ evolving model predicts too many red galaxies.

The blueness of the faint galaxy population is of course the cause of the discrepancy between the blue and IR number counts. What must be explained in any model is why there are so many very blue faint galaxies. However there is a more profound conclusion that can be drawn from Figure 4, namely that only those galaxies with the redder colors are being correctly understood in the

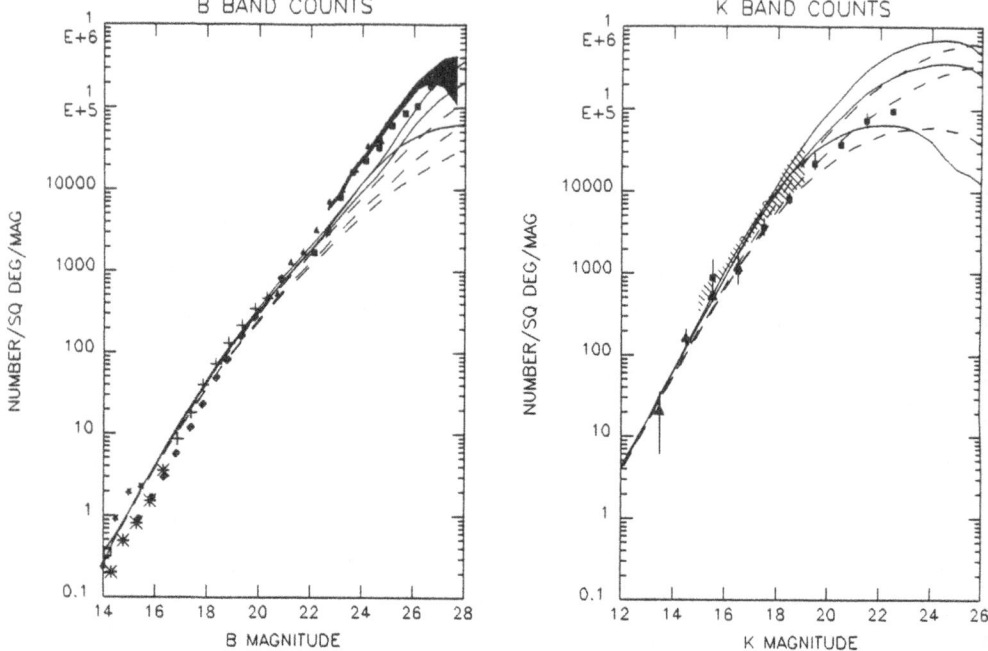

Figure 1 A comparison of blue and infrared counts with model predictions. The blue number counts are taken from the compilation of Metcalfe et al. (1990) in addition to those of Maddox et al. (1990; diamonds), Tyson (1988; shaded region) and Lilly et al. (1990; solid squares). The infrared counts are from Glazebrook et al. (1990; triangles), Jenkins and Reid (1990; shaded region) and Cowie et al. (1990; squares and diamonds). For the IR data I have shown error limits. Two classes of model are shown. The dashed lines show models in which galaxies do not change at all in luminosity or type. The bottom curve has $q_0 = 0.5$, the middle $q_0 = 0.02$, and the top a zero-curvature model with $\Omega_0 = 0.1$ and a cosmological constant. The three solid lines show models with the same three geometries as above but in which the galaxies evolve with z roughly following the prescription of Yoshii and Takahara (1988). The galaxies are assumed to form at $z_f = 5$ in all cases. The K band counts fit best to a $q_0 = 0.5$ cosmological geometry with evolution or a $q_0 = 0.02$ model with no evolution but the latter model is quite physically implausible. All the models which fit the K band counts grossly underpredict the faint-end blue counts.

data from a number of complete spectroscopic samples in Figure 2. To $B = 24$ there is only one possible object in these samples which could even conceivably be at $z > 1$; the majority have $z < 0.6$.

Until the faint K-band counts became available there were two possible explanations for the combined faint blue counts and blue selected spectroscopy, one

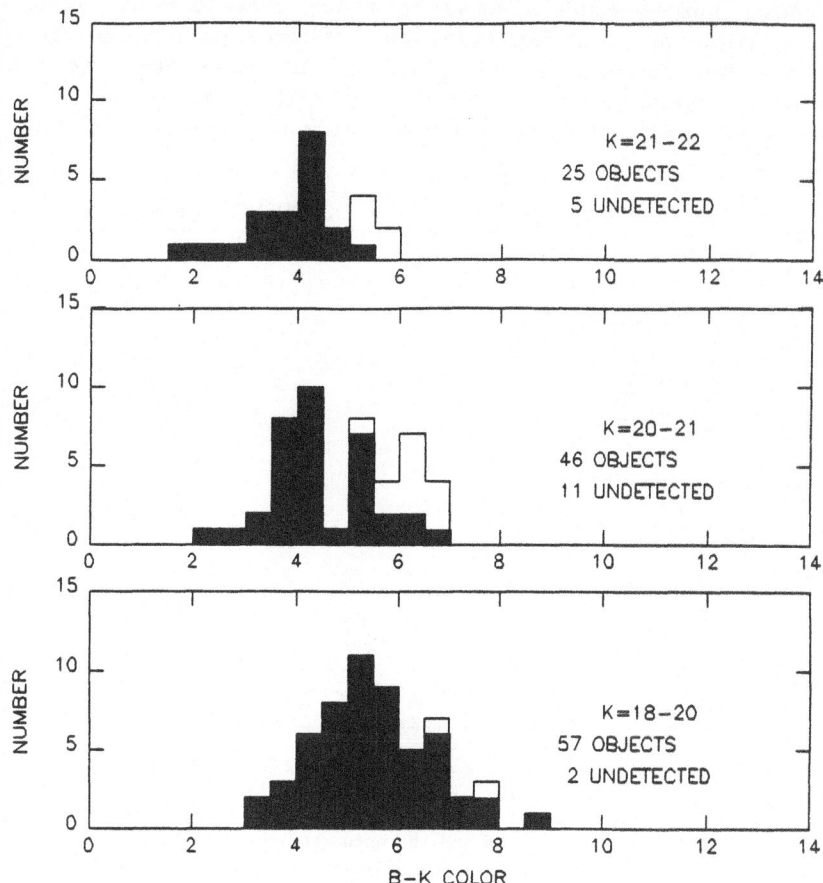

Figure 4 Histograms of the (B-K) distribution of the K sample as a function of magnitude. Galaxies which are not detected at the one sigma level in B are shown as the open area. They could lie anywhere to the red of their position in the graph.

conventional smoothly evolving models; the blue galaxies must correspond to some other population or to episodic star formation in the fainter end of the normal galaxy population (e.g. Broadhurst et al. 1988).

We can combine the redshift data with the K-band magnitudes to determine what the population producing the faint B counts actually is. That is, the absolute K magnitudes should give a good estimate of the galaxy mass except for the most extreme star-bursting cases where it constitutes an upper limit. (They can be roughly used without a K correction because of the flatness of galaxy SEDs in the near infrared.) The absolute K magnitudes for the galaxies giving the $B < 24$ counts are shown in Figure 5. At brighter B magnitudes most of the galaxies are at or near the K_* of -25.3 typical of the most luminous elliptical galaxies ($H_0 = 50$ km s^{-1} Mpc^{-1}). At the faint end, which is also the position at which the counts begin to rise rapidly above predictions, we begin to see many much smaller galxies that are typically about 4 magnitudes fainter in K. This is an unexpected effect in a magnitude-limited sample where we expect the counts to be dominated by the near-L_* galaxies that can be seen to the limits of largest volumes. Roughly two thirds of the faintest galaxies appear to be dwarf galaxies of this type. If we try to turn this into a luminosity function we find that the total K luminosity density is $8.1 \times 10^8 (H_0/50) L_\odot/\text{Mpc}^3$ with roughly half coming from the dwarfs and half from normal galaxies.

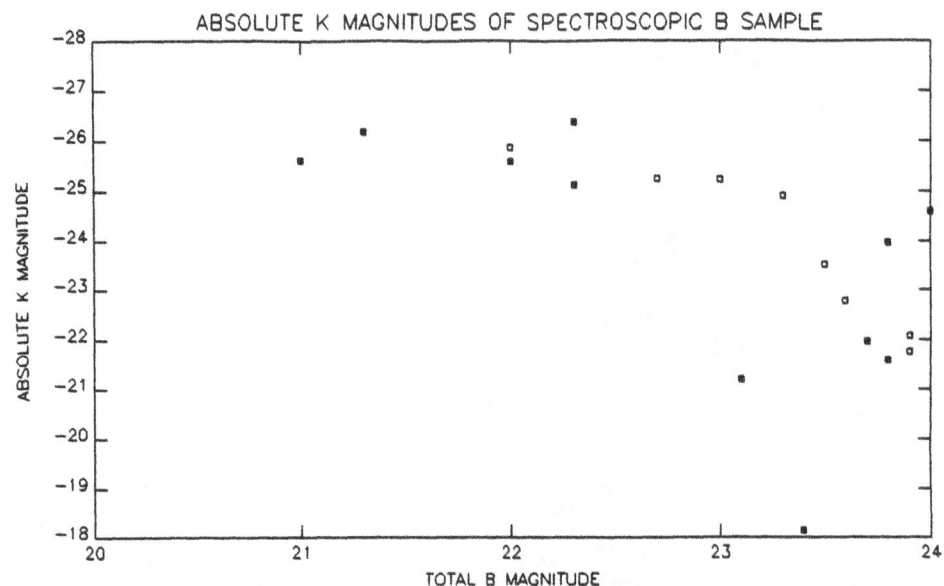

Figure 5   Absolute K magnitudes for the $B < 24$ sample. The solid symbols show objects with spectroscopic redshifts and the open symbols those where the redshifts have been estimated from colors. The dashed line shows the approximate magnitude of a typical giant galaxy.

Since the amount of K light in the dwarfs is essentially equal to that in the normal galaxies the dwarfs must contain a comparable amount of baryonic matter in low mass stars. Given that they are likely to have considerably more gas than stars, they are almost certain to dominate the baryonic mass. In particular if thay contain approximately four times as much mass in gas as in stars, typical of local blue dwarf irregular galaxies, then the baryonic density would be roughly 1.4×10^{-31} g cm^{-3}

An alternative way to estimate the baryonic mass in the dwarfs is to consider the amount they contribute on average to the sky surface brightness in the blue or ultraviolet. This quantity is a direct measure of the average density of the metals produced by the population *regardless of cosmology or the details of the star formation process* (Songaila et al. 1990). The reason for this is fairly clear: the ultraviolet light density is an integrated history of massive star formation in a given local volume and these massive stars also generate the released metals. The detailed relationship is

$$S_\nu = 3.6 \times 10^{-25} \left(\frac{\rho Z}{10^{-34}\,\text{g cm}^{-3}} \right) \text{ergs cm}^{-2}\,\text{s}^{-1}\,\text{Hz}^{-1}\,\text{deg}^{-2}, \qquad (1)$$

where S_ν is the average sky surface brightness at frequency ν, and ρZ is the local density of metals.

From the spectroscopic data discussed above we know that the bulk of the sky light (probably at least two thirds) at $B = 22 - 24$ comes from the dwarf galaxies at $z \sim 0.3 - 0.4$. This 'blue sky' contribution dominates the ultraviolet sky light with $S_\nu \approx 10^{-24}$ ergs cm^{-2} s^{-1} Hz^{-1} and corresponds to $\rho Z \approx 3 \times 10^{-34}$ g cm^{-3}. Even if the fractional metal content Z were as high as high as 0.02, typical of disks or the centers of massive elliptical galaxies, the density of material in this new population would be $\sim 10^{-32}$ g cm^{-3}, nearly equal to that in all known galaxies. On the other hand, it is much more likely that Z is considerably lower in these systems, which may be closer in properties to local giant extragalactic HII regions and blue irregular galaxies; in this case their average baryon density would be around 10^{-31} g cm^{-3}, consistent with the more direct estimate.

These estimates are quite approximate but can be compared directly with homogeneous Big Bang nucleosynthesis which predicts a baryon density of $2 - 9 \times 10^{-31}$ g cm^{-3} (e.g. Yang et al. 1984) and it is apparent that the dwarfs can provide the required baryon density without any problem. The interesting thing is that it is entirely plausible, physically, for the blue galaxy population to contain the bulk of the baryons. The dwarfs could also be more uniformly distributed than, or even distributed preferentially away from, normal galaxies at the same epoch, which would result in quite significant biasing of baryonic matter from present-day luminous material.

4. Summary

The data is still fluid and has yet to be confirmed by other groups, and the model results are still subject to interpretation. However, to my mind the most natural conclusion at this time is that we live in a $q_0 = 0.5$, $\Lambda = 0$ universe, which allows us to understand the K band counts naturally, and that there is a population of dwarf galaxies, responsible for the large number of faint blue galaxies observed, that flourished at $z \sim 0.4$ but has now burned out and either destroyed itself or in some other way (perhaps by being too faint) and is no longer counted in the present galaxy population.

If we accept that the blue number counts and spectroscopy imply the existence of the dwarf population then it contains at least as much baryonic mass as (and probably considerably more than) the present-day 'normal' galaxies. This could account for the missing baryon problem which is suggested by primordial nucleosynthesis models.

Finally since the dwarfs may not be distributed as the obseved normal galaxies are it is possible that they may imply substantial biasing of the present luminous material from the baryonic matter. Actually determining whether this is in fact the case will require large faint galaxy samples that can be cross correlated with brighter normal galaxies at the same redshift.

I am very grateful to Toni Songaila for many interesting discussions which greatly clarified my thinking on these problems. I would also like to thank John Kormendy and Ken Freeman for extremely illuminating dicussions about dwarf galaxies and my student Jon Gardner and colleagues Klaus Hodapp, Esther Hu, Simon Lilly and Richard Wainscoat on whose work this talk is partly based. This work was supported in part by NASA grant NAGW 959.

References

Broadhurst, T. J., Ellis, R. S. and Shanks, T. 1988, *M.N.R.A.S.*, **235**, 827.

Colless, M. M., Ellis, R. S., Taylor, K. and Hook, R. N. 1990, *M.N.R.A.S.*, **244**, 408.

Cowie, L. L., Gardner, J. P., Wainscoat, R. J., and Hodapp, K. 1991, preprint.

Efstathiou, G., Ellis, R. S. and Peterson, B. A. 1988, *M.N.R.A.S.*, **232**, 431.

Ellis, R. S. 1983, in *The Origin and Evolution of Galaxies* ed. B. T. Jones and J. Jones (Dordrecht: Reidel), p. 255.

Fukugita, M., Takahara, F., Yamashita, K. and Yoshii, Y. 1990, preprint.

Glazebrook, K., Peacock, J. A., Collins, C. A., and Miller, L. 1990, preprint.

Jenkins, C. and Reid, J. 1990, *A. J.*, in press.

Lilly, S. J., Cowie, L. L., and Gardner, J. P. 1990, *Ap. J. Suppl.*, in press.

Maddox, S. J., Sutherland, W. J., Efstathiou, G., Loveday, J., and Peterson., B. A. 1990, *M.N.R.A.S.*, **247**, 1p.

Metcalfe N., Shanks, T., and Fong, R. 1990, preprint.

Songaila, A., Cowie, L. L. and Lilly, S. J. 1990, *Ap. J.*, **348**, 371.

Yang, J., Turner, M. S., Steigman, G., Schramm, D. N. and Olive, K. A. 1984, *Ap. J.*, **281**, 493.

Yoshii, Y. and Takahara, F. 1988, *Ap. J.*, **326**, 1.

ANALYSIS OF THE LARGE SCALE STRUCTURE WITH DEEP PENCIL BEAM SURVEYS

A.S. SZALAY[1,2], R.S. ELLIS[3], D.C. KOO[4] and T.J. BROADHURST[5]
1. *Department of Physics and Astronomy,*
   The Johns Hopkins University, Baltimore MD 21218
2. *Department of Physics, Eötvös University, Budapest, H-1088*
3. *Department of Physics, University of Durham, DH1 3LE*
4. *Lick Observatory, University of California, Santa Cruz, CA 95064*
5. *Mathematics Department, Queen Mary & Westfield College,*
   London E1 4NS

ABSTRACT. Recent observations of the large scale structure of the universe suggest inhomogeneities on scales larger than previously thought. A deep redshift survey with a 'pencil beam' geometry of galaxies at the Galactic Poles indicates strong clustering, with a provocative regularity on a scale exceeding $100\ h^{-1}$ Mpc. We examine in detail the role of such deep 'pencil beam' surveys in studies of the large scale structure using the 1-D power spectrum of the distribution to illustrate the statistical significance of the feature claimed by Broadhurst *etal* (1990). We demonstrate how deep probes complement sparse-sampled wide angle local surveys providing new information on large scales. To test the results found from our original survey, we examine future sampling strategies that would delineate the topology more clearly and briefly discuss the implications of such results in theories accounting for the origin of structure. We quantify the sensitivity of pencil beam surveys to the topology of large-scale structures and compare them to sparse sampled wide-angle local surveys. In particular, we also present a more detailed statistical analysis for the Galactic Pole data set, and show that it remains significant at the 2×10^{-4} level.

1. Introduction

Many of the most fundamental cosmological questions relate to the formation of the large scale structure; these connect the smallest and largest physical scales known. Those processes responsible for small quantum fluctuations in the early universe occur at energies and length scales inaccessible to modern accelerators. Recent results concerning the large scales over which galaxies may be organized suggest a new tool for examining processes in the early Universe.

To model the structure seen, cosmologists consider a variety of initial conditions and follow subsequent evolution under the assumption of gravitational instability, with combined analytic and numerical techniques, making certain extra assumptions. The predictions are compared with the distribution of galaxies and fluctuation limits for the cosmic background radiation (CBR).

The most popular galaxy formation theory is the biased Cold Dark Matter (CDM) model (Blumenthal *et al* 1984), in which most of the mass density is in the form of noninteracting dark particles. Together with a scale invariant Zeldovich–Harrison spectrum for the initial density fluctuations, the theory satisfies many observational constraints on small (< 100 h^{-1} Mpc) scales (White *et al* 1987). However, to explain the absence of CBR fluctuations, its proponents invoked the concept of 'biasing' whereby galaxies only form at high peaks of the mass fluctuations. Much stronger correlations in the distribution of visible galaxies are predicted than for the underlying mass (Kaiser 1984, Bardeen *et al* 1986).

CDM predicts the universe should be relatively homogeneous on scales above 30-40 h^{-1} Mpc and it is thus important to test this observationally. Surprisingly little accurate data exists. The apparent strong clustering for rich clusters (Bahcall and Soneira 1983) is in marginal conflict with theory. Large scale motions of galaxy samples offer an alternative test since they are related to density fluctuations. Recent results indicate coherent streaming within 80 h^{-1} Mpc scales (Rubin *et al* 1976, Burstein *et al* 1986, Lynden-Bell *et al* 1988) but these analyses rely on environment-independent distance indicators whose robustness is largely untested.

Recent galaxy redshift surveys suggest significant deviations from uniformity on large scales. Kirshner *et al* (1983) found a 60 h^{-1} Mpc sphere with a large underdensity in the galaxy distribution. deLapparent, Geller and Huchra (1986) delineated the 'Great Wall' in their CfA surveys - a structure connecting several known Abell clusters over a spatial extent of 100 by 50 h^{-1} Mpc. Chincarini *et al* (1983) and Giovanelli *et al* (1986) had earlier shown the Perseus-Pisces supercluster extends also to large distances in the transverse directions.

Are such 'walls' representative features of the universe? N-body simulations are able to generate apparently similar structures from CDM initial conditions (Park 1990), but few quantitative tests have been done so far. Statistical tests are readily constructed when the topology is understood, but that is evidently not yet the case.

Our task in this paper is to examine the ways in which evidence for large scale power might be assembled. In §2 we begin by contrasting two rather different observing strategies – the sparse-sampled local survey and the deep pencil beam survey demonstrating the unique features of each. In §3 we examine the latter strategy quantitatively, then in §4 we discuss the statistical consequences of the one-dimensional nature of the pencil beams. In §5 we present a more detailed statistical analysis of the recent survey of Broadhurst *et al* (1990, BEKS), then in §6 we discuss ongoing research and the extension of our strategy to answer further questions.

2. Strategies for Mapping Large Scale Structure

First we address the question of the best strategy for delineating the large scale distribution of galaxies, given that its precise topology remains unclear. With the exception of dedicated telescopes, a typical redshift survey can only measure a few thousand redshifts in a few years' observation. Indeed, there is probably no single strategy optimal in all situations since the choice is strongly influenced by what presumptions hold about the statistical properties and the topology of the large scale galaxy distribution. There is a fundamental difference between the strategies and goals of the sparse-sampled wide angle surveys, and the deep pencil beam surveys. Their correlation functions measure different statistical aspects of the same distribution. In this section, aimed towards studies of the large scale structure, we will concentrate our attention on these two strategies, demonstrating that for a full picture both are required.

In general, answers depend on what questions were asked. When discussing large scale structure, one should first specify what do these words mean. For some, large scale structure is equal to the large scale behaviour of the galaxy two point correlation function. Others like to draw maps of the galaxy distribution, and call that the large scale structure. These data sets contain different information, of course, and therein lies the fundamental difficulty in presenting a coherent view of the same universe. We will elaborate on these differences below, and focus on finding large coherent structures: our use of the words Large Scale Structure...

There are two competing sources of statistical noise in redshift surveys. The 'sampling noise' is determined by the number of redshifts obtained, and the 'clustering noise' is from the observed small scale correlation function of galaxies; the latter depends on the shape and size of the survey volume. For a given geometry, there is an optimum number of galaxies to observe, at which point the 'sampling noise' equals the 'clustering noise'. Beyond this critical number, the signal to noise for large scale correlations is not significantly improved. Such an argument has been used, e.g. to determine the optimum 'sparse sampling' rate for wide angle or all-sky surveys (Kaiser 1986).

If fluctuations in the universe are strictly Gaussian, their full statistical description is contained in the two-point correlation function. In this case the 'sparse sampled' survey is indeed the best way to obtain this information. If the universe, however, contains some very sharp large scale features, like the 'Great Wall', a sparsely sampled survey may fail to identify those and it may thus not be possible to test whether the sample is strictly 'fair', i.e. representative.

Another strategy that can be adopted uses a smaller solid angle, sampling fainter galaxies in the same amount of telescope time. There is some freedom in how the shape of the volume is chosen. Restricting the survey to a thin slice increases the dynamic range of relative distances compared to a wide angle, symmetric cone. Choosing a pencil beam enables us to go even deeper, at the expense of becoming much narrower. Because of this, the CfA slice delineated well the transverse spatial coherence of the 'Great Wall' (de Lapparent, Geller and Huchra 1986).

The slices or pencil beams cannot be arbitrarily thin, due to the well-known small scale clustering of galaxies

$$\xi(r) = \left(\frac{r}{r_o}\right)^{-1.8} \tag{1}$$

where $r_o = 5 \ h^{-1}$ Mpc is the correlation length (Peebles 1980). Any linear dimension chosen to be less than r_o will ensure noise from small scales will dominate any signal from large scale structure (see §3). Thus there is a practical limit on how thin such slices should be; the CfA slice is fairly optimal.

The presence of structures like the 'Great Wall' also means, that either higher order correlations are present, or the phases of the fluctuations are correlated: clearly different from a homogenous isotropic Gaussian random field. We can see this using a simple model, due to Peebles (1989). Consider an extreme distribution with all galaxies on sheets, spaced at separations λ, to form a three dimensional cubic lattice. The 3-D correlation function can be calculated analytically

$$\xi(r) = \frac{1}{6}\left[\frac{\lambda}{r} + 2\frac{\lambda}{r}\text{Int}\frac{r}{\lambda} - 2\right]. \tag{2}$$

The correlation function obviously reveals a feature at $r = \lambda$, but with a fairly small amplitude, $\xi(\lambda) \approx 0.16$, since in different directions the sheets are seen under different inclination angles, smoothing out the correlations. However, if we observe the same structure with pencil beam surveys, every beam would still go through a sequence of sheets, yielding very strong features in each 1-D correlation function. An infinite number of pencil beams would provide a much weaker signal, in a similar but not equivalent manner to the 3-D correlations. The key difference is that sharp walls are non-Gaussian structures whose phases in the Fourier decomposition are strongly correlated. A 3-D correlation function – equivalent to the power spectrum i.e. no phase information – will give a poor characterization of such a distribution.

From these it is obvious, that sharp sheets are well detected in pencil beams, and the one dimensional correlation functions can have a very large amplitude at their typical separation, while the three dimensional correlations – not to mention the projected angular correlations – would be quite small. This, of course does not guarantee that the detected features are all walls.

If 'Great Walls' are numerous, i.e. if they are typical of the large scale structure, from the surface density of galaxies seen, we can estimate what would constitute a 'fair sample'. Suppose a fraction f of the bright galaxies are on walls with a surface density equal to that observed, $\mu = 0.3$ galaxies Mpc^{-2} (deLapparent et al 1991). The characteristic 'cell' size can then be estimated from the corresponding average volume density of bright galaxies, $n = 0.02$ galaxies Mpc^{-3} (deLapparent et al 1991). We estimate the fraction f from the CfA survey itself, the wall contains 906 galaxies out of 2536 (Ramella et al 1991) thus $f = 906/2536 \approx 0.36$. Given a topology, we can determine the mean surface-to-volume ratio. This number is remarkably robust, thus for simplicity we can consider a spherical topology and count only half the surface area, since walls separate two volumes. With these numbers the typical size of the voids would be

$$\lambda = 2R = \frac{3\mu}{nf} = 126 \ h^{-1} \, \text{Mpc} \tag{3}$$

This number is an order of magnitude estimator, of course, but shows that if the 'Great Wall' is a typical structure of the universe, it cannot be replicated much more often than λ. In this case, a strategy penetrating depths several times larger than this scale is required. For a fixed number of redshifts, ignoring practical considerations of multiple object techniques for the moment, it is then advantageous to narrow the slice further into a pencil beam of a small solid angle. The beam size constraint can be understood intuitively as follows: if the radius is small ($< r_o$), one cannot be certain whether the detected overdensities in the galaxy distribution indicate large coherent structures, or small scale gravitational clustering with a typical size $\simeq r_o$. Larger radii overcome this problem, but then the depth of the survey is sacrificed since the sampling rate cannot be decreased significantly.

An optimum geometry can be defined using exactly the same considerations as developed by Kaiser (1986) for the wide-angle sparse-sampled case viz. balancing 'sampling noise' and 'clustering noise' under certain topological assumptions. For example, if the large scale structure is dominated by walls with a well defined surface density μ, separated by large voids of diameter λ, how can we best determine if such structures are common? Evidently, the sparse-sampled local surveys are *far from optimum* in this case. We will consider a 1D correlation function along a pencil beam, and show, that in the presence of walls, it can be very effective in detecting them.

3. Optimum Choice of the Pencil Beam Geometry

To quantify these arguments in a simple analytical model we approximate such a pencil beam survey as a cylinder of length D, cross-sectional radius R, and calculate the clustering noise assuming a correlation function measured in bins of length L. The number of independent cells contributing to a given bin in the large scale correlation function is then $M = D/L$. If we assume that there is no large-scale clustering other than that of Eq.(1) truncated on scales over 30 h^{-1} Mpc, then the relative variance of the one-dimensional correlation function (the pair count N_p) along the line of sight can be written as

$$\epsilon_\xi^2 = \frac{\langle \Delta N_p^2 \rangle}{\langle N_p \rangle^2} = \frac{4}{M} \left(\frac{1}{n_o} + \xi_o \right) + \frac{1}{M} \left(\frac{1}{n_o} + \xi_o \right)^2 \approx 4 \left(\frac{1}{N_T} + \frac{\xi_o}{M} \right), \qquad (4)$$

where n_o is the mean number of galaxies in a cell, and $N_T = M n_o$ is the total number of galaxies observed. For larger radii the first term is the important one, but for small R the quadratic term dominates. This can be considered as a 1-D equivalent of Kaiser's (1986) ε_ξ. Here ξ_o is the mean of the small scale correlation function averaged over a cell of radius R and length L, well approximated by an expression depending only on the geometry, obtained from numerical integration:

$$\xi_o = 2.24 \, r_o^{1.8} L^{-0.75} R^{-1.05}. \qquad (5)$$

The optimum number of galaxies to be observed is given by balancing the two criteria: $N_{crit} \approx M/\xi_o$, since oversampling will not reduce ϵ_ξ considerably further. In practice, if there are only a few pencil beams, it may be advisable to oversample slightly, by a factor of three or so, to make evident, that the data are not affected by sampling, but for a large scale program in the future this may not be necessary.

Now we address the choice of the best geometry, which is further complicated by constraints imposed by the field of view of the telescope, and other considerations, like the detection of individual walls. Substituting the appropriate numerical values for the number density of galaxies and the geometry relevant for the pencil beams, and assuming a sampling rate f, we find that ϵ_ξ scales inversely with D,

$$\epsilon_\xi^2 = \frac{1}{M} F(R, f) \propto D^{-1} F(R, f) \qquad (6)$$

where $F(R, f)$ is the dimensionless noise per cell in the survey: the more independent cells there are, the smaller the total statistical noise becomes. For example, if the large scale topology is dominated by walls separated by $\lambda \approx 120$ h^{-1} Mpc, then the number of walls found by a pencil beam is $N_w = D/\lambda$. If the walls are clustered on scales $\ll D$, the signal to noise in the detection of a nonzero clustering amplitude is increasing with the depth of the survey in the usual manner

$$\Delta_\xi \propto N_w^{1/2} \propto D^{1/2} \qquad (7)$$

Such a detection can be compared with the null-hypothesis of no large scale correlations quantitatively, e.g. in order to reject the null hypothesis at some significance level, or to compare with the results of N-body simulations. Any pencil beam redshift survey should then be optimized for achieving this using the smallest possible amount of telescope time. The parameter to maximize is the ratio $\eta = \Delta_\xi / \epsilon_\xi$.

$$\eta = \frac{\Delta_\xi}{\epsilon_\xi} \propto D \ F(R, f)^{-1/2} \tag{8}$$

Assuming Euclidean geometry for simplicity at this stage, the number of galaxies and the total observing time depend on R and D as

$$\begin{aligned} N_T &\propto f R^2 D, \\ T_o &\propto f N_T D^2 \propto f R^2 D^3 \end{aligned} \tag{9}$$

We can express the effective depth that can be reached, given T_o and R, scaling as

$$D \propto T_o^{1/3} R^{-2/3} f^{-1/3} \tag{10}$$

Substituting this into Eq.(8) yields:

$$\eta \propto T_o^{1/3} f^{-1/3} R^{-2/3} F^{-1/2} \tag{11}$$

For detection at a given significance level, we can ask what amount of observing time is required for different radii. Expressing T_o as a function of R, at a fixed value of η

$$T_o \propto \eta^3 f R^2 F^{3/2}. \tag{12}$$

Results of a more complete numerical analysis, taking realistic values, and assuming a sampling rate close the complete $f = 1$ are presented in Figure 1. This curve has a well defined minimum at $R \approx r_o$, confirming our intuitive estimate very well. In a real situation, i.e. that our spectra are sky limited, and taking cosmological effects into account would further strengthen the point, that one should not go *arbitrarily* deep and narrow.

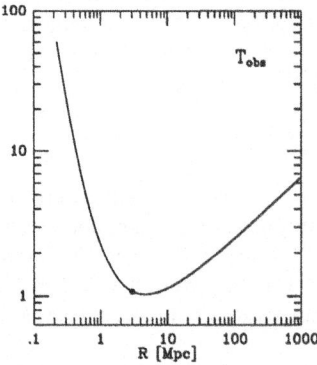 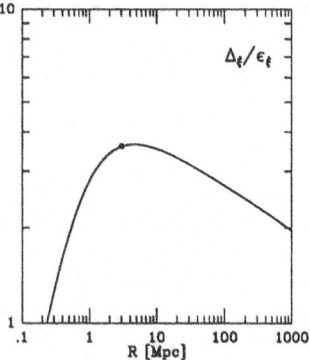

Figure 1. The dependence of the observing time T_o and the efficiency η on the radius of the cylindrical pencil beam, when detecting an existing signal in the large scale correlation function at a given significance. The small dot indicates the strategy of the current project, close to the optimal.

We have to consider another question: how frequently does one detect the individual walls. This problem has been recently addressed by deLapparent *et al* (1991), and Ramella *et al* (1991), and discussed by Geller (1990). They consider the 'Great Wall' data set, and place the wall at different redshifts. Since the wall does not have a uniform surface density, rather it is quite 'patchy', they infer that the BEKS pencil beams would typically detect more than 5 galaxies in 50 % of the walls in the redshift range of 0.1 - 0.4. This number is very sensitive on the diameter of the pencil beam, for a 3 h^{-1} Mpc diameter the probability is 20 %, while for 10 h^{-1} Mpc it is 90 %. They also note – entirely correctly – that there may be a bias in the surface densities from the pencil beams, since 'misses' are not counted, but detections are from the higher than average surface areas.

We can raise one additional point here: in these simulations the walls were assumed to be almost perpendicular to the line-of-sight, as the 'Great Wall' itself. If we consider a random inclination angle, the surface area increases as $1/\cos i$. The mean of this quantity is divergent, we have to truncate at the aspect ratio of the wall: $\ln(100/5) \approx 3$. Thus the typical diameter is increased by about $\sqrt{3} = 1.73$, from 6 to about 10 h^{-1} Mpc, making it more likely that the pencil beams would detect almost every 'Great Wall'-like structure.

Above a somewhat arbitrary threshold of 5 galaxies has been set. This can be further quantified: in order to detect a wall, the number of galaxies detected is compared to the small scale variance of the galaxy counts at that redshift. Consider cells of length $\Delta L \approx 5$ h^{-1} Mpc along the survey, equal to the approximate width of the walls (deLapparent *et al* 1986), and close to the resolution of our redshifts (300 km/s^{-1}). Those cells containing a wall should have a typical galaxy count N. The expected count n_o and its variance σ_N for the same cell are

$$N = f R^2 \pi \mu / \cos i$$
$$n_o = f n R^2 \pi \Delta L$$
$$\sigma_N^2 = n_o^2 \left(\frac{1}{n_o} + \xi_o \right) \approx n R^2 \pi \Delta L \xi_o \tag{13}$$

for small ΔL. If we set the detection threshold to $N > n_o + \alpha \sigma_N$, in the first order the detection is independent of f, assuming that the clustering noise dominates. For a realistic survey, with $\alpha = 3$ and our canonical numbers we get $f > 0.1$, and $N > 2.75$. The previously used $N = 5$ corresponds to a marginally better detection at the 3.3σ level. One can see now the origin of this quantity. If $N \approx 1$, there is little use for this argument, since we will never detect fractional galaxies, this is another reason to consider $N = 5$ as a very reasonable detection threshold. So generally, $f \ll 0.1$ seems to exclude the detection of the individual walls, leaving only little freedom in choosing our sampling by statistical considerations, unless we are ready to give up the individual detections.

To consider this problem in its completeness can be quite complicated, involving higher order statistics. If we wanted to set the optimal threshold for separating walls from small scale clumps, we should study the distribution P_N, the probability of finding N galaxies in a given cell. The null hypothesis of the small scale gravitational clustering can yield such a distribution, and this should be compared to the similar one derived from the cell counts within the 'Great Wall', as discussed in Ramella *et al* (1991). Then one should chose the threshold for the optimal separation of the two. However, the P_N distribution contains contributions from all N-point correlation functions.

4. Statistics of One-Dimensional Projections

As it is well known in statistics, one-dimensional subsets or projections of three-dimensional random fields have rather different statistical properties from the original one. If we have a density fluctuation field $\delta(\mathbf{x})$, and an observational selection function $W(\mathbf{x})$, we measure $\Delta(\mathbf{x}) = \delta(\mathbf{x})W(\mathbf{x})$, their product, excluding sampling noise, etc. If $\delta(\mathbf{x})$ is a random field with a known power spectrum $P_3(\mathbf{k})$, the power spectrum of the observed field Δ can be written using the convolution theorem:

$$\langle|\Delta(\mathbf{k'})|^2\rangle = \langle|\delta * w|^2\rangle = \frac{1}{(2\pi)^3}\int d^3\mathbf{k}\, P_3(\mathbf{k})|w(\mathbf{k'} - \mathbf{k})|^2 \tag{14}$$

where $w(\mathbf{k})$ is the window function, the Fourier transform of the selection function $W(\mathbf{x})$. In obtaining the above relation, we have only used the translational invariance of the ensemble average, requiring that

$$\langle\delta(\mathbf{k'})\delta^*(\mathbf{k})\rangle = (2\pi)^3\delta^{(3)}(\mathbf{k'} - \mathbf{k})\, P_3(\mathbf{k}) \tag{15}$$

where $\delta^{(3)}$ is the three dimensional Dirac-delta. For simplicity let us consider an extreme window, an infinitely thin pencil along the z-axis, extending to infinity. The window function in this case is $w(k_1, k_2, k_3) = 2\pi\delta^{(1)}(k_3)$. It is easy to show, that in this case the ensemble averaged one-dimensional power spectrum $P_1(k)$ becomes

$$P_1(k_3) = \frac{1}{(2\pi)^2}\int dk_1 dk_2\, P_3(k_1, k_2, k_3) \tag{16}$$

The integral describes a projection: in a given point k_3 along the survey axis in Fourier-space we project all power in the (k_1, k_2) plane, perpendicular to the axis, intersecting it at k_3, into a single value $P_1(k_3)$, a lot of power is aliased from small to large scales. Now we can consider various options for the three-dimensional power spectrum, from the most general case to the homogenous isotropic process. Until now, our treatment is fully general. If $\delta(\mathbf{x})$ is not a homogenous random process, like the cubic lattice discussed in §2, generally $P_3(\mathbf{k})$ can have sharp spikes, that can show up in $P_1(k_3)$ as well. On the other hand if $\delta(\mathbf{x})$ is a homogenous and isotropic random process, the power spectrum $P_3(\mathbf{k})$ will depend on $k = |\mathbf{k}|$ only, and $P_1(k)$ can be written as

$$P_1(k) = \frac{1}{2\pi}\int_k^\infty d\ell\, \ell\, P_3(\ell) = P_1(0) - \frac{1}{2\pi}\int_0^k d\ell\, \ell\, P_3(\ell). \tag{17}$$

In this latter case there is an other implication: the one dimensional power spectrum originating from a three dimensional homogenous isotropic random process (not necessarily Gaussian) is always monotonically decreasing. As the simplest example, plane waves with their wave vectors perpendicular to the line of sight will contribute to fluctuations in the DC level of the 1D power.

As one can see in the form of Eq.(17) if there is a large scale (low frequency) cutoff in P_3 at the wave number k_0, then P_1 is constant from 0 to k_0 : it is a 'white noise', with the level $P_1(0)$. The noise level can be physically interpreted as the three dimensional variance $\sigma_3^2 = \langle|\delta(\mathbf{x})|^2\rangle$ multiplied with the expectation value $\langle 1/\ell\rangle/\pi$, the inverse wave number where the maximum contribution to P_1 comes from

$$P_1(0) = \frac{1}{2\pi} \int_0^\infty d\ell\, \ell\, P_3(\ell) = \frac{1}{\pi} \langle \frac{1}{\ell} \rangle \left[\frac{1}{2\pi^2} \int_0^\infty d\ell\, \ell^2\, P_3(\ell) \right] \tag{18}$$

The values of $\Delta(k)$ at different k are not independent, they correlate due to the convolution with the window function:

$$\langle \Delta(\mathbf{k}'')\Delta^*(\mathbf{k}') \rangle = \frac{1}{(2\pi)^3} \int d^3\mathbf{k}\, P_3(k)w(\mathbf{k}'' - \mathbf{k})w^*(\mathbf{k}' - \mathbf{k}) \tag{19}$$

For realistic survey geometries we have to calculate the appropriate window function, and with a basic model for $P_3(k)$, like CDM or a minimal truncated power law correlation function we can predict the 1D power spectrum. If we approximate the survey as a cylinder of fixed radius R and effective depth D, there are two separable window functions, a radial and a transverse one. For more complicated geometries generally the transverse window function depends on the z coordinate, for a fixed solid angle the z dependence is relatively simple. There is a natural coherence scale in k-space due to the large D, roughly $k_* = 1/D$. The projection in k-space takes place then in disks of this thickness, and 3D power is projected only from within a transverse window of conjugate radius of approximately $2\pi/R$. For a cylindrical survey the transverse window function is independent of the z coordinate

$$w(\ell R) = 2\frac{J_1(\ell R)}{\ell R}, \tag{20}$$

where ℓ is the magnitude of the wave vector in the plane of the projection. The physical importance of these two length scales can be easily understood, the depth D will enter through the number of independent cells along the line of sight, while the transverse scale R determines the level of the small scale clustering noise entering the problem. Interestingly enough, if the survey geometry consists of multiple pencilbeams, sparsely sampling a larger surface area with a radius B, the important scale will be this larger B, rather than the small R (see §6 for more). Let us consider two such transverse windows (both two dimensional), the first describing the individual pencil beams with radius R, $W_R(\mathbf{x})$, and the wider one with the scale B, $W_B(\mathbf{x})$, describing the pattern of the sparsely distributed pencils. The total selection window becomes the convolution,

$$W(\mathbf{x}) = \int d^2\mathbf{x}'\, W_B(\mathbf{x}') W_R(\mathbf{x} - \mathbf{x}'). \tag{21}$$

Due to the convolution theorem, the window functions in Fourier space will be the products, thus $w(\mathbf{k}) = w_B(\mathbf{k}) w_R(\mathbf{k})$. The total width of $w(\mathbf{k})$ will determine how much small scale (large k) power will be included in the integrals. The window function narrower in k-space will dominate, thus the relevant scale is the wider radius B. If W_B consists of N points over the area size B, $\langle |w_B(\mathbf{k})|^2 \rangle$ will have a DC component of $1/N$, thus the correlation noise due to the width of a single pencil beam will be attenuated by this amount.

5. Application to the BEKS Survey

The advent of efficient multi-object spectrographs on large optical telescopes has considerably accelerated the progress towards completing redshift surveys of faint galaxies. The instrumentation used in such surveys typically sample 10-50 galaxies within fields of 10-40 arcmin diameter and thus 'pencil beam' geometries are produced which are characteristically different from strategies used to map the galaxy distribution locally. With 40' fields on 4m telescopes, the diameter of the survey at the median redshift of about $z \approx 0.3$ is $6~h^{-1}$ Mpc; such geometries are close to optimal for detecting wall-like topologies on scales comparable to those revealed in the CfA surveys (§2).

The deep surveys (Koo, Kron and Szalay 1986, Broadhurst, Ellis and Shanks 1988) were originally motivated by the need to determine the redshift distribution of magnitude-limited samples as a means of discriminating between various models explaining the excess numbers of faint counts (see Ellis 1991 for a review). However, since such surveys reach considerably deeper than earlier work, the redshift distributions have also been used to quantify the large scale distribution of galaxies (BEKS 1990).

BEKS provided evidence from a combined sample of galaxies in the North and South Galactic poles for structures on scales >100 h^{-1} Mpc with a provocative regularity. The BEKS survey consisted of two deep surveys spanning 2000 h^{-1} Mpc – the deepest so far – and two previous brighter surveys by others (Kirshner et al 1983, Peterson et al 1986), which together cover a volume well approximated by a cylinder of a constant comoving radius (Figure 2). The two Northern fields lie within the CfA slice, and the 'Great Wall' is readily detected. Surprisingly, however, at large radial distances most galaxies lie in a few discrete 'spikes' separated typically by 130 h^{-1} Mpc. The spread of spike separations is, in fact, surprisingly small $\simeq 23\%$. This is revealed using the 1-D pair counts (Figure 3) and the 1D power spectrum (Figure 4) which has a very sharp peak at the wave-number corresponding to 130 h^{-1} Mpc.

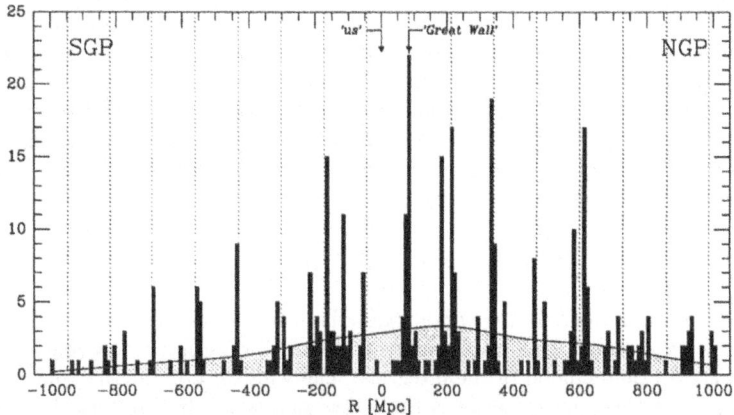

Figure 2. The combined redshift distribution of the four data sets, consisting of a deep and narrow (30 arc mins), and a bright and wide, survey at each of the Galactic Poles (South to the left, North to the right). The full selection function (smoothed from the data) is shown as a shaded curve. We used $\Omega = 1$ and $h = 1$ to convert redshifts to comoving distance. Both our position and that of the 'Great Wall' is indicated with small arrows.

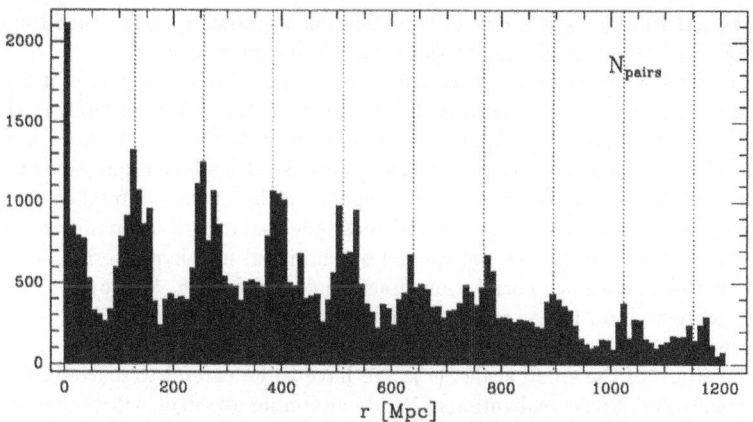

Figure 3. One-dimensional pair count correlation for all the redshift data available in the NGP-SGP axis. The dashed line indicates the multiples of 128 h^{-1} Mpc. The relative excess of the pair counts over the smooth background is at least 1.5 at the first peak. Note the anticorrelation between 50 and 100 h^{-1} Mpc.

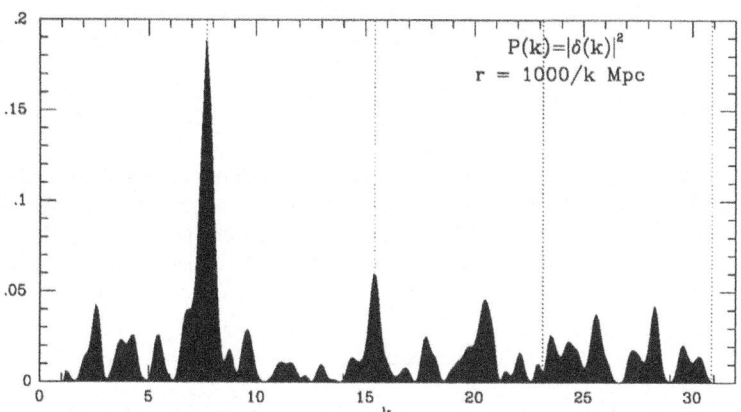

Figure 4. One-dimensional power spectrum of the BEKS sample. The dashed line corresponds to a separation of 128 h^{-1} Mpc, the best fit to the first two harmonics. The one-dimensional power spectrum is a non-trivial projection from three dimensions. The observed peak has a formal probability of 2.2×10^{-4}. The physical scale corresponding to the wave number k is $r(k) = 1000/k$ h^{-1} Mpc.

We can approximate our radial selection function roughly with a Gaussian, this multiplies the true galaxy distribution. In Fourier space the true power spectrum of the galaxy distribution is convolved with the square of the window function, also a Gaussian with an approximate conjugate width of $k_* = 0.001$ Mpc^{-1}. This is the natural line width of our power spectrum, the coherence length in k-space. For this reason we chose this as our units for the wave numbers. The width of the spectral feature at $k = 7.7$ is close to 1, hardly larger than the natural line width due to the selection window, implying that the intrinsic coherence length in r space is similar to the range of our observations!

The first question we have to ask, how likely is this just due to chance, is it an upward fluctuation, arising from a simple underlying null hypothesis? When estimating this probability, we have to be aware, that this is an *a posteriori* statistics, i.e. we have first chosen the highest peak in the observed power spectrum, and asked statistical questions later. The probability of finding such peaks in the power spectrum by chance is further complicated by the fact that the small scale correlations increase the variance of the number of galaxies. Thus before we proceed any further we specify our minimal null hypothesis:

- galaxies have the well-known correlation function $\xi(r) = (r/5\ h^{-1}\,\mathrm{Mpc})^{-1.8}$
- they are uncorrelated on larger scales ($> 30\ h^{-1}\,\mathrm{Mpc}$)

As we have seen above, this null hypothesis implies a white noise power spectrum on scales above the truncation of the small scale $\xi(r)$. We have to be careful to distinguish between ensemble averages and single realizations. While ensemble averages will be monotonically decreasing, or remain white noise, the individual realizations can have peaks in various places. Also, if our survey is not deep enough to be statistically fair, maybe the hypothesis of a homogenous isotropic process is not valid (e.g. we sample only the few nearest cells).

We denote our Fourier transform of the galaxy distribution as f_k, and the single realization of the 1D power spectrum from the data as A_k, calculated by treating each galaxy as a Dirac delta at its observed position r_n, neglecting redshift distortions,

$$f_k = \frac{1}{N} \sum_1^N e^{2\pi i k r_n} \tag{22}$$

$$A_k = |f_k|^2.$$

Under our null hypothesis, the power spectrum amplitudes have an exponential distribution, since the real and complex parts of the Fourier transform are distributed as a Gaussian. Even if parts of the signal in r-space are very non-Gaussian, the power spectrum amplitudes satisfy this property quite well. In this case the probability distribution is described by the noise level A_0:

$$dP(A_k) = \exp(-A_k/A_0)\frac{dA_k}{A_0} \tag{23}$$

All models for the small scale clustering or a more detailed influence of the survey geometry enter through this single number. One can have external and internal estimators of this quantity. All those estimators using a certain two point function are external, in the sense that little is known about the fine details of the two-point correlation function on scales much beyond $15\ h^{-1}$ Mpc, we only rely upon our assumptions.

The amplitude for this white noise is easily calculated analytically from our null hypothesis with the truncated $\xi(r)$, if we assume a simple cylindrical geometry. The noise level is the sum of the sampling and clustering noise:

$$A_0 = \frac{1}{N_T} + \frac{\xi_o}{M}, \tag{24}$$

where N_T is the total number of galaxies in the survey, ξ_o is the mean value of the small scale correlation function over a cell of length 30 h^{-1} Mpc along the pencil beam, and M

is the number of independent cells along the line of sight, as discussed in §2. For the BEKS data $N_T = 400$, and $M/\xi_o = 80$, depending only on the geometry. The expected level of white noise is thus:

$$A_0 = \frac{1}{400} + \frac{1}{80} = \frac{1}{67} = 0.015 \tag{25}$$

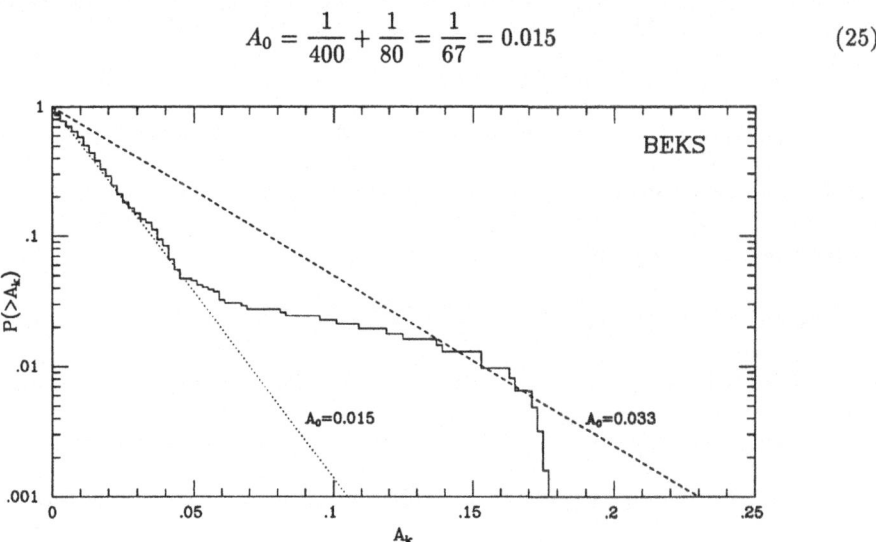

Figure 5. The cumulative distribution of the power spectrum amplitudes A_k from Fig.4. The shoulder is the effect of the large peak, while the small amplitude part shows the exponential behaviour. The lines show various levels of the white noise, 0.015 is the best fit to the low amplitude part.

There is a way to obtain this number independently of any hypotheses, using an internal estimator of the same quantity. We can calculate the cumulative distribution of the actual – numerically determined – power spectrum amplitudes, and fit an exponential to this distribution. The slope of the exponential at small amplitudes gives us the required noise level, which turns out to be

$$A_0 = 0.015, \tag{26}$$

in excellent agreement with the estimate in Eq. (25). Of course, our analytical model is somewhat oversimplified, a proper treatment of the window shapes coming from the radial and angular selection criteria yields a 25 % higher number. This small difference may arise from the assumption that $\xi(r)$ extends out to 30 h^{-1} Mpc, whereas empirical values of J_3, the integral of ξ, imply convergence by 15 h^{-1} Mpc; this would somewhat overestimate the strength of the small scale correlations. Also, a similar variance in the intrinsic estimator is expected from higher (4th) moments of the galaxy distribution.

Such estimates for the general noise level have to be confronted with the observed BEKS peak amplitude $P_1(7.7) = 0.177$. Since under our null hypothesis the amplitude of the power spectrum has an exponential distribution, we can model our a posteriori observation as determining the power spectrum in N_k independent bins, then picking the one with the largest amplitude. One may wonder, whether the formalism developed by Bardeen *et al* (1986) on peaks of random Gaussain fields should be used. However, since we pick the largest peak in the Fourier spectrum, it is also the maximum value of the observed spectrum. The cumulative distribution of the maximum of N samples drawn from an exponential distribution is well known (Kendall and Stuart 1977):

$$G(x|N) = 1 - \exp(-Ne^{-x}) \approx Ne^{-x}, \tag{27}$$

where $x = A_k/A_0 = 11.8$ is the relative height of the peak over the noise. In our case we set the truncation scale at 30 h^{-1} Mpc, corresponding to $k \approx 30$ in our units of the natural line width. This is equal to the number of independent k-bins over the range where the white noise behaviour is expected to hold. Our final probability of finding such a peak in the power spectrum becomes

$$G(11.8|30) = 2.2 \times 10^{-4}. \tag{28}$$

suggesting a statistically significant clustering of galaxies on very large scales. The estimate is not sensitive to the truncation of $\xi(r)$ at 30 h^{-1} Mpc. On the other hand, it is very sensitive on the noise level A_0 : if A_0 turned out to be a factor of 2 larger, 0.030, the corresponding probability would grow to 0.1. This very strong dependence is due to the appearence of A_0 in the double exponential of G. It is easy to see, how a slight mistake or a somewhat incorrect assumption in calculating the external estimator may lead to a very different probability than the one quoted here. This is the reason, why we strongly advocate the use of the intrinsically determined noise level only, and discard the use of the external estimations.

6. Discussion

As BEKS stressed in their paper, further data is urgently needed to address many questions posed by their striking result.
- How sharp is the detected 130 h^{-1} Mpc scale?
- Does the galaxy distribution show similar structures in other directions?
- Do the observed structures have a large transverse extent as would be expected if the large scale topology is cellular?

In order to get a better handle on the errors, and to determine which part of the data contributes most of the signal, we have successively excluded individual spikes in the redshift histogram, as well as whole catalogs, which are combined into our survey. After these eliminations the characteristic scale still shows up in every part of the data set, although the statistical significance decreases according to the expectations.

We have to be aware that the derived radial distances arise directly from the raw redshifts, no velocity corrections have been used. On the other hand, on the relevant scales (> 100 h^{-1} Mpc), only very substantial velocities can have any appreciable effect. Since the survey volume is still small enough to contain only about 0.2 Abell cluster per cone, it is rather unlikely that objects with cluster-like velocity dispersions are present. We know of one Zwicky cluster in one of the fields, and of course our pencil beam at the NGP passes near Coma.

When we convert from redshifts to comoving distance, the conversion is sensitive to the curvature of the universe. By taking various values of Ω one can shift the high redshift tail of the distribution relative to the near part. This is reflected in the pair correlations. Tests indicate, that to first order the sharpness of the peak in the Fourier spectrum is rather insensitive to our choice of Ω, although it may split up into a doublet, containing the same amount of power for very low values of Ω.

We have observations under way in several other directions. Preliminary analyses of these directions indicate, that those other fields by themselves are at least as clustered on >100 h^{-1} Mpc scales as the SA57 (NGP) field. Adding the SGP, the NGP and the two bright surveys effectively tripled the dynamical range of the survey, which resulted in a similar improvement in the signal-to-noise ratio of the correlation function. We are currently taking data in the complementary directions to these new surveys, expecting a similar trend in the S/N ratio.

In order to assess the transverse extent of the spikes detected in the data we can use the angular positions within the spikes. This already indicates that some of the spikes extend almost uniformly across the field. As discussed in §4, it is rather advantageous to use a sparsely distributed pattern over a 25-100 square degree area of 6-10 well sampled pencil beams of 0.3 square degrees each. With this we gain on several grounds, without much sacrifice:

- the effective radius grows to several degrees, lowering the correlation noise
- the wall detection is enhanced by the use of multiple beams
- it is well suited for the multiobject spectrographs
- it improves the detection of walls compared to the same number of galaxies from a uniformly sparse survey

Using the Nessie fiber spectrograph at the Kitt Peak 4m Mayall telescope we are in the process of obtaining redshifts (up to 70 each) in several small (40 arc min) pencil beams within 5 degrees of our well sampled areas. These transverse separations correspond to distances of about 30 h^{-1} Mpc. The cross-correlation of these sparse surveys with the well sampled central ones enables us to derive statistically meaningful information about the transverse extent of the walls, even if the individual surveys are undersampled. Already 160 redshifts were obtained around the SA57 field, in agreement with our expectations. More than 400 new redshifts were obtained in the SGP region recently, confirming the previously found Southern spikes over a 10×10 degree area rather dramatically.

We expect that within a year we will have at least four independent directions with over 200 galaxies each, around some of those we will have sparsely sampled with a few adjacent pencil beams, and altogether have a database of over 2000 faint galaxies total. This should be large enough to answer quite a few of the still outstanding questions.

Acknowledgements

We would like to acknowledge useful conversations with M. Geller and M. Ramella. We would like to thank R. Kron and J. Munn, for their permission to present the unpublished NGP faint data set, obtained in collaboration with them.

References

Bahcall, N. and Soneira, R. (1983) *Ap. J.*, **270**, 20.

Bardeen,J., Bond, J.R., Kaiser, N. and Szalay,A.S. (1986) *Ap. J.*, **304**, 15.

Blumenthal,G.R., Faber, S.M., Primack, J.R. and Rees,M.J. (1984) *Nature*, **311**, 517.

Broadhurst,T.J., Ellis, R.S., and Shanks, T. (1988) *M.N.R.A.S.*, **235**, 827.

Broadhurst,T.J., Ellis, R.S., Koo, D.C. and Szalay, A.S. (1990) *Nature*, **343**, 726.

Burstein, D., Davies, R.L., Dressler, A., Faber, S.M., Lynden-Bell, D., Terlevich, R.J. and Wegner, G. (1986) In *Galaxy Distances and Deviations from the Hubble Expansion*, ed. B. Madore and B. Tully, (Boston:Reidel), p.123

Chincarini, G.L., Giovanelli, R., Haynes, M.P. (1983) *Astron. Astrophys.*, **121**, 5.

deLapparent, V., Geller, M.J. and Huchra, J.P. (1986) *Ap. J.*, **302**, L1.

deLapparent, V., Geller, M.J. and Huchra, J.P. (1991) In preparation.

Geller, M.J. (1990) Private communication.

Giovanelli, R., Haynes, M.P. and Chincarini, G.L. (1986) *Ap. J.*, **300**, 77.

Kaiser, N. (1984) *Ap. J.*, **273**, L17.

Kaiser, N. (1986) *M.N.R.A.S.*, **219**, 785.

Kendall, M. and Stuart, A. (1977) *The Advanced Theory of Statistics*, (Macmillan Publishing: New York).

Kirshner, R.P., Oemler, A., Schecter, P.L. and Schectman, S.A. (1983) *Astron. J.*, **88**, 1285.

Koo, D.C., Kron, R.G. and Szalay, A.S. (1987) *Proc. of the 13th Texas Symposium on Rel. Astrophys.*, ed. M.P.Ulmer, (World Scientific) p. 227.

Koo, D.C., Kron, R.G., Munn, J. and Szalay, A.S. (1991) In preparation.

Lynden-Bell, D., Faber, S.M., Burstein, D., Davies, R.L., Dressler, A., Terlevich, R.J. and Wegner, G. (1988) *Ap. J.*, **326**, 19.

Park, C. (1990) *M.N.R.A.S.*, **242**, 59p.

Peebles, P.J.E. (1980) *The Large Scale Structure of the Universe*, (Princeton University Press).

Peebles, P.J.E. (1989) In *Fractals in Physics*, eds. A. Aharony and J. Feder (North Holland).

Peterson, B.A., Ellis, R.S., Bean, A., Efstathiou, G.P., Shanks, T. and Zou, Z-L. (1985) *M.N.R.A.S.*, **221**, 233.

Ramella, M., Geller, M.J. and Huchra, J. (1991) In preparation.

Rubin, V. Thonnard, N., Ford, W.K. and Roberts, M. (1976) *Astron. J.*, **81**, 719.

Szalay, A.S., Koo, D.C., Ellis, R.S. and Broadhurst, T.J., (1991) In preparation.

Voronoi, G. (1908) *Reine Angew. Math.*, **134**, 198.

White, S.D.M., Frenk, C.S., Davis, M. and Efstathiou, G. (1987) *Ap. J.*, **313**, 505.

DISTANCE TO THE COMA CLUSTER AND THE VALUE OF H₀

S. OKAMURA
Kiso Observatory
Institute of Astronomy, University of Tokyo
Mitake-mura, Kiso-gun, Nagano-ken, 397-01 Japan

M. FUKUGITA
Yukawa Institute for Theoretical Physics
Kyoto University, Kyoto, 606 Japan

1. Introduction

The Hubble constant H_0 is one of the most fundamental cosmological parameters. In spite of astronomers' efforts over 60 years, H_0 is still uncertain by a factor of two (H_0=50–100 km s^{-1}Mpc^{-1}) as shown in Fig. 1. A conspicuous feature seen in this figure is the shortness of error bars compared with the range of distribution. This may be ascribed to the fact that only the random errors were taken into account in most of the previous studies with no allowance made for probable systematic errors. Figure 1 demonstrates the need for better error estimates in the art of distance estimation.

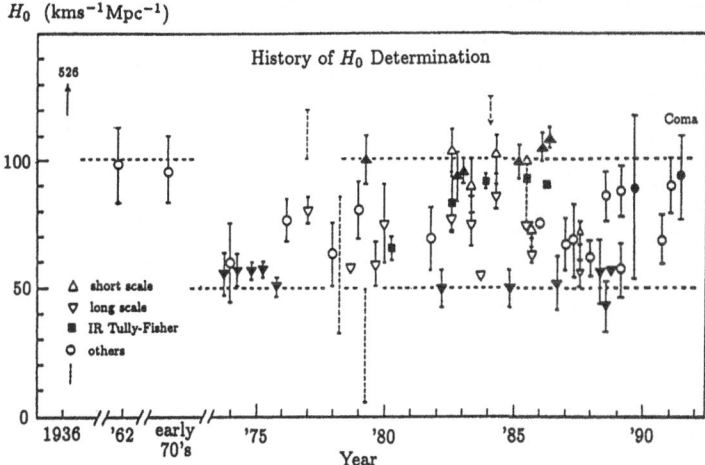

Fig. 1. Representative determinations of the Hubble constant. Different symbols show different methods and/or different zero points. Filled upward triangles come from de Vaucouleurs and coworkers (short scale), and filled downward triangles from Sandage, Tammann and their coworkers (long scale). Filled squares are from Aaronson and coworkers. The last but three data point with the large error bar indicates H_0 from the Virgo cluster as summarized by Huchra (1988). The point labelled as Coma represents our present result for the Coma cluster.

451

It has been widely admitted that the Virgo cluster is the fundamental first step to the determination of the cosmic distance scale, hence of H_0 as well. The Virgo cluster is the nearest cluster located at $v_H \sim 1300$ km s^{-1} (v_H is the Hubble recession velocity). The cluster is close enough to apply a variety of distance-determination methods, yet at least sufficiently far so that the Hubble flow is considerably larger than the peculiar velocity. The Virgo cluster, however, shows complex features in both spatial and velocity structures (e.g., Binggeli et al. 1987; Huchra 1985). The definition of the cluster center is controversial, and the radial velocity of galaxies which are supposed to lie near the center exhibits quite a wide distribution ranging from -350 km s^{-1}(e.g., NGC4406) to more than 2000 km s^{-1}. This causes substantial uncertainties not only in the infall velocity towards the center but also in the Hubble recession velocity to be used as an average value. The depth of the cluster is also substantial (van den Bergh et al. 1990; Burstein and Raychaudhury 1989) and there is a difficulty in identifying galaxies located close to the 'center' (Pierce and Tully 1988). This makes it controversial to use the Virgo cluster to directly extract H_0 or to use the average value of the distance for the cluster 'center' as a first step.

In this sense, the use of clusters beyond the Virgo located in the velocity range of $v_H \sim 3000$-10000 km s^{-1} are preferable to the determination of H_0. The limit of 10000 km s^{-1} is the current limit of dynamical observations. Clusters in this range, however, suffer from the paucity of both accurate photometric data and dynamical data, which are necessary to determine the parameters of empirical distance-indicator relations such as the Tully-Fisher or the $D_n - \sigma$ relation. Previous studies on the distance determination of clusters beyond the Virgo distance included only 10 galaxies or so per cluster (e.g., Aaronson et al. 1986; Bottinelli et al. 1987; Kraan-Korteweg et al. 1988: KCT). Such small samples have left the suspicion that the resulting distance may suffer from a large error due to a sample incompleteness bias (e.g., Teerikorpi 1987; in particular KCT). The magnitude of the error depends upon the intrinsic dispersion around the distance-indicator relation, which is also controversial (cf. Pierce and Tully 1988 versus KCT).

We report here the determination of the Coma cluster distance using the B-band Tully-Fisher relation with the sample whose sample incompleteness is under our control. Surface photometry was carried out to estimate the total magnitude unambiguously. We circumvent the Virgo cluster by combining the Coma galaxies directly with the local calibrators. Coma cluster is one of the best clusters for the determination of H_0, since it is located as far as at $v_H \sim 7000$ km s^{-1} and in the direction perpendicular to the alleged large-scale streaming motion (Dressler et al. 1987a).

Our analysis resulted in $H_0 = 92 \pm 16$ km s^{-1}Mpc^{-1}(the last data point in Fig. 1). The error estimate includes both random and systematic errors. We believe that the present study is one of the most extensive Tully-Fisher studies to date of a cluster beyond the Virgo distance.

2. Key Points of the Present Study

Since the present study will be published elsewhere in detail (Fukugita et al. 1991), we only stress the key points here.

2.1. LARGE SAMPLE

The present sample consists of all spiral galaxies for which HI line width data are available in a circle of a 4° radius centered on the Coma cluster. The total sample consists of 48 galaxies including 18 for which new HI line widths were obtained with the 305m Arecibo telescope by Williams and Rood (1989). Selection criteria on the cluster membership and on the inclination angle left 30 galaxies for the Tully-Fisher analysis. The size of this final sample is more than twice as large as that used in previous studies. (Bottinelli et al. (1987) used 24 galaxies but they imposed much looser selection criteria that made their result quite inaccurate.) The average recession velocity of our sample is 6926 km s^{-1}, in good agreement with the velocity given by other authors (e.g., 6925 km s^{-1} is obtained from the comprehensive catalogue of Kent and Gunn (1982)), indicating that the sampling is not too strongly biased towards fore or background.

2.2. ACCURATE SURFACE PHOTOMETRY

The photometric data used in most previous Tully-Fisher studies in the B band relied on very limited number of sources. The total magnitude B_T were essentially taken from aperture photometry by Bothun et al. (1985). They adopted a common growth curve for all galaxies in order to extrapolate their aperture magnitudes to B_T. For most of the Coma galaxies, a magnitude using only a single aperture was used to estimate B_T. The axial ratio of a galaxy was generally computed using the major and the minor axis lengths given in the visual estimate by Nilson (1973: UGC).

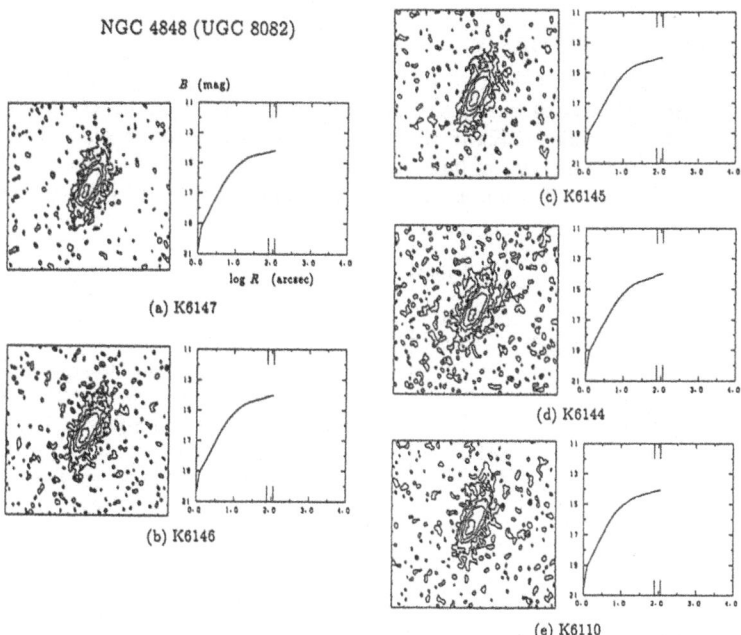

Fig. 2. Surface photometry using the Kiso Schmidt plates. NGC4848 was included in all the five plates. Left panels show isophotes and the right panels show growth curves. The isophote interval is 1 mag with the outermost 25 mag arcsec^{-2}. Note the good consistency among different plates.

In order to improve the photometric data, we took five 6° × 6° Schmidt plates in the B band with sufficient overlaps for the 10° × 10° region centered on the Coma cluster using the 105cm Schmidt telescope at the Kiso Observatory. Detailed surface photometry was carried out for all the 48 galaxies as well as some 10 calibrating galaxies. Internal accuracy of B_T was estimted to be 0.04 mag (1σ dispersion) in random errors and 0.08 mag (1σ dispersion) in the zero point on the basis of the data of the same galaxy exposed on the different plates (*cf.* Fig. 2). External accuracy was investigated by comparing the aperture magnitudes of Bothun *et al.* (1985) and Cornell *et al.* (1987) with the aperture magnitudes from our photometry for the same apertures. A systematic zero point difference of 0.13 mag was found with our magnitude brighter. All of the possible internal and external errors are included in the error estimate.

2.3. SAMPLE INCOMPLETENESS

The sample incompleteness was estimated as a function of the magnitude by comparing the magnitude histogram of the present sample with that of galaxies listed in Zwicky *et al.* (1961-68: CGCG) and in UGC in the circle of a 4° radius centered on NGC 4889. CGCG is claimed to be nominally complete down to 15.5 mag while UGC to 14.5 mag. The resulting incompleteness function $f(m)$ was used to estimate the bias quantitatively. The correction for this bias to the nominal distance modulus was estimated to be $\sim$ +0.3 mag.

2.4. LOCAL CALIBRATORS

The choice of local calibrators and their distances used to be one of the critical factors in the Tully-Fisher studies made in the past. For example, Aaronson and Mould (1983) found the zero point difference of 0.65 mag between the RSA zero point and the value by de Vaucouleurs. By now this uncertainty was essentially removed by a significant amount

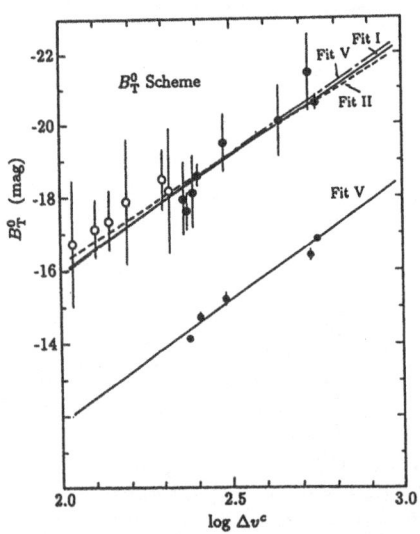

Fig. 3. Local galaxy calibration of the B-band Tully-Fisher relation for the B_T^0 scheme. Fit I (solid line; based on 8 filled circles) and Fit II (broken line; based on all 14 data points) are based on the compilation by Tammann (1987), and Fit V is based on the new Cepheid distances by Freedman (1990). Fit V is shown separately for clarity shifted by 4 mag in the ordinate and only the regression line (dot-dashed line) is superposed on Fits I and II for the sake of comparison.

of work on Cepheid variables for local calibrators made by Madore, McAlary, Welch, Visvanathan, and more recently, by Freedman and their collaborators. We confirmed that the compilation given by Tammann (1987) is basically correct (except for the distance to M81).

In our work we primarily used the Tammann's compilation but attached weight to local calibrators with Cepheid distances (Fig. 3). The calibration was confirmed by more systematic new Cepheid observations by Freedman (1990), and also by new methods using the planetary nebula luminosity function (Ciardullo et al. 1989; Jacoby et al. 1989, 1990). We do not expect an uncertainty more than ±0.1 mag in the local distance calibration.

2.5. DEPENDENCE ON CORRECTION PRESCRIPTIONS

It has often been stated that different correction prescriptions would lead to different answers. There are several prescriptions for the correction of observed total magnitude for the Galactic and the internal absorptions to estimate the intrinsic value. One of them is the B_T^0 scheme by de Vaucouleurs et al. (1976: RC 2) and the other extreme is the $B_T^{0,i}$ scheme by Sandage and Tammann (1981: RSA). We derived the distance to the Coma cluster based on six different combinations, i.e., two different prescriptions for the absorption correction as well as three different choices of local calibrators and their distances. We found that most uncertainties cancel between the cluster galaxies and the local calibrators. The maximum difference in the resulting distance modulus of the Coma cluster was found to be 0.11 mag.

2.6. ERROR ESTIMATE

We estimated carefully possible errors which could creep in many steps to determine the parameters of the Tully-Fisher relation. The errors were divided into systematic and random errors. Most of the systematic errors arising from the correction prescriptions cancel and have little effects on the final distance modulus as noted above (residual 0.06 mag). The major systematic errors that we cannot control are those due to the absolute zero point of photometry (+0.13 mag) and those due to the local calibrators (±0.08 mag).

Systematic errors were summed up and the net effect was found to be ∼0.2 mag (+0.24/−0.13 mag). Random errors were added in quadrature and the sum was 0.2 mag. Our error estimate is then ∼ ±0.4 mag in the distance modulus, including both systematic and random errors.

3. Results and Discussion

The nominal distance modulus that we obtained was $(m - M)_0 = 34.07$ (B_T^0 scheme) and 34.16 ($B_T^{0,i}$ scheme). After correcting for the effect of the sample incompleteness bias, we have 34.3 mag and 34.5 mag, respectively. Our best estimate is then

$$(m - M)_0 = 34.47 \ (+0.44, -0.33) \ \text{mag}, \qquad \Delta = 78 \ (+18, -11) \ \text{Mpc}.$$

The radial velocity of the Coma cluster with respect to the centroid of the Local

Group was computed to be $v_0 = 6925$ kms^{-1} (dispersion$=933$ km s^{-1}) as the mean of the observed velocities of 314 member galaxies within the $4°$ circle on the basis of the compilation by Kent and Gunn (1982). After correcting this value for the assumed Virgo infall velocity of 300 ± 100km s^{-1}, we have

$$v_{\rm H} = 7212 \pm 100 \quad \text{kms}^{-1}$$

for the Hubble recession velocity of the Coma cluster. This leads to the Hubble constant

$$H_0 = v_{\rm H}/\Delta = 92 \pm 16 \quad \text{kms}^{-1}\text{Mpc}^{-1},$$

which is shown in Fig. 1.

Table 1. The Coma distance modulus (μ_0) and the difference in the distance moduli between the Coma and the Virgo clusters ($\Delta\mu_0$)

Method	μ_0 (mag)	$\Delta\mu_0$ (mag)	ref.
qualified indicators			
TF(H-band)	34.51 ± 0.10	3.69 ± 0.12:	Aaronson et al. (1986)
TF(B-band)	33.8 ± 0.9?[1]		Bottinelli et al. (1987)
TF(H-band)	35.70		KCT
TF(B-band)	35.50		idem
TF($B+H$)		3.70 ± 0.17	Giraud (1986)
TF(B-band)	$34.47^{+0.44}_{-0.33}$		this work
$D_n - \sigma$		3.65	Dressler et al. (1987b)
	35.29		calib.: Dressler (1987)
	34.74		calib.: Fukugita & Hogan (1991)
SNe Ia(E)		3.89 ± 0.21	Tammann (1978)
SNe Ia (all types)		3.75 ± 0.20	Capaccioli et al. (1990)
	35.05 ± 0.45		calib.: idem
	34.58 ± 0.31		calib.: Fukugita & Hogan (1991)
SNe Ia (E+S0)		3.60 ± 0.30	Capaccioli et al. (1990)
	34.90 ± 0.45		calib.: idem
	34.46 ± 0.32		calib.: Fukugita & Hogan (1991)
other indicators			
$L - \sigma$		3.76 ± 0.12	Lucey (1986)
CM relation $u - V$		3.89 ± 0.20	Visvanathan & Sandage (1977)
$u - V$		3.5 ± 0.2	Aaronson et al. (1981)
$u - K$		3.0 ± 0.2	idem
$U - B$		3.66 ± 0.14	Sandage (1972)
$V - K$		3.66 ± 0.35	Persson et al. (1979)
$V - K$		2.6 ± 0.3	Aaronson et al. (1981)
Mg$_2$ index		3.87	Dressler (1984)
Mg$_2$ index		3.99	Dressler et al. (1987b)
Brightest cluster member		3.37 ± 0.40	Sandage & Hardy (1973)
D-μ_V diagram		3.2 ± 0.2	Kodaira et al. (1983)
Nuclear mag. of E's		4.2	Weedman (1976)
Universal M/L for E's		3.80 ± 0.06	Vader (1986)

1) computed from H_0 values for Coma galaxies given in their Table 2.

We expect that the error induced by the large-scale peculiar velocity field is not significant. An analysis of the velocity field was carried out by Bertschinger *et al.* (1990). The Coma cluster is just outside of the region that they analyzed, but their analysis indicates a quiet velocity field ($\sim$–200 km s^{-1}) near the Coma cluster. This may increase H_0 only by 3%.

Finally, we should perhaps comment on the distances obtained by previous authors (Table 1), in particular, on those by Sandage, Tammann and collaborators, which show the largest discrepancy with the present estimate. Let us take the work of KCT who obtained with the use of of the B-band Tully-Fisher method $(m - M)_0$=35.5, 1 mag larger than ours. The key is their '2σ upper envelope fit', where they assumed a large intrinsic scatter ($\sigma \sim$0.7 mag) for the Tully-Fisher relation and further presumed that the employed galaxies are located just below the 2σ line due to the sample incompleteness (Fig. 4a). In our work, however, we have explicitly demonstarated that the incompleteness bias cannot be that strong. We extracted a subsample consisting of 11 of 13 bright galaxies used by Aaronson *et al.* (1986) and by KCT. This subsample gave the distance modulus which is different by only less than 0.07 mag from that for the total sample, for which the sample incompleteness bias is known (Fig. 4b). In our analysis we also found that the dispersion is unlikely to be so large; we suspect that the apparent large dispersion that KCT obtained from their Virgo sample is largely due to the contamination from galaxies away from the Virgo center as mentioned in section 1 (see also Burstein and Raychaudhury 1989; van den Bergh *et al.* 1990).

The distance by Aaronson *et al.* (1986) agrees nominally with our value. We suspect that the correction for the sample incompleteness bias would increase their value slightly. The distance by Bottinelli *et al.* (1987) has a large error because of loose sample selection criteria as mentioned in section 2.1 and also of errors in photometry data.

Most of distance estimates to the Coma cluster to date are based on the determination of the difference of the distance moduli between Virgo and Coma (Table 1). The full

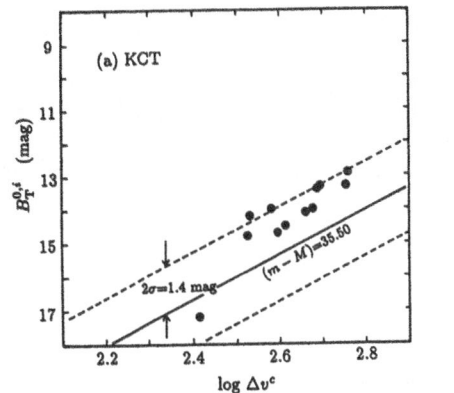
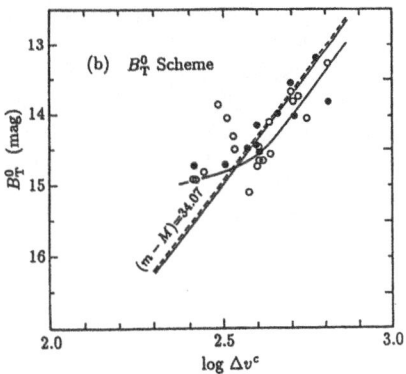

Fig. 4. (a) A diagram showing the '2σ upper envelope fit' of KCT for Coma cluster. (b) Tully-Fisher relation for a subsample. The 11 of 13 galaxies used in Aaronson *et al.* (1986) and in KCT are denoted by filled circles. (Note that the scales of the two figures are different.) The solid line is the best fit to the subsample and the broken line to the total sample. Note only a slight difference of < 0.1 mag. The curve indicates the bias (+0.27 mag) due to the sample incompleteness.

distance to the Coma then depends upon what value is used for the distance to the Virgo 'center', and such determinations suffer from the problem stressed in section 1.

We would like to thank H.J.Rood, K.Tarusawa, and B.Williams for collaboration on this work. We are also indebted to G.D.Bothun and J.Ostriker for useful comments.

References

Aaronson, M., and Mould, J. 1983, *Ap. J.*, **265**, 1.

Aaronson, M., Bothun, G., Mould, J., Huchra, J., Schommer, R.A., and Cornell, M.E. 1986, *Ap. J.*, **302**, 536.

Aaronson, M., Persson, S.E., and Frogel, J.A. 1981, *Ap. J.*, **245**, 18.

Bertschinger, E., Dekel, A., Faber, S.M., Dressler, A., and Burstein, D. 1990, *Ap. J.*, **364**, 370.

Binggeli, B., Tammann, G.A., and Sandage, A. 1987, *A. J.*, **94**, 251.

Bothun, G.D., Aaronson, M., Schommer, B., Mould, J., Huchra, J., and Sullivan III, W.T. 1985, *Ap. J. Suppl.*, **57**, 423.

Bottinelli, L., Fouque, P., Gouguenheim, L., Paturel, G., and Teerikorpi, P. 1987, *Astr. Ap.*, **181**, 1.

Burstein, D., and Raychaudhury, S. 1989, *Ap. J.*, **343**, 18.

Capaccioli, M., Cappellaro, E., Della Valle, M., D'Onofrio, M., Rosina, L., and Turatto, M. 1990, *Ap. J.*, **350**, 110.

Ciardullo, R., Jacoby, G.H., Ford, H.C., and Neill, J.D. 1989, *Ap. J.*, **339**, 53.

Cornell, M.E., Aaronson, M., Bothun, G., and Mould, J. 1987, *Ap. J. Suppl.*, **64**, 507.

de Vaucouleurs, G., de Vaucouleurs, A., and Corwin, H.G. 1976, *Second Reference Catalogue of Bright Galaxies* (Austin: Univ. Texas Press) (RC2)

Dressler, A. 1984, *Ap. J.*, **281**, 512.

Dressler, A. 1987, *Ap. J.*, **317**, 1.

Dressler, A., Faber, S.M., Burstein, D., Davies, R.L., Lynden-Bell, D., Terlevich, R.J., and Wegner, G. 1987a, *Ap. J. (Letters)*, **313**, L37.

Dressler, A., Lynden-Bell, D., Burstein, D., Davies, R., Faber, S.M., Terlevich, R., and Wegner, G. 1987b, *Ap. J.*, **313**, 42.

Freedman, W.L. 1990, *Ap. J. (Letters)*, **355**, L35.

Fukugita, M., and Hogan, C.J. 1991, *Ap. J. (Letters)*, (*in press*)

Fukugita, M., Okamura, S., Tarusawa, K., Rood, H.J., and Williams, B. 1991, *Ap. J.* (in press)

Giraud, E. 1986, *Astr. Ap.*, **164**, 17.

Huchra, J.P. 1985, in *The Virgo Cluster of Galaxies*, eds. O.-G.Richter and B.Binggeli (Garching: ESO), p.181.

Huchra, J.P. 1988, in *The Extragalactic Distance Scale*, eds. S. van den Bergh and C.J.Pritchet (Provo: Brigham Young Univ.), p.257.

Jacoby, G.H., Ciardullo, R., Ford, H.C., and Booth, J. 1989, *Ap. J.*, **344**, 704.

Jacoby, G.H., Walker, A., and Ciardullo, R. 1990, *Ap. J.*, **365**, 471.

Kent, S.M., and Gunn, J.E. 1982, *A. J.*, **87**, 945.

Kodaira, K., Okamura, S., and Watanabe, M. 1983, *Ap. J. (Letters)*, **274**, L49.

Kraan-Korteweg, R.C., Cameron, L.M., and Tamann, G.A. 1988, *Ap. J.*, **331**, 620. (KCT)

Lucey, J.R. 1986, *M. N. R. A. S.*, **222**, 417.

Nilson, P. 1973, *Uppsala Genral Catalogue of Galaxies*, Acta Uppsala Univ., Ser. V:A, Vol.1 (UGC)

Persson, S.E, Frogel, J.A., and Aaronson, M. 1979, *Ap. J. Suppl.*, **39**, 61.

Pierce, M.J., and Tully, R.B. 1988, *Ap. J.*, **330**, 579.

Sandage, A. 1972, *Ap. J.*, **176**, 21.

Sandage, A., and Hardy, E. 1973, *Ap. J.*, **183**, 743.

Sandage, A., and Tamann, G.A. 1981, *A Revised Shapley-Ames Catalog of Bright Galaxies*, (Washington: Carnegie Institution) (RSA)

Tammann, G.A. 1978, *Mem. Soc. Astron. Ital.*, **49**, 315.

Tammann, G.A. 1987, in *Observational Cosmology*, IAU Symp. No. 124, eds. A.Hewitt *et al.* (Dordrecht: Reidel), p.151.

Teerikorpi, P. 1987, *Astr. Ap.*, **173**, 39.

Vader, J.P. 1986, *Ap. J.*, **306**, 390.

van den Bergh, S., Pierce, M.J., and Tully, R.B. 1990, *Ap. J.*, **359**, 4.

Visvanathan, N., and Sandage, A. 1977, *Ap. J.*, **216**, 214.

Weedman, D.W. 1976, *Ap. J.*, **203**, 6.

Williams, B., and Rood, H.J. 1989, *private communication*

Zwicky, F., Herzog, E., Wild, P. Karpowicz, M., and Kowal, C.T. 1961-1968, *Catalogue of Galaxies and Clusters of Galaxies* (Zurich: L.Speich) (CGCG)

COSMOLOGICAL IMPLICATIONS OF HI ABSORPTION SYSTEMS

S. IKEUCHI
National Astronomical Observatory
Theoretical Astrophysics Division
Mitaka, Tokyo 181
Japan

ABSTRACT: As a unified model for HI absorption systems of quasars such as the Lyman α forest, Lyman limit systems and damped Lyman α line systems, we examine the evolution of intergalactic clouds confined by the gravity of cold dark matter, the so called minihalos. For reproducing three independent observed results, the HI column density distribution, the number density evolution and the continuum depression of high redshift quasars, the minihalos have to expand due to the increase of diffuse UV flux with decreasing redshift. From this result, we discuss the cosmological implications of absorption systems of quasars in relation to galaxy formation.

1. Observations

As is well known (Blades *et al.* 1988), many absorption line systems havde been detected in the continua of quasars (QSOs). Here, we focus to the HI absorption systems which are divided to three kinds in the characteristics of absorption profiles. They are (1) the Lyman α forest with HI column density $N_{HI} = 10^{13} - 10^{17} \mathrm{cm}^{-2}$, numerously observed as sharp, narrow lines at the blueside of Lyman α emission lines of QSOs, (2) the Lyman limit systems (LLS) with $N_{HI} = 10^{17} - 10^{20} \mathrm{cm}^{-2}$, observed as continuum absorptions at the Lyman edges, and (3) the damped Lyman α line systems with $N_{HI} > 10^{20} \mathrm{cm}^{-2}$, observed as broad saturated lines. All of these absorption systems have the redshifts z_a far less than the emission redshifts z_e of QSOs. Therefore, these are thought to be intervening systems not associated with QSOs. Since LLS and damped Lyman α systems are frequently associated with metallic line systems, it is highly probable that the star formation has already occurred in these systems.

For these HI absorption systems, several important observational results are reported, which suggest their origin and evolutionary features.

(I) HI column density dsitribution : As for three different HI absorption systems in the range $10^{14} \sim 10^{22} \mathrm{cm}^{-2}$, the HI column density distribution which is the observed number of absorbers per unit HI column density, $d\mathcal{N}/dN_{HI}$, normalized at an appropriate redshift is expressed as a power law form of a single index β, *i.e.*, $d\mathcal{N}/dN_{HI} \propto N_{HI}^{-\beta}$ with $\beta = 1.6 \sim 1.8$ (Tytler 1987, Sargent *et al.* 1989). Bechtold (1987) claimed that the power index at $N_{HI} > 10^{18} \mathrm{cm}^{-2}$ is as small as $\beta \sim 0.8$. Moreover, Wolfe (1990) reported that the number of damped Lyman α lines with $N_{HI} > 10^{21} \mathrm{cm}^{-2}$ increases by a factor of 3 with $\beta \simeq 1.6$.

This simple distribution law suggests that these three absorption systems may be common in origin.

(II) Continuum depression of high z QSOs :

The continuum depression at the blueside of Lyman α emission line is defined as

$$D = 1 - \frac{f(\text{obs})}{f(\text{cont})}, \tag{1}$$

where $f(\text{obs})$ and $f(\text{cont})$ are, respectively, the observed intensity and the extrapolated intensity from the redside of Lyman α emission line. Schneider et al. (1989) indicated that this continuum depression increases rapidly at $z > 4$. This may occur due to the overlapped absorptions of HI systems.

(III) Number density evolution :

The observed number of absorption systems per unit redshift, $d\mathcal{N}/dz$, is simply approximated by a power law form of (1+z) as $d\mathcal{N}/dz \propto (1+z)^{\gamma}, i.e.,$

Lyman α forest	$\gamma = 2.17 \pm 0.14,$
LLS	$\gamma = 0.59 \pm 0.8.$

For the LLS, Lanzetta (1990) indicated that they evolve rapidly, $\gamma = 5.7 \pm 1.9$, at $z > 2.5$. Moreover, if we sum the number of LLS and CIV absorption systems the power $\gamma \simeq 0$ for $z > 2.0$. In the standard Friedman universe, the power index γ should be in the range $0.5 \sim 1.0$ if the number density in the comoving coordinate is conserved. This means the rapid evolution of Lyman α forest and LLS.

In the present paper, we attempt to reproduce the above three observational results by means of intergalactic clouds confined by the gravity of cold dark matter(CDM), the so called minihalos. We consider that three kinds of absorption systems correspond to different places of the same minihalo (section 2) and that the evolution of minihalos are driven by the time variation of diffuse UV flux (section 3). All the above three observational results can be reproduced if the minhalos are expanding at $z > 2$. The implications of this result are discussed in section 4.

2. Mninihalo Model

The original idea of minihalos has been proposed by Rees (1986) and Ikeuchi (1986) based upon the CDM cosmogony. The intergalactic cloud with gas mass $10^6 \sim 10^{10} M_\odot$ confined by the CDM gravity is irradiated by the diffuse UV flux $J(z)$. The parameters characterizing the minhalos are $C = \rho_b(0)/\rho_d(0)$ (the central density ratio of baryons to CDM), $X = \sigma_d^2/c_s^2$ (the ratio of velocity dispersion of CDM to gas sound velocity), and $D = \rho_d(0)$ (the central CDM density normalized appropriately). As is shown by Ikeuchi et al. (1988), the minhalos are stable for $C \leq 10$ and $X \geq 0.8$. Depending upon the intensity of diffuse UV flux, the gas temperature and density of clouds are $T_b \sim (2-4) \times 10^4 K$ and $n_b \simeq 10^{-4} \sim 10^{-1} cm^{-3}$, and the HI column density ranges 10^{10} to 10^{18} cm^{-2} corresponding to the impact parameter $p \sim R_b$ to $p \sim 0$. Therefore, one static minhalo has all corresponding HI column density regions from Lyman α forest to LLS.

The noticeable characteristic of equilibrium minihalos is that the gas density distribution is almost isothermal, $n_b(r) \propto r^{-n}$ with $n \simeq 2$, irrespective of C when $X \leq 1$. This means that the HI column density depends upon the impact parameter, p, as $N_{HI} \propto$

$p^{-n-1} = p^{-3}$ and the expected area of N_{HI} is expressed as $p\,dp \propto N_{HI}^{-(n+3)/(n+1)}dN_{HI} = N_{HI}^{-5/3}dN_{HI}$. Since this area is proportional to the expected number, we obtain $d\mathcal{N}/dN_{HI} \propto N_{HI}^{-(n+3)/(n+1)} = N_{HI}^{-5/3}$. Therefore, the HI column density distribution from the Lyman α forest to LLS is naturally reproduced as $\beta = 5/3 \sim 1.7$.

We calculate the structure and evolution of minihalos when the diffuse UV flux changes as $J(z) = J(0)(1+z)^\alpha$. For each epoch we calculate $n_b(r)$ and $T_b(r)$ from which we obtain $n_{HI}(r)$ and $N_{HI}(p)$. Then, we have

$$\frac{d^2\mathcal{N}}{dz\,dN_{HI}} = \int dM \frac{c\pi}{H_0} n_A(0)(\frac{M}{M_*})^{-\delta} \frac{p\,dp}{dN_{HI}}(1+z)^{1-q_0}, \tag{2}$$

where $n_A(0), H_0$, and q_0 are the present space density of minihalos, the Hubble constant and deceleration parameter. We assume the mass function of them to be $n_A(z, M) = n_A(0)(1+z)^3(M/M_*)^{-\delta}$. From equation (2), we easily obtain

$$\frac{d^2\mathcal{N}}{dz\,dN_{HI}} \propto N_{HI}^{-\beta}(1+z)^\gamma, \tag{3}$$

for each epoch. This should be compared with the observations of HI column density distribution and number density evolution.

The continuum depression is simply expressed

$$D(z) = 1 - \exp(-\tau(z)), \tag{4}$$

where the optical depth due to HI components of minihalos is calculated as

$$\tau(z) = \int_{N_1}^{N_2} \frac{d^2\mathcal{N}}{dz\,dN_{HI}} N_{HI}\,dN_{HI}/N_{HI,crit}, \tag{5}$$

where N_1 and N_2 are the minimum and maximum column density of minihalos and $N_{HI,crit}$ is the critical HI column density for $\tau = 1$. By inserting equation (2) into (5) we can calculate the z-dependence of continuum depression.

As is seen in the above, we follow the evolution of minihalos with a set of parameters C, X and D when the diffuse UV flux changes with z. For the evolution law of $J(z)$ at $z > 2$, there is no definite answer. If $\alpha > 0$ for $J(z) = J(0)(1+z)^\alpha$, the UV flux decreases with decreasing z and the gas cloud contracts because the temperature decreases. If $\alpha < 0$, the gas cloud expands. Here, we examine two cases and compare the results with observations.

3. Evolution of Minihalos

3.1 CONTRACTING MINIHALOS

In the fist place, we examine the case of
$$J(z) = 10^{-21}(1 + z/3.5)^\alpha \text{ erg cm}^{-2} \text{ s}^{-1} \text{ str}^{-1} \text{ Hz}^{-1},$$

$$\begin{aligned} \alpha &= 1.0 \quad \text{at} \quad z > 2, \\ \alpha &= 4.0 \quad \text{at} \quad z < 2. \end{aligned} \tag{6}$$

This case nearly corresponds to the result by Bajtlik *et al* (1988) in consideration of the proximity effect of Lyman α forest.

The initial condition is set at $z_i = 5$ in a hydrostatic equilibrium. With decreasing z, the gas temperature decreases which leads to the contraction of a cloud. As a result, the gas density increases and the recombination of ionized gas makes the increase of HI. Therefore, the UV flux is absorbed in the outer layer of a cloud, self-schielding. Meanwhile, the neutral core is formed because of complete attenuation of UV flux. The HI column density of this neutral core increases from $N_{HI} \sim 10^{17.5} \mathrm{cm}^{-2}$ ($\tau = 1$) to $10^{22} \mathrm{cm}^{-2}$, which corresponds to the damped Lyman α line systems. Finally, the neutral core collapses.

Along this evolution, the HI column density distribution approximately follows a power law form and well reproduces the observational result as seen in Figure 1.

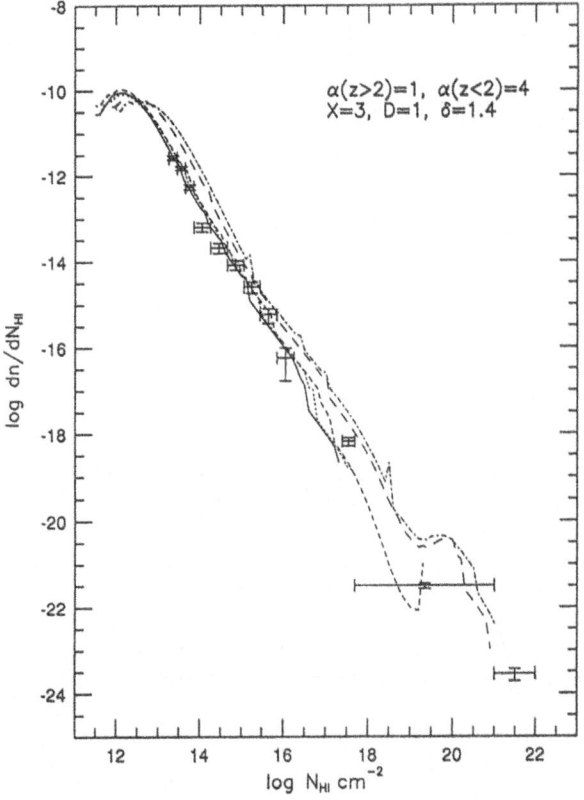

Figure 1. The calculated and observed HI column density distributions for a contracting minihalo. The observational results are taken from Sargent *et al.* (1989). Each line corresponds to the different epoch.

However, the continuum depression and number density evolution show contrary behaviors to observations, *i.e.*, $dD/dz < 0$ at $z > 4$ and $d\mathcal{N}/dz \propto (1+z)^\gamma$, $\gamma < 0$. This result is naturally expected. Since the outer envelope of a cloud does not much contract and smaller gas clouds can arise detectable due to the increase of N_{HI}, the expected number of observable Lyman α forest and their optical depth increase with decreasing z. Then, we have to abandon this contracting minihalos.

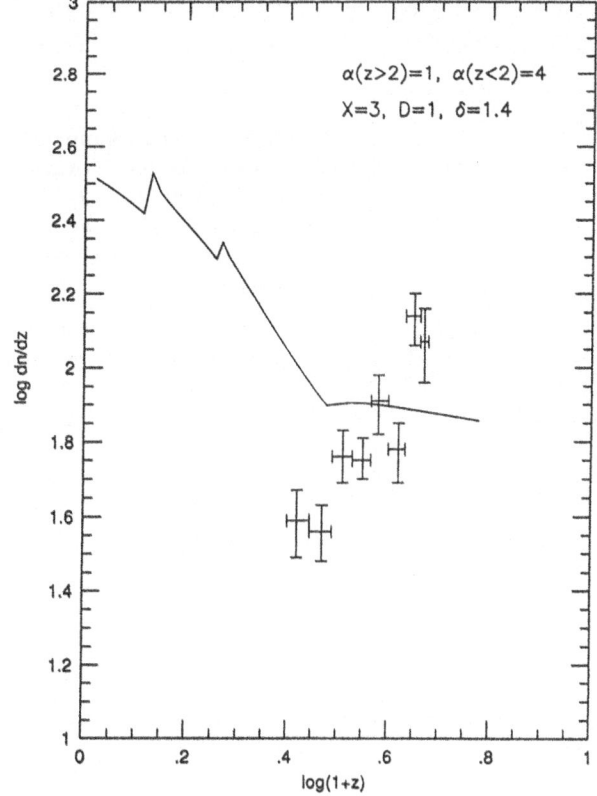

Figure 2. The number density evolution of the Lyman α forest. The calculated result shown by solid lines contradicts with observations.

3.2 EXPANDING MINIHALOS

We examine the case

$$J(z) = 10^{-21}(1 + z/3.5)^\alpha \text{ erg cm}^{-2} \text{ s}^{-1} \text{ str}^{-1} \text{ Hz}^{-1},$$

$$\alpha = -4 \quad \text{at} \quad z > 2,$$
$$\alpha = 4 \quad \text{at} \quad z < 2. \tag{7}$$

We assume the initial cloud to be isothermal, $T_b = 10^4$K, at $z_i = 5$. In this stage, the inner part of a cloud is optically thick and is to be observed as damped Lyman α line systems, while the outer part is optically thin.

With decreasing z, the diffuse UV flux increases and the cloud is heated up. As a result the temperature increases to expand the cloud, and the density decreases, so that the HI column density decreases. In Figures 3, 4 and 5 we show the HI column density distribution $d\mathcal{N}/dN_{HI}$, the number density evolution $d\mathcal{N}/dz$ for the Lyman α forest and LLS, and the continuum depression D. As is seen, all the observed results are well reproduced. With expansion, the lower mass end of minihalos cannot be detectable because of too small HI column density. Therefore, the number of absorption systems rapidly decreases which leads to the quick decrease of continuum depression. Therefore, this expanding minihalo model looks well for reproducing the observed characteristics of all HI absorption systems.

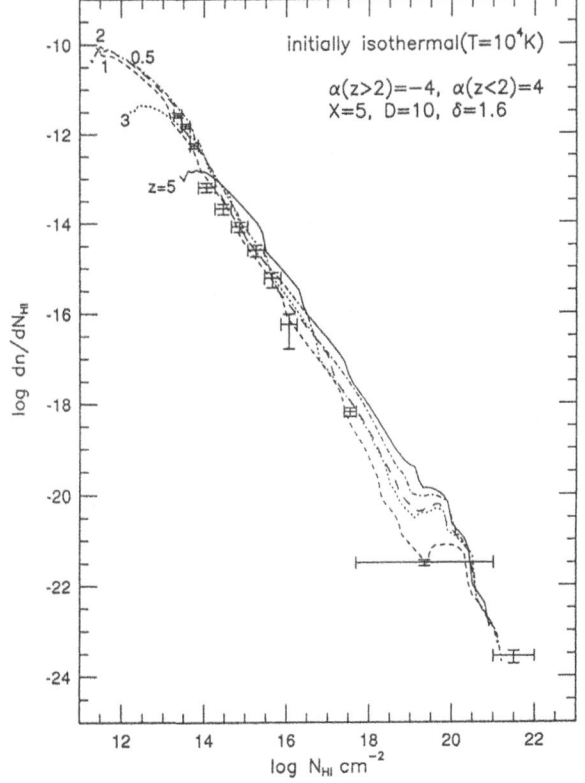

Figure 3. The HI column density distribution for an expanding minihalo at $z > 2$. The observations of HI column density distribution are well reproduced as well.

4. Implications

From the above results, we can obtain the following implications in relation to the galaxy formation.

(I) Expanding minihalos well reproduce three independent observational results. This implies that the diffuse UV flux changes as $J \propto (1+z)^\gamma$, $\gamma < 0$ at $z > 2$. This behavior is expected when the UV flux is proportional to the observed number of QSOs. However, in this case it is difficult to explain the Gunn-Peterson test of high z QSOs because the UV flux is too small to ionize completely the intergalactic gas at $z > 4$. We should examine more carefully this problem.

(II) Metallic line systems like CIV and SiIV absorptions will be intercorporated with this expanding minhalo model. If the optically thick minihalos at $z_i = 5$ might collapse to form stars at the central part, the high HI column density regions with $N_{HI} > 10^{18}$cm^{-2} will be associated with metallic line absorptions. Furthermore, the core contraction of minihalos after expansion will occur at $z < 2$ because $\alpha \sim 4$ at this stage. The star formation in this core will also contribute to the metallic line absorptions. This means the rapid increase of metallic line systems at $z < 2$, of which behavior is exactly confirmed in CIV systems (Steidel *et al.* 1988). The constancy of the sum of LLS and CIV systems just indicates that LLS changes to CIV systems.

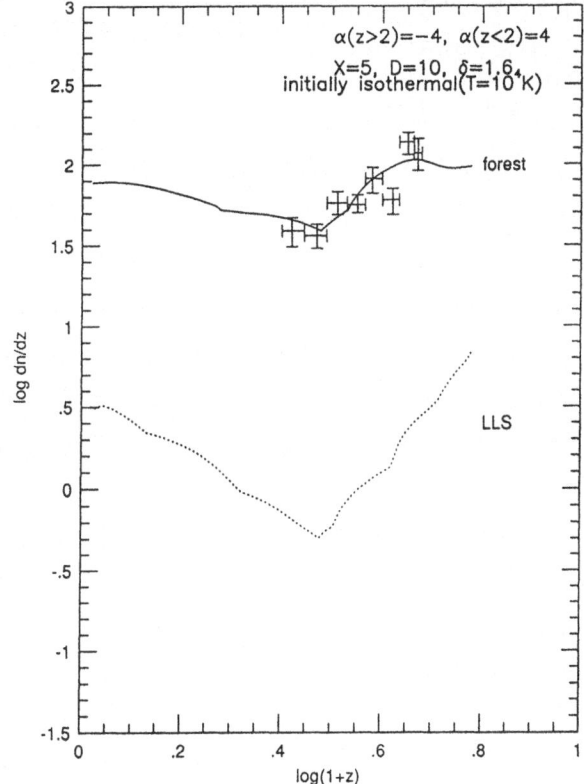

Figure 4. The calculated number density evolutions of the Lyman α forest and LLS. The observations for the former are well reproduced.

(III) Suppose that protogalactic clouds are formed at $z > 5$ with mass function $N(M) \propto M^{-\delta}$, $\delta = 1.2 \sim 1.8$ within the CDM cosgomony. We can divide them into three cosmic objects according to their evolution. (1) Massive clouds with gas mass greater than $10^{10} M_{\odot}$ would collapse at $z > 5$ to galaxies. (2) Clouds with gas mass between 10^8 and $10^{10} M_{\odot}$ would once expand during the epoch $z = 5$ to $z = 2$ because of the increase of UV flux, and their central regions are observed as damped Lyman α line systems and LLS. Sometimes, they are observed as metallic line systems if star formation occurred at the central region. At $z \leq 2$, these massive clouds would quickly change to contraction because of the rapid decrease of UV flux and finally become to less massive galaxies, which are detected metallic line systems. (3) Less massive clouds with gas mass less than $10^8 M_{\odot}$ would slowly contract at $z < 2$ but do not attain the final collapse till the present epoch because of weak gravity. They might be cooled and distribute numerously. We can expect such clouds everywhere.

The above behaviors are essentially determined by the intensity and evolution law of diffuse UV flux. If we schematically describe it as

$$J(z) = J(z_*)(1 + z/1 + z_0)^{\alpha(z)} \tag{8}$$

the critical masses of the above three cases are determined by $J(z_*)$ and the evolutionary features are controlled by the sign and z-dependence of $\alpha(z)$. Therefore, it is the most important work to exactly determine $J(z)$ by using other informations of high z universe.

468

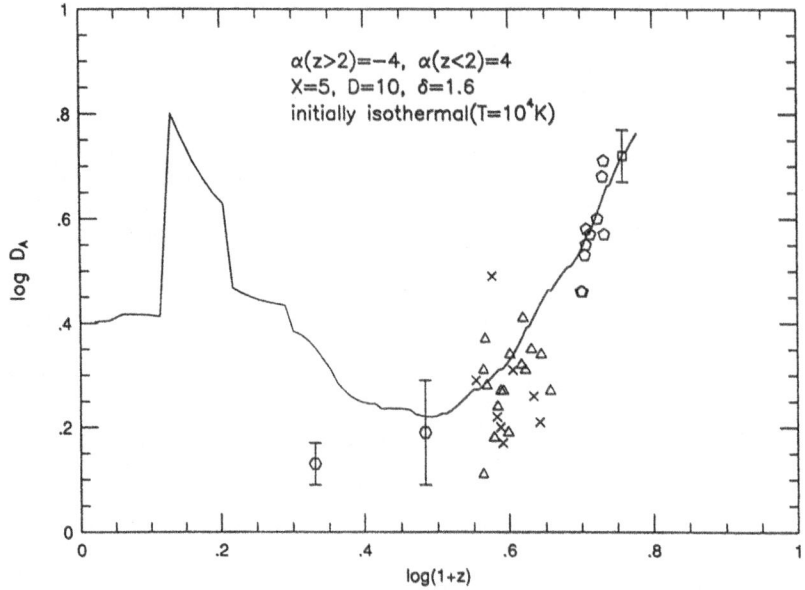

Figure 5. The continuum depression for an expanding minihalos. The observational results are taken from Schneider *et al.* (1989).

References

Bajtlik, S., Duncan, R.C., and Ostriker, J.P. 1988, *Astrophys. J.* <u>327</u>, 570.

Bechtold, J. 1987, in *High Redshift and Priemeval Galaxies* ed. by J. Bergeron *et al.* (Editions Frontieres, Paris), p.397.

Blades, J.C., Turnshek, D., and Norman, C. 1988, *Quasar Absorption Lines : Probing the Universe* (Cambridge Univ. Press, Cambridge).

Ikeuchi, S. 1986, *Astrophys. Space Sci.* <u>118</u>, 509.

Ikeuchi, S., Murakami, I., and Rees, M.J. 1988, *Month. Not. Roy. Astron. Soc.* <u>236</u>, 21p.

Lanzetta, K.M. 1990, Submitted to *Astrophys. J.*

Rees, M.J. 1986, *Month. Not. Roy. Astron. Soc.* <u>218</u>, 25p.

Sargent, W.L.W., Steidel, C.C., and Boksenberg, A. 1989, *Astrophys. J. Suppl.* <u>69</u>, 703.

Schneider, D.P., Schmidt, M., and Gunn, J.E. 1989, *Astron. J.* <u>98</u>, 1507.

Steidel, C.C., Sargent, W.L.W., and Boksenberg, A. 1988, *Astrophys. J. Letters*, <u>333</u>, L5.

Tytler, D. 1987, *Astrophys. J.* <u>321</u>, 69.

Wolfe, A. 1990, Talk at the Oxford Workshop on High Redshift Galaxies.

GRAVITATIONAL LENS EFFECT ON COSMOLOGICAL DISTANCE

MISAO SASAKI
Uji Research Center
Yukawa Institute for Theoretical Physics
Kyoto University
Uji 611
Japan

ABSTRACT. Based on the cosmological post Newtonian expansion, a general formula for the distance-redshift relation in a realistic inhomogeneous universe is presented. Then, focusing on the gravitational lens effect, the validity of using the Friedmann formula for the distance-redshift relation in an inhomogeneous universe is discussed.

1. Introduction

Owing to extensive observational efforts, various surprising features of the large-scale structure of the universe have been discovered in recent years and a new era of the observational cosmology has opened.

On the other hand, theoretical bases of the observational cosmology have not seemed to be sufficiently developed. Namely, actual observed data are inherently those on the past null cone of an observer and are necessarily contaminated by various inhomogeneous structures along the light path, while it is a common practice to tacitly assume as if observations were made on a homogeneous Friedmann background when observational data are to be interpreted. However, the very existence of the large-scale structure implies the universe is highly inhomogeneous at least on scales much smaller than the horizon scale. Hence it is very important to investigate the effect of inhomogeneities on cosmological observations.

In this talk, we report on some recent progress in this direction, based on work in collaboration with Futamase [1] and with Watanabe [2].

2. Cosmological post Newtonian metric

In order to discuss the light propagation in an inhomogeneous universe, it is first necessary to know the metric for a realistic inhomogeneous universe model. For this purpose, we resort to the cosmological post Newtonian expansion of the Einstein

469

equations formulated by Futamase [3]. In this formalism, the existence of the two small parameters ϵ and κ is assumed; ϵ represents the amplitude of the gravitational potential (ϕ) of inhomogeneities, $\phi \sim \epsilon^2$, and κ is the ratio of the typical scale of the inhomogeneities (ℓ) to the horizon scale (L), $\kappa = \ell/L$.

The expansion makes sense if $\epsilon^2 \ll \kappa \ll 1$ and the metric reduces to the ordinary Newtonian form,

$$ds^2 = a^2(\eta) \left[-(1 + 2\Psi)d\eta^2 + (1 - 2\Psi)\gamma_{ij}dx^i dx^j \right] , \tag{1}$$

where

$$\left(\frac{a'}{a}\right)^2 = \frac{8\pi G}{3}\rho_b a^2 - K ; \quad \rho_b = \langle \rho \rangle , \; K = \pm 1, 0 ,$$

$$\overset{(3)}{\Delta} \Psi = 4\pi G a^2 \delta\rho ; \quad \delta\rho = \rho - \rho_b .$$

For galactic scales, $\epsilon \sim 10^{-3}$ and $\kappa \sim 10^{-4.5}$. Hence $\epsilon^2/\kappa \sim 10^{-1.5} \ll 1$ and the use of the above Newtonian form metric is justified. Note that the density contrast $\delta\rho/\rho$ is $O(\epsilon^2/\kappa^2)$ which can be much larger than unity for $\kappa \ll \epsilon$ even if $\epsilon^2/\kappa \ll 1$ as in the case of galactic scale inhomogeneities.

3. Distance-redshift relation

To investigate the light propagation in an inhomogeneous universe, it is convenient to make use of the conformal invariant property of a light path (*i.e.*, null geodesic) and analyze the path on the conformally transformed metric,

$$\begin{aligned} ds^2 \to d\tilde{s}^2 = a^{-2}ds^2 &= \tilde{g}_{\mu\nu}dx^\mu dx^\nu \\ &= -(1 + 2\Psi)d\eta^2 + (1 - 2\Psi)\gamma_{ij}dx^i dx^j , \\ dv \to d\lambda = a^{-2}dv ; &\quad \tilde{k}^\mu = \frac{dx^\mu}{d\lambda} , \end{aligned} \tag{2}$$

where v and λ are the affine parameters on the physical and conformally transformed metric, respectively.

The optical equations on the metric $\tilde{g}_{\mu\nu}$ are

$$\begin{aligned} \frac{d}{d\lambda}\tilde{\theta} &= -\tilde{R}_{\alpha\beta}\tilde{k}^\alpha \tilde{k}^\beta - \frac{1}{2}\tilde{\theta}^2 - 2\tilde{\sigma}^2 , \\ \frac{d}{d\lambda}\tilde{\sigma} &= -\tilde{C}_{\alpha\rho\beta\sigma}\tilde{k}^\alpha \tilde{k}^\beta \tilde{t}^\rho \tilde{t}^\sigma - \tilde{\theta}\tilde{\sigma} . \end{aligned} \tag{3}$$

Under the condition $\epsilon^2/\kappa \ll 1$, one can show that the effect of inhomogeneities can be treated perturbatively [1]:

$$\delta\tilde{\theta} \ll \tilde{\theta}, \qquad |\tilde{\sigma}| \ll \tilde{\theta}. \tag{4}$$

That is, the linear approximation is valid when investigating the light propagation in an inhomogeneous universe.

Then it has been shown that the distance-redshift relation in an inhomogeneous universe takes the form [1,4],

$$\frac{\delta d_L(z,\gamma^i)}{d_L(z)} = \sqrt{-K}\delta\lambda_s \coth(\sqrt{-K}\lambda_s) + I,$$

$$d_L(z) = a_0 \frac{\sinh(\sqrt{-K}\lambda_s)}{\sqrt{-K}}(1+z), \qquad \lambda_s = \eta_0 - \eta_s(z), \tag{5}$$

where $\delta\lambda_s$ represents fluctuations in the affine parameter distance and I the gravitational lens effect. Although $\delta d_L/d_L$ in the above is for the luminosity distance, it coincides with $\delta d_A/d_A$ where d_A is the angular-diameter distance because of the reciprocity theorem [5]. The explicit expression for $\delta\lambda_s$ can be found in Ref.[1]. Here we concentrate on the term I:

$$I = \frac{-1}{\sqrt{-K}} \int_0^{\lambda_s} d\lambda \, [\coth(\sqrt{-K}\lambda) - \coth(\sqrt{-K}\lambda_s)]$$

$$\times \sinh^2(\sqrt{-K}\lambda) \left\{ \frac{1}{2}\delta(R_{\mu\nu}k^\mu k^\nu)_\lambda + \tilde{\sigma}^2(\lambda) \right\} ; \tag{6}$$

$$\frac{1}{2}\delta(R_{\mu\nu}k^\mu k^\nu)_\lambda = \overset{(3)}{\Delta}\Psi - (\Psi'' + 2\frac{d}{d\lambda}\Psi')$$

$$+ 2K \left\{ (\Psi - v_{;i}\gamma^i - \frac{1}{2}v^2)_s + 2\int_\lambda^{\lambda_s} d\lambda_1 \, \Psi'(\lambda_1) \right\}.$$

A simple order estimate shows that $\overset{(3)}{\Delta}\Psi$-term in I is dominant for $z \gtrsim z_{cr} \equiv \epsilon^{1/3}$ (for galaxies, $z_{cr} \sim 0.1$). We also note that shear $\tilde{\sigma}^2$ itself may be negligible, but may not be negligible in its effect to the distance, since its contribution adds up secularly. The effect of shear will be discussed in the next section.

Consider observations of objects at $z \gtrsim z_{cr} \sim 0.1$ in the Einstein-de Sitter (ED) background ($K = 0$):

$$I \equiv \left(\frac{\delta d}{d}\right) \approx \frac{1}{\lambda_s} \int_0^{\lambda_s} d\lambda (\lambda - \lambda_s)\lambda \left[\overset{(3)}{\triangle}\Psi + \tilde{\sigma}^2\right]. \tag{7}$$

Further, we assume:
(1) All lens galaxies have equal mass and radius and distributed randomly.
(2) No matter in the intergalactic space.
Let R_* be the radius of a lens galaxy and R_0 be the mean separation distance between galaxies. Then ϵ^2/κ is expressed as

$$\frac{\epsilon^2}{\kappa} \equiv \frac{3H_0 R_0^3}{8\pi R_*^2} \approx 1.6 \times 10^{-2} h^{-2} \left(\frac{50\,\text{kpc}}{R_*}\right)^2 \left(\frac{R_0}{1\text{Mpc}h^{-1}}\right)^3. \tag{8}$$

Neglecting the contribution of shear, a statistical consideration leads to the probability distribution of I as [1]

$$P(I) \approx (1 - e^{-g(z)}) \frac{1}{\sqrt{2\pi f(z)}} \exp\left[-\frac{(I - I_1(z))^2}{2f(z)}\right] + e^{-g(z)}\delta\left(I - I_0(z)\right), \tag{9}$$

where

$$I_1(z) = -\frac{1}{e^{g(z)} - 1} I_0(z), \qquad I_0(z) = 3\left[\frac{\sqrt{1+z}+1}{\sqrt{1+z}-1}\ln(1+z) - 4\right],$$

$$f(z) = \frac{8}{5}\left(\frac{\epsilon^2}{\kappa}\right)\left(\frac{\sqrt{1+z}-1}{\sqrt{1+z}}\right)^3, \quad g(z) = \frac{1}{4}\left(\frac{\epsilon^2}{\kappa}\right)^{-1}\left[(1+z)^{3/2} - 1\right].$$

The meaning of each term in Eq.(9) should be apparent. $I_0(z)$ corresponds to the Dyer-Roeder(DR) distance [6] which holds for rays propagated only through intergalactic space. The appearance of a delta function is due to over-simplification of the present model and in reality density fluctuations and/or non-vanishing shear in the intergalactic space will yield finite dispersion. $I_1(z)$ is the mean deviation of d_L from the ED value, $g(z)$ is the optical depth for gravitational lensing, and $f(z)$ is the dispersion of $\delta d/d$ for rays passed through at least one galaxy. For any z, $f(z) = O(\epsilon^2/\kappa) \ll 1$ and $I_1(z) \ll 1$. This supports the validity of linear approximation. The fact that $g(z) \gg 1$ for $z \gtrsim 1$ justifies the use of the distance-redshift relation of a Friedmann model for most light rays.

4. Effect of shear

In the previous section, we have neglected the contribution of shear. Here let us estimate its effect on the distance.

The equation for the shear is

$$\sigma(z) = -d_A^{-2}(z) \int_0^{v(z)} dv' \left(R_{abcd} k^a k^c \bar{t}^b \bar{t}^d d_A^2 \right) (v').$$ (10)

Assume that d_A can be approximated by that of a parametrized DR distance,

$$d_{DR}(z; \alpha) = \frac{2}{\beta H_0} \left((1+z)^{(\beta-5)/4} - (1+z)^{-(\beta+5)/4} \right); \quad \beta = \sqrt{25 - 24\alpha},$$

where $\alpha = 1$ for the ED model and $\alpha = 0$ for the completely clumpy DR model. Then, one can show that [2]

$$\langle |\sigma|^2 \rangle (z; \alpha) = 6 \left(\frac{\epsilon^2}{\kappa} \right) H_0^2 I(z; \alpha);$$

$$I(z; \alpha) = d_{DR}^{-4}(z; \alpha) \int_0^z (1+z')^{5/2} d_{DR}^4(z'; \alpha) dz'.$$ (11)

We found the above formula with $\alpha = 1$ is in good agreement with the results of a numerical simulation [2], provided $\epsilon^2/\kappa \lesssim 1$.

Using the formula (11), we can estimate a correction to the distance due to shear:

$$I_\sigma \equiv \left(\frac{\delta d}{d} \right)_\sigma = -\frac{1}{\lambda_s} \int_0^{\lambda_s} d\lambda (\lambda_s - \lambda) \lambda \, \tilde{\sigma}^2,$$ (12)

where $\tilde{\sigma}(\lambda) = a^2 \sigma(v)$. Then,

$$\langle I_\sigma \rangle = -\frac{4}{5} \left(\frac{\epsilon^2}{\kappa} \right) \left(\frac{\sqrt{1+z} - 1}{\sqrt{1+z}} \right)^3.$$ (13)

In particular, the DR part of the probability distribution will have a finite dispersion,

$$\Delta I_\sigma^2 = \langle I_\sigma^2 \rangle - \langle I_\sigma \rangle^2 \approx \frac{2^7 \cdot 3}{5^3 \cdot 7} \left(\frac{\epsilon^2}{\kappa} \right)^2 H_0 R_0 \left(\frac{\sqrt{1+z} - 1}{\sqrt{1+z}} \right)^5.$$ (14)

Thus, as long as $\epsilon^2/\kappa \lesssim 1$, the main effect of the shear is to reduce the DR distance by an amount of $\langle I_\sigma \rangle \times d_{DR} (= O(\epsilon^2/\kappa))$.

Hence, together with the result of section 3, the use of the Friedmann formula in an inhomogeneous universe is approximately justified if the statistics of observations is sufficiently good and the statistical average is considered. However, it is also important to note that if $\epsilon^2/\kappa \gtrsim 1$, neither Friedmann distance nor DR distance will lose their significance.

5. Conclusion

In this talk, I have presented the distance formula for a realistic inhomogeneous universe, based on the cosmological post Newtonian expansion, and discussed the effect of gravitational lensing on the distance. It was shown that provided the condition $\epsilon^2/\kappa \ll 1$ is satisfied, where κ is the ratio of a characteristic scale of inhomogeneities to the horizon size and ϵ^2 is the magnitude of the gravitational potential, the Friedmann formula for the distance-redshift relation is approximately correct even in an inhomogeneous universe. Further, it was shown that the main effect of shear along the light path is to reduce the so-called Dyer-Roeder distance, which holds for rays passed only through vacuum space, by a fraction of $O(\epsilon^2/\kappa)$.

Of course, the above conclusion is based on some rather idealistic assumptions; (1) the lens galaxies are assumed to be transparent, (2) galaxies are completely randomly distributed, and (3) all the galaxies have the same mass. Although, the conclusion will not change qualitatively, it will be interesting to see how quantitatively the result changes by relaxing some or all of the above assumptions. In particular, it is left as a future issue to clarify the effect of spatial correlation and mass distribution of lens galaxies.

I am very much grateful to T. Futamase and K. Watanabe for fruitful collaboration on which this talk is based.

References

1. T. Futamase and M. Sasaki, *Phys. Rev.* **D40** (1989) 2502.
2. K. Watanabe and M. Sasaki, *Publ. Astron. Soc. Japan* (1990) L33.
3. T. Futamase, *Phys. Rev. Lett.* **61** (1988) 2175.
4. M. Sasaki, *Mon. Not. R. astr. Soc.* **228** (1987) 653.
5. see *e.g.*, G. F. R. Ellis, in *Proceedings of the International School of Physics, Course XLVII, General Relativity and Cosmology*, ed. B. K. Sachs (Academic Press, New York, 1971), p.104.
6. C. C. Dyer and R. C. Roeder, *Astrophys. J.* **172** (1972) L115.

Apparent and Biased Effects in Oscillating Universe

Masahiro MORIKAWA *

Department of Physics, University of British Columbia
Vancouver, B. C. Canada V6T 2A6

Abstract

We study a cosmological model in which the expansion rate H and the gravitational constant G are oscillating. This model is proposed to explain the recently observed periodicity in the redshift distribution of galaxies with a characteristic scale of $128h^{-1}$Mpc. Apparent and biased enhancement of the galaxy distribution in the concentric spherical shells with periodically spaced radii centered on the Milky Way is derived. The apparent effect comes from the distorsion of the ordinary distance-redshift relation due to the H change. The biased effect comes from glow of a galaxy due to the G change. Our main subjects in this paper are, 1) Doublet peak structure of the periodic distribution of galaxies. 2) Difference in the strength of quasar and Lyman-α cloud clusterings. 3) Bending of the magnitude-redshift curve in the Hubble diagram.

1. Introduction

In previous papers (Morikawa 1990a,b), we proposed a model of the Universe with oscillating Hubble parameter in order to explain the remarkable coherent periodic distribution of galaxies over thousands of Mpc (Broadhurst et al. 1990:=BEKS). The oscillation of the expansion rate of the Universe produces an apparent concentration of galaxies on concentric spherical shells centered on the Milky Way with equal separation. This oscillation of the Hubble parameter H is naturally induced by the nonconformal scalar field model. This scalar field couples to the scalar curvature, and this coupling is the essence of the strong H oscillation. On the other hand, this coupling inevitably produces the oscillation of the gravitational constant G as well. Since the luminosity of a star and a galaxy is very sensitive to the change of G, we expect real luminosity oscillation of galaxies in redshift, and biased enhancement of the number count observation (Morikawa 1990b, Hill et al. 1990, Rees 1990). G and H-oscillation effects should be mixed in the actual number count observations. We study how they show up in the number count observation. We also argue how we can check our model.

Suppose that the Hubble parameter is oscillating in cosmic time with a period of order 10^8 years on top of its standard evolution. Then even if the matter distribution is uniform in constant cosmic time t slices and its temporal evolution

---

* Bitnet address: USERHIRO@UBCMTSG

is monotonic, there appears an apparent inhomogeneity and a periodicity in the matter distribution when plotted against redshift z. Actually, from the relation $dz = -(1+z)H\,dt$, the oscillation of H distorts a smooth observable plotted against t into an oscillatory one when differentially plotted against z. This transmutation of the temporal structure into the spatial one is caused simply by the fact that we always observe the past when we look at distant objects. Since this oscillation is spatially uniform and isotropic, it never distorts the isotropy of the cosmic microwave radiation. The model of the oscillating H is realized by introducing a scalar field $\phi(t)$ with an extra curvature coupling $\xi\phi^2 R$. We study the oscillating H based on this model.

The effect is not only an apparent one, but there also appears a real effect in our model. The curvature coupling makes the effective gravitational constant G ϕ-dependent, and G becomes periodic as H. Since the luminosity of a star and a galaxy is quite sensitive to the change of G, we expect observable real periodic glow of galaxies. This luminosity change affects the direct luminosity observation. Moreover, it affects the number count observation because the overall brightening of all the galaxies makes the detectable number of galaxies increase.

These two effects should be mixed in the actual observations. Previously, when we consider only the oscillation of H, we came across several difficulties in the explanations: Why there is an asymmetry between the north and south data of the number count observation (BEKS)?; Why the redshift distribution of the Ly-α coulds is so smooth?; Why the redshift distribution of quasars is so clampy than expected?; Why the temporal change of G is so small than predicted? $\cdots$ We will try to explain them in this paper.

2. Apparent and Biased Number Count

As a model of a universe with oscillating Hubble parameter, we consider a spatially uniform nonconformal scalar field $\phi(t)$ in a flat Friedmann-Robertson-Walker Universe model whose line element is given by $ds^2 = dt^2 - a^2(t)[d\chi^2 + \chi^2 d\Omega^2]$. In this model, we have four parameters: the mass m, the curvature coupling constant ξ for ϕ, present Hubble parameter H_0, and the deceleration parameter q_0. Let us first calculate the evolution analytically as much as possible, though all the numerical calculations shown in figures are done exactly. In order to do so, we adopt a power law expansion of the universe with a small oscillatory part (osc.):

$$a(\tau) = a_0(t/t_0)^c + \text{osc.}, \qquad H(t) = ct^{-1} + \text{osc.}, \qquad (1)$$

with some constants c, a_0 and t_0. If we neglect the oscillatory part osc. in the equation of motion for the scalar field, then it becomes approximately the Bessel equation with the solution:

$$\phi(t) = J_\nu(mt)/\sqrt{6Gmt}, \qquad (2)$$

where $\nu^2 = -(4/3)\xi + (1/4)$, (and $c = 2/3$). From this, we get an approximate expression for the Hubble parameter:

$$\frac{H(t)}{m} \approx \sqrt{\dot{\phi}^2 + m^2 \phi^2} + 6m\xi\phi\dot{\phi} \approx \frac{2}{3mt} - \frac{4\xi}{3(mt)^2} \cos(2mt - \pi\nu), \qquad (3)$$

for large times ($mt \gg 1$). The second term in the RHS represents the small oscillating part and the first term the slowly varying part. It is obvious from this expression that the period of the oscillation (π/m in terms of t) is determined by the mass m and the strength of the oscillation is determined by the parameter ξ.

In the similar way, the effective gravitational constant G becomes a function of the scalar field and changes with time:

$$G(t) = \frac{1}{G_0^{-1} - 8\pi\xi\phi^2} \approx \frac{G_0}{1 - (4\xi/3)(mt)^{-2} \sin(2mt - \pi\nu)}. \qquad (4)$$

Now we apply our model to explain the coherent galaxy correlations (BEKS). The differential number count of galaxies is given by

$$dN = n_0 a_0^2 \chi^2 H^{-1} d\Omega dz, \qquad (5)$$

where n_0 is the galaxy density at present. Therefore, the oscillation in H directly induces that in the differential number count as a function of z.

We consider the effect of G-change on the luminosity of a star and a galaxy. The temperature of the center of a star T is roughly given by $T \approx GM/R$, where M is a mass of the star and R is its radius. The luminosity of the star L is proportional to the radiation energy gradient T^4/R, to the surface area R^2, and to the mean free path of a photon $T^3(M/R^3)^{-2}$. Therefore, we get $L \approx G^7 M^5$ (Teller, 1948). Numerical works (Chin and Strothers, 1976) support this strong non-linearity of the G dependence of L. The luminosity change of an individual star directly affect the galaxy luminosity. Moreover, when G increases, we expect, a) the galaxy becomes more compact and induces more frequent interactions. b) the galactic nuclei will become more active if the activity is gravitational. These effects also enhance the luminosity of the galaxy. The net effect of G change on the luminosity, therefore, seems to be strongly nonlinear. Unfortunately, we cannot exactly estimate this net effect at present because we don't know the mechanism of the interaction of galaxies and the activity mechanism of the galactic nuclei. Therefore, we adopt the simplest procedure here; we assume the luminosity of a galaxy L is given by a power of G: $L/L_0 = (G/G_0)^{7\beta}$ with β some constant. The power seems to be very large (at least $\beta = 1$) reflecting nonlinear effects.

The increase of the individual galactic luminosity strongly enhances the number count observation of galaxies. This enhancement is calculable, in principle, if we get the absolute luminosity distribution of galaxies (luminosity function). For example, if the limiting magnitude of detection is located at the steep part of the luminosity function, the steepness enhances the number count. However, the estimate of the luminosity function depends on the real evolution of galaxies, whose estimate is difficult. We assume here that the number count is simply proportional to the luminosity. This linear relation is expected when the observation is deep enough to detect the most faint galaxies. We take care of the actual nonlinearity by the parameter β in the present preliminary argument.

G and H oscillations have exactly the same periodicity since these oscillations are from the single field ϕ. The relative phase of G and H is shifted 25% as is seen from Eqs.(3,4). This shows up in the periodic number count observation as a double peak structure with the separation one quarter of the complete period. This separation amounts to $32h^{-1}$Mpc or to the redshift 0.011 (for $z \ll 1$).

In Figure 1, we show a typical numerical calculation of the number count as a function of redshift. The solid line is a normalized number count (Eq(5)) and the broken line is the modulated number count due to the G change. This periodic number count is the one recently reported in the measurement BEKS. The measurement is performed at both Galactic poles within a narrow angle $0.3°$. The most prominent fact in this observation is the existence of a single periodicity $128h^{-1}$Mpc in the comoving distance on a large scale ($2000h^{-1}$Mpc where $H_0 \equiv h$ 100km/sec/Mpc) in both directions. Our model explains this observation. We superimpose there data.

From this graph, we see that some number count peaks in north data coincide with the G peaks (broken line) while those in south data coincide with the H^{-1} peaks (solid line). This may be the reason why the proper redshift zero-point seemed to be shifted, in BEKS data, about $\Delta z = 0.011$, though a question still remains: Why the north data picked up G peaks and the south data H^{-1} peaks. I think the sharpness of the peaks in the BEKS data is artifact of the one dimensional pencil beam survey, and therefore, we should not take the precise position of the peaks seriously. Actually, the cluster/void structure of galaxies enhances the sharpness: If a supercluster happen to be located on the H^{-1} or G peaks, the number count of the pencil beam survey gives a sharp peak. The true distribution can be given only from the three dimensional complete survey, in which, each sharp peaks are averaged out to give much smooth distribution. In this sense, I agree with the recent claims (Kaiser 1990, Rhie 1990) that the apparent periodicity is easily produced in a pencil beam survey. However, I consider this easiness should be understood as the mildness of the contrast in the periodic galaxy distribution.

Our oscillating universe model predicts a concentric spherical shell structure of the galaxy distribution centered at the Milky Way. This structure and the location of peaks do not conflict with several other observations: Szalay 1990, Kopylov et al. 1988, Kirshner et al. 1990, Geller and Huchra 1989.

3. Difference in the Quasar and Ly-α Cloud Clustering

The present model also predicts the clustering of quasars and Ly-α clouds in higher redshift regions. This clustering mechanism is partially the same as the previous argument (Morikawa 1990a,b), however if we take into account the real clustering due to the G change, then the clustering strength becomes different for quasars and Ly-α clouds. The apparent effect due to the H oscillation affects the redshift distribution irrespective of the object, while the real effect due to the G oscillation selectively affects the redshift distribution of the emitting objects such as quasars but not that of the absorbing objects such as Ly-α clouds. Since the amplitude of G oscillation is much larger than that of H in higher redshift, we expect a strong quasar clustering and a weak Ly-α cloud clustering.

In Figure 2, we plot the prediction of the global number count, which is exactly the extension of the Figure 1. The bars show the quasar distribution (Hewitt and Burbidge 1987). We observe that the width of the apparent distribution (solid line) is smaller than the real distribution (broken line). This means that if the Ly-α data are bined in finite redshift intervals, the apparent density contrast becomes much smaller. Observationally the clustering of quasars is clear, and the phase of enhancement seems not to conflict with our prediction. However, we cannot predict the amount of the enhancement precisely because at present we have no systematic way to derive the luminosity change from G change, especially for the unknown system such as quasars. On the other hand, the clustering of Ly-α clouds is a touchy problem (Chu and Zhu 1989, Bartlett et al. 1990), however, it is clear that the clustering is at least weaker than that of quasars. In the subsequent study, we will quantitatively compare the prediction and the observations.

4. Magnitude-Redshift Relation

If the absolute luminosity is periodically changing, we have to check the direct observation of magnitude-redshift relation. The apparent magnitude m is given by

$$m = -5 + 5\log(d_L/\mathrm{pc}) - (5/2)\log(L/L_0), \tag{6}$$

with possible corrections. Here, d_L is the luminosity distance.

In Figure 3, the prediction of the apparent magnitude-redshift relation is plotted. On top of it, data of radio galaxies (dots) and quasars (squares) are superimposed (Wampler 1987). We observe that the relation is smooth enough at the low redshift regions ($\Delta m < 0.3$ for $z < 0.3$) and well within the observed scatter. On the other hand in the higher redshift regions, the curve tends to bend to the luminous direction, which is consistent with the observation. If we did not consider the G change, we would get a curve given by the upper envelop of the solid line, which systematically deviates from the observation. Also in the number count magnitude relation, we observe the bending of the curve toward the luminous direction. Ordinarily, these bendings are interpreted as the evolutionary effect of the galaxy. In the subsequent publications, we will clarify the contribution of this effect.

5. Conclusions

We have studied the apparent effect due to the oscillation of H and the biased effect due to the oscillation of G on few cosmological observations. Our predictions are: 1) Doublet peak structure of the periodic distribution of galaxies. The separation of the doublet is one quarter of the complete period. Therefore, the shift in the proper zero-point of redshift in BEKS data does not conflict with our model. 2) Quasar clustering is much stronger than Lyman-α cloud clustering. Therefore, the analysis Bartlett et al. 1990 does not conflict with our model. 3) The magnitude-redshift curve in the Hubble diagram tends to bend toward the luminous direction. These must be carefully compared with observations to get the constraints on the parameters of our model.

In general, the spatial change of the gravitational constant G, possibly caused by the condensation of much heavier nonconformal scalar fields, provides a bias mechanism for the baryon distribution. We will quantitatively argue this bias effect in wider contexts such as the dark matter in the halo, variation of the age of the globular cluster, metalicity distribution in the galaxy and in the cluster of galaxies.

Acknowledgement

I am grateful to Salman Habib, Paul Hickson, and Bill Unruh for various discussions. I also would like to thank The Japan Society for the Promotion of Science for financial support.

References

Bartlett, J. G., R. Esmailzadeh, R. and Hall, L. J. 1990, preprint CfPA-TH-90-25.

Broadhurst, T. J., Ellis, R. S., Koo, D. C., and Szalay, A. S. 1990, *Nature*, **343**, 726.

Chin, C. W., and Stothers, R. 1976, *Phys. Rev. Lett.* **36**, 833.

Chu, Y., and Zhu, X. 1989, *Astron. Astrophys.* **222**, 1.

Geller, M. J. and Huchra, J. P. 1989, *Science* **246**, 897.

Hellings, R. W., Adams, P. J., Anderson, J. D., Keesey, E. L., Lau, E. L., Standish, E. M., Canuto, V. M. and Goldman, I. 1983, *Phys. Rev. Lett.*, **51**, 1609.

Hewitt, A. and Burbidge, G. R., 1987, *Astrophys. J. Suppl.* **63**, 1.

Hill, C. T., Steinhardt, P. J., and Turner, M. S., 1990, preprint FERMI-PUB-90/129-T.

Kaiser, N. 1990, preprint

Kirshner et al. 1990. *Astronomical J.* **100**, 1409.

Kopylov et al. 1988, in *Large Scale Structure of the Universe*, (Proceedings of the Symposium of the IAU) 129.

Morikawa. M., 1990a, *Astrophys. J. Lett.* **362**, L37.

Morikawa. M., 1990b, *Astrophys. J.* to appear.

Rees, M., 1990, private communication.

Rhie, S. H. 1990, in the proceedings of the workshop *After the First Three minutes* (Maryland).

Szalay, A. S., 1990, talk at the conference *Inflation and Exotic Cosmic Structure Formation*, UBC Vancouver.

Teller, E. 1948, *Phys. Rev.* **73**, 801.

Wampler, E. J. 1987, *Astron. and Astrophys.*, **178**, 1.

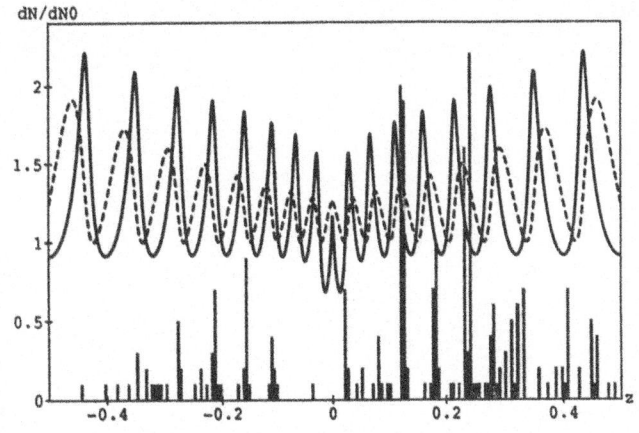

Figure 1
A numerical calculation of the number count as a function of redshift (normalized by the standard model). The solid line is a normalized number count (Eq(5)) and the broken line is the modulated number count due to the G change. We superpose BEKS data (The vertical axis: ×10). The positive and negative redshifts mean the north and south directions, respectively. To avoid a conflict with the Viking radar echo experiment (Hellings et al. 1983), we simply take $\dot{G}/G = 0$ at present. Another requirement that H increases with redshift completely determines the present phase ($\phi_0 = 0.0103$, $\dot{\phi}_0 = 0$). We take $\xi = 10$, $\beta = 5$. We do not argue the contrast of the peaks, which is sensitive to the parameters ξ, β. However, the oscillation period is almost unique.

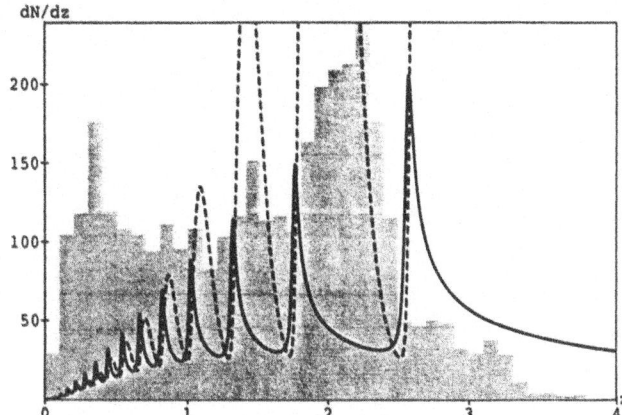

Figure 2
Same as Figure 1, but on a global scale without normalization. Data are from Hewitt and Burbidge 1987 (The vertical axis: ×20). Again, the contrast is not important. The phase of the oscillation is a direct continuation from the Figure 1.

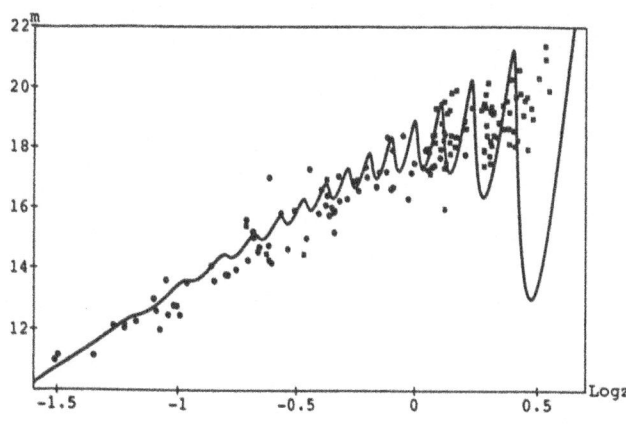

Figure 3
A numerical calculation of the apparent magnitude redshift relation. Observational data (Wampler 1987) is superimposed (dots for 3CR radio galaxies, squares for quasars).

OSCILLATING PHYSICS AND PERIODIC UNIVERSES[1]

PAUL J. STEINHARDT
Department of Physics
University of Pennsylvania
Philadelphia, PA 19104

ABSTRACT. The periodicity in the pencil-beam red shift survey of galaxies reported by Broadhurst, *et al.* may be an illusion caused by the periodic oscillations of physical constants. We analyze scenarios entailing gravitational, electromagnetic, and other interactions.

Broadhurst, Ellis, Koo and Szalay[2] (BEKS) have reported on a pencil-beam red shift survey in which they find an apparent periodicity of 128 h^{-1} Mpc in the galaxy distribution. We have recently considered whether the periodicity might really be an observational effect induced by spatially uniform but temporally oscillating physical "constants" of nature. Normally, it is assumed that the red shift distribution of observable (luminous) galaxies, $dL_{galaxy}(z)/dz$, is directly proportional to the spatial distribution of all galaxies, $dN_{galaxy}(r)/dr$; this would be correct if luminous and non-luminous galaxies are distributed equally and if the red shift z is directly proportional to r. However, with oscillating Newton's constant G or oscillating Rydberg, the measured redshifts vary periodicially in time; if oscillating masses or couplings cause galaxy luminosity to vary (independent of the real spatial density), L_{galaxy} varies periodically. Since different red shifts correspond to different times, the consequence in each case would be a periodic modulation of the red shift distribution independent of angle, akin to what BEKS observe. We find that only a relatively tiny oscillation amplitude, as small as .3% in some scenarios, is sufficient to mimic the BEKS effect.

A natural method of inducing oscillations of a fundamental physical "constant" is to suppose that it is really a function of a fundamental scalar field, ϕ, whose average value $\langle\phi\rangle$ is fixed by the minimum of its potential and which oscillates about $\langle\phi\rangle$. To reproduce the BEKS data, ϕ must be a field with an ultra-low mass (a soft boson) coherently oscillating with a period $c\tau \simeq L = 128\ h^{-1}$ Mpc corresponding to a mass $m = \mathcal{O}(10^{-31})$ eV), where the present Hubble parameter is $H_0 = 100\ h$ km sec^{-1} Mpc^{-1}. Below, we summarize scenarios in which an oscillating soft-boson can create the BEKS effect:[1]

OSCILLATING DARK MATTER

The Idea: Suppose the energy density in the oscillating soft-boson field, ϕ, contributes nearly 100 per cent of Ω, the ratio of the total mass density of the universe to the critical density. Then, oscillations in the field induce an oscillation in the Hubble constant with amplitude, $\mathcal{A} = 3H/m$ (assuming $\Omega = 1$).

483

Problems: To obtain the desired period (128 Mpc), $m \approx 80H$ or $\mathcal{A} = 3H/4m \approx$ .01, at least a factor of 50 too small to produce the observed effect $(\mathcal{A} \gtrsim .50)$. The amplitude would be enhanced by assuming a larger Ω, but this is unacceptable. The effect can also be enhanced by coupling the field non-minimally to gravity, as Morikawa has proposed.[3] However, a second problem results in any event: The soft-boson does not begin to oscillate until $H \approx m$; given the small m dictated by the observed period, the universe would be required to remain radiation-dominated until very late times, $z = 10$! We conclude that this approach is ruled out.

OSCILLATING G

The Idea: If the soft-boson field is non-minimally coupled to gravity, G will oscillate $(G = G_0 + \delta G \cos{(mt)})$ and induce an oscillation in H. Unlike Morikawa, we require that the soft-boson energy does *not* dominate Ω, so the universe becomes matter-dominated at the standard epoch, $z \approx 10^4$. Since $H^2 = \dot{G}H + H^2$, the oscillation amplitude is $(m/H)(\delta G/G)$. For $m \approx 80H$, a rather tiny $\delta G/G \gtrsim .003$ suffices to obtain an oscillation amplitude in H, $\mathcal{A} \gtrsim .50$, that can mimic the BEKS effect.

Problems: Limits from nucleosynthesis or stellar evolution are not sensitive to such small $\delta G/G$. However, Viking radar ranging data constrains $\dot{G}/GH < .3h^{-1}$, The disagreement is marginal (a factor of $3.3h$) at worst, and may be eliminated altogether in alternative models where $\dot{G}/GH$ decreases with time.[4] We conclude that this model may be viable.

OSCILLATING ATOMIC LINES

The Idea: If the soft-boson couples to the electromagnetic field $(\phi F_{\mu\nu}F^{\mu\nu})$ or to the electron mass $(\phi\bar{\psi}_e\psi_e)$, it induces oscillations in α or the electron mass, m_e. Atomic line spacings, proportional to the Rydberg, oscillate periodically in time, producing an effect similar to oscillating the Hubble constant. The requisite oscillation amplitude is $\delta\alpha/\alpha \gtrsim .003$ or $\delta m_e/m_e > .003$.

Problems: Limits on $\delta\alpha/\alpha$ based on analysis of samarium isotopes in the Oklo natural nuclear reactor $(\delta\alpha/\alpha < 10^{-8})$ rule out the oscillating α scenario. Limits from rhenium isotopes in meteorites and from comparisons of 21 cm lines to fine structure in distant absorption clouds in front of QSO's are less stringent, but also rule out the requisite amplitude. The limits on m_e are weaker, though. Varying m_e induces radiative corrections to α. Using the Oklo limit above, we have discovered a new, world-record constraint: $\delta m_e/m_e < 6.5 \times 10^{-6}$, more than 60 times more stringent than previously quoted, direct limits based on the meteorite or QSO measurements cited above. A priori, the limit easily rules out the oscillating m_e scenario. However, it is conceivable that a special symmetry between leptons causes the radiative corrections to α from varying m_e to be exactly cancelled by corrections from other leptons. Then, one must fall back on the direct limits, which are inconsistent with the requisite amplitude by less than an order of magnitude — perhaps only a a marginal failure.

OSCILLATING LUMINOSITY

The Idea: The BEKS sample is chosen largely from the exponential, high-luminosity tail of the galaxy distribution. An oscillating soft-boson which somehow varies galaxy luminosity by twofold, say, could explain the BEKS observations. The soft-boson may change masses or couplings of weakly interacting massive particles and, thereby, alter the luminosity of certain types of stars; or, the soft-boson may result in the periodic decay of of dark matter in halos, causing galaxies to "glow" periodically.

Problems: We do not know observations that rule out this scenario. However, the scenario is less well-defined than the others and depends sensitively on the BEKS selection function, which is not well-understood at present.

The oscillating physics scenarios described above are subject to two criticisms: (1) they produce effects that are periodic in time, rather than z, in disagreement with BEKS at larger ($z = .4$) red shifts; (2) BEKS suggest that we are presently off-center with respect to the north-south survey and that the periods in two directions orthogonal to north-south may differ by as much as 10 per cent. These criticisms may be obviated if an anharmonic potential ($\lambda\phi^4$) is assumed for the soft-boson field.[4] The period becomes time-dependent, more closely approximating periodicity in z; also, the value of the period depends on the initial value of ϕ, so that a small initial inhomogeneity in ϕ would distort the shapes and periods of the concentric surfaces of high dL_{galaxy}/dz.

The key test for oscillating physics will be if further observations show peaks in the red shift distribution on roughly concentric surfaces. Distinguishing which oscillating physics is achievable with only modest improvements in present measurement techniques. Conversely, should the BEKS periodicity prove to be a chimera, the absence of peaks can be used to derive new and stringent limits on the variations of the constants.

This work was supported in part by U.S. DOE Grant No. DOE-EY-76-C-02-3071.

References

[1] This report summarizes the collaborative work of C. T. Hill, P. J. Steinhardt, and M. S. Turner, Fermilab Preprint FNAL-PUB-129-T (1990); see this paper for a complete list of references.

[2] Broadhurst, T. J., Ellis, R. S., Koo, D. C., and Szalay, A. S., *Nature* **343**, 726-728 (1990).

[3] Morikawa, M., Univ. of British Columbia preprints 90-0208 and 90-0380 (1990).

[4] R. Crittendon, P. J. Steinhardt, C. T. Hill, M. S. Turner, Penn preprint (1990).

PROBES OF THE HIGH-REDSHIFT UNIVERSE

M.J. REES
Institute of Astronomy
Madingley Road
Cambridge, CB3 0HA, England

ABSTRACT. Some lines of evidence on physical conditions at redshifts $z \gtrsim 3$ are briefly reviewed. The 21 cm line emission from inhomogeneous neutral gas could yield information on redshifts preceeding the epoch of galaxy formation and reheating.

1. INTRODUCTION

Many fundamental features of the present-day Universe are probably direct consequences of exotic physics during very early eras. Among these are the observed inhomogeneities. Indeed, on sufficiently large scales nonlinearities have not yet developed, so we can in principle (by studying the microwave background isotropy, for instance, or by investigating the present mass distribution on supercluster scales) directly probe the imprint of a possible inflationary phase. On the scale of individual galaxies, however, the effects of gas dynamics, dissipation etc. are likely to have been important. The detailed scenarios of galaxy and cluster formation can now be tested by various observations, and are the subject of other talks at this conference. Let me note only that the choice of the correct theory is closely linked to a decision on what the dark matter actually is. It is embarrassing that more than 90 per cent of the Universe is unaccounted for, and clearly we cannot expect a definitive galaxy formation theory until this basic issue is settled.

2. THE EPOCH OF GALAXY FORMATION

We can infer on quite general grounds that the formation of galaxies, especially disc-like systems, cannot have been completed before the Universe was about 2 billion years old, corresponding to $z \simeq 2$. This argument is based on the angular momentum of disc galaxies. Neighbouring protogalaxies will exert tidal torques on each other, the result being that a typical protogalactic gas cloud will have acquired some angular momentum by its turnaround time. However, analytic and numerical work suggests that this angular momentum is only a few per cent of what is needed for rotational support. In consequence, a large collapse factor would be needed before centrifugal effects could balance gravity (Fall and Efstathiou 1980; Gunn, 1982). For the case of disc formation in a nonbaryonic halo with an isothermal ($\rho \propto r^{-2}$) density distribution, the typical collapse factor is of order 10.

This implies that the material lying 10 kpc from the centre of a disc like our Milky Way must have fallen in from a radius of order 100 kpc. The free-fall time from that radius is of order 10^9 years, so the formation of 10 kpc discs could not have been completed until the Universe was at least 2 billion years old. This argument can be formalised, but essentially requires that protogalaxies should have been large and diffuse. [The angular momentum problem, incidentally, would become even more acute if dark matter and extensive halos did not exist.]

We certainly, on the basis of the above argument, expect that galaxies should look very different at $z = 2$. In particular, a galaxy should be surrounded by a large amount of gas, falling in from radii as large as 100 kpc. If a quasar lights up in the centre of a high redshift galaxy, this circumgalactic gas may become conspicuous via scattering or reprocessing of the quasar light. (This topic is relevant to some results discussed in Dr Cowie's presentation.) Nor should we be surprised about the evidence from the quasar distribution, and from their absorption spectra, that conditions at $z \simeq 2$ were very different from those at the present epoch. Note that this argument does not tell us when galaxy formation *starts* – other considerations are required in order to decide whether the first bound systems formed at (say) $z = 5$ or $z = 50$.

3. HIGH-REDSHIFT QUASARS

It has been known for 20 years that bright quasars were much more common at $z \simeq 2$ than at the present epoch. In the last few years, the has been progress in identifying statistically significant samples of quasars at much larger redshifts (see Warren and Hewett, 1990, for a review). It obviously becomes progressively more difficult to discover objects as the redshift increases; however there is now a consensus that the quasars genuinely 'thin out'. At $z \simeq 4$, the comoving density of quasars with $M_B \simeq -26$ is 5–10 times lower than at $z \simeq 2$.

There are now 19 known quasars with $z > 4$, the current record holder being $z = 4.73$ (Schneider, Schmidt and Gunn, 1989). These seem to continue the same trend, even though there is actually no evidence that the most exceptionally powerful objects, with $M_B < -28$, participate in the decline. We are still a long way from a proper theory for quasar 'demography' (Rees, 1990; Warren and Hewett 1990).

The very existence of these high redshift quasars means that some galaxies must already have formed, and evolved to the stage where they have well-defined nuclei in which runaway quasar activity can occur, when the Universe was only $\sim 10^9$ years old. The existence of any quasars with redshift of order 5 is fatal for the classic adiabatic 'pancake' model in which the first bound systems have cluster masses and must have 'turned around' relatively recently. The mass involved in a quasar of luminosity $L = 10^{47}h_{50}^{-2}$ erg s^{-1} can be expressed straightforwardly in terms of its luminosity, lifetime and efficiency as

$$M_Q = 2 \times 10^9 L_{47} h_{50}^{-2} \varepsilon_{0.1}^{-1} t_{Q8}.$$

In this expression, ε is the efficiency, in units of 10%, and t_{Q8} the quasar lifetime in units of 10^8 years. The number of active quasars per comoving volume in the redshift range $z \simeq 3-4$ is $2 \times 10^{-7} h_{50}^{-3}$ Mpc^{-3}. The time corresponding to that redshift range is $5 \times 10^8 h_{50}^{-1}$ yrs. We can therefore ask whether there are enough appropriate host galaxies for the $\sim 5 h_{50}^{-1} t_{Q8}^{-1}$ generations of quasars implied by the observations. The evidence is becoming a marginal embarrassment for the cold dark matter model (Efstathiou and Rees 1988, Rees 1988). In that cosmogony, most large galaxies assembled relatively recently. The characteristic scale of the bound CDM halos at $z = 4$ is only $10^9 M_\odot$. It would need a fluctuation way out on the high amplitude tail of a gaussian distribution to have already produced a galactic mass halo by that time. Quantifying this argument further is stymied by uncertainty about what sort of galaxy is needed to host a quasar, how long quasars live, etc. It would certainly be easier to explain high-z quasars if the initial fluctuations were non-gaussian: one would then envisage rare locations where deep and massive potential wells formed early, even though more typical fluctuations on a given mass scale condensed out more recently.

Another clue comes from studies of how quasars are clustered over the sky. Bright quasars are sufficiently thinly spread that one would not expect to see evidence of clustering, unless this occurred on a vastly larger scale than is indicated by normal galaxies. However, the surveys of fainter quasars, particularly by Boyle and his collaborators (Mitchell *et al.* 1990), show clustering both at high and low redshifts. What is surprising is that the degree of clustering does not appear to change much with redshift. This is compatible with a model in which quasars form in a biased fashion at high amplitude peaks in the cluster scale density distribution.

Quasars can provide a wealth of information on the process of galaxy formation in a third way, namely by serving as probes for diffuse matter along the line of sight. The complex absorption spectra of quasars reveal systems probably identifiable with protogalaxies, along with a much greater number believed to involve clouds of subgalactic mass.

4. THE EPOCH Z > 5

One important immediate inference can be drawn from the highest-redshift quasars: the absence of a Gunn-Peterson (1965) trough shortward of Lyman α implies that the intergalactic gas has already been ionized. It has either been photoionized by a UV background, or else heated to $> 10^6$ K so that collisions maintain $ne/n_H \gtrsim 10^6$. It seems unlikely that the thinly-spread powerful quasars could themselves have provided their requisite energy; a numerous population of weak (Seyfert-level) galactic nuclei cannot, however, be excluded, since these could not be detected individually at $z \gtrsim 4$. An alternative possibility – one that would be specially attractive in 'hierarchical' models – is that the intergalactic medium is ionized by high-mass stars forming in aggregates of subgalactic scale. In the CDM model, for instance, many such systems would form as early as $z \simeq 10$; even if only $\sim 10^{-3}$ of the baryons condensed into O and B stars within such systems, sufficient UV would be emitted to photoionize all the remaining baryons in the intergalactic medium (Couchman and Rees 1986).

Unless quasars are discovered at vastly greater redshifts than the current record, information will remain sadly lacking on the cosmic dark age, extending from the epoch of recombination to the stage when the Universe was as much as 10^9 years old. I should like to briefly mention a new technique (Scott and Rees, 1990; Subramanian 1990) which offers, at least in principle, the chance of probing incipient large-scale structure at high redshifts, perhaps even before reheating occurred. This technique depends on studying the 21cm line expected from diffuse neutral hydrogen. The contribution from this line, in terms of brightness temperature, would be much less than the 2.7° of the microwave background, and also much less than the nonthermal background at radio frequencies due to synchrotron emission

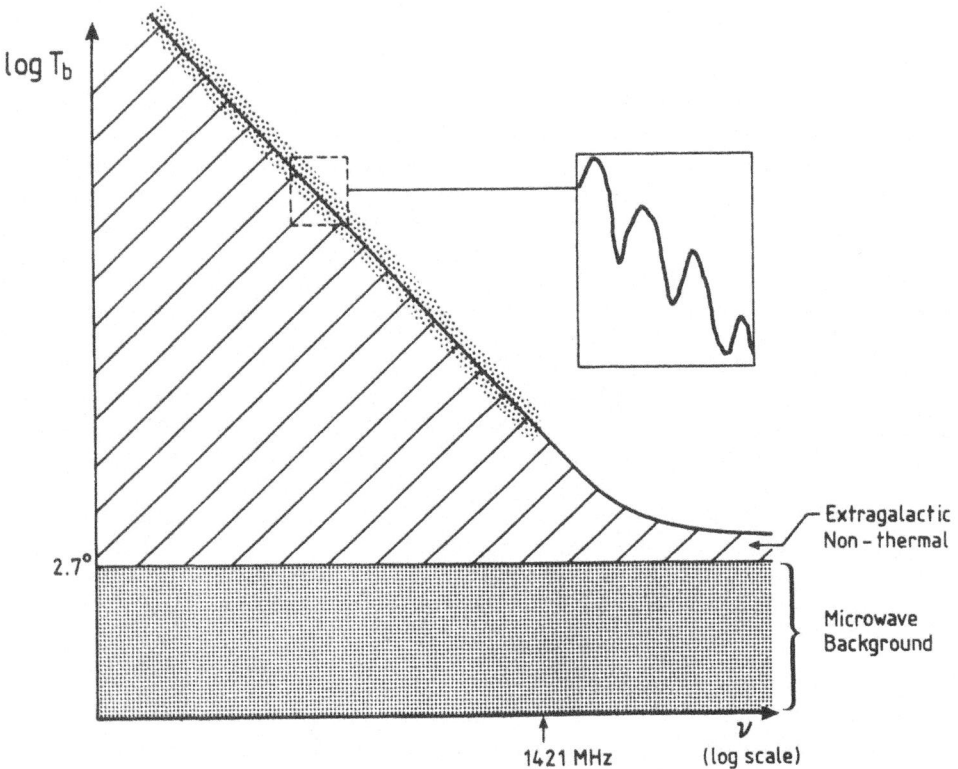

Figure 1. The dominant extragalactic backgrounds in the radio bands are the primordial 2.7 K black body radiation, and the non-thermal synchrotron background, whose brightness temperature goes as $\sim \nu^{-2.7}$. Intergalactic HI emits and/or absorbs via the 21 cm transition, and in consequence changes the background temperature. Although this effect would be undetectably small if the HI were smoothly distributed, any 'clumping' of the gas into incipient clusters would create spectral and angular structure in the background. By scanning in angle using a narrow bandwidth of frequencies, structures in the high-z neutral hydrogen could be detected. By comparing the angular structures seen in two 'maps' made at slightly different frequencies, one could distinguish between effects due to discrete non-thermal sources (for which the two maps would correlate) and those due to HI (where the maps would not correlate).

from extragalactic sources. However, it may be possible to pick out the 21cm contribution, because of its characteristic angular structure, combined with fine structure in frequency space. (see Figure 1 and caption)

The contribution to the radio background temperature at $1420\,(1+z)^{-1}$ MHz due to uniformly distributed HI at redshift z is easily calculated to be

$$\Delta T_b \simeq 0.1 h\, \Omega_B \,(1+z)^{\frac{1}{2}}\, F\delta^\circ\mathrm{K}.$$

The factor F is unity if the spin temperature $T_s >> T_{rad}$. If there has been no heat input into the primordial gas before the relevant epoch, then T_s may be less than T_{rad}; in the opposite limit when $T_s << T_{rad}$, $F = -(T_{rad}/T_s)$. (In the latter case, in other words, the 21cm transition *reduces* the background temperature, a given column density giving an effect larger by T_{rad}/T_s than in the $T_s >> T_{rad}$ case.) If we observe a region where the density is higher than average, or the cosmic expansion slower than average, the effect will be enhanced by a factor δ. For a linear fluctuation, δ is enhanced above unity by $\frac{5}{3}$ the amplitude of the density perturbation, the extra $\frac{2}{3}$ coming from the reduced expansion rate in an overdense growing perturbation (which further increases the HI column density per unit redshift interval).

So a smooth distribution of pregalactic HI would contribute a background temperature T_b, which is smaller than the microwave background, and very much smaller than the non-thermal background that dominates the radio sky. However, if the HI is spatially clustered, or the velocities are perturbed from the mean Hubble flow, there will be spatial and spectral structure in its emission (which can be deleted by difference measurements, using a scheme involving switching between nearby patches of sky, the observations being made at a range of frequencies). The expected fine structure in frequency space allows the 'signal' from non-uniform HI to be distinguished from patchiness in the non-thermal synchrotron background. One of the standard radio-frequency bands is 151 MHz. This is specially appropriate because it corresponds to $z \simeq 8.4$, a redshift where the primordial hydrogen may still be neutral.

The most hopeful possibility for carrying out this kind of 'tomography' on protoclusters involves the proposed GMRT in India (Swarup 1984). This instrument is planned to comprise an array of 34 dishes, each 45 metres in diameter. The dishes will not be sufficiently well surfaced to be effective at high frequencies. However, the array will be 8 times more sensitive than the VLA at 327 MHz. It will also operate at 151 MHz, where its resolution is 20 arc seconds for the entire array, and 7 minutes if one uses only the central array where the dishes are more closely packed. The expected sensitivity is such that incipient clusters will certainly be detectable if they resemble the predictions of the explosion model. In this model (Ikeuchi 1981, Ostriker and Cowie 1981) the gas would have been shock heated in a huge shell, but would have undergone Compton cooling and recombination at $z \gtrsim 8$ before fragmenting into galaxies. The gas would then be neutral, but on an adiabat with $T_s >> R_{rad}$. Cluster-scale perturbations, perhaps still with $T_s < T_{rad}$, would also be detectable in a high density baryon-dominated universe. There is less chance of a positive result in the case of, for instance, the standard cold dark matter scenario. However, the precursors of specially large systems resembling the 'Great Attractor' (Lynden-Bell *et al.* 1988) or the 'Shapley Concentration' (Scaramella *et al.* 1989; Lahav *et al.* 1989; Raychaudhury 1989) should reveal themselves in this way.

5. A POSSIBLE COSMOLOGICAL TEST USING HIGH-Z HI

High-z 21 cm observations would also offer the possibility of an important cosmological test if structure were to be seen. We can assume that the perturbations, even if not all spherical, are at least statistically isotropic. The angular diameter and velocity difference across a comoving spherical region are related by

$$2\theta = \frac{v\Omega_0^2(1+z)}{2c\left(1+\Omega_0 z\right)^{\frac{1}{2}}\left[\Omega_0 z - (2-\Omega_0)\left\{(1+\Omega_0 z)^{\frac{1}{2}} - 1\right\}\right]}.$$

Hence correlating the characteristic size with the characteristic velocity difference associated with any inhomogeneities that are detected would tell us something about the value of Ω_0. Putting some numbers in we obtain $2\theta \simeq 0.242\,v/c$ for $\Omega_0 = 1$, and $2\theta \simeq 0.213\,v/c$ for $\Omega_0 = 0.1$ (which turns out to be very close to the minimum value for this factor). So we can see that the test is not particularly sensitive; observations of > 100 independent regions would certainly be required to yield a significant result. However, the GMRT (a synthesis instrument), would automatically yield ~ 5 arcmin resolution over a field of about 4 square degrees so it should be feasible to tell between these two extreme values of Ω_0.

The shells in explosion models would be expanding somewhat faster than the Hubble flow, the extra factor depending on which explosive phase they were in, on whether the explosion was impulsive or had energy input over some period of time and on other details of the model. Hence relating the velocity difference and angular size would give information about both the value of Ω_0 and the evolution of the explosion together, but it would not be possible to separate these effects by observing individual shells.

6. SOME CONCLUDING COMMENTS

The most impressive calculations against which one can confront the data on present-day clustering are the results of N-body simulations. However, it is important to bear in mind that these tell us, at best, the distribution of the postulated non-dissipative dark matter. The next step is to include gas dynamics. However, even when this has been done, it will not tell us how efficiently this gas turns into stars, and the extent to which the star formation depends on cosmic epoch, galactic environment, etc. For a long time, I believe we shall have to depend on parameter fitting, using empirical luminosity functions, etc., when we try to compare these dynamical models with the observations.

Nonetheless, a great deal can be learnt from the inclusion of gas dynamics in the simulations (which is at least a well-posed phenomenon), even if neither star formation nor realistic radiative cooling is allowed for. Although cold gas tracks the distribution of the collisionless dark matter in the linear domain, once shell crossing and shocks occur there is no reason to believe that the distributions will remain similar. In particular, there is the possibility that some gas will remain on low adiabats, because it manages to escape passage through a strong shock. Conversely, there is the possibility that in some contexts the dark matter may remain colder than the gas. For instance, if two protogalaxies crash into one another, the gas is shock-heated to a temperature depending on their relative velocities, but the dark matter from each galaxy can interpenetrate the other, and it is not obvious how much the velocity distribution for the collisionless matter is raised.

We have heard at this meeting about the difficulties faced by those who wish to reconcile the observations with the predictions of the cold dark matter model. I think it is important to be realistic about how much one can expect of the model that theorists have so far explored. These models allow one to calculate the large-scale distribution of the dark matter, and therefore allow one to predict the gravitationally-induced velocities of galaxies, and the primary angular fluctuations in the microwave background. On the other hand, the complicated processes, involving different kinds of feedback, whereby the gas in a potential well turns into a galaxy, are not the kind of thing one can expect to be described by a simple few-parameter system. For this reason, I do not think the supporters of cold dark matter can be accused of evasiveness or special pleading if they feel the need to tinker with the biasing parameter, and make this more complicated than a simple threshold.

In general, astrophysicists must distinguish between those branches of their subject which they expect to be described exactly by a few-parameter system, analogous to what one hopes for in particle physics, and those areas where we cannot expect to progress beyond parameter fitting. Perhaps the paradigm science here is geophysics rather than physics. Plate tectonics was an important contribution to geophysics, and a great unifying idea, but it is obviously no disparagement of that idea to point out that it cannot explain the exact shape of each continent. In cosmology, the amplitude and spectrum of initial fluctuations, the baryon/photon ratio, etc. may be explicable by the kind of theory physicists like; and one may be able to compute the present *mass* distribution with some confidence if it evolves via purely gravitational effects. But the interactive gas-dynamical processes are more like geophysics. The relation between the distribution of mass and light may be as complicated as the shape of the continents – it may be just as unrealistic to describe it by a single parameter. There are some features of the Universe, and of large-scale structure in particular, which it is too much to expect a deductive theory to account for. The knack is to distinguish those features from the ones which can in principle provide a decisive discriminant between the embarrassingly large number of alternative theories that remain in play.

I am grateful to Douglas Scott for collaboration on the work summarised in sections 4 and 5.

REFERENCES

Couchman, H.M.P. and Rees, M.J. 1986, *Mon. Not. R. astr. Soc.* **221**, 53.
Efstathiou, G.P. and Rees, M.J. 1988, *Mon. Not. R. astr. Soc.* **230**, 5P.
Fall, S.M. and Efstathiou, G.P. 1980, *Mon. Not. R. astr. Soc.* **193**, 189.
Gunn, J.E. 1982 in *'Astrophysical Cosmology'* ed. H.A. Bruck *et al.* (Vatican Publications).
Gunn, J.E. and Peterson, B. 1965, *Astrophys. J.* **142**, 1633.
Hogan, C. and Rees, M.J. 1979, *Mon. Not. R. astr. Soc.* **188**, 791.
Ikeuchi, S. 1981, *Publ. Astr. Soc. Japan* **33**, 211.
Lahav, O., Edge, A.C., Fabian, A.C. and Putney, A. 1989, *Mon. Not. R. astr. Soc.* **238**, 881.
Lynden-Bell, D., Faber, S.M., Burstein, D., Davies, R.L., Dressler, A. Terlevich, R.J. and Wegner, G. 1988, *Astrophys. J.* **326**, 19.
Mitchell, L., Miller, P.S. and Boyle, B.J. 1990. in preparation
Ostriker, J.P. and Cowie, L. 1981, *Astrophys. J. (Letters)* **243**, L127.
Raychaudhury, S. 1989, *Nature* **342**, 251.
Rees, M.J. 1988 in *'Big Bang, Active Galactic Nuclei and Supernovae'* ed. S. Hayakawa and K. Sato (Universal Acad. Press Inc.) p121.
Rees, M.J. 1990, *Science* **247**, 817.

Scaramella, R., Baiesi-Pillastrini, G., Chincarini, G., Vettolani, G. and Zamorani, G. 1989, *Nature* **338**, 562.

Schneider, D., Schmidt, M. and Gunn, J.E. 1989, *Astron J.* **98**, 1951.

Scott, D. and Rees, M.J. 1990, *Mon. Not. R. astr. Soc.* **247**, 510.

Subramanian, R. 1990, *Journal of Astron. & Astrophys.* (in press).

Swarup, G. 1984 *'Giant Metre-Wavelength Radio Telescope - a Proposal'*, Radio Astronomy Centre, Tata Institute, Ootacamund, India.

Warren, S.J. and Hewett, P.C. 1990, *Rep. Prog. Phys.* **53**, 1095.

LARGE SCALE COHERENT STRUCTURES IN THE UNIVERSE

LEV KOFMAN
CITA University of Toronto
M5S 1A1, Canada

ABSTRACT. In the first part of the talk, the non-linear dynamics of the large scale structure of the Universe due to gravitational instability is considered in the framework of the Adhesion Model. I argue that pancaking should occur on the all scales, forming large scale coherent structures – 'superpancakes', due to large-scale coherent motions of clumps of matter formed before. In the second part I consider a simple statistics of nonlinear matter distribution: the one-point probability distribution functions of the smoothed peculiar velocity field and density field. In the Zel'dovich approximation, the velocity distribution function remains constant even through nonlinear evolution, while the density distribution becomes non Gaussian very rapidly, and depends on the linear *rms* with the given filtered scale only. One can expect a different shape of distribution function in other cosmological scenarious besides gravitational instability.

1. Basic Equations

I will assume that:

i) Inflation produces a flat universe $\Omega \simeq 1$,

ii)at the present epoch, most of the mass is in the form of dark matter of relic origin, primarily Cold Dark Matter (CDM),

iii)Nonlinear LSS originated by gravitational instability of small primordial inhomogeneities arising from vacuum fluctuations during Inflation.

Let $\vec{x}, \vec{v} = a\frac{d\vec{x}}{dt}, \rho(t, \vec{x})$ and $\phi(t, \vec{x})$ be, respectively, the comoving coordinates, peculiar velocity and density of dark matter (neglecting baryons), and peculiar Newtonian gravitational potential. The motion of dust-like dark matter before the particle orbits cross obeys a nonlinear system of equations

$$\frac{\partial \rho}{\partial t} + 3H\rho + \frac{1}{a}\nabla_x(\rho\vec{v}) = 0, \tag{1}$$

$$\frac{\partial \vec{v}}{\partial t} + \frac{1}{a}(\vec{v}\nabla_x)\vec{v} + H\vec{v} = -\frac{1}{a}\nabla_x\phi, \tag{2}$$

$$\nabla_x{}^2\phi = 4\pi Ga^2(\rho - \bar{\rho}), \tag{3}$$

495

where a is a scalar factor of an expanding Universe, $H = \dot{a}/a$ and $\bar{\rho}$ is a mean density. In the linear regime the evolution of density perturbations $\delta\rho = \rho - \bar{\rho}$ is defined by the growing mode of adiabatic perturbations and is given by

$$\frac{\delta\rho}{\rho} = Aa(t)\nabla_x{}^2\phi, \qquad (4)$$

and for the velocity

$$\vec{v} = -\frac{2c^2}{3H}\left(\frac{1}{a}\nabla_x\frac{\phi}{c^2}\right). \qquad (5)$$

In (4) the combination $A = (\frac{3}{2}H^2a^3)^{-1}$ does not depend on time for a flat Universe with dust-like matter. Thus the initial condition of the basic equations is defined by the space distribution of the linear gravitational potential ϕ, which practically does not change in time in the linear regime. Usually ρ is taken as realization of a random (Gaussian) field, which statistical characteristics are constructed from the dispertion measures σ_j^2. For the field smoothed by the window function $W(R)$ with filtered scale R:

$$\sigma_j^2(R) = \frac{1}{2\pi^2}\int_0^\infty dk k^2 k^{2j}(k^4|\phi_{0k}|^2 c_k^2)W_k^2(R). \qquad (6)$$

The properties of the field depends strongly on the particuliar cosmological model, both through the initial spectrum of perturbations $|\phi_{0k}|^2$ from Inflation, and the transfer function of the dark matter c_k.

In this contribution I focus on approximated methods of treatment with nonlinear dynamics and statistics of nonlinear matter distribution which obeys eqs.(1)-(3). In Sec.2 the Zel'dovich approximation is formulated in terms of an equation for gravitational potential. It turns out that the time evolution of the 3D-hypersurface of gravitational potential is completely similar to the changing of the wave front according to the Huygens principle in the geometrical optics, till the epoch of the caustic (pancake) formation. In Sec. 3 we review the Adhesion Model which allows us to use the Zel'dovich approximation beyond the caustics. The resulting matter distribution is a random packed cellular structure with mass concentrations succesively in pancakes, filaments and knots. In Sec.4 we follow the further motion of these knots under the influence of the long wave perturbations. This leads to the formation of the 'superpancakes' and 'superfilaments', which are properly large scale coherent structures. In Sec.5 we consider the probability distribution functions of smoothed fields of density and velocity in the Zel'dovich approximation. The result for $P(\vec{v})$ is remarkable simple: it does not change in time. The density distribution, on the contrary, departs very rapidly from the initial Gaussian form. Finally, in Sec.6, we discuss the properties of probability distribution functions.

2. Zel'dovich approximation via gravitational potential

Let us use a new time variable $a(t)$ instead of t, and introduce a comoving velocity $\vec{u} = \frac{d\vec{x}}{da} = \vec{v}/a\dot{a}$. Then eq.(2) has the form

$$\frac{\partial\vec{u}}{\partial a} + (\vec{u}\nabla_x)\vec{u} = -\frac{3}{2a}(\vec{u} + A\nabla_x\phi) \qquad (7)$$

If we start with a potential motion (5), then, from (2), according to the Kelvin's theorem, it remains a potential field forever (no baryons). Let Φ be a velocity potential so $\vec{u} = \nabla_x \Phi$. Then from (7) we find a general relation between the velocity potential and the Newtonian gravitational potential

$$\frac{\partial \Phi}{\partial a} + \frac{1}{2}(\nabla_x \Phi)^2 = -\frac{3}{2a}(\Phi + A\phi). \tag{8}$$

Substituting density ρ from eq.(3) into (1) we have

$$\nabla_x \left[\frac{\partial}{\partial a}(a\nabla_x \phi) + (A^{-1} + a\nabla_x^2 \phi)\nabla_x \Phi \right] = 0. \tag{9}$$

Thus, we have reduced three eqs.(1)-(3) to two eqs. (8), (9), for two scalar fields ϕ and Φ, or, substituting ϕ from (8) in (9), even to a single equation for field Φ.

In the one dimensional case the exact solution of eqs. (8) and (9) is

$$\phi = -A^{-1}\Phi \tag{10}$$

and

$$\frac{\partial \Phi}{\partial a} + \frac{1}{2}(\nabla_x \Phi)^2 = 0. \tag{11}$$

In the three dimentional case, in the linear regime, this solution also satisfies the basic equations. The ansatz, which leads to the Zel'dovich approximation, consists of using equations (10) and (11) in nonlinear regime also.

Now we review the Zel'dovich ansatz in terms of the potential Φ. Eq.(11) is similar to the eikonal equations in geometrical optics, where Φ plays the role of phase of wave front, and a – the role of transversal coordinate (for more about the similarity see [1]). Let the (deformation) tensor $D_{ij} = \nabla_{x_i}\nabla_{x_j}\Phi$ has the eigenvalues λ_i, and let their symmetric combinations be

$$J_1 = \lambda_1 + \lambda_2 + \lambda_3, \quad J_2 = \lambda_1\lambda_2 + \lambda_1\lambda_3 + \lambda_2\lambda_3, \quad J_3 = \lambda_1\lambda_2\lambda_3. \tag{12}$$

Taking the derivatives of Φ in respect with ∇_x, we get the following set of equations

$$\frac{\partial \vec{u}}{\partial a} + (\vec{u}\nabla_x)\vec{u} = 0, \tag{13}$$

$$\frac{DJ_1}{Da} + J_1^2 - 2J_2 = 0, \quad \frac{DJ_2}{Da} + J_1 J_2 - 3J_3 = 0, \quad \frac{DJ_3}{Da} + J_1 J_2 = 0, \tag{14}$$

where $\frac{D}{Da} = \frac{\partial}{\partial a} + \vec{u}\nabla_x$. Eq.(13) describes the free streaming of particles in a comoving coordinates. The solution of eq.(14) in terms of principal radii of curvature of 3D-hypersurfaces of the potential $R_i = \lambda_i^{-1}$ is extremely simple:

$$R_i = R_{0,i} + a. \tag{16}$$

Additionally, the solution of eq.(11) is

$$\Phi(a, \vec{x}) = \Phi_0(\vec{q}) + \frac{(\vec{x} - \vec{q})^2}{2a}. \tag{17}$$

From (17) we get

$$\vec{x} = \vec{q} + a\nabla_q \Phi_0(\vec{q}) \tag{18}$$

and

$$\vec{u} = \nabla_x \Phi(a, \vec{x}) = \nabla_q \Phi_0(\vec{q}). \tag{19}$$

Eq.(18) is the usual form of the Zel'dovich approximation [2], which describes the displacement of a particle with initial Lagrangian coordinate $\vec{q}$ to current Eulerian coordinate $\vec{x}$, under the initial perturbation $\Phi_0(\vec{q})$. The solution (16) and (17) describes the deformations of 3D-hypersurface of the potential Φ. Eq.(16) is nothing but the Huygens principle of geometrical optics. Eq.(17) gives us another but equivalent geometrical prescription: the insertion of the osculating paraboloid $P = \frac{(\vec{x}-\vec{q})^2}{2a} - \Phi(\vec{x}, a)$ tangential to the hypersurface of the initial linear gravitational potential $\phi_0 = -A^{-1}\Phi_0(\vec{q})$ at the point $\vec{q}$. Then the Eulerian coordinate $\vec{x}$ at a chosen time $a(t)$ is just the projection of the apex of P. For this the hypersurface $\Phi(\vec{x}, a)$ crosses over the apexes of paraboloids.

In particular, at the nonlinear stage, negative peaks of ϕ are sharpened, and positive peaks are flattened [3]. The solutions (16)-(19) are valid till the formation of folds of ϕ-hypersurface, which corresponds to caustics (pancakes). The illustration of the time evolution of the gravitational potential according to the Huygens principle is drawn in Fig.1a.

3. Beyond the Zel'dovich solution: Into The Adhesion Model

The well known advantage of the Zel'dovich approximation is a correct prediction of pancake formation. The pancake model is the basis for the old fashioned HDM model, where pancakes have the sizes of superclusters, and form very recently. Then, ten years ago, one could think that pancakes enlarge and form a connected network of sheets and filaments, which exist as an intermediate asymptote and later decay into isolated (randomly) distributed clumps.

Actually, in the basic eqs.(1)-(3), no intrinsic scale is present, and in every model the first non-linear objects are very asymmetric pancakes. Any initial shape of potential Φ_0, according to (17), leads to singularities – caustics. Pancakes inevitably arise in the more popular CDM model, but here they have cosmologically negligible sizes and form very early. Thus traces of pancakes of first generation have already disappeared in the present structure. The direct application of the Zel'dovich approximation becomes invalid soon after the formation of the pancakes, hence it is widely believed that the advantages of the pancake scenario are disadvantages for the models like CDM. Here LSS formation has features of hierarchical clustering. Direct extrapolation of the linear growth of density field peaks (4) inspires the Press-Schechter [4] prescription in which structure evolves hierarchically. There the smallest mass scales collapse first, and then merge to form structures on larger and larger scales.

The Press-Schechter and the Zel'dovich models seemed, at first glance, to be competing and probably even mutually incompatible. However the metamorphosis which has taken place with the pancakes in an Adhesion Model provides compatibility with both scenarios.

The Adhesion Model [1,5-7] is an approximate description of the large scale gravitating cosmological system by the Burgers equation

$$\frac{\partial \vec{u}}{\partial a} + (\vec{u}\nabla_x)\vec{u} = \nu\nabla_x^2\vec{u}. \tag{20}$$

The new term (compare with eq.(13)) on the right side mimics the adhesion of particles inside pancakes just after their orbits cross. The introduction of the artificial viscosity term is based on the fact that in the real system the thickness of pancakes l remains much smaller than their sizes or distances D between them. The dimensional coefficient of viscosity $\nu \approx lD$ must be much smaller then D^2, in which case the viscosity term is negligible outside of pancakes.

For the potential motion under consideration (the Kelvin's theorem is also valid here) eq.(20) permits an analytical solution [10], in terms of the potential

$$\Phi(\vec{x}, a) = -2\nu\ln\left[\frac{1}{(4\pi\nu a)^{3/2}} \int d^3\vec{q}\, exp\left[-\frac{1}{2\nu}(\Phi_0(\vec{q}) + \frac{(\vec{x} - \vec{q})^2}{2a})\right]\right]. \quad (21)$$

For the most interesting case $\nu \to 0$ the steepest descent method can be applied, resulting in

$$\Phi(\vec{x}, a) = -2\nu\ln\left[\sum_\alpha \sqrt{\delta_{ij} + aD_{ij}(\vec{q}_\alpha)}\, exp\left[-\frac{1}{2\nu}(\Phi_0(\vec{q}_\alpha) + \frac{(\vec{x} - \vec{q}_\alpha)^2}{2a})\right]\right]. \quad (22)$$

Here $\vec{q}_\alpha$ are the extremal points of (17), which are nothing but the Lagrangian positions of particles at Eulerian position $\vec{x}$ under the Zel'dovich approximation. In the case of a one to one correspondence between Lagrangian and Eulerian spaces, eq.(22) reverts to (17). In the more general case, one can generalize to (22) the geometrical interpretation of the Zel'dovich solution. Contrary to the Zel'dovich solution, adhesion of particles into the pancake corresponds to the case of a paraboloid having two points of contact with the ϕ_0-hypersurface but without crossing at any point. The projection of the apex of this paraboloid gives us the position $\vec{x}$ of the 'stuck pancake', which actually lies inside the real pancake. If P touches the hypersurface in three or even four points simultaneously, then its apex indicates the position of a filament and knot, respectively.

In the geometrical language of the ϕ-deformation the solution (22) corresponds to the advanced Zel'dovich approximation but with excision of the piece of the hypersurface below the cusps, see Fig.1b.

Eventually the Lagrangian region have run into cellular structure formed by pancakes, filaments and knots. The cellular structure is the skeleton of the matter distribution, which is governed by the initial linear random field of the gravitational potential. Let us note that the parameter ν drops out of the final geometrical interpretation. The main feature of such an approach is that the analysis may be carried out for an arbitrary moment independent of previous evolution.

The reliability of the Adhesion Model (20) has been checked by direct comparision with N-body simulation with the same initial condition [9-12]. Fig.2 illustrates the agreement between the results from high-resolution N-body simulations [13] and numerical code based straightforwardly on the insertion of paraboloids [9,14].

4. Large Scale Coherent Structures

In the Adhesion Model, at the beginning, the first generation pancakes have appeared, associated with maximum of λ_1. Pancakes grow, merge or intersect [15] creating filaments, and in turn knots form at the intersection of the filaments. Thus, a randomly packed cellular structure arises [5,9]. Then some cells of this structure grow with the time, some cells disappear, and meanwhile knots can merge to form knots of larger mass. Thus the dynamics of cellular structure provides the hierarchical clustering of knots.

At very large scales (large filtered scale R) the rms fluctuations (6) are small and the linear approximation is valid. At the smaller but still fairly large scales, the fluctuations are quasilinear, and it is reasonable to assume that the Zel'dovich approximation is valid here. In order to combine of 'bottom-up' gravitational clustering and large-scale pancaking we need the new qualitative concept, which reflects the way of development of structure.

If we consider the deformation of developed cellular structure due to long-wave perturbations of the gravitational potential, we immediately recognize that knots tend to form large-scale substructure as a result of the large-scale coherent motion of matter. In various numerical codes one can see coherent objects resembling filaments [13,16,17]. However, statements about the formation of the network of structures in CDM-type models is author-dependent. In the Adhesion Model at the latest stages, only the large-scale inhomogeneities of the gravitational potential drive the skeleton of the matter distribution. Let us now filter the initial potential and construct the skeleton. Fig.3 demonstrates that coherent structures tend to coincide with the skeleton from smoothed potential. The alignment of clumps with the skeleton is the result of the action of the smoothed velocity field. We therefore call these coherent structures 'superpancakes' and 'superfilaments'. Thus, behind the seeming coherent structures in N-body simulation, we have 'superpancakes' in the Adhesion Model. Actually 'superpancakes' could be treated by means of the Zel'dovich approximation applied to the smoothed gravitational potential.

'Superpancakes', contrary to the former first generation pancakes, are builted not from dark matter particles themself, but from clumps which form before. Prominent 'superpancakes' consists of a number of clumps. However, one expects that at the very late epoch 'superpancakes' could form from very few large 'superclumps'. Meanwhile there is still large scale coherent motion of 'superclumps'.

The formation of the 'superpancakes' due to large scale inhomogeneities of the gravitational potential and the dynamical washing out of the small scale initial inhomogeneities in terms of gravitational potential is sketched in Fig.1c.

5. Statistics of Nonlinear Matter Distribution

The statistics of a continuous density function $\rho(t, \vec{x})$ are entirely described by the set of correlation functions $\xi^{(p)}(\vec{x}_1, ..., \vec{x}_p) = <\rho(\vec{x}_1)...\rho(\vec{x}_p)>$. The dynamical equations for correlation functions are a non-closed BBGKY hierarchy of equations [18]. Another more direct way to define the statistics is based on n-point probability distribution function

$$P(\rho_1, ...\rho_n) = \int \frac{d^n\theta}{(2\pi)^n} e^{i \sum_{k=1}^{n} \rho_k \theta_k} exp\left[\sum_{p=1}^{\infty} \frac{i^p}{p!} \sum_{j_1=1}^{n} \sum_{j_p=1}^{n} \theta_{j_1}...\theta_{j_p} \xi^{(p)}(\vec{x}_{j_1}, ..., \vec{x}_{j_p})\right].$$

(23)

Here we consider the one-point probability distribution functions of smoothed density and velocity fields in the Zel'dovich approximation. The density following a particle is

$$\rho(\vec{q}, a) = \frac{\bar{\rho}(a)}{|(1 - a\lambda_1)(1 - a\lambda_2)(1 - a\lambda_3)|}.$$

(24)

In the most interesting case, the initial ρ in the linear regime (4) is a random Gaussian field, for which the point probability is normal

$$P_0(\rho) = \frac{1}{\sqrt{2\pi}\sigma} exp[-\frac{(\rho - \bar{\rho})^2}{2\sigma^2}], \qquad (25)$$

where $\sigma = a\sigma_0$ is the linear rms density. As non-linearity develops, the density distribution becomes non-Gaussian. The former symmetric distribution (25) of small fluctuations is broken, because of the positive density contrast can reach any large value while the negative density contrast is restricted by $\rho \geq 0$. Eq.(24) describes the evacuation of matter from regions around positive peaks of the initial gravitational potential ϕ, resulting in more or less spherical voids, which expand in time and tend to occupy a largest fraction of volume with $\rho < \bar{\rho}$, as well as formation of very anisotropic collapsed dense pancakes which tend to occupy a smaller fraction of volume. Hence the probability function $P(\rho)$, meaning the fraction of volume with a given value of density, is expected to be very non Gaussian even in the quasilinear stage.

For an initially Gaussian Φ-field, the joint distribution function of ordered eigenvalues λ_i involving in (24) is (see [19])

$$W(\lambda_1, \lambda_2, \lambda_3) = 3!\frac{5^3 27}{8\sqrt{5}\pi\sigma_0^6}(\lambda_1 - \lambda_2)(\lambda_1 - \lambda_3)(\lambda_2 - \lambda_3)exp\Big[-\frac{1}{\sigma_0^2}(3J_1^2 - \frac{15}{2}J_2)\Big]. \quad (26)$$

Our goal is to find the distribution function $P(\rho, a)$ basing on (24) and(26). Let us introduce auxillary variables depending on λ_i

$$z = 1 - aJ_1 + a^2 J_2 - a^3 J_3,$$

$$y = a^2 J_2 - a^3 J_3,$$

$$x = a^3 J_3, \qquad (27)$$

where J_i are defined in (12). Their joint distribution function is

$$G(x, y, z) = \frac{P(\lambda_1, \lambda_2, \lambda_3)}{|\frac{\partial(x, y, z)}{\partial(\lambda_1, \lambda_2, \lambda_3)}|}, \qquad (28)$$

where λ_i are expressed through x, y, z. From (27) we get a cubic equation for λ_i

$$\lambda^3 - \frac{1 - z + y}{a}\lambda^2 + \frac{x + y}{a^2}\lambda - \frac{x}{a^3} = 0, \qquad (29)$$

only three real ordered roots of which $\lambda_i(x, y, z)$ we have to substitute in (28). For the complex roots one puts $G = 0$. Then the distribution fuction of the denominator z of the right side eq.(24) is

$$F(z) = \int dy \int dx G(x, y, z). \qquad (30)$$

Finally, for the distribution function of density $\rho = \bar{\rho}/|z|$ in the Eulerian space $P(\rho(\vec{x}), a)$, we have

$$P(\rho) = \left(\frac{c}{\rho}\right)\frac{\bar{\rho}}{\rho^2}\left[F(\frac{\bar{\rho}}{\rho}) + F(-\frac{\bar{\rho}}{\rho})\right]. \tag{31}$$

Let us draw attention to the extra factor $\frac{c}{\rho}$ that appears here due to the mapping from the Lagrangian space into Eulerian space as $|\frac{\partial \vec{x}}{\partial \vec{q}}| = \frac{\bar{\rho}}{\rho}$. The normalization factor c guarantees mass conservation

$$\int_0^\infty d\rho \rho P(\rho) = \bar{\rho}, \tag{32}$$

while the normalization of $P(\rho)$ deviates from unity and corresponds to the mean number of folds of multistream flows

$$\int d\rho P(\rho) = <N>. \tag{33}$$

Thus the problem is reduced to finding the region in the (x,y,z)-space for which the all three roots of eq.(29) are real and which gives the appropriate integration limit in (30). After that, collecting together (26),(28), (30) and (31), we find the probability function of density

$$P(\rho, a) = \frac{C}{\rho^3 a^4}\int_{3\bar{\rho}^{1/3}}^\infty d\alpha \; e^{-\frac{(\alpha-3)^2}{2\sigma^2}}\left(1 + e^{-\frac{6\alpha}{\sigma^2}}\right)\left(e^{-\frac{5\beta_1^2}{2\sigma^2}} + e^{-\frac{5\beta_2^2}{2\sigma^2}} - e^{-\frac{5\beta_3^2}{2\sigma^2}}\right), \tag{34}$$

$$\beta_n(\alpha) \equiv \alpha\left(\frac{1}{2} + \cos\left[\frac{2}{3}(n-1)\pi + \frac{1}{3}\mathrm{arcos}\left(\frac{54\bar{\rho}}{\alpha^3} - 1\right)\right]\right), \quad \tilde{\rho} \equiv \bar{\rho}/\rho, \quad \sigma^2 \equiv (a\sigma_i)^2/5$$

with σ_o the standard deviation of the initial Gaussian distribution (25). The numerical factors we collect in the unique factor C, which is defined by the condition (32). The function $P(\rho)$ depends only on a single parameter – the linear rms density fluctuations $a\sigma_0(R)$, which is varying either in time, or as a result of changing of filter radius R. In the limit of very small $a\sigma_0(R)$ from (34) we recover to the correct result (25). The family of $P(\rho)$ for different parameters is plotted in Fig.4. In the next Section we will discuss the properties and physical meaning of $P(\rho)$.

There is another way (for more details see [20]) to get the probability distribution functions in the Zel'dovich approximation, which is based on closed kinetic equation for the joint probability distribution function of the potential Φ, velocity $\vec{u}$, invariants of deformation tensor J_i and density ρ. This kinetic equation is derived from eqs.(1), (11),(13),(14) which describe an inertial motion of collisionless particles in comoving coordinates and are valid for the generic case including multistream flows after the formation of caustics. The kinetic equation can be solved by the method characteristics, and for density probability distribution function the result is reduced to (34). For the velocity probability distribution function $Q(\vec{u}, a)$ one obtains the equation

$$\frac{DQ}{Da} = 0. \tag{35}$$

Using (19), we get the solution

$$Q(\vec{u}, a) = Q_0[\vec{u}_0(\vec{x} - a\vec{u})]. \tag{36}$$

If the equation

$$\vec{u}(\vec{x}, a) = \vec{u}_0(\vec{q}), \vec{q} = \vec{x} - a\vec{u} \tag{37}$$

has a single root – no orbit crossing – then the probability distribution function of the velocity remains its initial form!(see also [21]). In the conclusion, we will discuss this point in more details. Meanwhile, solution (36) is valid for all times.

Thus, for Gaussian initial fluctuations, the one-point probability distribution function of the velocity (in the non linear regime but with no orbit crossing) is

$$Q(u_i) = \frac{1}{\sqrt{2\pi\sigma_u^2}} exp\left(-\frac{u_i^2}{2\sigma_u^2}\right), \tag{38}$$

where $\sigma_u(R)$ is initial rms of smoothed fluctuations of velocity component. This property only applies to the one-point distribution; the two-point and higher order probability distribution functions are non Gaussian in the non-linear regime.

6. The Signature of Gravitational Instability in the Density Distribution Function

The physical meaning of the probability distribution functions obtained above is based on the following results established in [21]. The density and velocity distribution functions in the Zel'dovich approximation seem to be in a good agreement with those computed in cosmological N-body simulations, while the *rms* value σ is restricted to satisfy this approximation. Also, there may be evidence for Gaussian initial fluctuations on the scales more than $5h^{-1}Mpc$, since the one-point distribution of observed velocities, as processed throught the POTENT analysis and IRAS analysis, is indistinguishable from a normal curve, in agreement with (38). A preliminary study of the observed deviations from Gaussian distribution shows that they are consistent with being entirely due to non linear effects of gravitational instability.

Thus, the appropriateness of the density probability distribution function from the Zel'dovich approximation applying for the smoothed initial fluctuations provides the concept of the successive pancaking on the all scales.

In conclusion, we discuss some properties of the probability distribution function (34). At large densities $\rho \gg \bar{\rho}$ the asymptotic behavior of the expression (34) is

$$P(\rho, a) \propto \rho^{-3}. \tag{39}$$

It is easy to recognize that this asymptotic is provided by pancakes. The density fall-off in the vicinity of pancake is $\rho \propto \Delta x^{-1/2}$, where Δx is the distance from caustic [1]. The fraction of volume occupied by large density in such a region $dx \propto P(\rho)d\rho$, and therefore we get (39). Thus the asymptotic (39) is consistent with the formation of the pancakes. Actually one can expect that in the real system $P(\rho)$ has the asymptotic (39) around the first caustics to form, and only tends towards (39) for smoothed field, because 'superpancakes' are not strictly caustics. For small $|\rho - \bar{\rho}|$ the distribution function $P(\rho)$ is close to a log-normal distribution. The continuity eq.(1) reads as $d(ln\rho)/da = -\nabla \vec{u}$. The more Gaussian is the velocity field, the more $P(\rho)$ approaches a log-normal distribution. The asymptote at small ρ corresponds to the shape of voids.

The time evolution of $P(\rho)$, especially its asymptotes, is a tool to follow the dynamics of the large scale structure. The shape and asymptotes of the density and velocity distribution functions in other cosmological scenarios (explosions, textures, strings) should be different from those in the gravitational instability picture. It could be very interesting to distinguish them and to apply this comparatively simple test to observations.

Acknowledgements

I thank A.Melott and S.Shandarin for realization of the N-body simulation to compare with the Adhesion Model. I thank D.Pogosyan, S.Shandarin, E.Bertschinger, J.Gelb, A.Nusser and A.Dekel for useful discussions and collaboration.

References
1. Shandarin, S.F. and Zel'dovich, Ya.B., *Rev.Mod.Phys.*, **61**(1989)185.
2. Zel'dovich, Ya.B.,*Astron.Astrophys.* **5**(1970)84.
3. Kofman, L.A. In:*'Morphological Cosmology'*, Springer Lect.Not.in Physics, **332**(1989)154.
4. Press, W., and Schechter, P., *Astrophys.J.* **187**(1974)425.
5. Gurbatov, S.N., Saichev, A.I., and Shandarin, S.F., *M.N.R.A.S.* **236**(1989)385; *Sov. Phys. Doklady* **30**(1985)921.
6. Shandarin, S.F., in: *Large Scale Structures of the Universe*, eds. J.Audouze et al.,Kluwer Academic, Dortreht (1988), p.273.
7. Kofman, L.A. and Shandarin, S.F., *Nature* **334**(1988)129.
8. Burgers, J.M., *The Non-linear Diffusion Equation*, Reidel, Dortreht (1974).
9. Kofman, L.A., Pogosyan, D.Yu. and Shandarin, S.F., *M.N.R.A.S.* **242**(1990)200.
10. Kofman, L.A.,Melott, A., Pogosyan, D,Yu. and Shandarin, S.F, in preparation.
11. Weinberg, D., and Gunn, J., *Astrophys.J. (Lett.)* **352**(1989)25.
12. Nusser, A. and Dekel, A., *Astrophys.J.* **362**(1990)14.
13. Melott, A. and Shandarin, S.F., *Astrophys.J.* **343**(1989)26.
14. Pogosyan, D.Yu., Tartu Preprint (1989).
15. Arnold, V.I., Shandarin, S.F. and Zel'dovich Ya.B., Geophys. Astrophys. Fluid Dynamics, **20**(1982)111.
16. Bertschinger, E., *Ann. Ny. Acad. Sci.***57**(1989)151.
17. Kates, R.,Kotok, E. and Klypin, A.A., *Astron.Astrophys.* **241**(1991).
18. Peebles P.J.E. *The Large-scale Structure of the Universe*, Princeton University Press 1980.
19. Doroschkevich, A.G., *Astrofizica*, **6**(1970)581.
20. Kofman, L.A., CITA Preprint (1991).
21. Kofman, L.A., Gelb, J., Bertschinger, E., Nusser, A. and Dekel, A., in the preparation.

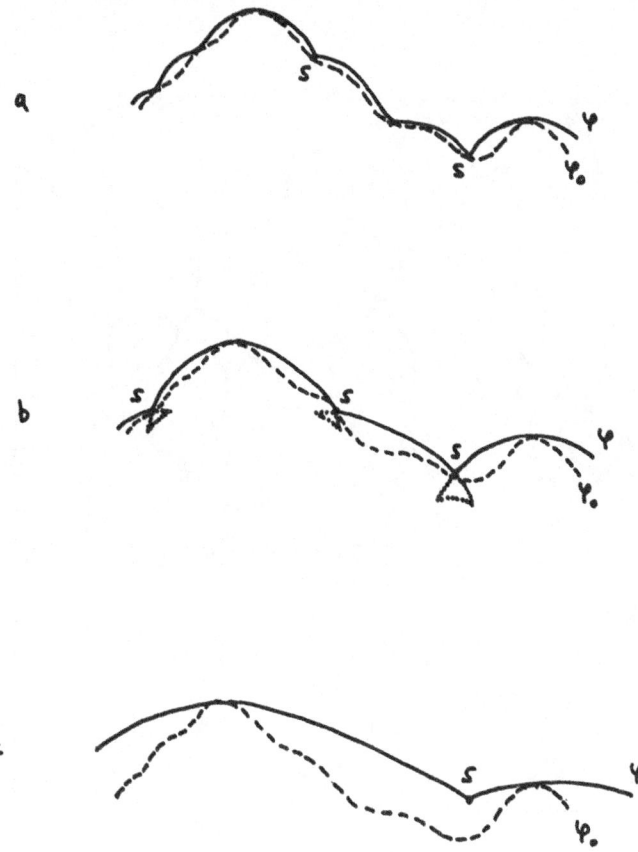

Figure 1.　1D sketch of the time evolution of the hypersurface of the gravitational potential ϕ:

a) in the Zel'dovich approximation, according to the Huygens principle. The dashed line is the initial profile, the solid line is the evolved profile, and the singularities S corresponds to the pancakes.

b) in the Adhesion Model. The dotted line below the cusp is droped off, the points S corresponds to the 'stuck' pancakes.

c) the advanced stage in the Adhesion Model. The final profile of ϕ is defined by the large scale initial inhomogeneities, meanwhile the small scale details are dynamically forgotten

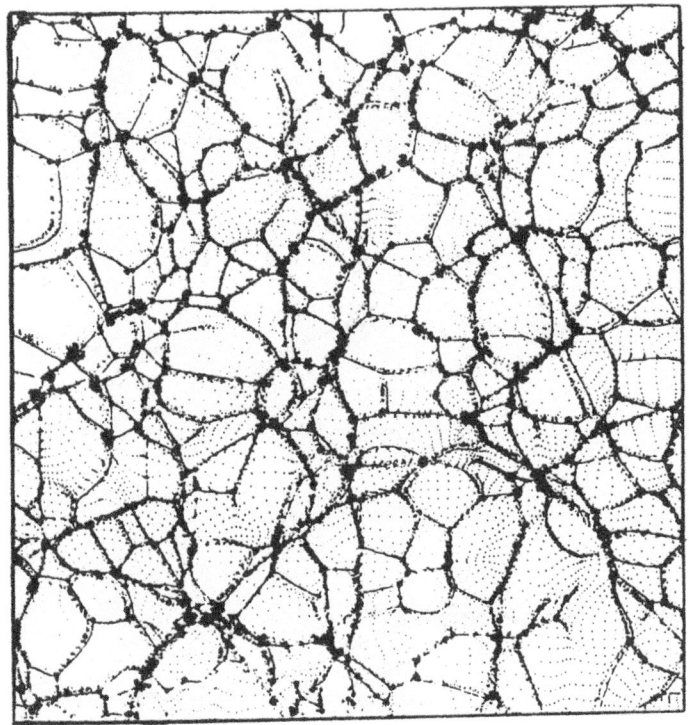

Figure 2. The comparison of the 2D N-body simulation [13](points) with the Adhesion Model (solid line) for the same initial condition (from [10]). The structure for the moment $\sigma = 4$ is shown.

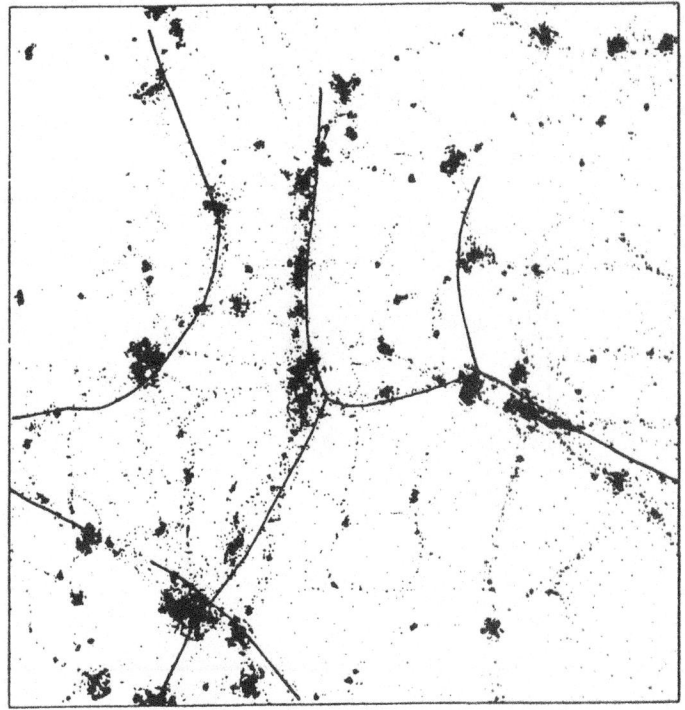

Figure 3. Same as Fig.2, but for the later time $\sigma = 16$, and with smoothed initial potential in the Adhesion Model.

508

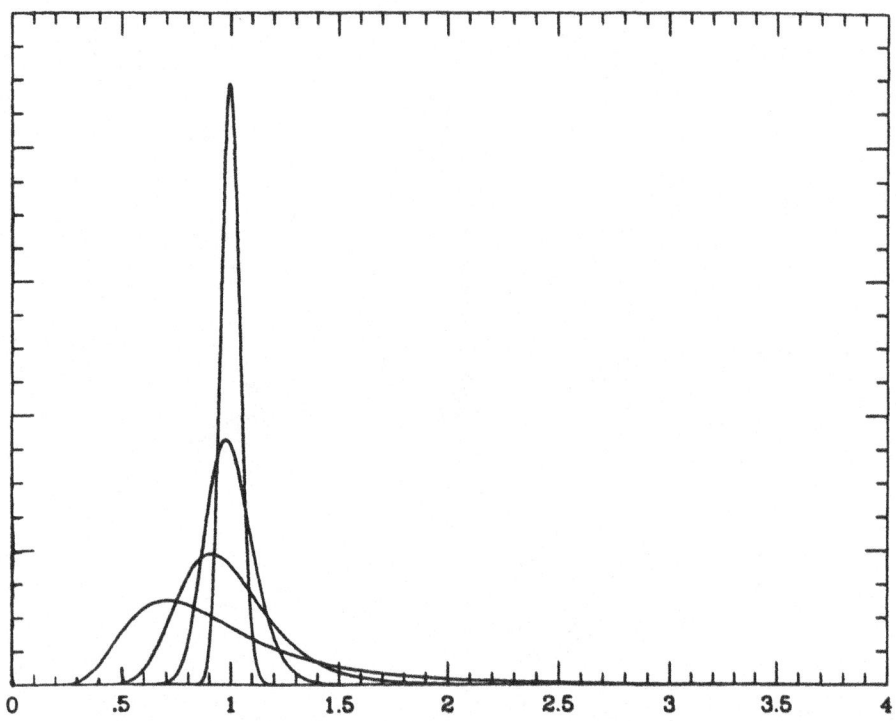

Figure 4. The density probability distribution function $P(\rho)$ in the Zel'dovich approximation (34) for $\sigma = 0.05; 0.1; 0.2; 0.3$.

NON-GAUSSIAN DENSITY FIELD
IN NONLINEAR GRAVITATIONAL CLUSTERING

Yasushi Suto
Uji Research Center
Yukawa Institute for Theoretical Physics
Kyoto University
Uji 611
Japan

ABSTRACT. Nonlinear gravitational clustering is one of the most important and fundamental processes to link the theoretical models with the observed cosmic structures. In this talk, I address two aspects of the nonlinear gravitational effects, mainly on the basis of the recent cosmological N-body simulations; one is the growth rate of density fluctuations in nonlinear regime, and the other is the generation of the non-Gaussian density field.

1. Introduction

The present paradigm in cosmology assumes that the structure in the universe was originated from the primordial adiabatic random-Gaussian density fluctuations, partly motivated by the inflationary theories (see, *e.g.*, Blau and Guth 1987). To be more specific, the Fourier transform of the density contrast:

$$\delta(\mathbf{k}) \equiv \frac{1}{L^3} \int \frac{\delta\rho(\mathbf{r})}{\rho} \exp(i\mathbf{k} \cdot \mathbf{r}) d^3r \equiv |\delta_\mathbf{k}| \exp(i\phi_\mathbf{k}), \tag{1}$$

is assumed to have the following properties (L is the size of the periodic box) ; (i) the density field is isotropic and $\delta(\mathbf{k})$ depends only on $k = |\mathbf{k}|$, (ii) the power spectrum $P(k) = \langle|\delta_\mathbf{k}|^2\rangle$ is in proportion to k, *i.e.*, scale-invariant Harrison-Zel'dovich spectrum (Harrison 1970; Zel'dovich 1972), (iii) the phases of each mode $\phi_\mathbf{k}$ are randomly distributed and thus there is no correlation among them, and (iv) the amplitudes $|\delta_\mathbf{k}|^2$ are Gaussian distributed with variance $P(k)/2$. The above statements are summarized in terms of the probability distribution function of the primordial density field as

$$\text{Prob}(|\delta_\mathbf{k}|^2, \phi_\mathbf{k}) d|\delta_\mathbf{k}|^2 d\phi_\mathbf{k} = \frac{1}{P(k)} \exp\left(-\frac{|\delta_\mathbf{k}|^2}{P(k)}\right) d|\delta_\mathbf{k}|^2 \frac{d\phi_\mathbf{k}}{2\pi}, \tag{2}$$

with $P(k) \propto k$.

Since the cosmological evolution of density fluctuations is a manifestation of the long-range correlation via the Newtonian gravitational force, the above properties, supposedly valid in the primordial density field, will be substantially modified in the course of purely gravitational nonlinear evolution. The resulting statistical properties of the nonlinear density fields, such as the abundances of density peaks, correlation functions of peaks and the probability distribution function of the Fourier components, are essential in exploring the initial condition of the universe through the present distribution of galaxies.

The present talk is based on the recent results of the cosmological N-body simulations (Suto 1989; Suginohara *et al* 1991; Suginohara and Suto 1991a), together with the analytical work (Suto and Sasaki 1991). The simulations adopt the tree method (Barnes and Hut 1986) and were carried out with $N = 64^3$ particles in a periodic cube L^3. The softening parameter is $\epsilon = L/1280$, representing fully three orders of magnitude in linear scale. The initial conditions are generated so as to realize random Gaussian density field with the scale-free fluctuation spectra

$$P(k) \propto k^n, \qquad (n = 1, \ 0, \ -1, \ \text{and} \ -2). \qquad (3)$$

These spectra were chosen so as to study the general properties of the nonlinear gravitational clustering, rather than to examine the specific astrophysical models. The Zel'dovich approximation is used to single out the growing mode of the density flucuations. The mass of each particle is identical and given by M/N (M is the total mass of the system). In practice, we use the units in which $L = M = 1$. For simplicity, we analysed the data simulated in the Einstein – de Sitter universes. Results for cold dark matter scenarios were presented by Tatsushi Suginohara in this conference, and will be described elsewhere (Suto and Suginohara 1991; Suginohara and Suto1991b).

2. Growth of Fluctuations — Quasi-Nonlinear Theory —

Linear perturbation theory is a useful approximation in describing the evolution of density fluctuations in an expanding universe as long as their amplitude is small (Peebles 1980). In fact, this approach provides the basis in considering various cosmological consequences of the theoretical models. On the other hand, the observed distribution of galaxies is strongly inhomogeneous on $r \lesssim 5h^{-1}$Mpc scales, where h is the Hubble constant in units of 100km·sec^{-1}· Mpc^{-1}. Thus nonlinear gravitational effects on the growth of density fluctuations definitely play an important role in the formation of cosmic structures. Does nonlinearity enhance the growth rate of fluctuation power spectrum $P(k)$? Surprisingly, no definite answer was given to such a basic question; some argued on the basis of the spherical model that nonlinearity should enhance the growth rate, while several numerical simulations starting from the Poisson initial conditions suggested the opposite. In developing a cosmo-logical N-body code, I found that the nonlinear growth rate sensitively depends on the spectral shape of the fluctuations (Suto 1989); for scale-free power-spectrum (eq.[3]), $n \gtrsim 0$ models exhibit slower growth in nonlinear regime relative to linear theory, while the growth is enhanced for $n \lesssim -2$ models. This result was con-firmed later by the more comprehensive numerical work (Suginohara et al. 1991) with higher resolution. Is it a physical effect ? or does it suffer from any numer-ical artifacts, for example, due to the discretization effect inherent in the particle method ? In order to answer the question, we considered the perturbation analysis and showed that the simulation results are in good agreement with the analytical argument (Suto and Sasaki 1991).

Let us expand the Fourier transform of density fluctuations (1) as a perturbation series (Juszkiewicz 1981; Vishniac 1983) :

$$\delta(\mathbf{k}, t) \equiv \delta_1(\mathbf{k}, t) + \delta_2(\mathbf{k}, t) + \delta_3(\mathbf{k}, t) + \cdots. \tag{4}$$

Then one obtains to second order,

$$|\delta(\mathbf{k}, t)|^2 = |\delta_1(\mathbf{k}, t)|^2 + 2\mathrm{Re}[\delta_1^*(\mathbf{k}, t)\delta_2(\mathbf{k}, t)] + |\delta_2(\mathbf{k}, t)|^2 + 2\mathrm{Re}[\delta_1^*(\mathbf{k}, t)\delta_3(\mathbf{k}, t)]. \tag{5}$$

The linear perturbation δ_1 is assumed to be given by a random Gaussian field. Then the second term in equation (5) vanishes after taking ensemble average, and the power-spectrum of density fluctuations reduces to

$$P(k, t) \equiv P_{11}(k)\,(t/t_i)^{4/3} + [P_{22}(k) + P_{13}(k)]\,(t/t_i)^{8/3}, \tag{6}$$

where $P_{11}(k)$, $P_{22}(k)$, and $P_{13}(k)$ are given at the initial time t_i. In fact, $P_{22}(k)$ and $P_{13}(k)$ are generated from $P_{11}(k)$, as a source term, in a perturbative sense through the continuity, Poisson and Euler equations (Juszkiewicz 1981; Vishniac 1983).

After a lengthy, but straightforward calculation, we found that equation (15) in Vishniac (1983) can be simply written as

$$P_{22}(k) = \frac{k^3}{98(2\pi)^2} \int_0^\infty P_{11}(kr)dr \int_{-1}^1 dx P_{11}(k\sqrt{1+r^2-2rx}) \frac{(3r+7x-10rx^2)^2}{(1+r^2-2rx)^2}, \tag{7}$$

$$P_{13}(k) = \frac{k^3 P_{11}(k)}{252(2\pi)^2} \int_0^\infty dr\, P_{11}(kr)$$

$$\times \left[\frac{12}{r^2} - 158 + 100r^2 - 42r^4 + \frac{3}{r^3}(r^2-1)^3(7r^2+2)\ln\left|\frac{1+r}{1-r}\right| \right].$$

Furthermore we found that equation (7) can be integrated analytically for the power-law fluctuation spectrum with a cutoff at large wavenumber:

$$P_{11}(k) = \begin{cases} A(k/k_c)^n, & \text{for } 0 < k < k_c \ ; \\ 0, & \text{for } k > k_c \ . \end{cases} \tag{8}$$

Since we are mainly interested in the bahavior for $k \ll k_c$, the presence of the small-scale cutoff k_c does not significantly affect the dynamics in such a mildly nonlinear regime.

The complicated expressions after integrating equation (7) are explicitly given in Suto and Sasaki (1991). Here we only show the plots of the resulting evolution of power-spectra with $n = 1$, 0, and -1 (Figure 1). In Figure 1, we did not plot points where $P(k)$ becomes negative, which simply implies that our quasi-nonlinear approach breaks down and higher-order terms should be taken into account. The results, however, are valid at least in mildly nonlinear regime of fluctuations. For $n \gtrsim 0$, nonlinear effect tends to *suppress* the growth of fluctuations relative to linear theory, while the growth is enhanced for $n \lesssim -1$; the power transfer due to the mode-coupling sensitively depends on the shape of power-spectrum. In addition to the scale-free models, we consider the following spectrum:

$$P_{11}(k) = \begin{cases} A(k/k_*), & \text{for } 0 < k < k_* \ ; \\ A(k/k_*)^{-3}, & \text{for } k > k_* \ , \end{cases} \tag{9}$$

which in equation (7) can be also integrated analytically . This model is intended to represent the asymptotic behavior of the CDM spectrum both on small and large scales. The above spectrum evolution is plotted in Fig. 1(d). As in the previous examples, the growth is suppressed for $n = 1$ region, and enhanced for $n = -3$ region. The secondary peak emerges around $k = 2k_*$, reflecting that the perturbation series are considered here only to second order.

In summary, perturbation analysis gives a physical explanation to the spectral dependent growth in mildly nonlinear regime which was found earlier in numerical simulations (Suto 1989; Suginohara et al. 1991). It is interesting to note that the power-spectrum in the regime approaches $P(k) \propto k^{-1}$ which yields the correlation function of $\xi(r) \propto r^{-2}$.

3. Distribution Function of Amplitudes and Phases of Fluctuations

As density fluctuations evolve, their statistical properties are supposed to deviate from those of the initial random-Gaussian field. The nonlinear clustering should definitely introduce the phase correlation in some way; the high density clumps, for example, would attract the nearby particles and produce a (coherent) wave packet in a sense. Our first attempt is simply to take the Fourier transforms of the

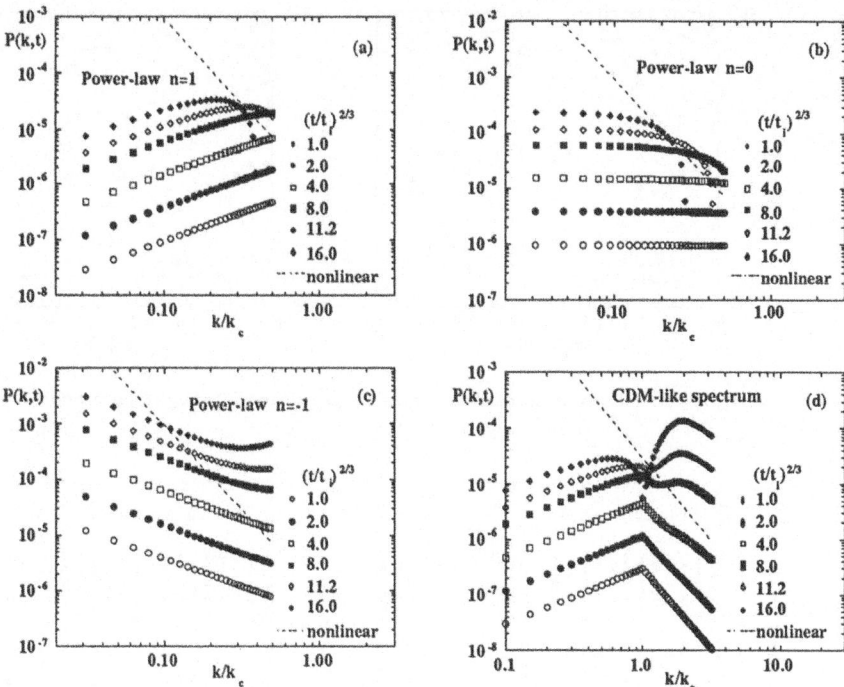

Figure 1: Quasi-nonlinear evolution of power spectrum (Suto and Sasaki 1991)

density on a 128^3 grid. Then after averaging over the direction of $\mathbf{k}$, we obtained the distribution function of the amplitudes and phases for each k-mode (Figures 2 and 3).

The rms density fluctuation smoothed over the filtering scale R_f :

$$\sigma^2(R_f) = \frac{L^3}{2\pi^2} \int_0^\infty \langle|\delta(\mathbf{k})|^2\rangle\exp(-k^2 R_f^2)k^2 dk. \tag{10}$$

reaches 5.0 for $R_f = L/96$ at the expansion factor $a_f = 37.6$ in the above example. It is a bit surprising that the distribution function even at fully nonlinear epochs looks almost random-Gaussian. Although Figures 2 and 3 are for $n = 0$ model, the behavior is essentially independent of the spectral index n and of the epoch of the simulations. We suspect, however, that the above result is somewhat misleading; the Gaussian distribution would be simply the consequence of the central limit theorem and the apparent random phase distribution which we obtained might result from the averaging over the direction of $|\mathbf{k}|$. Without further theoretical guidance, it would be difficult to find the non random-Gaussian nature in Fourier space; the analysis in real space would be more suitable for the purpose, as will be shown in the next section.

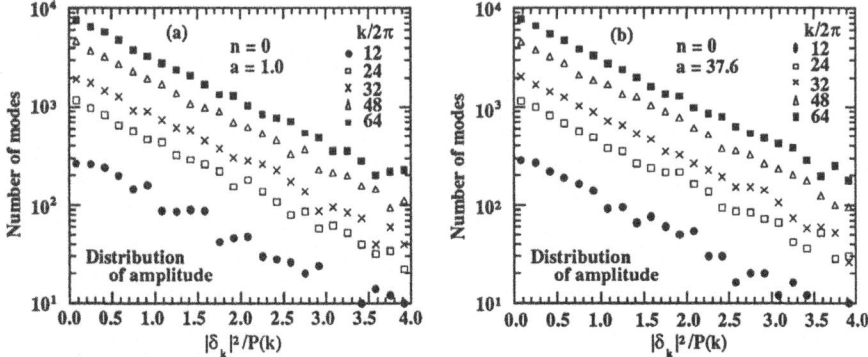

Figure 2: Distribution of amplitudes for $n = 0$ model (Suginohara and Suto 1991a)

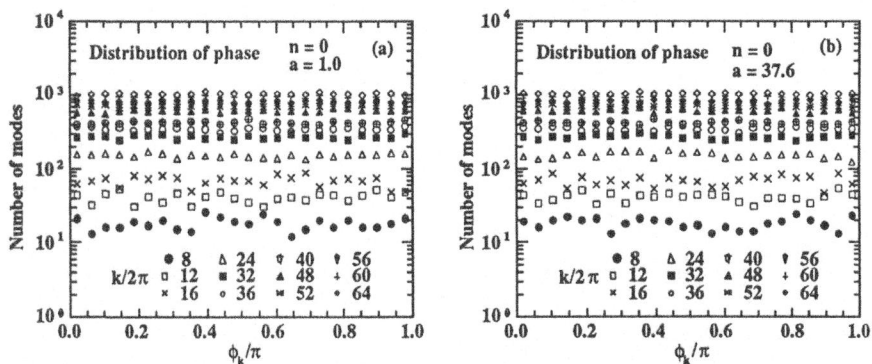

Figure 3: Distribution of phases for $n = 0$ model (Suginohara and Suto 1991a)

4. Non-Gaussian Features due to the Gravitational Clustering

In order to exhibit the non random-Gaussianity of the underlying density field, we plot the distribution of density contrast $(\delta\rho/\rho)_g$ on the 128^3 grid (Figure 4a) after Gaussian smoothing with $R_f = L/96$ (Suginohara and Suto 1991a). The initial distribution $(a = 1)$ is very close to Gaussian and the tails at large $(\delta\rho/\rho)_g$ would be a consequence of the initial condition setup. The final distribution, on the other hand, is highly non-Gaussian and has a similar shape independently of n. While the particle conservation guarantees $\langle(\delta\rho/\rho)_g\rangle = 0$, the higher moments are generated in the course of clustering (see *e.g.*, Peebles 1980). Figure 4b displays the probability distribution of the local density on the grid on a log-log plot. Now it is clear that the final distribution is very close to the power-law; we found that $n(\rho) \sim 0.1(\rho/\bar{\rho})^{-\alpha}$ with $\alpha \sim (1.7 - 2.0)$, almost independently of the initial spectral index. Kofman (1990) presented a theory to derive this distribution function using the Zel'dovich approximation, and in fact our result looks in reasonable agreement with his model for a range of $\sigma(R_f)$.

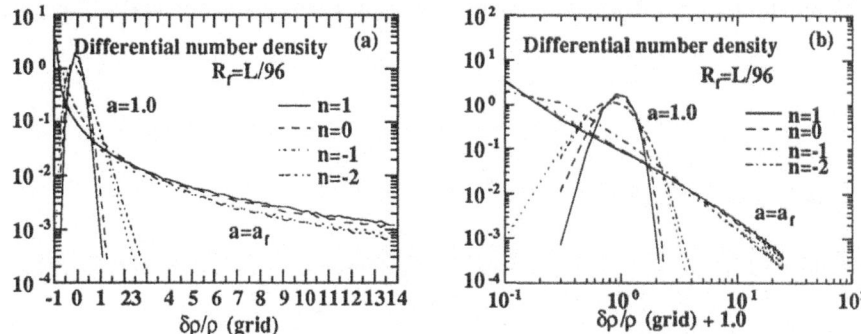

Figure 4: Probability distribution for local density on grids (Suginohara and Suto 1991a)

5. Summary

In the present talk, I focused on the two aspects of the nonlinear gravitational clustering. It is shown that the growth rate in the mildly nonlinear regime sensitively depends on the shape of the underlying fluctuation spectrum; if the initial power spectrum is given as a power-law $\propto k^n$, the growth is enhanced for $n \gtrsim -1$, while suppressed for $n \lesssim 0$. It is interesting to compare the behavior with the similarity solution (*e.g.*, Peebles 1980); in fully nonlinear regime, two-point correlation function is expected to grow $\propto a(t)^{6/(5+n)}$, where $a(t)$ is the cosmic scale factor. Thus the solution predicts the slower growth rate for $n > -2$ compared with linear theory (Suto 1989). Therefore the growth in mildly and fully nonlinear regimes is different even qualitatively, and sensitive to the spectral index n. On the contrary, the statistical properties generated in the course of nonlinear gravitational clustering are fairly independent of the spectral index n. We found that the distribution of local density approaches a power-law, from the initial Gaussian distribution. The distribution of velocity field, however, is shown to be very close to Gaussian even in nonlinear regime, again nearly independent of n (Suto and Fujita 1990).

Both results are important not only in understanding the basic physical processes of nonlinear evolution of gravitating systems, but also in the quantitative confrontation of theoretical models with observations. Certainly there remains ample room for the improvement and further study of the nonlinear effects. I hope that the above two examples will shed some light on the future works in this direction.

It is my pleasure to thank my collaborators, Misao Sasaki and Tatsushi Suginohara for many interesting discussions and contributions. I am also grateful to Jeremiah P. Ostriker and L.A. Kofman for useful conversations during the conference. I thank Katsuhiko Sato for hospitality at Department of Physics, the University of Tokyo. The numerical simulations presented in my talk were carried out on the Hitac S-820/80 at KEK (National Laboratory for High Energy Physics, Japan).

References

Bardeen, J.M., Bond, J.R., Kaiser, N. and Szalay, A.S. 1986, *Ap.J.*, **304**, 15.

Barnes, J. and Hut, P. 1986, *Nature*, **324**, 446.

Blau, S.K. and Guth, A.H. 1987, *in 300 Years of Gravitation* eds. S.W.Hawking and W.Israel, (Cambridge: Cambridge University Press), 521.

Harrison, E.R. 1970, *Phys.Rev.*, **D1**, 2726.

Juszkiewicz, R. 1981, *M.N.R.A.S.*, **197**, 931.

Kofman, L. 1990, talk in this conference.

Peebles, P.J.E. 1980,   *The Large Scale Structure of the Universe* (Princeton University Press: Princeton).

Suginohara, T. and Suto, Y. 1991a, *Ap.J.*, **371**, April 20 issue, in press.

Suginohara, T., and Suto, Y. 1991b, *Publ. Astron. Soc. Japan (Letters)*, in press.

Suginohara, T. , Suto, Y., Bouchet, F.R., and Hernquist, L. 1991, *Ap.J. Suppl.*, March issue in press.

Suto, Y. 1989, in *Proceedings of the Workshop on "Dark Matter and Structure in the Universe"*, ed. M.Sasaki (RITP;Hiroshima), p.129.

Suto, Y. and Fujita, M. 1990, *Ap.J*, **360**, 7.

Suto, Y. and Sasaki, M. 1991, *Phys.Rev.Lett.*, **66**, 264.

Suto, Y. and Suginohara, T. 1991, *Ap.J.(Letters)*, in press.

Vishniac, E.T. 1983, *M.N.R.A.S.*, **203**, 345.

Zel'dovich, Ya.B. 1972, *M.N.R.A.S.*, **160**, 1p.

UNITED WAYS OF INFLATION IN GENERALIZED GRAVITIES.

L. AMENDOLA, S. CAPOZZIELLO, M. LITTERIO, and F. OCCHIONERO.
Osservatorio Astronomico di Roma
Viale del Parco Mellini 84
00136 Roma, Italy

Admovere oculis distantia sidera nostris,
aetheraque ingenio supposuere suo [1] .

ABSTRACT. Generalized theories of gravity can be viewed in a unified way when the trajectories of cosmological models can be projected onto a two dimensional phase space. This tool is applied here to the case of a non-minimally coupled scalar field, to the compactification of multidimensional cosmology, and to fourth order gravity. In conclusion, as a natural follow-up of these models, we outline the possibility of a scalar coupling with the curvature squared term. In all cases, we have in mind a spatially flat Friedmann-Robertson-Walker metric with scale factor $a(t)$.

1 The Non-Minimally Coupled Case.

Generalized theories of gravity, alternative to canonical General Relativity (GR), have been a very active area of research, in the last decade, both on the particle physics side and on the cosmological side. The reasons of this interest trace to the fact that any theory which tries to put together gravity and quantized matter fields seems to require, as its low-energy limit, some sort of correction of the Hilbert Lagrangian. In a series of papers [2] , which we summarize here and where complete reference to earlier work is given, we have discussed in the greatest possible generality the common features of inflation arising in different kinds of theories. Here we report briefly about the Non-Minimal Coupling (NMC) between matter and gravity (in this Section), about the effect of adding spatial dimensions to the four dimensional manifold (in Sec.2), and about adding terms of higher order in the Ricci scalar generating Fourth Order Gravity (FOG) (in Sec.3). The last section is devoted to a model which, to our knowledge, is new and contains a coupling of a scalar field to the square of the Ricci scalar.

The unifying concept concerns the method and is provided by the Phase Space (PS) portrait of the dynamical properties of the models (i.e. we span a PS for each case and then search for common features). The PS portrait is a convenient tool, because it allows a global perception of stability of the trajectories of cosmological models deduced from the particular theory and an immediate identification of their relevant eras (in particular, it is interesting to single out Friedmannian and inflationary stages). We investigate the following main features in each model: the existence and the local stability of critical points; the existence and the stability of singularity-free trajectories; the existence and stability of inflationary trajectories; the basin of attraction of the Friedmannian region; the semiclassical stability of vacuum states; the existence of classically not allowed region of PS. Representations of the PS are worked out in the Poincaré (PPS) projection, which shrinks all the $\{x, y\}$ PS into the unitary $\{r, \theta\}$ circle.

A seminal paper in this area is the work by Belinsky et al. [4] who analyzed and classified cosmological trajectories for a minimally coupled massive scalar field with emphasis on evaluating the likelihood of inflation.

The very first scalar-tensor models proposed by Jordan, Brans and Dicke, has received much attention in recent years as a simple way to shape and to investigate inflationary scenarios. These models acquire a large interest in the context of Quantum Cosmology. In Ref.[2] we generalize the picture presented in Ref.[4] by exploring the $(\phi, \dot{\phi})$ phase space of the NMC Lagrangian

$$\mathcal{L} = -\frac{R}{16\pi} + \frac{1}{2}\xi\phi^2 R + \frac{1}{2}g^{\mu\nu}\phi_{;\mu}\phi_{;\nu} - V(\phi) , \qquad (1)$$

where V is a chaotic potential, which drives the evolution of ϕ in the Klein-Gordon equation

$$\Box\phi + V'(\phi) - \xi\phi R = 0 . \qquad (2)$$

We span a wide class of polynomial potentials and a wide range for the parameter ξ. In almost all cases we find inflationary attractors, with a scale factor either power law or exponential in time, and we give analytical expressions for the asymptotic behaviors. A central oscillating Friedmann region is always recovered. Recently this oscillating pattern has been advocated [8] to explain the apparent redshift periodicity in deep observational surveys. In our case, a narrow range of ξ just above the conformal value $1/6$ turns out to be the most favorable in the sense that the corresponding solutions admit quasi-de Sitter inflation, semiclassical stability and singularity-free trajectories, [2] .

2 Multidimensional Cosmology.

Another application comes from multidimensional cosmology [3] : the extra dimensions are assumed to be compactified to a D-sphere of radius $b(t)$, of present value b_0. The metric then reads: $\bar{g}_{MN} = \text{diag}[\hat{g}_{\mu\nu}(x) ; b^2(x)h_{ij}(y)]$. The dimensionally reduced (from $N = D + 4$ to 4 dimensions) Lagrangian density is

$$\mathcal{L} = -\frac{1}{16\pi G_N} \left(\frac{b}{b_0}\right)^D \left\{\hat{R} + \frac{D(D-1)}{b^2} \left(\hat{g}^{\mu\nu}\partial_\mu b \partial_\nu b - 1\right) - \right.$$
$$\left. - 16\pi G_N V_D^0 \left[\hat{\mathcal{L}}(\hat{\psi}) - 2\bar{\Lambda}\right]\right\}, \tag{3}$$

where $\hat{\mathcal{L}}(\hat{\psi})$ stands for other fields, $\bar{\Lambda}$ is a cosmological constant, V_D^0 is the present value of the volume of the internal space and G_N is the Newtonian gravitational constant. Then, the ordinary Einstein–Hilbert action is recovered by assuming

$$\hat{g}_{\mu\nu} = (b/b_0)^{-D} g_{\mu\nu}, \tag{4}$$

with the dilaton field defined by

$$\sigma = \sigma_0 \ln\left(\frac{b}{b_0}\right), \qquad \sigma_0 = \left[\frac{D(D+2)}{16\pi G_N}\right]^{1/2}. \tag{5}$$

In a PS portrait, $(b,\dot{b})$, a static internal space shows up (as a point like attractor) only when quantum correction [11] are taken into account.

In the same cases one dimensional attractors also exist, along which the internal space expands exponentially. If no other fields are included, the static solution is a global minimum of the potential and the dilaton has zero energy. However, during cosmological inflation, this minimum is displaced above zero and a semiclassical tunnelling becomes possible from the static to the expanding phase. In [3] by comparing the tunnelling rates of the inflaton ϕ and the dilaton σ under the respective barriers, we show that this is not the case for any of the known inflationary models. In particular in Linde's chaotic inflation we derive an upper bound on b_0.

3   Fourth Order Gravity.

The model proposed by Starobinsky [9] , derived from the one-loop approximation to Quantum Gravity (QG), belongs to FOG theories and deserves special attention: in this report, for lack of space, we concentrate on models which work like Starobinsky's, but are obtained in a schematic way from a variational principle and do not contain the trace anomaly. Let us remind here that already the simplest generalization

$$\mathcal{L} = -\frac{1}{16\pi}\left(R - \frac{1}{6M^2}R^2\right), \tag{6}$$

yields, through the trace of the field equations, the fundamental equation of FOG,

$$\Box R + M^2(R + 8\pi T) = 0, \tag{7}$$

where T either refers to ordinary matter (not explicitly included above) or simulates the terms coming from the trace anomaly. With the latter, Eq.(7) yields a non-singular de Sitter phase which decays, when $H \simeq M$, into an oscillating regime with frequency M; after matter creation, canonical GR is recovered in the familiar Friedmannian eras. The (00) field equation yields a first integral of (7):

$$H\dot{R} = -R\left(H^2 + \frac{R}{12}\right) + 3M^2\left(H^2 - \frac{8\pi}{3}\rho\right) , \tag{8}$$

Here we add to (6) cubic or higher order terms in the Ricci scalar, while the trace anomaly contribution will be studied with similar method in future work [2] . In particular, the Lagrangian density

$$\mathcal{L} = -\frac{1}{16\pi}\left(R + \alpha R^2 + \beta R^N\right) , \tag{9}$$

where the identification of α is obvious and β is a free parameter the sign of which depends on N, yields

$$\Box R + \frac{\mathcal{L}'R - 2\mathcal{L} + 3\mathcal{L}'''\dot{R}^2}{3\mathcal{L}''} = 0 . \tag{10}$$

In Fig.1, we use $(R, \dot{R})$ as dynamical variables to keep the analogy with $(\phi, \dot{\phi})$ of [4] and [2] and we plot on the corresponding PPS a sample of the trajectories. The PS is a two dimensional surface $H = H(R, \dot{R})$ embedded in the tridimensional space $\{H, R, \dot{R}\}$, because of a constraint like (8). We project the trajectories onto the plane $\{R, \dot{R}\}$ exactly as in [2] we projected $\{H, \phi, \dot{\phi}\}$ onto the plane $\{\phi, \dot{\phi}\}$. Projections onto $\{H, \dot{H}\}$, [10], and onto $\{H, R\}$ are also possible and physically equivalent. In every case, the central damping region has a Friedmannian interpretation; an inflationary attractor always exists. The inflationary phase is the desired one provided, after staying sufficient time upon the attractor, the solutions fall toward the origin (Friedmann). Alternatively, in fact, they can also evolve away from the origin and undergo an eternal inflationary expansion toward infinity (North or South Pole on the PPS). We stress again the strict analogy of representation and features in NMC and FOG models even if the physics is completely different; the double inflationary scenarios, [5] , where the first inflation is driven by R in FOG and the second by ϕ in GR, can be explored sequentially on the $\{R, \dot{R}\}$ and $\{\phi, \dot{\phi}\}$ PS's, the second being a sort of zoomed up view of the central region of Fig.1.

4   Scalar Coupling to Curvature Squared.

As a final example we consider the Lagrangian density

$$\mathcal{L} = -\frac{1}{16\pi}\left(R - \frac{1}{6M^2}e^{-2\tau\psi}R^2\right) \tag{11}$$

which generalizes (6): here $\tau > 0$ is a free parameter and ψ is a scalar field. The basic motivation of (11) is that it follows directly from the compactification of a multidimensional theory containing quadratic terms: in that case ψ is to be identified as the dilaton; strictly speaking, in this operation other terms arise too, but they are neglected for simplicity. It is remarked that for $\tau \to 0$ this case goes into that of the previous Sec., for $\psi \to +\infty$ the model goes to canonical GR and for $\psi \to -\infty$ the case is totally new. Very schematically one can say that the mass M of the Starobinsky scalaron is replaced by a reduced mass $m(\psi) = M \exp \tau \psi$ which either increases or decreases depending on the evolution of ψ i.e. on the features which are built in the potential $V(\psi)$. In fact, the corresponding field equation obtained by varying the action with respect to ψ reads:

$$\Box \psi + V'(\psi) + \frac{\tau}{48\pi m^2(\psi)} R^2 = 0 \,. \tag{12}$$

which is a simple generalization of (2). Once again Einstein's equations are needed only in contracted form: they yield

$$\Box R + m^2(\psi)(R + 8\pi T) = \tau(\ldots) \,, \tag{13}$$

in place of (7). The main novelty with respect to the above is the occurence of the reduced mass $m(\psi)$ in place of M, which signals that the oscillating phases will have a secularly changing frequency; the dots on the r.h.s. stand for terms bilinear in the gradients of R and ψ, and for terms arising from (12). As before, the (00) field equation yields a first integral of (13),

$$H\dot{R} = -R\left(H^2 + \frac{R}{12}\right) + 3m^2(\psi)\left(H^2 - \frac{8\pi}{3}\rho\right) + 2\tau H R\dot{\psi} \,, \tag{14}$$

which is used as a check in the numerical work. A typical evolution for the scale factor $a(t)$ is seen numerically to contain an initial R-driven inflation down to $H \simeq m(\psi)$, followed by an oscillating regime with frequency $m(\psi)$ the average scale factor of which obeys power laws as in the Starobinsky case.

An intuitive view of the early evolution is obtained by conformally transforming the theory into canonical GR, [7] ,

$$\tilde{g}_{\alpha\beta} = e^{2\omega} g_{\alpha\beta}, \qquad 2\omega = \ln\left|\frac{\partial \mathcal{L}}{\partial R}\right| \,. \tag{15}$$

The theory contains then two coupled scalar fields, ψ and ω, the potential of which is shown in Fig.2 for the case where $V(\psi)$ is a quartic with a false vacuum at the origin and a true vacuum at some $\psi_0 < 0$. In the case shown, a channel connects the false vacuum with the true vacuum along which the model rolls down to become eventually Friedmann. There are also cases, however, where the false vacuum develops into a real minimum where the model remains trapped in perpetual inflation.

The evolution sketched here has first a phase of superinflation ($\dot{H} > 0$) down the ω-axis (perpendicular to the plane of the figure) followed by a turnaround and a phase of ordinary inflation ($\dot{H} < 0$) into the true vacuum. In many ways this reminds of extended inflation [6] in the sense that in this model also the bubble nucleation rate is time dependent due to the false vacuum background evolution. However this model escapes the constraints from the condition of bubble space filling, because all the space ends up into the final true vacuum phase. Other aspects of this model as the possibility of introducing features in the perturbation spectrum, will be studied.

References

1. P. Ovidius Naso, *Fastorum Libri I v.305*, ca. 770 ab U.c..

2. L. Amendola, F. Occhionero, *Astron. Nachr.* **311** 161 (1990); L. Amendola, M. Litterio, F. Occhionero, *Int. J. of Mod. Phys. A* **20** 3861 (1990); F. Occhionero, S. Capozziello, L. Amendola, *in preparation*.

3. E.W. Kolb, L. Amendola, M. Litterio, F. Occhionero, *unpublished*; L. Amendola, E.W. Kolb, M. Litterio, F. Occhionero *Phys. Rev.* **D42**, 1944 (1990).

4. Belinsky, V.A., L.P., Grishchuck, I.M., Khalatnikov, Ya.B., Zel'dovich, *Phys. Lett.* **B 155**, 232 (1985).

5. L.A. Kofman, A.D. Linde, A.A. Starobinsky *Phys. Lett.* **157B** 361 (1985); L. Amendola, F., Occhionero, D., Saez *Ap. J.* **349**, 399 (1990); S. Gottlöber, V. Müller, A.A. Starobinsky, NORDITA preprint (1990).

6. D. La and P.J. Steinhardt *Phys. Rev. Lett.* **62**, 376 (1989).

7. B. Whitt, *Phys. Lett.* **145B**, 176 (1984); J.D. Barrow and S. Cotsakis, *Phys. Lett.* **214B**, 515 (1988); K. Maeda, UTAP preprint (1988).

8. M. Morikawa preprint (1990) and *Ap. J.* (in press) (1991).

9. A.A. Starobinsky, *Sov. Phys. JETP Letters* **30**, 682 (1979); *Phys. Lett.* **91B**, 99 (1980).

10. L.A. Kofman, V.F. Mukhanov, D.Yu. Pogosyan *Sov. Phys. JETP* **66**, 433 (1987); S. Gottlöber and V. Müller, *Class. Quantum Grav.* **4**, 1427 (1987).

11. S. Randjbar-Daemi, A. Salam and J. Strathdee, *Phys. Lett.* **135B**, 388(1984); P.G.O. Freund and M. Rubin, *Phys. Lett.* **97B**, 233 (1980).

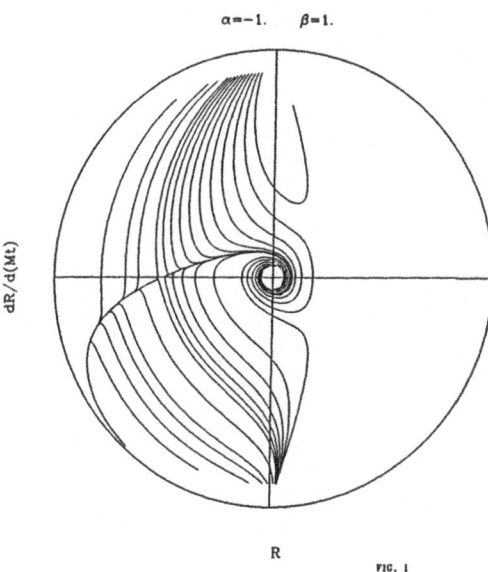

N=3

α=−1.   β=1.

dR/d(Mt)

R

FIG. 1

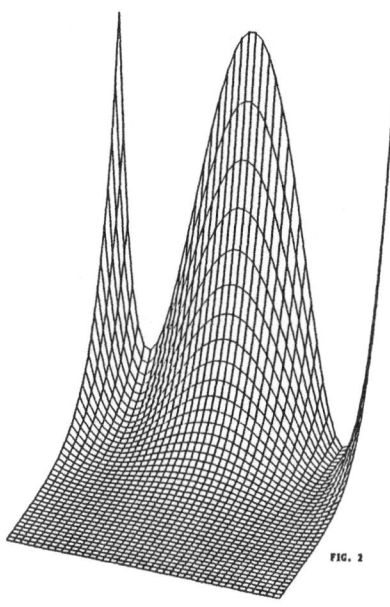

FIG. 2

REVIVING MASSIVE NEUTRINOS FOR LARGE-SCALE STRUCTURE

ROBERT J. SCHERRER
Department of Physics
The Ohio State University
Columbus, OH 43210

ABSTRACT. Although cold dark matter is currently favored for the formation of large-scale structure, neutrinos are the only presently observed particle which could serve as dark matter, and massive neutrinos act as hot dark matter, rather than cold dark matter. We present the results of our seeded hot dark matter model, in which the universe is dominated by massive neutrinos ($m_\nu \sim 25$ eV), and relic compact objects produce the density perturbations. The simplest such model, with randomly-distributed static point seed masses, produces results in good agreement with the observations, suggesting that massive neutrinos can be revived as a candidate for the dark matter. When combined with inflation, seeded hot dark matter produces a model in which perturbations on large and small scales are completely decoupled.

1. Introduction

In the study of the large-scale structure of the universe, there has arisen a "standard" model, in which the density perturbations are Gaussian curvature fluctuations with a Zel'dovich power spectrum; such perturbations, for example, can be produced by inflation. Within the context of such models, cold dark matter models produce results in much better agreement with the observations than hot dark matter models (White, Frenk, and Davis 1983; Kaiser 1983; White, Davis, and Frenk 1984; Davis, et al. 1985; White, et al. 1987; although see Melott 1985 for an opposing point of view). On the other hand, the only presently-known particle which could serve as dark matter is the neutrino, and the mass necessary for a neutrino to give $\Omega = 1$ is $m_\nu \sim 100h^2$, where $h = 0.5 - 1$ is the Hubble parameter in units of 100 km sec^{-1} Mpc^{-1}. Neutrinos with a mass in this range act as hot dark matter. Therefore, it is reasonable to ask whether it is possible to construct a hot dark matter model which produces results for large-scale structure in agreement with the observations.

The present work is motivated by the recent interest in cosmological relic objects such as cosmic strings (Zel'dovich 1980; Vilenkin 1981; Turok 1984; Scherrer, Melott and Bertschinger 1989 and references therein), soliton objects (Frieman et al. 1988; Griest and Kolb 1989; Frieman et al. 1989) and global texture (Turok and Spergel 1990). Such models provide an alternative to the inflationary model for the generation of density perturbations. While inflation produces gaussian curvature

525

fluctuations, relic seed objects produce non-gaussian isocurvature fluctuations. The spectrum of density fluctuations in these models depends on the mass spectrum and spatial distribution of the seed objects (Scherrer and Bertschinger 1991), but there exists a unique simplest seed model, in which all of the seeds have the same mass and are static uncorrelated point masses; it is this model which we examine here. We do not pretend that this model provides an accurate description of any of the specific seed models discussed above; our aim is simply to investigate the simplest seeded hot dark matter model in order to understand the general features of such models.

2. The Seeded Hot Dark Matter Model

In the standard adiabatic hot dark matter model, the perturbations are due entirely to clustering of the neutrinos (and whatever other components are present). Because the neutrinos are hot, they have a large free-streaming length. This free-streaming length represents the smallest possible scale on which neutrino perturbations can exist, since the neutrinos can stream out of smaller-sized regions and effectively erase all fluctuations on those scales. This free-streaming length increases as $(1 + z)^{-1}$ while the neutrinos are relativistic, reaching a maximum comoving length of roughly 13 Mpc/h^2 (we take $\Omega = 1$ throughout). At this point, all perturbations on smaller scales have been wiped out. After the neutrinos become non-relativistic, the comoving free-streaming scale drops as $(1 + z)^{1/2}$. At this point, density perturbations on smaller scales can begin growing again, but in the adiabatic scenario, all of these perturbations have been erased, so there can be no further perturbation growth on these scales. The power spectrum has a characteristic sharp cutoff at about 13 Mpc/h^2, and galaxy formation occurs very late via pancake fragmentation.

Now, however, suppose that we have condensed seed objects present. The neutrino perturbations are erased in the usual way as the streaming length increases, but the seed objects remain unaffected. When the streaming length decreases again, the neutrino perturbations on small scales are gone, but the seeds will induce new perturbations on small scales which can begin growing again. This process strongly suppresses the small-scale power due to the seed perturbations, but this model still has significantly more power on small scales than the standard adiabatic hot dark matter model.

To be more specific, suppose we have an uncorrelated distribution of static point masses, all with the same mass m_s. In this context, a point mass is any mass with a size smaller than the smallest comoving scale of interest at the epoch of equal matter and radiation, in this case, about 100 pc. The density perturbation induced around a single seed is spherically symmetric and is given by

$$\delta(r) = m_s \, g(r), \tag{1}$$

where $g(r)$ gives the spatial behavior of the accretion pattern. It is actually more convenient to work with $\hat{g}(k)$, the fourier transform of $g(r)$. For static point masses in a pure hot dark matter universe, the asymptotic behavior of $\hat{g}(k)$ is easy to calculate exactly; $\hat{g}(k) \propto k^{-2}$ at large k, due to free-streaming of the neutrinos, and $\hat{g}(k) \propto k^{2}$ at small k, due to the isocurvature suppression of the density perturbations. The power spectrum is a simple function of m_{s} and $\hat{g}(k)$ (Villumsen, Scherrer, and Bertschinger 1991, Scherrer and Bertschinger 1991):

$$P(k) = (2\pi)^{3} n_{s} m_{s}^{2} \hat{g}^{2}(k) \tag{2}$$

Note that the shape of the power spectrum is independent of the seed mass m_{s} and number density n_{s}; these simply fix the overall amplitude. Our arguments above tell us that the asymptotic shape of the power spectrum will be $P(k) \propto k^{4}$ for small k and $P(k) \propto k^{-4}$ for large k. Including baryons enhances the power on small scales (since they do not undergo free streaming) and the resulting power spectrum has a slope between -3 and -4 on small scales. A graph of the power spectrum is given in figure 1. Note that this power spectrum differs slightly from that given in Villumsen et al. (1991) because it includes a fully relativistic treatment of the horizon-crossing of the perturbations (Bertschinger 1990); Villumsen et al. used a quasi-Newtonian approximation.

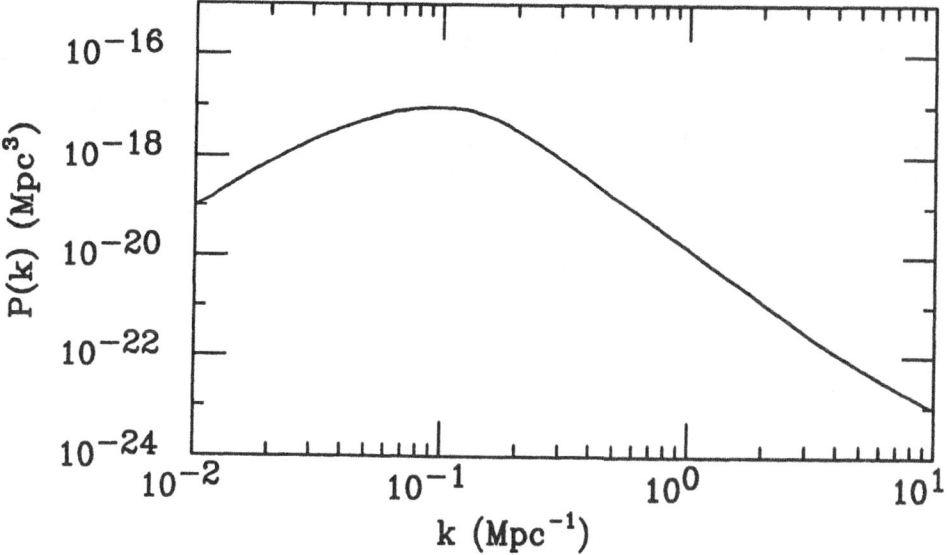

Figure 1: The power spectrum $P(k)$ for uncorrelated, static point seed masses with $m_{s} = 1\ M_{\odot}$ and $n_{s} = 1\ \mathrm{Mpc}^{-3}$ and a universe dominated by massive neutrinos, with $h = 1/2$.

The seeded hot dark matter model has much more power on small scales than the standard adiabatic hot dark matter model. It more closely resembles adiabatic cold dark matter, although the seeded hot dark matter power spectrum decreases more rapidly on both large and small scales.

3. Results

We have produced a linear density field by convolving a random poisson distribution of seeds, all having the same mass m_s and mean number density n_s, with the accretion pattern appropriate for hot dark matter (Bertschinger and Watts 1988; Scherrer and Bertschinger 1991). We take $h = 1/2$ and a baryon density of $\Omega_b = 0.08$. We have examined a family of models with fixed $m_s^2 n_s$, since this fixes the amplitude of the power spectrum. By varying m_s and n_s within this constraint, we obtain a family of models with identical power spectra; the models are different from each other because the fluctuations are non-gaussian.

We evolve our density field forward to the present using a PM code with 64^3 particles in a 128^3 grid (Villumsen 1989). The location of the galaxies is determined using a friends-of-friends algorithm. Our galaxy-galaxy two-point correlation function evolves very little at late times, so we use the pairwise radial galaxy velocities to fix the normalization of the power spectrum. By requiring the pairwise radial velocity to be ~ 300 km/sec on a scale of a few Mpc, we find that

$$m_s^2 n_s = 7 \times 10^{16} M_\odot^2 \text{Mpc}^{-3}. \tag{3}$$

The three models we examined have $n_s = 1/64$ Mpc^{-3}, $m_s = 2 \times 10^9 M_\odot$, $n_s = 1/4$ Mpc^{-3}, $m_s = 5 \times 10^8 M_\odot$ and $n_s = 8$ Mpc^{-3}, $m_s = 9 \times 10^7 M_\odot$. We find that all three models produce a two-point galaxy-galaxy correlation function of $\xi(r) = (r/r_0)^{-1.8}$ with $r_0 = 5h^{-1}$ Mpc. Remarkably, this correlation function shows very little evolution between $1 + z = 1.2$ and $1 + z = 0.87$; in effect the correlation function reaches the observed value and just sits there. The correlation function does evolve at higher redshift, but in a direction opposite from that expected in the Gaussian linear bias approximation. At high redshift, the correlation function has higher amplitude and is steeper. This effect occurs because high-σ peaks are more numerous in this model than in a Gaussian density field; the first galaxies form in these peaks and so have a large-amplitude correlation function at short distances. This provides a major prediction of this model: high-redshift galaxies should be strongly correlated.

Galaxy formation in this model takes place much earlier than in adiabatic hot dark matter models because of the enhanced small-scale power. More than half of our galaxies form between $z = 2$ and $z = 1$. The models with more massive seeds produce more galaxies at higher redshift; this is the only significant difference we have found between the three seed models we have examined. The total number of galaxies, in contrast, is essentially independent of m_s and n_s, as long as the amplitude of the power spectrum is held constant. This indicates that the number of galaxies which form is a function of the amplitude of the fluctuations, rather than the total number of seeds. We also find that the low density (void) regions in our simulation are much smoother than in the standard cold dark matter model, suggesting that dwarf galaxies will not form in such regions.

Turner (1990) has noted that the existence of high-redshift quasars requires significant evolution at high redshift to produce the accretion seeds for the quasars. In our model, the seed masses themselves can serve as the quasar accretion seeds. They have the necessary mass ($m_s > 10^7 M_\odot$) and exist primordially, so they can begin to form quasars as early as necessary.

4. Concluding Remarks

The main problem with this model is the necessity of accounting for rather massive seeds. If we fix the seed mass and number density to satisfy $m_s^2 n_s = 7 \times 10^{16} M_\odot^2$ Mpc^{-3}, then the constraint that the seeds themselves not dominate the density of the universe gives a lower bound on the seed mass: $m_s > 10^7 M_\odot$. Although we have not examined the formation of individual galaxies in this model, another potential problem for all massive neutrino models is the Tremaine-Gunn limit (Tremaine and Gunn 1979). For a discussion of this problem in connection with seeded hot dark matter, see van Dalen (this volume).

Despite these problems, our results suggest that it is possible to construct a massive neutrino model for large-scale structure which is physically motivated and produces results in agreement with the observations. The interesting coincidence is that the convolution of an uncorrelated distribution of seeds with hot dark matter accretion produces a power spectrum with a reasonable slope for structure formation. Randomly-distributed seeds convolved with cold dark matter would produce too much small scale power (Melott and Scherrer 1987), while introducing seed correlations with hot dark matter would change the power spectrum and alter our results. On the other hand, a distribution for the masses of the seeds, rather than a single mass, would not alter our results very much, as long as the seeds remained spatially uncorrelated (Villumsen, et al. 1991; Scherrer and Bertschinger 1991).

Finally, we mention the interesting model which is produced if seeded hot dark matter is combined with inflation (Steigman 1990). (It is possible to construct models in which seed masses can survive inflation; see Vishniac, Olive, and Seckel 1987; Copeland, Kolb, and Liddle 1990). The seeded hot dark matter model has a power spectrum $P(k) \propto k^4$ on large scales and $P(k) \propto k^{-3 \sim -4}$ on small scales. Inflation will induce adiabatic perturbations in the neutrinos with a power spectrum $P(k) \propto k^1$ on large scales and a sharp cutoff at the same point at which the seeded hot dark matter power spectrum turns over. In the linear regime, we can simply add these perturbations to derive a total power spectrum. The result is a power spectrum in which the inflationary perturbations dominate on large scales, and the seed perturbations dominate on small scales; we obtain $P(k) \propto k^1$ on large scales and $P(k) \propto k^{-3 \sim -4}$ on small scales. More importantly, the amplitudes of the large and small scale perturbations are decoupled, since the inflationary perturbations have no effect on small scales and the seed perturbations are completely dominated by the inflationary perturbations on large scales. One could then retain all of the desirable results obtained here, while producing any desired amplitude of fluctuations on very large scales.

5. Acknowledgments

This work is part of a collaboration with J. Villumsen and E. Bertschinger, whom I thank for allowing me to draw on this jointly-generated work. This work was supported in part by a University Seed Grant from the Ohio State University Office of Research and Graduate Studies, and by the Department of Energy.

References

Bertschinger, E. 1990, private communication.

Bertschinger, E., and Watts, P.N. 1988, Ap.J., **328**, 23.

Copeland, E.J., Kolb, E.W., and Liddle, A.R. 1990, Phys. Rev. D, **42**, 2911.

Davis, M., Efstathiou, G., Frenk, C.S., White, S.D.M. 1985, Ap. J., **292**, 371.

Frieman, J.A., Gelmini, G.B., Gleiser, M., and Kolb, E.W. 1988, Phys. Rev. Lett., **60**, 2101.

Frieman, J.A., Olinto, A.V., Gleiser, M., and Alcock, C. 1989, Phys. Rev. D, **40**, 3241.

Griest, K., and Kolb, E.W. 1989, Phys. Rev. D, **40**, 3231.

Kaiser, N. 1983, Ap. J. (Letters), **273**, L17.

Melott, A.L. 1985, Ap.J., **289**, 2.

Melott, A.L., and Scherrer, R.J. 1987, Nature, **328**, 691.

Scherrer, R.J., and Bertschinger, E. 1991, in preparation.

Scherrer, R.J., Melott, A.L., and Bertschinger, E. 1989, Phys. Rev. Lett., **62**, 379.

Steigman, G. 1990, private communication.

Tremaine, S., and Gunn, J.E. 1979, Phys. Rev. Lett., **42**, 407.

Turner, E.L. 1990, A.J., submitted.

Turok, N. 1984, Nucl. Phys. B, **242**, 520.

Turok, N., and Spergel, D.N. 1990, Phys. Rev. Lett., **64**, 2736.

Vilenkin, A. 1981, Phys. Rev. Lett., **46**, 1169, 1496(E).

Villumsen, J.V. 1989, Ap.J. Suppl., **71**, 407.

Villumsen, J.V., Scherrer, R.J., and Bertschinger, E. 1991, Ap.J., in press.

Vishniac, E.T., Olive, K.A., and Seckel, D. 1987, Nucl. Phys. B, **289**, 717.
White, S.D.M., Davis, M., and Frenk, C.S. 1984, M.N.R.A.S., **209**, 27P.
White, S.D.M., Frenk, C.S., and Davis, M. 1983, Ap.J. (Letters), **274**, L1.
White, S.D.M, Frenk, C.S, Davis, M., Efstathiou, G. 1987, Ap.J., **313**, 505.
Zel'dovich, Ya. B. 1980, M.N.R.A.S., **192**, 663.

CELLULAR DISTRIBUTION OF THE UNIVERSE AND DARK MATTER

D. Calzetti, M. Giavalisco, R. Ruffini
I.C.R.A. - Dipartimento di Fisica, Università di Roma,
Piazzale A. Moro, 2,
I-00185 Roma, Italy

ABSTRACT.
The medium-scale inhomogeneities in the distribution of structures in the Universe can be explained within a massive "inos" dark matter model. In this environment a cell-fractal distribution arises, giving at present time a general structure of fractal cells randomly distributed in space. In order to determine the compatiblity of the available data with the model proposed and the average size of the cells, three different analyses have been performed on the Zwicky and CfAI catalogues, respectively. All of them refer to the standard way of determining the two point correlation function $w(\theta)$ and $\xi(r)$, but we have analyzed the behaviour of $1 + w$ and $1 + \xi$, as requested by the model. In particular, one of the three tests is independent of the shape and the parameters of the luminosity function. The result is that the data are in agreement with a cell-fractal structure of cell diameter $\simeq 60 - 70h^{-1}$ Mpc ($H_o = 100$ km/s/Mpc).

1. Introduction

In the last years, many works on large-scale structure (see, for a review, [1]) has suggested that the distribution of structures themselves (galaxies, clusters, ...) cannot be considered homogeneous at least up to scales about $50h^{-1}$ Mpc ($H_o = 100$ km/s/Mpc). On the other hand, the success in explaining the recession velocity of galaxies and the Cosmic Background Radiation of Friedmann-Gamow theory gives evidence to the necessity of matching two different models for the distribution of matter: inhomogeneous on medium scales, homogeneous and isotropic (Friedmann-Gamow-like) on large scales.

Great importance has then assumed the study of the possible "regularities" existing in the inhomogeneous distribution. The inverse power law behaviour of the two-point correlation function, both for galaxies and clusters, with the same index γ [18; 2] and the linear growing of the correlation length with the sample size [11] has suggested the presence of a fractal structure [16; 17].

It can be summarized that galaxies, clusters and higher order clustering are distributed in fractal cells randomly distributed in space, where the cell size R_{co} must be determined from observations and explained by physical theories [23].

In section 2 we will present a model for the formation of the cell-fractal Universe and in section 3 we will show that, once a proper statistical indicator is chosen, the evidences desumed from optical surveys as the Zwicky [26] and the CfAI [15] catalogues are fairly in agreement with the model presented here, with a cell radius about $30 - 35h^{-1}$ Mpc in size.

2. The Model

The basic assumptions for constructing a cell fractal model are the following [23]:

- visible matter is a good tracer of dark matter;

- the dark matter component dominates the matter content of the Universe and is composed of massive "inos" obeyng Fermi statistics [22, 21];

- the basic mechanism for fragmentation of the initially homogeneous "ino" distribution in an expanding Universe is the Jeans gravitational instability [12];

- galaxies and clusters of galaxies are envisaged as different hierarchical order of a cellular fractal structured Universe.

- each "elementary cell" has a characteristic radius R_{co} and is formed of anisotropic and inhomogeneous fractal distribution;

- the "elementary cell" is expected to form at the era where the "inos" become non relativistic

$$1 + z_{nr} \approx 1.7 \times 10^4 \, (m_\nu/10eV) \, A(\xi_\nu).$$

As a consequence of the above assumptions, the "elementary cells" should be formed homogeneously in the Universe, so that the Friedmann-like Universe is reached for sizes $r \gg R_{co}$ by averaging the matter distribution on many fractal cells.

The size $R_{co}(z_{nr})$ is assumed to coincide with the Jeans length

$$\lambda_J(z_{nr}) \approx 1.3 \times 10^4 \, Kpc \, (m_\nu/10eV)^{-5/2} \, B(\xi_\nu);$$

then, the mass of "inos" interested to the instability processes is given by

$$M_J(z_{nr}) \approx 1.5 \times 10^{17} \, M_\odot \, (m_\nu/10eV)^{-2} \, C(\xi_\nu).$$

The functions $A(\xi_\nu)$, $B(\xi_\nu)$, and $C(\xi_\nu)$ depend on the chemical potential, number of species and helicity paramenter of the "inos" and on the ratio $\Omega_{dark}/\Omega_{tot}$.

For instance, if $m_\nu \approx 10eV$ and $\Omega_{inos} \approx 1$ we have $z_{nr} \approx 10^4$, $\xi_\nu \approx 5$, $R_{co} \approx 10^2 \, Mpc$ nowadays and $M_J \approx 10^{18} \, M_\odot$.

In few words, this cosmological model is able to connect the medium scale inhomogeneous density distribution with the large scale Friedmann-like density behaviour.

In addition this model naturally leads to the explanation of the observed absence of large scale inhomogeneities in the black body radiation: at the time of z_{rec} the density contrasts given by the model are still very small $(\delta\rho/\rho \sim 10^{-4})$, showing no difficulty in explaining inhomogeneities of $\delta T/T \sim 3 \, 10^{-5}$ on scales of a few degrees

[13]. It is, however, very interesting that some of the inhomogeneities predicted may be compared with the forthcoming data of COBE observations.

The formation of the fractal structure inside the cell by a cascade model of successive fragmentations due to the decrease of the Jeans mass has been explored by a direct numerical simulation [24].

One of the most distinctive features of this cellular model is the fact that it predicts a non zero chemical potential for the dark matter component, varying in the range $\xi_o \sim 5 - 10$. Recently, the consequences of this non null chemical potential on the process of cosmological nucleosyntesis has been probed [3] and they have found that a possible balance between the electron neutrino chemical potential and the one of the τ and μ neutrinos is:

$$\xi_{\nu_{\mu+\tau}} \sim 10\xi_{\nu_e},$$

leading to results compatible with the observational data.

3. Self-similarity in the distribution: determination of the cut-off in galaxy clustering

The cellular fractal model provides two observables on the angular two-point correlation function, which can be compared with the available data on galaxies [6]:

- the angular normalized density $1 + w(\theta)$ behaves as a pure power law for $\theta \ll 1$ and for $R_{co} \simeq R_*$ (being R_* the equivalent depth of the apparent magnitude limited sample considered):

$$1 + w(\theta) = B\theta^{-\beta}. \tag{1}$$

Here $w(\theta)$ is the two-point angular correlation function.

- The amplitude B is a function of the index $\gamma = 1+\beta$ and of the ratio $x = R_{co}/R_*$:

$$B \propto x^\gamma \frac{\dfrac{3-\gamma}{3}x^\gamma D_1(x) + D_2(x)}{\left(\dfrac{3-\gamma}{3}x^\gamma M_1(x) + M_2(x)\right)^2}, \tag{2}$$

where D_1, D_2, M_1, M_2 contain the integrals of the galaxy luminosity function.

Eq.(2) has been obtained supposing for the cell-fractal structure a density behaviour:

$$n(r) = \begin{cases} \frac{3-\gamma}{3} R_{co}^\gamma < n > r^{-\gamma} & r \leq R_{co} \\ < n > & r > R_{co} \end{cases} \tag{3}$$

We have firstly chosen to work with angular data, in order to avoid any problem connected with the third dimension.

The above two points have been recently analyzed by Calzetti et al. [7]. The Zwicky catalogue [26] has been cutted in five sub-samples of different apparent magnitude limits (red-shift unlimited sets), each every $\Delta m = 0.5$, ranging from $m = 13.5$ to $m = 15.5$. Then, for each sub-sample, the $1 + w(\theta)$ has been calculated and it resulted to be compatible with a power law form (Eq.(1)), with index $\beta \simeq 0.3$, for *all the sub-samples*. R_{co}/R_* has been estimated from Eq.(2), giving for the upper cut-off of the cell structure: $R_{co} \simeq 30h^{-1}$ Mpc.

By the way, the test is not self-consistent , since Eq.(2) has been used for calculating R_{co} and not for comparing it with an independently determined value.

Then, an alternative test has been performed [8] on the data. Let us consider a tridimensional catalogue, which a given number $i = 1, \ldots, h$ of subsamples has been selected from, ordered by increasing apparent magnitude limit, as before. It is possible to impose a new restriction on the sub-sets: the farest galaxy considered must be such that its distance R_t:

1. $R_t \leq R_{co}$; thus, defining $y = R_t/R_*$, the normalized density become:

$$1 + w(\theta) = B(y,\gamma) \, \theta^{-\beta},$$

so that taking:

2. $y = $ const, B depends only on γ.

These two conditions imply that only the hth sample has the largest R_t possible, i.e. $R_t = R_{co}$ (see Tab. 1). Condition 1 guarantees that we are working inside the fractal cell and 2 means that no spurious effects due to different apparent magnitude limits are present. As a consequence, the structure analyzed with our samples is purely fractal, if the cell model is correct. We expect, in this case, again Eq.(1) holds, but now $B = const$. In particular, if we are interested only in the behaviour of B and not in its absolute value, the analysis is independent of the luminosity function shape and parameters (and, obviously, independent also of the absolute value of R_*).

Needing a tridimensional catalogue, we have considered the CfAI [15], which is complete in the Northern Sky ($b \geq 40°$, for avoiding galactic obscuration effect) up to $m = 14.5$. Since now the range in apparent magnitude is smaller than in the Zwicky, the sub-samples have been chosen with magnitude interval $\Delta m = 0.1$, giving 10 sets ($13.6 \leq m \leq 14.5$). The choice of the red-shift cuts is shown in Tab. 1 (*red-shift limited sets*).

From the previous analysis, $R_{co} \sim 30h^{-1}$ Mpc, not large in value. In order to avoid the heavy presence of crowded clusters in the final data, Virgo has been removed.

Repeating on the new samples the same tests as before, $1+w(\theta)$ is again compatible with a power law form (Eq.(1)) and B is fairly constant, as expected (see Fig. 1 and Tab. 1).

As far as this analysis suggest, the Universe is structured in fractal cells, each of diameter $\sim 60h^{-1}$ Mpc, being the result independent of the particular equivalent depth or of the luminosity function shape chosen. In any case, the determined R_{co} can be even a lower limit on the actual cut-off and new tests, also with larger R_{co}, are needed for drawing a more complete picture.

A different analysis, which has given up to now preliminary results, has been applied to tridimensional data from the CfAI, divided, in this case, in 16 sub-samples ($13.0 \leq m \leq 14.5$, $\Delta m = 0.1$).

Again, in order to compare the results from this test with the previous ones, the same observational costraints of the angular analysis have been used ($\delta \geq 40°$ and corrections for galactic obscuration).

For each sub-sample we have estimated the correlation functions from the relation $1 + \xi(r) = N_{oo}(r)/N_{or}(r)$, where $N_{oo}(r)$ is twice the number of pairs of galaxies having separations in the range r, $r + dr$ and $N_{or}(r)$ is an analogue quantity where now the second galaxy of the pair belongs to a properly generated random sample [10].

The analysis has shown: 1) again the use of $1 + \xi$ as an indicator of the correlation seems to be more appropriate to describe the galaxy clustering than the usual ξ; 2) the correlation function expressed by $1 + \xi$ exhibits a double behaviour with the transition occuring for a separation of the order of 3 Mpc (see Fig. 2). For $r > r_t$ it is well described by a power law with a slope which is a slowly increasing function of the sample depth in the range ($0.6 \div 0.8$), while for $r < r_t$ the power law model is a bit worse determided (because of the large uncertainties for small separations), but the average value of its slope, $\gamma = 1.9 \pm 0.1$, is more confident with the canonical result.

Let us now focus on the intermediate scale regime $r > r_t$. A discussion on the small scale and on the connection between the two regimes (small and intermediate scales) will be presented in a forthcoming work (Calzetti, Giavalisco, Ruffini, 1990, in prep.). This is the region usually analyzed by correlation function studies. As pointed out by Chincarini and Guzzo [9] (but see also [5, 6] and Calzetti et al., 1990 in prep.), the use of $1 + \xi$ has permitted to evidenciate a self-similar clustering extending at least up to 35 Mpc. Actually this is the largest value of the effective sample radius for our data (for the definition of effective radius, see [4]). This means that the region within which self-similarity holds is 70 Mpc in diameter. This is consistent with the observed flattening of correlation function for separation values of the order of 40 Mpc.

In Fig. 3 the observed scaling relation for the amplitude A with the sample depth (in both cases of small and intermediate separations) is reported in double logarithmic scale and its trend, as a power law with the sample depth, is consistent with a self-similar organization of the galaxy clustering.

It should be noted that in the statistics used for fitting the data in both cases of angular and spatial normalized density, two kind of problems appear: a) the sub-samples are not statistically independent, since the larger includes the smaller; b) there may be a correlation effect between the amplitudes and the power law index, because the first depends on the second. On the ground of the large uncertainties got from the data, we think this should not influence too much our conclusions, but verifications are in progress.

For what concern the transition to the large scale homogeneity our data do not permit to draw any conclusion. In the intermediate regime the self-similar scaling seems to continue to the deepest survey ($R_{eff} \sim 35$ Mpc) and the correlation functions sensibly flatten for a separation of the order of 40 Mpc in each of the 16 samples. This suggests that for such a value the self-similarity fades away into an homogeneous large scale distribution, but no details of such a transition can by now be described.

References.

1. Bahcall, N. A.: 1988, Ann. Rev. of A. A., **26**, 631

2. Bahcall, N. A., Soneira, R. M.: 1983, Ap. J., **270**, 20

3. Bianconi, A., Lee, H.W., Ruffini, R.: 1990, A.A., in press

4. Calzetti, D., Einasto, J., Giavalisco, M., Ruffini, R., Saar, E.: 1987, Astrophys. and Space Sci., **137**, 101

5. Calzetti, D., Giavalisco, M., Ruffini, R.: 1988, A. A., **198** , 1

6. Calzetti, D., Giavalisco, M., Ruffini, R.: 1989, A. A., **226**, 1

7. Calzetti, D., Giavalisco, M., Ruffini, R.: 1990, in prep.

8. Calzetti, D., Giavalisco, M., Ruffini, R., Taraglio, S., Bahcall, N.A.: 1990, submitted to A. A.

9. Chincarini, G., Guzzo, G.: 1990, Nature, in press

10. Davis, M., Peebles, P.J.E.,: 1983, Ap. J., **267**, 465

11. Einasto, J., Klypin, A.A., Saar, E.: 1986, M.N.R.A.S., **219**, 457

12. Fabbri, R., Jantzen, R.T., Ruffini, R.: 1982, A. A., **114**, 219

13. Fabbri, R., Ruffini, R.: 1990, A. A., **228**, 1

14. Groth, E.J., Peebles, P.J.E.: 1977, Ap. J. **217**, 385

15. Huchra, J.P., Davis, M., Lathan, D., Tonry, J.: 1983, Ap. J. Suppl., **52**, 89

16. Mandelbrot, B.B.: 1975, C. R. Acad. Sc. Paris **CCLXXX A** , 1551

17. Mandelbrot, B.B.: 1979, C. R. Acad. Sc. Paris **CCLXXXVIII A** , 81

18. Peebles P.J.E.: 1980, *The Large-Scale Structure of the Universe*, Princeton Un. Press

19. Peebles, P.J.E., Hauser, M.G.: 1974, Ap. J. Suppl., **28**, 19

20. Peebles, P.J.E., Groth, E.J.: 1975, Ap. J. **196**, 1

21. Ruffini, R., Song, D.J.: 1987, A. A., **179**, 3

22. Ruffini, R., Song, D.J., Stella, L.: 1983, A. A., **125**, 265

23. Ruffini, R., Song, D.J., Taraglio S.: 1988, A. A., **190** , 1

24. Ruffini, R., Song, D.J., Taraglio S.: 1990, A. A., in press

25. Totsuji, M., Kihara, T.,: 1969, Publ. Astr. Soc. Japan, **21**, 221

26. Zwicky, F., Herzog, E., Wild, P., Karpowicz, M., Kowal, C.T.: 1961-68, *Catalog of galaxies and of clusters of galaxies* , Pasadena: CIT

Table 1: for each redshift limited sub-sample, characterized by a different m_c, the values of the red-shift cut R_t, of the logarithm of the normalized density amplitude $\log B$ and of its uncertainty $d \log B$ are reported.

m_c	$R_t(km/s)$	$\log B_{1+w}$	$d \log B_{1+w}$
13.6	1980	0.53	0.05
13.7	2074	0.54	0.04
13.8	2171	0.53	0.04
13.9	2274	0.53	0.04
14.0	2381	0.53	0.04
14.1	2403	0.52	0.03
14.2	2611	0.51	0.03
14.3	2734	0.49	0.03
14.4	2863	0.49	0.03
14.5	2997	0.47	0.03

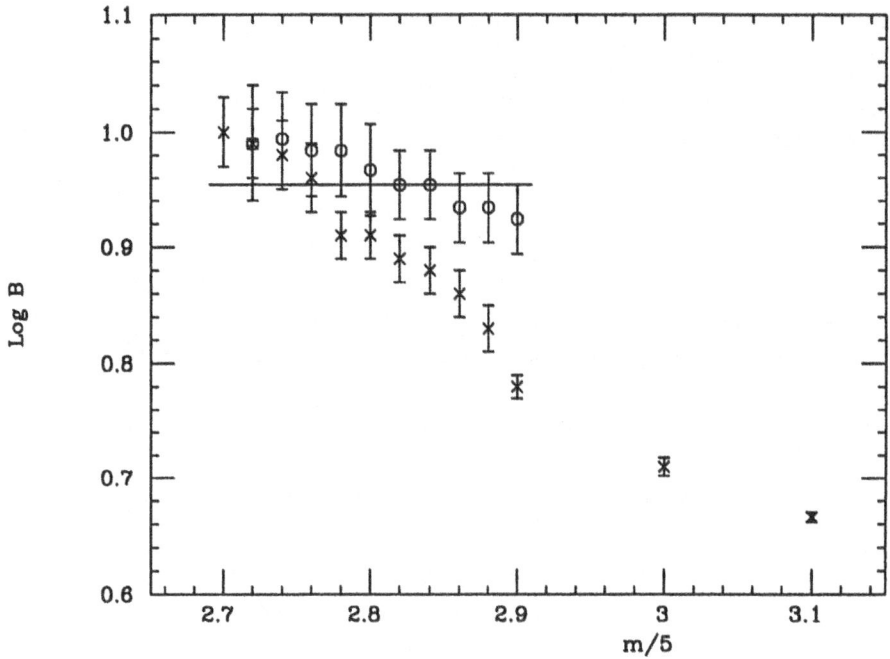

Fig. 1: The behaviour of $\log B$ vs. $m/5 \propto \log R_*$ is shown for the two different cases, respectively, of the red-shift unlimited samples (crosses) and of the red-shift limited samples (circles). It is evident that the second set of data gives amplitudes fairly consistent with a constant line (drawn for comparison). To better evidenciate the differences, the first set of sub-samples has been analyzed with $\Delta m = 0.1$ in the range $13.5 \leq m \leq 14.5$. The $1 - \sigma$ error bars are also reported. The scale of the plot is arbitrary.

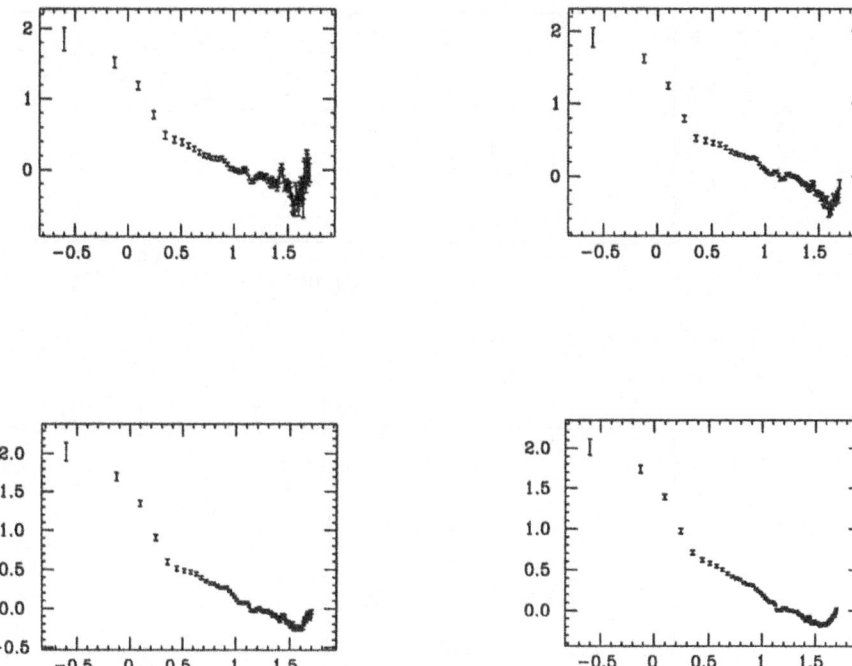

Fig. 2: The plots of $\log(1+\xi(r))$ vs. $\log r$ are shown for four of the 16 tridimensional sub-samples of the CfAI (m_c =13.0, 13.5, 14.0, 14.5). The transition region at $r_t \approx 3$ Mpc is here evident.

542

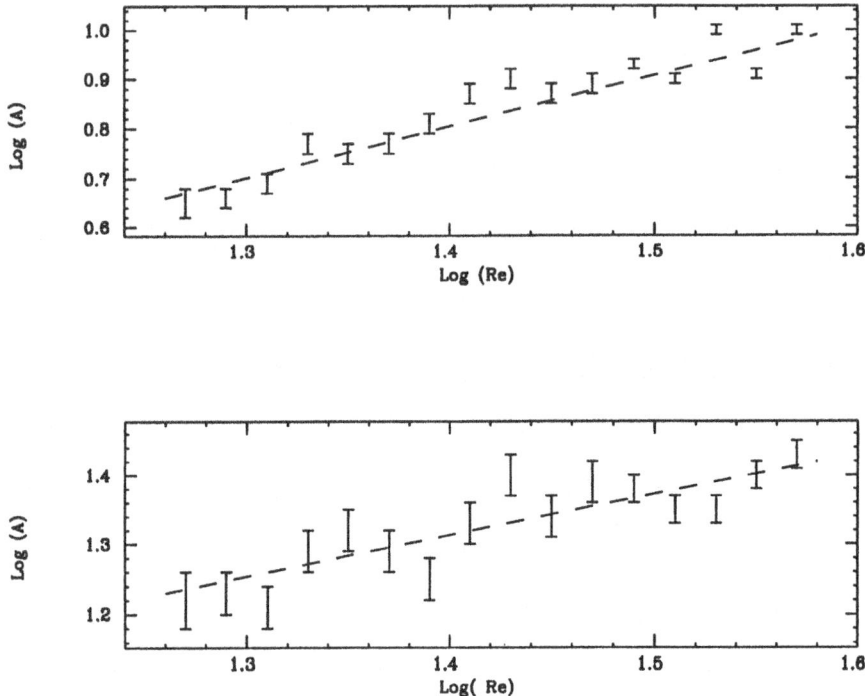

Fig. 3: The amplitudes A of the quantity $1 + \xi(r)$ for the 16 sub-samples vs. the effective radius R_{eff} are reported in double logarithmic scale for both cases of intermediate scales (upper panel) and small scales (lower panel). The fit with a straight line, as predicted by the model, is also shown.

THE LARGE-SCALE PECULIAR VELOCITY AT HIGH REDSHIFT

Xingfen Zhu and Yaoquan Chu
Center for Astrophysics
University of Sci. & Tech. of China
Hefei, Anhui 230026, China
and
CCAST (World Lab.) Beijing, China

The peculiar velocity of extragalactic objects has become one of the focal points of observational cosmology. The information about the distribution of peculiar velocity of extragalactic objects is as important as that about their spatial distribution to understanding the formation of large-scale structure in the universe. Up to now we know little about the peculiar velocity of quasar. The difficulty of dertermaing the quasar peculiar velocity is that there is no methods independent of redshift to determine the distance of quasar. The purpose of this paper is to discuss the possibility of deriving large scale peculiar velocity information from quasar pairs.

Let us conside a sample of quasars pairs, if the distribution of quasars is randomly homogeneous in space, then the orientation of the quasar pairs should also be isotropic. We have

$$N(D_r)\, dD_r = \frac{3}{2} N_T \left(1 - \frac{D_r^2}{D_{max}^2} \right) d\left(\frac{D_r}{D_{max}} \right) \qquad (1)$$

$$\overline{D_r} = \frac{1}{N_T} \int_0^{D_{max}} D_r N(D_r)\, dD_r = \frac{3}{8} D_{max} \qquad (2)$$

where the D_r is the radial projections of the distance between two quasars in a pairs and N_T is the total number of pairs in the sample. The quasar clustering and the peculiar velocity of quasars would lead to deviation from the above distribution.

As we know, the clustering of quasars at high redshifts is very weak(1). Here we conside a simple model in which we assume that the two quasars in every pair are approching each other at average speed of V. For a sample consisting all pairs with $D < D_{max}$, we have

$$N(D_r)\, dD_r = \frac{3}{2} N_T \left(1 + \frac{v}{c} \cdot \frac{c}{H_0} \cdot \frac{1}{D_{max}} \right)^{-1} \left[1 - \frac{D_r^2}{D_{max}^2} + \frac{v}{c} \cdot \frac{c}{H_0} \cdot \frac{2}{D_{max}} \left(1 - \frac{2D_r}{D_{max}} + \frac{D_r^2}{D_{max}^2} \right) \right] d\left(\frac{D_r}{D_{max}} \right) \quad (3)$$

$$\overline{D}_r \sim \frac{3}{8} D_{max} \left(1 - \frac{1}{3} \frac{v}{c} \cdot \frac{c}{H_0} \cdot \frac{1}{D_{max}} \right) \tag{4}$$

Equ. (3) and (4) show the deviation of $N(D_r)$ and $\overline{D}_r$ from (1) and (2) are due to the peculiar velocity of quasars.

For statistical analysise, we use the quasar sample given by Savage & Bolton(2), which includes quasars in two $5° \times 5°$ regions around respectively ($02^h, -50°$) and ($22^h, -18°$) in the southern hemisphere. For every pairs of quasar in the sample, we calculate the distance between two quasars, D, and its transversal and radial projections D_t and D_r using their redshifts and angular distance on the sky. For the sake of comparison we also calculate the distribution of the Monte Carlo sample which is given by the avarage of 10 Monte Carlo results of randomizations of angular coordinates α and δ for each quasar. We consider two subsamples with $D_{max}=50$ and 100 Mpc for each of the two regions. Statistical results show that the deviations of $\overline{D}_r$ from those given by Eq. (2) are all less than 6% for four subsamples. We also get the upper limit of peculiar velocity V_{up} from:

$$V_{up} = 8 H_0 \left| \overline{D}_r - \frac{3}{8} D_{max} \right| \tag{5}$$

All results of V_{up} for the four subsamples are around a thousand km/s. Therefore, if all deviations of $\overline{D}_r$ from that given by Eq.(2) are due to the motion of two quasars in a pair to approach each other, such peculiar velocity should then have the order of one thousand km/s.

Our conclusion are:
1) Quasars are clustered weakly on the scale of 50 - 100 Mpc and just in the initial stage of formating clusters.
2) One thousand kilometers per second is an upper limit to peculiar velocity of quasars.

REFERENCE
(1) Y.-Q. Chu and L.-Z. Fang, in "Observational Cosmology" eds. A. Hewitt, G. Burbidge and L.-Z. Fang, Reidel 1987, p. 627.
(2) Savage, A. and Bolton, J.G., Mon. Not. R. Astr. Soc., vol. 188, (1979), p.599.

AN EVIDENCE OF MULTIPLY-CONNECTED TOPOLOGY MODEL OF THE UNIVERSE

Xingfen Zhu and Yaoquan Chu
Center for Astrophysics
University of Sci. & Tech. of China
Hefei, Anhui 230026, China
            and
CCAST (World Lab.), Beijing, China

Recently many observational results of the spacial distribution of extragalactic objects challenge the standard view of cosmological model. For example, it is well known that there are two main arguments which are against the cosmological origin of quasar redshift. That is, the associations of quasars with low redshift galaxies (see: Chu, Zhu, Burbidge and Hewitt (1); Arp (2)) and the redshift periodic distribution of quasars and other extragalactic objects. (see: Fang, Chu et. al.(3), Chu and Zhu (4)). The aim of this work is to point out that these two phenomena maybe related to each other and is not unacceptable under the scheme of cosmological redshift and cosmological principle, only if the universe is multiply connected. In a multiply-connected universe, we can see the original image of an object (low redshift galaxy) and its ghost images (high redshift quasars) in the same direction. For many types of topologies, the image redshifts would show periodicity with respect to a variable which depends on redshift Z, such as:

$$F(Z) = \ln (1+Z) = An + B$$

where A, B are constant, n is positive integer. n=0 correspond to the galaxy and n > 0 to the associated quasars. Therefore we have:

$$\ln (1+Z_g) - \ln (1+Z_q) = An$$

where Z_g, Z_q are, respectivily, the redshifts of galaxy and quasars in an association.

In order to test the quasar-galaxy redshifts periodicity, we collected all samples of the multiple quasars which close to a bright galaxy from published literature. There are total 14 associations, which include 57 quasars. Multiple quasars close to a galaxy are more unlikely to be formed by projection.

We adopted the power-spectrum method to analyse the periodicity of redshift distribution. The statistical results are listed in Table 1.

545

Table 1: Power (P_n) in redshift distribution peaks of
multiple quasars near bright galaxy

Argument	P(5)	Probability
$\ln(1+Z_q)$	6.83	1-97.2%
$\ln(1+Z_q)-\ln(1+Z_g)$	7.21	1-98.1%

P(5) corresponds to the length of period $\ln(1+Z)=0.206$.
The probability that any peak in a random spectrum should
reach a power $P > P_0$ is:
$$p(>P_0) = 1 - (1-\exp(-P_0))^m.$$
It should be point out that the peak at n=5 is a broad peak
with considerable peak at n=4 and n=6 as well. The confidence
level of 97% to 98% derived for n=5 peak alone must therefore be a considerable underestimation of the true significance of the peak. A rough estimated the accidental probability of the peak in the power-spectrum should be 1-99.99%.

Up to now the periodic distribution of redshift is the
subject of controversy. Here we present a new evidence with
high statistical significance and the sample of multiple-
quasars associated with bright galaxy should be less of influence from selected effect such as emission identification,

To conclude, we would say that an cosmological model with
multiple-connectivity of topology seems good to explaination
both of the quasar-galaxy association and the periodic distribution of redshift of extragalactic objects.

References:
(1) Chu Y., Zhu, X., Burbidge G. and Hewitt A., 1984, Astro.
    & Astrophys., vol. 138, p.408.
(2) Arp H., 1987, "Quasars, Redshifts, and Controversies".
    Interstellar Media.
(3) Fang L., Chu Y., et. al., 1982, Astro. & Astrophys.,
    vol. 106, p.287
(4) Chu Y. and Zhu X., 1989, Astro. & Astrophs., vol.222, p.1

NON–GAUSSIAN INITIAL CONDITIONS
IN CDM MODEL SIMULATIONS

F. LUCCHIN
Dipartimento di Astronomia
vicolo dell'Osservatorio
I–35100 Padova
Italy

ABSTRACT. Results are reported on N–body simulations of the large–scale structure of the universe starting from non–Gaussian initial conditions in a Cold Dark Matter scenario.

The model for galaxy formation which has been most investigated is the standard Cold Dark Matter scenario based on the assumptions: i) the universe is flat; ii) the primordial power–spectrum of adiabatic fluctuations is scale–invariant (primordial spectral index $n = 1$); iii) the statistics of initial perturbations is Gaussian; such a specific choices are motivated by the prediction of the typical inflationary model. The model proved to be quite satisfactory in explaining most of the basic features of galaxy clustering without serious conflicts with the observed limits on the Cosmic Background Radiation. One difficulty of the model is however the general lack of structure on large scales (large–scale streaming motions, cluster–cluster correlation function, "bubbly" appearance of the slice of the CfA catalog, ...).

A CDM model with $n < 1$ and/or $\Omega_0 < 1$ seems to be the most conservative change to the standard scenario able to solve the large–scale problem. Such a possibility, investigated by Vittorio, Matarrese and Lucchin (1988), is however restricted by CBR anisotropy limits. Another possibility is to introduce ad–hoc characteristic scales in the primordial perturbations (see, e.g., Bardeen, Bond and Efstathiou 1987); this is possible in particular inflationary models, which however do not always fulfill a simplicity criterion.

One can consider instead the unexplored possibility that the statistics of primordial perturbations is non–Gaussian. Along this line Messina *et al.* (1990) investigated a class of non–correlated non–Gaussian models by N–body simulations, mainly to test the sensitivity of the clustering to initial non–random phases. In a second work Moscardini *et al.* (1990) performed simulations starting from non–Gaussian initial conditions in the frame of the CDM scenario: the present contribution reports the main results of this paper. The natural difficulty one has to face is the wide indeterminacy in assuming non–Gaussian statistics. Some sort of scale–invariance can be considered as a guiding principle, in the same spirit as this requirement allowed to select the Zeldovich spectrum as the most natural one. Another criterion can be that of building up the perturbation process by performing simple non–linear transformations, for instance of multiplicative type, on a Gaussian random field: it is interesting to note that inflation–generated non–Gaussian perturbations are expected to be of multiplicative type (see, e. g., Matarrese, Ortolan and Lucchin, 1990). In particular, we have considered three types of multiplicative non–Gaussian statistics for the peculiar

547

gravitational potential before the inclusion of the CDM transfer function: the *Convolution* model is obtained by convolving two independent Gaussian processes; the *Lognormal* statistics is the extreme case of multiplicative distribution and it formally splits in two different models, characterized respectively by positive or negative skewness for the linear mass fluctuation; our last model is a *Chi-squared* distribution with one degree of freedom; also in this case two possibilities appear according to the previous criterion. In general all these models are obtained by performing non–linear transformations on a Gaussian random field. The distributions are built up in such a way that the power–spectrum for the density perturbations is $\mathcal{P}(k) = \mathcal{P}_0 k T^2(k)$, where the CDM transfer function $T(k)$ is taken from Davis *et al.* (1985). This choice for the spectrum allows a comparison with the evolution of a standard, i.e. Gaussian, CDM model. A particle–mesh code with $N_p = 64^3$ particles on $N_g = 64^3$ grid–points was used. The box–size of the simulations is $L = 65\ h^{-2}$ Mpc, so that each particle carries a mass $m = 2.94 \times 10^{11}\ h^{-4}$ M$_\odot$ (h is the Hubble constant in units 100 Km/sec/Mpc). After evolving all the models starting from the same amplitude at the initial time, the clustering properties are analysed from a reference non–linearity time, fixed when in the Gaussian model non–linear events, up to the present time, which occurs when the particle two–point function is best fitted by a power–law $\xi(r) = (r/r_0)^{-\gamma}$, with $\gamma = 1.8$, in a suitable interval.

The main result will be that both the dynamics of clustering and the present texture are mostly sensitive to the sign of the primordial skewness. Both the structure and the dynamics of the Convolution model, which has vanishing initial skewness, are quite similar to the standard Gaussian CDM. Positive models cluster more rapidly showing a lumpy structure with small correlation length; the resulting peculiar velocities are also quite small. Large–scale structures in negative models form late, by the merging of shells surrounding primordial underdense regions, and give rise to a cellular structure with filaments, sheets and large voids; the resulting correlation length r_0 and *rms* bulk motion V_{bulk} turn out to be very large: f.i. with $h = 0.5$ one obtain the recently favoured value $r_0 \simeq 7.5 h^{-1}$ Mpc and a $V_{bulk} \sim 900$ Km/sec on patches of radius 500 Km/sec.

References

Bardeen, J.M., Bond, J.R. and Efstathiou, G. (1987) 'Cosmic fluctuations spectra with large–scale power' Astrophys. J. 321, 28–35.

Davis, M., Efstathiou, G., Frenk, C.S. and White, S.D.M. (1985) 'The evolution of large-scale structure in a universe dominated by cold dark matter', Astrophys. J. 292, 371–394.

Matarrese, S., Ortolan, A., and Lucchin, F. (1989) 'Inflation in the scaling limit' Phys. Rew. D40, 290–298.

Messina, A., Moscardini, L., Lucchin, F. and Matarrese, S. (1990) 'Non–Gaussian initial conditions in cosmological N–body simulations–I.Space–uncorrelated models', Mon. Not. R. astr. Soc. 245, 244–254.

Moscardini, L., Matarrese, S. Lucchin, F. and Messina, A. (1990) "Non–Gaussian initial conditions in cosmological N–body simulations–II.Cold dark matter models', Mon. Not. R. astr. Soc. (in press).

Vittorio, N., Matarrese, S. and Lucchin, F. (1988) 'Cold dark matter dominated inflationary model with $\Omega_0 < 1$ and $n < 1$', Astrophys. J. 328, 69–76.

EVOLUTION OF GALAXY CLUSTER AND PRIMORDIAL NUCLEOSYNTHESIS

V.V.CHUVENKOV, A.YU.GLUKHOV, B.V.VAINER
Department of Astrophysics
Rostov State University
5 Zorge
344104 Rostov-on-Don
U S S R

ABSTRACT. The model of galaxy cluster evolution is presented, matter exchange between galaxies and intergalactic medium and time dependence of initial mass function being taken into account. In all the model versions corresponding to available observed data primordial deuterium abundance in a galaxy decreases by a factor 2 to the present time, that enables to explain the results of primordial nucleosynthesis in the standart Big Bang model with baryon density parameter $\Omega_b = 0.1$.

1. INTRODUCTION

Evolution of galaxies is an important part of cosmological evolution and it should be taken into account for comparison of primordial yields of chemical elements with their present observed abundances. The complication and ambiguity in determination of the factor of chanding the element abundances during the period of galaxy evolution lead to the following way to solve this problem: it is necessary to design a model, in which as much as possible of physical characteristics compatible with observed data should be included into model scheme. In this work we shall consider evolution of chemical composition and luminosity of galaxies and emission of galactic dust.

2. BASIC EQUATIONS

Evolution of relatiye content of gas G and abundances of chemical elements X^i in a galaxy is described by the following equations (Tinsley (1980)):

$$\frac{dG_g}{dt} = -\Psi + I_1 - E + A \qquad (1)$$

549

$$\frac{d}{dt} X_g^i G_g = -X_g^i \Psi + I_2 - X_g^i E + X_{IGM}^i A, \tag{2}$$

where Ψ is the star formation rate (SFR), E is the rate of matter ejection from galaxy, A is the rate of accretion on to a galaxy;

$$I_1 = \int E_s(m) \; \varphi(m,t) \; \Psi(t-\tau_m) dm \qquad \text{and}$$

$$I_2 = \int E_{is}(m) \; \varphi(m,t) \; \Psi(t-\tau_m) dm$$

are, accordingly, the rate of return of stellar matter into interstellar medium (ISM) and rate of ISM enrichment by element i; φ is the initial mass function (IMF), E_s and E_{is} are the shares of total stellar mass and mass of element i returned into ISM, τ_m is the lifetime of star with mass m.

Chemical evolution of intergalactic medium (IGM) is determined by the equations:

$$\frac{dG_{IGM}}{dt} = E - A \tag{3}$$

$$\frac{d}{dt} X_{IGM}^i = X_g^i E - X_{IGM}^i A \tag{4}$$

Galaxy luminosity at the moment t is presented as:

$$L_g(t) = \int L_s(m) \int_{t-\tau_m}^{t} \Psi(t') \; \varphi(m,t') dt' dm \left[\frac{M_\odot}{L_\odot}\right]^{-1}, \tag{5}$$

where $L_s(m)$ is the bolometric luminosity of star with mass m.

At last, the temperature of galactic dust heated by stellar emission is:

$$T_d(t) = 9.6 \cdot 10^{-7} \left[\rho_d/X_d \int (1-A_d) Q_d I_\lambda d\lambda\right]^{1/5} [K], \tag{6}$$

where ρ_d and X_d are, accordingly, the density and the concentration of dust in galaxy, A_d and Q_d are the dust albedo and absorption factor, and I_λ is the spectral intensity of galaxy emission. In eqs. (5) and (6) we take into account only main sequence stars.

3. RESULTS

In calculations of galaxy chemical evolution final results depend on the assumed IMF and SFR. In the present work we have used the time-dependent form of IMF described by the equation:

$$\varphi(m,t) = Bm^{-1}\exp\left[2.3f(t) - (\lg m + f(t))^2\right], \qquad (7)$$

where $f(t) = -3.08\exp(-2t) + 1.02$ determines law of removing its maximum from $m=10$ $M_\odot$ at $t=0$ to the established
value $m=10^{-5}$ $M_\odot$ at present time $t=13.5$ Gyrs (coefficient
B is driven from normalization).

SFR is set in two forms (Fig.1). First one is proportional to galaxy gas fraction and decreases by time monoto-
neously that is characteristics of elliptical galaxies and
second one has two outbursts corresponding to the formation of spiral galaxy subsystems. Galaxy ejection is due
to the activity of II-type supernova and accretion on to
galaxy is $A=1$ $M_\odot/yr$.

The results are presented at Figs.2-4. It is seen
that they are in a good agreement with observed data.

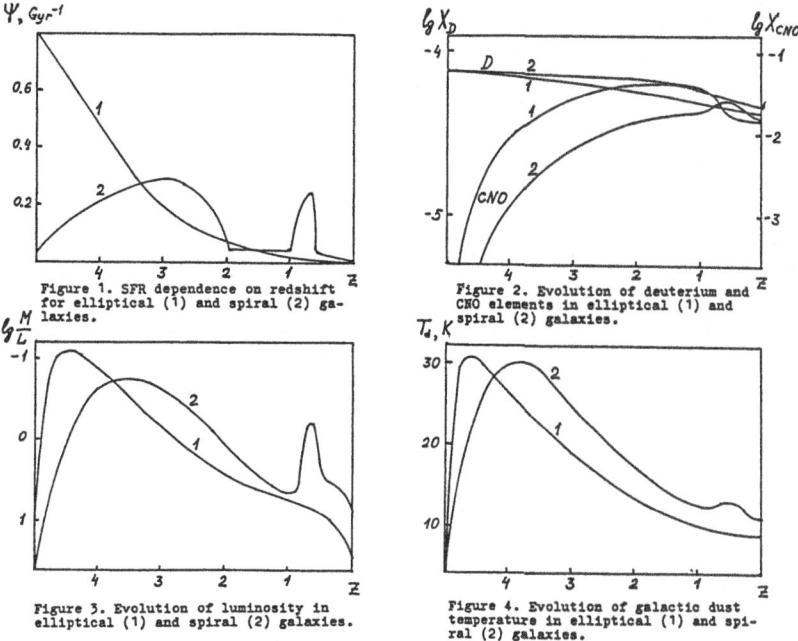

Figure 1. SFR dependence on redshift for elliptical (1) and spiral (2) galaxies.

Figure 2. Evolution of deuterium and CNO elements in elliptical (1) and spiral (2) galaxies.

Figure 3. Evolution of luminosity in elliptical (1) and spiral (2) galaxies.

Figure 4. Evolution of galactic dust temperature in elliptical (1) and spiral (2) galaxies.

The main cosmological result of the work is concerned
with evolution of deuterium abundance (See Fig.2). It decreases by factor no more than 2 during the stage of galaxy evolution, that leads to the conclusion about the value
of baryon density parameter $\Omega_b=0.1$.

4. REFERENCES

Tinsley B.M.(1980). Fundam. Cosmic Phys. 5, 287.

CONSTRAINTS ON UNIVERSE MODELS
WITH COSMOLOGICAL CONSTANT
FROM COSMIC MICROWAVE BACKGROUND ANISOTROPY

NAOSHI SUGIYAMA

Department of Physics, Kyoto University, Kyoto 606, Japan

with NAOTERU GOUDA AND MISAO SASAKI

ABSTRACT. Thorough numerical calculations of the fluctuations in the cosmic microwave background radiation using the gauge invariant formalism are carried out for various cosmological models with the cosmological constant.

1.Introduction

One of the most important constraints on cosmological models has been provided by the isotropy of the cosmic microwave background (CMB) radiation. It was found that it is very difficult to construct a model which satisfies constraints from observations of the CMB isotropy, large scale structures of the present universe and the age of the universe at once. We can resolve the paradoxical situation, *i.e.*, low density $\Omega_0 \simeq 0.1 \sim 0.3$, flat(inflation) and long lives universe $t_0 \gtrsim 13$, if one introduces a positive cosmological constant Λ since the existence of the Λ-term extends the cosmic age and permits a spatially flat universe with low density.

We also expect that a positive Λ will loosen the constraint from the CMB isotropy (Vittorio and Silk 1985 and Holtzmann 1989). In this paper, we numerically study CMB anisotropies for various cosmological models with the cosmological constant. And we investigate systematically the effect of the cosmological constant on CMB anisotropies and compare the results with recent observations by Readhead *et al.* (1989) and Davies *et al.* (1987). We consider almost all representative cosmological scenarios; pure baryonic, cold dark matter dominated and hot dark matter dominated universes with both adiabatic and isocurvature initial perturbations. However, we focus on spatially flat universes to be consistent with the inflationary scenario. For evolution equations we adopt the gauge invariant formalism of Bardeen (1980) which is further developed by Kodama and Sasaki (1984).

2.Results and Conclusions

Since the cosmological constant plays an important role only at late epochs, the evolution of density perturbations is not influenced by it before the universe becomes optically thin. Hence the cosmological constant does not affect physical processes before recombination such as Silk damping. There are two main effects of the cosmological constant. One is the change in the growth factor of the total matter perturbation from the recombination to the present. The other is a modification of the temperature correlation due to the spatial flatness. We find that both of two effects work to reduce the CMB anisotropy. The first effect occurs as follows. In an open universe, the growth of matter density perturbation is suppressed after the redshift of $z \simeq 1/\Omega_0 - 2$ due to the curvature effect. On the other hand, the suppression occurs at a later stage $z \simeq (1/\Omega_0 - 1)^{1/3} - 1$ in the flat universe with the cosmological constant. As a result, the growth factor of the latter becomes larger than that of the former.

The modification of the temperature correlation occurs as follows. The CMB anisotropy at an observational angle θ is essentially the difference between temperatures in two points separated by $\alpha\theta$ on the last scattering surface, where $\alpha \simeq (-K)^{-1/2}\sinh[(-K)^{1/2}\delta\eta]$ for a small angular separation with K given by $K = -a^2 H(1-\Omega)$ and $\delta\eta$ being the difference of the conformal time between present and the last scattering surface. Then the difference, $i.e.$, the CMB anisotropy increases as we see farther separated points.

We use J_3 at $25h^{-1}$Mpc as the normalization method for perturbation valuables. Our final constraints from the cosmic microwave background isotropy are as follows. As for pure baryonic models with initial adiabatic perturbations, unfortunately, $\delta T/T$ is slightly above the observational limits. On the other hand, as for isocurvature perturbations, we have found that the model $(\Omega_0, h) = (0.4, 1.0)$ is marginally allowed. Thus the existence of the cosmological constant seems to revive the pure baryonic universe, though the cosmic age of this model $\sim$ 9Gyr may be too short. For dark matter dominated models on Ω_0 are listed in the following. We also list the corresponding constraint on the age of the universe t_0 from that on Ω_0. For cold dark matter dominated models with adiabatic perturbations,

$$\Omega_0 \geq 0.10, \ t_0 \leq 12.5\text{Gyr} \ (h = 1.0), \tag{1}$$

$$\Omega_0 \geq 0.36, \ t_0 \leq 17.8\text{Gyr} \ (h = 0.5), \tag{2}$$

and for those with isocurvature perturbations,

$$\Omega_0 \gtrsim 0.31, \ t_0 \lesssim 9.4\text{Gyr} \ (h = 1.0). \tag{3}$$

For hot dark matter dominated models with adiabatic perturbations,

$$\Omega_0 \geq 0.25, \ t_0 \leq 9.9\text{Gyr} \ (h = 1.0), \tag{4}$$

$$\Omega_0 \geq 0.85, \ t_0 \leq 13.7\text{Gyr} \ (h = 0.5), \tag{5}$$

and for those with isocurvature perturbations,

$$\Omega_0 \gtrsim 0.57, \ t_0 \lesssim 7.8\text{Gyr} \ (h = 1.0). \tag{6}$$

In the case of $h = 0.5$, both cold and hot dark matter models with isocurvature perturbations are rejected.

Reference

Vittorio, N., and Silk, J., 1985, *Ap. J.*, (*Letters*), **297**, L1.

Holtzmann, J. A., 1989, *Ap. J. Suppl*, **71**, 1.

Readhead, A. C. S., Lawrence, C. R., Myers, S. T., Sargent, W. L. W., Hardebeck, H. E., and Moffet, A. T., 1989, *Ap. J.*, **346**, 566.

Davies, R. D., *et al.*, 1987, *Nature*, **326**, 462.

Bardeen, J. M., 1980, *Phys. Rev.*, **D22**, 1882.

Kodama, H., and Sasaki, M., 1984, *Prog. Theor. Phys. Suppl.*, **78**, 1.

GRAVITATIONAL LENS EFFECT AND DISTANCE MEASURE

K. WATANABE, M. SASAKI, and K. TOMITA
Yukawa Institute for Theoretical Physics
Gokasho Uji 611
Japan

ABSTRACT. The distance measure and time delays of images of a strongly gravitational lensed source are numerically calculated in a simple cosmological model.

1. Introduction

It is very important to investigate the light propagations in a realistic inhomogeneous universe because all of cosmological observations are done by light and light paths are necessarily affected by local inhomogeneities like galaxies before they reach us. This topic has been investigated by many authors, *e.g.*, Sachs and Wolfe (1967), Kantowski (1969), Dyer and Roeder (1973), Futamase and Sasaki (1988). The gravitational lens effect is one of the typical effects on the observations due to inhomogeneities (see *e.g.*, Refsdal 1964), and is sometimes discussed in the context of determination of cosmological parameters. Though the distance measure and time delays of images play an important role there (see *e.g.*, Cooke and Kantowski 1975, Blandford and Narayan 1986), it is not so evident what kind of distances must be used in the theory of the gravitational lens effect, *e.g.*, the lens equation and the time delay formula, because the gravitational lens effect may much affect the cosmological distance, *i.e.*, the angular diameter distance and so on. We therefore investigate the angular diameter distance and time delay by the numerical ray-shooting method.

2. Model Universe and Simulations

The ray-shooting are performed in a model universe with an average density that yields a spatially flat dust universe, *i.e.*, the Einstein-de Sitter (ED) universe. As for the matter distribution, we do not take into account the correlation of lens objects and the matter is assumed to be condensed into lens objects. The lens size is fixed as $100h^{-1}$ kpc and the redshift, z_S, and angular size, α, of a source are $z_S = 2$ and $\alpha = 2$ arcsec, respectively. To specify the model, we have to fix the mass of lenses. Instead of fixing it, we choose another parameter, ϵ^2/κ, to characterize the model, where ϵ^2 is the typical strength of the Newtonian potential, Ψ, generated by inhomogeneities: $\epsilon^2/\kappa \sim \Psi$, and κ is the ratio of the typical scale of inhomogeneities, ℓ, to the horizon scale, L: $\kappa \sim \ell/L$. As shown by Futamase and Sasaki(1988) and other authors, the parameter, ϵ^2/κ. is the measure of strength of the gravitational lens effect. In our model, we investigated in the case of $\epsilon^2/\kappa = 1$, which is the critical value.

We first simulate the ray-shootings to make a map of a ray bundle on the source plane. We then regard a small region (2 × 2 arcsec) as a source position and map this region back to the image plane to get multiple images of a single source. We next perform further ray-shootings in the directions of images to calculate the average of the distance measure and the arrival time for each images.

3. Results

In our simulation, we found four images for a single source. The typical separations of images were about 30 arcsec, which is rather larger than observed cases (~ 5 arcsec). Since the angular separation of images is expected to be in proportion to the lens size (see Refsdal 1964), this large separation is due to our choice of lens size. The distance measure of these images were much smaller ($\lesssim 0.05 H_0^{-1}$) than the standard one in the Friedmann model but the dispersion was quite small. The time delays between images were also calculated and we found that $\Delta t \sim 10^{-8} H_0^{-1}$, as was expected from the simple estimate of time delay due to the potential term in the time delay formula. Since these are rather preliminary results, the more detail analysis must be done.

4. Discussions

The Einstein ring is observed in an ideal situation that both (spherical) lens and source align on the line of sight of an observer. Using the lens equation, Refsdal (1964) derived the angular diameter of the Einstein ring, α_0, as $\alpha_0 = 4\sqrt{GMD_{LS}/D_L D_S} = 4 \times \sqrt{\epsilon^2/\kappa} \times \kappa \times \sqrt{D_{LS}/D_L D_S}$. If the scaling law on two parameters, ϵ^2/κ and κ, is actually found in numerical simulations of the Einstein ring, we can determine what kind of distances must be used in the lens equation. We therefore carried out the ray-shooting simulations but the authors would like to wait a chance to discuss the details.

References

Blandford, R., and Narayan, R., 1986, *Astrophys. J.*, **310**, 568.

Cooke, J. H., and Kantowski, R., 1975, *Astrophys. J. (Letters)*, **195**, L11.

Dyer, C. C., and Roeder, R. C., 1973, *Astrophys. J. (Letters)*, **173**, L31.

Futamase, T., and Sasaki, M., 1989, *Phys. Rev.*, **D172**, 2502.

Kantowski, R., 1969, *Astrophys. J.*, **155**, 89.

Refsdal, S.. 1964, *Mon. Not. R. astr. Soc.*, **128**, 295.

Sachs, R. K., and Wolfe, A. M. 1967, *Astrophys. J.*, **147**, 73.

Tomita, K. and Watanabe, K., 1989, *Prog. Theor. Phys.*, **82**, 563.

Tomita, K. and Watanabe, K., 1990, *Prog. Theor. Phys.*, **83**, 467.

Watanabe, K., and Sasaki, M., 1990, *Publ. Astr. Soc. Japan (Letters)*, **42**, L33.

Watanabe, K., and Tomita, K., 1990, *Astrophys. J.*, **355**, 1.

Watanabe, K., and Tomita, K., 1991, *Astrophys. J.*, April 1 issue (in press).

CONSTRAINTS ON THE COSMIC STRING SCENARIO FROM THE LARGE-SCALE DISTRIBUTION OF GALAXIES

S. MIYOSHI[1,2] and T. HARA[1]
[1]Department of Physics, Kyoto Sangyo University
Kamigamo-Motoyama, Kita-ku, Kyoto 603, Japan
[2]Institute of Astronomy, University of Cambridge
Madingley Road, Cambridge CB3 OHA, UK

ABSTRACT. From the observed large-scale distribution of galaxies we can get some constraints on the cosmic string scenario of galaxy formation.

Recently very deep pencil-beam surveys of galaxies in the directions of north and south Galactic poles have been made to study the large-scale clustering of galaxies. The analysis by Broadhurst et al. (1990) shows that there is an apparent regularity in the galaxy distribution with a characteristic scale of 128 h^{-1} Mpc, where h is Hubble's constant in units of 100 km s^{-1} Mpc^{-1}, over a linear scale extending to $\sim$2000 h^{-1} Mpc.

We can see a striking periodicity in Fig. 2 of Broadhurst et al. (1990), which plots the pair-count correlation for the observed redshift data. However, it does not necessarily mean the existence of the (same) periodicity in the histogram of observed redshifts. That is, a distinct periodicity could appear in the pair-count correlation even when the histogram of observed redshifts is clumpy in the manner that peak distances between neighbouring clumps are scattered around a fixed value. Indeed we can see some amount of fluctuation of peak distance in Fig. 1 of Broadhurst et al. (1990), which plots histograms of observed redshifts. The redshift distributions of 148 and 134 galaxies within two narrow (0.3 deg^2) fields of high galactic latitude (Koo and Kron 1987) also show considerable irregularities.

Thus the most important point of Broadhurst et al.'s (1990) result is not the periodicity in the pair-count correlation but the existence of the typical scale in the large-scale distribution of galaxies.

In the cosmic string scenario of galaxy formation (Rees 1986; Hara and Miyoshi 1987a,b, 1989), the typical separation L between the wakes of open cosmic strings at present is of the order of the comoving horizon scale at t_1, the epoch when the wakes started to form, i.e. L$\sim$4000$\times$ $(1+z_1)^{-1/2}$ h^{-1} Mpc in the flat (Ω=1) universe, where z_1 is the redshift corresponding to t_1. The equality of L with the scale 128 h^{-1} Mpc found by Broadhurst et al. (1990) yields $1+z_1\sim$1000. This means that the formation of the wakes relevant to the galaxy formation started after the recombination $(1+z_{REC} \simeq 1300)$. Taking into account the possible variation of observed characteristic scale of galaxy distribution with direction, it is most probable that the wake formation started just after the recombination time t_{REC}. In order that the baryonic matter can accumulate on to wakes, its sound velocity should be less than the infall velocity. It gives the following restriction on the string line mass density:

$$G\mu/c^2 \gtrsim 2.4 \times 10^{-6}(1-v/c)^{1/2}([1+z_i]/1300)^{1/2}$$

in case of $v/c \sim 1$, where μ is the string line mass density, G the gravitational constant, c the light velocity, and v the string velocity (perpendicular to its length).

If the universe is dominated by baryons or hot dark matter, no significant wakes could form until the recombination time t_{REC}. However, if the universe is dominated by cold dark matter, wakes of dark matter start to form at the earlier time t_{EQ} when the universe becomes matter dominant, thereafter evolving as shown by Rees (1986) and Hara and Miyoshi (1987a), and baryons can fall on to these wakes after t_{REC}. In this case there should be smaller subsidiary structures with comoving sizes or separations of ~ 20 h^{-2} Mpc $(c\Delta z \sim 2000$ h^{-1} km s$^{-1})$.

References

Broadhurst, T. J., Ellis, R. S., Koo, D. C., and Szalay, A. S. (1990) 'Large-scale distribution of galaxies at the Galactic poles', Nature, 343, 726-728.

Hara, T. and Miyoshi, S. (1987a) 'Formation of the first systems in the wakes of moving cosmic strings', Prog. Theor. Phys., 77, 1152-1162.

Hara, T. and Miyoshi, S. (1987b) 'Flare-up of the universe after $z \simeq 10^2$ for cosmic string model', Prog. Theor. Phys., 78, 1081-1098.

Hara, T. and Miyoshi, S. (1989) 'Large-scale structures and streaming velocities due to open cosmic strings', Prog. Theor. Phys., 81, 1187-1197.

Koo, D. C. and Kron, R. G. (1987) 'Evolution of very faint galaxies and quasars', in A. Hewitt, G. R. Burbidge and L. Z. Fang (eds.), Observational Cosmology (IAU Sympo., No. 124), Reidel, Dordrecht, pp. 383-388.

Rees, M. J. (1986) 'Baryon concentration in string wakes at $z \gtrsim 200$: implications for galaxy formation and large-scale structure', Mon. Not. R. astr. Soc., 222, 27p-32p.

CLUSTERING AND VELOCITY FIELD OF GALAXIES IN COLD DARK MATTER SCENARIO

TATSUSHI SUGINOHARA[1] AND YASUSHI SUTO[2]

[1] *Department of Physics, The University of Tokyo, Bunkyo 113, Japan*
[2] *Uji Research Center, YITP, Kyoto University, Uji 611, Japan*

Cold dark matter (CDM) scenario with biased galaxy formation in an Einstein – de Sitter universe has been attracting much attention since it was reported to be successful in reproducing observations on various scales using N-body simulations (*e.g.*, Davis *et al.* 1985). In the present talk we will show the results of a new simulation which aims at re-examining the previous results with $N = 128^3$ particles. Details will be described elsewhere (Suginohara *et al.* 1991; Suto and Suginohara 1991).

The simulation is carried out in a periodic cube whose size L is set to $65h^{-2}$Mpc, where h is the hubble constant in units of $100\,\mathrm{km\,s^{-1}Mpc^{-1}}$. In comparing the results with observations, we consider two specific possibilities in which (i) 1.5σ peaks, or (ii) 2.5σ peaks in the initial density field are assumed to preferentially form luminous galaxies, where σ is the rms fluctuation for the density field smoothed over the filtering length $R_f = 0.34h^{-2}\,\mathrm{Mpc}$.

We use the redshift correlation function of galaxies, $\xi_s(s)$ (de Lapparent, Geller and Huchra 1988), to identify the present epoch of the simulation data, rather than the real space correlation function $\xi_r(x)$ (*e.g.*, Davis and Peebles 1983). It turns out that both procedures predict quite different fluctuation amplitude to be matched with the present universe, implying that the previous results (*e.g.*, Davis *et al.* 1985) should be carefully interpreted. If the present epoch is identified using $\xi_s(s)$, the present model predicts an unacceptably large value for the one-dimensional rms velocity dispersion $\sim 1000\,\mathrm{km\,s^{-1}}$ (Table 1). Therefore to the extent that the observed $\xi_s(s)$ is reliable as a statistical measure of the fair sample of our universe, the conventional CDM scenario in an Einstein – de Sitter universe with the primordial scale-invariant, random-Gaussian density field should be ruled out.

References

Davis, M., Efstathiou, G., Frenk, C. S., and White, S. D. M. 1985, *Ap. J.*, **292**, 371.
Davis, M., and Peebles, P. J. E. 1983, *Ap. J.*, **267**, 465.
de Lapparent, V., Geller, M. J., and Huchra, J. P. 1988, *Ap. J.*, **332**, 44.
Suginohara, T., Suto, Y., Bouchet, F. R., and Hernquist, L. 1991, *Ap. J. Suppl.*, **75**, March issue, in press.
Suto, Y., and Suginohara, T. 1991, *Ap. J. (Letters)*, in press.

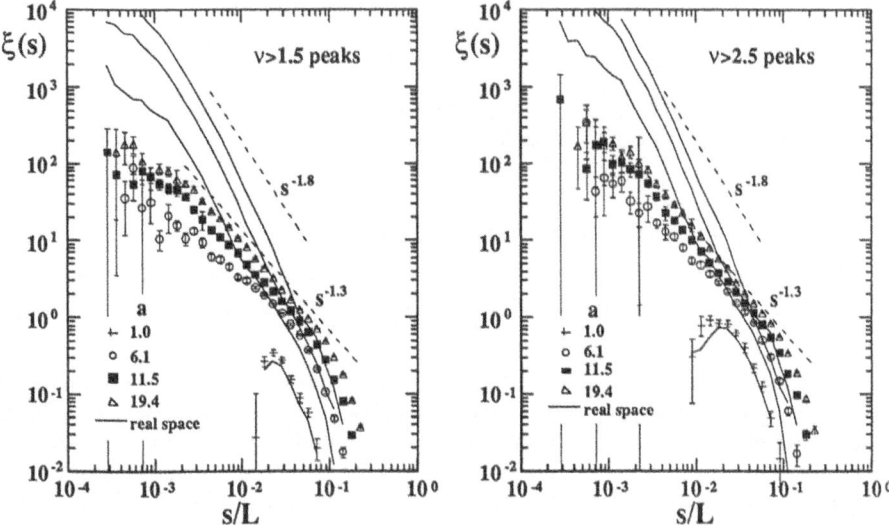

Figure 1 (*left*): Correlation functions for 1.5σ peaks in real space (solid curves) and in redshift space (symbols). The symbols are labelled by the expansion factor a, which is normalized to unity at the start of the simulation.

Figure 2 (*right*): The same as Fig.1 for 2.5σ peaks.

sample	a	ξ a)	h b)	v_p c)
1.5σ	6.1	$s^{-1.2}(x^{-1.8})$	$0.26^{+0.08}_{-0.05}(0.36 \pm 0.02)$	$1000 \pm 250(\ 700 \pm 50)$
1.5σ	19.4	$s^{-1.4}(x^{-2.6})$	$0.52^{+0.15}_{-0.09}(0.6\ \pm 0.03)$	$1400 \pm 300(1200 \pm 50)$
2.5σ	6.1	$s^{-1.3}(x^{-1.8})$	$0.34^{+0.1}_{-0.06}(0.42 \pm 0.02)$	$850 \pm 200(\ 700 \pm 50)$
2.5σ	19.4	$s^{-1.4}(x^{-2.7})$	$0.52^{+0.15}_{-0.09}(0.6\ \pm 0.03)$	$1400 \pm 300(1200 \pm 50)$

Table 1: Fitted parameters using the correlation functions.

a) Slope of the redshift-space correlation functions over the range $1 < \xi < 10$. The numbers in parentheses indicate the slope of the real-space correlation functions.

b) The Hubble constant derived by matching the correlation length in redshift space to the observed value. The numbers in parentheses are derived using the correlation length in real space.

c) Dispersion of the relative peculiar velocities in radial direction for particle pairs with separation $\sim 1h^{-1}\text{Mpc}$, in units of $\text{km}\,\text{s}^{-1}$. The velocities are scaled using the value of h given in the previous column.

FORMATION OF GALACTIC HALOS FROM SEEDED HOT DARK MATTER

Anthony van Dalen
Department of Physics

The Ohio State University, Columbus, OH 43210

Abstract: I summarize work done with Jens V. Villumsen[1] on the formation of galactic halos in a flat universe dominated by massive neutrinos. We find that the halos generated are well approximated by isothermal spheres, generating flat rotation curves, in agreement with galactic rotational velocity observations. Further, it affirms the assumption used in the calculation of the Tremaine-Gunn limit[2]. However, we find that with a massive object dominating the central region the core radius becomes an ill-defined quantity, so with a large fraction of the inner region made up of some other material, such as baryons, it is possible to produce small galaxies with low velocity dispersions.

1. The Model: The standard hot dark matter model for structure formation via gaussian initial fluctuations has the severe problem of virtually no primordial structure on galactic scales. This makes the formation of galaxies problematic. If one accepts the theoretical prejudice of $\Omega = 1$, it is tempting to try to save the hot dark matter universe, since it has a natural candidate in the neutrino. One way of forming small scale structure in such a universe is to suppose that small, very massive objects are the source of the perturbations[3]. Candidates for these seeds include primordial black holes, non-topological solitons, and cosmic texture.

Another issue, assuming the hot dark matter is massive neutrinos, comes from Liouville's theorem. Since the initial distribution of neutrinos has a peak phase space density, one can limit the characteristics of the final distribution. Assuming an isothermal sphere solution, one finds

$$m_{TG} > 120 eV (\frac{100 \text{km s}^{-1}}{\sigma})^{1/4} (\frac{1\text{kpc}}{r_c})^{1/2} (\frac{2}{g_\nu})^{1/4}, \tag{1}$$

which is the Tremaine-Gunn limit on the mass of the neutrino for a givien core radius, r_c, and one dimensional velocity dispersion, σ. Conventionally, one reverses this inequality to put a limit on the core radius. The neutrino mass is given by $98 \Omega_\nu h^2 eV$. For a given system, σ is derivable from the observed rotation curve. The "problem" then, is that some small galaxies seem to have observed core radii that are too small for the infered velocity dispersions.

We examined the evolution of a single seed perturbation in detail to determine the characteristics of the system formed, with particular emphasis on the relevance of the Tremaine-Gunn limit.

2. The Calculation: The collapse phase was evolved using a 3-D N-body code[4], which solves Poisson's equation by an nth order expansion of the potential in spherical harmonics. In this calculation the expansion was cut off at 2. Going to higher order was not found to be necessary. The calculation was begun at $1 + z = 20$ and continued down to 1.7. The input was derived via a separate, fully relativistic code for evolving linear perturbations[5]. The universe was assumed to have $h = 0.5$ and a closure density in massive neutrinos. No baryonic material was included. The seed used was $10^{10} M_\odot$.

3. Results: Figure 1 shows the density profile of the system at $1 + z = 2.6$ and 1.7. The plotted line is the average of 5 different random realizations of the same initial conditions. The error bars are the rms deviations. Figure 2 shows the rotation curve, the radial component of the velocity dispersion, and the mean radial velocity as a function of radius, at $1 + z = 1.7$. The system is isothermal and istropic in the core, but the tangential velocity components drop to 60% of the radial in the outer parts of the system. Overall, the system is well modeled by an isothermal sphere. Further, the amplitude of the density profile is proportional to the scale factor. It is possible to derive scaling relations for the velocity dispersion, size, mass, and density of the system that are consistent with this. We find that no perturbation less than roughly $10^9 M_\odot$ will accrete more than its own mass by $1 + z = 1$.

The seed at the center produces a weak density singularity, which becomes less important as the system becomes halo dominated. This demonstrates a significant weakness in the Tremaine-Gunn limit: A large mass concentration at the center of the system, which should form naturally in a universe with baryons, makes the "core radius" a less meaningful quantity. It is often suggested that "dark baryons" are necessary to form the inner halo of small galaxies. Figure 1 indicates that this extra populatation may not be necessary. The visible part of a given galaxy may make the halo suficiently singular so that it and the visible system alone will produce a flat rotation curve.

While the core radius limit may not be relevant, the density of a thermalized distribution of neutrinos will always be limited:

$$\rho_{TG} < 3 \times 10^6 M_\odot \text{kpc}^{-3} (\frac{\sigma}{100 \text{km s}^{-1}})^3 (\frac{m_\nu}{25 eV})^4. \qquad (2)$$

The systems modeled were found to be consistent with this limit, although the galaxy DDO 154[6] was not. However, this latter analysis is based on the parameters derived by Carignan and Freeman, who do not take into account the perturbing effect of the visible part of the galaxy on the halo, nor the possibility that the galaxy is further away than they estimate.

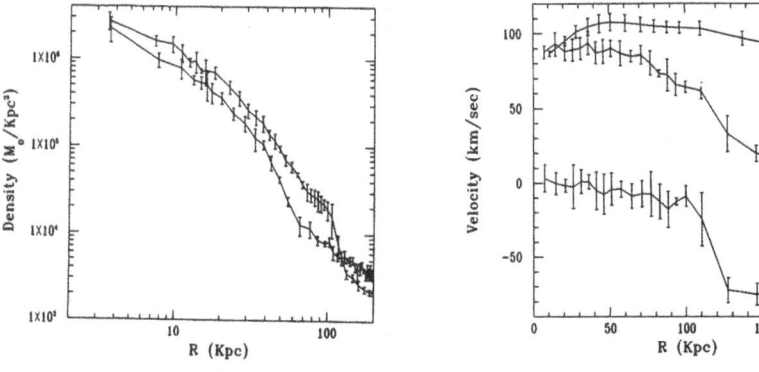

Figure 1: Density profile at $1 + z = 2.6$ (lower curve) and $1 + z = 1.7$.
Figure 2: From top to bottom: averaged circular velocity, radial velocity dispersion, and mean radial velocity vs radius at $1 + z = 1.7$.

1. van Dalen, A. and Villumsen, J.V. 1990, Ap. J., (submitted).
2. Tremaine, S. and Gunn, J.E. 1979, Phys. Rev. Letters, **42**, 407.
3. Villumsen, J.V., Scherrer, R.J., Bertschinger, E. 1990, Ap. J., (in press).
4. Villumsen, J.V. 1984, Ap. J., **295**, 388.
5. van Dalen, A. 1990, Ap. J., **351**, 356.
6. Carignan, C. and Freeman, K.C. 1988, Ap. J. (Letters), **332**, L33.

GRAVITATIONAL CLUSTERING OF GALAXIES: COMPARISON BETWEEN THERMODYNAMIC THEORY AND N-BODY SIMULATIONS

MAKOTO ITOH
Department of Astronomy, Kyoto University,
Sakyou-ku, Kyoto 606,
Japan

ABSTRACT. $f(N)$ statistics are very useful for describing the distribution of galaxies. Thermodynamic theory gives the functional form of $f(N)$ and agrees with the results of N-body simulations very well. It can also give the good measure of clustering and initial conditions.

1. $f(N)$ Statistics and Thermodynamic Theory

One of the important problems in cosmology is a quantitative description of the distribution of galaxies. One useful description is a two-point correlation function, $\xi(r)$. However, two-point correlation function contains very limited information. Fortunately, there is more effective approach which is $f(N)$ statistics for finding N galaxies in the volume of size V.

Saslaw and Hamilton(1984) derived a formula of $f(N)$ based on gravitational thermodynamics:

$$f(N) = \frac{\overline{N}(1-b)}{N!} [\overline{N}(1-b) + Nb]^{N-1} e^{-\overline{N}(1-b)-Nb},$$

where $\overline{N} = \overline{n}V$, $\overline{n}$ is a number density of galaxies and b is the ratio of the gravitational correlation energy to the kinetic energy of galaxies.

2. Comparison with N-Body Simulations

We carried out 4000-body simulations using Aarseth's COMOVEV code in order to examine the evolution of b and dependence on density parameter (Itoh et al. 1988). Initial distribution is Poisson. We calculated three models which have different density parameters; Ω_0 =1.0, 0.1 and 0.01 at $a/a_0 =$ 32, where a is the expansion factor with initial value a_0.

Figure 1. shows the results of the flat universe at $a/a_0 = 7.86$. We calculated $f(N)$ for four different sizes of sampling sphere whose radius is R. Solid histogram is the distribution functions calculated from simulations and dashed curve is the thermodynamic distribution functions. We used b as a fitting parameter in least-square fitting between experimental and theoretical distribution function. Agreements between them are very well. Fitted values of b is almost constant about 0.75.

Figure 2. shows the evolution of b and dependence on density parameter. Fitted values of b increase as clustering proceeds. Open universe evolves a little more slowly than flat universe. The two dashed lines show the range of observational fitted value of b ($b_{\text{obs}} = 0.70 \pm 0.05$). For the case of $\Omega_0 = 0.01$, fitted value of b saturates at about 0.25. From these figures, we can conclude that thermodynamic distribution function can describe the distribution of galaxies very well. Fitted value of b is a good measure of clustering. Comparing with an observational result, density parameter should be larger than 0.1.

In order to investigate the effects of primordial density fluctuations, we calculated other N-body simulations with primordial density fluctuations by using tree-code with 32768 particles (Suto et al. 1990). Power spectrum of primordial density fluctuations is $|\delta_k|^2 \propto k^n$ ($n = -2, -1, 0, 1$). It is thought that even in the evolved stage, initial conditions still remain within the large-scale linear regime. In fact, the fitted value of b depend on the index of initial power spectrum. We can derive the scale dependence of b for the large-scale using thermodynamic theory and linear density perturbation theory:

$$1 - b(R) \propto R^{n/2} \qquad (n = -2, -1, 0, 1) \qquad \text{for large } R,$$

where R is a radius of sampling sphere. Therefore scale dependence of b will give a clue to the primordial density fluctuations.

Figure 3. shows the scale dependence of the fitted value of b. We can see that the fitted value of b for large scale evolves keeping this relation. This means that initial conditions remain in large-scale linear regime.

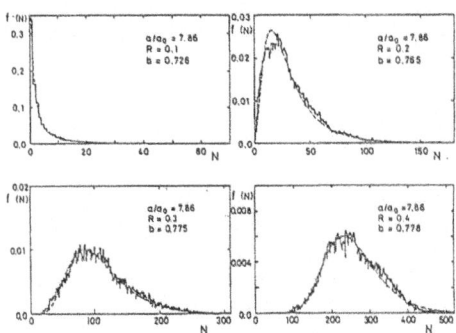

Figure 1. Distribution function of $f(N)$.

3. Summary

Our main results are as follows: (1) thermodynamic distribution function can describe the distribution of galaxies very well. (2) fitted value of b is a good measure of clustering. (3) density parameter should be larger than 0.1. (4) if we analyze observational distribution of galaxies more detail, we may get a clue to the primordial density fluctuations of the universe from the scale dependence of fitted value of b.

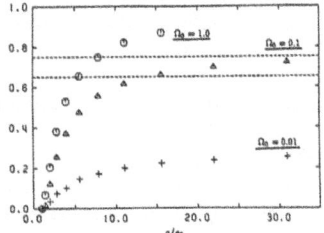

Figure 2. Evolution of b.

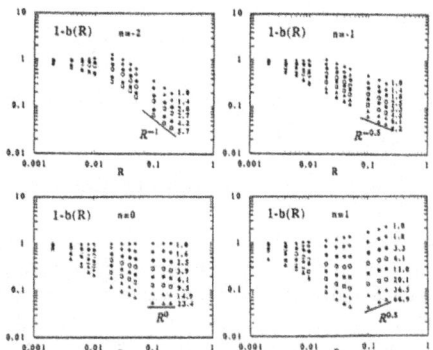

Figure 3. Scale dependence of b.

References

Itoh, M., Inagaki, S., and Saslaw, W.C. 1988, *Astrophys. J.*, **331**, 45.
Saslaw, W.C., and Hamilton, A.J.S. 1984, *Astrophys. J.*, **276**, 13.
Suto, Y., Itoh, M., and Inagaki, S. 1990, *Astrophys. J.*, **350**, 492.

THE HYDRODYNAMICAL INTERACTION OF GALAXIES

M. UMEMURA
National Astronomical Observatory
Mitaka, Tokyo 181
Japan

S. YOSHIOKA
Tokyo Mercantile Marine College
Koutou, Tokyo 135
Japan

ABSTRACT. We present 3D hydrodynamical calculations on interacting/merging galaxies. We find that nearly head-on collisions of galaxies lead to new galaxy formation. The dissipative effect of gaseous component significantly affects on merging process of interacting galaxies.

1. Introduction

The galaxy interaction attracts great attention in relation to galaxy formation and evolution in dark matter dominated universes and recent radio observations on interacting/merging galaxies as well. The calculations for the galaxy interaction has been so far performed mainly by the N-body simulations of collisionless particles. The gasdynamical effects, which have been ignored, are expected to play a significant role in galaxy collisions.

We perform 3D hydrodynamical calculations using an SPH scheme combined with N-body scheme, and pursue the dynamical evolution of both collisionless stellar component and gaseous matter.

2. Spherical Galaxy Encounters

First we adopt the spherical isothermal model, $\rho(r) = \sigma^2/2\pi G r^2$, for galaxies, with $M_* = M_g = 5 \times 10^8 M_\odot$, $R_G = 10$kpc, $\sigma_* = 10$km s^{-1}, and $T_g = (2\times)10^4$K. The calculations are parameterized by the relative distance r_p and velocity v_p between two galaxies at the pericenter.

Case 1 (nearly head-on): $r_p = 0.15$kpc, $v_p = 388$km s^{-1}

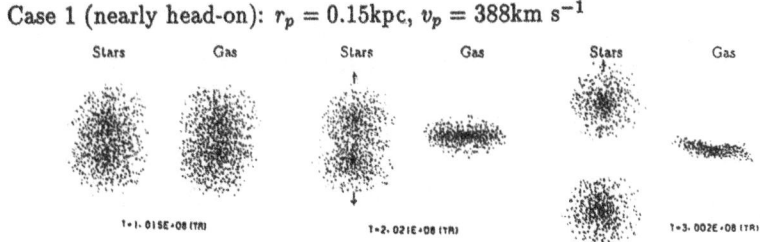

Case 2 (parabolic): $r_p = 4.6$kpc, $v_p = 63$km s^{-1}

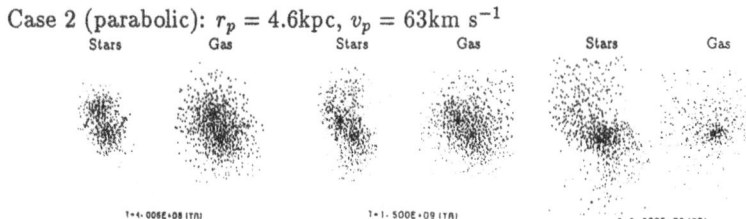

The results for a nearly head-on collision (Case 1) show that stellar components go through each other and a gaseous disk is left which undergoes gravitational instability. This leads to the third galaxy formation. In a parabolic encounter case (Case 2), two galaxies consequently merge, though this case doesn't satisfy the merging criterion if there is just collisionless component. This shows the dissipative effect is very important in understanding the merging process.

3. Disk Encounters

Next, we adopt exponential disks, $I(r) = I_0 \exp(-r/r_d)$, $r_d = 5$kpc, with total mass of $10^{10} M_\odot$. The gas is isothermal with $T_g = 2 \times 10^4$K. As for the kinematics, the rotational balance is assumed against the gravitational field at each point.

Case 3 (edge-to-edge collision): $r_p = 15$kpc, $v_p = 400$km s^{-1}, $f_g = 0.1$

Case 4 (edge-to-face collision): $r_p = 15$kpc, $v_p = 400$km s^{-1}, $f_g = 0.1$

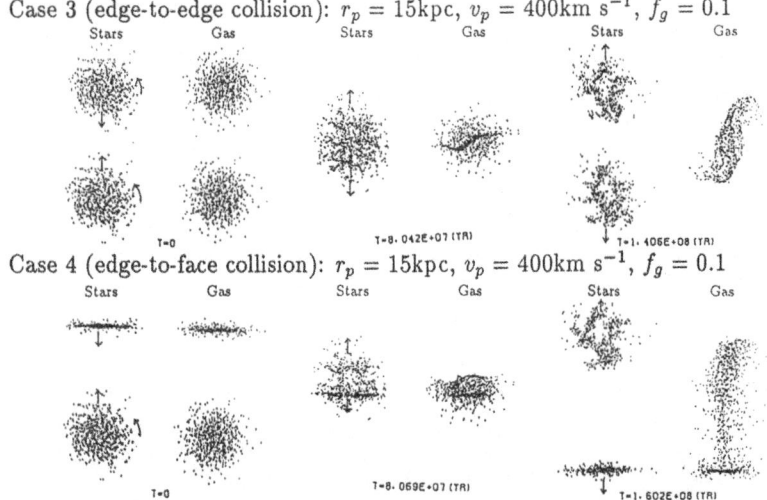

In Case 3, after the stellar components go through each other, the gaseous components merge into a disk, where a spiral shock occurs due to the angular momenta of parent galaxies. In Case 4, in the galaxy which undergoes edge-on collision, the bulk of gaseous component is stripped to form a gas streaming. On the other hand, in the galaxy which undergoes face-on collision, the gaseous disk survives and a spiral pattern appear due to the disturbance by the penetration of the edge-on colliding galaxy.

Finally, we consider parabolic encounters as merging cases. As the results, galaxies with gas fraction $f_g = 0.5$ merge three times faster than those with $f_g = 0.1$. In this process, the central phase space density of the merger remnant is raised by two order due to dissipative effect, which is just comparable to those of elliptical galaxies.

VELOCITY FIELDS AROUND SUPERCLUSTERS :
THE EFFECT OF ASPHERICITY

T. WATANABE
Department of Astronomy,
Faculty of Science,
Kyoto University,
Kyoto 606, Japan

ABSTRACT. Velocity fields in some aspherical configurations are investigated. They tend to deviate from that of a spherical configuration due to initial shapes of systems. The value of the cosmological density parameter may be affected by the flatness of the Local Supercluster.

1. Introduction

In theoretical approach, peculiar velocity fields are modeled in some special cases — linear perturbation theory [Peebles(1980)], spherical nonlinear model [Silk(1974) and Shechter(1980)] or an approximation found by Yahil(1985):

$$\frac{v_p}{Hr} = \frac{1}{3}\Omega^{0.6}\delta_r(1 + \delta_r)^{-1/4}. \tag{1}$$

where v_p is a radial peculiar infall velocity and δ_r is a mean density contrast within a shell of radius r. But in the real universe some objects like the Local Supercluster (the LSC) have appriciable flatness. So we are interested in some questions:

Is the spherical model also valid to aspherical systems?

If the answer is 'no', how their velocity fields depend on their asphericity?

To answere them, we consider the simplest case, the evolution of homogeneous oblate spheroid in an expanding background. Historically evolution of such figures has already been investigated by White and Silk(1979), but their velocity field during the evolution and deviation from spherical model were not considered in detail. So we focus on them.

2. Homogeneous Spheroid Model

We integrated numerically the equation of motion of homogeneous oblate spheroid found by White and Silk, assuming that the homogeniety of matter is conserved inside and outside of the spheroid in some cosmological models. In figure 1, we show the axial velocities in units of Hubble expansion as a function of the density contrast. Apparently each velocity component is greatly affected by its initial flatness.

We also found that a velocity component on each axis is approximated to following forms

$$\frac{v_i}{Ha_i} = \left[1 + \left(\frac{2}{3\alpha_i(t_i)} - 1\right)\left(\frac{1 + n\delta(t)}{1 + n\delta(t_i)}\right)^{1/3n}\right]^{-1}\left(\frac{H_R}{H} - 1\right)_{\text{spheroid}}, \tag{2}$$

where

$$\alpha_i = \begin{cases} \dfrac{\sqrt{1-e^2}}{e^2}\left(\dfrac{\sin^{-1}e}{e} - \sqrt{1-e^2}\right) & i = 1, 2 \\ \dfrac{2\sqrt{1-e^2}}{e^2}\left(\dfrac{1}{\sqrt{1-e^2}} - \dfrac{\sin^{-1}e}{e}\right) & i = 3 \end{cases} \quad , \quad n \sim \begin{cases} 1 & i = 1, 2 \\ q_i^{-0.75}, & i = 3 \end{cases}$$

$q_i = (a_3/a_1)$ at initial time, and $(H_R/\dot{H} - 1)_{\text{spheroid}}$ is substituted by eq.(1) δ of the spheroid in place of δ_r. This is available within the range of $0 \lesssim \delta \lesssim 10$ and almost independent of cosmological models.

3. The Cosmological Density Parameter

Using the results of calculation, we can talk about the value of Ω determined dynamically by application of the theoretical models on some flattend systems such as the LSC. Now we define two Ω's, Ω_R derived from the spherical model and Ω_S from our spheroid model. For simplicity we assume that (i) our model describes the dynamics of the LSC with sufficient accuracy and that (ii) our Galaxy is on the fundamental plane.

We can calculate the ratio Ω_S/Ω_R on some bases.

(i) the relation between δ of the spheroid itself and δ_r within our distance to the center is derived only by geometrical parameters p and q. Here p is the ratio of the distance of our Galaxy to the center of the LSC to the semi-major axis of the LSC, and q is the axial ratio.

(ii) the q - δ relation is uniquely determined by the initial axial ratio q_i almost independently of cosmological models and

(iii) when we have model independent Virgocentric velocity, the ratio is derived using the approximation of axial velocities we found.

Here we show the contour of this ratio in the case of $\delta = 3$ in figure 2. A figure on a contour is value of the ratio. You see that Ω derived from our spheroidal model can be systematically larger than that derived from the spherical model when the system has appriciable flatness and the distance of our Galaxy to the center is sufficiently small.

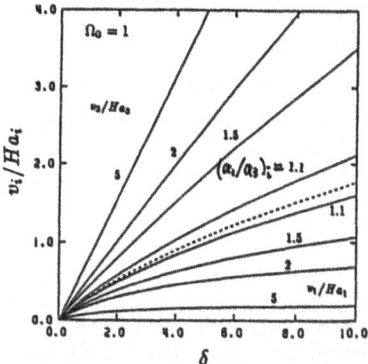

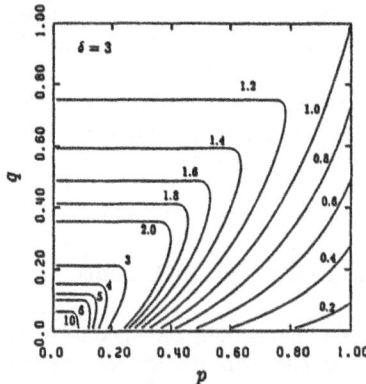

Fig.1. The axial peculiar velocities as a function of the density contrast. The radial peculiar velocity of the spherical configuration with the same initial density contrast is shown by broken curves.

Fig.2. Contours of the ratio Ω_S/Ω_R as a function of p and q. The case of $\delta = 3$ is shown.

Evolution and Statistics from the Number Counts of X-Ray Cluster and Intergalactic Medium

Hitoshi HANAMI

Laboratory of physics, College of Humanities and Social Sciences
Iwate University, 18-34, 3-Chome Ueda, Morioka, 020 JAPAN

ABSTRACT: We discuss the distribution and the evolution of X-ray clusters of galaxies. Hot Intra-Cluster Plasma (ICP) is the source of X-ray emission. We calculate the mass function and the temperature distribution by galaxy formation scenarios. We compared the computed results with the observational data. According the cold dark matter scenario, we need the biasing effects for the X-ray Clusters.

1. Introduction

From recent X-ray observations, the intergalactic medium in many clusters (cf. Coma, Perseus) have been found to be in the state of diffuse, thin and hot plasma, which may be trapped in the gravitational potential of clusters (e.g. Sarazin [1988]). These X-ray emitters are diffuse one which is opposed to the Active Galactic Nuclei (AGNs), and assocites with not only rich clusters but also poor clusters and even the groups of galaxies. We can estimate the total gravitational mass of the clusters, if the systems can be approximated to be in hydrostatic equilibrium state.

2. Relation between Mass and X-ray Temperature for Clusters

The total mass of a cluster is not directly measurable. If the mass of baryonic matter is proportional to the total dark matter, the optical luminosity of the clusters may be related to the total mass. More observable quantities related to the total mass, however, are the temperature of the hot gas. The typical virialized density is $\rho_c = 180\rho_0(1 + z)^3$. From this relation, we can get the typical radius of a cluster $R_c = \left(\frac{3M_c}{4\pi\rho_c}\right)^{(1/3)}$. If infalling and virializing compression process is the origin of the hot gas heating, we can estimate the temperature from the mass and the radius. If the gas is heated during violent relaxtion at the same time that a cluster is collapsed, it have the same energy per unit mass as the matter in galaxies, $kT_c \approx \sigma^2\mu m \approx \frac{\alpha G\mu m}{3}\frac{M_c}{R_c}$, where σ and α is the line-of sight velocity dispersion of the cluster and a factor of order unity which is found to be 0.8 in typical clusters from N-body simulations (e.g. Evrard [1989]). With above estimation of the radius of cluster the temperature of mass is related to the total mass of a cluster.

3. Theoretical Mass and X-ray Tempareture Function

Theoretical and simulation analysis for the cluster evolution reports the statistical property is represented by the multiplicity mass function which was started with Press-Schechter ([1974]) formula (e.g. Efstathiou et al.[1988]).

$$N(M,z)d\ln M = \sqrt{\frac{2}{\pi}}\frac{\rho_0}{M}\frac{d\ln\sigma_0(M)}{d\ln M}\nu(M,z)\exp\left(\frac{-\nu(M,z)^2}{2}\right)d\ln M.$$

$$\nu(M,z) = \frac{(1+\Omega^{0.6}z)}{(1+\Omega^{0.6}z_i)}\frac{\delta}{\sigma_0(M)}$$

$$\sigma_0(M)^2 = \int W(kR)P(k)d^3k$$

569

$$M = f\rho_0(1 + z)^3 R^3.$$

Using these equations we can calculate the mass function with a given initial perturbation. We had considered two cases: HCM (Hot Dark Matter) Scenario and CDM (Cold Dark Matter) Scenario (e.g. Bardeen et al. [1986]). The temperature distribution function can be obtained from the mass functions, if these collapsed systems are relaxed to hydrostatic equilibrium states as shown above.

4. Results

Figure Shows the cumulative temperature distribution function according to the above galaxy formation scenarios. We compared the computed results with the observational data by Edge et al. ([1990]). According the cold dark matter scenario, we need the biasing effects for the X-ray Clusters. On the other hand, the naive HDM scenario can not explain the X-ray cumulative temperature function. These are strong constraint for the galaxy formation theories.

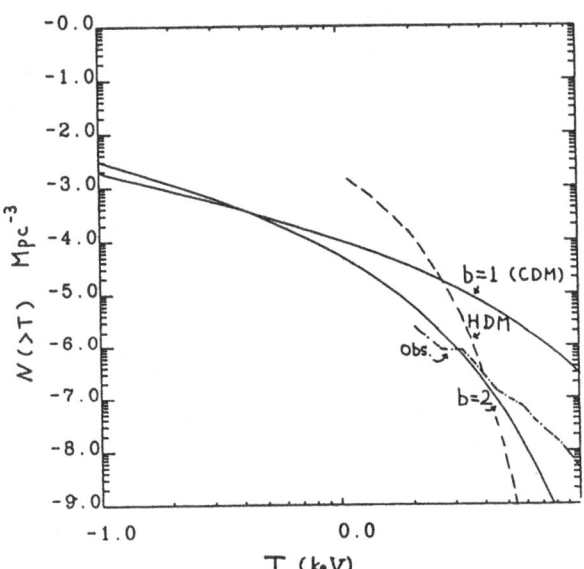

Figure The cumulative temperature function. The solid lines, the dotted line and the dush-dotted line is from the CDM ($b = 1, 2$), the HDM and the observation data, respectively.

References

Bardeen, J.M., Bond, J.R., Kaiser, N. and Szalay, A.S. 1986, Ap.J., 304, 15

Edge, A.C., Stewart, G.C., Fabian, A.C. and Arnaud, K.A. 1990, MNRAS, 245, 559

Efstathiou, G., Frenk, C.S., White, S.D.M. and Davis, M. 1988, MNRAS, 235, 715

Evrard, A.E., 1989, Ap.J., 341, L71

Press, W.H. and Schechter, P. 1974, Ap.J., 187, 425

Sarazin, C. L. 1988, X-Ray Emission from Clusters of Galaxies (Cambridge: Cambridge University Press)

A SAMPLE OF ULTRALUMINOUS IRAS GALAXIES IN SOUTHERN SKY

Z.-L., Zou[1], X.-Y., Xia[1,2], Z.-G., Deng[1,3]
1. Beijing Astron. Obs., Academy of Sciences, Beijing, China
2. Dept. of Phys., Tianjin Normal Univ., Tianjing, China
3. Graduate School, Academy of Sciences, Beijing, China

Infrared Astronomical Satellite (IRAS) revealed a class of luminous infrared galaxies which emit most (95%) of the energy in the infrared waveband. Much evidence has accumulated showing that they are all disturbed or interacting systems. Sanders et al. (1988) have examined deep CCD images of a complete sample of ten galaxies with far infrared luminosity larger than 10^{12} solar luminosities, selected from the bright galaxy survey (S(60)> 5.4Jy). They find that all of them are mergers. Lawrence et al.(1989) studied a larger sample which included 60 high luminosity IR galaxies, selected from QMC-GRO survey with S60 > 2Jy. They discovered that the proportion of low luminosity (L60 < 10^{11} solar luminosity) IRAS galaxies which are interacting or merging, 11±8%, is consistent with that in an optically selected control sample (18±5%). For high luminosity IR galaxies (L(60) > 10^{11} solar luminosity), the proportion is much higher (46±12%), but still well short of the 100% found by sanders et al.. It seems that the proportion of interacting or merging systems increases with IR luminosity in the sample.

In order to resolve this problem, we have examined a deepest sample so far, which is taken from QCD survey carried out by QMC (A. lawrence, J.Crawford, W.Saunders and M.Rowan-Robinson), Cambridge (G.Efstathiou and N.Kaiser) and Durham (R.Ellis, C.Frenk, I.Parry and Xia Xiaoyang). It covers the whole sky, selecting one in six IRAS galaxies with S(60)> 0.60 Jy. The total number of QCD sources is 2387.

The QCD team has measured their redshifts on the INT, WHT and AAT since 1987, and culculated the far infrared luminosity of each object as follows.

Far infrared flux:
$$FIR=1.75(2.55S(60)+1.01S(100))10^{-14} \text{ W/m}^2 \qquad (1)$$

Cosmological constants:
$$H_o=50 \text{ km/s·Mpc} \quad , \quad q_o=0.5 \qquad (2)$$

Luminosity distance:
$$d=2c[z+1-(z+1)^{0.5}]/H_o \qquad (3)$$

Far infrared luminosity:
$$\log(LIR)=\log 4\pi +2\log d+\log(FIR) \qquad (4)$$

In the Southern sky (dec.<-17°.5), we take 41 ultraluminous IR galaxies with $LIR>10^{12}$ solar luminosities as our sample, and 41 low-luminosity IR galaxies with $LIR<10^{9.7}$ solar luminosities as control sample. Their images on the film copies of SRC Sky Survey have been scanned by PDS machine in Purple Mountain Observatory, with scan parameters, 10 micron square aperture and 5 micron step. The data were reduced on sun workstation with IRAF software in Nanjing University, and VAX11/780 computer with ASPIC software in Beijing Astronomical Observatory as well.

The images of the galaxies were examined both in the dencity contour maps and the Sky Survey material. The contour maps are useful to reveal double nuclei or disturbed structure. The Sky Survey material is needed for the nearer objects to search for companions. Following Lawrence et al., we divided objects into seven classes: (0) isolated, (1) distant faint companion, (2) distant bright companion, (3) near-faint companion, (4) near bright companion, (5) interacting pair, (6) merger or highly peculiar.

Table 1 gives the number in each interaction class for four subsample-QMC-RGO sample with $L(60)<10^{11}$ solar luminosity, with $L(60) > 10^{11}$ solar luminosity, QCD sample with $LIR<10^{9.7}$ solar luminosity and with $LIR>10^{12}$ solar luminosity.

Table 1. Distribution with interaction class
Number in interaction class:

	0	1	2	3	4	5	6	fraction in 5+6
(a) QMC-RGO								
log(L60)<11.0	13	2	2	0	0	2	0	11%+/-8%
(b) QMC-RGO								
log(L60)>11.0	13	2	2	3	2	13	6	46%+/-13%
(c) QCD								
log (LIR)<9.7	28	5	4	0	0	3	1	10%+/-5%
(d) QCD								
log(LIR)>12.0	3	2	1	6	4	14	11	61%+/-12%

Close pair and interaction seem to be more common at higher luminosity. Statistical tests show that the low luminosity IR galaxies in subsamples (a)and (c) are consistent with each other. High luminosity IR galaxies are significantly different, at less than 1% level, from the low luminosity IR galaxies in QMC-RGO and QCD samples individually.

These results support Lawrence et al.'s conclusion that there is a clear statistical link between interactions and IR activity, but interaction does not seem to be the only cause of activity.

References:

1. Lawrence, A., Rowan-Robinson, M., Leech, K., Jones, D.H.P. and Wall, J.V., (1989) 'High-luminosity IRAS galaxies-I. The proportion of IRAS galaxies in interacting systems', Mon.Not.R.astr.Soc., Vol. 240, 329.
2. Sanders, D.B., Soifer, B.T., Elias, J.H., Madore, B.F., Matthews, K., Neugebauer, G., and Scoville,N.Z., (1986)'Ultraluminous infrared galaxies and the origin of quasars', Astrophys.J., Vol.325, 74.

BACKREACTIONS OF INHOMOGENEITIES ON EXPANDING UNIVERSES

S. BILDHAUER* and T. FUTAMASE
Department of Physics, Faculty of Science
Hirosaki University, Hirosaki
Aomori-ken 036, Japan

ABSTRACT. In a post-Newtonian approximation scheme in general relativity, an inhomogeneous metric is constructed for the non-linear regime. After spatially averaging, the universe is nearly homogeneous and isotropic. For simple solutions of the local Newtonian equations, it is shown that the expansion in the direction where structure preferentially forms is less rapid than the average expansion. The resulting anisotropy of the CMBR depends strongly on the time at which perturbations get non-linear. A special model is constructed in which the observed dipole component may be due to large-scale inhomogeneities.

1. The approximation scheme

The approximation scheme has been developed to construct a model of a fully inhomogeneous universe, (Futamase 1988,1989). It makes use of a spatial average to determine the backreactions of the inhomogeneities on the global behaviour of spacetime. It includes the linear approximation and a post-Newtonian approximation in general relativity to treat the non-linear regime. We, here, are interested in the situation, where the effect of self gravity is more important than the expansion of the universe, i.e., where the density contrast is larger than one. By an expansion of the Ricci tensor and spatial averaging of the Einstein equations, equations for the scale factor and the averaged metric are obtained, which describe the backreactions of the inhomogeneities. The equations which govern the local fluctuations are derived by subtracting the averaged equations from the original Einstein equations. The knowledge of the solution of these equations at lowest order allows one in the approximation scheme to calculate first non-trivial corrections to the expansion and to the averaged metric. We take the one-dimensional, self-gravitating dust motion described by Zeldovich's solution and consider two examples.

2. A universe which is homogeneous in the average

The situation we here have in mind is a universe which locally has a large density contrast, but which is homogeneous on large scales. The scale of structure considered is d, say 5-10 Mpc for clusters, 100 Mpc for superclusters. For this solution, the coupling of the equations for local inhomogeneities and global expansion has the following consequences, (Bildhauer 1990) : The expansion in the direction in which structure preferentially forms is less rapid than the average expansion. The resulting Bianchi-type I anisotropy generates an anisotropy of the CMBR. This anisotropy depends strongly on the time, at which perturbations get non-linear, z_i. Observational upper limits on anisotropies of the CMBR on large angular scales restrict in this simple model z_i to be smaller

---

* On leave of absence from Max-Planck-Institute for Physics and Astrophysics, Karl-Schwarzschild-Str.1, 8046 Garching bei München, FRG

than 2.5 (0.8) for a Hubble parameter of $H_0 = 50 \frac{km}{sMpc}$ or 1.75 (0.5) for $H_0 = 100 \frac{km}{sMpc}$ for $d = 10(100) Mpc$.

3. A globally inhomogeneous universe

We construct a globally as well as locally inhomogeneous universe. The density contrast is assumed very large in the local scale of the order, say d_1, but it varies rather smoothly in the horizon scale, d_2. On an intermediate scale, the universe is regarded as nearly periodic. In order to approximate such a spacetime, we spatially average the Einstein equations over the intermediate scale. In each intermediate region, we have different values of the averaged Einstein equations and therefore each region has a different expansion rate. For explicit calculation, we take a one-dimensional Zeldovich solution, involving the two scales d_1 and d_2. The horizon scale inhomogeneity causes a dipole moment in the CMBR. By an integration along the light paths, analytic expressions for the dipole and the quadrupole component in dependence on z_i and the initial conditions for the perturbations on the two scales are obtained, (Bildhauer & Futamase, 1990). For a choice of parameters, which is consistent with the observational quadrupole data, the observed dipole moment can be reproduced. Thus, according to this scheme, the CMBR data do not exclude a very large scale inhomogeneity, and the observed dipole component may be partially of cosmological origin.

Acknowledgement

S.B. thanks Hirosaki University for their kind hospitality and the Japan Society for the Promotion of Science and the Alexander von Humboldt-Stiftung (Feodor Lynen-programme) for their support. Our work was supported in part by the Japanese Grant-in-Aid for Science Research fund of the Ministry of Education, Science and Culture No. 01540226 and No. 02962003.

References

Bildhauer, S. (1990) 'Remarks on possible backreactions of inhomogeneities on expanding universes', to appear in Prog. Theor. Phys., **84**.

Bildhauer, S., Futamase, T. (1990) 'The Cosmic Microwave Background in a globally inhomogeneous universe', submitted to Mon. Not. R. astro. Soc. .

Futamase, T. (1988) 'Approximation Scheme for Constructing a Clumpy Universe in General Relativity', Phys. Rev. Lett., **61**, 2175.

Futamase, T. (1989) 'An approximation scheme for constructing inhomogeneous universes in general relativity', Mon. Not. R. astro. Soc., **237**, 187.

MULTI-LEVEL FEATURE IN THE LARGE SCALE DISTRIBUTION OF GALAXIES

Z.-G. Deng[1,2], X.-Y. Xia[2,3], Y.-Z. Liu[1], and Z.-L. Zou[2]

1 Graduate School, Academia Sinica, Beijing 100039, China
2 Beijing Astron. Obs., Academia Sinica, Beijing 100080, China
3 Dept. of Phys., Tianjin Normal Univ., Tianjin 300074, China

Discoveries of the super-large scale structures in the distribution of galaxies (e.g. the giant void in Bootes, the 'Great Wall' of galaxies and so-called 'great attractor' etc.) have raised the problem of how to determine the characteristic of distributions on very large scale. Two-point correlation function can not meet this requirement. To calculate two-point correlation function from a flux-limited sample we need to weight each pair count by a selection function which depends on the two-point correlation function itself. Besides, it has too large error bars for a volume-limited sample sorted out from flux-limited sample. These restrict the maximum scale of the analysis. Usually, two-point correlation function can only give us the characteristics of a distribution on scales about $10 \ h^{-1}$ Mpc.

We substitute the two-point correlation analysis with the analysis of function $N(r)$. $N(r)$ is the mean number of galaxies which have distances from any given galaxy less than r. $N(r)$ is a cumulative quantity and would have much smaller error bars, especially for larger r. So, it may be used to investigate the distribution feature on much larger scales. Function $N(r)$ has quite clear meaning. Fractal dimension of a distribution can be defined by (Wen et al., 1989)

$$D = d(\log(N(r)))/d(\log(r)). \tag{1}$$

To reduce the edge effect we have adopted the same measures as Wen et al. That is, by comparing the $\log(N_r(r))$-$\log(r)$ curve from Monte Carlo with straight line with slope 3, we give the restriction of the maximum scale r_{max} in the analysis, and substitute the Eq.(1) by

$$D = 3 + d(\log(N(r)/N_r(r)))/d(\log(r)). \tag{2}$$

Samples of our analysis are taken from the CfA redshift survey and the redshift surveys of IRAS galaxies in fields F15 and NGW. Four absolute magnitude-limited subsamples are sorted out from each of the northern and southern regions in the CfA survey, and five volume-limited subsamples are from each of the sample in fields F15 and NGW.

Resulting $\log(N(r)/N_r(r))$ vs. $\log(r)$ plots show that all these dia-

576

grams give fractal dimensions significantly different from 3 on their maximum analysed scales. It means that there must be structures in the distribution of galaxies on scale larger than 50 h_{50}^{-1} Mpc. Structures on this scale can be described quantitatively by the analysis of function $N(r)$ but can not be determined by two-point correlation function.

We can also see from these plots that all $\log(N(r)/N_x(r))$ vs. $\log(r)$ curves can not be fitted by a single straight line. It means that the distribution of galaxies in our samples is not simply fractal. However, they can be fitted by two or three straight lines very well. Two of these $\log(N(r)/N_x(r))$-$\log(r)$ plots are given in Fig.1 as samples. Straight lines in Fig.1 are obtained from linear regression. The piece-like structure of these plots shows that the distribution of galaxies has multi-level structure. That is, it is fractal with definite fractal dimension in certain scale ranges, but transits into fractal structure with different fractal dimension in another scale range.

Our analysis reveals another feature of the distribution. That is, fractal dimension of a fractal level on larger scale is generally larger than that of fractal level on smaller scale. On level with largest scale in the analysis, almost all subsamples show fractal dimensions larger than 2, and less than 2 significantly on the level with smallest scale. This systematic change in fractal dimensions with scales of levels implies that the principle feature of the distrbution of galaxies is clustering of galaxies on smaller scales and becomes to be a distribution with voids (or cellar structure) on larger scales.

There are scale ranges in which the distribution of galaxies transits from one fractal level to another. These scale ranges may be considered as some kind of typical scale and may be called turning scale. Existence of the turning scales shows that the distribution of galaxies does have typical scales. It implies that the formation and evolution of galaxy can not be a pure gravitational process.

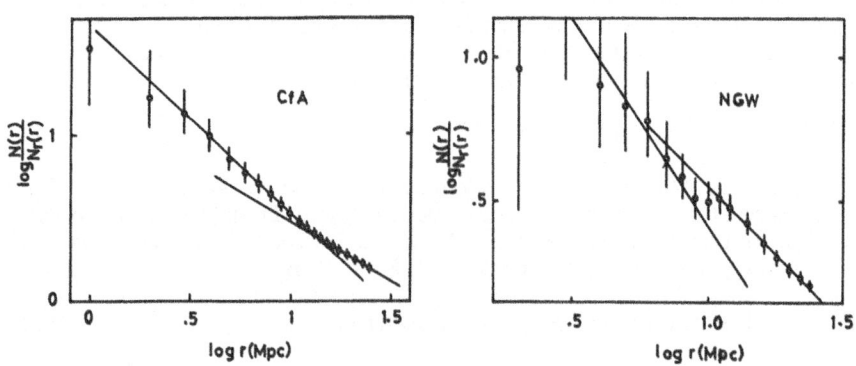

Fig.1

References:
1.   Z. Wen, Z.-G. Deng, Y.-Z. Liu, and X.-Y. Xia (1989), 'The fractal dimension in the large-scale distribution of galaxies with different luminosities', Astron. & Astrophys., 219, 1-6.

FRAGMENTATION AND CLUSTERING IN THE WAKE FORMED BY COSMIC STRING

Tetsuya HARA and Shigeru MIYOSHI
Department of physics, Kyoto Sangyo University, Kyoto 603 ,JAPAN

ABSTRACT. Numerical calculations are performed for the fragmentation, clustering and merging of dark matter under the open cosmic string model. Both the fragmentation in one dimensional collapse of pancake scenario and clustering theory from small scale to large scale within a wake are realized.

1 Introduction

Sheet-like structures in a large scale distribution of galaxies seem to be simply explained by open cosmic strings, for matter accumulates the traces of the string. According to this scenario, the characteristic size and the typical surface density of sheets are $\sim 10^2 \text{Mpc} \times 10^2 \text{Mpc}$ and $\sim 10^{12} (G\mu\beta\gamma/2 \cdot 10^{-6}) M_\odot/\text{Mpc}^2$,[1] where μ and β are the line density and the velocity of the string in unit of c=1, and $\gamma=(1-\beta^2)^{-1/2}$.

The accumulated dark matter in the plane may be fragmented due to the plane instability. The representative mass fragmented at 1+z is[1]

$$M \simeq 2 \cdot 10^{14} (G\mu\beta\gamma/2 \cdot 10^{-6})^3 \cdot ((1+z_i)/6.10^3)^{3/2}/(1+z)^3 M_\odot .$$

Then the mass sacle of galaxies is formed around $1+z \simeq 10$ and the typical mass at present is the order of group of galaxies. Numerical calculations are performed with 32^3 particles and 32^3 cells in a cube with comoving length $L=45.6(M/2.10^{11} M_\odot)^{1/3} h_{50}^{2/3} \text{Mpc}$.

2. Results of numerical calculations and discussions

The scale expansion of the factor e^4 ($\simeq 54.6$) is computed by 1000 time steps. At the first time, the fragmentation occurrs randomly, and small density peaks appears almost every where within the wake. As time has passed, the number of the density peak has decreased due to clustering and merging. The clustering and merging occur within the wake and, in the end, almost the regular stable configuration of finite numbers of fragments is realized, which may be due to the boundary condition of periodicity.

577

578

Under the cosmic string scenario, it could be explained the mass
scale of group of galaxies through clustering of dark matter. However,
mass scale of galaxies must be investigated through thermal process.
The characteristic stage for thermal process is at $1+z \simeq 8$ when Compton
cooling time due to 3K backgound radiation becomes longer than the
expansion time. The Jeans mass at this stage is as

$$M_J \simeq 5 \cdot 10^{10} (T/10^4)^{1/2} (\Omega_{bm}/0.1)^{-1/2} ((1+z)/8)^{-3/2} M_\odot \quad .$$

The detail thermal process for the formation of galaxies must be
pursued.

References
1)   T. Hara and S. Miyoshi, Prog. Theor. Phys. 77(1987),1152.
       T. Hara and S. Miyoshi, Prog. Theor. Phys. 81(1989),1187.

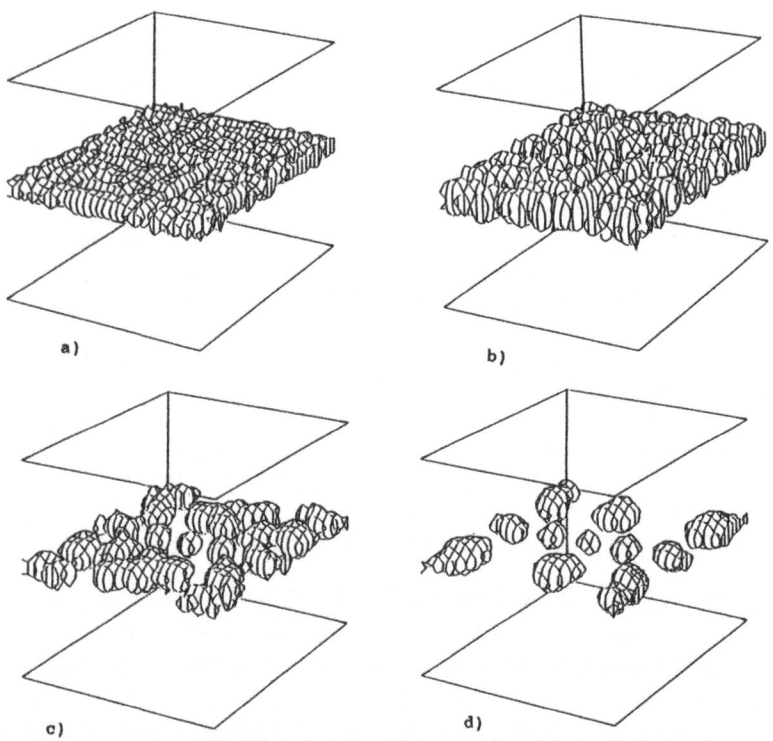

Fig. 1 Comparison of the morphology of the same density
       contrast surface ($\rho = 4 \times \rho_b$) to the background density.
       The universe has expanded from initial stage e times for
       a), e^2 times for b), e^3 times for c) and e^4 times for d).

PERTURBATIONS OF THE ONE-DIMENSIONAL STRUCTURE IN THE UNIVERSE

S. BILDHAUER, O. ASO, M. KASAI and T. FUTAMASE
Department of Physics
Hirosaki University
Hirosaki, Aomori-ken 036
Japan

ABSTRACT. Linear perturbation theory on the background of a one-dimensional, non-linear inhomogeneous universe is considered. For an Einstein-de Sitter behaviour of the scale factor, the exact solution of the perturbation equations is given. The non-linear effect of the density inhomogeneity on the evolution of the peculiar velocity is written in form of a relation between the peculiar velocity and the density parameter Ω including the correction due to the density contrast δ_0 of the Zeldovich solution. It is shown that the component parallel to the walls is characterized by $\Omega^{0.6}(1 + \delta_0)^{-0.57}$.

1. Introduction

Recently it was reported by Broadhurst *et al.* (1990) that the large-scale distribution of galaxies had an apparent regularity with a characteristic size of $128\,h^{-1}$ Mpc. This suggests that the Universe has a coherent one-dimensional structure which consists of at least 13 "megawalls". If the periodicity is genuine, the explanation of the origin of the unique regular separation is urgently needed, but it also has great significance to study the stability of the large-scale one-dimensional structure from the viewpoint of observational cosmology. For analysis of Virgo infall to determine the density parameter Ω, people mostly use the formula which was derived under the assumption that the density contrast is much less than unity or sometimes take the spherical model with non-linear correction. In reality, however, the Virgo infall is neither quite a linear phenomena nor appropriately described by a spherically symmetric flow. Thus, studies of non-linear models may be quite useful (Shinohe & Futamase 1990).

2. Evolution of Perturbations

We investigate the effect of inhomogeneity and anisotropy on the evolution of the density perturbations and the peculiar velocities on an one-dimensional inhomogeneous background. To study perturbations in the universe with coherent large-scale structure, we consider an exact solution for the inhomogeneous universe, the Zeldovich solution in Newtonian cosmology as a background and derive the equations for the perturbed quantities. For an Einstein-de Sitter behavior of the scale factor, *i.e.*, $a(t) \propto t^{(2/3)}$, the exact solutions of the density perturbation are found; one corresponds to the growing mode solution δ_1 and the other is the decaying mode solution $\delta_2 \propto 1/t$. At early epochs, the evolution of δ_1 coincides with the usual linear perturbations $\delta_1 \propto a(t)$, but at later stages, the local density drives the evolution such that the growth is faster in high density regions and slower in low density regions. Numerical integrations ensure that the same result generally holds for $\Omega \neq 1$ cases.

579

3. The Density Parameter-Peculiar Velocity Relation

The relation between the peculiar velocity v_α parallel to the "megawalls" and the gravitational acceleration g_α is given by

$$v_\alpha = \frac{2 g_\alpha}{3 H \Omega} f.$$

Employing the growing mode solution obtained for $\Omega = 1$, the perturbation parameter f can be written in terms of δ_0. A useful analytic approximation is $f = (1 + \delta_0)^{-0.57}$. Compared with the famous $\Omega^{0.6}$ formula in the homogeneous and isotropic background (Peebles 1980), this can be viewed as a non-linear effect of the density inhomogeneity of the Zeldovich background. The behavior of the peculiar velocity in general cases can be obtained numerically and is well fitted by the formula $f = \Omega^{0.6}(1 + \delta_0)^{-0.57}$ for $0 < \Omega < 1$. In the case of an axially symmetric mass excess, the peculiar velocity inside the wall at a distance of r from the center of the mass excess is

$$v = \frac{1}{2} H_0 r \delta_1 \Omega^{0.6} (1 + \delta_0)^{-0.57},$$

whereas in perturbation theory on a homogeneous background it is given by

$$v = \frac{1}{2} H_0 r \delta_1 \Omega_{\text{hom}}^{0.6}.$$

For given measured peculiar velocity and density contrast, the relation between Ω and Ω_{hom} is approximately

$$\Omega \simeq \Omega_{\text{hom}}(1 + \delta_0).$$

Thus, the perturbation theory on a homogeneous background tends to underestimate Ω in dense regions. (Note that in the case of a spherically symmetric mass excess, the factor $1/2$ is replaced by $1/3$.)

Acknowledgement

This work was supported in part by Hirosaki University Special Research Grant, and by the Japanese Grant-in-Aid for Science Research fund of the Ministry of Education, Science and Culture No. 01540226, No. 02740127 and No. 02962003.

References

Bildhauer, S., Aso, O., Kasai, M. & Futamase, T., 1990. Hirosaki University preprint.
Broadhurst, T. J., Ellis, R. S., Koo, D. C. & Szalay, A. S., 1990. Nature, 343, 726.
Peebles, P. J. E., 1980. The Large Scale Structure of The Universe, Princeton University Press, Princeton.
Shinohe, H. & Futamase, T., 1990. Prog. Theor. Phys., 83, 350.
Zeldovich, Ya. B., 1970. Astron. Astrophys., 5, 84.

Three Points Spatial Correlation Function on the Self Similar Observer Homogeneous Structure

Hyung-Won Lee and Remo Ruffini
ICRA - International Center for Relativistic Astrophysics
and
Dipartimento di Fisica, Università di Roma "La Sapienza"
Piazzale Aldo Moro 2, 00185 Roma, Italia.

ABSTRACT. We presented an explicit form of 2 and 3 points spatial correlation functions for observer homogeneous and self similar structure. Crucial difference from the usual hierarchical form is the fact that it increase with separation after a certain value. This behaviour seems to be consistent with a rough estimation with CfA II catalog.

1. Introduction

One of the best way to study large scale structure of Universe is measuring correlation functions(Peebles (1980)). The observational determination for various catalog gives strong evidence for power law dependence of correlation functions(Groth and Peebles (1977), Davis and Peebles (1983), Einasto, et al. (1984), Bahcall (1986)). But they have different amplitude for galaxy and cluster of galaxies. These differences remain a challenge to theoretical models(Bahcall (1988), White, et al. (1988)).

2. Various Density Functions

Usually we treat galaxies as a point like objects in the catalog. Hence continuous density function can not be defined rather various density functions we can define are related to the number counts in various shape. But these number counts can be approximated by appropriate integrations over the various shape. Thus various density functions can be related to corresponding integrations as following.

$$< n > n_d^V(r) = \frac{1}{V_s} \int d^3 r_0 n(r_0) \frac{1}{4\pi} \int d\Omega_r n(r_0 + r) \qquad (1)$$

581

and

$$< n > n_d^V(r_1) n_d^S(r_1, r_{12}, \theta)$$
$$= \frac{1}{V_s} \int d^3 r_0 n(r_0) \cdot \frac{1}{4\pi} \int d\Omega_{r_1} n(r_0 + r_1)$$
$$\cdot \frac{1}{2\pi} \int dC(r_1, r_{12}, \theta) n(r_0 + r_2), \tag{2}$$

where $n_d^V(r)$ is the differential volume density with the volume $V(r) = 4\pi r^3/3$ and $n_d^S(r_1, r_{12}, \theta)$ the differential sphere density with two fixed objects separated by r_1 and subtanded angle θ.

3. Explicit form of Correlation Functions

By the virtue of the self similar structure, we can obtain

$$n_d^S(r_1, r_{12}, \theta) = n_d^V(r_{12}). \tag{3}$$

Since correlation functions can be represented by appropriate integrations, we obtain explicit forms of 2 and 3 points correaltion functions as following by relating those to eqs. (1) and (2).

$$\xi(r) = (1 - \frac{\gamma}{3})R_s^\gamma r^{-\gamma} - 1 \tag{4}$$

and

$$\xi(r_1, r_2, r_{12}) = \frac{1}{3}\Big\{ \xi(r_1)\xi(r_{12}) + \xi(r_{12})\xi(r_2) + \xi(r_2)\xi(r_1)$$
$$- \xi(r_1) - \xi(r_2) - \xi(r_{12}) \Big\}, \tag{5}$$

where R_s is the sample radius.

References

Bahcall,N.A.:1986, *Astrophys. J.* , 302,L41

Bahcall,N.A.:1988,*Ann. Rev. Astro. Ap.*,in press

Davis,M., Peebles,P.J.E.:1983, *Astrophys. J.* , **267**,465

Einasto,J.,Klypin,A.A.,Saar,E.,Shandarin,S.F.:1984,*Monthly Notices Roy. Astron. Soc.* **206**, 529

Groth,E.J.,Peebles,P.J.E.:1977, *Astrophys. J.* , **217**,38

Mandelbrot, B.B.: 1982, *The Fractal Geometry of Nature*, Freeman, San Francisco

Peebles,P.J.E.:1980,*The Large-Scale Structure of the Universe*, Princeton University press

Limits from cosmological nucleosynthesis
on the leptonic numbers of the universe

A. Bianconi, H.W. Lee, R. Ruffini
ICRA – International Center for Relativistic Astrophysics
and
Dipartimento di Fisica, Università di Roma "La Sapienza"
Piazzale Aldo Moro 2, 00185 Roma, Italia.

ABSTRACT. Constraints on chemical potentials and masses of "inos" are calculated using cosmological standard nucleosynthesis processes. It is shown that the electron neutrino chemical potential, ξ_{ν_e}, should not be grater than a value of order of 1 and the possible effective chemical potential of the other neutrino species , ξ, and electron neutrino chemical potential are related approximately $10\xi_{\nu_e} \approx \xi$ in order not to conflict to observational data. The allowed region consistent with the ^{4}He abundance observations are insensitive to the baryon to proton ratio, η, while those imposed by other light elements (D, ^{3}He, ^{7}Li) strongly depend on η.

1. Introduction

In the classical work on cosmological nucleosynthesis, see e.g. Fermi and Turkevich(1950), Alpher and Herman(1950) and Hayashi(1950), the presently observable ratio between the neutron and proton abundances, n_n/n_p, was mainly established in the early stages of the universe, at temperatures T $\gtrsim$ 1 MeV, when the weak interactions kept neutrons and protons in chemical equilibrium. Since neutrinos and antineutrinos do enter into reactions, a non zero electron neutrino chemical potential $\xi_{\nu_e} \equiv \mu_{\nu_e}/T$, causes an asymmetry between the neutrinos and the antineutrinos and, consequently, modifies the resulting n_n/n_p equilibrium ratio, leading to observable effects on the ^{4}He relative abundance.

The aim of our work is to point out that *larger* values of the leptonic number and, therefore, larger values of the chemical potential for the neutrinos are indeed viable, if we duly account for the effect in the expansion rate of the cosmological matter.

Similar analysis were already attempted by Beaudet and Goret 1976, Yahil and Beaudet(1976), and David and Reeves(1980). In the present paper we explicitly show how the large values of the chemical potential and the masses of "inos" species in Ruffini, et al.(1988) are indeed in agreement with possible values to be expected from cosmological nucleosynthesis.

2. Results of theoretical and numerical analysis

The observational constraints on ^{4}He,D,^{3}He,^{7}Li abundances are taken by Yang et al.(1984) or Boesgaard and Steigman(1985), and we refer these articles for a detailed discussions. In particular we remark that D and ^{3}He abundances are not used separately, but via their number sum $(D + {}^3He)/H$, as there are good reasons to think that this sum has been conserved throughout the evolution of the universe.

The main results of our analysis, done with the Kawano(1988) code, can be summarized as following:

1) The chemical potential of the electron neutrinos are constrained by the abundance of ^{4}He $\simeq$ 25 % to vary in the range

$$0 < \xi_{\nu_e} < a, \ a \approx 1 \div 1.5.$$

2) Present limits on the observed abundances of D and ^{3}He define a "safe belt" of admitted values of ξ_{ν_e}, ξ, η. It depends strongly on the value of ξ and η and to a lower extent from the value of ξ_{ν_e}. The allowed region is given by

$$\xi \approx 10\xi_{\nu_e}.$$

3. Conclusions

From the computations performed in the proceeding paragraphs we can conclude that electron neutrino and μ and τ neutrinos can play a very different role in cosmological nucleosynthesis and, correspondingly, the observed abundances allow values of the chemical potential in range

$$0 < \xi_{\nu_e} < 1.2 \tag{1}$$

$$0 < \xi < 12. \tag{2}$$

It is interesting that additional constraints on the value of the chemical potential of the neutrinos can be obtained if one turns from the epoch of cosmological nucleosynthesis to the current average values of the neutrinos in cosmology. The evolution of the statistical distribution and thermodynamical parameters of a fermion (or boson) species after decoupling ("redshifted statistics") has been extensively studied by Ruffini, et al.(1983) with general conditions ($\mu \neq 0, m \neq 0$). Applying the "redshifted statistics" to a species of light neutrinos we can see that a neutrino species with $m_\nu \sim$ some eV and $\xi_\nu \sim$ some units is compatible with $\Omega_{\nu+\bar{\nu}} \approx 1$.

We can therefore conclude that an *asymmetry* between the matter and anti-matter component in the dark–matter component of the universe is allowed by nucleosynthesis argument and leads to the constraints expressed by Eqs. (1), (2).

In this light, the explanation of the existence of a baryon number *asymmetry* in the universe

$$n = \frac{n_B - n_{\bar{B}}}{n_\gamma}$$

should be addressed, in our opinion, in concomitance of the explanation of possible lepton numbers in the range:

$$0 < |L_{\nu_e}| \equiv \frac{n_{\nu_e} - n_{\bar{\nu}_e}}{n_\gamma} \lesssim 0.94$$

$$0 < |L_{\nu_\mu}| \equiv \frac{n_{\nu_\mu} - n_{\bar{\nu}_\mu}}{n_\gamma} \lesssim 120.0$$

$$0 < |L_{\nu_\tau}| \equiv \frac{n_{\nu_\tau} - n_{\bar{\nu}_\tau}}{n_\gamma} \lesssim 120.0.$$

References

Alpher, R.A., Herman, R.C.: 1950, *Rev. Mod. Phys.* **22**, 153

Beaudet, G., Goret, P.: 1976, *Astr. Ap.* **49**, 415

Boesgaard, A.M., Steigmann, G.: 1985, *Ann. Rev. Astron. Astroph.* **23**, 319

David, Y., Reeves, H.: 1980, *Phil. Trans. R. Soc. Lond.* **A296**, 415

Fermi, E., Turkevich, A.: 1950, unpublished. Results published in Alpher, Herman : 1950

Hayashi, G.: 1950, *Prog. Theor. Phys.(Japan)* **5**, 224

Kawano, L.: 1988, *FERMILAB–PUB–88/34–A*

Ruffini, R., Song, D.J., Stella, L.: 1983, *Astron. Astrophy.* **125**, 265

Ruffini, R., Song, D.J., Taraglio, S.: 1988, *Astron. Astrophys.* **190**, 1

Yahil, A., Beaudet, G.: 1976, *Ap. J.* **206**, 26

Yang, Y., Turner, M.S., Steigman, G., Schramm, D.N., Olive, K.A.: 1984, *Ap. J.* **281**, 493

INHOMOGENEITIES AND NON-FRIEDMANN LIGHT PROPAGATION

E. V. Linder
Steward Observatory
University of Arizona
Tucson, Arizona, USA 85721

ABSTRACT. Driven by the need to reconcile large scale inhomogeneity with global isotropy, we investigate the rigorous development of a metric realistically describing our cosmology. Using the result we then estimate the order of magnitudes for various observational effects due to the presence of inhomogeneities affecting light propagation.

1. Doubts About Friedmann

Observations of the universe reveal two strong, disparate cosmological characteristics. The microwave background radiation and galaxy number counts show a high degree of isotropy but recent observations reveal structure on larger and larger scales, to such an extent that we must question whether the Friedmann-Robertson-Walker cosmological model is a reasonable approximation. Here we take a rigorous, general relativistic approach to the question.

2. Averaging Is Bad

One's first inclination, however, might be to attempt to define some sort of spatial averaging procedure that smooths out the inhomogeneities on some very large scale. After all, averaging works well in physics, from thermodynamics to bulk electrodynamics. Aside from mathematical objections, the following intuitive argument expresses my doubts.

The binding energy of a large region of the universe, being just the sum of kinetic and potential energies, is roughly

$$B \sim (1/2)m(HR)^2 - GmM/R$$
$$\sim (4\pi/3)GmR^2[3H^2/(8\pi G) - \rho]. \tag{1}$$

When the density is far below the critical density the binding energy is positive and we have an analogy to a gaseous phase, where thermodynamic averaging is useful. For high density, $B < 0$, and the analogy is to a solid phase. The large regions of the universe over which we would like to average, though, have $\rho \approx 3H^2/(8\pi G)$, giving a "liquid phase" for which simple averaging procedures fail. Thus I am pessimistic about this approach.

3. Relativity to the Rescue

Within general relativity we can write down a guess for the metric:

$$ds^2 = a^2(\eta)[\gamma_{ab} + h_{ab}]dx^a dx^b, \tag{2}$$

where a is the expansion factor, η the conformal time, γ_{ab} the conformal Robertson-Walker (perfectly smooth) metric, and h_{ab} perturbations representing the contribution of inhomogeneities. Using two ingredients — post-Newtonian formalism (Chandrasekhar 1965) and the two length scale approach (also known as mean field theory, cf. Misner, Thorne, and Wheeler 1973, §22) we parametrize the order of magnitude of h_{ab} as ϵ^2 and its characteristic length scale (inverse logarithmic derivative) as κ (see Futamase 1989).

Expanding the Einstein equations in ϵ, κ we find the consistency condition for truncating the series at, say, order $\epsilon^4 \kappa^{-2}$ (pseudotensor order) is simply $\epsilon, \kappa \ll 1$. The density perturbations are found to be $\epsilon^2 \kappa^{-2}$, so the nonlinear regime corresponds to $\epsilon \gg \kappa$; pressure perturbations are ϵ^2, and the expansion factor $a \sim a_{RW}(1+\epsilon^4 \kappa^{-2})$. If $\epsilon^2 \ll \kappa$ then the expansion follows the Robertson-Walker behavior and the metric also takes a particularly simple post-Newtonian form:

$$ds^2 = a_{RW}^2(\eta)[-(1+2\phi)d\eta^2 + (1-2\phi)\gamma_{ij}dx^i dx^j], \tag{3}$$

where ϕ is the gravitational potential of the inhomogeneities. If $\epsilon^2 \ll \kappa$ is violated, the metric does not take any such simple form and the global expansion is not Friedmann. Velocity perturbations are order $\epsilon^2 \kappa^{-1}$ and ϵ in the two cases.

4. How Messed Up Are Observations?

Use of the more realistic, approximate metric (3) allows calculation of inhomogeneity effects on light propagation and hence observations. Some of these effects are summarized below.

Small Effects	Order	Observable Effects	Order
Expansion Backreaction	$\epsilon^4 \kappa^{-2}$	Rms Light Deviation/Separation	$\epsilon^2 \kappa^{-2}$
Shear	$\epsilon^2 \kappa^{-1}$	Image Distortion	$\epsilon^2 \kappa^{-2}$
Photon Momentum Change	ϵ^2	Image Amplification	$\epsilon^2 \kappa^{-2}$
Distance-Redshift Relation	$\epsilon^2 \kappa^{-1}$	Multiple Imaging	$\epsilon^2 \kappa^{-2}$
Rms Light Deflection	$\epsilon^2 \kappa^{-1}$	Microwave Background Anisotropy	$\epsilon^2 \kappa^{-2}$

Typical numbers for galaxy (cluster) scales are $\epsilon^2 \sim 10^{-6.5}(10^{-5})$ and $\kappa \sim 10^{-5}(10^{-2.5})$ so $\epsilon^2 \kappa^{-1} \sim 10^{-1.5}(10^{-2.5})$. Thus, on these scales the proper metric to describe our cosmology is equation (3). This will break down on smaller (e.g. stellar) scales, but here the density is high enough that perhaps some averaging procedure is legitimate (see §2 above).

I gratefully acknowledge my collaborators Bob Wagoner and Mark Jacobs, and thank the conference organizers and the astrophysics groups at Hirosaki University and the Yukawa Institute for Theoretical Physics for hospitality and support. This work was supported in part by NASA grant NAGW-763 at Steward Observatory and NAGW-299 and NSF PHY86-03273 at Stanford University; conference attendance was made possible by an NSF International Travel Grant through the American Astronomical Society.

References

Chandrasekhar, S. (1965) *The post-Newtonian equations of hydrodynamics in general relativity,* Ap. J. 142, 1488-1512.

Futamase, T. (1989) *An approximation scheme for constructing inhomogeneous universes in general relativity,* MNRAS 237, 187-200.

Misner, C.W., Thorne, K.S., and Wheeler, J.A. (1973) *Gravitation,* W.H. Freeman, San Francisco.

ON THE GENERATION OF NON-GAUSSIAN FLUCTUATIONS DURING CHAOTIC INFLATION

I. YI, E. T. VISHNIAC, and S. MINESHIGE
Astronomy Department
University of Texas, Austin, TX 78712
USA

ABSTRACT. We solve the stochastic evolution equations for the inflaton field during the slow-roll period of chaotic inflation. We find exact analytic solutions for the $V(\Phi) = \lambda\Phi^4/4$ potential. The resulting probability distributions are sharply peaked around the classical deterministic trajectory and are approximately gaussian, but the exact probability distributions show that skewness is very sensitively dependent on λ and also weakly dependent on initial and final conditions. The non-Gaussian statistics induced by the nonlinear stochasticity effectively measures the expected level of non-Gaussian deviation in the inflation-generated density fluctuations.

1. Quantum Fluctuations and Exact Solutions

In the slow-roll limit of the chaotic inflation, the classical equation of motion for the inflation-driving scalar field becomes

$$\dot{\Phi} = -V'(\Phi)/3H(\Phi), \tag{1}$$

with the Hubble parameter, H, given by $H^2(\Phi) = 8\pi V(\Phi)/3M_p^2$, where $M_p = G^{-1/2}$ is the Planck mass. The classical trajectory of the field Φ_d is then

$$\Phi_d = \Phi_0 \exp\left(-\sqrt{\lambda/6\pi}M_p(t - t_0)\right), \tag{2}$$

for the quartic potential $V(\Phi) = \lambda\Phi^4/4$ where $\Phi(t = t_0) = \Phi_0$.

In the stochastic approach to chaotic inflation, quantum fluctuations generated inside the apparent horizon ($\sim 1/H$) during the de Sitter stage are multiplicatively added to the coarse-grained field. We find

$$\dot{\Phi} = -\frac{V'(\Phi)}{3H(\Phi)} + \frac{H^{3/2}(\Phi)}{\sqrt{8\pi^2}}\Gamma(t) \tag{3}$$

with the Gaussian random noise defined by $< \Gamma(t) >= 0$, $< \Gamma(t_1)\Gamma(t_2) >= 2\delta(t_1 - t_2)$. The Fokker-Planck equation for the probability $P(x = \Phi/M_p; \tau = tM_p)$ corresponding to equation (3) is

$$\partial_\tau P(x;\tau) = \sqrt{\frac{\lambda}{6\pi}}\partial_x(xP) + \sqrt{\frac{\lambda^3}{216\pi}}\partial_x(x^k\partial_x x^{6-k}P). \tag{4}$$

Here k is a constant, depending on the interpretation rule. $k = 0$ corresponds to Ito's interpretation and $k = 3$ corresponds to the Stratonovich interpretation, where the former rule is formally causal but the latter is not.

The exact solution of equation (4) with $P(x;0) \propto \delta(x - x_0)$ is (see, Yi et al. 1990);

$$P(x;\tau) \approx \frac{(k-1)^2}{4} \left(\frac{24}{\lambda(1-k)^2}\right)^{\frac{k+1}{4}} x_0^{\frac{7-3k}{2}} x^{\frac{k-11}{2}} \frac{\exp(-\sqrt{\frac{\lambda}{24\pi}}(1-k)\tau)}{\exp(\sqrt{\frac{8\lambda}{3\pi}}\tau) - 1}$$

$$\times \exp\left(-\frac{3}{2\lambda}\frac{x^{-4} + x_d^{-4}}{\exp(\sqrt{\frac{8\lambda}{3\pi}}\tau) - 1}\right) I_{\frac{|k-1|}{4}}\left(\frac{3}{\lambda}\frac{(xx_d)^{-2}}{\exp(\sqrt{\frac{8\lambda}{3\pi}}\tau) - 1}\right), \tag{5}$$

where $I_l(z)$ is the modified Bessel function and x_d is the dimensionless deterministic trajectory $x_d = x_0 \exp(-\sqrt{\lambda/6\pi}\tau)$. In the limit $(xx_d) \ll \sqrt{3/\lambda}/\sqrt{(x_0/x_d)^4 - 1}$, we find

$$P(x;\tau) \approx \sqrt{\frac{\lambda}{6\pi}} \left(\frac{24}{\lambda(1-k)^2}\right)^{\frac{k+1}{4}} x^{\frac{k-9}{2}} x_0^{\frac{9-3k}{2}} \frac{\exp(-\sqrt{\frac{\lambda}{24\pi}}(3-k)\tau)}{\sqrt{\exp(\sqrt{\frac{8\lambda}{3\pi}}\tau) - 1}}$$

$$\times \exp\left(-\frac{3}{2\lambda}\frac{(x^{-2} - x_d^{-2})^2}{\exp(\sqrt{\frac{8\lambda}{3\pi}}\tau) - 1}\right). \tag{6}$$

The *effective standard deviation* is

$$\sigma_{NG} \approx |\sqrt{\lambda/12}x_d(x_0^4 - x_d^4)^{1/2} \pm (\lambda/8)x_d(x_0^4 - x_d^4)|. \tag{7}$$

The ratio

$$\frac{P(x_d + N\sigma_{NG})}{P(x_d - N\sigma_{NG})} \approx \exp\left(-N^3\sqrt{3\lambda/4}(x_0^4 - x_d^4)^{1/2}\right) \tag{8}$$

measures the local skewness near the deterministic trajectory x_d. The number of *effective* standard deviations for which non-Gaussian deviation becomes significant is

$$N \approx (-\ln s)^{1/3}\lambda^{-1/6}(x_0^4 - x_d^4)^{-1/6}, \tag{9}$$

where s is a constant smaller than unity which defines the significance of non-Gaussian deviation. Near the end of inflation, $x_d \sim 0.4$, $N \approx 28$ for $\lambda = 5 \times 10^{-14}$, $N \approx 18$ for $\lambda = 10^{-12}$, and $N \approx 8$ for $\lambda = 10^{-10}$ (with $x_0 = 4.4$ and 10% deviaton). For an initial condition set at an earlier epoch $x_0 \gg 4.4 m_p$, skewness becomes very significant.

2. Discussion and Conclusions

Uncertainties come from a) we have assumed that $P(x)$ is a δ function initially, b) we have assumed that the quantum noise inside the horizon is purely Gaussian, and c) we have neglected the role of spatial gradients. A rigorous solution to these difficulties would require including some spatial degrees of freedom in the field and introducing a measure of inhomogeneity which is neither local nor global, but is a measure of astronomically significant inhomogeneities around a given point in space.

The statistical importance of events which are rare in pure gaussian statistics might not be negligible. The very large-scale distribution of galaxies or the large-scale mapping of CMBR, which might be only very weakly influenced by the nonlinear gravitational evolution, may reveal this type of nonrandom phase initial conditions. Since the adiabatic fluctuations (and λ) are tightly constrianed by CMBR, possibilities of nongaussian statistics are more likely in isothermal density fluctuations.

Reference

Yi, I., Vishniac, E. T., and Mineshige, S. (1990) 'On the generation of nongaussian fluctuations during chaotic inflation', *Phys. Rev.* **D**, in press.

A UNIFIED PICTURE OF HI ABSORPTION-LINE SYSTEMS BY MINI-HALOS

I. MURAKAMI and S. IKEUCHI
National Astronomical Observatory
Mitaka, Tokyo 181
Japan

ABSTRUCT. We consider an evolution of intergalactic clouds confined by the gravity of cold dark matter (CDM), the so called minihalo, to reproduce the observed results of neutral hydrogen absorption line systems of quasars. Assuming a simplified evolution law of diffuse UV flux and mass function of gas clouds, we can reproduce several observational properties by expanding minihalos which experience an expansion caused by the increase of diffuse UV flux at $z > 2$.

1. Introduction

Neutral hydrogen absorption systems are observed in high redshift quasars (QSOs), that tell us informations about the high redshift universe. Several characteristic properties of these systems are reported: i) a single power-law form of HI column density distribution from Lyman α forest to damped Lyman α systems (Sargent *et al.* 1989), ii) rapid increase of number density evolution for Lyman α forest (Murdoch *et al.* 1986), iii) rapid increase of continuum depression of QSOs at $z \gtrsim 3$ (Schneider *et al.* 1989) indicating the cumulative absorptions by Lyman α forest.

Minihalo model is proposed from the CDM cosmogony (Rees 1986, Ikeuchi 1986). In this model, gas is confined by the gravity of CDM. HI column density distribution is naturally reproduced by the isothermal nature of clouds (Ikeuchi, Murakami, and Rees 1988).

Here, we examine the evolution of spherical gas clouds confined by dark halos driven by the time variation of diffuse UV flux and study several statistcal properties.

2. Minihalo Model

We consider a system in which CDM and gas with primordial abundance are distributed spherically. Keeping the potential of isothermal CDM unchanged, which is initially in a gravitational equilibrium with gas, we calculate the evolution of a gas cloud, considering radiative cooling and UV heating. Here, we take $\Omega = 1$, $q_0 = 0.5$, and $H_0 = 100 \mathrm{km\ sec^{-1}\ Mpc^{-1}}$.

We assume an intensity of diffuse UV flux including an effect of absorption due to gas: $J(\nu; z) = 10^{-21}(\nu/\nu_{LL})^{-1}((1 + z)/3)^\alpha \exp(- \int_r^R n_{HI}dr/10^{17.5}\mathrm{cm^{-2}})\mathrm{erg\ s^{-1}cm^{-2}Hz^{-1}}$, where R is cloud radius. We fix $\alpha = 4$ for $z \leq 2$. For $z \geq 2$, α is taken as a parameter.

In order to compare models with observations, we calculate HI column density as a function of an impact parameter, p, which is a distance of a line of sight from a cloud center. Assuming that the comoving number density of minihalos is constant and cloud mass function is represented by a power law, we obtain column density distribution at z;

$$\frac{d^2\tilde{N}}{dzdN_{HI}} = \frac{c\pi}{H_0}n_*(0)\int d\left(\frac{M_c}{M_*}\right)\left(\frac{M_c}{M_*}\right)^{-\delta}\frac{pdp}{dN_{HI}}(1+z)^{1-q_0} \qquad (1)$$

where q_0 is 0.5 here. Integrating this equation with N_{HI}, we obtain number density evolution of Lyman α forest. We consider continuum depression as the cumulative absorption by gas clouds and calculate the optical depth using distribution of clouds derived above. In order to determine $n_*(0)$, we set this value to fit equation (1) for observed HI column density distribution at $z = 3.6$ and $N_{HI} = 10^{13.75}\text{cm}^{-2}$.

3. Results

The increase of diffuse UV flux ($\alpha < 0$) causes an expansion of gas clouds and the decrease of it ($\alpha > 0$) causes a contraction of them to produce nuetral core. HI column density distribution at every z can be roughly approximated to a power-law form with same index β. Number density evolution and evolution of continuum depression can be reproduced by expanding minihalos, which experience an expansion, that causes the decrease of amplitude of N_{HI} distribution for $z \gtrsim 2$ (figure 1).

As is seen in the above, we can naturally reproduce three statistical properties of absorption lines of QSOs by expanding minihalos, and in this picture, highly ionized envelopes and/or less massive one-phase minihalos correspond to Lyman α forest and a neutral core in a massive minihalo is detected as a high column density system. More detailed results are discussed by Ikeuchi (1991) and Murakami and Ikeuchi (1990).

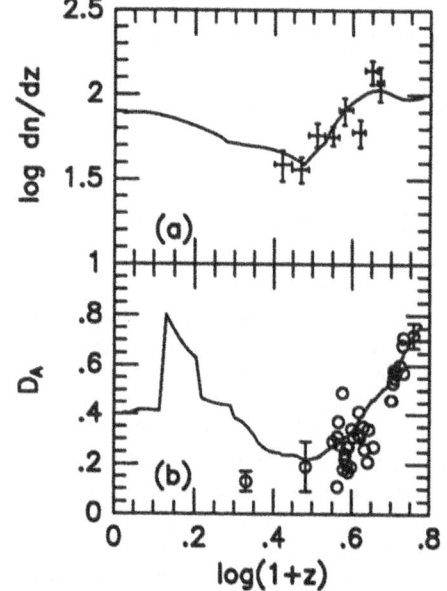

Fig.1 Statistical properties for expanding mini-halos for initially isothermal case. Parameters are $\alpha = -4$ ($z > 2$), $\rho_{CDM}(r = 0) = 10\rho_{crit}(z = 10)$, $\sigma^2_{CDM} = 5c_s(T = 10^4 K)$, and $\delta = 1.6$. (a) Number density evolution: Crosses are estimated by Murdoch et al. (1986). (b) Continuum depression. Circles are data estimated by Schneider et al. (1989).

References

Ikeuchi, S. 1986, *Ap. Sp. Sci.*, **118**, 509-514.
Ikeuchi, S. 1991, in *this volumn*.
Ikeuchi, S., Murakami, I., and Rees, M. J. 1988, *M.N.R.A.S.*, **236**, 21p-28p.
Murakami, I. and Ikeuchi, S. 1990, in preperation.
Murdoch, H.S., Hunstead, R.W., Pettini, M., and Blades, J.C. 1986, *Ap. J.*, **309**, 19-32.
Rees, M. J. 1986, *M.N.R.A.S.*, **218**, 25p-30p.
Sargent, W. L. W., Steidel, C. C., and Boksenberg, A. 1989, *Ap. J. Suppl.*, **69**, 703-761.
Schneider, D.P., Schmidt M., and Gunn, J.E. 1989, *A. J.*, **98**, 1951-1958.

GRAVITATIONAL LENS EFFECT ON THE IMAGES OF HIGH REDSHIFT OBJECTS

K. TOMITA and K. WATANABE

Uji Research Center
Yukawa Institute for Theoretical Physics
Kyoto University, Uji 611
Japan

ABSTRACT. In order to study the gravitational lens effect of galaxies on the images of high redshift objects, we solved numerically the null geodesic equations in an inhomogeneous model universe consisting of galaxies. Morphological studies were done as well as quantitative analyses for the image deformation.

1. Light propagation in an inhomogeneous model universe

The light rays emitted by high redshift objects like quasars propagate through the intergalactic space, the neighbourhood and sometimes the inside of galaxies before they reach us. Their paths are more or less deflected by the gravitational forces of galaxies. In the recent observational studies on the morphologies of high redshift quasars, many interesting distorted or warping images were found (Bartel and Miley (1988) and Bartel et al (1988)). It seems to us that they originate in gravitational lensing by galaxies. To clarify their origin, we studied in our recent work (Tomita and Watanabe (1990)) about what deformation was aroused by the galactic lens effect on the images of high redshift objects.

The method we used is as follows:

a. First we make numerically an inhomogeneous model universe which consists of galaxies and in which the galactic distribution satisfies the periodic condition with the spatial period of 33.3 h^{-1} Mpc. The time evolution of the distribution is determined by use of the N body simulation method in which the gravitational forces are derived in the Newtonian approximation.

b. Next we study the light propagation by solving the geodesic equation given in the post-Newtonian approximation. Then all galaxies play the role of lenses and the images of high-redshift objects are deformed by them.

The deflection angles depend on the galactic masses and radii, the number density, the birth time of galaxies and the background cosmological model. If we take into account the dark matter in the galactic halo region, the galactic masses are by a factor 10 - 100 larger than the standard mass ($\sim 10^{11} M_{\odot}$). In our work we adopted the rather large value for the galactic mass and the Einstein-de Sitter background model. For the galactic distribution we assumed that the present correlation length is about $5h^{-1}$ Mpc (the Hubble constant $H_0 = 100h$ km s^{-1}Mpc^{-1}).

2. Image deformation

We consider an observer who is at the origin of a periodic box and receives the ray bundles incoming from many arbitrary directions. Each bundle consists of 900 rays which are put in the same separation angle (= 1 arcsec) in a square region of 30 arcsec × 30 arcsec at the present epoch. If we go back in time, the cross sections of ray bundles are deformed from the square to complicated forms. Many figures showing the deformation were derived. It was found from their figures that often ray bundles focus and have caustic points or planes at epochs $z \sim 1$ and reexpand thereafter. Moreover, we derived the deformed images which were brought by the lens effect to originally circular objects (with various layers) put at epochs $z_e = 1 - 5$. For this purpose we used the correspondence between the present and past positions of rays. From the images in the ten ray bundles we found as a general properties that the circular images are deformed to long and narrow shapes in most cases. These morphological changes are interesting when we compare them with the warping structure of high redshift quasars.

In order to represent quatitatively the characteristic behaviors of image deformation, moreover, we introduced the following two quantities. One of them is the root mean squares of the difference of deflection angles of two rays. Another one is the average angle between two deviation vectors. These vectors represent the deviations among three neighbouring rays. They are chozen so as to be perpendicular each other at present epoch. These two quantities were calculated for ten ray bundles. The details are shown in our cited paper.

References

Bartel, P.D. and Miley, G.K. (1988) 'Evolution of radio structure in quasars', Nature **333**, 319-325.

Bartel, P.D., Miley, G.K., Schilizzi, R.K. and Lonsdale, C. J. (1988) 'Observations of the large scale radio structure in high redshift quasars', Astron. Astrophys. Suppl. **73**, 515-547.

Tomita, K. and Watanabe, K. (1990) 'Gravitational lens effect on the images of high redshift objects', Prog. Theor. Phys. **83**, 467-490.

GRAVITATIONAL LENS DETERMINATIONS OF COSMOLOGICAL PARAMETERS

EDWIN L. TURNER
National Astronomical Observatory
Mitaka, Tokyo 181, Japan
and
Princeton University Observatory
Princeton, NJ 08544-1001, USA

ABSTRACT. Currently available attempts to determine the three classical cosmological parameters (H_0, Ω_0, and Λ) using gravitational lens techniques and arguments are reviewed. These suggest that $H_0 \gtrsim 75$ km/s/Mpc, that contributions to Ω_0 by certain hypothetical types of dark matter are < 1, and that Λ is considerably smaller than $3H_0^2$ if a flat cosmological model is assumed. Unfortunately, none of these three conclusions is satisfactorily free of model dependence, possible systematic error, caveats, etc. There is, however, good reason to expect that the situation can be improved in this regard.

1. Introduction

Gravitational lensing phenomena have a deep and direct connection to cosmology because they involve geometric optics over distances of order the cosmic scale (c/H_0). One may think of them as natural optical benches with dimensions of order the size of the Universe. It is thus natural to attempt to employ them in our quest for the values of various cosmological parameters.

The current status of some of these attempts is briefly described and reviewed below. The conventional definitions are adopted

$$H_0 = [\dot{a}/a]_{z=0} \tag{1}$$

$$\Omega_0 = \frac{8\pi G\overline{\rho_0}}{3H_0^2} \tag{2}$$

$$\lambda = \Lambda/3H_0^2 \tag{3}$$

where the symbols all have their standard cosmological definitions (Weinberg 1972). Each parameter is discussed separately and in the reverse of the above order.

2. The Cosmological Constant

For various reasons (see the article by M. Turner in this volume for a review) considerable attention has recently been focused on cosmological models which have

$$\Omega_0 + \lambda = 1 \tag{4}$$

and which have $\lambda \gtrsim 0.8$. Such models exhibit much higher strong lensing event frequencies for a fixed comoving population of lens objects than do models which also obey equation (4) but have $\lambda \ll 1$ (Fukugita, Futamase, and Kasai 1990; Turner 1991). For typical quasars, the effect is approximately an order-of-magnitude and hence provides a potentially quite sensitive measure of λ. Comparison of actually observed galaxy-quasar lensing rates with those expected theoretically (Turner, Ostriker, and Gott 1984) appears to rule out large, in the above context, values of λ (Turner 1991). However, given a variety of systematic uncertainties in the theoretical predictions and the possibility of incompleteness and bias in existing lens samples, the conclusion must be regarded as provisional at the moment (Kochanek 1991a, Fukugita and Turner 1991, Mao 1991). Eventually, more statistically suitable lens samples and more detailed and realistic theoretical calculations may give us the best available empirical measures of the cosmological constant using this technique.

3. The Density Parameter

3.1. DARK GALAXIES

Press and Gunn (1973) long ago pointed out that the frequency of lensing events caused by a population of objects measures their contribution to Ω_0. Their argument may be used in a somewhat different way by noting that lensing galaxies (individual or in clusters) have now been identified for 10 of the 15 known (see Turner 1989 for a review) gravitational lens candidates with image splittings $1'' \lesssim \Delta\theta \lesssim 10''$, roughly corresponding to $10^{11} M_\odot$ to $10^{13} M_\odot$ lenses. Among the other 5 cases, it is easily possible that some are not actually lens systems at all or that the lensing galaxy has not yet been detected because it is faint and/or lost in the "glare" of the quasar images.

In any case, this fact already carries an interesting cosmological conclusion. It implies that dark objects with mass distributions and small scale clustering just like those of known galaxies *cannot* make a larger contribution to Ω_0 than the visible galaxy population. For example, the voids in the large scale galaxy distribution (Geller and Huchra 1989) cannot contain a dark matter distribution statistically like matter in the regions containing galaxies. Furthermore, the extreme opposite view that their are no dark galaxies is consistent with presently available data.

Unfortunately, lensing cross sections can be quite sensitive to details of lens object mass distributions (Hinshaw and Krauss 1987) so that this limit on dark galaxies cannot be naively applied to any sort of dark objects of similar mass. Nevertheless, for any specific model of dark objects (*e.g.*, CDM objects below the "bias threshold"), it will be possible to predict a frequency of non-galaxy lens (Babul and Lee 1990) systems and thus test it. The steadily improving sensitivity and angular resolution of optical observations should provide either more compelling cases of dark lensing objects or better limits on their occurrence

3.2. MICRO-LENSING LIMITS

Canizares (1985) suggested that we might e able to detect or limit the existence of a population of point masses carrying a significant cosmological density via micro-lensing effects. The idea is based on two points: 1) the observed range of quasar redshifts spans the domain of very small to moderate (~ 1/2) probabilities of significant micro-lens point source amplification (or magnification), and 2) quasar emission that arises from compact regions will be more strongly affected by micro-lensing than extended emission. Such considerations lead to the general *qualitative* expectation that quantities such as x-ray to optical flux ratios, optical to radio power ratios, continuum to line ratios, and emission line equivalent width distribution dispersions would increase with redshift if micro-lensing were common in quasar samples. In fact, no such trends have been observed; in all cases, there is either no significant trend or the opposite one.

Taken at face value, this suggests that there is no cosmic population of point masses with a total mass density close to the critical value. Unfortunately, our ignorance of the details of quasar sizes and emission structure prevent us from knowing precisely what lens mass range this test probes. Even more seriously, the possibility of source evolution masking micro-lensing trends introduces a major systematic uncertainty. Nevertheless, it is clear that this clever test has not yet been pursued very vigorously and that it may have much it can yet tell us.

4. Hubble's Constant

Since Refsdal's (1964) pioneering work brought discussions of gravitational lensing into modern astrophysics, the possibility of measuring large absolute cosmic distances (and hence H_0) using a single step, physically simple, direct technique has excited great interest. Unhappily, this excitement has not translated into the same sort of rapid advances which other areas of lens investigations have witnessed. In fact, the situation has not changed very much from that reviewed two years ago (Turner 1989).

The only lens system so far much studied for the purposes of determining H_0 is 0957+561A,B. The observational evidence for a time delay of somewhat more than 400 days continues to accumulate (Schild and Choflin 1987, Vanderriest *et al.* 1989, Lehar *et al.* 1989, Shild 1990). If the most straightforward models (Falco, Gorenstein, and Shapiro 1990) are adopted, the indicated value of H_0 is near 80 km/s/Mpc. Nevertheless, this intriguing result remains seriously model dependent (Kochanek 1991b) and thus systematically uncertain.

One interesting new result in this area is Yoshida's (1990) analysis of the Vanderriest *et al.* (1986) limits on the time delays in 1115+080 which gives $H_0 \gtrsim 75$ km/s/Mpc for a reasonably general model.

It is not yet clear how well H_0 can be determined from gravitational lens studies; however, it is completely clear that much more can and should be done. Probably too much effort has been devoted to 0957+561A,B compared to other lens systems; its primary virtue would seem to be that it was the first discovered. It does however have several serious defects for these purposes, among them that the source is not very strongly variable (as quasars go), that the lens mass distribution is a complex composite of galaxy and cluster, and that it has only two bright images. It is perfectly reasonable to imagine that we could have clear and well determined delays measured among the several images of two or more lens systems. If relatively simple and natural models (Kochanek

1991c) not only explained the image positions and brightness ratios (which can be well determined once the delay is measured) of all images but also gave a consistent H_0 value from each of the measured delays in all of the different lens systems, then we would have reason to confidently regard the H_0 issue as settled.

5. Discussion

So far, study of gravitational lenses has not yielded a decisive determination of any fundamental cosmological parameter. Of course, the same may be said of all techniques available for investigating their values. More positively, it should be noted that now, only a decade after their observational discovery, gravitational lenses are "in the game" in a serious way. In other words, they are giving us indications and hints which are arguably as interesting and reliable as those available from other sources. Moreover, there is every reason to expect that much more can be learned from them.

ACKNOWLEDGEMENTS. This review has benefitted from conversations with many colleagues and was supported in part by NASA grant NAGW-2173 and by a Visiting Professorship at the National Astronomical Observatory of Japan.

REFERENCES

Babul, A., and Lee, M.H. (1990) preprint.
Canizares, C.R. (1985) *Ap.J., 263*, 508.
Falco, E.E., Gorenstein, M.V., and Shapiro, I.I. (1990) *Ap.J.*, in press.
Fukugita, M., Futamase, T., and Kasai, M. (1990) *MNRAS, 246*, 24p.
Fukugita, M., and Turner, E.L. (1991) in preparation.
Geller, M.J., and Huchra, J. (1989) *Science, 246*, 897.
Hinshaw, G., and Krauss, L.M. (1987), *Ap.J.* 320, 468.
Kochanek, C.S. (1991a) *Ap.J.*, submitted.
Kochanek, C.S. (1991b) *Ap.J.*, submitted.
Kochanek, C.S. (1991c) *Ap.J.*, submitted.
Lehar, J., Hewitt, J.N., and Roberts, D.H. (1989) in *Gravitational Lenses*, eds. J. Moran, J.N. Hewitt, and K.Y. Lo (Dordrecht: Reidel) 84.
Mao, S. (1991) preprint.
Press, W.H., and Gunn, J.E. (1973) *Ap.J., 185*, 397.
Refsdal, S. (1964) *MNRAS, 128*, 295.
Schild, R. (1990) in *Gravitational Lensing*, eds. Y. Mellier, B. Fort, and G. Soucail (Berlin: Springer-Verlag), 102.
Schild, R., and Cholfin, B. (1986) *Ap.J., 300*, 209.
Turner, E.L. (1989) in *Proceedings of the Fourteenth Texas Symposium on Relativcistic Astrophysics*, ed. E.J. Fenyves (New York: NY Academy of Sciences), 319.
Turner, E.L. (1991) *Ap.J. Lett.*, in press.
Turner, E.L., Ostriker, J.P., and Gott, J.R. (1984) *Ap.J., 284*, 1.
Vanderriest, C., Schneider, J., Herpe, G., Chevreton, M., Moles, M., and Wlerick, G. (1989) *Astr. Ap., 215*, 1.
Vanderriest, C., Wlerick, G., Lelievre, G., Schneider, J., Sol, H., Horville, D., Renard, L., and Servan, B. (1986) *Astr. Ap., 158*, L5.
Weinberg, S. (1972) *Gravitation and Cosmology*, (New York: Wiley), 405.
Yoshida, H. (1990) Ph.D. Thesis, University of Tsukuba.

Global Cosmological Parameters

M. FUKUGITA

Yukawa Institute for Theoretical Physics

Kyoto University, Kyoto, 606 Japan

The determination of the global cosmological parameters is discussed. It is argued that the Hubble constant is likely to lie between 75 and 100km s^{-1}Mpc^{-1}, and a low value ($H_0 \approx$ 50km s^{-1}Mpc^{-1}) is quite unlikely. It is also shown that the current observational evidence supports a low density (typically $\Omega \approx 0.1$) universe. Combining these parameters with the accepted value for the age of the universe, it is concluded that a sizable amount of the cosmological constant is probably necessary. We discuss that the same conclusion is also inferred from the number count of faint galaxies.

1. Introduction

In this report I shall discuss the status for the determination of the cosmological parameters, with emphasis on the work that I have been carrying out over the last few years. I assume that the Universe is of the Friedmann-type (i.e., matter + possible cosmological constant) [1], and examine whether the parameters determined from various observations are consistent with each other.

The Freedmann universe is described by three fundamental parameters. The Hubble constant H_0 is the only parameter that has the dimension and determines the physical scale of the universe. Two others are the mass density of the universe $\Omega_0 = \rho/\rho_{crit}$ and the cosmological constant Λ, for which we use a dimensionless quantity λ_0. The motto of this Symposium is perhaps characterized by $H_0 = 50$km s^{-1}Mpc^{-1}, $\Omega_0 = 1$ and $\Lambda = 0$, as most of speakers assumed explicitly or implicitly in their talks. My conclusion is that the observation indicates neither $H_0 \approx 50$km s^{-1}Mpc^{-1} nor $\Omega_0 \approx 0.1$, and that $\Lambda \neq 0$ is probably necessary. I use the normalization

$$H_0 = 100h \text{ km s}^{-1}\text{Mpc}^{-1},$$
$$\Omega_0 = \rho/\rho_{crit}, \tag{1}$$
$$\lambda_0 = \Lambda/3H_0^2,$$

so that $\Omega_0 + \lambda_0 = 1$ corresponds to the flat universe.

2. Hubble Constant

It has customarily been quoted that H_0 lies between 50 and 100km s^{-1}Mpc^{-1}. The statistics of published literature on H_0, however, indicates that the reported value shows a bimodal distribution and people are divided into two groups; one group of people claims H_0 close to 100km s^{-1}Mpc, while the other insists on $H_0 \approx 50$km s^{-1} Mpc^{-1}. In this talk I would like to discuss whether the observational determination of H_0 really leaves such a large uncertainly.

The difficulty has always been to find the reliable distance to galaxies sufficiently far away for local "peculiar" velocities, as deviation from a pure Hubble expansion are called, to be negligible. The use of Cepheid variables with ground based telescopes is limited to M81 (NGC 3031), which is at about 3 Mpc away. Up to this distance there is currently little disagreement among authors, thanks to the extensive Cepheid work with infrared and multicolour photometry [2-4]. The compilation by Tammann in 1987 [5] gives basically correct distances to within M81 groups, except for M81 itself, for which he gave a distance 1 mag fainter. On the other hand, another useful indicator of type Ia supernovae (SNe Ia) can apply only from the Virgo distance onward due to its rare occurrence. A reliable indicator has been lacking between M81 and the Virgo cluster (see Fig. 1).

The only "qualified" indicator which links these two objects is the Tully-Fisher method [6], which utilizes the empirical relation between the rotation velocity as measured by the HI 21cm line width and the luminosity of spiral galaxies. The basic drawback of this method is that the galaxy itself is used as the indicator. The statistical ensemble of galaxies located at the same distance should be introduced to obtain the distance to the required accuracy. This is contrary to the case of star indicators, with which one can, in principle, measure the distance to a galaxy with a desired precision, provided that systematic errors are under control. Another drawback of using a galaxy as an indicator concerns theoretical aspects. The evolution of stars (or objects associated with a star) is believed to be determined solely by the mass and the composition. On the other hand, galaxies are complicated dynamical systems with history, and their evolution may depend on such aspects; the distance indicator relation is purely empirical and one has to rely on the relation, blindly hoping that no hidden systematic errors accumulate. There are also proposed other indicators that reach the distance as far as the Virgo cluster (globular cluster luminosity functions, novae etc.), but they do not seem to be calibrated so well and often give divergent distances, as we discuss below.

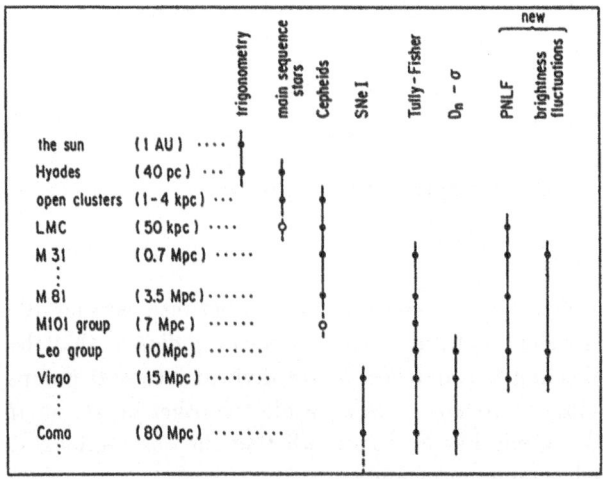

Fig. 1. "Qualified" distance indicators.

Schematically we can say that people who believe in the Tully-Fisher distance are likely to conclude $H_0 \approx 80 - 100$km s^{-1}Mpc^{-1} mostly from the analysis beyond the Virgo cluster. Other people who insist $H_0 \approx 50$km s^{-1}Mpc^{-1} presume that the naive application of the Tully-Fisher relation does not yield the correct distance, and tends to believe in the distance from supernovae, which apparently give a convergent value around $H_0 \approx 50$km s^{-1}Mpc^{-1}.

In 1983 Aaronson and Mould [7] claimed that the zero point error for the Tully-Fisher relation biases the distance as large as 0.65 mag depending upon whether one adopts the distance by Sandage and Tammann [8] or that by de Vaucouleurs [9]. Now we have five calibrating galaxies (M31, M33, N2403, N300 and M81) with a good Cepheid distance. The uncertainty as claimed in 1983 is basically removed, and the overall uncertainty for the zero point is now at a 10% level ($\pm 0.1 - 0.15$ mag). For instance, with the distance determined by Freedman [3], the five points are nicely on a straight line with a dispersion $\lesssim 0.2 - 0.25$ mag both for the H-band and the B-band Tully-Fisher relations (Fig. 2). The Tully-Fisher analysis for the Virgo cluster, however, is still controversial as we shall see below.

One of the most important progress made recently is the discovery of the planetary nebula luminosity function (PNLF) as an excellent standard candle by Jacoby and collaborators [10-15]. The PNLF has a sharp cut off at the brightest end, and it is suggested to be insensitive to metallicity of host galaxies. The important point is that this indicator fills the gap between M81 and the Virgo cluster (see Fig. 1), and it gives a distance to individual galaxies placed near the Virgo centre. We particularly note that the distance by PNLF accurately ($\lesssim 10\%$) agrees with the Cepheid distance [14,11] for LMC $-$M31$-$M81. The distance to Leo I group [12] shows a good agreement ($\approx 10\%$) with the value given by the Tully-Fisher relation. The importance of the accurate measurement to the Leo distance is that it makes possible to calibrate the zero point of the $D_n - \sigma$ relation for ellipticals [16].

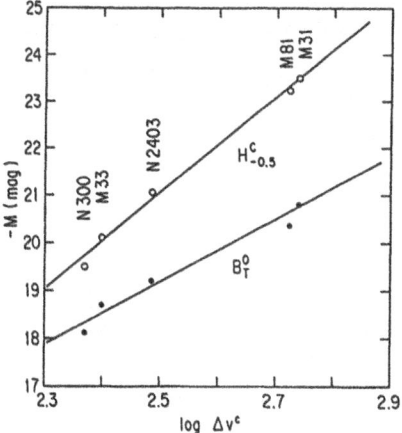

Fig. 2. Tully-Fisher relation for local calibrators with the Cepheid distance. $H^c_{-0.5}$ stands for the scheme by Aaronson et al. [35], and B^0_T for the scheme in the *Second Reference Catalogue of Bright Galaxies*.

With this calibration the distance given by PNLF agrees with that by $D_n - \sigma$ for individual galaxies in the Virgo cluster within 0.25 mag (12 %), a dispersion expected for the $D_n - \sigma$ relation. Let us also note that the PNLF distance agrees with that by surface brightness fluctuation (SFB) technique [17] to within 0.2−0.3 mag for 3 common galaxies in Virgo. The average distance to six galaxies near the Virgo centre is 14.7 ± 1 Mpc $[(m - M)_0 = 30.8 \pm 0.11]$ according to Jacoby et al.[14]. This distance also agrees well with that for the "Virgo centre" from the Tully-Fisher analysis with the sample by Pierce and Tully [18]. For a detailed comparison see Table I of ref. [19].

Another important point is that three galaxies with the PNLF distance have SNe Ia, which allows a reliable calibration of maximum light of SN Ia [19]. We obtained the absolute magnitude of the maximum brightness $M_{\max}(B) = -18.5$ ($\sigma = 0.26$). A similar value is also derived from the distance using $D_n - \sigma$, the Tully-Fisher and brightness fluctuations for 8 SNe I near the Virgo centre. Note that this value is 1−1.5 mag fainter than has been adopted [20,21]. We emphasize the advantage of the method adopted here that the calibration is made on the basis of the distance to individual galaxies, rather than the use of the distance to the "Virgo cluster" as has been done traditionally [20]. With this method the errors due to uncertainties for the Virgo membership or the line-of-sight effect does not enter in the final result; we need to care only the statistical (dispersion) error.

The Hubble diagram for more distant supernovae is obtained by a number of authors, and it is summarized to be [21]

$$M_{\max} = -(18.13 - 18.5) + 5 \log h. \qquad (2)$$

Combining this with the above zero point we are led to $75 \leq H_0 \leq 105$ km s^{-1} Mpc^{-1}.

With this calibration we may estimate the distance to the Coma cluster using 5 type I SNe observed in Coma (photographic magnitudes)[22]: $d = 81 \pm 15$Mpc. This is compared with the value from the Tully-Fisher analysis 78 (+18/−11) Mpc, which includes a 13% upward shift from the cluster population incompleteness bias [23]. Application of the $D_n - \sigma$ relation calibrated by PNLF for the Leo group yields ~ 88 Mpc. These values all correspond to $H_0 \approx 80 - 90$km s^{-1}Mpc^{-1}. In Fig. 3 we show indicators which give convergent values ($\lesssim 20$ %) up to the Coma distance.

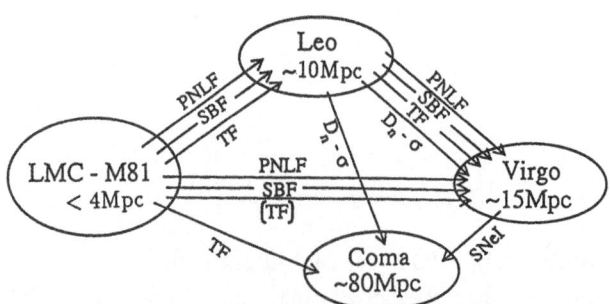

Fig. 3.   Distance indicators which give a convergent answer.

We should remark here on the arguments which support the long distance leading to $H_0 \approx 50 \mathrm{km\ s^{-1}Mpc^{-1}}$. The most often quoted is the Hubble constant derived from SNeI, just discussed above. The traditional calibrations have depended on 5 methods [21]; (i) calibration of the maximum luminosity by one SN in IC4182 and two SNe in N5253, (ii) the calibration using the Virgo distance, (iii) historical SNe, (iv) expansion parallax using the Baade-Wesselink method, and (v) theory. For (i) we just say that both galaxies lack reliable distance estimates with qualified indicators; the distances are estimated using the brightest stars as a standard candle. As for (ii) we perhaps need a little detailed arguments, which are deferred to below. It is clear that (iii) cannot be accurate enough to determine the absolute magnitude within 1 mag. The method of (iv) usually assumes that the flux has the black-body spectrum. If, for example, electron scattering (for SNe II) or bound-bound opacity (for SNe I) dominates over absorption opacity, the flux may be diluted [24,25]. For a type II supernova an example of such uncertainty is seen best for SN1987A. While Branch [26] deduced 55 kpc with the black-body spectrum, some other authors [24,27] obtained a value varying from 43 to 49kpc; the extreme values represent almost the longest and the shortest among the estimates of the LMC distance. The calibration of (v) uses M_{max} calculated theoretically in a simple model of SNe I [28]. This argument hinges on the assumption that the bolometric luminosity at maximum light is given by the instantaneous decay luminosity of ^{56}Ni, the mass of which is computed in a deflagration model. Although the model seems successful enough to serve as a convincing description of how SNe I work, it has not been tested quantitatively at the level required of detail for it to give a calibration for a standard candle. It seems at least possible, in the absence of any evidence to the contrary, that the allocation of the energy budget between kinetic energy, blue light and other forms of radiation is not quite as predicted by the simple model. Indeed there are some indications that the kinetic energy of type Ia supernovae is not universal [29].

The issue of the distance to Virgo deserves more discussion. The best representation of the long distance to the Virgo cluster may be that by Sandage and Tammann [30]. For example, they have listed 5 independent observations that agree with each other to give $d = 22$ Mpc to the Virgo centre: (i) The first indicator is globular cluster luminosity function (GCLF) by Harris [31]. The method assumes that the peak luminosity of globular cluster is universal. It is known, however, that the shape of GCLF is not quite universal. For instance, GCLF for M87 has a considerably longer tail in the luminous side compared with those of M31 and of our Galaxy. GCLF of LMC has no peak. The assumption is not really calibrated. In fact the correlation with the distance by other methods is poor; GCLF applied to NGC3379 (Leo I) gives a distance 30-45% shorter than that by PNLF, the brightness fluctuation technique and the Tully-Fisher method, while the same method gives a distance to M87 1.5 times larger. (ii) The nova study for Virgo galaxies [32] requires a very difficult observation. The plot for a maximum brightness–light declining rate for Virgo novae shows a wild scatter; the dispersion amounts to >1 mag and χ^2 is larger than 10/DF (see Fig. 7 of ref. 32). The fit by Pritchet and van den Bergh, which gives $(m - M)_0 = 31.45 \pm 0.44$, is constrained essentially by one point, nova no. 2 in NGC4472 (east). We feel that we can not determine the distance with a good precision. In addition Sandage and Tammann increased their distance modulus by 0.2mag by taking a larger absorption of $A_B = 0.51$ for novae in M31 used for calibration (cf. $A_B = 0.31$ is adopted by Pritchet and van den Bergh). This large absorption would bring the nova distance to M31 (Cohen [33]) into $(m - M)_0 = 23.84$ in a gross disagreement with the Cepheid distance (24.25 ± 0.15). Here is also an example of poor correlation with other indicators. (iii) The $D_n - \sigma$ relation is certainly calibrated better by elliptical galaxies, rather than the bulge of M31 [34] (Sandage and Tammann here also increased the Virgo distance modulus by 0.2mag compared with the original value by Dressler). (iv) To the Baade-Wesselink distance to type II SNe in the Virgo galaxies the criticism as discussed above applies. (v) There is an apparent disagreement in the Tully-Fisher distance to the Virgo cluster. Aaronson et al. [35] and Pierce and Tully [18] obtained $d \simeq 15$ Mpc $[(m - M)_0 \simeq 30.9]$. On the other hand, Kraan-Korteweg, Cameron and Tammann (KCT) [36] and Fouqué et al. [37] obtained $d = 19 - 22$ Mpc $[(m - M)_0 = 31.4 - 31.7]$. The latter authors claim that the short distance by the former authors is due to the sample incompleteness bias and that the use of the "complete sample" should lead to the long distance. The fact that KCT and Fouqué et al. obtained a dispersion, as large as 0.7 mag, which is much larger than was found anywhere else (0.2-0.4mag), however, seems to indicate that some of galaxies in their sample are located behind the centre. (For example, Freedman [3] showed from the dispersion for nearby galaxies that the probability of the dispersion as large as 0.7 mag is less than 2%, provided that the dispersion does not depend on the environment.) The agreement of the Pierce-Tully distance with the distance from elliptical galaxies suggest that bright galaxies selected by Pierce and Tully (and by Aaronson et al.) are located close to M87, but the samples by KCT and Fouqué et al. contains a significant number of galaxies in the background. Evidence in support for this interpretation is also discussed by van den Bergh et al. [38] using the luminosity class classification based on CCD imaging (see also Burstein and Raychaudhury [39]). I feel, however, that the issue is not quite settled, and that one cannot use the Tully-Fisher method to establish the distance to "the Virgo cluster", unless the issue with the line-of-sight problem is settled. Let me stress again that the calibration of SNeI using individual galaxies as discussed above circumvents this problem.

From the discussion given above we may conclude that the Hubble constant takes a value between 75 and 100 km s^{-1}Mpc^{-1}. Solid observational evidence lacks for $H_0 \approx 50$. The fact that the Hubble diagram for first rank ellipticals [40] exhibits a straight line up to $z \approx 0.5$ suggests that we may accept this Hubble constant as a global value.

3. Mass Density

The observational determination of the mass density uses one of the following three methods; (i) classical test for the space-time geometry [41], (ii) estimation of the luminosity density and of the average mass-to-light ratio $\langle M/L \rangle$ of galaxies [42], and (iii) dynamical test for the mass distribution of galaxies [43].

The geometry test is one of the most classical subjects in cosmology, yet it is rather recent that it started to give useful information on the cosmological parameters (the status as of 1986–7 is summarized by Sandage [44]). The best-known test is the magnitude-redshift relation for distant bright galaxies. This method, however, seems to be not so useful because it uses the brightest galaxy as a standard candle and thus suffers from strong selection effects; in particular the sample for high redshifts $z \gtrsim 0.5$ is taken from radio galaxies which would be particularly luminous. For such galaxies a particularly strong evolution effect may also be expected. On the other hand, the magnitude-number count $N(m)$ employs more normal galaxies and the evolution effect, at least to some extent, is under control. In order to discriminate among cosmological models, the survey is necessary beyond 24 B_J mag, and such data became available only in 1988 by Tyson [45] and then by Lilly et al.[46]. I describe in some detail our analysis in the next section, and here only mention the result that the number count strongly favours a low density ($\Omega_0 \approx 0.1$) universe.

For the estimation of (ii) the luminosity density $\mathcal{L}$ is obtained by integrating the luminosity function which is usually taken to be of the Schechter form,

$$\phi(L)dL = \phi_*(L/L_*)^\alpha \exp(-L/L_*)dL/L_*. \tag{3}$$

The luminosity density is determined up to a typical uncertainty of a factor of 2, $\mathcal{L}_B = 1 - 2.5 \times 10^8 L_\odot h/(\text{Mpc})^3$ [47]. The resulting Ω_0 with this method largely depends on the adopted value of $\langle M/L \rangle$. If we take $\langle M/L_B \rangle \approx 15h$ [48] we obtain $\Omega_0 \approx \mathcal{L}_B \langle M/L_B \rangle / \rho_{\text{crit}} \approx 0.01$[42]. This value, however, increases to 0.1 if we adopt $\langle M/L_B \rangle \approx 100h$. Let us note that $\langle M/L_B \rangle \approx 15h$ is a value obtained from analyses for the mass inside a limited radius (typically the Holmberg radius) [48] and M/L_B becomes as large as ≈ 100 even for the Galaxy if the radius is taken to be sufficiently large [48]. Therefore, it seems perhaps more reasonable to accept this larger value, which leads to $\Omega_0 \approx 0.1$.

Dynamical estimates of Ω_0 uses the relation between the mass distribution and the peculiar velocity. It is supposed that the Galaxy is located at about the edge of the Virgo cluster, and falls towards its centre due to the peculiar acceleration by the mass concentration. With the linear perturbation theory the peculiar velocity is given by [43]

$$v_p = \frac{1}{4\pi} H_0 \Omega_0^{0.6} \frac{1}{\rho} \int d^3x \frac{\hat{x}}{x^2} \rho(x) \tag{4}$$

$$\approx \frac{1}{3} H_0 R \Omega_0^{0.6} \langle \delta\rho/\rho \rangle. \tag{5}$$

For the infall velocity $|v_p| \approx 200\text{-}400\text{km/s}$[49] and the Hubble velocity for the Virgo centre $H_0 R = |v_{\text{obs}} + v_p| \approx 1200\pm200\text{km/s}$, we obtain $\Omega_0 \approx 0.05 - 0.3$ from the average density enhancement $\langle \delta\rho/\rho \rangle \approx 2 - 3$[49,50]. For collapsed clusters of galaxies such as Coma estimates were made by applying the cosmic virial theorem [43]. The potential energy of the system is estimated by an integral of the two-point correlation function of galaxies and the kinetic energy by the relative velocity of galaxy pairs. The statistical stability condition that the core ($R < 1h^{-1}\text{Mpc}$) not dissolve within a Hubble time results in $0.1 \leq \Omega_0 \leq 0.3$ [43,51]. A similar value was also derived from the cosmic energy equation with the aid of a similarity solution of the BBGKY hierarchy [43].

There have been several attempts to apply eq.(4) to the large-scale flow beyond the Virgo cluster. The dipole pattern in the cosmic background radiation (CBR) [52] has been interpreted as a result of a local-group motion of 600km/s towards $\ell = 268°$, $b = 27°$. This apex differs from the centre of the Virgo cluster ($\ell = 284°, b = 74°$) by $45°$. A simple interpretation for this effect is that the Virgo cluster itself is also moving towards the direction of the Hydra-Centaurus supercluster. Several authors [53-56] estimated the mass-density dipole that appears in the right-hand side of eq.(4) using the galaxy sample obtained by the Infrared Astronomical Satellite (IRAS)[57]. Its merit lies in the homogeneous coverage and calibration over almost the entire sky, and a small Galactic extinction effect close to the Galactic plane, which allows estimation of the dipole integral. Since the flux f_i of a galaxy i is proportional to L_i/r_i^2, the density dipole integral $\int (\hat{r}/r^2)\delta(r)$ is essentially given by simply summing up the flux over all the galaxy sample, provided that far infrared light traces mass. The dipoles thus obtained agree among the authors, and point towards rather close to the CBR velocity dipole ($\sim 25°$ away from its apex). By identifying the mass density dipole with the CBR dipole with the assumption that the latter is the peculiar velocity of the local group due to the mass distribution around the local group, Yahil et al. [53] and Villumsen and Strauss [55] obtained $\Omega_0 \approx 1 \pm 0.2$. This value differs largely from those obtained from the optical means.

The basic assumption in dynamical methods is always that light (optical or infrared) traces mass at least at a scale of clusters. If this assumption is not tenable, the resulting answer would differ considerably from the true one. We have examined [58] this assumption for the Pisces-Perseus region using the HI survey data given by Giovanelli and Haynes [59]. By comparing the luminosity distribution with the mass distribution inferred from the HI line width, we found that optical light traces very well mass, at least that associated with galaxies, but far infrared light does not unless the selection effect, which is strongly distance dependent, is properly estimated, say by an optical catalogue; if the fraction of galaxies not visible in far infrared is not corrected for, the density enhancement in the Pisces-Perseus region is not correctly obtained with IRAS galaxies.

4. Cosmological Parameters from Deep Surveys

Under the assumption that the comoving number density of galaxies is conserved, the number of galaxies in the angular area ω in the magnitude range $m - m + dm$ and the redshift range $z - z + dz$ is given by

$$N(m, z)dm\, dz = \frac{\omega}{4\pi} \frac{dV}{dz} \sum_i \phi_i(M(m, z))\, dm\, dz, \qquad (6)$$

where the summation is taken over morphological types of galaxies, ϕ_i is the luminosity function for type i given by (3), and V is the comoving volume that depends on Ω_0 and λ_0. The absolute magnitude in the band pass λ, M_λ is written

$$m_\lambda = M_\lambda + 5\log(d_L/10\text{pc}) + E_\lambda + K_\lambda \qquad (7)$$

where d_L is the luminosity distance, K_λ the correction for the frequency shift and E_λ the correction for luminosity evolution of galaxies.

To evaluate the E correction we adopt the evolution model by Arimoto and Yoshii [60], which is a Tinsley-Bruzual type population synthesis model [61] but is improved to account for chemical evolution. In this model the galaxies evolve only through $L_i^*(z)$. The model is controlled by three parameters, the initial mass function, the star formation rate and the duration of star formation. The parameters are fixed, so that the model gives a correct present day spectral energy distribution (SED). In this model ellipticals are characterized by a large star formation rate ($\sim 50 - 100\times$ that in spirals) and an early termination epoch of $t_c \simeq 0.7$ Gyr (t_c is taken to be infinity for spirals).

1. Result from a simple treatment [62]

The galaxy number count as a function of magnitude is given by the integral of (6) over the redshift. In Fig. 4 the comparison between the prediction and the observation in the B-band (more precisely B_J band as defined by the GG385 filter and the Kodak IIIa-J plate) is shown. For simplicity we plotted only the data by Tyson [45]. The prediction is presented for three typical cosmological models (a) $\Omega_0 = 1$, $\lambda_0 = 0$, (b) $\Omega_0 = 0.1$, $\lambda_0 = 0$ and (c) $\Omega_0 = 0.1$, $\lambda_0 = 0.9$. Dashed curves are the prediction with a no-evolution model ($E_\lambda = 0$) for comparison.

It is clear that the prediction of a no-evolution model gives $n(m)$ substantially smaller than the observation for $B_J \gtrsim 22$ mag for all sets of the cosmological parameters [63-65]. The difference among different models becomes clear for $B_J \gtrsim 24$ mag. It is also apparent that the $\Omega_0 = 1$ model strongly deviates from the observation; most conspicuous is the fact that the predicted curve reaches a maximum at around $B_J = 27$ mag and the observed count exceeds this maximum value by a factor of 5 or more. We emphasize that the predicted maximum number of $N(m)$ is almost independent of the detail of the luminosity evolution model, since evolution shifts the curve only horizontally. Therefore, we conclude that the $\Omega_0 = 1$ flat model is strongly disfavoured (see also ref. [45,66]). The open universe model (b) does not yield the "best fit". We consider, however, that this case is probably acceptable if we allow uncertainties in the evolution model. The "best fit" to the observation is given by models with a substantial amount of the cosmological constant $\lambda_0 = 0.5 - 1.2$. This includes the case for the flat universe $\lambda_0 + \Omega_0 = 1$ as in our model (c). The features discussed here persist in both R and I bands, and basically the same conclusion can be drawn from the comparison in each band.

Let us mention here some doubts concerning the arguments given above: (i) The excess of $N(m)$ at faint magnitudes may be due to an excess of faint galaxies located rather nearby. (ii) Merging of galaxies may disturb the conclusion against the $\Omega_0 = 1$ model [67]. (iii) The shape of predicted curves of $N(m)$ does not agree with the observation at faint magnitudes [68]; the prediction is quite flat for $B_J \gtrsim 28 - 30$ mag, while the observation by Tyson indicates a rather sharp decrease beyond $B_J \gtrsim 27 - 28$ mag. (iv) The absence of high redshift galaxies in the redshift distribution $N(z)$ of the Durham redshift survey [69] favours no-evolution model; the prediction with evolution models has a conspicuous high redshift tails compared with the observation.

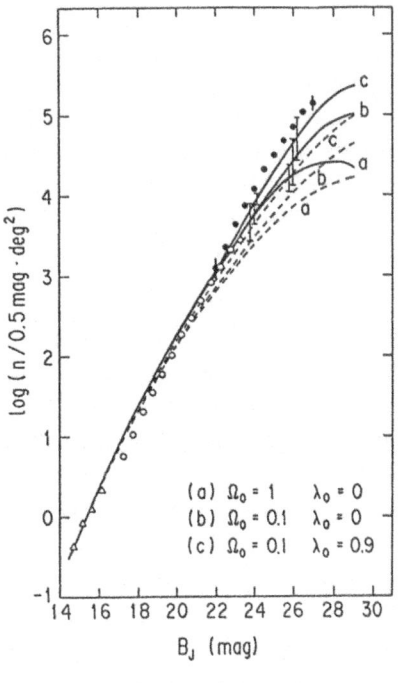

Fig. 4. Number count of galaxies as a function of the magnitude in the B_J band [62]. Model predictions are shown for (a) $\Omega_0 = 1$, $\lambda_0 = 0$ (h=0.5), (b) $\Omega_0 = 0.1$ (h=0.7), $\lambda_0 = 0$ and (c) $\Omega_0 = 0.1$, $\lambda_0 = 0.9$ (h=1.0). Dashed curves represent the corresponding no-evolution model predictions. Data points ($B_J > 17$mag) are taken from Tyson [45].

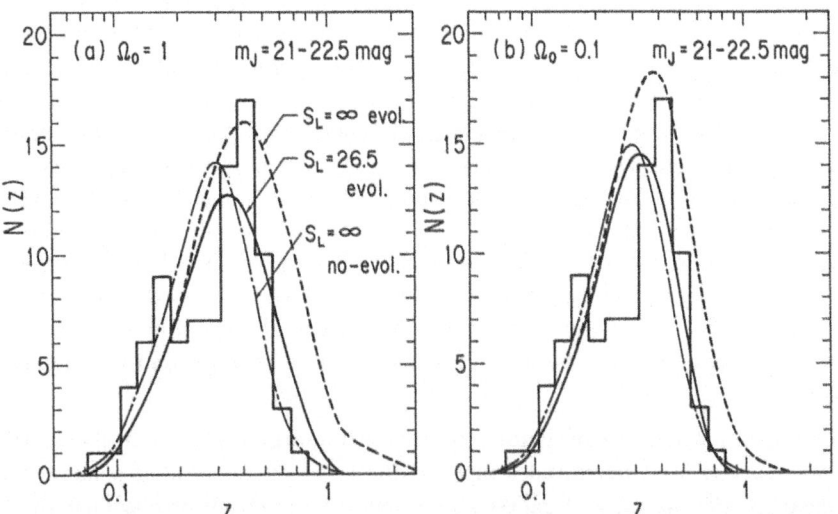

Fig. 5. The redshift distribution of galaxies for $B_J = 21 - 22.5$ mag [70]. The Durham data [69] are compared with the model predictions for (a) $\Omega_0 = 1$ and (b) $\Omega_0 = 0.1$. The solid curve shows a model with canonical galaxy evolution and with the isophotal selection effect taken into account. The model is normalized to the observed number of galaxies. Dashed (dot-dashed) curve represents the models with (without) galaxy evolution, for which the total magnitude scheme is employed. The same normalization is taken for the model with galaxy evolution. The model without galaxy evolution is renormalized to the observed number of galaxies.

In order to examine point (i), we artificially changed the slope of the Schechter function beyond the acceptable range. The change $\alpha = -1.1 \pm 0.2 \rightarrow -2$ in (3), in fact, makes the predicted curve in good agreement with the observation irrespective of the choice of cosmological models. This change, however, results in a gross disagreement between the prediction and the observed $N(z)$[69]; the change causes a pronounced peak at a redshift much lower than is seen in the observation. Points (ii)-(iv) are answered below.

2. *Importance of the selection effects* [70]

Observations generally include a number of selection procedures. Here we consider effects of (i) the procedure to define the magnitude of galaxies, (ii) finite seeing, and (iii) the selection of a finite-size objects as a galaxy candidate. Some of such effects have already been discussed in the literature [71].

The most common procedure to define the magnitude is to use the isophotal magnitude scheme. For example, Tyson [45] measured the isophote down to 29 mag $(\text{arcsec})^{-2}$ and assumed it to give correct total magnitudes to 27 mag. The Durham group [69] chose galaxies for their redshift survey from Schmidt plates using the $s_L = 26.5$ mag $(\text{arcsec})^{-2}$ isophote.

With the use of the isophotal magnitude eq.(7) receives an additional term that refers to the galaxy profile $G(r)$

$$m_\lambda = M_\lambda + 5 \log (d_L/10 \text{ pc}) + E_\lambda + K_\lambda - 2.5 \log G(r_{\text{cut}})/G(\infty) \tag{8}$$

where r_{cut} is the radius corresponding to the isophotal cutoff. Since the surface brightness decreases as $(1 + z)^{-4}$, high-redshift galaxies may receive a large selection effect from this cutoff.

We found indeed that this effect is very important in $N(z)$ even for galaxies as bright as $20-22.5$ mag. Fig. 5 compares $N(z)$ defined by the total magnitude system to that with the isophotal magnitude system with the cutoff at $s_L = 26.5$ mag$(\text{arcsec})^{-2}$ for the $\Omega_0 = 1$ model and the $\Omega_0 = 0.1$ model. With the isophotal cutoff a substantial fraction of high-redshift galaxies is missed. The effect is especially pronounced for $z \gtrsim 0.6$. That the observation lacks high-redshift galaxies is thus regarded as an artifact of the selection effect. In this figure it is seen that the no-evolution model (dash-dotted curve) mimics this effect, which led the Durham group [70] to conclude that the observation favours no-evolution model. This observation concerns not too faint galaxies and the effects of (ii) and (iii) are found to be unimportant.

On the other hand, the effect of isophotal cutoff is often not too serious in the integrated quantity like $N(m)$; other effects are more important. Fig. 6 shows the effect of isophotal cutoff $s_L = 29$ mag, and the finite-size selection effect corresponding to the minimum diameter of $D_{\text{min}} = 1.7$ arcsec [45] with a finite seeing effect also taken into account. The effect of the isophotal cutoff is gradual but is visible even at $B_J = 22$ mag. The effect of finite seeing is visible only at the faintest magnitude. The effect of the finite-size cutoff, however, becomes very important beyond $B_J > 27$ mag; the effect is quite sudden and sharp. It is interesting to note that we miss nominally

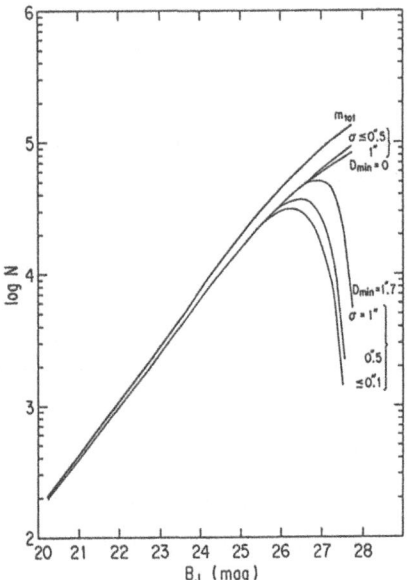

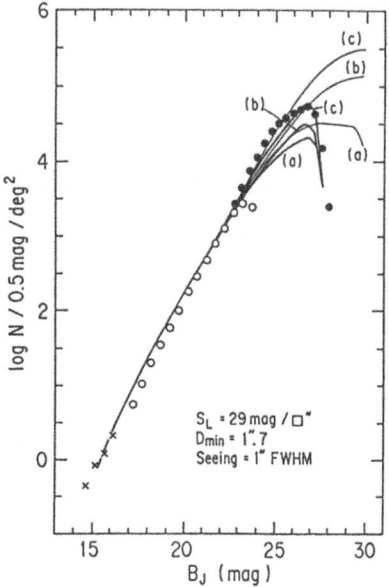

Fig. 6. Selection effects for the number count of galaxies [70]. A thick curve is for the total magnitude and thin curves are for the isophotal magnitude with $s_L = 29$ mag $(\text{arcsec})^{-2}$ with and without other selection effects.

Fig. 7. The raw data by Tyson [45] and predictions for the number count with the selection effects taken into account ($s_L = 29$ mag $(\text{arcsec})^{-2}$, $D_{min} = 1.7$, $\sigma = 1$.) [70]. Thin curves are predictions without the selection effects.

more faint galaxies with better seeing! As an example for a realistic case we exhibit in Fig. 7 the prediction both with and without selection effects in case of the Tyson's observation. An excellent fit is achieved with the $\Omega_0 = 0.1$, $\lambda_0 = 0.9$ model with the selection effects taken into account. This figure shows that a sharp decrease seen in $N(m)$ at the faintest magnitude is due to the selection effect.

Whether merging of galaxies saves the problem of the deficiency of a predicted number of faint galaxies for the $\Omega_0 = 1$ universe becomes subtle, if the selection effect is taken into account. If the mass density of premergers is equal to or higher than postmergers (cf. at the present day a galaxy with a small mass has a larger mass density), premergers mostly do not satisfy the minimum size criterion set by Tyson and should be missed in his count. Hense even if the number density of galaxies increases as $\sim (1 + z)^{3+n}$, $n > 0$ as postulated by Guiderdoni and Rocca-Volmerange [67], the increase of $N(m)$ by merging is at most 30 %, and it does not saves the $\Omega_0 = 1$ universe.

5. Conclusion

We have argued that the observationally favoured cosmological parameters are

$$H_0 = 75 - 100 \text{ km s}^{-1} \text{ Mpc}^{-1}$$
$$\Omega_0 \approx 0.1.$$

This leads to the upper limit of cosmic age as small as 10 Gyr, if $\Lambda=0$. On the other hand, the age inferred from evolution of globular clusters is 13-18 Gyr [72]. This rather long age is also corroborated by nucleochronology. In the framework of the Friedmann cosmology, this inconsistency implies a non-vanishing cosmological constant.

We have shown that a non-zero cosmological constant is also favoured by the number count of faint galaxies; the observed number of galaxies at faint magnitudes is significantly more than that expected from the $\Lambda=0$ geometry and the canonical galaxy evolution model.

The cosmological constant that concerns us here corresponds to the vacuum energy of 3±1 meV. This is not ridiculously small a value. It may be interesting to note a coincidence with the muon neutrino mass $\sim 10^{-4}$ eV as inferred from the solar neutrino problem [73]; the numbers suggest $m_\nu = g\langle\phi\rangle$ with $g \sim 0.1$, yet we do not know any model which realizes this relation.

One of the ways in which one can actively test for non-zero Λ is to utilize statistics of gravitational lensing for quasars. It has been shown by Futamase, Kasai and myself [74], and by Turner [75] independently that the lensing optical depth is strongly boosted by the presence of non-zero Λ. In fact, Turner [76] has estimated that about 60 lenses should be detected for $\Omega_0 = 0.1$ and $\lambda_0 = 0.9$ in the 4250 quasars listed in the Hewitt-Burbidge catalogue, whereas lensing candidates count only 9, and suggested that this model is already excluded. The estimate of Turner, however, turned out to be an overestimate, and this predicted number is reduced to be 9-20 in a recent improved estimate [76]. On the other hand, the properties of some of the candidates are not quite expected in a simple lensing model, and it is not clear whether this number is taken literally for lenses. In this view the $\Omega_0 = 0.9$, $\lambda_0 = 0.1$ model is just marginally allowed. We may hope the observation of lensing in the near future will settle the issue about the cosmological constant.

I would like to thank S. Okamura for useful comments.

References

1. A. Freedmann, Z. Phys. **10**, 377 (1922); **21**, 326 (1924).

2. D.L. Welch, C.W. McAlary, R.A. McLaren, B.F. Madore, Ap. J. **305**, 385 (1986); B.F. Madore et al., Ap. J. **294**, 560 (1985); C.W. McAlary and B.F. Madore, Ap. J. **282**, 101 (1984).

3. W.L. Freedman, Ap. J. (Letters) **355**, L35 (1990).

4. For a review, B.F. Madore (ed.), *Cepheids : Theory and Observations* (Cambridge Univ. Press, Cambridge, 1985).

5. G. A. Tammann, in *IAU Symposium 124, Observational Cosmology*, eds. A. Hewitt et al. (Reidel, Dordrecht, 1987), p.151.

6. R.B. Tully and J. R. Fisher, Astr. Ap **54**, 661 (1977).

7. M. Aaronson and J. Mould. Ap. J. **265**, 1 (1983).

8. A. Sandage and G.A. Tammann, *A Revised Shapley-Ames Catalog of Bright Galaxies* (Carnegie Institute, Washington) (RSA.)

9. G. de Vaucouleurs, Ap. J. **223**, 730; **224**, 710 (1978).

10. G.H. Jacoby, Ap. J. **339**, 39 (1989); R. Ciardullo, G.H. Jacoby, H.C. Ford and J.D. Neill, Ap. J. **339**, 53 (1989).

11. G.H. Jacoby, R. Ciardullo, H.C. Ford and J. Booth, Ap. J. **344**, 704 (1989).

12. R. Ciardullo, G.H. Jacoby and H.C. Ford, Ap. J. **344**, 715 (1989).

13. G.H. Jacoby, R. Ciardullo and H.C. Ford, Ap. J. **356**, 332 (1990).

14. G.H. Jacoby, A.R. Walker and R. Ciardullo, Ap. J. **365**, 471 (1990).

15. S.R. Pottasch, Astr. Ap. **236**, 231 (1990); see also M. Fukugita, and C.J. Hogan, Nature **347**, 120 (1990).

16. A. Dressler et al., Ap. J. **313**, 42 (1987).

17. J.L. Tonry, E.A. Ajhar and G.A. Luppino, Ap. J. **346**, L57 (1990); MIT preprint (1990).

18. M.J. Pierce and R.B. Tully, Ap. J. **330**, 579 (1988).

19. M. Fukugita and C.J. Hogan, Ap. J. (Letters) **368**, L11 (1991).

20. E.g., B. Leibundgut and G.A. Tammann, Basel preprint 35 (1989).

21. C.T. Kowal, A.J. **73**, 1021 (1968); R. Barbon, M. Capaccioli and F. Ciatti, Astr. Ap. **44**, 267 (1975); D. Branch and C. Bettis, A.J. **83**, 224 (1978); G.A. Tammann, in *Supernovae : A Survey of Current Research*, ed. M.J. Rees and R.J. Stoneham (Reidel, Dordredrt), p.371; W.D. Arnett, D. Branch and J.C. Wheeler, Nature **314**, 337 (1985); R. Cadonau, A. Sandage and G.A. Tammann, in *Supernovae as Distance Indicators*, ed. N. Bartel (Springer, Berlin 1985), p.151; D. Branch, ibid, p.138; G.A. Tammann and B. Leibundgut, Astr. Ap. **236**, 9 (1990); D.L. Miller and D. Branch, A.J. **100**, 530 (1990).

22. M. Capaccioli et al., Ap. J. **350**, 110 (1990).

23. M. Fukugita, S. Okamura, K. Tarusawa, H.J. Rood and B.A. Williams, Kyoto preprint YITP/K-874 (Ap. J. in press).

24. M. Chilukuri and R. V. Wagoner, in *Atmospheric Diagnostics of Stellar Evolution*, ed. K. Nomoto (Springer Berlin), p.295.

25. D. Branch, ref. 21.

26. D. Branch, Ap. J. (Letters) **320**, L23 (1987).

27. R.V. Wagoner, in *Supernovae*, Proceedings of the Tenth Santa Cruz Workshop ed. S.E. Woosley (Springer, New York 1989); P.Höflich, in *Atomospheric Diagnosties of Stellar Evolution*, ed. K. Nomoto (Springer Berlin). R.G. Eastman and R.P. Kirshner, Ap. J. **347**, 771 (1989); W. Schmutz et al., Ap. J. **355**, 255 (1990).

28. W.D. Arnett, D. Branch and J.C. Wheeler, ref. 21.

29. D. Branch, W. Drucker and D.J. Jeffery, Ap. J. **330**, L117 (1988).

30. A. Sandage and G.A. Tammann, Ap. J. **365**, 1 (1990); G.A. Tammann, in *The Extragalactic Distance Scale : Proceedings of the ASP 100th Anniversary Program*, ed. S. van den Bergh ed C.J. Pritchet (ASP Conf. Ser. 4) p.282.

31. W.E. Harris, in *The Extragalactic Distance scale : Proceedings of the ASP 100th Anniversary Program*, ed. S. van den Bergh ed C.J. Pritchet (ASP Conf. Ser. 4) p.231

32. C.J. Pritchet and S. van den Bergh, Ap. J. **318**, 507 (1987).

33. J.G. Cohen, Ap. J. **292**, 90 (1985).

34. A. Dressler, Ap. J. **317**, 1 (1987).

35. M. Aaronson et al., Ap. J. **302**, 536 (1986).

36. R.C. Kraan-Korteweg, L.M. Cameron and G. A. Tammann, Ap. J. **331**, 620 (1988).

37. P. Fouqué, L. Bottinelli, L. Gouguenheim. and G. Paturel, Ap. J. **349**, 1 (1990).

38. S. van den Bergh, M.J. Pierce and R.B. Tully, Ap. J. **359**, 4 (1990).

39. D. Burstein and S. Raychaudhury, Ap. J. **343**, 18 (1985).

40. A. Sandage, J. Kristian and J.A. Westphal, Ap. J. **205**, 688 (1976).

41. E. Hubble, *Realm of the Nebulae*, 1936 (Yale University Press); H.P. Robertson, Pub. Astr. Soc. Pacific **67**, 82 (1955); A. Sandage, Ap. J. **133**, 355 (1961).

42. J.R. Gott et al., Ap. J. **194**, 543 (1974).

43. P.J.E. Peebles, *The Large-Scale Structure of the Universe* (Princeton Univ. Press, Princeton, 1980); M. Davis and P.J.E. Peebles, Ap. J. **267**, 465 (1983).

44. A. Sandage, in *Observational Cosmology, Proceedings of the IAU Symposium 124*, Beijing, 1987 (Reidel, Dordrecht) p.1; Ann. Rev. Astr. Ap. **26**, 561 (1988).

45. J.A. Tyson, A.J. **96**, 1 (1988).

46. S.J. Lilly, L.L. Cowie and J.P. Gardner, University of Hawaii preprint (1990) (Ap. J. in press).

47. R.P. Kirshner, A. Oemler, Jr. and P.L. Schechter, A.J. **83**, 1549 (1978); **84**, 951 (1979); R.P. Kirshner, A. Oemler, Jr., P.L. Schechter and S.A. Shectman, A.J. **88**, 1285 (1983); M. Davis and J. Huchra, Ap. J. **254**, 437 (1982); J.E. Felten, A.J. **82**, 861 (1977); Comm. Ap. **11**, 53 (1985); G. Efstathiou, R.S. Ellis and B.A. Peterson, MNRAS **232**, 431 (1988).

48. S.M. Faber and J.S. Gallagher, Ann. Rev. Astr. Ap. **17**, 135 (1979).

49. M. Davis and P.J.E. Peebles, Ann. Rev. Astr. Ap. **21**, 109 (1983); A. Yahil, in *The Virgo Cluster of Galaxies* 1985, (ESO, Garching), p.359; M. Aaronson et al., Ap. J. **258**, 64 (1982).

50. P.L. Schechter, A.J. **85**, 801 (1980); J.L. Tonry and M. Davis, Ap. J. **246**, 680 (1981).

51. A.J. Bean et al., MNRAS **205**, 605 (1983).

52. P.M. Lubin, G.L. Epstein and G.F. Smoot, Phys. Rev. Lett. **50**, 616 (1983); D.J. Fixsen, E.S. Cheng and D.T. Wilkinson, Phys. Rev. Lett. **50**, 620 (1983); G. Smoot, this proceedings.

53. A. Yahil, D. Walker and M. Rowan-Robinson, Ap. J. **301**, L1 (1986).

54. A. Meiksin and M. Davis, A.J. **91**, 191 (1986).

55. J.V. Villumsen and M.A. Strauss, Ap. J. **322**, 37 (1987).

56. O. Lahav, M. Rowan-Robinson and D. Lynden-Bell, MNRAS 234, 677 (1988).

57. *IRAS Point Source Catalogue*, 1985 (U.S. Government Printing Office, Washington D.C.).

58. M. Fukugita and T. Ichikawa, Kyoto University report RIFP-767 (1988) and preprint in preparation.

59. R. Giovanelli and M.P. Haynes, A.J. **90**, 2445 (1985); R. Giovanelli et al., A.J. **92**, 250 (1986).

60. N. Arimoto and Y. Yoshii, Astr. Ap. **164**, 260 (1980); **173**,23 (1987).

61. B.M. Tinsley, Ap. J. **241**, 41 (1980); G. Bruzual A., Ap. J. **273**, 105 (1983).

62. M. Fukugita, F. Takahara, K. Yamashita and Y. Yoshii, Ap. J. (Letters) **361**, L1 (1990); preprint in preparation.

63. A.G. Bruzual and R.G. Kron, Ap. J. **241**, 25 (1980).

64. T. Shanks et al., MNRAS **206**, 767 (1984).

65. Y. Yoshii and F. Takahara, Ap. J. **326**, 1 (1988).

66. B. Guiderdoni and B. Rocca-Volmerange, Astr. Ap. **227**, 362 (1990); Y. Yoshii and B.A. Peterson, Ap. J. in press.

67. B. Rocca-Volmerange, and B. Guiderdoni, MNRAS **247**, 166 (1990).

68. T. Shanks, Durham preprint (1990).

69. T.J. Broadhurst, R.S. Ellis and T. Shanks, MNRAS **235**, 827 (1988); M. Colless et al., MNRAS **244**, 408 (1990).

70. Y. Yoshii and M. Fukugita, Kyoto preprint and preprint in preparation.

71. E.g., G.F.R. Ellis, J.J. Perry and A.W. Sievers, A.J. **89**, 1124 (1981); S. Phillips, J.I. Davies and M.J. Disney, MNRAS **242**, 235 (1990).

72. D. van den Berg, Ap. J. (suppl.) **51**, 29 (1983); I. Iben, Jr. and A. Renzini, Phys. Repts. **105**, 331 (1984); J.E. Hesser et al., Publ. Astr. Soc. Pacific **99**, 739 (1987); G. Alcaino, W, Liller and F. Alvarado, Ap. J. **330**, 569 (1988).

73. E.g. J.N.Bahcall, Princeton preprint IASSNS-AST 90/30 (1990); M. Fukugita and T. Yanagida, Santa Barbara preprint NSF-ITP-91-07 (1991).

74. M. Fukugita, T. Futamase and M. Kasai, MNRAS **246**, 24p (1990).

75. E.L. Turner, Ap. J. (Letters) **365**, L43 (1990).

76. A. Hewitt and G. Burbidge, Ap. J. Suppl. **63**, 1 (1987); **69**, 1 (1989).

77. M. Fukugita and E.L. Turner, Princeton University preprint POP-391 (1991).

GRAVITATIONAL LENS EFFECTS
ON THE REDSHIFT-VOLUME TEST OF Ω_0

M. OMOTE

Institute of Physics,
University of Tsukuba,
Tsukuba 305, JAPAN
and

H. YOSHIDA

Department of Physics,
Fukushima Medical College,
Fukushima City 960-12, JAPAN

ABSTRACT.Gravitational lens effects on the redshift-volume test are investigated. We reanalyze Loh-Spillar's data of this test by taking into account the gravitational lens effects caused by inhomogeneities of the universe, and show that the effects decrease the value of the density parameter Ω_0 to be 0.38 ± 0.44.

1. Introduction

Loh and Spillar (1986, hereafter LS) measured the number and total flux for a flux-limited sample of galaxies, and showed $\Omega_0 \simeq 1.0$. They assumed that all galaxies evolve with time by the same amount in luminosity, and that the universe is described by the Friedmann universe. Some authors discussed the validity of the first assumption and concluded that the evolutionary effect of galaxies cannot be ignored. In this paper, we discuss the effects of local inhomogeneities of the universe on the redshift-volume test.

2. The k-th Flux Moment in the Clumpy universe

We assume that the universe is globally described by the Friedmann universe, but locally by the clumpy universe (Dyer and Roeder 1972,1973,1974). In the clumpy universe, all the inhomogeneities are concentrated into the clumps, which are uniformly distributed act as gravitational lenses. For a sample of galaxies with observed flux $l \geq l_0$, the k-th flux moment in the clumpy universe, in the interval dz (z: redshift), dL (L: intrinsic luminosity), and dA (A: gravitational amplification factor), can be written as:

$$dM_k(z, l, A) = n(z, L)\left(\frac{l}{l_0}\right)^k dA p(z; A) dL \frac{dV}{dz} dz, \tag{1}$$

where $n(z, L)$ is the luminosity function, and $p(z; A)$ is the probability in the case when lights emitted from a source at z are amplified by factor A due to the gravitational lens effect. By assuming that all the clumps have point structures, $p(z; A)$ can be expressed as a simple form (Omote and Yoshida 1990). If the light is amplified by the factor A, the relation l and L is given by

$$l = \frac{AL}{4\pi(\frac{c}{H_0})^2(1 + z)^3 d_{DR}(z)}, \tag{2}$$

where $d_{DR}(z)$ is the angular diameter distance in the clumpy universe (in unit of $\frac{c}{H_0}$).

3. Results and Conclusion

We used the Schechter form with $\alpha = -1.07$ (Efstathiou *et al.* 1988) as the luminosity function in eq.(1), and assume the vanishing cosmological constant. In order to investigate the influence of the gravitational lens effects on the value of Ω_0, we have estimated the values of Ω_0 and ϕ_* in the clumpy universe and in the Friedmann universe, by virtue of eqs.(1) and (2), and found $\Omega_0 = 1.77 \pm 0.87$, $\phi_* = (1.59 \pm 0.44) \times 10^{-3} h^3 [\text{Mpc}]^{-3}$ in the Friedmann universe, and $\Omega_0 = 0.38 \pm 0.44$, $\phi_* = (1.04 \pm 0.31) \times 10^{-3} h^3 [\text{Mpc}]^{-3}$ in the clumpy universe.

Thus it is found that the gravitational lens effects are not negligible in analyzing the data of LS's test, and that the effects decrease the value of Ω_0.

References

Dyer, C.C., and Roeder, R.C. (1972), *Ap.J.(Letters)*,**174**,L115.
Dyer, C.C., and Roeder, R.C. (1973), *Ap.J.(Letters)*,**180**,L31.
Dyer, C.C., and Roeder, R.C. (1974), *Ap.J.*,**189**,167.
Efstathiou, G., Ellis, R.S., and Peterson, B.A. (1988), *M.N.R.A.S.*,**232**,431.
Loh, E.D., and Spillar, E.J. (1986), *Ap.J.(Letters)*,**307**, L1.
Omote, M., and Yoshida, H. (1990), *Ap.J.*,**361**,27.

Summary of the Conference

Satio Hayakawa
Nagoya University
P.O.Box 464-01
Furo-cho Chikusa-ku Nagoya
Japan

1. Retrospect

In planning the present conference the Local Organizing Committee attempted to celebrate 70 years of Professor Chushiro Hayashi and commemorate his pioneering work on primordial nucleosynthesis and the evolution of early universe published 40 years ago. To our greatest regret, however, he seems to have suspected our plan and shyed to be present at this conference. I would, therefore, like to tell you how he initiated a work on primordial nucleosynthesis 40 years ago and what he thought 25 years ago when the cosmic background radiation was discovered.

Having been stimulated by Gamow's idea on the primordial origin of elements, he asked himself if primordial matter could consist only of neutrons as assumed in the $\alpha\beta\gamma$ theory. He noticed that protons should coexist with neutrons due to induced β-processes and the β-decay of neutrons, and that this would modify the element formation processes proposed by Gamow. His paper titled "Proton-Neutron Concentration Ratio in the Expanding Universe at the Stages Preceding the Formation of the Elements" was read at a meeting of the Physical Society of Japan held in fall of 1949 and was submitted to Progress of Theoretical Physics on January 12, 1950 (Prog. Theor. Phys. $\underline{5}$, 224 (1950)). In this paper he correctly pointed out that the proton/neutron ratio should be almost independent of the initial condition, though the ratio of about 4:1 he obtained was smaller than the present value because of different values of parameters adopted. He then estimated the H/He number ratio to be 6:1, consistent with the He abundance known in those days. He further discussed that the presence of protons could overcome the crevasses at A=5 and 8 for building up heavier elements and that nuclear reactions starting with this mixture of protons and neutrons would explain the abundances of light elements up to ^{12}C including deuterons.

In 1950s thereafter, primordial nucleosynthesis was hidden behind nucleosynthesis in stars, as developed by Fowler and Hoyle. If the expansion of the universe would hold, however, Hayashi thought that nuclear reactions which took place in stars should occur in the early universe as well. He took into account the 3α reaction and tried to

617

explain the formation of carbon and other elements in collaboration with M.Nishida. For high densities of matter, primordial nucleosynthesis would form not only ^{2}H, ^{3}He and ^{4}He but also ^{12}C and ^{16}O (C.Hayashi and M.Nishida, Prog. Theor. Phys. 16, 613 (1956)). Now everybody knows that the matter density is not high enough for the 3α reaction to operate.

As soon as the microwave background was discovered, Hayashi suggested to his student, H.Sato, to calculate the primordial nucleosynthesis using this information on temperature. The result was published (H. Sato, Prog. Theor. Phys. 38, 1083 (1967)) nearly at the same time as independent works by Peebles and by Wagoner, Fowler and Hoyle. In a review paper by H.Sato, T.Matsuda, and H.Takeda, (Prog. Theor. Phys. Suppl. No.49 (1971)), they wrote a foot note" Discussion in this subsection is due to C. Hayashi (unpublished, 1965)", in which the time evolution of the abundances of nuclides H to B and the final He abundance were given.

2. Primordial Nucleosynthesis

Owing to its long history over 40 years, the results of primordial nucleosynthesis are almost established. If papers presented at the present conference are compared with those at similar ones, the Yamada Conference XX on Big Bang, Active Galactic Nuclei and Supernove (Proceedings edited by S.Hayakawa and K.Sato, Universal Academy Press, Tokyo, 1989)) held two and a half years ago here and the NATO Advanced Study Institute on Baryonic Dark Matter (Proceedings edited by D.Lynden-Bell and G.Gilmore, Kluwer Academic Publ. Dordrecht, 1990)) held a year ago at Cambridge University, one can find no essential changes in recent years. The primordial abundances derived from various observations and astration corrections seem to have reached general agreement. The abundance of ^{4}He is Y=0.24±0.1 which favors the number of neutrino families N_ν=3. Our astrophysicists are gratified by its confirmation by accelerator experiment. A problem may, however, arise if Y<0.23. Both the abundances of D+^{3}He and ^{7}Li have narrowed down the baryon-photon number ratio to be η=(3-5)x10^{-10}. The baryon density relative to the critical density, $\Omega_B \simeq 0.1$, provides a basis of discussing the dark matter problem.

Attemps have been made at increasing the baryon density to Ω=1 favored by the inflation model. However, the introduction of inhomogeneity can hardly increase Ω_B beyond 0.4 because of diffusion effects, as shown by quantitative numerical studies. The possibility that baryonic matter is partly locked in quark nuggets without participating in nucleosynthesis meets a difficulty that relic quarks would take part in observable effects.

3. Inflation Models

In comparison with primordial nucleosynthesis the history of inflation theory is very short, only about ten years since the proposal of Guth and Katsu Sato. Within the last ten years, however, the theory has rapidly inflated, starting with prehistoric, passing through old, new, chaotic and bubbling, and then is extended to post-modern, as artistically presented by Kolb. In the course of development some "constants" have been made time dependent. The ratio of the tunneling rate to the expansion rate, $\epsilon(t)$, has been assumed to increase in such a way that the initial inflation is followed by the phase transition. Likewise other "constants" such as the cosmological constant and the gravitational constant may also be time variable.

The inflation models thus modified result in interesting consequences which will be subject to observational test. Some topological defects such as domain walls may be inflated away, while the others such as texture may be created thereafter and survive. Gravitational waves produced by bubble collisions may form background radiation which is to be detectable through the pulsar period fluctuations. Inhomogeneities are produced in the course of inflation by rapid variations of fields and phase transitions. Their identification by observation may be possible only after the inhomogeneities due to late time transitions are distinguished from the effects of inflation.

4. Cosmic Background Radiation

The results obtained by COBE showed a Planckian spectrum over a wide wavelength range from submillimeter to centimeter. The Wien part restricts the Sunyaev-Zeldovich parameter to $y<10^{-3}$, while the Rayleigh-Jeans part gives an upper limit of the chemical potential, $\mu<5\times10^{-3}$. I am sorry that our result for submillimeter excess has disturbed the cosmology community. The origin of the excess appears to be due to a change in the zero level caused by ground line interference, which happened only after the installation of the payload for launching. It has been emphasized that more improved technology is required to find out theoretically expected values, $y\sim10^{-4}$ and $\mu\sim10^{-4}$.

COBE also obtained the dipole anisotropy with an amplitude of $T_{DP}=3.3\pm0.2$mK, while only upper limits of 10^{-4} were obtained for the quadruple amplitude and the 7° temperature fluctuation. The temperature distribution can be measured with better sensitivity in the Wien region. In this region, however, the contribution of foreground sources may be appreciable and have to be subtracted. In order for this to be possible, our Japanese group is preparing for a multiband observation covering a wide sky region with an orbiting, cooled telescope which will be launched in 1994.

An interesting possibility has been suggested that the Ly α line trapped by resonance scattering after recombination may cause instability which possibly forms a large density contrast. In this connection efforts for the observation of Ly α radiation redshifted to an infrared continuum shall be encouraged.

X-ray astronomy satellite Ginga has observed X-ray background and many AGNs. AGNs give the largest contribution to X-ray background, which is estimated to be about 30%. The rest is suspected to be due to distant, faint sources including obscured AGNs. The confirmation is expected with the next satellite ASTRO-D which will be launched in early 1993.

5. Dark Matter and Large Scale Structure

The expansion of the universe is dictated by three quantities, the expansion rate or the Hubble constant H_0, the matter density in units of the critical density Ω, and the cosmological constant Λ, the latter two being combined to give the deceleration parameter q_0. Despite much effort of astronomical observations, none of them have been determined with sufficient reliability.

Among them the Hubble constant is regarded as most fundamental, since the critical density is proportional to H_0^2, thus affecting the value of Ω, and the age of the universe is inversely proportional to H_0. As is well known, the value of H_0 has changed considerably from time to time. In late 1940s $1/H_0$ was shorter than the age of the earth, and this motivated the invention of steady state cosmology. After an increase of the distance scale due to the distinction between the period-amplitude relations of Populations I and II variables, the value of $1/H_0$ has gradually increased, being superposed by oscillations. In recent years the value of H_0 has still been debated; astronomers of high reputation tell us different values between 50 and 100 $kms^{-1}Mpc^{-1}$.

At this conference a cute idea has been proposed that H_0 oscillates not with historical time but with cosmological time in the course of the formation of galaxies. This would result in the periodicity in the distribution of galaxies as observed in the northern and southern pole directions. The pencil beam survey of galaxies has been extended to other directions and shows similar periodic distributions. Since the periodic distribution seems to hold over a wide region, this feature is called the crystallization of galaxies.

The difficulty in getting the reliable value of H_0 is due to the fact that the distance-redshift relation can be obtained for nearby galaxies, most of which fall towards the great attractor. In order to avoid such motion, more distant galaxies have to be observed. The observation of galaxies in the Coma cluster has been performed with a Schmidt telescope at Kiso Observatory and has given $H_0=95(+20, -17)kms^{-1} Mpc^{-1}$.

Such a large value of H_0 would result in too short an age of the universe if $\Lambda=0$. However, the number-magnitude relation of galaxies based on a recent deep survey gives $\Omega \simeq 0.1$ and $\lambda=c^2\Lambda/3H_0^2 \simeq 0.9$. Large scale structures also favor $\Omega \simeq 0.1$, and an infrared deep survey suggests finite Λ. A set $\Omega+\lambda=1$ is consisted with the inflation model and does not contradict the age of the Universe. A small value of Ω and a finite value of Λ require the reexamination of cosmological models, particularly of the dark matter problem.

If $\Omega=1$ is adopted for the matter density, this is much larger than the density of baryonic matter, $\Omega_B \simeq 0.1$, derived from primordial nucleosynthesis. A large difference $\Omega-\Omega_B$ should be attributed to matter consisting of weakly interacting particles. Their existence has been favored with hope that they play an efficient role in the formation of galaxies within a relatively short period up to $z \simeq 5$ starting with an extremely uniform distribution of matter at $z \sim 10^3$. Since light particles such as neutrinos quickly diffuse away to form a uniform distribution, massive particles are supposed to take major part in the formation of clumps which attract baryonic matter to grow into galaxies. Candidates of such weakly interacting massive particles, WIMPs, have been proposed on the particle theoretical ground and searched for by various means.

Results thus far obtained do not positively indicate their existence but rule out quite a few of possible candidates. Now we wonder if any of the candidates would survive. However, $\Omega \simeq 0.1$ as indicated at this conference seems to relieve us from this worry, so that no WIMPs may exist. Since the values of Ω and Ω_B can be claimed at best within uncertainty of factor two, searches for exotic particles shall not be discouraged, but negative results will no longer discourage us.

The density of baryonic matter is higher than that of luminous matter, and the difference between them can be attributed to dark matter of baryonic nature. We know that dark matter exists in the galactic halo and in the intracluster space, and its amount is estimated from the dynamical mass which considerably exceeds the luminous mass. Such dark matter may consist mainly of baryonic matter which escapes out of the observation. The baryonic dark matter is less exotic than the nonbaryonic one, and some of its constituents may be astronomical objects which can be detected as faint objects and through the background radiation. A number of papers presented at this conference are suggestive of disclosing the nature of dark matter in connection with large scale structures, dynamics and physics of stellar and gaseous systems, gravitational lensing, and so forth.

In conclusion, I would like to express my feeling that cosmology is coming close to the dawn, so that the structure of the universe and dark matter are becoming visible. We owe the dawning to efforts of increasing the collecting power of faint light in a wide wavelength range, of refining the analysis of data, and of developing ideas. However, it is still dark, especially for oversea participants who still keep their night time at home in this afternoon, but they may now feel some daylight after their stay here for a week.

On behalf of the Local Organizing Committee, I would like to thank all participants for their enlightening contributions which have made this Conference exceedingly stimulating. It is my great pleasure, as the Chairman of Commission 19 for Astrophysics of IUPAP, that the activity of our Commission will be appreciated by the physics community owing to the success of this conference. Finally, on behalf of participants, I thank Dr. Katsuhiko Sato and his associates, excepting me, an idle member, for their excellent organization which has made this conference fruitful.

IUPAP CONFERENCE

Primordial Nucleosynthesis and Evolution of Early Universe

Participant List

September 4 - 8, 1990
Sanjo Conference Hall,
University of Tokyo
Bunkyo-ku, Tokyo 113, JAPAN

Aizu Ko

3-24-3 Katahira, Asao-ku,
Kawasaki 215, Japan

Arafune Jiro
ICR, Univ. of Tokyo
ICR, Univ.of Tokyo, Midori-cho,
Tanashi,Tokyo 188, Japan

Arai Kenzo
Kumamoto University
Dept. of Physics, Kumamoto Univ.
Kurokami 2-39-1, Kumamoto, Japan

Arima Akito
University of Tokyo
University of Tokyo, Bunkyo-ku,
Tokyo 113, Japan

Audouze J
Inst. d'Astrophysique
Presidence de la Republique, SS rue
du Faubourg Saint-Honore 75008
Paris, France

Baba Kazuo
Nara Saho-Jogakuin Junior College
Nara Saho-Jogakuin Junior College,
Rokuyaon-cho, Nara, 630, Japan

Banerjee B.
Tata Inst.
Theoretical Physics Group, Tata
Institute of Fundamental Research,
Homi Bhabha Road, Colaba,
Bombay 400 005, India

Beckman John
Inst. de Astrophys. de Canarias
Instituto de Astrophysica de
Canarias
38200 La Laguna, Tenerife,
Spain

Berkin Andrew
Waseda University
Dept.of Physics,Waseda Univ.
Okubo 3-4-1,Shinjuku-ku,Tokyo
169, Japan

Bildhauer Stefan
Hirosaki University
Dept. of Physics, Hirosaki
University, 3 Bunkyo-cho,
Hirosaki Aomori 036, Japan

Bond J. R.
CITA
Canadian Institute for Theoretical
Astrophysics University of Tronto,
Canada

Buchmann Alfons
University of Tokyo
Dept. of Physics, Fac. of Sciences,
University of Tokyo, Bunkyo-ku,
Tokyo 113, Japan

Calzetti Daniela
University of Rome
G9, Dipartimento Difisica,
Universita 'La Sapienza', Piazzale A.
Moro, 2-00185 Roma, Italy

Cao Shenglin
Beijing Normal University
Department of Astronomy, Beijing
National University, Beijing
100875, China

Cho Y. M.
Seoul National University
Dept. of Physics, Seoul National
University
Seoul, 151-742, Korea

Chuvenkov Vladimir
Rostov State University
Department of Astrophysics
Rostov State University
5 Zorge, 344104 Rostov-on-Don,
USSR

Cowie L.
University of Hawaii
Institute for Astronomy, Univ.
Hawaii at Manoa, 2680 Woodlawn
Drive, Honolulu
Hawaii 96822, USA

Den Mitsue
Kyoto University
Department of Physics, Kyoto
University
Kitashirakawa Oiwake-cho,
Sakyo-ku, Kyoto 606, Japan

Deng Zugan
Academia Sinica
Dept. of Physics, Graduate School,
Academia Sinica, P. O. Box 3908,
Beijing 100039, China

Doi Mamoru
University of Tokyo
Institute of Astronomy, University
of Tokyo, Mitaka, Tokyo 181,
Japan

Dolgov Alexandre D.
Inst. for Theor. & Exp. Phys.
Institute for Theoretical and
Experimental Physics
Moscow 117259, USSR

624

Esmailzadeh Rahim
University of California
Department of Astronomy
University of California, Berkeley
Berkeley, CA 94720, USA

Fowler William A.
Caltech
Kellogg Rad. Lab. Caltech 106-38
Pasadena, CA 91125, USA

Fujii Yasunori
Univ. of Tokyo-Komaba
Inst. of Physics, Univ.of
Tokyo-Komaba, Komaba,
Meguro-ku, Tokyo 153, Japan

Fukuda Tomokazu
INS, Univ. of Tokyo
INS, University of Tokyo,
Midori-cho 3-2-1, Tanashi, Tokyo
188, Japan

Fukugita Masataka
YITP
YITP, Kyoto University,
Kitashirakawa sakyou-ku, Kyoto
606, Japan

Futamase Toshifumi
Hirosaki University
Dept. of Physics, Hirosaki
University,
3 Bunkyo-cho, Hirosaki Aomori
036, Japan

Giavalisco Mauro
University of Rome
Dipartimento di
Fisica-G9-Universita 'La Sapienza',
Piazzale Aldo Moro 2, 00185,
Rome, Italy

Gouda Naoteru
Kyoto University
Department of Physics, Kyoto
University
Kitashirakawa Oiwake-cho,
Sakyo-ku Kyoto 606, Japan

Hagio Fumihiko
Kumamoto Inst. of Technology
Dept. Phys., The Kumamoto
Institute of Technology
Ikeda 4, Kumamoto, Japan

Hanami Hitoshi
Iwate University
Laboratory of Physics, College of
Humanities and Social Sciences
Iwate Univ., Morioka 020, Japan

Hara Tetsuya
Kyoto Sangyo University
Kyoto Sangyo University, Kitaku,
Kamigamo, Kyoto 603, Japan

Hasegawa Takashi
University of Tokyo
National Astronomical
Observatory, Mitaka, Tokyo 181,
Japan

Hawking Stephen W.
Cambridge Universitry
Department of Applied
Mathematics and Theoretical
Physics, University of Cambridge,
Silver Str. UK

Hayakawa Satio
Nagoya University
Nagoya University, Furo-cho,
Chikusa-ku, Nagoya 464-01, Japan

Hayashi Hirofumi
Shizuoka University
Faculty of Education, Shizuoka
University, Ooya 836, Shizuoka,
Japan

Hayashi Koichi
Kinki University
Dept. of Physics, Kinki University,
Higashi-Osaka 577, Japan

Hobbs Lewis
University of Chicago
Univ. of Chicago, Yerkes
Observatory
Williams Bay, WI 53191-0258,
USA

Hogan Craig
University of Arizona/YITP
YITP, Kyoto University, Kyoto
606, Japan

Holman Richard
Carnegie Mellon University
Dept. of Physics, Carnegie Mellon
University,
Pittsburgh, PA 15213, USA

Hoshi Reiun
Rikkyo University
Depertment of Physics, Rikkyo
University, Nishi-Ikebukuro 3,
Toshima-ku, Tokyo 171,
Japan

Hosoya Akio
Tokyo Inst. of Technology
Dept of Physics, Tokyo Inst. of
Technology,Ohokayama,
Meguro-ku, Tokyo, Japan

Ikeuchi Satoru
NAO
National Astronomical
Observatory, Mitaka, Tokyo 181,
Japan

Inagaki Shogo
Kyoto University
Dept. of Astronomy, Fac ulty of
Science,Kyoto University,
Kirashirakawa-Oiwake-cho,
Sakyo-ku, Kyoto 606, Japan

Ino Toshio

Takachiho Univ.
Ohmiya 2-19-1, Suginami-ku,
Tokyo 168,
Japan

Inoue Hajime
ISAS
ISAS, Yoshinodai, Sagamihara-shi,
Kanagawa 229, Japan

Ishihara Hideki
Kyoto University
Department of Physics, Kyoto
University
Kitashirakawa Oiwake-cho,
Sakyo-ku, Kyoto 606, Japan

Ito Makoto
Kyoto University
Department of Astronomy,
Kyoto University,
Kyoto 606, Japan

Iwamoto Naoki
University of Toledo
Department of Physics &
Astronomy
University of Toledo, 2801 W.
Bancroft St. Toledo, OH43606,

Iye Masarori
NAO
National Astronomical Observatory
Mitaka, Tokyo 181, Japan

Izawa Mizuo
Shimonoseki Univ. of Fisheries
Shimonoseki Univ.of Fisheries,
P.O.Box 3 Yoshimi, Shimonoseki
759-65, Japan

625

Jugaku Jun
Tokai University
Institute for Civilization, Tokai
University,
Hiratuka, Kitakaname 1117
259-12, Japan

Kabe Seiji
KEK
KEK, Oho 1-1, Tsukuba, Ibaraki
305, Japan

Kajino Toshitaka
Tokyo Metropolitan Univ.
Depertment of Physics,Tpkyo
Metropolitan University, Setagaya,
Tokyo 158, Japan

Kang Ho-Shik
Ohio State University
174 W. 18th Ave.
Dept. of Phys., The Ohio State
University Columbus,
OH43210-1106, USA

Kasai Masumi
Hirosaki University
Dept of Physics, Hirosaki
University, 3 Bunkyo-cho,
Hirosaki, Aomori 036, Japan

Kawano Lawrence
Caltech
W.K.Kellogg Radiation Lab,
106-38 California Institute of
Technology Pasadena, CA 91125,
USA

Kawasaki Masahiro
Tohoku University
Department of Physics, Tohoku
University, Sendai 980, Japan

Khlopov M.
Keldysh Inst. Applied Math.
Space Research Inst., USSR Acad.
of Sci., Moscow, USSR

Kiguchi Masayoshi
Kinki University
Res. Inst. for Science and
Technology, Kinki University,
3-4-1 Kowakae, Higashi-Osaka,
Osaka 577, Japan

Kim Sung-Won
Ewha Womans University
Dept. of Science Education, Ewha
Womans University, Seoul
120-750, Korea

Kirilova Daniela Petrova
Joint Inst. of Nuclear Research
Moskovskaja Oblast, G. Dubna, Ul.
Stroiteli, 8, 609, 141980, USSR

Kitching Peter
INS, Univ. of Tokyo
INS, University of Tokyo,
Midori-cho 3-2-1, Tanashi, Tokyo
188, Japan

Kodaira Keiichi
NAO
National Astronomical
Observatory, Mitaka, Tokyo 181,
Japan

Kofman Lev
Tartu Astropys. Obs.
Tartu Astrophys. Obs. , Toravere,
Estonia, USSR

Kohama Akihisa
University of Tokyo
Dept. Phys. Fac. of Sci.
Univ. of Tokyo
Bunkyo-ku, Tokyo 113, Japan

Kokado Akira
Kwansei Gakuin University
Dept. of Physics, Kwansei Gakuin
University, 1-1-155 Uegahara,
Nishinomiya, Hyogo, Japan

Kolb Edward
Fermilab
Theoretical Astrophysics,
MS209, Fermilab, box 500
Batavia, IL 60510, USA

Konuma Michiji
Keio University
Dept. of Physics, Keio University,
Kohoku-ku, Yokoyama 223, Japan

Koyama Katsuji
Nagoya University
Dept. of Physics, Nagoya Univ.,
Furoh-cho, Chikusa-ku, Nagoya
464, Japan

Krauss Lawrence
Yale University
Center for Theoretical Physics,
Yale Univ.Slowne Lab, 217
Prospect St., New Haven, CT
06511, USA

Kubo Ryogo
Keio University
Keio University, Hiyoshi 3-14-1,
Kouhoku-ku, yokohama 223, Japan

Kubono Shigeru
INS, Univ. of Tokyo
Institute for Nuclear Study, Univ. of
Tokyo,
3-2-1 Midori-cho, Tanashi, Tokyo,
Japan

Kubotani Hiroto
Kyoto University
Department of Physics, Kyoto
University
Kitashirakawa Oiwake-cho,
Sakyo-ku , Kyoto 606, Japan

Lee Hyung Won
University of Rome
GRUPPO G9, Dipartimento di
Fisica, Universita di Roma 'La
Sapienza', Piazzala Aldo Moro 2,
00185 Roma, Italy

Lee C.H.
Hanyang University
Department of Physics, Hanyang
University
Seoul 133-791, Korea

Linder Eric
University of Arizona
Steward Observatory Univ. of
Arizona
Tucson, AZ85721,
USA

Lucchin Francesco
University of Padova
Dipartimento di Fisica
Univ. Padova
Via Marzolo 8, 35131 Padova, Italy

Madsen Jes
University of Aarhus
Institute of Astronomy, University
of Aarhus DK-8000 Aarhus C,
Denmark

Maeda Kei-ichi
Waseda University
Dept. Phys. Waseda Univ.
Okubo 3-4-1, Shinjuku-ku,
Tokyo 169, Japan

Malaney Robert
Lawrence Livermore Lab.
Lawrence Livermore National Lab.
L-413
Livermore, CA94550, USA

626

Mathews Grant J.
University of California
Lawrence Livermore National
Laboratory
L-297, Livermore, CA 94550, USA

Matsuda Takuya
Kyoto University
Department of Aeronautical
Engineering, Kyoto University,
Yoshidahon-machi, Sakyo-ku,
Kyoto 606, Japan

Minakata Hisakazu
Tokyo Metropolitan Univ.
Department of Physics, Tokyo
Metropolitan University, Fukazawa
2-1-1, Setagaya, Tokyo 158, Japan

Mineshige Shin
University of Cambridge
Institute of Astronomy,
Madingley Road, Cambridge
CB3 OHA, England

Minn Hokee
Seoul National University
Dept. of mathematics, College of
Natural Sciencesn Seoul National
University,
Seoul, Korea

Miyama Shoken
NAO
National Astronomical
Obserbatory, Mitaka, Tokyo 181,
Japan

Miyoshi Shigeru
University of Cambridge
Institute of Astronomy, University
of Cambridge Madingley Road,
Cambridge CB3 OHA, UK

Mizutani Kohei
Saitama University
Depertment of Physics, Saitama
Univertsity,250 Shimo-Okubo,
Urawa 338, Japan

Morikawa Yoshitomi
INS, Univ. of Tokyo
INS Univ. of Tokyo, Midori-cho
Tanashi, Tokyo 188, Japan

Morikawa Masahiro
Univ. of British Columbia
Department of Physics,Univ. of
British Columbia, Vancouver, BC.
Canada V6T 2A6

Morita Syuji
Kyoto University
Department of Physics, Kyoto
University
Kitashirakawa Oiwake-cho,
Sakyo-ku Kyoto 606, Japan

Morita Reiko
Setsunan University
Setsunan University,
Ikeda-naka-machi Neyagawa, Osaka
572, Japan

Murakami Izumi
University of Tokyo
Division of Theoretical
Astrophysics, National
Astronomical Observatory, Mitaka,
Tokyo 181, Japan

Nagai Yasuki
Tokyo Inst. of Technology
Department of Applied Physics,
Tokyo Institute of Technology,
Ookayama, Meguro, Tokyo 152,
Japan

Nagasawa Michiyasu
University of Tokyo
Department of Physics, University
of Tokyo, Bunkyo-ku, Tokyo 113,
Japan

Nakamura Takashi
YITP
YITP Kyoto University,
Kyoto, Japan

Nakamura Seitaro
Tokai University
Kitakinmoku 1117, Hiratsuka
259-12, Japan

Nakao Ken-ichi
YITP
Uji Research Center Yukawa Inst.
for Theoretical Phys. Kyoto
University, Uji 611, Japan

Nambu Yasusada
Kyoto University
Dept. of Physics, Kyoto Univ.,
Kyoto 606,
Japan

Namiki Masatoshi
Takachiho University
Takachiho Univ.
Ohmiya 2-19-1, Suginami-ku,
Tokyo 168,
Japan

Oberhummer Heinz
TU Wien
Institut fur Kernphysik, TU Wien
Wiedner Hauptstrasse 8-10
A-1040 Vienna, Austria

Occhionero Franco
Astron. Obs. of Rome
Astorn. Observ. of Rome
V. Le Parco Mellini 84
00136 Roma, Italy

Oda Minoru
Riken
Riken, 2-1 Hirosawa, Wako,
Saitama
351-01, Japan

Okamura Sadanori
Kiso Obs., Univ. of Tokyo
Kiso Obs., Mitake, Kiso-gun,
Nagano
397-01, Japan

Okamura Takashi
Kyoto University
Department of Physics, Kyoto
University
Kitashirakawa Oiwake-cho,
Sakyo-ku, Kyoto 606, Japan

Ostriker J.P.
Priceton University
Princeton Univ.
Peyton Hall, Ivy Lane, Princeton,
NJ 08544-1001, USA

Pagel B.E.J.
Nordita
Nordita, Blegdamsvei 17,
Copenhagen, DK-2100, Denmark

Primack Joel R.
University of California
Physics Department, Universtiy of
California Santa Cruz, CA 95064,
USA

Rees Martin
University of Cambridge
Inst. of Astron., Madingley Road,
Cambridge, CB3 OHA, UK

Salopek David
Fermilab
NASA/Fermilab Astrophysics
Group
P. O. Box 500 MS-209
Batavia, Illinois 60510, USA

Sasaki Misao
YITP
Uji Research Center Yukawa Inst.
for Theoretical Phys. Kyoto
University, Uji 611, Japan

Sato Katsuhiko
University of Tokyo
Dept. of Phys., Faculty of Science,
Univ. of Tokyo, 7-3-1 Hongo,
Bunkyo-ku, Tokyo 113, Japan

Sato Humitaka
Kyoto University
Department of Physics, Kyoto
University
Kitashirakawa Oiwake-cho,
Sakyo-ku, Kyoto 606, Japan

Scherrer Robert
Ohio State University
Department of physics, Ohio State
University, Columbus, OH 43210,
USA

Schramm David N.
University of Chicago
Dept. of Ast. & Astro., University
of Chicago, 5640 S.Ellis Avenue -
AAC 140, Chicago, IL 60637, USA

Seriu Masashi
Kyoto University
Department of Physics, Kyoto
University
Kitashirakawa Oiwake-cho,
Sakyo-ku , Kyoto 606, Japan

Shibata Masaru
Kyoto University
Department of Physics, Kyoto
University
Kitashirakawa Oiwake-cho,
Sakyo-ku, Kyoto 606, Japan

Shirafuji Takeshi
Saitama University
Department of Physics, Saitama
University Shimo-ookubo 255,
Urawa, Saitama 388, Japan

Silk Joe
University of California
Mt. Stromlo & Siding Spring
Observatories Weston Creek P.O.
ACT 2611, Australia

Smoot George
University of California
Bldg 50-232, Lawrence Berkeley
Lab
1 Cycrotron Rd, Berkeley, CA
94720,

Starobinsky Alexey A.
Landau Inst. for Theor. Phys.
Landau Institute for Theoretical
Physics
USSR Academy of Sciences,
Kosigna 2, GSP-1, Moscow,

Steigman Gary
Ohio State University
Dept. of Physics, Ohio State Univ.,
174West18th Ave., Columbus, OH
43210, USA

Steinhardt Paul
University of Pennsylvania
Dept. physics, Univ. of
Pennsylvania
David Rittenhouse Laboratory
Philadelphia, PA 19104, USA

Suginohara Tatsushi
University of Tokyo
Dept. of Phys., Faculty of Science,
Univ. of Tokyo, 7-3-1 Hongo,
Bunkyo-ku, Tokyo 113, Japan

Sugiyama Naoshi
Kyoto University
Dept. of Physics, Faculty of
Science Kyoto University,
Kitashirakawa-Oiwake-cho
Sakyo-ku, Kyoto 606, Japan

Sunyaev R. A.
Space Research Inst.
Space Research Inst., USSR Acad.
of Sci., Moscow, USSR

Suto Yasushi
YITP
Uji Research Center Yukawa Inst.
for Theoretical Phys. Kyoto
University, Uji 611, Japan

Suzuki Hideyuki
KEK
Dept. of Phys., Faculty of Science,
Univ. of Tokyo, 7-3-1 Hongo,
Bunkyo-ku, Tokyo 113, Japan

Suzuki Atsuto
KEK
KEK, Oho 1-1, Tsukuba, Ibaraki
305, Japan

Suzuki Yoichiro
ICR, Univ. of Tokyo
ICR, Univ. of Tokyo, 3-2-1
Midori-cho, Tanashi, Tokyo 188,
Japan

Szalay A.
Johns Hopkins University
Dept. Phys.and Astron. The John
Hopkins Univ. Baltimore,
MD21218, USA

Tajima Toshiki
Univ. Texas at Austin
Inst. for fusion studies Univ. Texas
at Austin Austin TX 78712-1068,
USA

Takahara Fumio
Tokyo Metropolitan University
Department of Physics, Tokyo
Metropolitan University, Fukazawa
2-1-1, Setagaya, Tokyo 158, Japan

Takatsuka Tatsuyuki
Iwate University
College of Humanities and Social
Sciences,
Iwate University, Morioka 020,
Japan

Tanaka Yasuo
ISAS
ISAS, Yoshinodai, Sagamihara-shi,
Kanagawa 229, Japan

Terasawa Nobuo
Riken
Riken, 2-1 Hirosawa, Wako,
Saitama
351-01, Japan

Tomita Kenji
YITP
Yukawa Institute for Theoretical
Physics, Kyoto University,
Gokanoshou, Uji, 611, Japan

Tsuchiya Toshio
Kyoto University
Department of Physics, Kyoto
University
Kitashirakawa Oiwake-cho,
Sakyo-ku, Kyoto 606, Japan

Tsuruta Satiko
Montana State University
Dept. of Phys., Montana State
Univ.,Bozeman, MT 59717, USA

Turner Michael S.
Fermilab
Theoretical Astrophysics,
MS209, Fermilab, Box 500
Batavia, IL 60510-0500, USA

628

Turner Edwin L.
NAO
National Astronomical Observatory
Mitaka, Tokyo 181, Japan

Uchida Yutaka
University of Tokyo
Department of Astronomy, Univ. of
Tokyo
Bunkyo-ku, Tokyo 113, Japan

Umemura Masayuki
NAO
National Astronomical
Observatory, Mitaka, Tokyo 181,
Japan

Vagnetti Fausto
II Universita di Roma
Astrophysica Dip. Fisica, II
Universita di Roma "Tor Vergata"
Via E. Carnevale, I-00173 Roma,
Italy

Van Dalen Anthony
Ohio State University
Dept. of Physics, The Ohio State
University 174 West 18th Ave,
Columbus OH 43210-1106, USA

Wasserman Ira
Cornell University
202 Space Sciences Bldg.
Cornell University, Ithaca, NY
14853, USA

Watanabe Tadashi
Tokyo Univ. of Information Sci.
Dept. of Information Systems,
Tokyo Univ. of Information
Sciences, Yatoh-cho
1200-2, Chiba 280-01, Japan

Watanabe Kazuya
YITP
YITP Uji Research center
Gokashou, Uji 611, Japan

Watanabe Takuya
Kyoto University
Dept of Astronomy, Kyoto
University,
Sakyo-ku, Kyoto 606, Japan

Yamada Shoichi
University of Tokyo
Dept. of Phys., Faculty of Science,
Univ. of Tokyo, 7-3-1 Hongo,
Bunkyo-ku, Tokyo 113, Japan

Yamaguchi Yoshio
Tokai University
Kitakinmoku 1117, Hiratsuka
259-12, Japan

Yamamoto Kazuhiro
YITP
YITP Uji Research center
Gokashou, Uji 611, Japan

Yamashita Kazuyuki
Kyoto University
Department of Physics, Kyoto
University
Kitashirakawa Oiwake-cho,
Sakyo-ku , Kyoto 606, Japan

Yamazaki Toshimitsu
INS, Univ. of Tokyo
INS, University of Tokyo,
Midori-cho 3-2-1, Tanashi, Tokyo
188, Japan

Yanagida Tsutomu
Tohoku University
Department of Physics, Tohoku
University, Sendai 980, Japan

Yokosawa Masayoshi
Ibaraki University
Dept. of Physics, Faculty of
Science, Ibaraki University,
Bunkyo 2-1-1, Mito 310, Japan

Yokoyama Jun'ichi
University of Tokyo
Dept. Phys. Fac. of Sci.
Univ. of Tokyo
Bunkyo-ku, Tokyo 113, Japan

Yoshida Hiroshi
Fukushima Medical College
Dept. Phys. Fukushima Medical
College
Fukushima-City 960-12, Japan

Yoshimura Motohiko
Tohoku University
Department of Physics, Tohoku
University, Sendai 980, Japan

Yoshioka Satoshi
Tokyo Univ. of Mercantile Marine
Tokyo University of Mercantile
Marine, Etchujima, Koto-ku, Tokyo
135, Japan

Zhu Xingfen
Univ. of Sci. & Tech. of China
Center for Astrophysics
Univ. of Science and Technology
of China Hefei, Anhui 230026,
China

Zou Zhenlong
Beijing Astro. Obs.
Beijing Astro. Observatory,
Chinese Academy of Sciences,
Beijing, 100080, P. R. China

The manufacturer's authorised representative in the EU is Springer
Nature Customer Service Centre GmbH, Europaplatz 3, 69115 Heidelberg,
Germany. If you have any concerns regarding our products, please
contact ProductSafety@springernature.com

Printed and bound by CPI Group (UK) Ltd, Croydon, CR0 4YY

30/04/2026

02100146-0002